IN CASE OF ACCIDENT[1]

In case of accident notify the laboratory instructor **immediately**.

FIRE

Burning Clothing. Prevent the person from running and fanning the flames. Rolling the person on the floor will help extinguish the flames and prevent inhalation of the flames. If a safety shower is nearby hold the person under the shower until flames are extinguished and chemicals washed away. Do not use a fire blanket if a shower is nearby. The blanket does not cool and smoldering continues. Remove contaminated clothing. Wrap the person in a blanket to avoid shock. Get prompt medical attention.

Do not, under any circumstances, use a carbon tetrachloride (toxic) fire extinguisher and be very careful using a CO_2 extinguisher (the person may smother).

Burning Reagents. Extinguish all nearby burners and remove combustible material and solvents. Small fires in flasks and beakers can be extinguished by covering the container with a fiberglass-wire gauze square, a big beaker, or a watch glass. Use a dry chemical or carbon dioxide fire extinguisher directed at the base of the flames. **Do not use water.**

Burns, Either Thermal or Chemical. Flush the burned area with cold water for at least 15 min. Resume if pain returns. Wash off chemicals with a mild detergent and water. Current practice recommends that no neutralizing chemicals, unguents, creams, lotions, or salves be applied. If chemicals are spilled on a person over a large area quickly remove the contaminated clothing while under the safety shower. Seconds count, and time should not be wasted because of modesty. Get prompt medical attention.

CHEMICALS IN THE EYE: Flush the eye with copious amounts of water for 15 min using an eyewash fountain or bottle or by placing the injured person face up on the floor and pouring water in the open eye. Hold the eye open to wash behind the eyelids. After 15 min of washing obtain prompt medical attention, regardless of the severity of the injury.

CUTS: Minor Cuts. This type of cut is most common in the organic laboratory and usually arises from broken glass. Wash the cut, remove any pieces of glass, and apply pressure to stop the bleeding. Get medical attention.

Major Cuts. If blood is spurting place a pad directly on the wound, apply firm pressure, wrap the injured to avoid shock, and get **immediate** medical attention. Never use a tourniquet.

POISONS: Call 800 information (1-800-555-1212) for the telephone number of the nearest Poison Control Center, which is usually also an 800 number.

---

[1] Adapted from *Safety in Academic Chemistry Laboratories,* prepared by the American Chemical Society Committee on Chemical Safety, March 1974.

AND

Macroscale
Microscale

ORGANIC
EXPERIMENTS

AND

Macroscale
Microscale
ORGANIC
EXPERIMENTS

THIRD EDITION

KENNETH L. WILLIAMSON

Mount Holyoke College

HOUGHTON MIFFLIN COMPANY

BOSTON   NEW YORK

Editor-in-Chief: *Kathi Prancan*
Senior Sponsoring Editor: *Richard Stratton*
Associate Editor: *Marianne Stepanian*
Packaging Services Supervisor: *Charline Lake*
Manufacturing Manager: *Florence Cadran*
Marketing Manager: *Penelope Hoblyn*

Cover design: Walter Kopec
Cover image: Jim Sherer Photography

Printed in the U.S.A.

Library of Congress Catalog Card Number: 98-72094

ISBN: 0-395-90220-7

3456789-DC-02 01 00

Contents

Aromatic Substitution and Elimination

Reactions of Aldehydes and Ketones

Reactions of Carboxylic Acids, Esters, and Amines

Photochemistry

Natural Product Chemistry and Biochemistry

Preface

Innovation and exploration characterize this Third Edition of *Macroscale and Microscale Organic Experiments*. Building on the strength and success of prior editions, the Third Edition boasts a reorganization that more closely follows organic chemistry textbooks as well as innovative new techniques, features, and experiments. Additionally, throughout this edition one will find experiments and references that tie this course to biochemical and biomedical applications of organic chemistry.

New to This Edition

Techniques

A method has been devised, using the 105° adapter in every Williamson microscale kit, for each student to construct a gas phase IR cell at no cost. Now they can add the third state of matter to traditional solid and liquid phase samples and explore for the first time the gases that are produced in these experiments. In addition, they can explore the myriad gases found in aerosol cans, inhalants, refrigerants, and the like, as well as those from biological sources such as marsh gas and odors ranging from perfumes to putrefaction.

Another innovation, the Wilfilter, is a simple polypropylene adapter that converts every reaction tube into a superior Craig tube. This conversion allows the isolation of minute quantities of crystals without transfer losses, free of solvent and almost dry. We are now using this in many experiments; it is a fast and efficient technique.

Features

The opportunities for students to explore organic chemistry in a variety of ways is presented in the new feature entitled "For Further Investigation." These sections require running reactions and then deducing the nature of the products, usually by IR and NMR spectroscopy. Open-ended investigations include exploring the nature of the gases found in commercial products.

Exploration of another nature is included at the end of many experiments: "Surfing the Web." I have tried to locate a few relevant sites that add new dimensions to the topic at hand as the unwary student can be overwhelmed by the riches of the Web. "Surfing the Web" leads students to, for example, infrared spectra correlated to vibrations of specific atoms and groups of atoms in a molecule, very clear pictures of melting crystals for the melting point experiment, animations of reactions, and full-color, close-up photographs of many of the experiments presented in this text.

New and Updated Experiments

The "NMR Spectroscopy" chapter has been rewritten to guide students in the use of this powerful technique for structure identification based on the assumption they will be using high-field Fourier transform spectrometers. Most of the NMR spectra have been revised and expanded, and the elements of two-dimensional NMR spectra are also presented.

Similarly, the "Infrared Spectroscopy" chapter has been rewritten so that students can carry out a logical step-by-step analysis of their spectra. Fourier transform IR spectrometers are rapidly coming into general use, because not only do they allow spectra to be obtained in less time, but they also give peak frequencies in digital form. In this edition FTIR spectra have replaced the analog versions of the previous edition. Also new to this chapter is a section on "Gas Phase IR Spectroscopy"; six more gas phase spectra can be found in the *Instructor's Guide*.

New to this edition is an experiment that allows students to prove the stereochemistry of reduction of a diketone to a diol: In Chapter 57, "Synthesis of 2,2-Dimethyl-1,5-dioxolane," the acetonide derivative of hydrobenzoin is prepared and its NMR spectrum analyzed to prove whether the hydrobenzoin is the meso or d,l-compound. Chapter 67, "Isolation of Lycopene and β-Carotene," has been placed in a more appropriate place in the organization of the manual, giving emphasis to the biochemical nature of the experiment. Chapter 34, "1,2,3,4-Tetraphenylnaphthalene via Benzyne," is an experiment that has been reinstated from a previous edition.

Throughout the text the scale of the macroscale experiments has been reduced, in most cases to half the former size. This has been done because of the widespread use of 14/20 standard-taper apparatus, the savings in the cost of waste disposal and purchase of chemicals, and for safety reasons. In a few cases microscale experiments have been scaled up, simply to make it easier to isolate products.

The "Searching the Chemical Literature" chapter has been rewritten to reflect not only changes in the references but also the availability of computer databases of this material.

Innovative Techniques from Previous Editions

Innovations from the last edition of this text have proved to be just as valuable as when first introduced. Among the two most important are the use of *t*-butyl methyl ether in place of diethyl ether and the use of anhydrous calcium chloride pellets as a drying agent. *t*-Butyl methyl ether is one of the most common and cheapest solvents being produced because it has replaced tetraethyl lead as an antiknock additive and as an oxygenate, added to reduce pollution, in gasoline. Not only is it cheaper than diethyl ether, it does not easily form peroxides and therefore can be stored more than 30 days after the container is opened; it has a higher boiling point (55°C) than diethyl ether and is therefore somewhat less flammable; and it forms a 4% azeotrope with water, so as it is removed it also removes the last traces of water from a solution. All in all it should become the solvent of choice, replacing diethyl ether in all applications except the Grignard reaction.

We continue to be very enthusiastic about the use of anhydrous calcium chloride pellets as a drying agent, an innovation introduced in the last edition. These pellets do not fracture and powder and so are ideally suited to drying microscale quantities of solutions where the solvent can simply be drawn off with a Pasteur pipette, making conventional filtration unnecessary.

The unique chapter on computational chemistry and molecular mechanics is an introduction to a tool that, like spectroscopy, can give new insight into the structure of organic molecules. Procedures for the application of this tool are given to no fewer than 20 experiments.

Our students are very enthusiastic about the synthesis of a fluorescer using the Wittig reaction and the synthesis of a Cyalume in order to make their own light sticks. Transfer hydrogenation using cyclohexene as the source of hydrogen is another experiment unique to *Macroscale and Microscale Organic Experiments*. Olive oil is thus hydrogenated to a fat that can be converted to soap. The use of Multifiber Fabric in the experiment on dyes and dyeing, Norit decolorizing charcoal in the form of pellets, the microscale cracking of dicyclopentadiene, and the multistep syntheses starting with the benzoin condensation of benzaldehyde, benzyne experiments, and tetraphenylcyclopentadienone are among the innovative experiments that have appeared in previous editions of this text, which dates to 1935 when Louis Fieser was the author. Remember, you saw it first in Williamson.

The "Biomimetic Synthesis of Pseudopellitierene" following the classic work of Robert Robinson is a relatively new experiment, as is the "Conversion of Camphene to Camphor."

Waste Disposal and Safety

The section at the end of every experiment entitled "Cleaning Up" has been written with the intent of focusing students' attention on not just the desired product from a reaction but also on all of the other substances produced in a typical organic reaction.

As one of the coauthors of the first edition of *Prudent Practices for the Disposal of Chemicals from Laboratories*, I have continued to follow closely the rapidly evolving regulatory climate and changes in laboratory safety rules and regulations. The safety information in this text is about as current as possible, but this is a rapidly changing area of chemistry; local rules and regulations must be known and adhered to.

Supplements

Instructor's Guide

The *Instructor's Guide* is an important adjunct to this text. It contains discussions about the time needed to carry out each experiment and assessment of the relative difficulty of each experiment, problems that might be encountered, answers to end-of-chapter questions, a list of chemical and apparatus required for each experiment—both per student and per 24-student laboratory—sources of supply for unusual items, and a discussion of hardware and software needed for running calculational chemistry and molecular mechanics experiments.

Desk copies of this *Instructor's Guide* are available by contacting your local Houghton Mifflin sales representative or by contacting:

Houghton Mifflin Co.
222 Berkeley St.
Boston, MA 02116-3764
(800) 733-1717

Web Site

Please visit our Web sites at *http://www.mtholyoke.edu/courses/kwilliam/microscale.shtml* and *www.hmco.com* for resource information intended to support users of this lab manual.

The Chemistry Tutor Version 2.0 CD-ROM

This CD, authored by Adam Drury of the University of Liverpool, is an interactive introduction to topics in organic chemistry, with a review of topics in general chemistry. It includes a tutorial with text, animations, and interactive problems. It contains 18 content modules of computer-assisted learning material plus four tools, including a periodic table database and chemical calculator.

Acknowledgments

I would like to acknowledge the help of many classes of Chemistry 302 at Mount Holyoke College in developing and refining the experiments in this text, as well as the contributions sent to me by students and faculty at many other institutions. I am also indebted to Rick Danheiser and Scott Virgil at MIT, where I taught in 1996 and 1997.

I also wish to express my thanks to my editor at Houghton Mifflin, Marianne Stepanian, as well as the reviewers of this text, Edward Alexander (San Diego Mesa College), Roger Murray (University of Delaware), George Thyvelikakath (Oral Roberts University), Grace B. Borowitz (Ramapo College of New Jersey), Janet E. Nelson (Middlebury College), Linda A. Jacob (Yale University), Ray Lutgring (University of Evansville), Tracy A. Oriskovich (The Pennsylvania State University), Robert D. Minard (The Pennsylvania State University), J. W. Sam Stevenson (Northeast State Technical Community College), Mark Arant (Northeast Louisiana University), William L. "Hank" Mancini (Paradise Valley Community College), Robert A. Braga (Georgia Institute of Technology), Leslie Gunatilaka (Virginia Polytechnic Institute and State University), and, in particular, Alan Shusterman (Reed) who reviewed the material on computational chemistry.

And finally I note with sadness the passing of Mary Fieser in 1997. While never a coauthor of Louis Fieser's laboratory manuals, this great chemist and author was a perceptive critic and constant inspiration.

ORGANIC EXPERIMENTS AND WASTE DISPOSAL

An unusual feature of this book is the advice at the end of each experiment on how to dispose of its chemical waste. Waste disposal thus becomes part of the experiment, which is not considered finished until the waste products are appropriately taken care of. This is a valuable addition to the book for several reasons.

Although chemical waste from laboratories is less than 0.1% of that generated in the United States, its disposal is nevertheless subject to many of the same federal, state, and local regulations as is chemical waste from industry. Accordingly, there are both strong ethical and legal reasons for proper disposal of laboratory wastes. These reasons are backed up by a financial concern, because the cost of waste disposal can become a significant part of the cost of operating a laboratory.

There is yet another reason to include instructions for waste disposal in a teaching laboratory. Students will someday be among those conducting and regulating waste disposal operations and voting on appropriations for them. Learning the principles and methods of sound waste disposal early in their careers will benefit them and society later.

The basics of waste disposal are easy to grasp. Some innocuous water-soluble wastes are flushed down the drain with a large proportion of water. Common inorganic acids and bases are neutralized and flushed down the drain. Containers are provided for several classes of solvents, for example, combustible solvents and halogenated solvents. (The containers are subsequently removed for suitable disposal by licensed waste handlers.) Some toxic substances can be oxidized or reduced to innocuous substances that can then be flushed down the drain; for example, hydrazines, mercaptans, and inorganic cyanides can be thus oxidized by sodium hypochlorite solution, widely available as household bleach. Dilute solutions of highly toxic cations are expensive to dispose of because of their bulk; precipitation of the cation by a suitable reagent followed by its separation greatly reduces its bulk and cost. These and many other procedures can be found throughout this book.

One other principle of waste control lies at the heart of this book. Microscale experimentation, by minimizing the scale of chemical operations, also minimizes the volume of waste. Chromatography procedures to separate and purify products, spectroscopy methods to identify and characterize products, and well-designed small-scale equipment enable one to conduct experiments today on a tenth to a thousandth the scale commonly in use a generation ago.

Chemists often provide great detail in their directions for preparing chemicals so that the synthesis can be repeated, but they seldom say much about how to dispose of the hazardous byproducts. Yet the proper disposal of a chemical's byproducts is as important as its proper preparation. Dr. Williamson sets a good example by providing explicit directions for such disposal.

Blaine C. McKusick

CHAPTER

1   *Introduction*

Prelab Exercise: Study the glassware diagrams and be prepared to identify the reaction tube, fractionating column, distilling head, addition port, and Hirsch funnel.

Welcome to the organic chemistry laboratory! This laboratory manual presents a unique approach for carrying out organic experiments—they can be conducted on either a microscale or a macroscale. The latter is the traditional way of teaching the principles of experimental organic chemistry and is the basis for all the experiments in this book, a book that traces its history to 1934 when Louis Fieser was its author. Most teaching institutions are equipped to carry out macroscale experiments. Instructors are familiar with these techniques and experiments, and much research in industry and academe is carried out on this scale. These experiments typically involve the use of about 10 g of *starting material,* the chief reagent used in the reaction.

For reasons primarily of safety and cost, there is a growing trend toward carrying out work in the laboratory on a microscale, a scale one-tenth to one-thousandth of that previously used. Using smaller quantities of chemicals exposes the laboratory worker to smaller amounts of toxic, flammable, explosive, carcinogenic, and teratogenic material. Microscale experiments can be carried out much more rapidly than macroscale experiments because of rapid heat transfer, rapid filtration, and rapid drying. Since the apparatus advocated by the author is inexpensive, more than one reaction can be set up at once. The cost of chemicals is, of course, greatly reduced. A principal advantage of microscale experimentation is that the quantity of waste is reduced by one-tenth to one-thousandth of that formerly produced.

To allow maximum flexibility in the conduct of organic experiments, this book presents procedures for the vast majority of the experiments on both the microscale and the macroscale. As will be seen, some of the equipment and techniques are different. A careful reading of the two procedures will indicate what changes and precautions must be employed in going from one scale to the other.

Synthesis and structure determination

Synthesis and structure determination are two major concerns of the organic chemist, and both are dealt with in this book. The rational synthesis of an organic compound, whether it involves the transformation of one functional group into another or a carbon–carbon bond-forming reaction, starts with a *reaction.*

Organic reactions usually take place in the liquid phase and are *homogeneous,* in that the reactants are all in one phase. The reactants can be solids and/or liquids dissolved in an appropriate solvent to mediate the reaction. Some reactions are *heterogeneous*—that is, one of the reactants is in the solid phase—and thus require stirring or shaking to bring the reactants in contact with one another. A few heterogeneous reactions involve the reaction of a gas, such as oxygen, carbon dioxide, or hydrogen, with material in solution. Examples of all of these are found among the experiments in this book.

An *exothermic* organic reaction evolves heat. If it is highly exothermic with a low activation energy, one reactant is added slowly to the other and heat is removed by external cooling. Most organic reactions are, however, mildly *endothermic,* which means the reaction mixture must be heated to overcome the activation barrier and to increase the rate of the reaction. A very useful rule of thumb is that *the rate of an organic reaction doubles with a 10°C rise of temperature.* The late Louis Fieser, an outstanding organic chemist and professor at Harvard University, introduced the idea of changing the traditional solvents of many reactions to high-boiling solvents in order to reduce reaction times. Throughout this book we will use solvents such as triethylene glycol, with a boiling point (bp) of 290°C, to replace ethanol (bp 78°C) and triethylene glycol dimethyl ether (bp 222°C) to replace dimethoxyethane (bp 85°C). The use of these high-boiling solvents can greatly increase the rates of many reactions.

Effect of temperature

Running an organic reaction is usually the easiest part of a synthesis. The challenge lies in isolating and purifying the product from the reaction because organic reactions seldom give quantitative yields of one pure substance.

"Working up the reaction"
Chapter 3: Crystallization
Chapter 7: Vacuum Distillation and Sublimation

In some cases the solvent and concentrations of reactants are chosen so that, after the reaction mixture has been cooled, the product will *crystallize.* It is then collected by *filtration,* and the crystals are washed with an appropriate solvent. If sufficiently pure at that point, the product is dried and collected; otherwise, it is purified by the process of recrystallization or, less commonly, by *sublimation.*

Chapter 8: Extraction of Acids and Bases

If the product of reaction does not crystallize from the reaction mixture, it is often isolated by the process of *extraction.* This involves adding a solvent to the reaction mixture that dissolves the product and is immiscible with the solvent used in the reaction. Shaking the mixture causes the product to dissolve in the extracting solvent, after which the two layers of liquid are separated and the product isolated from the extraction solvent.

Chapter 5: Distillation
Chapter 6: Steam Distillation

If the product is a liquid, it is isolated by *distillation,* usually after extraction. Occasionally the product can be isolated by the process of *steam distillation* from the reaction mixture.

Organic reactions are usually carried out by dissolving the reactants in a solvent and then heating the mixture to boiling. To keep the solvent from boiling away, the vapor is condensed to a liquid, which is allowed to run back into the boiling solvent.

On a microscale, reactions are carried out in a *reaction tube* (Fig. 1.1a). The mass of the reaction tube is so small that a milliliter of nitrobenzene (bp 210°C) will boil in 10 s and a milliliter of benzene (mp 5°C) will crystallize in the same period of time. Cooling is effected by simply shaking the tube in a small beaker of ice water and heating by immersing the reaction tube to the appropriate depth in an electrically heated sand bath. On a larger scale, heat transfer is not so fast because of the smaller ratio of surface area to volume in a round-bottomed flask.

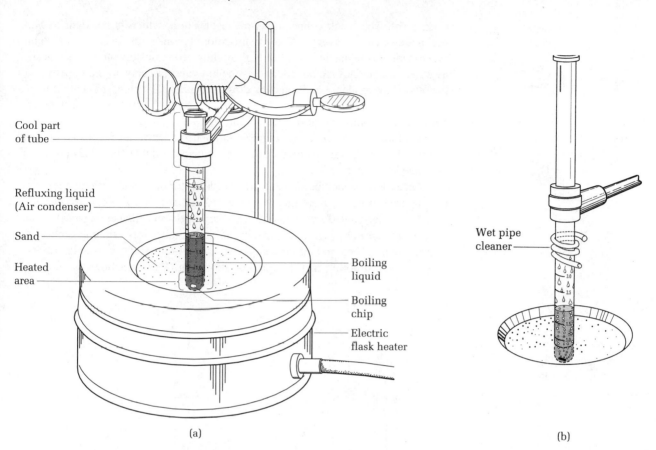

Cool part
of tube

Refluxing liquid
(Air condenser)

Sand

Heated
area

Boiling
liquid

Boiling
chip

Electric
flask heater

Wet pipe
cleaner

(a)

(b)

FIG. 1.1: (a) Reaction tube being heated on a hot sand bath in a flask heater. The area of the tube exposed to heat is small. The liquid boils and condenses on the cool upper portion of the tube, which functions as an air condenser. (b) The condensing area can be increased by adding the distilling column as an air condenser.

Cooling is again conducted using an ice bath, but heating is sometimes done on a steam bath for low-boiling liquids. Higher temperatures require electric *heating mantles* or *flask heaters*.

For microscale heating, a *sand bath* in an electric 100-mL flask heater filled with sand is a versatile heat source (Fig. 1.1a). The relatively poor heat conduction of sand results in a very large temperature difference between the top of the sand and the bottom. Thus, depending on the immersion depth in the sand, a similarly wide temperature range will be found in the reaction tube. Because the area of the tube exposed to heat is fairly small, it is difficult to transfer enough heat to the contents of the tube to cause solvents to boil away. The reaction tube is 100 mm long so that the upper part of the tube can function as an efficient *air condenser* (Fig. 1.1a), since the area of glass is large and the volume of vapor is comparatively small. The air condenser can be made even longer by attaching the empty *distilling column* to the reaction tube using the *connector with support rod*

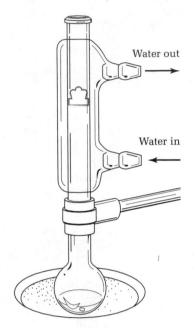

FIG. 1.2: Refluxing solvent in a 5-mL round-bottomed flask fitted with a water-cooled condenser.

(Fig. 1.1b). The black connector is made of Viton, which is resistant to high-boiling aromatic solvents. The cream-colored connector is made of Santo-prene, which is resistant to all but high-boiling aromatic solvents. Solvents such as water and ethanol are boiled, and as the hot vapor ascends to the upper part of the tube, they condense and run back down the tube. This process is called *refluxing* and is the most common method for conducting a reaction at a constant temperature, the boiling point of the solvent. For very low-boiling solvents such as diethyl ether (bp 35°C), a pipe cleaner dampened with water makes an efficient cooling device. A water-cooled condenser is also available (Fig. 1.2) but is seldom needed.

On a larger scale, the same electric flask heater or a sand bath on a hot plate can be used to heat a flask that is connected via a *standard-taper ground glass joint* to a water-cooled *reflux condenser,* where the water flows in a jacket around the central tube. The high heat capacity of water makes it possible to remove the large amount of heat put into the larger volume of refluxing vapor (Fig. 1.3).

It is worth noting that the reaction tube (Fig. 1.1a) functions as both flask

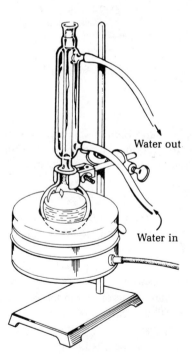

FIG. 1.3: Reflux apparatus for larger reactions. Liquid boils in the flask and condenses on the cold inner surface of the water-cooled condenser.

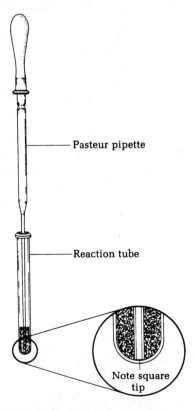

FIG. 1.4: Filtration using the Pasteur pipette and reaction tube.

and reflux condenser and is completely equivalent in function to the macroscale standard-taper flask and reflux condenser of Fig. 1.3 but costs about 1/40th as much.

The progress of a reaction can be followed by withdrawing tiny samples at intervals and analyzing them by *thin-layer chromatography.* If the product of a reaction crystallizes from the reaction mixture on cooling, it is isolated by *filtration.* On a microscale, this can be done in several ways. If the crystals are large enough and in the reaction tube, this is accomplished by inserting a *Pasteur pipette* to the bottom of the tube while expelling the air and withdrawing the solvent (Fig. 1.4). Very effective filtration occurs between the square tip of the pipette and the bottom of the tube. This method of filtration has several advantages over the alternatives. The mixture of crystals and solvent can be kept on ice during the entire process. This minimizes the solubility of the crystals in the solvent. There are no transfer losses of material because an external filtration device is not used. This technique allows several recrystallizations to be carried out in the same tube with final drying of the product under vacuum. Knowing the *tare* of the tube (the weight of the empty tube) allows the weight of the product to be determined without removing it from the tube. In this manner a compound can be synthesized, purified by crystallization, and dried without ever leaving the reaction tube. After removal of material for analysis, the compound in the tube can then be used for the next reaction. This technique is used in many of the microscale experiments in this book. When the crystals are dry, they are easily removed from the reaction tube. When they are wet, it is difficult to scrape them out. If the crystals are in more than about 2 mL of solvent, they can be isolated by filtration on the *Hirsch funnel.* The one that is in the microscale kit of apparatus is particularly easy to use because the funnel fits into the *filter flask* with no adapter, and it is equipped with a *polyethylene frit* for removal of the crystals (Fig. 1.5). The Wilfilter is especially good for collecting small quantities of crystals (Fig. 1.6).

Macroscale quantities of material are crystallized in *Erlenmeyer flasks;* the crystals are collected in porcelain or plastic *Büchner funnels* fitted with pieces of filter paper in the bottom of the funnel. A *filter adapter* (*Filtervac*) is used to form a vacuum-tight seal between the flask and funnel (Fig. 1.7).

Many solids can be purified by the process of *sublimation.* The solid is heated, and the vapor of the solid condenses on a cold surface to form crystals in an apparatus constructed from a *centrifuge tube* fitted with a rubber adapter (a *Pluro stopper*) and pushed into a *filter flask* (Fig. 1.8). Caffeine can be purified in this manner. This is primarily a microscale technique, although sublimers holding several grams of solid are available.

Mixtures of solids and, occasionally, of liquids can be separated and purified by *column chromatography.* The *chromatography column* for both microscale and macroscale work is very similar (Fig. 1.9).

Often the product of a reaction will not crystallize. It may be a liquid, it may be a mixture of compounds, or it may be too soluble in the solvent being used. In this case, an immiscible solvent is added, the two layers are shaken to effect *extraction,* and after the layers separate, one layer is removed. On a microscale, this can be done with a Pasteur pipette, and the process is repeated if necessary.

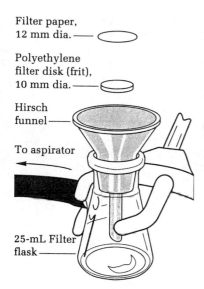

Filter paper,
12 mm dia.

Polyethylene
filter disk (frit),
10 mm dia.

Hirsch
funnel

To aspirator

25-mL Filter
flask

FIG. 1.5:   Hirsch funnel with an integral adapter, polyethylene frit, and 25-mL filter flask.

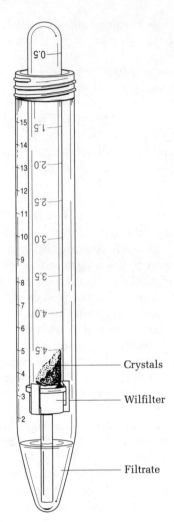

FIG. 1.6: Wilfilter filtration apparatus.

Crystals

Wilfilter

Filtrate

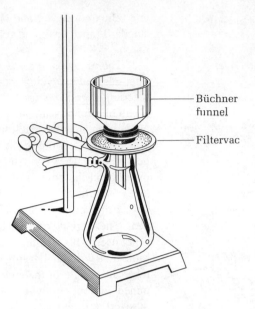

Büchner funnel

Filtervac

FIG. 1.7: Suction filter assembly.

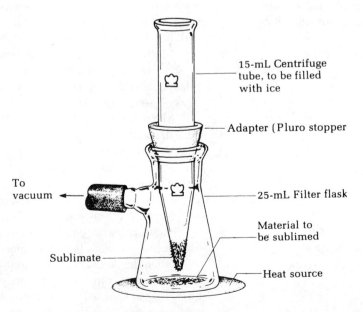

15-mL Centrifuge tube, to be filled with ice

Adapter (Pluro stopper

To vacuum

25-mL Filter flask

Material to be sublimed

Sublimate

Heat source

FIG. 1.8: Small-scale sublimation apparatus.

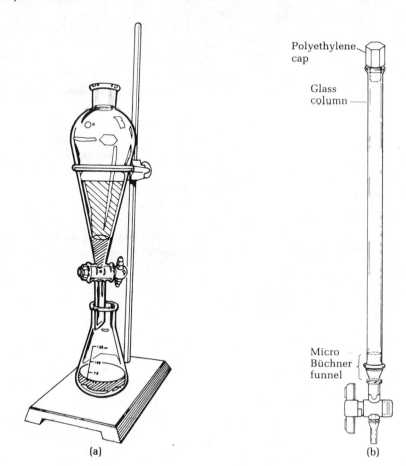

(a)   (b)

FIG. 1.10: (a) Separatory funnel with Teflon stopcock. (b) Microscale separatory funnel. Remove the polyethylene frit from the micro Büchner funnel before using.

A tall, thin column of liquid such as that produced in the reaction tube makes it easy to remove one layer selectively. This is much more difficult to do in the usual test tube because the height/diameter ratio is too small.

The chromatography column in the apparatus kit is also a *micro separatory funnel* (Fig. 1.10b). Remember to remove the frit at the column base of the micro Büchner funnel and to close the valve before adding liquid.

On a larger scale, a *separatory funnel* is used for extraction (Fig. 1.10a). The mixture can be shaken in the funnel and then the lower layer removed through the stopcock after the stopper is removed. These funnels are available in sizes from 10 to 5000 mL. The microscale version is seen in Fig. 1.10(b).

Some of the compounds to be synthesized in these experiments are liquids. On a very small scale, the best way to separate and purify a mixture of liquids is by *gas chromatography,* but this technique is limited to less than 100 mg of material on the usual gas chromatograph. For larger quantities of material, *distillation*

FIG. 1.9: Chromatography column consisting of funnel, tube, base fitted with a polyethylene frit, and Leur valve.

Chapter 5: Distillation

is used. For this purpose, small distilling flasks are used. These flasks have a large surface area to allow sufficient heat input to cause the liquid to vaporize rapidly so that it can be distilled and then condensed for collection in a *receiver*. The apparatus (Fig. 1.11) consists of a *distilling flask, distilling adapter* (which also functions as an air condenser on a microscale), a *thermometer adapter, thermometer,* and on a macroscale a water-cooled *condenser* and *distilling adapter* (Fig. 1.12). *Fractional distillation* is carried out using a small packed *fractionating column* (Fig. 1.13). The apparatus is very similar on both a microscale and macroscale. On a microscale, 2 to 4 mL of a liquid can be fractionally distilled, and 1 mL or more can be simply distilled. The usual scale in these experiments for macroscale distillation is about 25 mL. The individual components for microscale experimentation are shown in Fig. 1.14 and for macroscale work in Fig. 1.15.

Some liquids with a relatively high vapor pressure can be isolated and purified by *steam distillation,* a process in which the organic compound codistills

Chapter 6: Steam Distillation

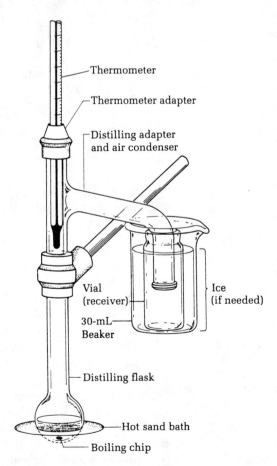

FIG. 1.11: Small-scale simple distillation apparatus.

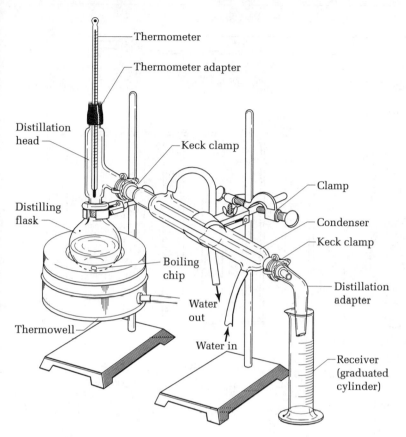

FIG. 1.12.   Apparatus for simple distillation.

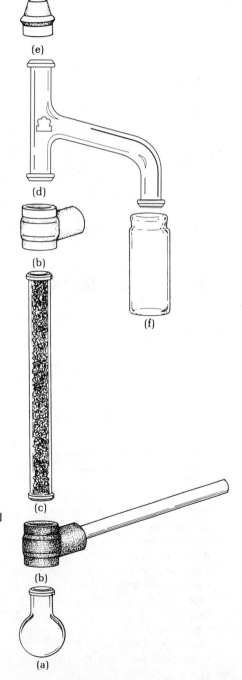

FIG. 1.13.   Microscale fractional distillation apparatus. (a) 5-mL round-bottomed flask.
(b) Elastomeric connector [Santoprene (white) or Viton (black)]. (c) Fractionating column packed with copper sponge. (d) Distilling head and air condenser. (e) Thermometer adapter (Santoprene).
(f) Receiver (1-dram vial).

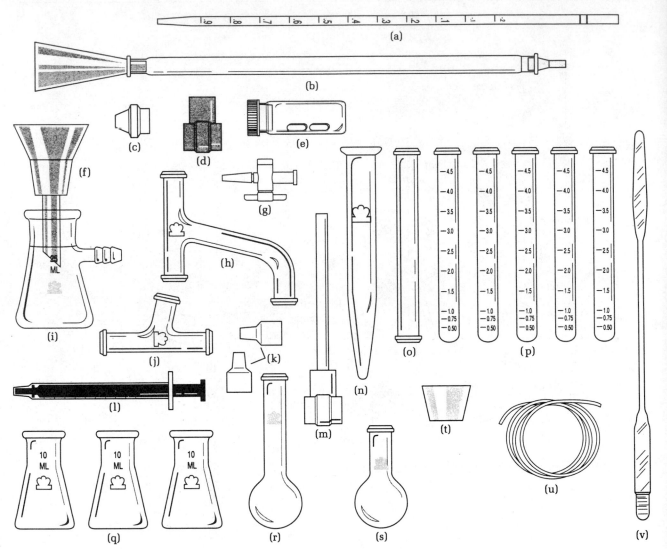

Fig. 1.14 Microscale apparatus kit.

(a) Pipette (1 mL), graduated in 1/100ths.

(b) Chromatography column (glass) with polypropylene funnel and 20-μm polyethylene frit in base, which doubles as a micro Büchner funnel. The column, base, and stopcock are also used as a separatory funnel.

(c) Thermometer adapter (Santoprene).

(d) Connector only (Viton).

(e) Magnetic stirring bars (4 × 12 mm) in distillation receiver vial.

(f) Hirsch funnel (polypropylene) with 20-μm fritted polyethylene disk.

(g) Stopcock for chromatography column and separatory funnel.

(h) Claisen adapter/distillation head with air condenser.

(i) Filter flask, 25 mL.

(j) Distillation head, 105° connecting adapter.

(k) Rubber septa/sleeve stoppers, 8 mm.

(l) Syringe (polypropylene).

(m) Connector (Santoprene) with support rod.

(n) Centrifuge tube (15 mL)/ sublimation receiver, with cap.

(o) Distillation column/air condenser.

(p) Reaction tube, calibrated, 10 × 100 mm.

(q) Erlenmeyer flasks, 10 mL.

(r) Long-necked flask, 5 mL.

(s) Short-necked flask, 5 mL.

(t) Filter adapter (Santoprene) for sublimation apparatus.

(u) Tubing (polyethylene), 1/16-in. diameter.

(v) Spatula (stainless steel) with scoop end.

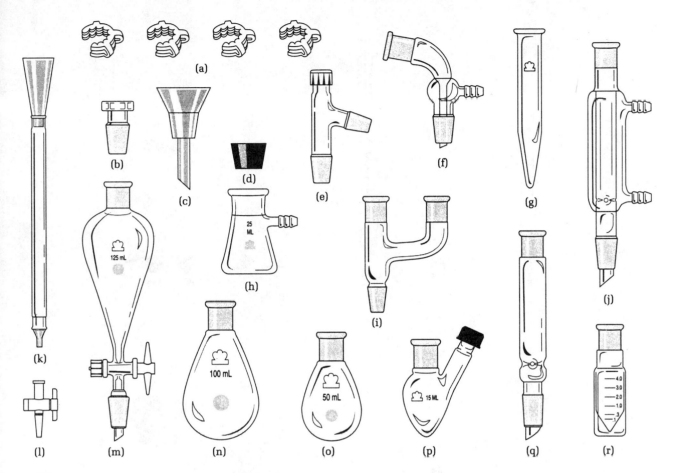

FIG. 1.15 Macroscale apparatus kit with 14/20 standard-taper ground-glass joints.

(a) Polyacetal Keck clamps, size 14.
(b) Hex-head glass stopper, 14/20 standard taper.
(c) Hirsch funnel (polypropylene) with 20-μm fritted polyethylene disk.
(d) Filter adapter (Santoprene) for sublimation apparatus.
(e) Distilling head with O-ring thermometer adapter.
(f) Vacuum adapter.

(g) Centrifuge tube (15 mL)/ sublimation receiver.
(h) Filter flask, 25 mL.
(i) Claisen adapter.
(j) Water-jacketed condenser.
(k) Chromatography column (glass) with polypropylene funnel and 20-μm polyethylene frit in base, which doubles as a micro Büchner funnel.

(l) Stopcock for chromatography column.
(m) Separatory funnel, 125 mL.
(n) Pear-shaped flask, 100 mL.
(o) Pear-shaped flask, 50 mL.
(p) Conical flask (15 mL) with side arm for inlet tube.
(q) Distilling column/air condenser.
(r) Conical reaction vial (5 mL)/distillation receiver.

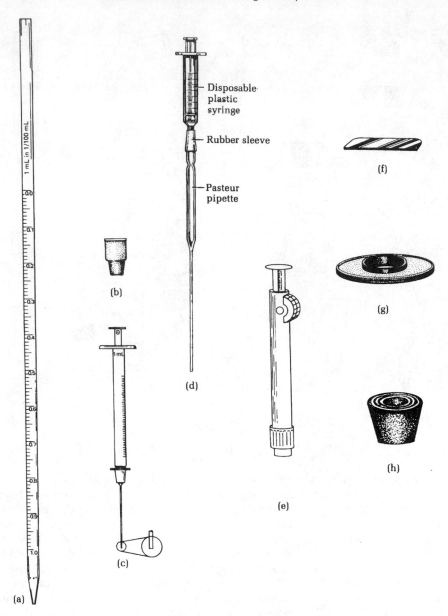

FIG. 1.16 Miscellaneous apparatus. (a) 1.0 ± 0.01 mL graduated pipette. (b) Septum. (c) 1.0-mL syringe with blunt needle. (d) Calibrated Pasteur pipette. (e) Pipette pump. (f) Glass scorer. (g) Filtervac. (h) Set of neoprene adapters.

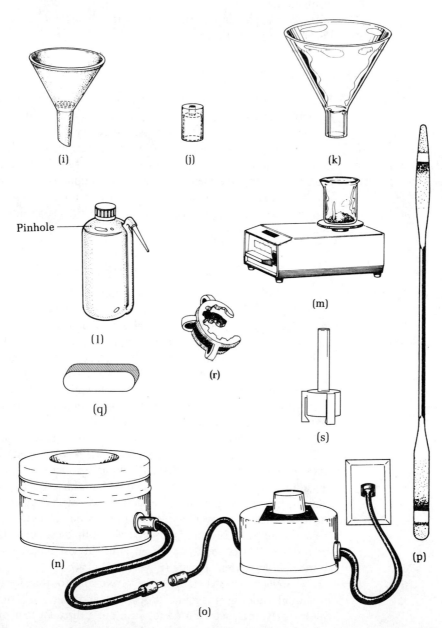

Pinhole

(i)   (j)   (k)   (l)   (m)   (q)   (r)   (s)   (n)   (o)   (p)

(i) Hirsch funnel with perforated plate in place. (j) Thermometer adapter. (k) Powder funnel. (l) Polyethylene wash bottle. (m) Single-pan electronic balance with automatic zeroing and digital readout, 100 g $\pm$ 0.001 g capacity. (n) Electric flask heater. (o) Solid-state control for electric flask heater. (p) Stainless steel spatula. (q) Stirring bar. (r) Keck clamp. (s) Wilfilter.

with water at a temperature below the boiling point of water. The microscale and macroscale apparatus for this process are shown in the chapter on steam distillation.

Other apparatus commonly used in the organic laboratory are shown in Fig. 1.16 on pages 12–13.

Check In

Your first duty will be to check in to your assigned desk. The identity of much of the apparatus should already be apparent from the preceding outline of the experimental processes used in the organic laboratory.

CAUTION: Notify your instructor immediately if you break a thermometer. Mercury is very toxic.

Check to see that your thermometer reads about 22–25°C (20°C = 68°F), normal room temperature. Examine the mercury column to see if the thread is broken—i.e., that the mercury column is continuous from the bulb up. Replace any flasks that have star-shaped cracks. Remember that apparatus with graduations and porcelain apparatus are expensive. Erlenmeyer flasks, beakers, and test tubes are, by comparison, fairly cheap.

Washing and Drying Laboratory Equipment

Clean apparatus immediately.

Considerable time can be saved by cleaning each piece of equipment soon after use, for you will know at that point what contaminant is present and be able to select the proper method for removal. A residue is easier to remove before it has dried and hardened. A small amount of organic residue usually can be dissolved with a few milliliters of an appropriate organic solvent. Acetone (bp 56.1°C) has great solvent power and is often effective, but it is extremely flammable and somewhat expensive. Because it is miscible with water and vaporizes readily, it is easy to remove from the vessel. Cleaning after an operation often can be carried out while another experiment is in process.

More experiments are ruined by damp or wet apparatus than for any other reason. Water is your enemy in the organic laboratory.

A *polyethylene bottle* [Fig. 1.16(l)] is a convenient wash bottle for acetone. The name, symbol, or formula of a solvent can be written on a bottle with a Magic Marker or wax pencil. For crystallizations, extractions, and quick cleaning of apparatus, it is convenient to have a bottle for each frequently used solvent—95% ethanol, ligroin, dichloromethane, and *t*-butyl methyl ether. A pinhole opposite the spout, which is covered with the finger when in use, will prevent the spout from dribbling the solvent. For microscale work, these solvents are best dispensed from 25- or 50-mL bottles with an attached test tube containing a graduated (1-mL) polypropylene pipette (Fig. 1.17).

Both ethanol and acetone are very flammable.

Pasteur pipettes (Fig. 1.18) are very useful for transferring small quantities of liquid, adding reagents dropwise, and carrying out crystallizations. Surprisingly, the acetone used to wash out a dirty Pasteur pipette usually costs more than the pipette itself. Discard used Pasteur pipettes in the special container for waste glass.

Sometimes a flask will not be clean after a washing with detergent and acetone. At that point try an abrasive household cleaner.

To dry a piece of apparatus rapidly, rinse with a few millimeters of acetone and invert over a beaker to drain. **Do not use compressed air,** which contains droplets of oil, water, and particles of rust. Instead draw a slow stream of air through the apparatus using the suction of your water aspirator.

Insertion of a glass tube into a rubber connector or adapter or hose is easy if the glass is lubricated with a very small drop of glycerol. Grasp the tube very close to the end to be inserted; if it is grasped at a distance, especially at the bend, the pressure applied for insertion may break the tube and result in a serious cut.

If a glass tube or thermometer should become stuck to a rubber connector, it can be removed by painting on glycerol and forcing the pointed tip of an 18-cm spatula between the rubber and glass. Another method is to select a cork borer that fits snugly over the glass tube, moisten it with glycerol, and slowly work it through the connector. When the stuck object is valuable, such as a thermometer, the best policy is to cut the rubber with a sharp knife.

Heat Source

A 10°C rise in temperature will approximately double the rate of an organic reaction. The processes of distillation, sublimation, and crystallization all require heat, which is most conveniently and safely applied from an electrically heated sand bath. A flask heater (see Fig. 1.16n, o) is two-thirds filled with sand, the temperature of which depends on the setting of the controller. This heater is small in diameter, giving good access to small apparatus, and the air above the heater is not hot. It is possible to hold a reaction tube containing refluxing solvents in the fingers of the hand. Because sand is a fairly poor conductor of heat, there is a very large variation in temperature in the sand bath depending on its depth. The temperature of a flask can be regulated by piling up or removing sand from near the flask with a spatula. The heater is easily capable of producing temperatures in excess of 300°C; therefore, never leave the controller at its maximum setting. Ordinarily, it is set at 20% of maximum. Because the flask heater can provide high temperatures, microscale equipment need not include a Bunsen burner, but on a macroscale, this has been a common way to heat reaction mixtures in the teaching laboratory. If at all possible, for safety reasons, use an electric flask heater. When the solvent boils below 90°C, the most common method for heating macroscale flasks is the *steam bath.*

A Petri dish containing sand and heated on a hot plate is not recommended for microscale experiments. It is too easy to burn oneself on the hot plate; too much heat wells up from the sand, so air condensers do not function well; the glass dishes break from thermal shock; and the ceramic coating on some hot plates chips and comes off.

Transfer of a Solid

A plastic funnel that fits the top of the reaction tube is most convenient for the transfer of solids to a reaction tube or to small Erlenmeyer flasks for microscale experiments (Fig. 1.19). It is also the top of the chromatography column (see Fig. 1.9). A special spatula with a scoop end is used to remove solid material from the

Wash acetone goes in the organic solvents waste container.

FIG. 1.17:   Recrystallization solvent bottle and dispenser.

Turn on the Thermowell about 20 minutes before you intend to use it. The sand heats slowly.

CAUTION: Never put a mercury thermometer in a sand bath! It will break, releasing highly toxic mercury vapor.

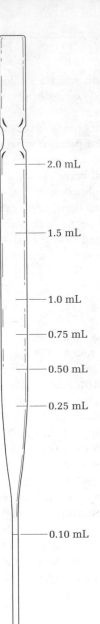

— 2.0 mL

— 1.5 mL

— 1.0 mL

— 0.75 mL

— 0.50 mL

— 0.25 mL

— 0.10 mL

Never pipette by mouth!

— 0.05 mL

FIG. 1.18 Approximate calibration of Kimble 9″ Pasteur pipette.

reaction tube (see Fig. 1.16p). On a large scale, a powder funnel is useful for adding solids to a flask (see Fig. 1.16k). A funnel also can be fashioned from a sheet of weighing paper.

Weighing and Measuring

The single-pan electronic balance (see Fig. 1.16m) capable of weighing to ±0.001 g and having a capacity of 100 g is the single most important instrument making small-scale organic experiments possible. Most of the quantitative measurements made in microscale experiments will use this balance. Weighing is a pleasure with these balances. For larger-scale experiments, a balance of such high accuracy is not necessary. Although the top-loading digital balances are the easiest to use, a triple-beam balance will work just as well. There should be one top-loading scale for every 12 students.

A container such as a reaction tube standing in a beaker or flask is placed on the pan. At the touch of a bar, the digital readout registers zero, and the desired quantity of the reagent can be added to the reaction tube as the weight is measured periodically to the nearest milligram. Even liquids are weighed when accuracy is needed. It is much easier to weigh a liquid to ±0.001 g than it is to measure it volumetrically to ±0.001 mL.

It is often convenient to weigh reagents on glossy weighing paper and then transfer the chemical to the reaction container. The success of an experiment often depends on using just the right amount of starting materials and reagents. Inexperienced workers might think that if one milliliter of a reagent will do the job, then two milliliters will do the job twice as well. Such assumptions are usually erroneous.

Liquids can be measured by either volume or weight according to the relationship

$$\text{Volume (mL)} = \frac{\text{weight (g)}}{\text{density (g/mL)}}$$

Modern Erlenmeyer flasks and beakers have approximate volume calibrations fused into the glass, but these are *very* approximate. Better graduations are found on the microscale *reaction tube*. Somewhat more accurate volumetric measurements are made in the 10-mL graduated cylinders. For volumes less than 4 mL, use a graduated pipette. **Never** apply suction to a pipette by mouth. The pipette can be fitted with a small rubber bulb. A Pasteur pipette can be converted into a calibrated pipette with the addition of a plastic syringe body (see Fig. 1.16d). Also see the calibration marks for a 9-in. Pasteur pipette in Fig. 1.18. You will find among your equipment a 1-mL pipette, calibrated in hundredths of a milliliter (see Fig. 1.16a). Determine whether it is designed to *deliver* 1 mL or *contain* 1 mL between the top and bottom calibration marks. For our purposes, the latter is the better pipette.

Because the viscosity, surface tension, vapor pressure, and wetting characteristics of organic liquids are different from those of water, the so-called automatic pipette (designed for aqueous solutions) gives poor accuracy in measuring organic liquids. *Syringes* (see Fig. 1.16c) and *pipette pumps* (Fig. 1.20), on the other hand, are quite useful, and frequent use will be made of them. Do not use a syringe that is equipped with a metal needle to measure corrosive reagents. Several reactions that require especially dry or oxygen-free atmospheres will be run in sealed systems. Reagents can be added to the system via syringe through a rubber *septum* (see Fig. 1.16b).

Careful measurements of weights and volumes take more time than less accurate measurements. Think carefully about which measurements need to be made with accuracy and which do not.

Tares

The tare of a container is its weight when empty. Throughout this laboratory course it will be necessary to know the tares of containers so that the weights of the compounds within can be calculated. If identifying marks can be placed on the containers (e.g., with a diamond stylus) you may want to record tares for frequently used containers in your laboratory notebook.

Tare = weight of empty container

To be strictly correct we should use the word *mass* instead of *weight*, because gravitational acceleration is not constant at all places on earth. But electronic balances record weights, unlike two-pan or triple-beam balances, which record masses.

The Laboratory Notebook

A complete, accurate record is an essential part of laboratory work. Failure to keep such a record means laboratory labor lost. An adequate record includes the procedure (what was done), observations (what happened), and conclusions (what the results mean).

The laboratory notebook:
  What you did.
  How you did it.
  What you observed.
  Your conclusions.

Never record anything on scraps of paper. Use a lined, $8\frac{1}{2}$- × 11-in. paperbound notebook, and record all data in ink. Allow space at the front for a table of contents, number the pages throughout, and date each day's work. Reserve the left-hand page for calculations and numerical data, and use the right-hand page for notes. Never record **anything** on scraps of paper to be recorded later in the notebook. Do not erase, remove, or obliterate notes; simply draw a single line through incorrect entries.

The notebook should contain a statement or title for each experiment and its purpose, followed by balanced equations for all principal and side reactions, and, where relevant, mechanisms of the reactions. Consult your textbook for supplementary information on the class of compounds or type of reaction involved. Give a reference to the procedure used; do not copy verbatim the procedure in the laboratory manual. Make particular note of safety precautions and the procedures for cleaning up at the end of the experiment.

Before coming to the lab to do preparative experiments, prepare a table (in your notebook) of reagents to be used and the products expected, with their physical properties. From your table, use the molar ratios of reactants to determine the limiting reagent, and then calculate the theoretical yield (in grams) of the desired product. Begin each experiment on a new page.

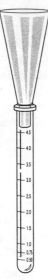

FIG. 1.19: Funnel for addition of solids and liquids to a reaction tube.

Include an outline of the procedure and the method of purification of the product in a flow sheet if this is the best way to organize the experiment (for example, an extraction; see Chapter 8). The flow sheet should list all possible products, byproducts, unused reagents, solvents, and so on that are expected to appear in the crude reaction mixture. On the flow sheet diagram indicate how each of these is removed, for example, by extraction, various washing procedures, distillation, or crystallization. With this information entered in the notebook before coming to the laboratory, you will be ready to carry out the experiments with the utmost efficiency. Plan your time before the laboratory period. Often two or three experiments can be run simultaneously.

When working in the laboratory, record everything you do and everything you observe **as it happens.** The recorded observations constitute the most important part of the laboratory record, since they form the basis for the conclusions you will draw at the end of each experiment. One way to do this is in a narrative form. Alternatively, the procedure can be written in outline form on the left-hand side of the page and the observations recorded on the right-hand side.

In some colleges and universities, you will be expected to have all the relevant information about the running of an experiment entered in your notebook *before coming to the laboratory* so that your textbook will not be needed when you are conducting experiments. In industrial laboratories, your notebook may be designed so that carbon copies of all entries are kept. These are signed and dated by your supervisor and removed from your notebook each day. Your notebook becomes a legal document in case you make a discovery worth hundreds of millions of dollars!

Record the physical properties of the product from your experiment, the yield in grams, and the percent yield. Analyze your results. When things do not turn out as expected, explain why. When your record of an experiment is complete, another chemist should be able to understand your account and determine what you did, how you did it, and what conclusions you reached. That is, from the information in your notebook, a chemist should be able to repeat your work.

Preparing a Laboratory Record

Use the following steps to prepare your laboratory record. The letters correspond to the completed laboratory records that appear at the end of this chapter. Because your laboratory notebook is so important, two examples, written in alternative forms, are presented.

A. Number each page. Allow space at the front of the notebook for a table of contents. Use a hardbound, lined notebook, and keep all notes in ink.
B. Date each entry.
C. Give a short title to the experiment, and enter it in the table of contents.
D. State the purpose of the experiment.
E. Write balanced equation(s) for the reaction(s).
F. Give a reference to the source of the experimental procedure.
G. Prepare a table of quantities and physical constants. Look up the needed data in the *Handbook of Chemistry and Physics*.

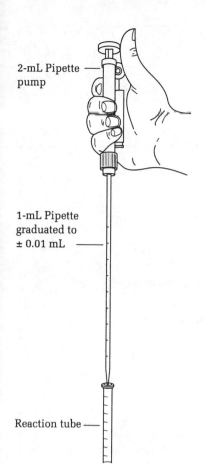

2-mL Pipette pump

1-mL Pipette graduated to ± 0.01 mL

Reaction tube

FIG. 1.20   Use of a pipette pump to measure liquids to ±0.01 mL.

H. Write equation(s) for the principal side reaction(s).

I. Write out the procedure with just enough information so that you can follow it easily. Do not simply copy the procedure from the text. Note any hazards and safety precautions. A very experienced chemist might write a procedure in a formal report as follows: "Dibenzalacetone was prepared by condensing at room temperature 1 mmol acetone with 2 mmol benzaldehyde in 1.6 mL of 95% ethanol to which was added 2 mL of aqueous 3 *M* sodium hydroxide solution. After 30 min the product was collected and crystallized from 70% ethanol to give 0.17 g (73%) of flat yellow plates of dibenzalacetone, mp 110.5–111.5°C." Note that in this formal report, no jargon (for example, EtOH for 95% ethanol, ΦCHO for benzaldehyde) is used and that the names of reagents are written out (sodium hydroxide, not NaOH). In this report the details of measuring, washing, drying, crystallizing, collecting the product, and so on are assumed to be understood by the reader.

J. Don't forget to note how to dispose of the byproducts from the experiment using the "Cleaning Up" section of the experimental procedure.

K. Record what you do as you do it. These observations are the most important part of the experiment. Note that conclusions do not appear among these observations.

L. Calculate the theoretical yield in grams. The experiment calls for exactly 2 mmol fluorobenzaldehyde and 1 mmol acetone, which will produce 1 mmol of product. The equation for the experiment also indicates that the product will be formed from exactly a 2:1 ratio of the reactants. Often, experiments are designed to have one reactant in great excess. In this experiment, a very slight excess of fluorobenzaldehyde was used inadvertently, so acetone becomes the limiting reagent.

M. Once the product is obtained, dried, and weighed, calculate the ratio of product actually isolated to the amount theoretically possible. Express this ratio as the percent yield.

N. Write out the mechanism of the reaction. If it is not given in the text of the experiment, look it up in your lecture text.

O. Draw conclusions from the observations. Write this part of the report in narrative form in complete English sentences. This part of the report can, of course, be written after leaving the laboratory.

P. Analyze TLC, IR, and NMR spectra if they are a part of the experiments. Rationalize observed versus reported melting points.

Q. Answer assigned questions from the end of the experiment.

R. This page presents an alternative method for entering the experimental procedure and observations in the notebook. Before coming to the laboratory, enter the procedure in outline form on the left side of the page. Then enter observations in brief form on the right side of the page as the experiment is carried out. Draw a single line through any words incorrectly entered. Do not erase or obliterate entries in the notebook, and never remove pages from the notebook.

Two samples of completed laboratory records follow. An alternative method for recording procedure, cleaning up, and observations is given on p. 24. The other parts of the report are the same.

(A)   p. 35
(B)   Sept. 23, 2001

(C)   **DIFLUORODIBENZALACETONE**

(D)   *Purpose:* To observe and carry out an aldol condensation reaction.

(E)   2 [4-Fluorobenzaldehyde] + CH$_3$CCH$_3$ $\xrightarrow{\text{3 N NaOH}}$ F—CH=CH—C—CH=CH—F

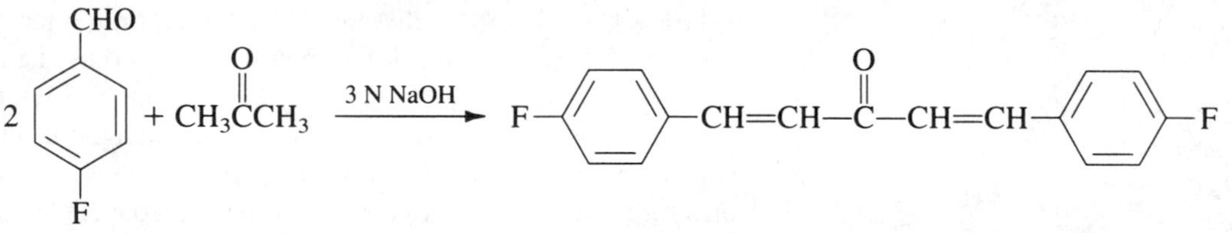

4-Fluorobenzaldehyde Acetone   **4,4′-Difluorodibenzalacetone**

(F)   *Reference:* Williamson, *Macroscale and Microscale Organic Experiments,* p. 792.

Table of Quantities and Physical Constants   (G)

Substance	Mol wt	G/mol Used or Produced	Mol Needed (from eq.)	Density	mp	bp	Solubility
4-Fluorobenzaldehyde	124.11	0.248/0.002	0.002	1.157		181°C	Slightly soluble in H$_2$O; soluble in ethanol, diethyl ether, acetone
Acetone	58.08	0.058/0.001	0.001	0.790		56°C	Soluble in H$_2$O, ethanol, diethyl ether, acetone
4,4′-Difluoro-dibenzalacetone	270.34	0.273/0.001			167°C		Insoluble in H$_2$O; slightly soluble in ethanol, diethyl ether, acetone
4-Fluoro-benzalacetone	164.19					54°C	Insoluble in H$_2$O; slightly soluble in ethanol, diethyl ether, acetone
Sodium hydroxide	40.01						Soluble in H$_2$O; insoluble in ethanol, diethyl ether, acetone

(H)   *Side Reactions:* No important side reactions, but if excess acetone is used, product will be contaminated with 4-fluorobenzalacetone.

p. 36

Ⓘ   *Procedure:* To a reaction tube, add 0.248 g (0.002 mole) of benzaldehyde and then 16 g of an ethanol solution that contains 0.58 mg of acetone. Mix the solution and observe it for 1 min. Remove sample for TLC. Then add 2 mL of 3 *M* NaOH(*aq*). Cap the tube with a septum and shake it at intervals over a 30-min period. Collect product on Hirsch funnel, wash crystals thoroughly with water, press as dry as possible, determine crude weight, and save sample for mp. Recrystallize product from 70% ethanol in a 10-mL Erlenmeyer flask. Cool solution slowly, scratch or add seed crystal, cool in ice, and collect product on Hirsch. Wash crystals with a few drops of ice-cold ethanol. Run TLC on silica gel plates using hexane to elute, develop in iodine chamber. Run NMR spectrum in deuterochloroform and IR spectrum as a mull.

Ⓙ   *Cleaning Up:* Neutralize filtrate with HCl and flush solution down the drain. Put recrystallization filtrate in organic solvents container.

Ⓚ   *Observations:* Into a reaction tube, 0.250 g of 4-fluorobenzaldehyde was weighed (0.248 called for). Using the 1.00-mL graduated pipette and pipette pump, 1.61 mL of the stock ethanol solution containing 0.58 mg of acetone was added (16 mL called for). The contents of the tube were mixed by flicking the tube, and then a drop of the reaction mixture was removed and diluted with 1 mL of hexane for later TLC. The water-clear solution did not change in appearance during 1 min.

     Then about 2 mL of 3 *M* NaOH(*aq*) was added using graduations on side of reaction tube. The tube was capped with a septum and shaken. The clear solution changed to a light yellow color immediately and got slightly warm. Then after about 50 sec, the entire solution became opaque. Then yellow oily drops collected on sides and bottom of tube. Shaken every 5 min for ½ hr. The oily drops crystallized, and more crystals formed. At the end of 30 min, the opaque solution became clear, and yellow crystals had formed. Tube filled with crystals.

     Product was collected on 12-mm filter paper on a Hirsch funnel; transfer of material to funnel was completed using the filtrate to wash out the tube. Crystals were washed with about 15 mL of water. The filtrate was slightly cloudy and yellow. It was poured into a beaker for later disposal. The crude product was pressed as dry as possible with a cork. Wt (damp) 270 mg. Sample saved for mp.

     Recrystallized from about 50 mL of 70% ethanol on sand bath. Almost forgot to add boiling stick! Clear, yellow solution cooled slowly by wrapping tube in cotton. A seed crystal from Julie added to start crystallization. After about 25 min, flask was placed in ice bath for 15 min, then product was collected on Hirsch funnel (filter paper), washed with a few drops of ice-cold solvent, pressed dry and then spread out on paper to dry. Wt 0.19 mg. Very nice flat, yellow crystals, mp 179.5–180.5°C. The crude material before recrystallization had mp 177.5–179°C.

     TLC on silica gel using hexane to elute gave only one spot, R_f 0.23, for the starting material and only one spot for the product, R_f 0.66. See attached plate.

The NMR spectrum that was run on the Bruker in deuterochloroform containing TMS showed a complex group of peaks centered at about 7 ppm with an integrated value of 17.19 and a sharp, clear quartet of peaks centered at 6.5 ppm with an integral of 8.62. The coupling constant of the AB quartet was 7.2 Hz. See attached spectrum.

The IR spectrum, run as a Nujol mull on the Mattson FT IR, showed intense peaks at 1590 and 1655 cm^{-1}, as well as small peaks at 3050, 3060, and 3072 cm^{-1}. See attached spectrum.

Cleaning Up: Recrystallization filtrate put in organic solvents container. First filtrate neutralized with HCl and poured down the drain.

(L) *Theoretical yield calculations:*

$$0.250 \text{ g fluorobenzaldehyde used } \frac{0.250 \text{ g}}{124.11 \text{ g/mole}} = 0.00201 \text{ mole}$$

$$0.058 \text{ g of acetone used } \frac{0.058 \text{ g}}{58.08 \text{ g/mole}} = 0.001 \text{ mole}$$

The moles needed (from equation) are 2 mmol of the benzaldehyde and 1 mmol of acetone. Therefore, acetone is the limiting reagent, and 1.00 mmol of product should result. The MW of the product is 270.34 g/mol.

$$0.001 \text{ mol} \times 270.34 \text{ g/mol}$$
$$= 0.270 \text{ g, theoretical yield of difluorodibenzalacetone}$$

(M) *Percent Yield of Product:*

$$\frac{0.19 \text{ g}}{0.270 \text{ g}} \times 100 = 70\%$$

(N) *The Mechanism of the Reaction:*

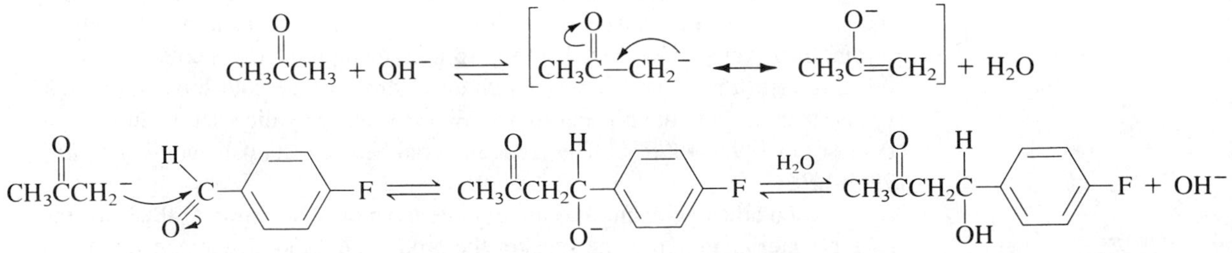

p. 38

(O) *Discussion and Conclusions:* Mixing the fluorobenzaldehyde, acetone, and
ethanol gave no apparent reaction because the solution did not change in
appearance, but less than a minute after adding the NaOH, a yellow oil
appeared and the reaction mixture got slightly warm, indicating that a reaction
was taking place. The mechanism given in the text (above) indicates that
hydroxide ion is a catalyst for the reaction, confirmed by this observation. The
reaction takes place spontaneously at room temperature. The reaction was
judged to be complete when the liquid surrounding the crystals became clear (it
was opaque) and the amount of crystals in the tube did not appear to change.
This took 30 min. The crude product was collected and washed with water.
Forgot to cool it in ice before filtering it off. The crude product weighed more
than the theoretical because it must still have been damp. However, this amount
of crude material indicates that the reaction probably went pretty well to
completion. The crude product was recrystallized from 5 mL of 70% ethanol.
This was probably too much, since it dissolved very rapidly in that amount.
This and the fact that the reaction mixture was not cooled in ice probably
accounts for the relatively low 70% yield. Probably could have obtained second
crop of crystals by concentrating filtrate. It turned very cloudy when water was
added to it.

Theoretically, it is possible for this reaction to give three different products
(*cis, cis-; cis, trans-;* and *trans, trans*-isomers). Because the mp of the crude
and recrystallized products were close to each other and rather sharp, and
because the TLC of the product gave only one spot, it is presumed that the
product is just one of these three possible isomers. The NMR spectrum shows a
sharp quartet of peaks from the vinyl protons, which also indicates only one
isomer. Since the coupling constant observed is 7.2 Hz, the protons must be *cis*
to each other. Therefore, the product is the *cis, cis*-isomer.

(P) The infrared spectrum shows two strong peaks at 1590 and 1655 cm^{-1},
indicative of an α-β-unsaturated ketone. The small peaks at 3050, 3060, and
3072 cm^{-1} are consistent with aromatic and vinyl protons. The yellow color
indicates that the molecule must have a long conjugated system.

The fact that the TLC of the starting material showed only one peak is
probably due to the evaporation of the acetone from the TLC plate. The spot
with R_f 0.23 must be from the 4-fluorobenzaldehyde.

(Q) Answers to assigned questions are written at the end of the report.

An alternative method for recording procedures, cleaning up, and observations:

p. 36

Ⓡ

Sept. 23, 2001 Procedure	Sept. 24, 2001 Observations
Weigh 0.248 g of 4-fluorobenz-aldehyde into reaction tube.	Actually used 0.250 g.
Add 1.6 mL of ethanol stock soln that contains 0.58 g of acetone.	Actually used 1.61 mL.
Mix, observe for 1 min.	Nothing seems to happen. Water clear soln.
Add 2 mL of 3 M NaOH(aq).	Clear, then faintly yellow but clear, then after about 50 sec, the entire soln very suddenly turned opaque. Got slightly warm. Light yellow color.
Cap tube with septum; shake at 5-min intervals for 30 min.	Oily drops separate on sides and bottom of tube. These crystallized; more crystals formed. Tube became filled with crystals. Liquid around crystals became clear.
Filter on Hirsch funnel.	Filtered (filter paper) on Hirsch.
Wash crystals with much water.	Washed with about 150 mL of H_2O. Transfer of crystals done using filtrate.
Press dry.	Crystals pressed dry with cork.
Weigh crude.	Crude (damp) wt 0.27 g.
Save sample for mp.	Crude mp 177.5–179°C.
Recrystallize from 70% ethanol; wash with a few drops of ice-cold solvent.	Used about 50 mL of 70% EtOH. Too much. Dissolved very rapidly.
Dry product.	Dried on paper for 30 min.
Weigh.	Wt 0.19 g.
Take mp.	Mp 179.5–180.5°C.
Run IR as mull.	Done.
Run NMR in $CDCl_3$.	Done.
Neutralize first filtrate with HCl, pour down drain.	Done.
Org. filtrate in waste organic bottle.	Done.

2

Laboratory Safety and Waste Disposal

Prelab Exercise: Read this chapter carefully. Locate the emergency eyewash station, safety shower, and fire extinguisher in your laboratory. Check your safety glasses or goggles for size and transparency. Learn which reactions must be carried out in the hood. Learn to use your laboratory fire extinguisher; learn how to summon help and how to put out a clothing fire. Learn first aid procedures for acid and alkali spills on the skin. Learn how to tell if your laboratory hood is working properly. Learn which operations under reduced pressure require special precautions. Check to see that compressed gas cylinders in your lab are firmly fastened to benches or walls. Learn the procedures for properly disposing of solid and liquid waste in your laboratory.

Small-scale organic experiments are much safer to conduct than their counterparts run on a scale 10 to 100 times larger. However, on either a microscale or a macroscale, the organic chemistry laboratory is an excellent place to learn and practice safety. Commonsense procedures practiced here also apply to other laboratories as well as the shop, kitchen, and studio.

General laboratory safety information particularly applicable to this organic chemistry laboratory course is presented in this chapter. It is not comprehensive. Throughout this text you will find specific cautions and safety information presented as margin notes printed in red. For a brief and thorough discussion of the topics in this chapter you should read the first 35 pages of *Safety in Academic Chemistry Laboratories,* American Chemical Society, Washington, DC, 1990, except for the admonition regarding contact lenses (see below).

Important General Rules

Know the safety rules of your particular laboratory. Know the locations of emergency eyewashes and safety showers. Never eat, drink, or smoke in the laboratory. Don't work alone. Perform no unauthorized experiments, and don't distract your fellow workers; horseplay has no place in the laboratory.

Eye protection

Eye protection is extremely important. Safety glasses of some type must be worn at all times. It has recently been determined "that contact lenses can be worn in most work environments provided the same approved eye protection is worn as required of other workers in the area" (*Chemical and Engineering News,* June 1, 1998, p. 6).

Ordinary prescription eyeglasses don't offer adequate protection. Laboratory safety glasses should be of plastic or tempered glass. If you do not have such glasses, wear goggles that afford protection from splashes and objects coming

FIG. 2.1   Safety goggles and safety glasses.

from the side as well as the front. If plastic safety glasses are permitted in your laboratory, they should have side shields (see Fig. 2.1).

Dress sensibly.

Dress sensibly in the laboratory. Wear shoes, not sandals or cloth-top sneakers. Confine long hair and loose clothes. Don't wear shorts. Don't use mouth suction to fill a pipette, and wash your hands before leaving the laboratory. Don't use a solvent to remove chemicals from skin. This will only hasten the absorption of the chemical through the skin.

Working with Flammable Substances

Relative flammability of organic solvents

Flammable substances are the most common hazard of the organic laboratory; two factors can make this laboratory much safer than its predecessor: making the scale of the experiments as small as possible and not using burners. Diethyl ether (bp 35°C), the most flammable substance you will usually work with in this course, has an ignition temperature of 160°C, which means that a hot plate at that temperature will cause it to burn. For comparison, *n*-hexane (bp 69°C), a constituent of gasoline, has an ignition temperature of 225°C. The flash points of these organic liquids—that is, the temperatures at which they will catch fire if exposed to a flame or spark—are below −20°C. These are very flammable liquids; however, if you are careful, they are not difficult to work with. Except for water, almost all of the liquids you will use in the laboratory will be flammable.

FIG. 2.2   Solvent safety can.

Bulk solvents should be stored in and dispensed from *safety cans* (see Fig. 2.2). These and other liquids will burn in the presence of the proper amount of their flammable vapors, oxygen, and a source of ignition (most commonly a flame or spark). It is usually difficult to remove oxygen from a fire, although it is possible to put out a fire in a beaker or a flask by simply covering the vessel with a flat object, thus cutting off the supply of air. Your lab notebook might do in an emergency. The best solution is to pay close attention to sources of ignition— open flames, sparks, and hot surfaces. Remember the vapors of flammable liquids are **always** heavier than air and thus will travel along bench tops and down drain troughs and will remain in sinks. For this reason all flames within the vicinity of a flammable liquid must be extinguished. Adequate ventilation is one of the best ways to prevent flammable vapors from accumulating. Work in an exhaust hood when manipulating large quantities of flammable liquids.

Flammable vapors travel along bench tops.

Should a person's clothing catch fire with a safety shower close at hand, shove the person under it. Otherwise, shove the person down and roll him or her over to extinguish the flames. It is extremely important to prevent the victim from running or standing because the greatest harm comes from breathing the hot vapors that rise past the mouth. The safety shower might then be used to extinguish glowing cloth that is no longer aflame. A so-called fire blanket should not

FIG. 2.3 Carbon dioxide fire extinguisher.

be used—it tends to funnel flames past the victim's mouth, and clothing continues to char beneath it. However, it is useful for retaining warmth to ward off shock after the flames are out.

An organic chemistry laboratory should be equipped with a carbon dioxide or dry chemical (monoammonium phosphate) *fire extinguisher* (see Fig. 2.3). To use this type of extinguisher, lift it from its support, pull the ring to break the seal, raise the horn, aim it at the base of the fire, and squeeze the handle. Do not hold onto the horn because it will become extremely cold. Do not replace the extinguisher; report the incident so the extinguisher can be refilled.

When disposing of certain chemicals, be alert for the possibility of *spontaneous combustion*. This may occur in oily rags; organic materials exposed to strong oxidizing agents such as nitric acid, permanganate ion, and peroxides; alkali metals such as sodium; or very finely divided metals such as zinc dust and platinum catalysts. Fires sometimes start when these chemicals are left in contact with filter paper.

Working with Hazardous Chemicals

If you do not know the properties of a chemical you will be working with, it is wise to regard the chemical as hazardous. The *flammability* of organic substances poses the most serious hazard in the organic laboratory. There is the possibility that storage containers in the laboratory may contribute to a fire. Large quantities of organic solvents should not be stored in glass bottles. Use safety cans. Do not store chemicals on the floor.

Flammable vapors plus air in a confined space are explosive.

A flammable liquid can often be vaporized to form, with air, a mixture that is *explosive* in a confined space. The beginning chemist is sometimes surprised to learn that diethyl ether is more likely to cause a laboratory fire or explosion than a worker's accidental anesthesia. The chances of being confined in a laboratory with a high enough concentration of ether to cause loss of consciousness are extremely small, but a spark in such a room would probably eradicate the building.

The probability of forming an explosive mixture of volatile organic liquids with air is much greater than that of producing an explosive solid or liquid. The chief functional groups that render compounds explosive are the *peroxide, acetylide, azide, diazonium, nitroso, nitro,* and *ozonide* groups (see Fig. 2.4). Not all

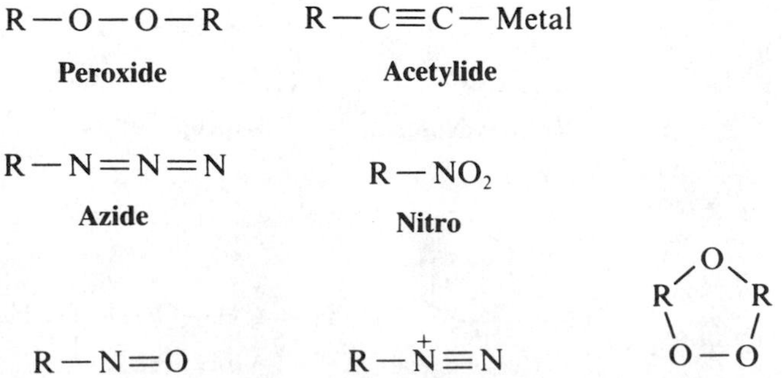

$$R-O-O-R \qquad R-C\equiv C-\text{Metal}$$
Peroxide \qquad\qquad **Acetylide**

$$R-N=N=N \qquad R-NO_2$$
Azide \qquad\qquad\qquad **Nitro**

FIG. 2.4 Functional groups that can be explosive in some compounds.

$$R-N=O \qquad R-\overset{+}{N}\equiv N \qquad \begin{array}{c} O \\ R \diagup \diagdown R \\ O-O \end{array}$$

members of these groups are equally sensitive to shock or heat. You would find it difficult to detonate trinitrotoluene (TNT) in the laboratory, but nitroglycerine is treacherously explosive. Peroxides present special problems that are dealt with below.

Safety glasses must be worn at all times.

You will need to contend with the corrosiveness of many of the reagents you will handle. The danger here is principally to the eyes. Proper eye protection is *mandatory,* and even small-scale experiments can be hazardous to the eyes. It takes only a single drop of a corrosive reagent to do lasting damage. Handling concentrated acids and alkalis, dehydrating agents, and oxidizing agents calls for commonsense care to avoid spills and splashes and to avoid breathing the often corrosive vapors.

Certain organic chemicals present problems with acute toxicity from short-duration exposure and chronic toxicity from long-term or repeated exposure. Exposure can come about through ingestion, contact with the skin, or, most commonly, inhalation. Currently, great attention is being focused on chemicals that are teratogens (chemicals that often have no effect on a pregnant woman but cause abnormalities in a fetus), mutagens (chemicals causing changes in the structure of the DNA, which can lead to mutations in offspring), and carcinogens (cancer-causing chemicals). Small-scale experiments reduce these hazards greatly but do not eliminate them.

Peroxides

Ethers form explosive peroxides.

Certain functional groups can make an organic molecule become sensitive to heat and shock, such that it will explode. Chemists work with these functional groups only when there are no good alternatives. One of these functional groups, the *peroxide* group, is particularly insidious because it can form spontaneously when oxygen and light are present (see Fig. 2.5). Ethers, especially *cyclic ethers* and those made from primary or secondary alcohols (such as tetrahydrofuran, diethyl ether, and diisopropyl ether), form peroxides. Other compounds that form peroxides are *aldehydes, alkenes* that have allylic hydrogen atoms (such as cyclohexene), compounds having benzylic hydrogens on a tertiary carbon atom (such as

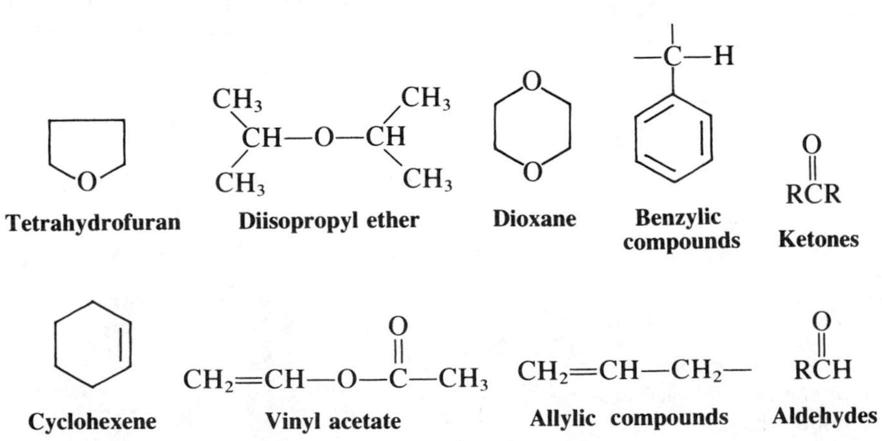

FIG. 2.5 Some compounds that form peroxides.

Tetrahydrofuran   **Diisopropyl ether**   **Dioxane**   **Benzylic compounds**   **Ketones**

Cyclohexene   **Vinyl acetate**   **Allylic compounds**   **Aldehydes**

isopropyl benzene), and vinyl compounds (such as vinyl acetate). Peroxides are low-power explosives but are extremely sensitive to shock, sparks, light, heat, friction, and impact. The biggest danger from peroxide impurities comes when the peroxide-forming compound is distilled. The peroxide has a higher boiling point than the parent compound and remains in the distilling flask as a residue that can become overheated and explode. This is one reason why it is very poor practice to distill anything to dryness.

Don't distill to dryness.

Detection of Peroxides	Removal of Peroxides
To a solution of 0.01 g of sodium iodide in 0.1 mL of glacial acetic acid, add 0.1 mL of the liquid suspected of containing a peroxide. If the mixture turns brown, a high concentration of peroxide is present; if it turns yellow, a low concentration of peroxide is present.	Pouring the solvent through a column of activated alumina will simultaneously remove peroxides and dry the solvent. Do not allow the column to dry out while in use. When the alumina column is no longer effective, wash the column with 5% aqueous ferrous sulfate, and discard it as nonhazardous waste.

Problems with peroxide formation are especially critical for ethers. Ethers form peroxides readily, and, because they are frequently used as solvents, they are often used in quantity and then removed to leave reaction products. Cans of diethyl ether should be dated when opened and if not used within one month should be treated for peroxides or disposed of.

tert-Butyl methyl ether with a primary carbon on one side of the oxygen and a tertiary carbon on the other does not form peroxides easily. It should be used in place of diethyl ether for extraction. See sidebar in Chapter 8.

You may have occasion to use *30% hydrogen peroxide*. This material causes severe burns if it contacts the skin, and it decomposes violently if contaminated with metals or their salts. Be particularly careful not to contaminate the reagent bottle.

Working with Corrosive Substances

Handle strong acids, alkalis, dehydrating agents, and oxidizing agents carefully so as to avoid contact with the skin and eyes and to avoid breathing the corrosive vapors that attack the respiratory tract. All strong concentrated acids attack the skin and eyes. *Concentrated sulfuric acid* is both a dehydrating agent and a strong acid and will cause very severe burns. *Nitric acid* and *chromic acid* (used in cleaning solutions) also cause bad burns. *Hydrofluoric acid* is especially harmful, causing deep, painful, and slow-healing wounds. It should be used only after thorough instruction.

Sodium, potassium, and *ammonium hydroxides* are common bases you will encounter. The first two are extremely damaging to the eye, and ammonium

Add H$_2$SO$_4$, P$_2$O$_5$, CaO, and NaOH to water, not the reverse.

hydroxide is a severe bronchial irritant. Like sulfuric acid, *sodium hydroxide, phosphorous pentoxide,* and *calcium oxide* are powerful dehydrating agents. Their great affinity for water will cause burns to the skin. Because they release a great deal of heat when they react with water, to avoid spattering they should always be added to water rather than water being added to them. That is, the heavier substance should always be added to the lighter one so that rapid mixing results as a consequence of the law of gravity.

You will receive special instruction when it comes time to handle *metallic sodium, lithium aluminum hydride,* and *sodium hydride,* substances that can react explosively with water.

HClO$_4$ perchloric acid

Among the strong oxidizing agents, *perchloric acid* is probably the most hazardous. It can form heavy metal and organic *perchlorates* that are *explosive,* and it can react explosively if it comes in contact with organic compounds.

Should one of these substances get on the skin or in the eyes, wash the affected area with very large quantities of water, using the safety shower and/or eyewash foundation (Fig. 2.6). Do not attempt to neutralize the reagent chemically. Remove contaminated clothing so that thorough washing can take place. Take care to wash the reagent from under the fingernails.

When you are using very small quantities of these reagents, no particular safety equipment is needed except for *safety glasses.* Take care not to let the reagents, such as sulfuric acid, run down the outside of a bottle or flask and come in contact with the fingers. Wipe up spills immediately with a very damp sponge, especially in the area around the balances. Pellets of sodium and potassium hydroxide are very hygroscopic and will dissolve in the water they pick up from the air; therefore, they should be wiped up very quickly. When working with larger quantities of these corrosive chemicals, wear protective gloves; with still larger quantities, use a face mask, gloves, and a neoprene apron. The corrosive vapors can be avoided by carrying out work in a good exhaust hood.

Wipe up spilled hydroxide pellets rapidly.

Do not use the plastic syringe with the metal needle to dispense corrosive inorganic reagents, such as concentrated acids or bases.

Working with Toxic Substances

Many chemicals have very specific toxic effects. They interfere with the body's metabolism in a known way. For example, the cyanide ion combines irreversibly with hemoglobin to form cyanometmyoglobin, which can no longer carry oxygen. Aniline acts in the same way. Carbon tetrachloride and some other halogenated compounds cause liver and kidney failure. Carcinogenic and mutagenic substances deserve special attention because of their long-term insidious effects. The ability of certain carcinogens to cause cancer is very great; for example, special precautions are needed in handling aflatoxin B$_1$. In other cases, such as with dioxane, the hazard is so low that no special precautions are needed beyond reasonable normal care in the laboratory.

Women of childbearing age should be careful when handling any substance of unknown properties. Certain substances are highly suspect teratogens and will cause abnormalities in an embryo or fetus. Among these are benzene, toluene, xylene, aniline, nitrobenzene, phenol, formaldehyde, dimethylformamide (DMF),

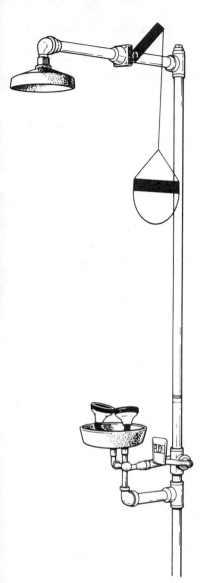

FIG. 2.6 Emergency shower and eyewash station.

dimethyl sulfoxide (DMSO), polychlorinated biphenyls (PCBs), estradiol, hydrogen sulfide, carbon disulfide, carbon monoxide, nitrites, nitrous oxide, organolead and mercury compounds, and the notorious sedative thalidomide. Some of these substances will be used in subsequent experiments. Use care. Of course, the leading known cause of embryotoxic effects is ethyl alcohol in the form of maternal alcoholism. The amount of ethanol vapor inhaled in the laboratory or absorbed through the skin is so small it is unlikely to have these morbid effects.

It is impossible to avoid handling every known or suspected toxic substance, so it is wise to know what measures should be taken. Because the eating of food or the consumption of beverages in the laboratory is strictly forbidden and because one should never taste material in the laboratory, the possibility of poisoning by mouth is remote. Be more careful than your predecessors—the hallucinogenic properties of LSD and **all** artificial sweeteners were discovered by accident. The two most important measures to be taken, then, are avoiding skin contact by wearing the *proper* type of protective gloves (see next section) and avoiding inhalation by working in a good exhaust hood.

Many of the chemicals used in this course will be unfamiliar to you. Their properties can be looked up in reference books, a very useful one being the *Aldrich Catalog Handbook of Fine Chemicals*. Note that 1,4-dichlorobenzene is listed as a "toxic irritant" and naphthalene is listed as an "irritant." Both are used as moth balls. Camphor, used in vaporizers, is classified as a "flammable solid irritant." Salicylic acid, which we will use to synthesize aspirin (Chapter 41) is listed as "moisture-sensitive toxic." Aspirin (acetylsalicylic acid) is classified as an "irritant." Caffeine, which we will isolate from tea or cola syrup (Chapter 8), is classified as "toxic." Substances not so familiar to you—1-naphthol and benzoic acid—are classified respectively as "toxic irritants" and "irritant." To put things in some perspective, nicotine is classified as "highly toxic." Pay attention to these health warnings. In laboratory quantities, common chemicals can be hazardous. Wash your hands carefully after coming in contact with laboratory chemicals. Consult Margaret-Ann Armour, *Hazardous Laboratory Chemicals Disposal Guide,* Lewis Publishers, Inc., 1996, for information on truly hazardous chemicals.

Because you have not had previous experience working with organic chemicals, most of the experiments you will carry out in this course will not involve the use of known carcinogens, although you will work routinely with flammable, corrosive, and toxic substances. A few experiments involve the use of substances that are suspected of being carcinogenic, such as hydrazine. If you pay proper attention to the rules of safety, you should find working with these substances no more hazardous than working with ammonia or nitric acid. The single, short-duration exposure you might receive from a suspected carcinogen, should an accident occur, would probably have no long-term consequences. The reason for taking the precautions noted in each experiment is to learn, from the beginning, good safety habits.

Gloves

Be aware that "protective gloves" in the organic laboratory may not offer much protection. Polyethylene and latex rubber gloves are very permeable to many organic liquids. An undetected pinhole can mean long-term contact with reagents. Disposable polyvinyl chloride (PVC) gloves offer reasonable protection from contact with aqueous solution of acids, bases, and dyes, but no one type of glove is useful as protection against all reagents. It is for this reason that no less than 25 different types of chemically resistant gloves are available from laboratory supply houses.

It is probably safer not to wear gloves and immediately wash your hands with soap and water after accidental contact with any harmful reagent or solvent than to wear inappropriate or defective gloves.

Using the Laboratory Hood

Modern practice dictates that in laboratories where workers spend most of their time working with chemicals, there should be one exhaust hood for every two people. This precaution is often not possible in the beginning organic chemistry laboratory, however. In this course you will find that for some experiments the hood must be used and for others it is adivsable; in these instances, it may be necessary to schedule experimental work around access to the hoods. However, many experiments formerly carried out in the hood can now be carried out at the desk because the concentration of vapors will be minimal when working at a microscale.

The hood offers a number of advantages for work with toxic and flammable substances. Not only does it draw off the toxic and flammable fumes, it also affords an excellent physical barrier on all four sides of a reacting system when the sash is pulled down. And should a chemical spill occur, it is nicely contained within the hood.

Keep the hood sash closed.

It is your responsibility each time you use a hood to see that it is working properly. You should find some type of indicating device that will give you this information on the hood itself. A simple propeller on a cork works well (Fig. 2.7). The hood is a backup device. Don't use it alone to dispose of chemicals by evaporation; use an aspirator tube or carry out a distillation. Toxic and flammable fumes should be trapped or condensed in some way and disposed of in the prescribed manner. Except when you are actually carrying out manipulations on the experimental apparatus, the sash should be pulled down. The water, gas, and elec-

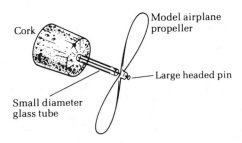

FIG. 2.7 Air flow indicator for hoods. The indicator should be permanently mounted in the hood and should be spinning whenever the hood is in operation.

trical controls should be on the outside of the hood so it is not necessary to open the hood to adjust them. The ability of the hood to remove vapors is greatly enhanced if the apparatus is kept as close to the back of the hood as possible. Everything should be at least 15 cm back from the hood sash. Chemicals should not be stored permanently in the hood but should be removed to ventilated storage areas. If the hood is cluttered with chemicals, you will not have good, smooth air flow or adequate room for experiments.

Working at Reduced Pressure

Implosion

Whenever a vessel or system is evacuated, an implosion could result from atmospheric pressure on the empty vessel. It makes little difference whether the vacuum is perfect or just 10 mm Hg; the pressure difference is almost the same (760 versus 750 mm Hg). An implosion may occur if there is a star crack in the flask or if the flask is scratched or etched. Only with heavy-walled flasks specifically designed for vacuum filtration is the use of a safety shield (Fig. 2.8) ordinarily unnecessary. The chances of implosion of the apparatus used for microscale experiments are very small.

Dewar flasks (thermos bottles) are often found in the laboratory without shielding. They should be wrapped with friction tape or covered with plastic net to prevent the glass from flying about in case of an implosion (Fig. 2.9). Similarly, vacuum desiccators should be wrapped with tape before being evacuated.

FIG. 2.8 Safety shield.

Working with Compressed Gas Cylinders

Many reactions are carried out under an inert atmosphere so that the reactants and/or products will not react with oxygen or moisture in the air. Nitrogen and argon are the inert gases most frequently used. Oxygen is widely used both as a reactant and to provide a hot flame for glassblowing and welding. It is used in the oxidative coupling of alkynes (Chapter 24). Helium is the carrier gas used in gas chromatography. Some other gases commonly used in the laboratory are ammonia, often used as a solvent; chlorine, used for chlorination reactions; acetylene, used in combination with oxygen for welding; and hydrogen used for high- and low-pressure hydrogenation reactions.

Always clamp gas cylinders.

The following rule applies to all compressed gases: Compressed gas cylinders should be firmly secured at all times. For temporary use, a clamp that attaches to the laboratory bench top and has a belt for the cylinder will suffice (Fig. 2.10). Eyebolts and chains should be used to secure cylinders in permanent installations.

A variety of outlet threads are used on gas cylinders to prevent incompatible gases from becoming mixed because of an interchange of connections. Both right- and left-handed external and internal threads are used. Left-handed nuts are notched to differentiate them from right-handed nuts. Right-handed threads are used on nonfuel and oxidizing gases, and left-handed threads are used on fuel gases, such as hydrogen.

FIG. 2.9 Dewar flask with safety net in place.

Cylinders come equipped with caps that should be left in place during storage and transportation. These caps can be removed by hand. Under these caps is

FIG. 2.10 Gas cylinder clamp.

a hand wheel valve. It can be opened by turning the wheel counterclockwise; however, because most compressed gases in full cylinders are under very high pressure (commonly up to 3000 lb/in.^{2}), a pressure regulator must be attached to the cylinder. This pressure regulator is almost always of the diaphragm type and has two gauges, one indicating the pressure in the cylinder, the other the outlet pressure (Fig. 2.11). On the outlet, low-pressure side of the regulator is located a

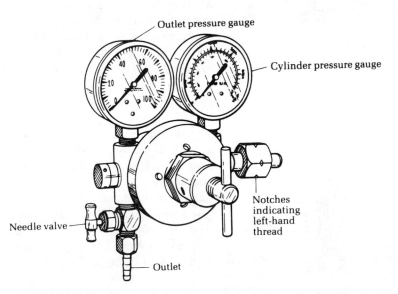

FIG. 2.11 Gas pressure regulator. Turn two-flanged diaphragm valve *clockwise* to increase outlet pressure.

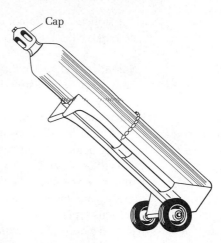

FIG. 2.12   Gas cylinder cart.

Clockwise movement of diaphragm valve handle increases pressure.

small needle valve and then the outlet connector. After connecting the regulator to the cylinder, unscrew the diaphragm valve (turn it counterclockwise) before opening the hand wheel valve on the top of the cylinder. This valve should be opened only as far as necessary. For most gas flow rates in the laboratory, this will be a very small amount. The gas flow and/or pressure is increased by turning the two-flanged diaphragm valve **clockwise**. When the apparatus is not being used, turn off the hand wheel valve (clockwise) on the top of the cylinder (Fig. 2.10). Before removing the regulator from the cylinder, reduce the flow or pressure to zero. Cylinders should never be emptied to zero pressure and left with the valve open because the residual contents will become contaminated with air. Empty cylinders should be labeled "empty," and their valves should be closed, capped, and returned to the storage area, separated from full cylinders. Gas cylinders should never be dragged or rolled from place to place but should be fastened into and moved in a cart designed for the purpose (Fig. 2.12).

Waste Disposal—Cleaning Up

Spilled solids should simply be swept up and placed in the appropriate solid waste container. This should be done promptly, because many solids are hygroscopic and become difficult if not impossible to sweep up in a short time. This is particularly true of sodium hydroxide and potassium hydroxide.

Spilled acids should be neutralized. Use sodium carbonate or, for larger spills, cement or limestone. For bases use sodium bisulfate. If the spilled material is very volatile, clear the area and let it evaporate, provided there is no chance of igniting flammable vapors. Other liquids can be taken up into such absorbents as vermiculite, diatomaceous earth, dry sand, kitty litter, or paper towels. Be particularly careful in wiping up spills with paper towels. If a strong oxidizer is present, the towels can later ignite. Bits of sodium metal will also cause paper towels to ignite. Sodium metal is best destroyed with *n*-butyl alcohol. Unless you are sure

Clean up spills rapidly.

Mercury requires special measures—see instructor.

the spilled liquid is not toxic, wear gloves when using paper towels or a sponge to remove the liquid.

Cleaning Up.

In the not-too-distant past it was common practice to wash all unwanted liquids from the organic laboratory down the drain and to place all solid waste in the trash basket. This was never a wise practice, and now, for environmental reasons, this is no longer allowed by law.

Organic reactions usually employ a solvent and often involve the use of a strong acid, a strong base, an oxidant, a reductant, or a catalyst. None of these should be washed down the drain or placed in the wastebasket. We will place the material we finally classify as waste in containers labeled for nonhazardous solid waste, organic solvents, halogenated organic solvents, and hazardous wastes of various types.

Nonhazardous waste encompasses such solids as paper, corks, sand, alumina, and sodium sulfate. These ultimately will end up in a sanitary landfill (the local dump). Any chemicals that are leached by rainwater from this landfill must not be harmful to the environment. In the *organic solvents* container are placed the solvents that are used for recrystallization and for running reactions, cleaning apparatus, etc. These solvents can contain dissolved, solid, nonhazardous organic solids. This solution will go to an incinerator where it will be burned. If the solvent is a halogenated one (e.g., dichloromethane) or contains halogenated material, it must go in the *halogenated organic solvents* container. Ultimately this will go to a special incinerator equipped with a scrubber to remove HCl from the combustion gases. The final container is for various *hazardous wastes*. Because hazardous wastes are often incompatible (oxidants with reductants, cyanides with acids, etc.), several different containers may be provided in the laboratory for these, e.g., for phosphorus compounds, heavy metal hydroxides, mercury salts, etc.

Some hazardous wastes are concentrated nitric acid, platinum catalysts, sodium hydrosulfite (a reducing agent), and Cr^{6+} (an oxidizing agent). To dispose of small quantities of a hazardous waste, e.g., concentrated sulfuric acid, the material must be carefully packed in bottles and placed in a 55-gal drum called a lab pack, to which is added an inert packing material. The lab pack is carefully documented and then hauled off by a bonded, licensed, and heavily regulated waste disposal company to a site where such waste is disposed of. Formerly, many hazardous wastes were disposed of by burial in a "secure landfill." The kinds of hazardous waste that can be thus disposed of have become extremely limited in recent years, and much of the waste undergoes various kinds of treatment at the disposal site (e.g., neutralization, incineration, reduction) to put it in a form that can be safely buried in a secure landfill or flushed to a sewer. There are relatively few places for approved disposal of hazardous waste. For example, there are none in New England, so most hazardous waste from this area is trucked to South Carolina! The charge to small generators of waste is usually based on the volume of waste. So, 1000 mL of a 2% cyanide solution would cost much more to dispose of than 20 g of solid cyanide, even though the total amount of this poisonous substance is the same. It now costs much more to dispose of most hazardous chemicals than it does to purchase them new.

Waste containers:
Nonhazardous solid waste
Organic solvents
Halogenated organic solvents
Hazardous waste (various types)

Waste disposal is very expensive.

The law: A waste is not a waste until the laboratory worker declares it a waste.

Cleaning up: reducing the volume of hazardous waste or converting hazardous waste to less hazardous or nonhazardous waste

American law states that a material is not a waste until the laboratory worker declares it a waste. So for pedagogical and practical reasons, we would like you to regard the chemical treatment of the byproducts of each reaction in this text as a part of the experiment.

In the section entitled "Cleaning Up" at the end of each experiment, the goal is to reduce the volume of hazardous waste, to convert hazardous waste to less hazardous waste, or to convert it to nonhazardous waste. The simplest example is concentrated sulfuric acid. As a byproduct from a reaction, it is obviously hazardous. But after careful dilution with water and neutralization with sodium carbonate, the sulfuric acid becomes a dilute solution of sodium sulfate, which in almost every locale can be flushed down the drain with a large excess of water. Anything flushed down the drain must be accompanied by a large excess of water. Similarly, concentrated base can be neutralized, oxidants such as Cr^{6+} can be reduced, and reductants such as hydrosulfite can be oxidized (by hypochlorite—household bleach). Dilute solutions of heavy metal ions can be precipitated as their insoluble sulfides or hydroxides. The precipitate may still be a hazardous waste, but it will have a much smaller volume.

One type of hazardous waste is unique: a harmless solid that is damp with an organic solvent. Alumina from a chromatography column and calcium chloride used to dry an ether solution are examples. Being solids, they obviously can't go in the organic solvents container, and being flammable they can't go in the nonhazardous waste container. A solution to this problem is to spread the solid out in the hood to let the solvent evaporate. You can then place the solid in the nonhazardous waste container. The saving in waste disposal costs by this operation is enormous.

Disposing of solids wet with organic solvents: alumina and anhydrous calcium chloride pellets

Our goal in "Cleaning Up" is to make you more aware of *all* aspects of an experiment. Waste disposal is now an extremely important aspect. Check to be sure the procedure you use is legal in your location. Three sources of information have been used as the basis of the procedures at the end of each experiment: the *Aldrich Catalog Handbook of Fine Chemicals,* which gives brief disposal procedures for every chemical in their catalog; *Prudent Practices in the Laboratory: Handling and Disposal of Chemicals,* National Academy Press, Washington, DC, 1995; and *Hazardous Laboratory Chemicals Disposal Guide,* M.-A. Armour, Lewis Publishers, Inc., 1996. The last title should be on the bookshelf of every laboratory. This 464-page book gives detailed information about several hundred hazardous substances, including physical properties, hazardous reactions, physiological properties and health hazards, spillage disposal, and waste disposal. Many of the treatment procedures in "Cleaning Up" are adaptations of these procedures. *Destruction of Hazardous Chemicals in the Laboratory*, G. Lunn and E. B. Sansone, Wiley, New York, NY, 1994, complements this book.

The area of waste disposal is changing rapidly. Many different laws apply—local, state, and federal. What may be permissible to wash down the drain or evaporate in the hood in one jurisdiction may be illegal in another, so before carrying out this part of the experiment check with your college or university waste disposal officer.

Questions

1. Write a balanced equation for the reaction between iodide ion, a peroxide, and hydrogen ion. What causes the orange or brown color?

2. Why does the horn of the carbon dioxide fire extinguisher become cold when the extinguisher is used?

3. Why is water not used to put out most fires in the organic laboratory?

3

Crystallization

Prelab Exercise: Write an expanded outline for the seven-step process of crystallization.

Crystallization: the most important purification method for solids, especially for small-scale experiments

On both a large and small scale, *crystallization* is the most important method for the purification of solid organic compounds. A crystalline organic substance is made up of a three-dimensional array of molecules held together primarily by van der Waals forces. These intramolecular attractions are fairly weak; most organic solids melt in the range of room temperature to 250°C.

Crystals can be grown from the molten state just as water is frozen into ice, but it is not easy to remove impurities from crystals made in this way. Thus most purifications in the laboratory involve dissolving the material to be purified in the appropriate hot solvent. As the solvent cools, the solution becomes saturated with respect to the substance, which then crystallizes. As the perfectly regular array of a crystal is formed, foreign molecules are excluded, and thus the crystal is one pure substance. Soluble impurities stay in solution because they are not concentrated enough to saturate the solution. The crystals are collected by *filtration,* the surface of the crystals is washed with cold solvent to remove the adhering impurities, and then the crystals are dried. This process is carried out on an enormous scale in the commercial purification of sugar.

In the organic laboratory, crystallization is usually the most rapid and convenient method for purifying the products of a reaction. Initially you will be told which solvent to use to crystallize a given substance and how much of it to use; later on you will judge how much solvent is needed; and finally the choice of both the solvent and its volume will be left to you. It takes both experience and knowledge to pick the correct solvent for a given purification.

The Seven Steps of Crystallization

The process of crystallization can be broken into seven discrete steps: choosing the solvent, dissolving the *solute, decolorizing* the solution, removing suspended solids, crystallizing the solute, collecting and washing the crystals, and drying the product. The process involves dissolving the impure substance in an appropriate hot solvent, removing some impurities by decolorizing and/or filtering the hot solution, allowing the substance to crystallize as the temperature of the solution falls, removing the crystallization solvent, and drying the resulting purified crystals.

How crystallization starts

Crystallization is initiated at a point of nucleation—a seed crystal, a speck of dust, or a scratch on the wall of the test tube if the solution is supersaturated

TABLE 3.1   Crystallization Solvents

Solvent	Boiling Point (°C)	Remarks
Water (H_2O)	100	The solvent of choice because it is cheap, nonflammable, and nontoxic and will dissolve a large variety of polar organic molecules. Its high boiling point and high heat of vaporization make it difficult to remove from crystals.
Acetic acid (CH_3COOH)	118	Will react with alcohols and amines. Difficult to remove from crystals. Not a common solvent for recrystallizations, although used as a solvent when carrying out oxidation reactions.
Dimethyl sulfoxide (DMSO), methyl sulfoxide (CH_3SOCH_3)	189	Also not a commonly used solvent for crystallization, but used for reactions.
Methanol (CH_3OH)	64	A very good solvent, used often for crystallization. Will dissolve molecules of higher polarity than will the other alcohols.
95% Ethanol (CH_3CH_2OH)	78	One of the most commonly used crystallization solvents. Its high boiling point makes it a better solvent for the less polar molecules than methanol. Evaporates readily from the crystals. Esters may undergo interchange of alcohol groups on recrystallization.
Acetone (CH_3COCH_3)	56	An excellent solvent, but its low boiling point means there is not much difference in solubility of a compound at its boiling point and at room temperature.
2-Butanone, methyl ethyl ketone, MEK ($CH_3COCH_2CH_3$)	80	An excellent solvent with many of the most desirable properties of a good crystallization solvent.
Ethyl acetate ($CH_3COOC_2H_5$)	78	Another excellent solvent that has about the right combination of moderately high boiling point and yet the volatility needed to remove it from crystals.
Dichloromethane, methylene chloride (CH_2Cl_2)	40	Although a common extraction solvent, dichloromethane boils too low to make it a good crystallization solvent. It is useful in a solvent pair with ligroin.

(continued)

Note: The solvents in this table are listed in decreasing order of polarity. Adjacent solvents in the list will in general be miscible with each other.

with respect to the substance being crystallized (the solute). Supersaturation will occur if a hot, saturated solution cools and crystals do not form. Large crystals, which are easy to isolate, are formed by nucleation and then slow cooling of the hot solution.

1. Choosing the Solvent and Solvent Pairs

Similia similibus solvuntur

In choosing the solvent the chemist is guided by the dictum "like dissolves like." Even the nonchemist knows that oil and water do not mix and that sugar and salt dissolve in water but not in oil. Hydrocarbon solvents such as hexane will dissolve hydrocarbons and other nonpolar compounds, and hydroxylic solvents such as water and ethanol will dissolve polar compounds. Often it is difficult to decide, simply by looking at the structure of a molecule, just how polar or nonpolar it is

TABLE 3.1   *continued*

Solvent	Boiling Point (°C)	Remarks
Diethyl ether, ether ($CH_3CH_2OCH_2CH_3$)	35	Its boiling point is too low for crystallization, although it is an extremely good solvent and fairly inert. Used in solvent pair with ligroin.
Methyl *t*-butyl ether ($CH_3OC(CH_3)_3$)	52	A relatively new solvent that is very inexpensive because of its large-scale use as an antiknock agent and oxygenate in gasoline. Does not easily form peroxides, and is less volatile than diethyl ether, but has the same solvent characteristics. See sidebar in Chapter 8.
Dioxane ($C_4H_8O_2$)	101	A very good solvent, not too difficult to remove from crystals; a mild carcinogen, forms peroxides.
Toluene ($C_6H_5CH_3$)	111	An excellent solvent that has replaced the formerly widely used benzene (a weak carcinogen) for crystallization of aryl compounds. Because of its boiling point it is not easily removed from crystals.
Pentane (C_5H_{12})	36	A widely used solvent for nonpolar substances. Not often used alone for crystallization, but good in combination with a number of other solvents as part of a solvent pair.
Hexane (C_6H_{14})	69	Frequently used to crystallize nonpolar substances. It is inert and has the correct balance between boiling point and volatility. Often used as part of a solvent pair. See ligroin.
Cyclohexane (C_6H_{12})	81	Similar in all respects to hexane. See ligroin.
Petroleum ether	30–60	A mixture of hydrocarbons of which pentane is a chief component. Used interchangeably with pentane because it is cheap. Unlike diethyl ether or *t*-butyl methyl ether, it is not an ether in the modern chemical sense.
Ligroin	60–90	A mixture of hydrocarbons with the properties of hexane and cyclohexane. A very commonly used crystallization solvent. Also sold as "hexanes."

and therefore which solvent would be best. Therefore, the solvent is often chosen by experimentation.

The ideal solvent

The best crystallization solvent (and none is ideal) will dissolve the solute when the solution is hot but not when the solution is cold; it will either not dissolve the impurities at all or it will dissolve them very well (so they won't crystallize out along with the solute); it will not react with the solute; and it will be nonflammable, nontoxic, inexpensive, and very volatile (so it can be removed from the crystals).

Miscible: capable of being mixed

Some common solvents and their properties are presented in Table 3.1 in order of decreasing polarity of the solvent. Solvents adjacent to each other in the list will dissolve in each other, i.e., they are miscible with each other, and each solvent will, in general, dissolve substances that are similar to it in chemical structure. These solvents are used both for crystallization and as solvents in which reactions are carried out.

Procedure

Picking a Solvent.

To pick a solvent for crystallization, put a few crystals of the impure solute in a small test tube or centrifuge tube, and add a very small drop of the solvent. Allow it to flow down the side of the tube and onto the crystals. If the crystals dissolve instantly at room temperature, that solvent cannot be used for crystallization because too much of the solute will remain in solution at low temperatures. If the crystals do not dissolve at room temperature, warm the tube on the hot sand bath, and observe the crystals. If they do not go into solution, add a drop more solvent. If the crystals go into solution at the boiling point of the solvent and then crystallize when the tube is cooled, you have found a good crystallization solvent. If not, remove the solvent by evaporation, and try another solvent. In this trial-and-error process it is easiest to try low-boiling solvents first, because they can be removed most easily. Occasionally no single satisfactory solvent can be found, so mixed solvents, or *solvent pairs,* are used.

Solvent Pairs. To use a mixed solvent, dissolve the crystals in the better solvent, and add the poorer solvent to the hot solution until it becomes cloudy and the solution is saturated with the solute. The two solvents must, of course, be miscible with each other. Some useful solvent pairs are given in Table 3.2.

2. Dissolving the Solute

Microscale Procedure

Once a crystallization solvent has been found, the impure crystals are placed in a reaction tube, solvent is added dropwise, the crystals are stirred with a microspatula or a small glass rod, and the tube is warmed on a steam bath or sand bath with the addition of more solvent until the crystals dissolve. A solution of this type can become *superheated,* that is, heated above its boiling point without actually boiling. When boiling does suddenly occur, it can happen with almost explosive violence. To prevent this from happening, a *wood applicator stick* can be added to the solution (Fig. 3.1). Air trapped in the wood comes out of the stick and forms the nuclei on which even boiling can occur. Porous porcelain *boiling chips* work in the same way. Never add a boiling chip or boiling stick to a hot solution, since it may be superheated and boil over.

    To remove solid impurities (insoluble byproducts of reaction, lint, dust, etc.), it is necessary to dilute the solution with excess solvent (the better solvent if a solvent mixture is being used), carry out the filtration near room temperature, and then evaporate the solvent to a point at which the hot solution is once more

Prevention of bumping

Do not use wood boiling sticks in place of boiling chips in a reaction. Use them only for crystallization.

TABLE 3.2 Solvent Pairs

Acetic acid–water	Ethyl acetate–cyclohexane
Ethanol–water	Acetone–ligroin
Acetone–water	Ethyl acetate–ligroin
Dioxane–water	*t*-Butyl methyl ether–ligroin
Acetone–ethanol	Dichloromethane–ligroin
Ethanol–*t*-butyl methyl ether	Toluene–ligroin

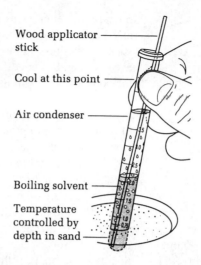

Wood applicator stick

Cool at this point

Air condenser

Boiling solvent

Temperature controlled by depth in sand

FIG. 3.1 The reaction tube being used for crystallization. The wood applicator stick ("boiling stick") promotes even boiling and is easier to remove than a boiling chip. The Thermowell sand is cool on top and hotter down deeper, so it provides a range of temperatures. The reaction tube is long and narrow; it can be held in the hand while the solvent refluxes. Do not use a boiling stick in place of a boiling chip in a reaction.

saturated with respect to the solute so that crystallization can take place in the usual way. This is described in Step 4.

Care must be exercised to use the correct amount of solvent. Observe the mixture carefully as solvent is being added. Allow sufficient time for the boiling solvent to dissolve the solute, and note the rate at which most of the material dissolves. When you believe most of the material has been dissolved, stop adding solvent. There is the possibility that your sample is contaminated with a small quantity of an insoluble impurity that never will dissolve. To hasten the solution process, crush large crystals with a stirring rod, taking care not to break the tube. On a microscale, there is a tendency to use too much solvent so that on cooling the hot solution little or no material crystallizes. This is not a hopeless situation. The remedy is to evaporate some of the solvent and repeat the cooling process. Inspect the hot solution. If it contains no undissolved impurities and is not colored from impurities, you can simply let it cool, allowing the solute to crystallize (Step 5), and then collect the crystals (Step 6). On the other hand, if the solution is colored, it must be treated with activated (decolorizing) charcoal and then filtered before crystallization (Step 3). If it contains solid impurities, it must be filtered before crystallization takes place (Step 4).

Don't use too much solvent.

 ## Macroscale Procedure

Place the substance to be crystallized in an Erlenmeyer flask (never use a beaker), add enough solvent to cover the crystals, and then heat the flask on a steam bath (if the solvent boils below 90°C) or a hot plate until the solvent boils. Stir the

FIG. 3.2 Swirling of a solution to mix contents and help dissolve material to be crystallized.

MICROSCALE AND MACROSCALE

Activated charcoal = decolorizing carbon = Norit

Pelletized Norit (also called active carbon, activated charcoal, decolorizing carbon)

mixture or, better, swirl it (Fig. 3.2) to promote dissolution. Add solvent gradually, keeping it at the boil, until all of the solute dissolves. Addition of a boiling stick or a boiling chip to the solution once most of the solid is gone will promote even boiling. It is not difficult to superheat the solution, i.e., heat it above the boiling point with no boiling taking place. Once the solution does boil it does so with explosive violence. Never add a boiling chip or boiling stick to a hot solution. A glass rod with a flattened end can sometimes be of use in crushing large particles of solute to speed up the dissolving process. Be sure no flames are nearby when working with flammable solvents.

Be careful not to add too much solvent. Note how rapidly most of the material dissolves, and then stop adding solvent when you suspect that almost all of the desired material has dissolved. It is best to err on the side of too little solvent rather than too much. Undissolved material noted at this point could be an insoluble impurity that never will dissolve. Allow the solvent to boil, and if no further material dissolves, proceed to Step 4 to remove suspended solids from the solution by filtration, or if the solution is colored, go to Step 3 to carry out the decolorization process. If the solution is clear, proceed to Step 5, Crystallizing the Solute.

3. Decolorizing the Solution; Use of Pelletized Norit

The vast majority of pure organic chemicals are colorless or a light shade of yellow. Occasionally a chemical reaction will produce high molecular weight by-products that are highly colored. The impurities can be adsorbed onto the surface of activated charcoal by simply boiling the solution with charcoal. Activated charcoal has an extremely large surface area per gram (several hundred square meters) and can bind a large number of molecules to this surface. On a commercial scale, the impurities in brown sugar are adsorbed onto charcoal in the process of refining sugar.

In the past, laboratory manuals have advocated the use of finely powdered activated charcoal for removal of colored impurities. This has two drawbacks. Because the charcoal is so finely divided it can only be separated from the solution by filtration through paper, and even then some of the finer particles pass through the filter paper. And the presence of the charcoal completely obscures the color of the solution, so that adding the correct amount of charcoal is mostly a matter of luck. If too little charcoal is added, the solution will still be colored after filtration, making repetition necessary; if too much is added, it will absorb some of the product in addition to the impurities. We have found that charcoal extruded as short cylindrical pieces measuring about 0.8×3 mm made by the Norit Company solves both of these problems. It works just as well as the finely divided powder, and it does not obscure the color of the solution. It can be added in small portions until the solution is decolorized and the size of the pieces makes it easy to remove from the solution.[1]

---

1. Available from Aldrich Chemical Co., 940 West St. Paul Ave., Milwaukee, WI 53233. Catalog number 32942-8 as Norit RO 0.8. This form of Norit is an extrudate 0.8 mm dia. It has a surface area of 1000 m^2/g, a total pore volume of 1.1 mL/g.

Add a small amount (0.1% of the solute weight is sufficient) of pelletized Norit to the colored solution, and then boil the solution for a few minutes. Be careful not to add the charcoal pieces to a superheated solution; the charcoal functions like hundreds of boiling chips and will cause the solution to boil over. Remove the Norit by filtration as described in Step 4.

4. Filtering Suspended Solids

The filtration of a hot, saturated solution to remove solid impurities or charcoal can be done in a number of ways. Processes include gravity filtration, pressure filtration, decantation, or removal of the solvent using a Pasteur pipette. Vacuum filtration is not used because the hot solvent will cool during the process and the product will crystallize in the filter.

Microscale Procedure

(A) Removal of Solution with a Pasteur Pipette.   If the solid impurities are large in size, they can be removed by filtration of the liquid through the small space between the square end of a Pasteur pipette and the bottom of a reaction tube (Fig. 3.3). Expel air from the pipette as it is being pushed to the bottom of the tube. Use a small additional quantity of solvent to rinse the tube and pipette. Anhydrous calcium chloride, a drying agent, is removed easily in this way. Removal of very fine material, such as traces of charcoal, is facilitated by filtration of the solution through a small piece of filter paper (3 mm^2) placed in the reaction tube. This process is even easier if the filter paper is the thick variety, such as that from which Soxhlet extraction thimbles are made.[2]

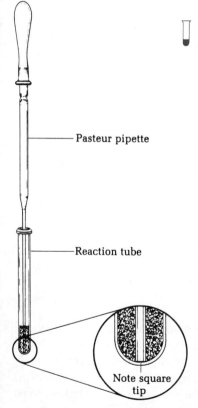

Pasteur pipette

Reaction tube

Note square tip

FIG. 3.3 Filtration using the Pasteur pipette and reaction.

(B) Filtration in a Pasteur Pipette.   To filter 0.1 to 2 mL of a solution, dilute the solution with enough solvent that the solute will not crystallize out at room temperature. Prepare a filter pipette by pushing a tiny bit of cotton into a Pasteur pipette, put the solution to be filtered into this filter pipette using another Pasteur pipette, and then force the liquid through the filter using air pressure from a pipette bulb (Fig. 3.4). Fresh solvent should be added to rinse the pipette and cotton. The filtered solution is then concentrated by evaporation. One problem encountered with this method is using too much cotton packed too tightly in the pipette so that the solution cannot be forced through it. To remove very fine impurities, such as traces of decolorizing charcoal, a 3- to 4-mm layer of Celite filter aid can be added to the top of the cotton.

(C) Removal of Fine Impurities by Centrifugation.   To remove fine solid impurities from up to 4 mL of solution, dilute the solution with enough solvent that the solute will not crystallize out at room temperature. Counterbalance the reaction tube, and centrifuge for about two min at high speed in a laboratory

2. J. L. Belletire and N. O. Mahmoodi, *J. Chem. Ed.* **66:**964, 1989.

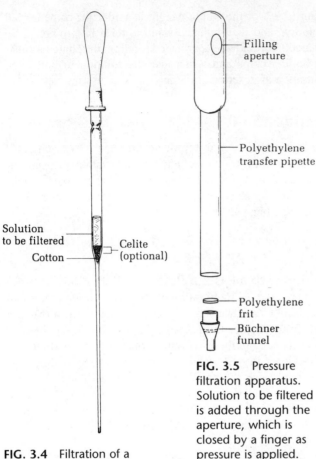

Solution
to be filtered

Celite
(optional)

Cotton

FIG. 3.4 Filtration of a
solution in a Pasteur
pipette.

Filling
aperture

Polyethylene
transfer pipette

Polyethylene
frit

Büchner
funnel

FIG. 3.5 Pressure
filtration apparatus.
Solution to be filtered
is added through the
aperture, which is
closed by a finger as
pressure is applied.

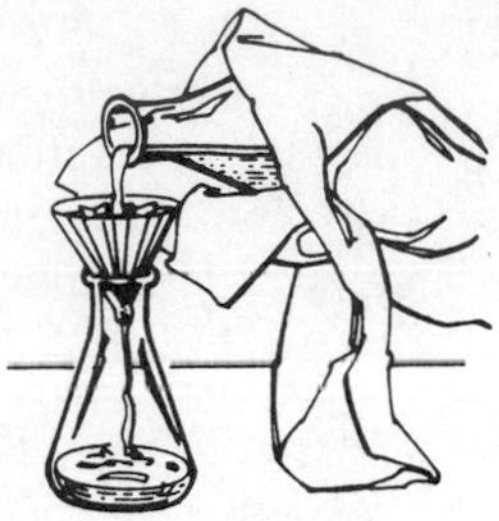

FIG. 3.6 Gravity filtration of
hot solution through fluted filter
paper.

centrifuge. The clear supernatant can be decanted (poured off) from the solid on
the bottom of the tube. Alternatively, with care, the solution can be removed with
a Pasteur pipette, leaving the solid behind.

(D) Pressure Filtration with Micro Büchner Funnel.

The technique
applicable to volumes from 0.1 to 5 mL is the use of the *micro Büchner funnel*. It
is made of polyethylene and is fitted with a porous *polyethylene frit* 6 mm in
diameter. This funnel fits in the bottom of an inexpensive disposable polyethylene pipette in which a hole is cut (Fig. 3.5). The solution to be filtered is placed
in the pipette using a Pasteur pipette. The thumb covers the hole in the plastic
pipette, and pressure is applied to filter the solution. It is good practice to place a
6-mm-diameter piece of filter paper over the frit, which can otherwise become
clogged with insoluble material.

Use filter paper on top of frit.

*Using the chromatography
column for pressure filtration*

The glass chromatography column can be used in the same way. A piece of
filter paper is placed over the frit. The solution to be filtered is placed in the chromatography column, and pressure is applied to the solution using a pipette bulb.
In both procedures, dilute the solution to be filtered so that it does not crystallize

out in the apparatus, and use a small amount of clean solvent to rinse the apparatus. The filtered solution is then concentrated by evaporation.

Macroscale Procedure

Decant: to pour off. A fast, easy separation procedure

(A) Decantation.

On a large scale, it is often possible to pour off (decant) the hot solution leaving the insoluble material behind. This is especially easy if the solid is granular like sodium sulfate. The solid remaining in the flask and the inside of the flask should be rinsed with a few milliliters of the solvent in order to recover as much of the product as possible.

(B) Gravity Filtration.

The most common method for the removal of insoluble solid material is gravity filtration through a fluted filter paper (Fig. 3.6). This is the method of choice for the removal of finely divided charcoal, dust, lint, etc. The following equipment is needed for this process: three Erlenmeyer flasks on a steam bath or hot plate—one to contain the solution to be filtered, one to contain a few milliliters of solvent and a stemless funnel, and the third to contain several milliliters of the crystallizing solvent to be used for rinsing purposes—a fluted piece of filter paper, a towel for holding the hot flask and drying out the stemless funnel, and boiling chips for all solutions.

A piece of filter paper is fluted as shown in Fig. 3.7 and is then placed in a stemless funnel. Appropriate sizes of Erlenmeyer flasks, stemless funnels, and filter paper are shown in Fig. 3.8. The funnel is stemless so that the saturated solution being filtered will not have a chance to cool and clog the stem with crystals. The filter paper should fit entirely inside the rim of the funnel; it is fluted to allow rapid filtration. Test to see that the funnel is stable in the neck of the Erlenmeyer flask. If it is not, support it with a ring attached to a ring stand. A few milliliters of solvent and a boiling chip should be placed in the flask into which the solution is to be filtered. This solvent is brought to a boil on the steam bath or hot plate along with the solution to be filtered.

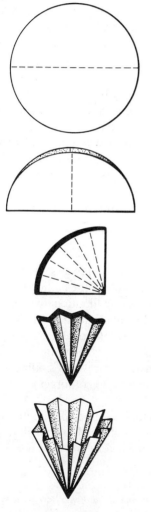

FIG. 3.7 Fluting a filter paper.

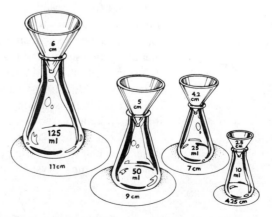

FIG. 3.8 Assemblies for gravity filtration. Stemless funnels have diameters of 2.5, 4.2, 5.0, and 6.0 cm.

The solution to be filtered should be saturated with the solute at the boiling point. Note the volume, and then add 10% more solvent. The resulting slightly dilute solution is not as likely to crystallize out in the funnel in the process of filtration. Bring the solution to be filtered to a boil, grasp the flask in a towel, and pour the solution into the filter paper in the stemless funnel (Fig. 3.6).

The funnel should be warm in order to prevent crystallization from occurring in the funnel. This can be accomplished in two ways: (1) Invert the funnel over a steam bath for a few seconds, then pick up the funnel with a towel, wipe it perfectly dry, place it on top of the Erlenmeyer flask, and add the fluted filter paper, or (2) place the stemless funnel in the neck of the Erlenmeyer flask and allow the solvent to reflux into the funnel, thereby warming it.

Pour the solution to be filtered at a steady rate into the fluted filter paper. Check to see whether crystallization is occurring in the filter. If it does, add boiling solvent (from the third Erlenmeyer flask heated on the steam bath or hot plate) until the crystals dissolve, dilute the solution being filtered, and carry on. Rinse the flask that contained the solution to be filtered with a few milliliters of boiling solvent, and rinse the fluted filter paper with this same solvent.

Be aware that the vapors of low-boiling solvents can ignite on an electric hot plate.

Because the filtrate has been diluted in order to prevent it from crystallizing during the filtration process, the excess solvent must now be removed by boiling the solution. The process can be speeded up somewhat by blowing a slow current of air into the flask in the hood or using an aspirator tube to pull vapors into the aspirator (Figs. 3.9 and 3.10). However, the fastest method is to heat the solvent in the filter flask on the sand bath while the flask is connected to the water aspirator. The vacuum is controlled with the thumb (Fig. 3.11).[3] If your thumb is not large enough, put a one-holed rubber stopper into the Hirsch funnel or the filter flask and again control the vacuum with the thumb. If the vacuum is not controlled, the solution may boil over and go out the vacuum hose.

5. Crystallizing the Solute

On both a macroscale and a microscale, the crystallization process should normally start from a solution that is saturated with the solute at the boiling point. If it has been necessary to remove impurities or charcoal by filtration, the solution has been diluted. To concentrate the solution, simply boil off the solvent under an aspirator tube as shown in Fig. 3.9 (macroscale) or blow off solvent using a gentle stream of air or, better, nitrogen in the hood as shown in Fig. 3.10 (microscale). Be sure to have a boiling chip (macroscale) or a boiling stick (microscale) in the solution during this process, and then do not forget to remove it before initiating crystallization.

A saturated solution

Once it has been ascertained that the hot solution is saturated with the compound just below the boiling point of the solvent, it is allowed to cool slowly to room temperature. Crystallization should begin immediately. If it does not, add a

---

3. D. W. Mayo, R. M. Pike, and S. M. Butcher, *Microscale Organic Laboratory* (New York, Wiley, 1986), p. 97.

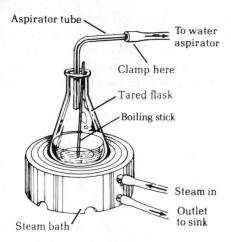

FIG. 3.9 Aspirator tube in use. A boiling stick may be necessary to promote even boiling.

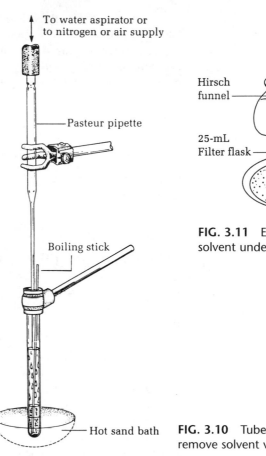

FIG. 3.10 Tube being used to remove solvent vapors.

FIG. 3.11 Evaporation of a solvent under a vacuum.

Add a seed crystal.

Slow cooling is important.

seed crystal or scratch the inside of the tube with a glass rod at the liquid–air interface. Crystallization must start on some nucleation center. A minute crystal of the desired compound saved from the crude material will suffice. If a seed crystal is not available, crystallization can be started on the rough surface of a fresh scratch on the inside of the container.

Once it is ascertained that crystallization has started, the solution must be cooled slowly without disturbing the container in order that large crystals can form. On a microscale, it is best to allow the reaction tube to cool in a beaker filled with cotton or paper towels, which act as insulation so cooling takes place slowly. Even insulated in this manner, the small reaction tube will cool to room temperature within a few minutes. Slow cooling will guarantee the formation of large crystals, which are easily separated by filtration and easily washed free of adhering impure solvent. On a small scale, it is difficult to obtain crystals that are too large and occlude impurities. Once the tube has cooled to room temperature without disturbance, it can be cooled in ice to maximize the amount of product that comes out of solution. The crystals are then separated from the *mother liquor* (the *filtrate*) by filtration.

On a macroscale, the Erlenmeyer flask is set atop a cork ring or other insulator and allowed to cool spontaneously to room temperature. If the flask is moved during crystallization, many nuclei will form and the crystals will be small and will have a large surface area. They will not be so easy to filter and wash clean of mother liquor. Once crystallization ceases at room temperature, the flask should be placed in ice to cool further. Take care to clamp the flask in the ice bath so that it does not tip over.

6. Collecting and Washing the Crystals

Once crystallization is complete, the crystals must be separated from the ice-cold mother liquor, washed with ice-cold solvent, and dried.

Microscale Procedure

(A) Filtration Using the Pasteur Pipette.
The most important filtration technique to be used in microscale organic experiments employs the Pasteur pipette (Fig. 3.12). About 70% of the crystalline products from the experiments in this text can be isolated in this way. The others will be isolated by filtration on the Hirsch funnel.

The ice-cold crystalline mixture is stirred with the Pasteur pipette, and while air is being expelled from the pipette, it is forced to the bottom of the reaction tube. The bulb is released, and the solvent is drawn into the pipette through the very small space between the square tip of the pipette and the curved bottom of the reaction tube. When all the solvent has been withdrawn, it is expelled into another reaction tube containing the crystals. It is sometimes useful to rap the tube containing the wet crystals against a hard surface to pack them so that more solvent can be removed. The tube is returned to the ice bath, and a few drops of cold solvent are added to the crystals. The mixture is stirred to wash the crystals, and the solvent is again removed. This process can be repeated as many times as necessary. Volatile solvents can be removed from the damp crystals under vacuum (Fig. 3.13). Alternatively, the last traces of solvent can be removed by centrifugation using the Wilfilter (C), (Fig. 3.15.)

(B) Filtration Using the Hirsch Funnel.
When the volume of material to be filtered is larger than about 1.5 mL, then the material is collected on the Hirsch funnel.

The Hirsch funnel in the Williamson/Kontes kit is unique. It is made of polypropylene and has an integral molded stopper that fits the 25-mL filter flask. It comes fitted with a 20-μm polyethylene fritted disk, which is not meant to be disposable, although it costs only about twice as much as an 11-cm piece of filter paper (Fig. 3.14). While products can be collected directly on this disk, it is good practice to place an 11- or 12-mm-diameter piece of no. 1 filter paper on the disk. In this way the frit will not become clogged with insoluble impurities. The disk of filter paper can be cut with a cork borer or leather punch. A piece of filter paper *must* be used on the old-style porcelain Hirsch funnels.

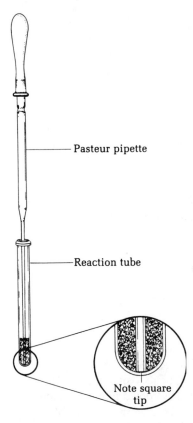

Pasteur pipette

Reaction tube

Note square tip

FIG. 3.12 Filtration using the Pasteur pipette and reaction tube.

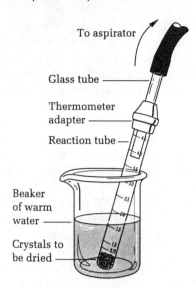

To aspirator

Glass tube ——

Thermometer
adapter ——

Reaction tube ——

Beaker
of warm
water ——

Crystals to
be dried ——

FIG. 3.13   Drying crystals under reduced pressure in a reaction tube.

Clamp the clean, dry 25-mL filter flask in an ice bath to prevent it from falling over, and place the Hirsch funnel with filter paper in the flask. Wet the filter paper with the solvent used in the crystallization, turn on the water aspirator (see below), and ascertain that the filter paper is pulled down onto the frit. Pour and scrape the crystals and mother liquor onto the Hirsch funnel, and as soon as the liquid is gone from the crystals, break the vacuum at the filter flask by removing the rubber hose. The filtrate can be used to rinse out the container that contained the crystals. Again, break the vacuum as soon as all the liquid has disappeared from the crystals; this prevents impurities from drying on the crystals. The

Break the vacuum, add a very small quantity of ice-cold wash solvent, reapply vacuum.

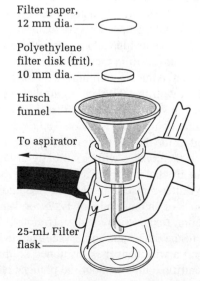

Filter paper,
12 mm dia. ——

Polyethylene
filter disk (frit),
10 mm dia. ——

Hirsch
funnel ——

To aspirator

25-mL Filter
flask ——

FIG. 3.14   Hirsch funnel used for vacuum filtration. This unique design has a removable and replaceable 20-μm polyethylene frit. No adapter is needed because there is a vacuum-tight fit to the filter flask. Always clamp the flask. Always use a piece of filter paper.

reason for cooling the filter flask is to keep the mother liquor cold so that it will not dissolve the crystals on the Hirsch funnel when the filtrate is used to wash crystals from the container onto the funnel. With a very few drops of ice-cold solvent, rinse the crystallization flask. This container should still be ice cold. Place the ice-cold solvent on the crystals, and then reapply the vacuum. As soon as the liquid is pulled from the crystals, break the vacuum. Repeat this washing process as many times as necessary to remove colored material or other impurities from the crystals. In some cases, only one very small wash will be needed. After the crystals have been washed with ice-cold solvent, the vacuum can be left on to dry the crystals. Sometimes it is useful to press solvent from the crystals using a cork.

(C) Filtration with the Wilfilter (Replacing the Craig Tube).

The isolation of less than 100 mg of recrystallized material from a reaction tube (or any other container) is not easy. If the amount of solvent is large enough (a milliliter or more) the material can be recovered by filtration on the Hirsch funnel (B). But when the volume of liquid is less than a milliliter, much product is left in the tube during the transfer to the Hirsch filter. The solvent can be removed with a Pasteur pipette pressed against the bottom of the tube, a very effective filtration technique, but scraping the damp crystals from the reaction tube results in major losses. If the solvent is relatively low-boiling, it can be evaporated by connecting the tube to a water aspirator (Fig. 3.13). Once the crystals are dry they are easily scraped from the tube with little or no loss. But some solvents, and water is the principal culprit, are not easily removed by evaporation. And even when removal of the solvent under vacuum is not terribly difficult, it takes time.

We have invented a filtration device that circumvents these problems, the Wilfilter. After crystallization has ceased most of the solvent is removed from the crystals using a Pasteur pipette in the usual way (Fig. 3.12). Then the polypropylene Wilfilter is placed on the top of the reaction tube followed by a 15 mL polypropylene centrifuge tube (Fig. 3.15). The assembly is inverted and placed in a centrifuge such as the International Clinical Centrifuge that holds 12 15-mL tubes. The assembly, properly counterbalanced, is centrifuged for about a minute at top speed. The centrifuge tube is removed from the centrifuge, and the reaction tube is then removed from the centrifuge tube. The three fingers on the Wilfilter keep it attached to the reaction tube. The filtrate is left in the centrifuge tube.

Filtration with the Wilfilter occurs between the top surface of the reaction tube and the flat surface of the Wilfilter. Liquid will pass through that space during centrifugation; crystals will not. The crystals will be found on the top of the Wilfilter and inside the reaction tube. The very large centrifugal forces remove all the liquid, so the crystals will be virtually dry and thus easily removed from the reaction tube by shaking or scraping with the metal spatula.

The Wilfilter replaces an older device known as the Craig tube (Fig. 3.16), which consists of an outer tube of 1-, 2-, or 3-mL capacity with an inner plunger made of Teflon (expensive) or glass (fragile). The material to be crystallized is transferred to the outer tube and crystallized in the usual way. The inner plunger is added, and a wire hanger is fashioned so that the assembly can be removed from the centrifuge tube without the plunger falling off. Filtration in this device

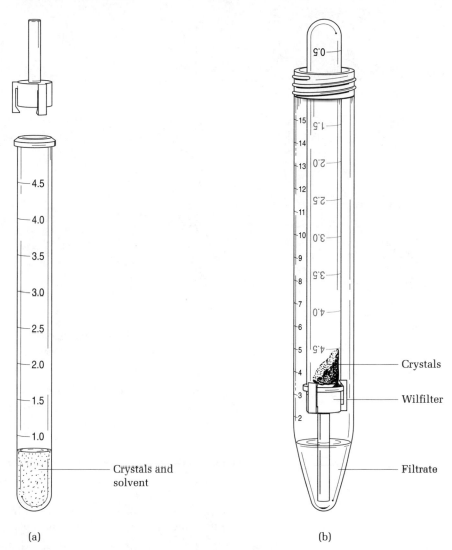

(a)   (b)

FIG. 3.15 The Wilfilter filtration apparatus. Filtration occurs between the flat face of the polypropylene Wilfilter and the top of the reaction tube.

occurs through the rough surface that has been ground into the shoulder of the outer tube.

The Wilfilter has the advantages that a special recrystallization device is not needed, no transfers of material are needed, it is not as limited in capacity (which is 4.5 mL), and its cost is one-fifth that of the Craig tube assembly.

(D) Filtration into a Reaction Tube on the Hirsch Funnel.

If it is desired to have the filtrate in a reaction tube instead of spread all over the bottom of the 25-mL filter flask, then the process described above can be carried out in

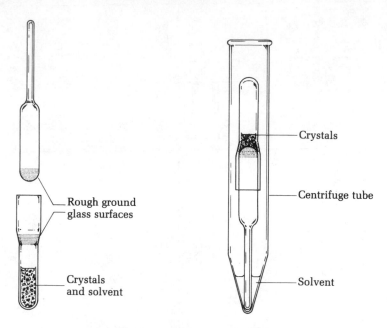

FIG. 3.16 Craig tube filtration apparatus. Filtration occurs between the rough ground glass surfaces when the apparatus is centrifuged.

the apparatus shown in Fig. 3.17. The vacuum hose is connected to the side arm using the thermometer adapter and a short length of glass tubing. Evaporate the filtrate in the reaction tube to collect a second crop of crystals.

(E) Filtration into a Reaction Tube on the Micro Büchner Funnel.

If the quantity of material being collected is very small, the bottom of the chromatography column is a micro Büchner funnel, which can be fitted into the top of the thermometer adapter as shown in Fig. 3.18. Again, it is good practice to cover the frit with a piece of 6-mm filter paper (cut with a cork borer).

(F) The Micro Büchner Funnel in an Enclosed Filtration Apparatus.

In the apparatus shown in Fig. 3.19, crystallization is carried out in the upper reaction tube in the normal way. The apparatus is then turned upside down, the crystals shaken down onto the micro Büchner funnel, and a vacuum applied through the side arm. In this apparatus, crystals can be collected in an oxygen-free atmosphere. This resembles a Schlenk tube.

 Macroscale Apparatus

Filtration on the Hirsch Funnel and the Büchner Funnel.   If the quantity of material is small (<2 g), the Hirsch funnel can be used in exactly the way described above. For larger quantities, the Büchner funnel is used. Properly

Hirsch funnel
with fritted disk

To
aspirator

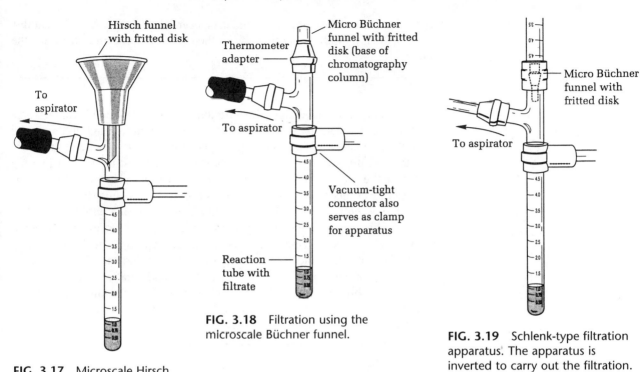

Thermometer
adapter

Micro Büchner
funnel with fritted
disk (base of
chromatography
column)

To aspirator

Vacuum-tight
connector also
serves as clamp
for apparatus

Reaction
tube with
filtrate

Micro Büchner
funnel with
fritted disk

To aspirator

FIG. 3.18 Filtration using the
microscale Büchner funnel.

FIG. 3.19 Schlenk-type filtration
apparatus. The apparatus is
inverted to carry out the filtration.

FIG. 3.17 Microscale Hirsch
filtration assembly. The Hirsch
funnel gives a vacuum-tight seal
to the 105° adapter.

Clamp the filter flask.

matched Büchner funnels, filter paper, and flasks are shown in Fig. 3.20. The
Hirsch funnel shown in the figure has a 5-cm bottom plate to accept 3.3-cm paper.

Place a piece of filter paper in the bottom of the Büchner funnel. Wet it with
solvent, and be sure it lies flat so that crystals cannot escape around the edge and
under the filter paper. Then with the vacuum off, pour the cold slurry of crystals
into the center of the filter paper. Apply the vacuum; as soon as the liquid disappears from the crystals break the vacuum to the flask by disconnecting the
hose. Rinse the Erlenmeyer flask with cold solvent. Add this to the crystals, and

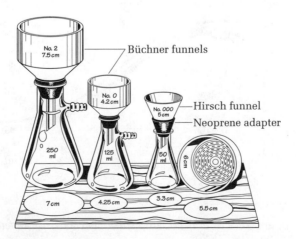

Büchner funnels

No. 2
7.5 cm

No. 0
4.2 cm

No. 000
5 cm

Hirsch funnel

Neoprene adapter

250
ml

125
ml

50
ml

6 cm

7 cm

4.25 cm

3.3 cm

5.5 cm

FIG. 3.20 Matching filter
assemblies. The 6.0-cm
polypropylene Büchner funnel
(*right*) resists breakage and can
be disassembled for cleaning.

reapply the vacuum just until the liquid disappears from the crystals. Repeat this process as many times as necessary, and then leave the vacuum on to dry the crystals.

The Water Aspirator and Trap.

The most common way to produce a vacuum in the organic laboratory for filtration purposes is by employing a *water aspirator*. Air is efficiently entrained in the water rushing through the aspirator so that it will produce a vacuum roughly equal to the vapor pressure of the water going through it (17 torr at 20°C, 5 torr at 4°C). A check valve is built into the aspirator, but even so when the water is turned off it may back into the evacuated system. For this reason a *trap* is always installed in the line (Fig. 3.21). *The water passing through the aspirator should always be turned on full force.* The system can be opened to the atmosphere by removing the hose from the small filter flask or by opening the screw clamp on the trap. Open the system, then turn off the water to avoid having water sucked back into the filter trap. Thin rubber tubing on the top of the trap will collapse and bend over when a good vacuum is established. You will, in time, learn to hear the difference in the sound of an aspirator when it is pulling a vacuum and when it is working on an open system.

Use a trap.

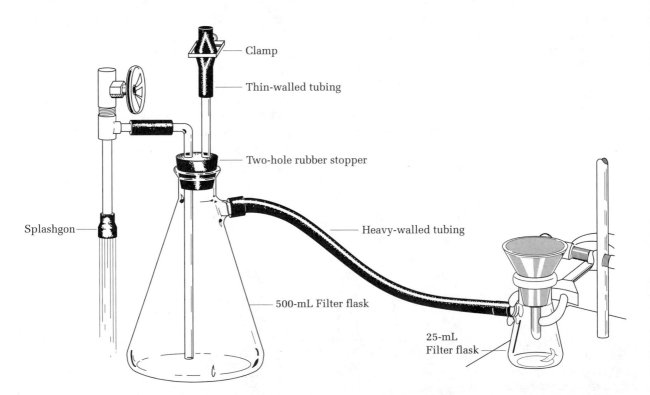

FIG. 3.21 Aspirator, filter trap, and Hirsch funnel. Clamp small filter flask to prevent its turning over.

Collecting a Second Crop of Crystals.   Regardless of the method used to collect the crystals on either a macroscale or a microscale, the filtrate and washings can be combined and evaporated to the point of saturation to obtain a second crop of crystals—hence the necessity for having a clean receptacle for the filtrate. This second crop will increase the overall yield, but the crystals will not usually be as pure as the first crop.

7. Drying the Product

Microscale Procedure

If possible, dry the product in the reaction tube after removal of the solvent using a Pasteur pipette. This can be done simply by connecting the tube to the water aspirator. If the tube is clamped in a beaker of hot water, the solvent will evaporate more rapidly under vacuum, but take care not to melt the product (Fig. 3.13). Water, which has a high heat of vaporization, is difficult to remove this way. Scrape the product out onto a watch glass, and allow it to dry to constant weight, which will indicate that all the solvent is gone. If the product is collected on the Hirsch funnel or the Wilfilter, the last bit of solvent can be removed by squeezing the crystals between sheets of filter paper before drying them on the watch glass.

Macroscale Procedure

Once the crystals have been washed on the Hirsch funnel or the Büchner funnel, press them down with a clean cork or other flat object and allow air to pass through them until they are substantially dry. Final drying can be done under reduced pressure (Fig. 3.22). The crystals can then be turned out of the funnel and squeezed between sheets of filter paper to remove the last bit of solvent before final drying on a watch glass.

Experiments

1. Solubility Tests

To test the solubility of a solid, transfer an amount roughly estimated to be about 10 mg (the amount that forms a symmetrical mound on the end of a stainless steel spatula) into a reaction tube and add about 0.25 mL of solvent from a calibrated dropper or pipette. Stir with a fire-polished stirring rod (4-mm), break up any lumps, and determine if the solid is readily soluble at room temperature. If the substance is readily soluble in methanol, ethanol, acetone, or acetic acid at room temperature, add a few drops of water from a wash bottle to see if a solid precipitates. If it does, heat the mixture, adjust the composition of the solvent pair to produce a hot solution saturated at the boiling point, let the solution stand undisturbed, and note the character of the crystals that form. If the substance fails to dissolve in a given solvent at room temperature, heat the suspension and see if

FIG. 3.22 Drying a solid by reduced air pressure.

Test Compounds:

Resorcinol

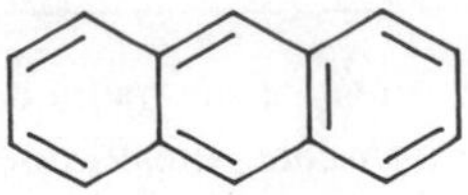

Anthracene

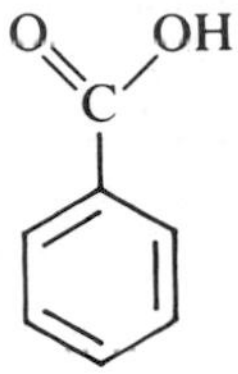

Benzoic acid

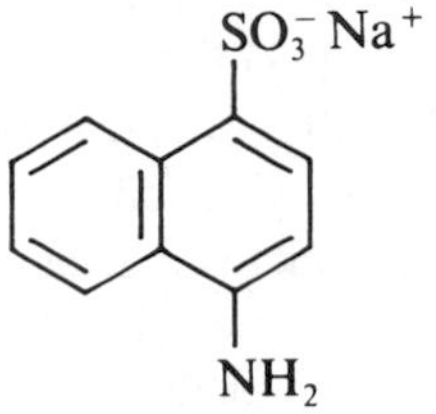

**4-Amino-1-
naphthalenesulfonic acid,
sodium salt**

solution occurs. If the solvent is flammable, heat the test tube on the steam bath or in a small beaker of water kept warm on the steam bath or a hot plate. If the solid completely dissolves, it can be declared readily soluble in the hot solvent; if some but not all dissolves, it is said to be moderately soluble, and further small amounts of solvent should then be added until solution is complete. When a substance has been dissolved in hot solvent, cool the solution by holding the flask under the tap and, if necessary, induce crystallization by rubbing the walls of the tube with a stirring rod to make sure that the concentration permits crystallization. Then reheat to dissolve the solid, let the solution stand undisturbed, and inspect the character of the ultimate crystals.

Make solubility tests on the test compounds shown to the left in each of the solvents listed below. Note the degree of solubility in the solvents, cold and hot, and suggest suitable solvents, solvent pairs, or other expedients for crystallization of each substance. Record the crystal form, at least to the extent of distinguishing between needles (pointed crystals), plates (flat and thin), and prisms. How do your observations conform to the generalization that like dissolves like?

Solvents:

    Water—hydroxylic, ionic
    Toluene—an aromatic hydrocarbon
    Ligroin—a mixture of aliphatic hydrocarbons

Cleaning Up[4]   Place organic solvents and solutions of the compounds in the organic solvents container. Dilute the aqueous solutions with water, and flush down the drain.

**MICROSCALE
AND MACROSCALE**

2. Crystallization of Pure Phthalic Acid, Naphthalene, and Anthracene

Phthalic acid   **Naphthalene**   **Anthracene**

The process of crystallization can be observed readily using phthalic acid. In the reference book *The Handbook of Chemistry and Physics,* in the table "Physical Constants of Organic Compounds," the entry for phthalic acid gives the following solubility data (in grams of solute per 100 mL of solvent). The superscripts refer to temperature in °C:

---

4. For this and all other "Cleaning Up" sections, see Chapter 2 for a complete discussion of waste disposal procedures.

Water	Alcohol	Ether, etc.
0.54^{14}	11.71^{18}	0.69^{15} eth., i. chl.
18^{99}		

The large difference in solubility in water as a function of temperature suggests this as the solvent of choice. The solubility in alcohol is high at room temperature. Ether is difficult to use because it is so volatile; the compound is insoluble in chloroform (i. chl.).

Microscale Procedure, Phthalic Acid

Set the heater control to about 20% of the maximum.

Crystallize 60 mg (0.060 g) of phthalic acid from the minimum volume of water, using the above data to calculate the required volume. First, turn on the electrically heated sand bath. Add the solid to a 10×100 mm reaction tube, and then, using a Pasteur pipette, add water dropwise. Use the calibration marks found in Fig. 1.18 to measure the volume of water in the pipette and the reaction tube. Add a boiling stick (a wooden applicator stick) to facilitate even boiling and prevent bumping. After a portion of the water has been added, gently heat the solution to boiling on a hot *sand bath* in the electric heater. The deeper the tube is placed in the sand, the hotter it will be. As soon as boiling begins, continue to add water dropwise until all the solid just dissolves. Cork the tube and clamp it as it cools, and observe the phenomenon of crystallization.

Alternate procedure: Dry the crystals under vacuum in a steam bath in the reaction tube.

After the tube reaches room temperature, cool it in ice, stir the crystals with a Pasteur pipette, and expel the air from the pipette as the tip is pushed to the bottom of the tube. When the tip is firmly and squarely seated in the bottom of the tube, release the bulb and withdraw the water. Rap the tube sharply on a wood surface to compress the crystals and remove as much of the water as possible with the pipette. Then cool the tube in ice and add a few drops of ice-cold ethanol to the tube in order to remove water from the crystals. Connect the tube to a water aspirator, and warm it in a beaker of hot water (see Fig. 3.13). Once all the solvent is removed, using the stainless steel spatula, scrape the crystals onto a piece of filter paper, fold the paper over the crystals, and squeeze out excess water before allowing the crystals to dry to constant weight. Weigh the dry crystals, and calculate the percent recovery of product.

Microscale Procedure, Naphthalene and Anthracene

These compounds can also be isolated using the Wilfilter.

Following the procedure outlined above, crystallize 40 mg of naphthalene from 80% aqueous methanol or 10 mg of anthracene from ethanol. These are more typical of compounds to be crystallized in later experiments in that they are soluble in organic solvents. It will be much easier to remove these solvents from the crystals under vacuum than it is to remove water from phthalic acid. You will seldom have occasion to crystallize less than 30 mg of a solid in these experiments.

Cleaning Up Dilute the aqueous filtrate with water, and flush the solution down the drain. Phthalic acid is not considered toxic to the environment. Methanol and ethanol filtrates go in the organic solvents container.

Macroscale Procedure

Crystallize 1.0 g of phthalic acid from the minimum volume of water, using the above data to calculate the required volume. Add the solid to the smallest practical Erlenmeyer flask, and then, using a Pasteur pipette, add water dropwise from a full 10-mL graduated cylinder. A boiling stick (a stick of wood) facilitates even boiling and will prevent bumping. After a portion of the water has been added, gently heat the solution to boiling on a hot plate. As soon as boiling begins, continue to add water dropwise until all the solid just dissolves. Place the flask on a cork ring or other insulator, and allow it to cool undisturbed to room temperature, during which time the crystallization process can be observed. Slow cooling favors large crystals. Then cool the flask in an ice bath, decant (pour off) the mother liquor (the liquid remaining with the crystals), and remove the last traces of liquid with a Pasteur pipette. Scrape the crystals onto a filter paper using a stainless steel spatula, squeeze the crystals between sheets of filter paper to remove traces of moisture, and allow the crystals to dry. Alternatively, the crystals can be collected on a Hirsch funnel. Compare the calculated volume of water with the volume of water actually used to dissolve the acid. Calculate the percent recovery of dry, recrystallized phthalic acid.

Cleaning Up Dilute the filtrate with water, and flush the solution down the drain. Phthalic acid is not considered toxic to the environment.

MICROSCALE

Decolorizing using pelletized Norit

3. Decolorizing a Solution with Decolorizing Charcoal

Into a reaction tube place 1.0 mL of a solution of methylene blue dye that has been made up at a concentration of 10 mg per 100 mL of water. Add to the tube a few pieces (10 or 12) of decolorizing charcoal, shake, and observe the color over a period of a minute or two. Heat the contents of the tube to boiling (reflux), and observe the color by holding the tube in front of a piece of white paper from time to time. How rapidly is the color removed? If the color is not removed in a minute or so, add more charcoal pellets.

Cleaning Up Place the Norit in the nonhazardous solid waste container.

MACROSCALE

4. Decolorization of Brown Sugar (Sucrose, $C_{12}H_{22}O_{11}$)

Raw sugar is refined commercially with the aid of decolorizing charcoal. The clarified solution is seeded generously with small sugar crystals and excess water removed under vacuum to facilitate crystallization. The pure white crystalline product is collected by centrifugation. Brown sugar is partially refined sugar and can be decolorized easily using charcoal.

Dissolve 15 g of dark brown sugar in 30 mL of water in a 50-mL Erlenmeyer flask by heating and stirring. Pour half the solution into another 50-mL flask. Heat one of the solutions nearly to the boiling point, allow it to cool slightly, and add to it 250 mg (0.25 g) of decolorizing charcoal (Norit pellets). Bring the solution back to near the boiling point for 2 minutes; then filter the hot solution into an Erlenmeyer flask through a fluted filter paper held in a previously heated funnel. Treat the other half of the sugar solution in exactly the same way, but use only 50 mg of decolorizing charcoal. In collaboration with a fellow student, try heating the solutions for only 15 s after addition of the charcoal. Compare your results.

Cleaning Up Decant (pour off) the aqueous layer. Place the Norit in the non-hazardous solid waste container. The sugar solution can be flushed down the drain.

MICROSCALE

Benzoic acid

5. Crystallization of Benzoic Acid from Water and a Solvent Pair

Crystallize 50 mg of benzoic acid from water in the same way phthalic acid was crystallized. Then in a dry reaction tube dissolve another 50-mg sample of benzoic acid in the minimum volume of hot toluene, and add cyclohexane to the hot solution dropwise. When the hot solution becomes cloudy and crystallization has started, allow the tube to cool slowly to room temperature; then cool it in ice and collect the crystals. Compare crystallization in water to that in the solvent pair.

Cleaning Up The aqueous solution, after dilution with water, can be flushed down the drain. The toluene and cyclohexane filtrates should be placed in the organic solvents container.

MACROSCALE

Naphthalene

Do not try to grasp Erlenmeyer flasks with a test tube holder.

Support the funnel in a ring stand.

6. Recrystallization of Naphthalene from a Mixed Solvent

Add 2.0 g of impure naphthalene[5] to a 50-mL Erlenmeyer flask along with 3 mL of methanol and a boiling stick to promote even boiling. Heat the mixture to boiling over a steam bath or hot plate, and then add methanol dropwise until the naphthalene just dissolves when the solvent is boiling. The total volume of methanol should be 4 mL. Remove the flask from the heat, and cool it rapidly in an ice bath. Note that the contents of the flask set to a solid mass, which would be impossible to handle. Add enough methanol to bring the total volume to 25 mL, heat the solution to the boiling point, remove the flask from the heat, allow it to cool slightly, and add 30 mg of decolorizing charcoal pellets to remove the colored impurity in the solution. Heat the solution to the boiling point for 2 min; if the color is not gone, add more Norit and boil again, then filter through a fluted filter paper in a previously warmed stemless funnel into a 50-mL Erlenmeyer flask. Sometimes

5. A mixture of 100 g of naphthalene, 0.3 g of a dye such as congo red, and perhaps sand, magnesium sulfate, dust, etc.

filtration is slow because the funnel fits so snugly into the mouth of the flask that a back pressure develops. If you note that raising the funnel increases the flow of filtrate, fold a small strip of paper two or three times and insert it between the funnel and flask. Wash the used flask with 2 mL of hot methanol, and use this liquid to wash the filter paper, transferring the solvent with a Pasteur pipette in a succession of drops around the upper rim of the filter paper. When the filtration is complete, the volume of methanol should be 15 mL. If it is not, evaporate excess methanol.

Because the filtrate is far from being saturated with naphthalene at this point, it will not yield crystals on cooling; however, the solubility of naphthalene in methanol can be greatly reduced by addition of water. Heat the solution to the boiling point, and add water dropwise from a 10-mL graduated cylinder, using a Pasteur pipette (or use a precalibrated pipette). After each addition of water the solution will turn cloudy for an instant. Swirl the contents of the flask, and heat to redissolve any precipitated naphthalene. After the addition of 3.5 mL of water the solution will be almost saturated with naphthalene at the boiling point of the solvent. Remove the flask from the heat, and place it on a cork ring or other insulating surface to cool, without being disturbed, to room temperature.

Immerse the flask in an ice bath along with another flask containing methanol and water in the ratio of 30:7. This cold solvent will be used for washing the crystals. The cold crystallization mixture is collected by vacuum filtration on a small Büchner funnel (50-mm) (Fig. 3.23). The water flowing through the aspirator should always be turned on full force. In collecting the product by suction filtration, use a spatula to dislodge crystals and ease them out of the flask. If crystals still remain in the flask, some filtrate can be poured back into the crystallization flask as a rinse for washing as often as desired, because it is saturated with solute. To free the crystals from contaminating mother liquor, break the suction, pour a few milliliters of the fresh cold solvent mixture into the Büchner funnel, and immediately reapply suction. Repeat this process until the crystals and the filtrate are free of color. Press the crystals with a clean cork to eliminate excess solvent, pull air through the filter cake for a few minutes, and then put the large flat platelike crystals out on a filter paper to dry. The yield of pure white crystalline naphthalene should be about 1.6 g. The mother liquor contains about 0.25 g, and about 0.15 g is retained in the charcoal and on the filter paper.

Cleaning Up   Place the Norit in the nonhazardous solid waste container. The methanol filtrate and washings are placed in the organic solvents container.

7. Purification of an Unknown

Bear in mind the seven-step crystallization procedure:

1. Choose the solvent.
2. Dissolve the solute.
3. Decolorize the solution (if necessary).
4. Filter suspended solids (if necessary).

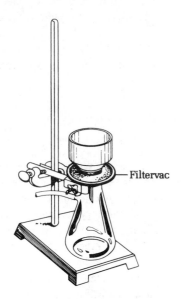

FIG. 3.23 Suction filter assembly clamped to provide firm support. The funnel must be pressed down on the Filtervac to establish reduced pressure in the flask.

— Filtervac

5. Crystallize the solute.
6. Collect and wash the crystals.
7. Dry the product.

You are to purify 2.0 g of an unknown provided by the instructor. Conduct tests for solubility and ability to crystallize in several organic solvents, solvent pairs, and water. Conserve your unknown by using very small quantities for solubility tests. If only a drop or two of solvent is used, the solvent can be evaporated by heating the test tube on the steam bath or sand bath, and the residue can be used for another test. Submit as much pure product as possible with evidence of its purity (i.e., the melting point). From the posted list identify the unknown.

Cleaning Up   Place decolorizing charcoal, if used, and filter paper in the non-hazardous solid waste container. Put organic solvents in the organic solvents container, and flush aqueous solutions down the drain.

Crystallization Problems and Their Solutions

Induction of Crystallization

Occasionally a sample will not crystallize from solution on cooling, even though the solution is saturated with the solute at elevated temperature. The easiest method for inducing crystallization is to add to the supersaturated solution a seed crystal that has been saved from the crude material (if it was crystalline before recrystallization was attempted). In a probably apocryphal tale, the great sugar chemist Emil Fischer merely had to wave his beard over a recalcitrant solution and the appropriate seed crystals would drop out, causing crystallization to occur. In the absence of seed crystals, crystallization can often be induced by scratching the inside of the flask with a stirring rod at the air/liquid interface. One theory holds that part of the freshly scratched glass surface has angles and planes corresponding to the crystal structure, and crystals start growing on these spots. Often crystallization is very slow to begin, and placing the sample in a refrigerator overnight will bring success. Other expedients are to change the solvent (usually to a poorer one) and to place the sample in an open container where slow evaporation and dust from the air may help induce crystallization.

Seeding

Scratching

Oils and "Oiling Out"

Some saturated solutions, especially those containing water, when they cool deposit not crystals but small droplets referred to as oils. Should these droplets subsequently crystallize and be collected, they will be found to be rather impure. Should the temperature of the saturated solution be above the melting point of the solute when it starts to come out of solution, the solute will, of necessity, be deposited as an oil. Similarly, the melting point of the desired compound may be depressed to a point such that a low-melting eutectic mixture of the solute and the solvent comes out of solution. The simplest remedy for this latter problem is to lower the temperature at which the solution becomes saturated with the solute by

Crystallize at a lower temperature

simply adding more solvent. In extreme cases it may be necessary to lower this temperature well below room temperature by cooling the solution with dry ice.

Crystallization Summary

1. **Choosing the solvent.** "Like dissolves like." Some common solvents are water, methanol, ethanol, ligroin, and toluene. When you use a solvent pair, dissolve the solute in the better solvent, and add the poorer solvent to the hot solution until saturation occurs. Some common solvent pairs are ethanol–water, *t*-butyl methyl ether–ligroin, and toluene–ligroin.

2. **Dissolving the solute.** To the crushed or ground solute in an Erlenmeyer flask or reaction tube add solvent; heat the mixture to boiling. Add more solvent as necessary to obtain a hot, saturated solution.

3. **Decolorizing the solution.** If it is necessary to remove colored impurities, cool the solution to near room temperature, and add more solvent to prevent crystallization from occurring. Add decolorizing charcoal in the form of pelletized Norit to the cooled solution, and then heat it to boiling for a few minutes, taking care to swirl the solution to prevent bumping. Remove the Norit by filtration, then concentrate the filtrate.

4. **Filtering suspended solids.** If it is necessary to remove suspended solids, dilute the hot solution slightly to prevent crystallization from occurring during filtration. Filter the hot solution. Add solvent if crystallization begins in the funnel. Concentrate the filtrate to obtain a saturated solution.

5. **Crystallizing the solute.** Let the hot saturated solution cool spontaneously to room temperature. Do not disturb the solution. Then cool it in ice. If crystallization does not occur, scratch the inside of the container or add seed crystals.

6. **Collecting and washing the crystals.** Collect the crystals using the Pasteur pipette method, the Wilfilter, or by vacuum filtration on a Hirsch funnel or a Büchner funnel. If the latter technique is employed, wet the filter paper with solvent, apply vacuum, break vacuum, add crystals and liquid, apply vacuum until solvent just disappears, break vacuum, add cold wash solvent, apply vacuum, and repeat until crystals are clean and filtrate comes through clear.

7. **Drying the product.** Press the product on the filter to remove solvent. Then remove it from the filter, squeeze it between sheets of filter paper to remove more solvent, and spread it on a watch glass to dry.

Questions

1. A sample of naphthalene, which should be pure white, was found to have a grayish color after the usual purification procedure. The melting point was correct and the melting point range small. Explain the gray color.

2. How many milliliters of boiling water are required to dissolve 25 g of phthalic acid? If the solution were cooled to 14°C, how many grams of phthalic acid would crystallize out?

3. What is the reason for using activated carbon during a crystallization?

4. If a little activated charcoal does a good job removing impurities in a crystallization, why not use a lot?

5. Under what circumstances is it wise to use a mixture of solvents to carry out a crystallization?

6. Why is gravity filtration and not suction filtration used to remove suspended impurities and charcoal from a hot solution?

7. Why is a fluted filter paper used in gravity filtration?

8. Why are stemless funnels used instead of long-stem funnels to filter hot solutions through fluted filter paper?

9. Why is the final product from the crystallization process isolated by vacuum filtration and not by gravity filtration?

4

Melting Points, Boiling Points, and Refractive Indices

Prelab Exercise: Predict what the melting points of the three urea–cinnamic acid mixtures will be.

Part 1. Melting Points

Melting points—a micro technique

The melting point of a pure solid organic compound is one of its characteristic physical properties, along with molecular weight, boiling point, refractive index, and density. A pure solid will melt reproducibly over a narrow range of temperatures, typically less than 1°C. The process of determining this melting "point" is done on a truly micro scale using less than 1 mg of material. The apparatus is very simple, consisting of a thermometer, a capillary tube to hold the sample, and a heating bath.

Characterization

An indication of purity

Melting points are determined for three reasons. If the compound is a known one, the melting point will help to characterize the sample in hand. If the compound is new, then the melting point is recorded in order to allow future characterization by others. And finally the range of the melting point is indicative of the purity of the compound; an impure compound will melt over a wide range of temperatures. Recrystallization of the compound will purify it, and the melting point range will decrease. In addition, the entire range will be displaced upward. For example, an impure sample might melt from 120–124°C and after recrystallization melt at 125–125.5°C. A solid is considered pure if the melting point does not rise after recrystallization.

A crystal is an orderly arrangement of molecules in a solid. As heat is added to the solid, the molecules will vibrate and perhaps rotate but still remain a solid. At a characteristic temperature it will suddenly acquire the necessary energy to overcome the forces that attract one molecule to another, and it will undergo translational motion—in other words, it will become a liquid.

Melting point generalizations

The forces by which one molecule is attracted to another include ionic attraction, van der Waals forces, hydrogen bonds, and dipole–dipole attraction. Most, but by no means all, organic molecules are covalent in nature and melt at temperatures below 300°C. Typical inorganic compounds are ionic and have much higher melting points; e.g., sodium chloride melts at 800°C. Ionic organic molecules often decompose before melting, as do compounds having strong hydrogen bonds such as sucrose.

Other factors being equal, larger molecules melt at higher temperatures than

FIG. 4.1   Melting point–composition diagram for mixtures of the solids X and Y.

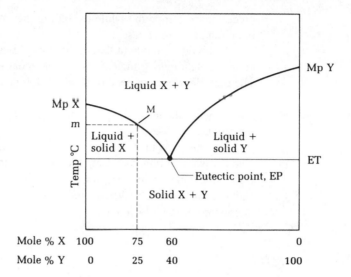

smaller ones. Among structural isomers the more symmetrical will have the higher melting point. Among optical isomers the R and S enantiomers will have the same melting points; but the racemate, the mixture of equal parts of R and S, will usually have a different melting point. Molecules that can form hydrogen bonds will usually have higher melting points than their counterparts of similar molecular weight.

A phase diagram

The melting point behavior of impure compounds is best understood by consideration of a simple binary mixture of compounds X and Y (Fig. 4.1). This melting point–composition diagram shows the melting point behavior as a function of composition. The melting point of a pure compound is the temperature at which the vapor pressures of the solid and liquid are equal. But in dealing with a mixture the situation is different. Consider the case of a mixture of 75% X and 25% Y. At a temperature below ET, the eutectic temperature, the mixture is solid Y and solid X. At the eutectic temperature the solid begins to melt. The melt is a solution of Y dissolved in liquid X. The vapor pressure of the solution of X and

Melting point depression

Y together is less than that of pure X at the melting point; therefore, the temperature at which X will melt is lower when mixed with Y. This is an application of Raoult's law (Chapter 5). As the temperature is raised, more and more of solid X melts until it is all gone at point *M* (temperature *m*). The melting point range is thus from ET to *m*. In practice it is very difficult to detect point ET when a melting point is determined in a capillary because it represents the point at which an infinitesimal amount of the liquid solution has started to melt.

In this hypothetical example the liquid solution becomes saturated with Y at point EP. This is the point at which X and Y and their liquid solutions are in equilibrium. A mixture of X and Y containing 60% X will appear to have a sharp melt-

The eutectic point

ing point at temperature ET. This point, EP, is the eutectic point.

In general the melting point range of a mixture of compounds is broad, and the breadth of the range is an indication of purity. The chances of accidentally coming on the eutectic composition are small. Recrystallization will enrich the predominant compound while excluding the impurity, and therefore the melting point range will decrease.

It should be apparent that the impurity must be soluble in the compound, so an insoluble impurity such as sand or charcoal will not depress the melting point. The impurity does not need to be a solid. It can be a liquid such as water (if it is soluble) or an organic solvent, such as the one used to recrystallize the compound; hence the necessity for drying the compound before determining the melting point.

Mixed melting points

Advantage is taken of the depression of melting points of mixtures to prove whether two compounds having the same melting points are identical. If X and Y are identical, then a mixture of the two will have the same melting point; but if X and Y are not identical, then a small amount of X in Y or of Y in X will cause the melting point to be lowered.

Apparatus

The apparatus needed for determining an accurate melting point need not be elaborate; the same results are obtained on the simplest as on the most complex devices.

Thomas–Hoover Uni-Melt

The Thomas–Hoover Uni-Melt apparatus (Fig. 4.2) will accommodate seven capillaries in a small, magnified, lighted beaker of high-boiling silicone oil that is stirred and heated electrically. The heating rate is controlled with a variable transformer that is part of the apparatus. The rising mercury column of the thermometer can be observed with an optional traveling periscope device so the eye need not move away from the capillary. For industrial analytical and control work there is even an apparatus (Mettler) that automatically determines the melting point and displays the result in digital form.

Mel-Temp

The Mel-Temp apparatus (Fig. 4.3) consists of an electrically heated aluminum block that accommodates three capillaries. The sample is illuminated through the lower port and observed with a six-power lens through the upper port. The heating rate can be controlled, and with a special thermometer the apparatus can be used up to 500°C, far above the useful limit of silicone oil (about 350°C).

In this melting point apparatus (Fig. 4.3) it is advisable to use a digital thermometer (Fig. 4.4) in place of the mercury-in-glass thermometer. The digital thermometer has a small heat capacity and fast response time. It is more robust than a glass thermometer and does not, of course, contain mercury, which is very toxic.

Capillaries can be obtained commercially or can be made by drawing out 12-mm soft-glass tubing. The tubing is rotated in the hottest part of the Bunsen burner flame until it is very soft and begins to sag. It should not be drawn out during heating, but is removed from the flame and after a moment's hesitation drawn steadily and not too rapidly to arm's length. With some practice it is possible to produce 10 to 15 good tubes in a single drawing. The long capillary tube can be

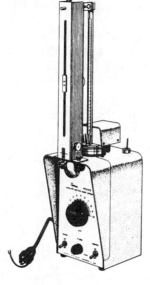

FIG. 4.2   Thomas–Hoover Uni-Melt melting point apparatus.

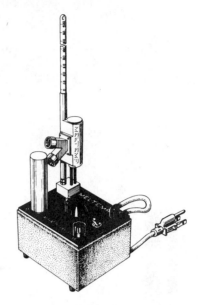

FIG. 4.3   Mel-Temp melting point apparatus.

Digital thermometer

°C

79

Surface probe

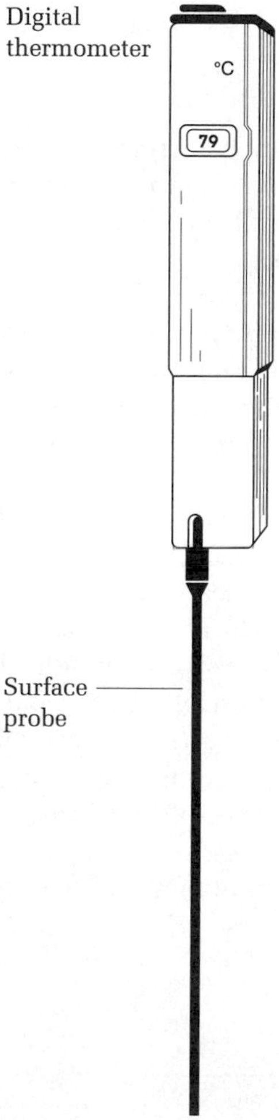

FIG. 4.4   Digital thermometer.

cut into 100-mm lengths with a glass scorer. Each tube is sealed by rotating the end in the edge of a small flame, as seen in Fig. 4.5.

Filling Melting Point Capillaries

The dry sample is ground to a fine powder on a watch glass or a piece of glassine paper on a hard surface using the flat portion of a spatula. It is formed into a small pile and the melting point capillary forced down into the pile. The sample is shaken into the closed end of the capillary by rapping sharply on a hard surface or by dropping it down a 2-ft length of glass tubing onto a hard surface. The height of the sample should be no more than 2–3 mm.

FIG. 4.5   Sealing a melting point capillary tube.

Sealed Capillaries

Samples that sublime

Some samples sublime (go from the solid directly to the vapor phase without appearing to melt) or undergo rapid air oxidation and decompose at the melting point. These samples should be sealed under vacuum. This can be accomplished by forcing a capillary through a hole previously made in a rubber septum and evacuating the capillary using the water aspirator or a mechanical vacuum pump (Fig. 4.6). Using the flame from a small micro burner, the tube is gently heated about 15 mm above the tightly packed sample. This will cause any material in this region to sublime away. It is then heated more strongly in the same place to collapse the tube, taking care that the tube is straight when it cools. It is also possible to seal the end of a Pasteur pipette, add the sample, pack it down, and seal off a sample under vacuum in the same way.

Handle flame with great care in the organic laboratory.

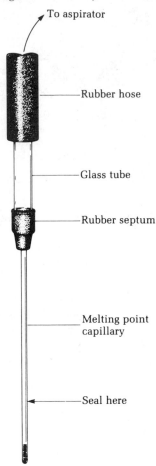

Determining the Melting Point

The accuracy of the melting point depends on the accuracy of the thermometer, so the first exercise in this experiment will be to calibrate the thermometer. Melting points of pure, known compounds will be determined and deviations recorded so that a correction can be applied to future melting points. Be forewarned, however, that thermometers are usually fairly accurate.

The most critical factor in determining an accurate melting point is the rate of heating. At the melting point the temperature rise should not be greater than 1°C per minute. This may seem extraordinarily slow, but it is necessary in order for heat from the bath to be transferred equally to the sample and to the glass and mercury of the thermometer.

From experience you know the rate at which ice melts. Consider doing a melting point experiment on an ice cube. Because water melts at 0°C, you would need to have a melting point bath a few degrees below zero. To observe the true melting point of the ice cube, you would need to raise the temperature extraordinarily slowly. The ice cube would appear to begin to melt at 0°C and, if you waited for temperature equilibrium to be established, it would all be melted at 0.5°C. If you were impatient and raised the temperature too rapidly, the ice might appear to melt over the range 0 to 20°C. Similarly, melting points determined in capillaries will not be accurate if the rate of heating is too fast.

The rate of heating is the most important factor in obtaining accurate melting points. Heat no faster than 1°C per minute.

FIG. 4.6 Evacuation of a melting point capillary prior to sealing.

 **MICROSCALE**

Experiments ———————————————————————

1. Calibration of the Thermometer

Determine the melting point of standard substances (Table 4.1) over the temperature range of interest. The difference between the values found and those expected constitutes the correction that must be applied to future temperature readings. If the thermometer has been calibrated previously, then determine one or more melting points of known substances to familiarize yourself with the technique. If the determinations do not agree within 1°C, then repeat the process. Both mercury-in-glass and digital thermometers will need to be calibrated.

MICROSCALE

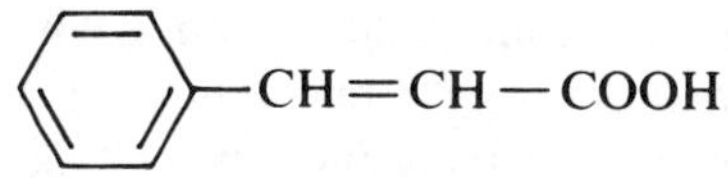

-CH=CH-COOH

Cinnamic acid

2. Melting Points of Pure Urea and Cinnamic Acid

Using a metal spatula, crush the sample to a fine powder on a hard surface such as a watch glass. Push a melting point capillary into the powder, and force the powder down in the capillary by tapping the capillary or by dropping it through a long glass tube held vertically and resting on a hard surface. The column of solid should be no more than 2–3 mm in height and should be tightly packed.

TABLE 4.1   Melting Point Standards

Compound	Structure	Melting Point (°C)
Naphthalene	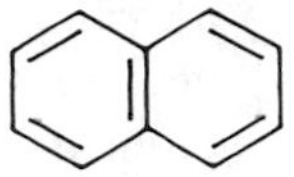	80–82
Urea	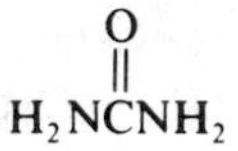	132.5–133
Sulfanilamide	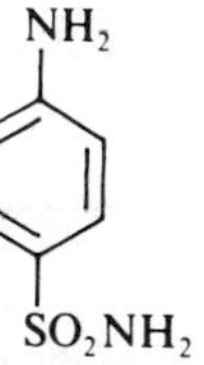	164–165
4-Toluic acid	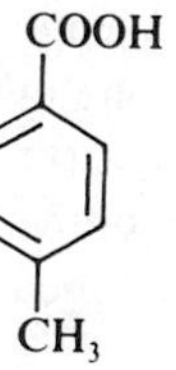	180–182
Anthracene	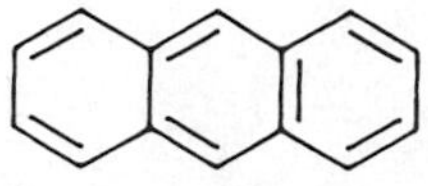	214–217
Caffeine (evacuated capillary)	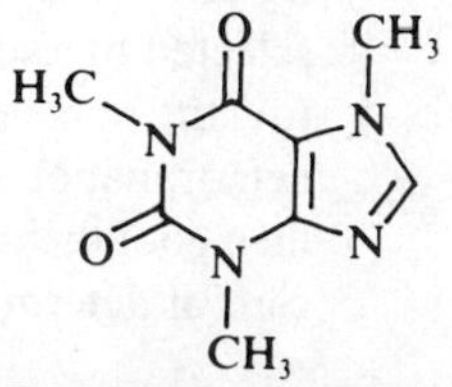	234–236.5

Except for the Thomas–Hoover and Mel-Temp apparatus, the capillary is held to the thermometer with a rubber band made by cutting a slice off the end of a piece of ³⁄₁₆-in. (5 mm) rubber tubing. This rubber band must be above the level of the oil bath; otherwise, it will break in the hot oil. Insertion of a fresh tube under the rubber band is facilitated by leaving the used tube in place. The sample should be close to and on a level with the center of the thermometer bulb.

Heat rapidly to within 20°C of the melting point.

If the approximate melting temperature is known, the bath can be heated rapidly until the temperature is about 20°C below this point, but the heating during the last 15–20°C should slow down considerably so that the rate of heating at the melting point is no more than 1°C per minute while the sample is melting. As the melting point is approached the sample may shrink because of crystal structure changes. However, the melting process begins when the first drops of liquid are seen in the capillary and ends when the last trace of solid disappears. For a pure compound this whole process may occur over a range of only 0.5°C; hence the necessity of having the temperature rise slowly during the determination.

Two or three melting points at once

If determinations are to be done on two or three samples that differ in melting point by as much as 10°C, two or three capillaries can be secured to the thermometer together and the melting points observed in succession without removal of the thermometer from the bath. As a precaution against interchange of tubes while they are being attached, use some system of identification, such as one, two, and three dots made with a marking pencil.

Determine the melting point of either urea (mp 132.5–133°C) or cinnamic acid (mp 132.5–133°C). Repeat the determination, and if the two determinations do not check within 1°C, do a third one.

☐ MICROSCALE

3. Melting Points of Urea–Cinnamic Acid Mixtures

Make mixtures of urea and cinnamic acid in the approximate proportions 1:4, 1:1, and 4:1 by putting side by side the correct number of equal-sized small piles of the two substances and then mixing them. Grind the mixture thoroughly for at least a minute on a watch glass using a metal spatula. Note the ranges of melting of the three mixtures, and use the temperatures of complete liquefaction to construct a rough diagram of mp versus composition.

☐ MICROSCALE

4. Unknowns

Determine the melting point of one or more of the following unknowns to be selected by the instructor (Table 4.2), and on the basis of the melting point identify the substance. Prepare two capillaries of each unknown. Run a very fast determination on the first sample to ascertain the approximate melting point, and then cool the melting point bath to just below the melting point and make a slow, careful determination using the other capillary.

TABLE 4.2 Melting Point Unknowns

Compound	Melting Point (°C)
Benzophenone	49–51
Maleic anhydride	52–54
4-Nitrotoluene	54–56
Naphthalene	80–82
Acetanilide	113.5–114
Benzoic acid	121.5–122
Urea	132.5–133
Salicylic acid	158.5–159
Sulfanilamide	165–166
Succinic acid	184.5–185
3,5-Dinitrobenzoic acid	205–207
p-Terphenyl	210–211

Part 2. Boiling Points

The boiling point of a pure organic liquid is one of its characteristic physical properties, just like its density, molecular weight, and refractive index, and the melting point of a solid. The boiling point is used to characterize a new organic liquid, and knowledge of the boiling point helps to compare one organic liquid with another, as in the process of identifying an unknown organic substance.

Comparison of boiling points with melting points is instructive. The process of determining the boiling point is more complex than that for the melting point: It requires more material, and because it is affected less by impurities, it is not as good an indication of purity. Boiling points can be determined on a few microliters of a liquid, but on a small scale it is difficult to determine the boiling point *range*. This requires enough material to distill—about 1 to 2 mL. Like the melting point, the boiling point of a liquid is affected by the forces that attract one molecule to another—ionic attraction, van der Waals forces, dipole–dipole interactions, and hydrogen bonding.

Structure and Boiling Point

In a homologous series of molecules the boiling point increases in a perfectly regular manner. The normal saturated hydrocarbons have boiling points ranging from $-162°C$ for methane to $330°C$ for $n\text{-}C_{19}H_{40}$, an increase of about $27°C$ for each CH_2 group. It is convenient to remember that n-heptane with a molecular weight of 100 has a boiling point near $100°C$ ($98.4°C$). A spherical molecule such as 2,2-dimethylpropane has a lower boiling point than n-pentane because it cannot have as many points of attraction to adjacent molecules. For molecules of the same molecular weight, those with dipoles, such as carbonyl groups, will have higher boiling points than those without, and molecules that can form hydrogen bonds will boil even higher. The boiling point of such molecules depends on the number of hydrogen bonds that can be formed, so that an alcohol with one hydroxyl group will boil lower than one with two if they both have the same molecular weight. A number of other generalizations can be made about boiling point behavior as a function of structure; you will learn about these throughout your study of organic chemistry.

Boiling Point as a Function of Pressure

Since the boiling point of a pure liquid is defined as the temperature at which the vapor pressure of the liquid exactly equals the pressure exerted on it, the boiling point will be a function of atmospheric pressure. At an altitude of 14,000 ft the boiling point of water is $81°C$. At pressures near that of the atmosphere at sea level (760 mm), the boiling point of most liquids decreases about $0.5°C$ for each 10-mm decrease in atmospheric pressure. This generalization does not hold for greatly reduced pressures because the boiling point decreases as a nonlinear function of pressure (see Fig. 5.1). Under these conditions a nomograph relating observed boiling point, boiling point at 760 mm, and pressure in millimeters should be consulted (see Fig. 7.10). This nomograph is not highly accurate; the change in boiling point as a function of pressure also depends on the type of compound (polar, nonpolar, hydrogen bonding, etc.). Consult the *Handbook of Chemistry and Physics* for the correction of boiling points to standard pressure.

The Laboratory Thermometer

CAUTION: Mercury is toxic. Report broken thermometers to your instructor.

Most mercury-in-glass laboratory thermometers have a mark around the stem that is three inches (76 mm) from the bottom of the bulb. This is the immersion line; the thermometer will record accurate temperatures if immersed to this line. Should you break a mercury thermometer, immediately inform your instructor, who will use special apparatus to clean up the mercury. Mercury vapor is very toxic.

Prevention of Superheating—Boiling Sticks and Boiling Stones

A very clean liquid in a very clean vessel will superheat and not boil when subjected to a temperature above its boiling point. This means that a thermometer placed in the liquid will register a temperature higher than the boiling point of the liquid. If boiling does occur under these conditions, it occurs with explosive violence. To avoid this problem, boiling stones are always added to liquids before heating them to boiling—whether to determine a boiling point or to carry out a reaction or distillation. The stones provide the nuclei on which the bubble of vapor indicative of a boiling liquid can form. Some boiling stones, also called boiling chips, are porous unglazed porcelain. This material is filled with air in numerous fine capillaries. Upon heating, this air expands to form the fine bubbles on which even boiling can take place. Once the liquid cools it will fill these capillaries and the boiling chip will become ineffective, so another must be added each time the liquid is heated to boiling. Sticks of wood—so-called applicator sticks about 1.5 mm in diameter—also promote even boiling and, unlike stones, are easy to remove from the solution. None of these work well for vacuum distillation.

Apparatus and Technique

By Distillation. When enough material is available, the best method for determining the boiling point of a liquid is to distill it (Chapter 5). Distillation allows the boiling range to be determined and thus gives an indication of purity. Bear in mind, however, that a constant boiling point is not a guarantee of homogeneity and thus purity. Constant-boiling azeotropes such as 95% ethanol abound.

Using a Digital Thermometer and a Reaction Tube. Boiling points can be measured rapidly and accurately using an electronic digital thermometer as depicted in Fig. 4.7. While digital thermometers are too expensive at present for each student to have one, two or three of these in the laboratory can greatly

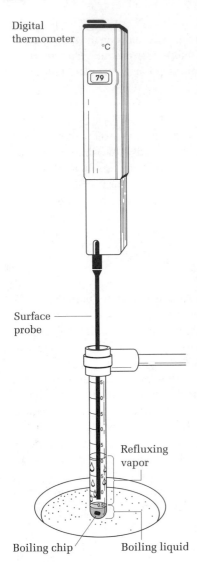

Digital
thermometer

°C

79

Surface
probe

Refluxing
vapor

Boiling chip   Boiling liquid

FIG. 4.7 Use of a digital
thermometer for determining
boiling points.

Smaller-scale boiling point
apparatus

speed up the determination of boiling points, and they are much safer to use because there is no danger from toxic mercury vapor if the thermometer is accidentally dropped.

A surface probe is the active element. Unlike the bulb of mercury at the end of a thermometer, this element has a very low heat capacity and a very fast response time, so boiling points can be determined very quickly in this apparatus. About 0.2–0.3 mL of the liquid and a boiling chip are heated on a sand bath until the liquid refluxes about 3 cm up the tube. The boiling point is the highest temperature recorded by the thermometer and maintained for about 1 min. Application of heat will drive tiny bubbles of air from the boiling chip. Do not mistake these tiny bubbles for real boiling. This can happen if the unknown has a very high boiling point. The probe should not touch the side of the reaction tube and should be about 5 mm above the liquid.

In a Reaction Tube.   If a digital thermometer is not available, use the apparatus shown in Fig. 4.8. Use of the distilling adapter on the top of the reaction tube allows access to the atmosphere. Place 0.3 mL of the liquid along with a boiling stone in a 10 × 100 mm reaction tube, clamp a thermometer so that the bulb is just above the level of the liquid, and then heat the liquid with a sand bath. It is *very important* that no part of the thermometer touch the reaction tube. Heating is regulated so that the boiling liquid refluxes (condenses the drips down) about 3 cm up the thermometer but does not boil out of the apparatus. If you cannot see the refluxing liquid, carefully run your finger down the side of the reaction tube until you feel heat. This indicates where the liquid is refluxing. Droplets of liquid must drip from the thermometer bulb in order to heat the mercury thoroughly. The boiling point is the highest temperature recorded by the thermometer and maintained over about a 1-min time interval.

Application of heat will drive tiny bubbles of air from the boiling chip. Do not mistake these tiny bubbles for real boiling. This can happen if the unknown has a very high boiling point. It may take several minutes to heat up the mercury in the thermometer bulb. True boiling is indicated by drops dripping from the thermometer and a constant temperature recorded on the thermometer. If the temperature is not constant, then you are probably not observing true boiling.

Using a 3- to 5-mm Tube.   For smaller quantities, the tube is attached to the side of the thermometer (Fig. 4.9) and heated with a liquid bath. The tube, which can be made from tubing 3 to 5 mm in diameter, contains a small inverted capillary. This is made by cutting a 6-mm piece from the sealed end of a melting point capillary, inverting it, and sealing it again to the capillary. A centimeter/millimeter ruler is printed on the inside cover of this book.

When the sample is heated in this device, the air in the inverted capillary will expand and an occasional bubble will escape. At the boiling point a continuous and rapid stream of bubbles will emerge from the inverted capillary. At this point the heating is stopped and the bath allowed to cool. A time will come when bubbling ceases and the liquid just begins to rise in the inverted capillary. The temperature at which this happens is recorded. The liquid is allowed to partially fill

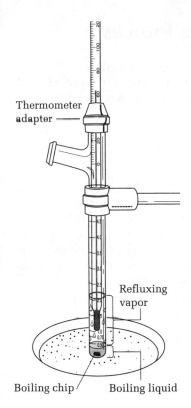

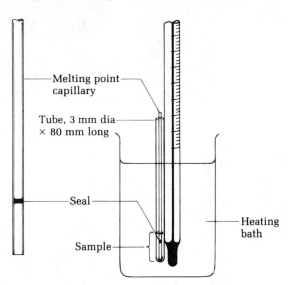

FIG. 4.9   Smaller-scale boiling point apparatus.

FIG. 4.8   Small-scale boiling-point apparatus. Be sure the thermometer does not touch the tube.

the small capillary, and the heat is applied carefully until the first bubble comes from the capillary. The temperature is recorded at that point. The two temperatures approximate the boiling point range for the liquid. The explanation: As the liquid was being heated, the air expanded in the inverted capillary and was replaced by vapor of the liquid. The liquid was actually slightly superheated when rapid bubbles emerged from the capillary, but on cooling the point was reached at which the pressure on the inside of the capillary matched the outside (atmospheric) pressure. This is, by definition, the boiling point.

Cleaning Up   Place the boiling point sample in either the halogenated or non-halogenated waste container. Do not pour it down the sink.

Part 3. Refractive Indices

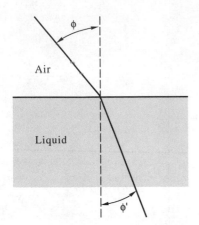

FIG. 4.10   Refraction of light.

The refractive index *n* is a physical constant that, like the boiling point, can be used to characterize liquids. It is the ratio of the velocity of light in air to the velocity of light in the liquid (Fig. 4.10). It is also equal to the ratio of the sine of the angle of incidence ϕ to the sine of the angle of refraction ϕ':

$$n = \frac{\text{Velocity in air}}{\text{Velocity in liquid}} = \frac{\sin \phi}{\sin \phi'}$$

The angle of refraction is also a function of temperature and the wavelength of light (consider the dispersion of white light by a prism). Because the velocity of light in air (strictly speaking, a vacuum) is always greater than that through a liquid, the refractive index is a number greater than 1; for example, hexane n_{D}^{20} 1.3751; diiodobenzene, n_{D}^{20} 1.7179. The superscript 20 indicates that the refractive index was measured at 20°C, and the subscript D refers to the yellow D-line from a sodium vapor lamp, light with a wavelength of 589 nm.

The measurement is made on a refractometer using a few drops of liquid. Compensation is made within the instrument for the fact that white light and not sodium vapor light is used, but a temperature correction must be applied to the observed reading by adding 0.00045 for each degree above 20°C:

$$n_{\mathrm{D}}^{20} = n_{\mathrm{D}}^{t} + 0.00045(t - 20°\text{C})$$

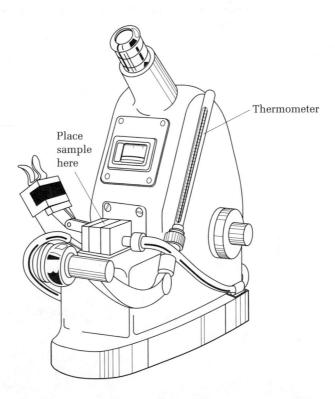

FIG. 4.11   Abbé refractometer. Sample block can be thermostatted.

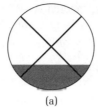
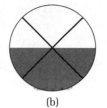
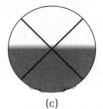

(a)   (b)   (c)

FIG. 4.12   (a) View into refractometer when index knob is out of adjustment. (b) View into refractometer when properly adjusted. (c) View when chromatic adjustment is incorrect.

The refractive index can be determined to 1 part in 10,000, but because the value is quite sensitive to impurities, there is not always very good agreement in the literature with regard to the last figure. For this reason, the refractive indices in this book have been rounded to the nearest part per thousand, as have the refractive indices reported in the Aldrich catalog of chemicals. To master the technique of using the refractometer, measure the refractive indices of several known, pure liquids before measuring an unknown.

Procedure

Refractometers come in many designs. In the most common, the Abbé design (Fig. 4.11), two or three drops of the sample are placed on the open prism using a polyethylene Beral pipette (to avoid scratching the prism face). The prism is closed, and the light is turned on and positioned for maximum brightness as seen through the eyepiece. If the refractometer is set to a nearly correct value, then a partially gray image will be seen, as in Fig. 4.12(a). Turn the knob so that the line separating the dark and light areas is at the crosshairs, as in Fig. 4.12(b). Sometimes the line separating the dark and light areas is fuzzy and colored. Turn the chromatic adjustment until the demarcation line is sharp and colorless. Then read the refractive index. On a newer instrument, press a button to light up the scale in the field of vision. On older models, read the refractive index through a separate eyepiece. Read the temperature on the thermometer attached to the refractometer, and make the appropriate temperature correction to the observed index of refraction.

Cleaning Up   When the measurement is completed, open the prism and wipe off the sample with lens paper using ethanol, acetone, or hexane as necessary.

Questions

1. What effect would poor circulation of the melting point bath liquid have on the observed melting point?

2. What is the effect of an insoluble impurity, such as sodium sulfate, on the observed melting point of a compound?

3. Three test tubes, labeled A, B, and C, contain substances with approximately the same melting points. How could you prove the test tubes contain three different chemical compounds?

4. One of the most common causes of inaccurate melting points is too rapid heating of the melting point bath. Under these circumstances, how will the observed melting point compare with the true melting point?

5. Strictly speaking, why is it incorrect to speak of a melting *point*?

6. What effect would the incomplete drying of a sample (for example, the incomplete removal of a recrystallization solvent) have on the melting point?

7. Why should the melting point sample be finely powdered?

8. You suspect that an unknown is acetanilide (mp 113.5–114°C). Give a qualitative estimation of the melting point when the acetanilide is mixed with 10% by weight of naphthalene.

9. You have an unknown with an observed melting point of 90–93°C. Is your unknown compound A with a reported melting point of 95.5–96°C or compound B with a reported melting point of 90.5–91°C? Explain.

10. Why is it important to heat the melting point bath or block slowly and steadily when the temperature gets close to the mp?

11. Why is it important to pack the sample tightly in the melting point capillary?

12. An unknown compound is suspected to be acetanilide (mp 113.5–114°C). What would happen to the mp if this unknown were mixed with (a) an equal quantity of pure acetanilide? (b) an equal quantity of benzoic acid?

13. Which would be expected to have the higher boiling point, *t*-butyl alcohol (2-methyl-2-propanol) or *n*-butyl alcohol (1-butanol)? Explain.

14. When borosilicate glass (Kimax, Pyrex), n^{20} 1.474, is immersed in a solution having the same refractive index, it is almost invisible. Soft glass with n^{20} 1.52 is quite visible. This is an easy way to distinguish between the two types of glass. Calculate the mole percents of toluene and heptane that will have a refractive index of 1.474 assuming a linear relationship between the refractive indices of the two.

Reference

A. Weissberger and B. W. Rossiter (eds.). *Physical Methods of Chemistry,* Vol. 1, Part V, Wiley-Interscience, New York, 1971.

Surfing the Web

http://ull.chemistry.uakron.edu/organic_lab/melting_point/

The melting points of pure urea and cinnamic acids and mixtures of the two, as well as the melting points of unknowns (Experiments 2, 3, and 4), are carried out on a Thomas–Hoover apparatus and illustrated with 12 extraordinarily clear color photos in this University of Akron site.

For updated information visit:

www.mtholyoke.edu/courses/kwilliam/microscale.shtml

or

www.hmco.com/amco/college/chemistry/Home.html

5

Distillation

Prelab Exercise: Predict what a plot of temperature vs. volume of distillate will look like for the simple distillation and the fractional distillation of (a) a cyclohexane–toluene mixture and (b) an ethanol–water mixture.

The origins of distillation are lost in antiquity as humans in their thirst for more potent beverages found that dilute solutions of alcohol from fermentation could be separated into alcohol-rich and water-rich portions by heating the solution to boiling and condensing the vapors above the boiling liquid—the process of distillation. Since ethyl alcohol, ethanol, boils at 78°C and water boils at 100°C, one might naively assume that heating a 50:50 mixture of ethanol and water to 78°C would cause the ethanol molecules to leave the solution as a vapor that could be condensed to give pure ethanol. Such is not the case. A mixture of 50:50 ethanol:water boils near 87°C, and the vapor above it is not 100% ethanol.

Consider a better-behaved mixture, cyclohexane and toluene. The vapor pressures as a function of temperature are plotted in Fig. 5.1. When the vapor pressure of the liquid equals the applied pressure the liquid boils, so this diagram shows that at 760 mm pressure, standard atmospheric pressure, these pure liquids boil at 78 and 111°C, respectively. If one of these pure liquids were to be distilled,

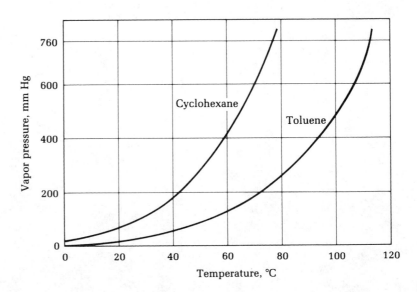

FIG. 5.1 Vapor pressure vs. temperature for cyclohexane and toluene.

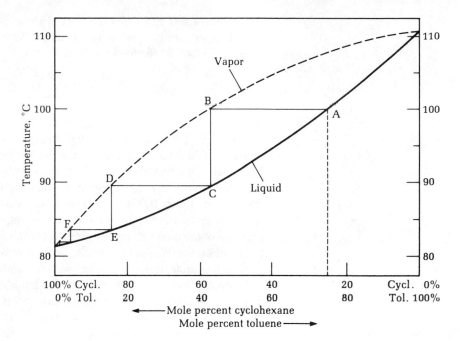

FIG. 5.2 Boiling point–composition curves for cyclohexane–toluene mixtures.

we would find that the boiling point of the liquid would equal the temperature of the vapor and that the temperature of the vapor would remain constant throughout the distillation.

Figure 5.2 is a boiling point–composition diagram for the cyclohexane–toluene system. If a mixture of 75 mole percent toluene and 25 mole percent cyclohexane is heated, we find from Fig. 5.2 that it boils at 100°C, or point A. Above a binary mixture of cyclohexane and toluene the vapor pressure has contributions from each component. Raoult's law states that the vapor pressure of the cyclohexane is equal to the product of the vapor pressure of pure cyclohexane and the mole fraction of cyclohexane in the liquid mixture:

$$P_c = P_c^\circ N_c$$

where P_c is the partial pressure of cyclohexane, P_c° is the vapor pressure of pure cyclohexane at the given temperature, and N_c is the mole fraction of cyclohexane in the mixture. Similarly for toluene:

$$P_t = P_t^\circ N_t$$

and the total vapor pressure above the solution, P_{Tot}, is given by the sum of the partial pressures due to cyclohexane and toluene:

$$P_{Tot} = P_c + P_t$$

Dalton's law states that the mole fraction of cyclohexane in the vapor at a given temperature is equal to the partial pressure of the cyclohexane at that temperature divided by the total pressure:

$$X_c = \frac{P_c}{\text{total vapor pressure}}$$

At 100°C cyclohexane has a partial pressure of 433 mm and toluene a partial pressure of 327 mm; the sum of the partial pressures is 760 mm, and so the liquid boils. If some of the liquid in equilibrium with this boiling mixture were condensed and analyzed, it would be found to be 433/760 or 57 mole percent cyclohexane (point B, Fig. 5.2). This is the best separation that can be achieved on simple distillation of this mixture. As the simple distillation proceeds, the boiling point of the mixture moves toward 110°C along the line from A, and the vapor composition becomes richer in toluene as it moves from B to 110°C. To obtain pure cyclohexane, it would be necessary to condense the liquid at B and redistill it. When this is done it is found that the liquid boils at 90°C (point C) and the vapor equilibrium with this liquid is about 85 mole percent cyclohexane (point D). So to separate a mixture of cyclohexane and toluene, a series of fractions would be collected and each of these partially redistilled. If this fractional distillation were done enough times the two components could be separated.

This series of redistillations can be done "automatically" in a fractionating column. Perhaps the easiest to understand is the bubble cap column used to distill crude oil fractionally. These columns dominate the skyline of oil refineries, some being 150 ft high and capable of distilling 200,000 barrels of crude oil per day. The crude oil enters the column as a hot vapor (Fig. 5.3). Some of this vapor with high-boiling components condenses on one of the plates. The more volatile substances travel through the bubble cap to the next higher plate where some of the less-volatile components condense. As high-boiling liquid material accumulates on a plate it descends through the overflow pipe to the next lower plate, and vapor rises through the bubble cap to the next higher plate. The temperature of the vapor that is rising through a cap is above the boiling point of the liquid on that plate. As bubbling takes place, heat is exchanged, and the less volatile components on that plate vaporize and go on to the next plate. The composition of the liquid on a plate is the same as that of the vapor coming from the plate below. So on each plate a simple distillation takes place. At equilibrium, vapor containing low-boiling material is ascending, and high-boiling liquid is descending through the column.

Figure 5.2 shows that the condensations and redistillations in a bubble cap column consisting of three plates correspond to moving on the boiling point–composition diagram from point A to point E.

In the laboratory the successive condensations and distillations that occur in the bubble cap column take place in a distilling column. The column is packed with some material on which heat exchange between ascending vapor and descending liquid can take place. A large surface area for this packing is desirable, but the packing cannot be so dense that pressure changes take place within

Fractional distillation

Overflow pipe

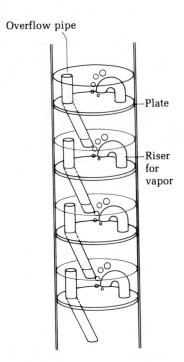

FIG. 5.3 Bubble plate distilling column.

Heat exchange between ascending vapor and descending liquid

Holdup: unrecoverable distillate that wets the column packing

Column packing

Height equivalent to a theoretical plate (HETP)

Equilibration is slow.

Good fractional distillation takes a long time.

the column causing nonequilibrium conditions. Also, if the column packing has a very large surface area, it will absorb (hold up) much of the material being distilled. A number of different packings for distilling columns have been tried—glass beads, glass helices, carborundum chips, etc. We find one of the best packings is a copper or steel sponge (Chore Boy). It is easy to put into the column, does not come out of the column as beads do, and has a large surface area, good heat transfer characteristics, and low holdup. It can be used in both microscale and macroscale apparatus.

The ability of different column packings to separate two materials of differing boiling points is evaluated by calculating the number of theoretical plates, each theoretical plate corresponding to one distillation and condensation as discussed above. Other things being equal, the number of theoretical plates is proportional to the height of the column, so various packings are evaluated according to the height equivalent to a theoretical plate (HETP); the smaller the HETP, the more plates the column will have and the more efficient it will be. The calculation is made by analyzing the proportion of lower- to higher-boiling material at the top of the column and in the distillation pot.[1]

Although not obvious, the most important variable contributing to a good fractional distillation is the rate at which the distillation is carried out. A series of simple distillations takes place within a fractionating column, and it is important that complete equilibrium be attained between the ascending vapors and the descending liquid. This process is not instantaneous. It should be an adiabatic process; that is, heat should be transferred from the ascending vapor to the descending liquid with no net loss or gain of heat. In larger, more complex distilling columns, a means is provided for adjusting the ratio between the amount of material that boils up and condenses (refluxes) and is returned to the column (thus allowing equilibrium to take place) and the amount that is removed as distillate. A reflux ratio of 30:1 or 50:1 would not be uncommon for a 40-plate column; distillation would take several hours.

Carrying out a fractional distillation on the truly microscale (<1 mg) is impossible, and it is even impossible on a small scale (10–400 mg). It can be carried out on a 4-mL scale. As seen in later chapters, various types of chromatography are employed for the separation of micro and semimicro quantities of material, while distillation is the best method for separating more than a few grams of material.

Azeotropes

Not all liquids form ideal solutions and conform to Raoult's law. Ethanol and water are such liquids. Because of molecular interaction, a mixture of 95.5% (by weight) of ethanol and 4.5% of water boils *below* (78.15°C) the boiling point of pure ethanol (78.3°C). Thus, no matter how efficient the distilling apparatus,

---

1. See A. Weissberger (ed.), *Technique of Organic Chemistry,* Vol. IV, "Distillation," Wiley-Interscience, New York, 1951.

The ethanol–water azeotrope

100% ethanol cannot be obtained by distillation of a mixture of, say, 75% water and 25% ethanol. A mixture of liquids of a certain definite composition that distills at a constant temperature without change in composition is called an *azeotrope;* 95% ethanol is such an azeotrope. The boiling point–composition curve for the ethanol–water mixture is seen in Fig. 5.4. To prepare 100% ethanol the water can be removed chemically (reaction with calcium oxide) or by removal of the water as an azeotrope (with still another liquid). An azeotropic mixture of 32.4% ethanol and 67.6% benzene (bp 80.1°C) boils at 68.2°C. A ternary azeotrope (bp 64.9°C) contains 74.1% benzene, 18.5% ethanol, and 7.4% water. Absolute alcohol (100% ethanol) is made by addition of benzene to 95% alcohol and removal of the water in the volatile benzene–water–alcohol azeotrope.

Ethanol and water form a minimum boiling azeotrope. Other substances, such as formic acid (bp 100.7°C) and water (bp 100°C), form maximum boiling azeotropes. For these two compounds the azeotrope boils at 107.3°C.

A pure liquid has a constant boiling point. A change in boiling point during distillation is an indication of impurity. The converse proposition, however, is not always true, and constancy of a boiling point does not necessarily mean that the liquid consists of only one compound. For instance, two miscible liquids of similar chemical structure that boil at the same temperature individually will have nearly the same boiling point as a mixture. And, as noted previously,

A constant bp on distillation does not guarantee *that the distillate is one pure compound.*

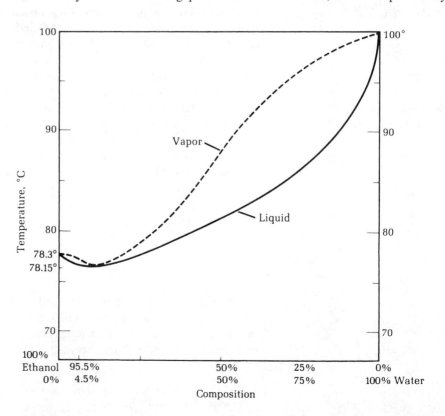

FIG. 5.4 Boiling point–composition curves for ethanol–water mixtures.

TABLE 5.1 Variation in Boiling Point with Pressure

Pressure (mm)	Water (°C)	Benzene (°C)
780	100.7	81.2
770	100.4	80.8
760	100.0	80.3
750	99.6	79.9
740	99.2	79.5
584*	92.8	71.2

*Instituto de Quimica, Mexico City, altitude 7,700 ft (2,310 m).

azeotropes have constant boiling points that can be either above or below the boiling points of the individual components.

Distilling a mixture of sugar and water

When a solution of sugar in water is distilled, the boiling point recorded on a thermometer located in the vapor phase is 100°C (at 760 torr) throughout the distillation, whereas the temperature of the boiling sugar solution itself is initially somewhat above 100°C and continues to rise as the concentration of sugar in the remaining solution increases. The vapor pressure of the solution is dependent on the number of water molecules present in a given volume; and hence with increasing concentration of nonvolatile sugar molecules and decreasing concentration of water, the vapor pressure at a given temperature decreases and a higher temperature is required for boiling. However, sugar molecules do not leave the solution, and the drop clinging to the thermometer is pure water in equilibrium with pure water vapor.

Bp changes with pressure.

When a distillation is carried out in a system open to the air and the boiling point is thus dependent on existing air pressure, the prevailing barometric pressure should be noted and allowance made for appreciable deviations from the accepted boiling point temperature (see Table 5.1). Distillation can also be done at the lower pressures that can be achieved by an oil pump or an aspirator with substantial reduction of boiling point.

Experiments

MICROSCALE

1. Calibration of Thermometer

If you have not previously carried out a calibration, test the 0°C point of your thermometer with a well-stirred mixture of crushed ice and distilled water. To check the 100°C point, put 2 mL of water in a test tube with a boiling chip to prevent bumping and boil the water gently over a hot sand bath with the thermometer in the vapor from the boiling water. Take care to see that the thermometer does not touch the side of the test tube. Then immerse the bulb of the thermometer in the liquid, and see if you can observe superheating. Check the atmospheric pressure to determine the true boiling point of the water.

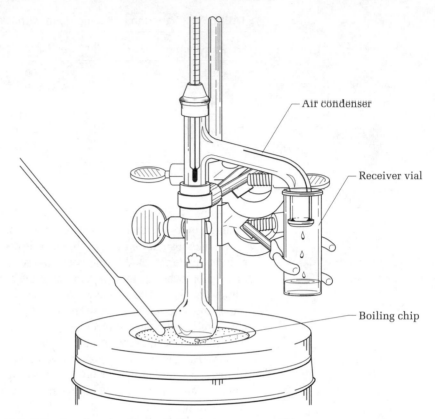

FIG. 5.5   Small-scale simple distillation apparatus. This apparatus can be adapted for fractional distillation by packing the long neck with a copper sponge. The temperature is regulated by either scraping sand away from or piling sand up around the flask.

▌ MICROSCALE

Apparatus for simple distillation.

Apparatus

In this experiment, the two liquids to be separated are placed in a 5-mL round-bottomed long-necked flask that is fitted to a distilling head (Fig. 5.5). The flask has a larger surface area exposed to heat than the reaction tube, so the necessary thermal energy can be put into the system to cause the materials to distill. The hot vapor rises and completely envelops the bulb of the thermometer before passing over it and down toward the receiver. The downward-sloping portion of the distilling head functions as an air condenser.

The rate of distillation is determined by the heat input to the apparatus. This is most easily controlled by using a spatula to pile up or scrape away hot sand from around the flask. This works very well. The effect of a column packing on separation efficiency can be evaluated by carrying out a careful distillation of cyclohexane and toluene in this apparatus and then packing the long neck of the flask with a copper sponge and repeating the distillation. Fractional distillation is best carried out using the 10-cm packed column (Fig. 5.6).

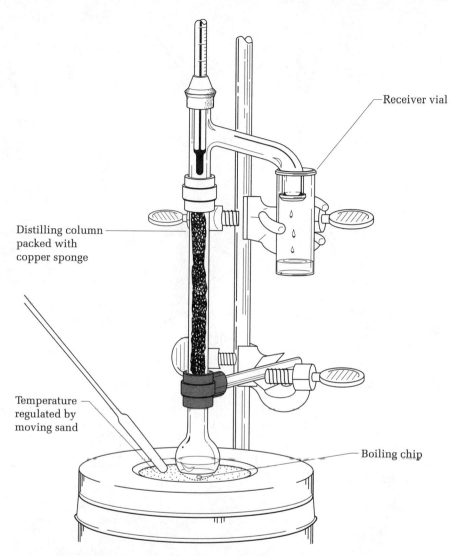

Receiver vial

Distilling column
packed with
copper sponge

Temperature
regulated by
moving sand

Boiling chip

FIG. 5.6   Small-scale fractional distillation apparatus. The 10-cm column is
packed with 1.5 g of copper sponge (Chore Boy).

*Another simple distillation
apparatus.*

Another design for a simple distillation apparatus is shown in Fig. 5.7. The
long air condenser will condense even low-boiling liquids, and the receiver is far
from the heat.

 **MICROSCALE**

2. Instant Microscale Distillation

Frequently, a very small quantity of freshly distilled material is needed in an
experiment. For example, two compounds that need to be distilled freshly are aniline, which turns black because of the formation of oxidation products, and
benzaldehyde, a liquid that easily oxides to solid benzoic acid. The impurities in

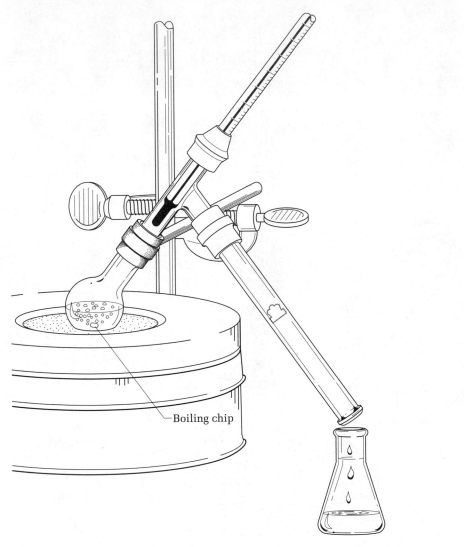

FIG. 5.7 Simple distillation apparatus.

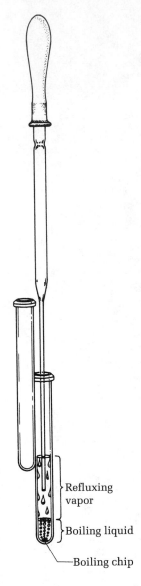

FIG. 5.8 Apparatus for instant microscale distillation.

both these compounds have much higher boiling points than the parent substance, so a very simple distillation suffices to separate them. This is accomplished as follows.

Place a few drops of the impure liquid in a reaction tube along with a boiling chip. Clamp the tube in a hot sand bath, and adjust the heat so that the liquid refluxes about halfway up the tube. Expel the air from a Pasteur pipette, thrust it down in the hot vapor, and then pull the hot vapor into the cold upper portion of the pipette. The vapor will immediately condense, and it can then be expelled into another reaction tube held adjacent to the hot one (Fig. 5.8). In this way enough pure material can be distilled to determine a boiling point, run a spectrum, make a derivative, or carry out a reaction. Sometimes the first drop or two will be cloudy, indicating the presence of water. This fraction should be discarded in order to obtain pure dry material.

Experiments

MICROSCALE AND MACROSCALE

1. Calibration of Thermometer

If you have not previously carried out a calibration, test the 0°C point of your thermometer with a well-stirred mixture of crushed ice and distilled water. To check the 100°C point, put 2 mL of water in a reaction tube with a boiling chip to prevent bumping, and boil the water gently over a hot sand bath with the thermometer in the vapor from the boiling water. Take care to see that the thermometer does not touch the side of the reaction tube. Then immerse the bulb of the thermometer in the liquid, and see if you can observe superheating. Check the atmospheric pressure to determine the true boiling point of the water.

▐ MICROSCALE

2. Simple Distillation

(A) Simple Distillation of a Cyclohexane–Toluene Mixture. To a 5-mL long-necked round-bottomed flask is added 2.0 mL of dry cyclohexane, 2.0 mL of dry toluene, and a boiling chip (see Fig. 5.5). This flask is joined by means of a Viton (black) connector to a distilling head fitted with a thermometer using a rubber connector. The thermometer bulb should be completely below the side arm of the Claisen head so that the mercury reaches the same temperature as the vapor that distills. The end of the distilling head dips well down into a vial, which rests on the bottom of a 30-mL beaker filled with ice. The distillation is started by piling up hot sand to heat the flask. As soon as boiling starts, the vapors can be seen to rise up the neck of the flask. Adjust the rate of heating by piling up or scraping away sand from the flask so that it takes *several minutes* for the vapor to rise to the thermometer: **The rate of distillation should be no faster than two drops per minute.**

Viton is resistant to hot aromatic vapors.

The thermometer bulb must be completely below the side arm.

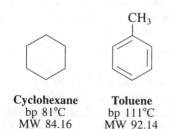

Cyclohexane
bp 81°C
MW 84.16
n_D^{20} 1.4260

Toluene
bp 111°C
MW 92.14
n_D^{20} 1.4960

There are 21 ± 3 drops per milliliter.

Record the temperature versus the number of drops during the entire distillation. If the rate of distillation is as slow as it should be, there will be sufficient time between drops to read and record the temperature. Continue the distillation until only about 0.4 mL remains in the distilling flask. At the end of the distillation, measure as accurately as possible, perhaps using a syringe, the volume of the distillate and, after it cools, the volume left in the pot; the difference is the holdup of the column if none has been lost by evaporation. Note the barometric pressure, make any thermometer corrections necessary, and make a plot of milliliters (drop number) versus temperature for the distillation.

Cleaning Up The pot residue should be placed in the organic solvents container. The distillate can be placed there, or it can be recycled.

(B) Simple Distillation of an Ethanol–Water Mixture. In a 5-mL round-bottomed long-necked flask place 4 mL of a 10% to 20% ethanol–water mixture. It could come from the fermentation of glucose (Chapter 59). Assemble the apparatus as described above, and carry out the distillation until you believe a representative sample of ethanol has collected in the receiver. In the hood place three drops of this sample on a watch glass, and try to ignite it with the blue cone of a microburner flame. Does it burn? Is any unburned residue observed? There was a time when alcohol–water mixtures were mixed with gunpowder and ignited to give proof that the alcohol had not been diluted. One hundred proof alcohol is 50% ethanol by volume.

Cleaning Up   The distillate and pot residue can be flushed down the drain.

MICROSCALE

Adjust the heat input to the flask by piling up or scraping away sand around the flask.

Never distill in an airtight system.

Insulate the distilling column but not the head.

21 ± 3 drops = 1 mL

3. Fractional Distillation

Apparatus

Assemble the apparatus shown in Fig. 5.6. The 10-cm column is packed with 1.5 g of copper sponge and connected to the 5-mL short-necked flask using the black (Viton) connector. The column should be vertical, and care should be taken to ensure that the bulb of the thermometer does not touch the side of the distilling head. The column, but not the distilling head, will be insulated with glass wool or cotton at the appropriate time to ensure that the process is adiabatic. Alternatively, the column can be insulated with a cut-off 15-mL polyethylene-centrifuge tube.

(A)   Fractional Distillation of a Cyclohexane–Toluene Mixture. To the short-necked flask is added 2.0 mL of cyclohexane, 2.0 mL of toluene, and a boiling chip. The distilling column is packed with 1.5 g of copper sponge (Fig. 5.6). The mixture is brought to a boil over a hot sand bath. Observe the ring of condensate that should rise slowly through the column; if you cannot at first see this ring, locate it by touching the column with the fingers. Reduce the heat by scraping sand away from the flask, and wrap the column, but not the distilling head, with glass wool or cotton if it is not already insulated.

The distilling head and the thermometer function as a small reflux condenser. Again, apply the heat, and as soon as the vapor reaches the thermometer bulb, reduce the heat by scraping away sand. **Distill the mixture at a rate no faster than two drops per minute,** and record the temperature as a function of the number of drops. If the heat input has been *very* carefully adjusted, the distillation will cease and the temperature reading will drop after the cyclohexane has distilled. Increase the heat input by piling up the sand around the flask in order to cause the toluene to distill. Stop the distillation when only about 0.4 mL remains in the flask, and measure the volume of distillate and the pot residue as before. Make a plot of boiling point versus milliliters of distillate (drops), and compare it to the simple distillation carried out in the same apparatus. Compare your results with those of Fig. 5.9.

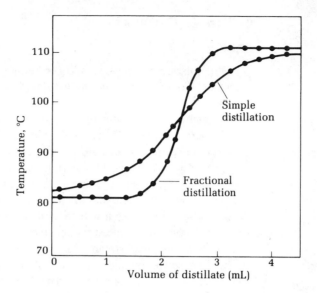

FIG. 5.9 Simple and fractional distillation curves for cyclohexane and toluene.

Cleaning Up   The pot residue should be placed in the organic solvents container. The distillate can be placed there or recycled.

(B) Fractional Distillation of an Ethanol–Water Mixture.   Distill 4 mL of the same ethanol–water mixture used in the simple distillation experiment following the procedure used for cyclohexane–toluene with either the short or the long distilling column. Remove what you regard to be the ethanol fraction, and repeat the ignition test. Is any difference noted?

Cleaning Up   The pot residue and distillate can be flushed down the drain.

(C) Fractional Distillation of an Unknown Mixture.   Distill 4 mL of a mixture of two of the compounds from Table 5.2 (at the end of the chapter). The two liquids will have boiling points at least 20°C apart. The composition of the mixture will be 20/80, 30/70, 40/60, 50/50, 60/40, 70/30, or 80/20 percents of the two components. Identify the two compounds, and determine the percent composition of each.

Once the mixture has been distilled and the volumes of the lower- and higher-boiling components determined, the mixture might be redistilled and separated into three fractions: the lower-boiling one, a small middle fraction, and the higher-boiling one. The composition and purity of the lower- and higher-boiling fractions can be determined by measuring the refractive indices of each.

Cleaning Up   Place all fractions and the pot residue in the organic solvents waste container.

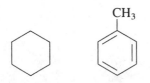 **MACROSCALE**

4. Simple Distillation

Apparatus

In any distillation, the flask should not be more than two-thirds full at the start. Great care should be taken not to distill to dryness because, in some cases, high-boiling explosive peroxides can become concentrated.

Assemble the apparatus for simple distillations shown in Fig. 5.10, starting with the support ring followed by the electric flask heater (Fig. 1.16n) and then the flask. One or two boiling stones are put in the flask to promote even boiling. Each ground joint is greased by putting three or four stripes of grease lengthwise around the male joint and pressing the joint firmly into the other without twisting. The air is thus eliminated, and the joint will appear almost transparent. (Do not use excess grease because it will contaminate the product.) Water enters the condenser at the tublature nearest the receiver. Because of the large heat capacity of water, only a very small stream (3 mm diameter) is needed; too much water pressure will cause the tubing to pop off. A heavy rubber band, or better, a Keck clamp, can be used to hold the condenser to the distillation head. Note that the bulb of the thermometer is below the opening into the side arm of the distillation head.

CAUTION: Cyclohexane and toluene are flammable; make sure distilling apparatus is tight.

CH₃

Cyclohexane
MW 84.16
bp 81.4°C
den 0.78
n_D^{20} 1.4260

Toluene
MW 92.14
bp 110.8°C
den 0.87
n_D^{20} 1.4960

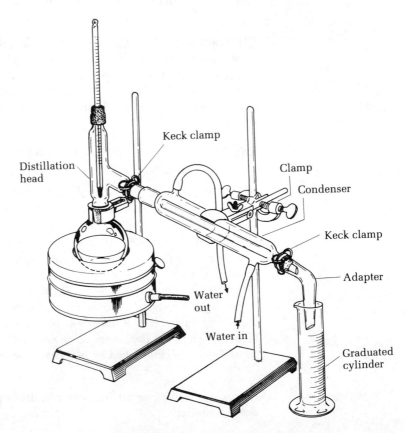

FIG. 5.10 Apparatus for simple distillation.

Do not add a boiling chip to a hot liquid. It may boil over.

Dispose of cyclohexane and toluene in the container provided. Do not pour them down the drain.

(A) Simple Distillation of a Cyclohexane–Toluene Mixture.

Place a mixture of 30 mL cyclohexane and 30 mL toluene and a boiling chip in a dry 100-mL round-bottomed flask, and assemble the apparatus for simple distillation. After making sure all connections are tight, heat the flask strongly until boiling starts. Then adjust the heat until the distillate drops at a regular rate of about one drop per second. Record both the temperature and the volume of distillate at regular intervals. After 50 mL of distillate is collected, discontinue the distillation. Record the barometric pressure, make any thermometer correction necessary, and plot boiling point versus volume of distillate. Save the distillate for fractional distillation.

Cleaning Up   The pot residue should be placed in the organic solvents container. The distillate also can be placed there or it can be recycled.

A 10% to 12% solution of ethanol in water is produced by fermentation. See Chapter 59.

(B) Simple Distillation of an Ethanol–Water Mixture.

In a 500-mL round-bottomed flask place 200 mL of a 20% aqueous solution of ethanol. Follow the procedure (above) for the distillation of a cyclohexane–toluene mixture. Discontinue the distillation after 50 mL of distillate has been collected. In the hood place three drops of distillate on a Pyrex watch glass, and try to ignite it with the blue cone of a microburner flame. Does it burn? Is any unburned residue observed?

Cleaning Up   The pot residue and distillate can be flushed down the drain.

 MACROSCALE

5. Fractional Distillation

Apparatus

Assemble the apparatus shown in Figs. 5.11 and 5.12. The fractionating column is packed with one-fourth to one-third of a metal sponge. The column should be perfectly vertical, and it should be insulated with glass wool covered with aluminum foil with the shiny side in. However, in order to observe what is taking place within the column, insulation is omitted for this experiment.

(A) Fractional Distillation of a Cyclohexane–Toluene Mixture.

After the flask from the simple distillation experiment has cooled, pour the 50 mL of distillate back into the distilling flask, add one or two new boiling chips, and assemble the apparatus for fractional distillation. The stillhead delivers into a short condenser fitted with a bent adapter leading into a 10-mL graduated cylinder. Gradually turn up the heat to the electric flask heater until the mixture of cyclohexane and toluene just begins to boil. As soon as boiling starts, turn down the power. Heat slowly at first. A ring of condensate will rise slowly through the column; if you cannot at first see this ring, locate it by cautiously touching the column with the fingers. The rise should be very gradual, in order that the column can acquire a uniform temperature gradient. Do not apply more heat until you are sure that the ring of condensate has stopped rising; then increase the heat gradu-

FIG. 5.11 Apparatus for fractional distillation. The position of the thermometer bulb is critical.

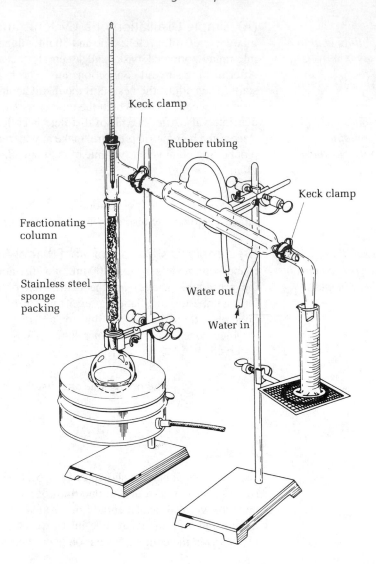

Keck clamp

Rubber tubing

Keck clamp

Fractionating column

Stainless steel sponge packing

Water out

Water in

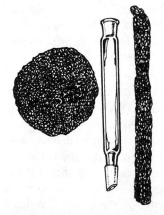

FIG. 5.12 Fractionating column and packing. Use one-third of sponge (Chore Boy).

ally. In a properly conducted operation, the vapor-condensate mixture reaches the top of the column only after several minutes. Once distillation has commenced, it should continue steadily without any drop in temperature at a rate not greater than 1 mL in 1.5–2 min. Observe the flow, and keep it steady by slight increases in heat as required. Protect the column from drafts by wrapping it with aluminum foil, glass wool, or even a towel. This insulation will help prevent flooding of the column, as will slow and steady distillation.

Record the temperature as each milliliter of distillate collects, and make more frequent readings when the temperature starts to rise abruptly. Each time the graduated cylinder fills, quickly empty it into a series of labeled 25-mL Erlenmeyer flasks. Stop the distillation when a second constant temperature is reached. Plot a distillation curve, and record what you observed inside the column in the course of the fractionation. Combine the fractions that you think are pure, and turn in the product in a bottle labeled with your name, desk number, the name of the product, the bp range, and the weight.

Cleaning Up   The pot residue should be placed in the organic solvents container. The cyclohexane and toluene fractions can be placed there also, or they can be recycled.

(B) Fractional Distillation of Ethanol–Water Mixtures. Place the 50 mL of distillate from the simple distillation experiment in a 100-mL round-bottomed flask, add one or two boiling chips, and assemble the apparatus for fractional distillation. Follow the procedure (above) for the fractional distillation of a cyclohexane–toluene mixture. Repeat the ignition test. Is any difference noted? Alternatively distill 60 mL of the 10–20% ethanol–water mixture that results from the fermentation of glucose (Chapter 65).

Cleaning Up   The pot residue and distillate can be flushed down the drain.

MICROSCALE AND MACROSCALE

6. Unknowns

You will be supplied with an unknown, prepared by the instructor, that is a mixture of two solvents from those listed in Table 5.2, only two of which form azeotropes. The solvents in the mixture will be mutually soluble and differ in boiling point by more than 20°C. Fractionate the unknown, and identify the components from the boiling points. Prepare a distillation curve. You may be directed to analyze your distillate by gas chromatography (Chapter 11) or refractive index (Chapter 4).

Cleaning Up   Organic material goes in the organic solvents container. Water and aqueous solutions can be flushed down the drain.

TABLE 5.2 Some Properties of Common Solvents

Solvent	Boiling Point (°C)
Acetone	56.5
Methanol	64.7
Hexane	68.8
1-Butanol	117.2
2-Methyl-2-propanol	82.2
Water	100.0
Toluene*	110.6

*Methanol and toluene form an azeotrope, bp 63.8°C (69% methanol).

Questions

1. In the simple distillation experiment (2), can you account for the boiling point of your product in terms of the known boiling points of the pure components of your mixture?

2. From the plot of boiling point versus volume of distillate in the simple distillation experiment, what can you conclude about the purity of your product?

3. From the boiling point versus volume of distillate plot in the fractional distillation of the cyclohexane–toluene mixture (3), what conclusion can you draw about the homogeneity of the distillate?

4. From the boiling point versus volume of distillate in the fractional distillation of the ethanol–water mixture (3), what conclusion can you draw about the homogeneity of the distillate? Does it have a constant boiling point? Is it a pure substance because it has a constant boiling point?

5. What is the effect on the boiling point of a solution (e.g., water) produced by a soluble nonvolatile substance (e.g., sodium chloride)? What is the effect of an insoluble substance such as sand or charcoal? What is the temperature of the vapor above these two boiling solutions?

6. In the distillation of a pure substance (e.g., water), why doesn't all the water vaporize at once when the boiling point is reached?

7. In fractional distillation, liquid can be seen running from the bottom of the distillation column back into the distilling flask. What effect does this returning condensate have on the fractional distillation?

8. Why is it extremely dangerous to attempt to carry out a distillation in a completely closed apparatus (one with no vent to the atmosphere)?

9. Why is better separation of two liquids achieved by slow rather than fast distillation?

10. Explain why a packed fractionating column is more efficient than an unpacked one.

11. In the distillation of the cyclohexane–toluene mixture, the first few drops of distillate may be cloudy. Explain.

12. What effect does the reduction of atmospheric pressure have on the boiling point? Can cyclohexane and toluene be separated if the external pressure is 350 mm Hg instead of 760 mm Hg?

13. When water-cooled condensers are used for distillation or for refluxing a liquid, the water enters the condenser at the lowest point and leaves at the highest. Why?

6

Steam Distillation

Prelab Exercise: If a mixture of toluene and water is distilled at 97°C, what weight of water would be necessary to carry over 1 g of toluene?

Each liquid exerts its own vapor pressure.

When a mixture of cyclohexane and toluene is distilled (Chapter 5) the boiling point of these two miscible liquids is *between* the boiling points of each of the pure components. By contrast, if a mixture of benzene and water (immiscible liquids) is distilled, the boiling point of the mixture will be *below* the boiling point of each pure component. Since the two liquids are essentially insoluble in each other, the benzene molecules in a droplet of benzene are not diluted by water molecules from nearby water droplets, and hence the vapor pressure exerted by the benzene is the same as that of benzene alone at the existing temperature. The same is true of the water present. Because they are immiscible, the two liquids independently exert pressures against the common external pressure, and when the sum of the two partial pressures equals the external pressure, boiling occurs. Benzene has a vapor pressure of 760 torr at 80.1°C, and if it is mixed with water the combined vapor pressure must equal 760 torr at some temperature below 80.1°C. This temperature, the boiling point of the mixture, can be calculated from known values of the vapor pressures of the separate liquids at that temperature. Vapor pressures found for water and benzene in the range 50–80°C are plotted in Fig. 6.1. The dotted line cuts the two curves at points where the sum of the vapor pressures is 760 torr; hence this temperature is the boiling point of the mixture (69.3°C).

The boiling point of immiscible liquids is below *the boiling point of either pure liquid.*

Practical use can sometimes be made of the fact that many water-insoluble liquids and solids behave as benzene does when mixed with water, volatilizing at temperatures below their boiling points. Thus, naphthalene, a solid, boils at 218°C but distills with water at a temperature below 100°C. Since naphthalene is not very volatile, considerable water is required to entrain it, and the conventional way of conducting the distillation is to pass steam into a boiling flask containing naphthalene and water. This process is called *steam distillation*. With more volatile compounds, or with a small amount of material, the substance can be heated with water in a simple distillation flask and the steam generated *in situ*.

Some high-boiling substances decompose before the boiling point is reached and, if impure, cannot be purified by ordinary distillation. However, they can be freed from contaminating substances by steam distillation at a lower temperature, at which they are stable. Steam distillation also offers the advantage of selectivity, because some water-insoluble substances are volatile with steam and others are not, and some volatilize so very slowly that sharp separation is possible. The technique is useful in processing natural oils and resins, which can be separated

Used for isolation of perfume and flavor oils.

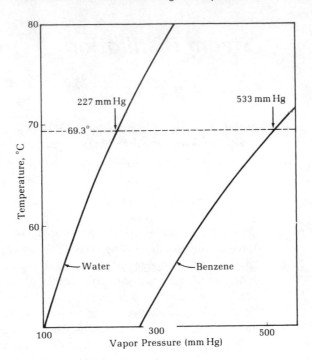

FIG. 6.1 Vapor pressure vs. temperature curves for water and benzene.

into steam-volatile and non–steam-volatile fractions. It is useful for recovery of a non–steam-volatile solid from its solution in a high-boiling solvent such as nitrobenzene, bp 210°C; all traces of the solvent can be eliminated and the temperature can be kept low.

The boiling point remains constant during a steam distillation as long as adequate amounts of both water and the organic component are present to saturate the vapor space. Determination of the boiling point and correction for any deviation from normal atmospheric pressure permits calculation of the amount of water required for distillation of a given amount of organic substance. According to Dalton's law, the molecular proportion of the two components in the distillate is equal to the ratio of their vapor pressures (p) in the boiling mixture; the more volatile component contributes the greater number of molecules to the vapor phase. Thus,

Dalton's law

$$\frac{\text{Moles of water}}{\text{Moles of substance}} = \frac{p_{\text{water}}}{p_{\text{substance}}}$$

The vapor pressure of water (p_{water}) at the boiling temperature in question can be found by interpolation of the data of Table 6.1, and that of the organic substance is, of course, equal to $760 - p_{\text{water}}$. Hence, the weight of water required per gram of substance is given by the expression

$$\frac{\text{Wt of water per}}{\text{g of substance}} = \frac{18 \times p_{\text{water}}}{\text{MW of substance} \times (760 - p_{\text{water}})}$$

TABLE 6.1   Vapor Pressure of Water in Millimeters of Mercury

t (°C)	p	t (°C)	p	t (°C)	p	t (°C)	p
60	149.3	70	233.7	80	355.1	90	525.8
61	156.4	71	243.9	81	369.7	91	546.0
62	163.8	72	254.6	82	384.9	92	567.0
63	171.4	73	265.7	83	400.6	93	588.6
64	179.3	74	277.2	84	416.8	94	610.9
65	187.5	75	289.1	85	433.6	95	633.9
66	196.1	76	301.4	86	450.9	96	657.6
67	205.0	77	314.1	87	468.7	97	682.1
68	214.2	78	327.3	88	487.1	98	707.3
69	223.7	79	341.0	89	506.1	99	733.2

From the data given in Fig. 6.1 for benzene–water, the fact that the mixture boils at 69.3°C, and the molecular weight of 78.11 for benzene, the water required for steam distillation of 1 g of benzene is only $227 \times \frac{18}{533} \times \frac{1}{78} = 0.10$ g. Nitrobenzene (bp 210°C, MW 123.11) steam distills at 99°C and requires 4.0 g of water per gram. The low molecular weight of water makes water a favorable liquid for two-phase distillation of organic compounds.

Steam distillation is employed in a number of experiments in this text. On a small scale steam is generated in the flask that contains the substance to be steam distilled by simply boiling a mixture of water and the immiscible substance.

Apparatus

The apparatus (Fig. 6.2) consists of a 5-mL short-necked boiling flask, addition port, and distilling head leading to an ice-cooled receiver. As the steam distillation proceeds, it may be necessary to add more water to the flask. This is done via syringe through the septum on the addition port.

Isolation of Citral

Terpenes

Citral is an example of a very large group of *natural products* called *terpenes*. They are responsible for the characteristic odors of plants such as eucalyptus, pine, mint, peppermint, and lemon. The odors of camphor, menthol, lavender, rose, and hundreds of other fragrances are due to terpenes, which have ten carbon atoms with double bonds, and aldehyde, ketone, or alcohol functional groups. (See Fig. 6.3.)

In nature these terpenes all arise from a common precursor, isopentenyl pyrophosphate. At one time they were thought to come from the simple diene, isoprene (2-methyl-1,3-butadiene), because the skeletons of terpenes can be dissected into isoprene units, having five carbon atoms arranged as in *The isoprene rule* 2-methylbutane. These isoprene units are almost always arranged in a "head-to-tail" fashion.

FIG. 6.2 Small-scale steam distillation apparatus. Add water via syringe at A.

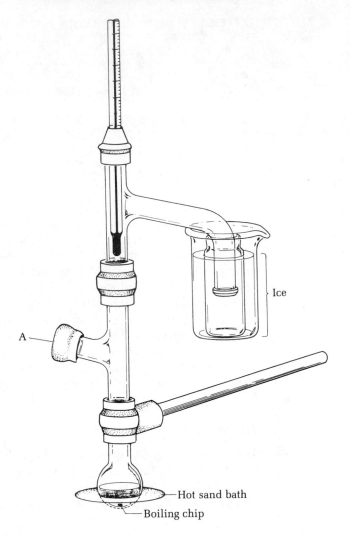

A

Ice

Hot sand bath

Boiling chip

In the present experiment citral is isolated by steam distillation of lemongrass oil, which is used to make lemongrass tea, a popular drink in Mexico. The distillate contains 90% citral and 10% neral, the isomer about the 2,3-bond. Citral is used in a commercial synthesis of vitamin A.

Lemongrass oil contains a number of substances; simple or fractional distillation would not be a practical method for obtaining pure citral. And because it boils at 229°C, it has a tendency to polymerize, oxidize, and decompose during distillation. For example, heating with potassium bisulfate, an acidic compound, converts citral to 1-methyl-4-isopropylbenzene (p-cymene).

Steam distillation: the isolation of heat-sensitive compounds

Steam distillation is thus a very gentle method for isolating citral. The distillation takes place below the boiling point of water. The distillate consists of a mixture of citral (and some neral) and water. It is isolated by shaking the mixture with *t*-butyl methyl ether. The citral dissolves in the *t*-butyl methyl ether, which

FIG. 6.3 Structures of some terpenes.

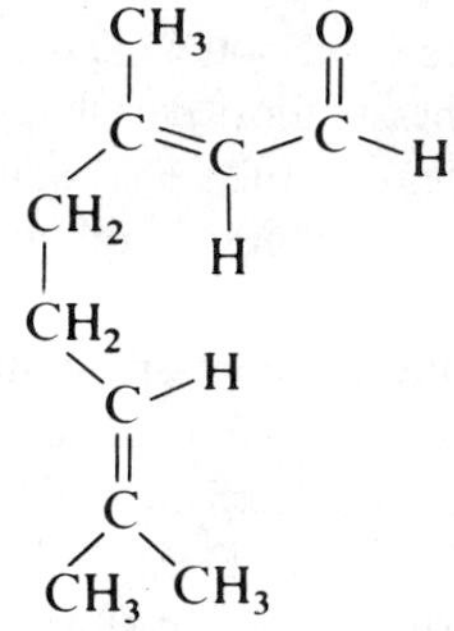

Citral

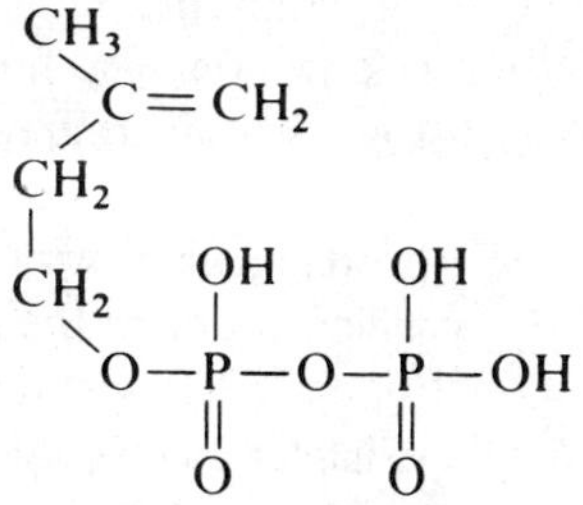

Isopentenyl pyrophosphate

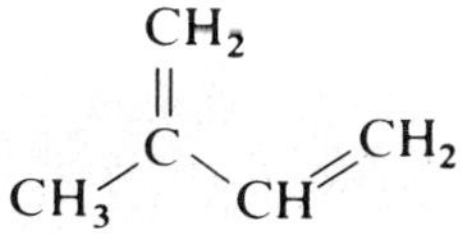

Isoprene
(2-Methyl-1,3-butadiene)

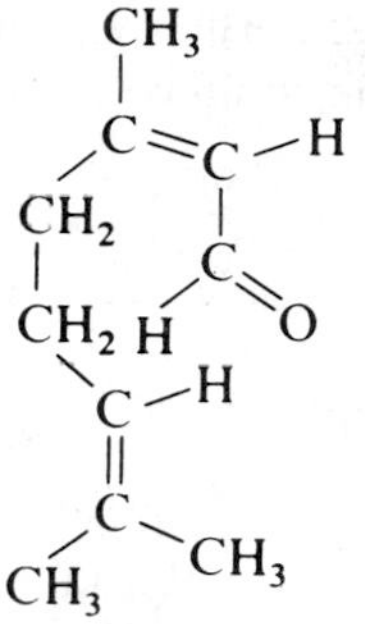

Neral

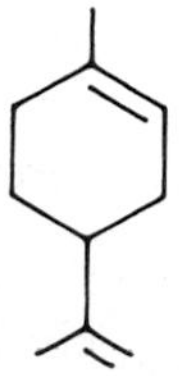

Limonene
(lemons)

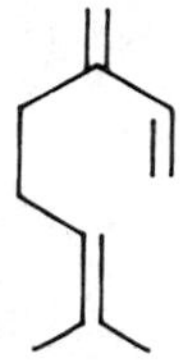

Myrcene
(bayberry)

α-Pinene
(turpentine)

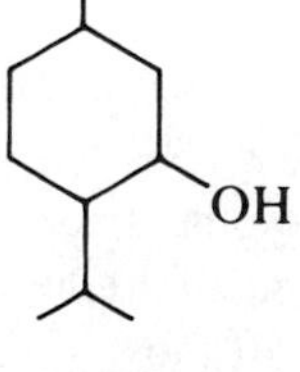

Menthol
(mint)

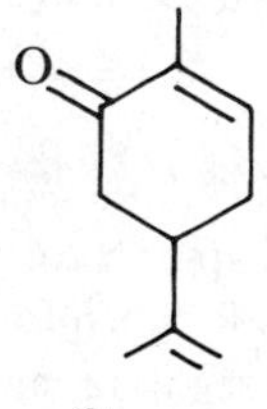

Carvone
(caraway seeds)

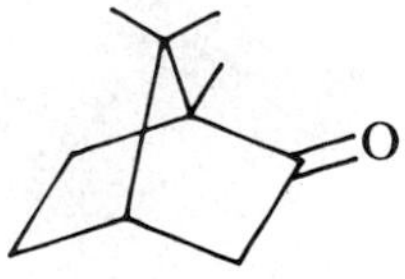

Camphor

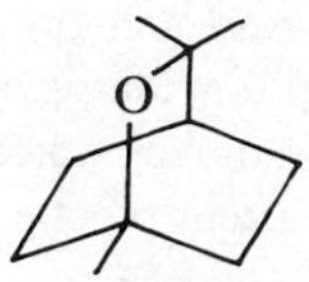

1,8-Cineole
(eucalyptus)

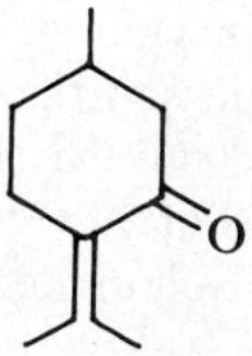

Pulegone
(pennyroyal oil)

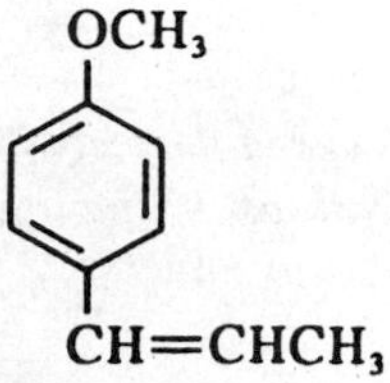

Anethole
(anise seed)

Citral and neral are geometric isomers.

is immiscible with water, and the two layers are separated. The *t*-butyl methyl ether is dried (it dissolves some water) and evaporated to leave citral.

Store the citral in the smallest possible container in the dark for later characterization. The homogeneity of the substance can be investigated using thin-layer chromatography (Chapter 9) or gas chromatography (Chapter 11). Chemically the molecule can be characterized by reaction with bromine and also permanganate, which shows it contains double bonds, and by the Tollens test, which confirms the presence of an aldehyde (Chapter 36). An infrared spectrum (Chapter 12) would confirm the presence of the aldehyde; an ultraviolet spectrum (Chapter 14) would indicate it is a conjugated aldehyde; and a nuclear magnetic resonance (NMR) spectrum (Chapter 13) would clearly show the aldehyde proton, the three methyl groups, and the two olefinic protons. Finally the molecule can be reacted with another substance to form a crystalline derivative for further identification (Chapter 70). You will probably not be required to carry out all of these analyses; however, these are the major tools available to the organic chemist for ascertaining purity and determining the structure of an organic molecule.

Experiments

Steam Distillation Apparatus

MICROSCALE

Into a 5-mL short-necked round-bottomed flask place a boiling chip, 0.5 mL of lemongrass oil[1] (*not* lemon oil), and 3 mL of water. Assemble the apparatus as depicted in Fig. 6.2, and distill as rapidly as possible, taking care that all the distillate condenses. Using the syringe, inject water dropwise through the septum to keep the volume in the flask constant. Distill into an ice-cooled vial, as shown, or, better, into the 15-mL centrifuge tube until no more oily drops can be detected, about 10 to 12 mL.

Wrap the vertical part of the apparatus in cotton or glass wool to speed the distillation.

MICROSCALE

1. Extraction of Citral

See Chapter 8 for information on the theory and practice of extraction.

Add 2.5 mL of *tert*-butyl methyl ether to the cold distillate. Shake, let the layers separate, and then draw off most of the ether with a Pasteur pipette, leaving the aqueous layer in the reaction tube. Continue this extraction process by again adding 1.5 mL of ether to the centrifuge tube, shaking, and separating to remove the citral from the steam distillate.

Place the ether in a reaction tube, and then add about 2 mL of saturated sodium chloride solution to the ether. Stopper the tube, shake it, and then remove and discard the aqueous layer. Add anhydrous calcium chloride pellets to the ether until they no longer clump together, stopper the tube, and shake it over a period of 5 to 10 minutes. This removes water adhering to the reaction tube and dissolved in the ether. Force a Pasteur pipette to the bottom of the reaction tube,

In Sweden they say, "Add anhydrous drying agent until it begins to snow."

---

1. Obtained from Pfaltz and Bauer, 172 E. Aurora St., Waterbury, CT 06708.

expel the air from the pipette, draw up the ether, and expel it into another tared reaction tube. Add fresh ether, which will serve to wash off the drying agent, and add that ether to the tared reaction tube. Add a boiling stick or boiling chip to the ether in the reaction tube, place it in a beaker of hot water, and evaporate the ether by boiling and drawing the ether vapors into the water aspirator or, better, by blowing a gentle stream of air or nitrogen into the tube (in the hood). The last traces of ether can be removed by connecting the reaction tube directly to the water aspirator. The residue should be a clear, fragrant oil. If it is cloudy, it is wet. Determine the weight of the citral, and calculate the percentage of citral recovered from the lemongrass oil, assuming that the density of the lemongrass oil is the same as that of citral (0.89). Transfer the product to the smallest possible airtight container, and store it in the dark for later analyses. Analyze the infrared spectrum in detail.

See page 137 *for the removal of water from ether solutions.*

Cleaning Up   The aqueous layer can be flushed down the drain. Any ether goes into the organic solvents container. Allow ether to evaporate from the calcium chloride in the hood, and then place it in the nonhazardous solid waste container.

🧪 MACROSCALE

Anthracene
mp 216°C

Toluene
bp 111°C
den 0.866

2. Recovery of a Dissolved Substance

Steam Distillation Apparatus

In the assembly shown in Fig. 6.4 steam is passed into a 250-mL, round-bottomed flask through a section of 6-mm glass tubing fitted into a stillhead with a piece of 5-mm rubber tubing connected to a trap, which in turn is connected to the steam line. The trap serves two purposes: It allows water, which is in the steam line, to be removed before it reaches the round-bottomed flask, and adjustment of the clamp on the hose at the bottom of the trap allows precise control of the steam flow. The stopper in the trap should be wired on, as shown, as a precaution. A bent adapter attached to a *long* condenser delivers the condensate into a 250-mL Erlenmeyer flask.

Measure 50 mL of a 0.2% solution of anthracene in toluene into the 250-mL round-bottomed flask, and add 100 mL of water. For an initial distillation to determine the boiling point and composition of the toluene–water azeotrope, fit the stillhead with a thermometer instead of the steam-inlet tube (see Fig. 5.5). Heat the mixture with a hot plate and sand bath or electric flask heater, distill about 50 mL of the azeotrope, and record a value for the boiling point. After removing the heat, pour the distillate into a graduated cylinder and measure the volumes of toluene and water. Calculate the weight of water per gram of toluene and compare the result with the theoretical value calculated from the vapor pressure of water at the observed boiling point (see Table 6.1).

Replace the thermometer with the steam-inlet tube. To start the steam distillation, heat the flask containing the mixture with a Thermowell to prevent water from condensing in the flask to the point where water and product splash over into the receiver. Then turn on the steam valve, making sure the screw clamp on

CAUTION: Live steam causes severe burns. If apparatus leaks steam, turn off steam valve immediately.

FIG. 6.4 Steam distillation apparatus.

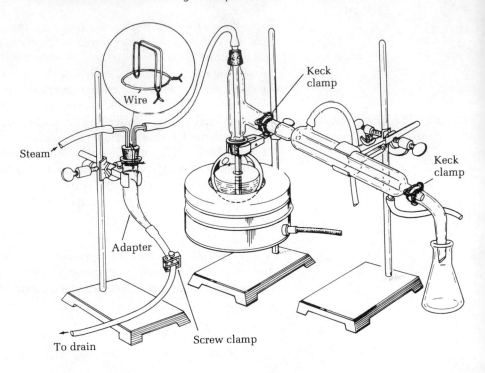

the bottom of the trap is open. Slowly close the clamp, and allow steam to pass into the flask. Try to adjust the rate of steam addition and the rate of heating so the water level in the flask remains constant. Unlike ordinary distillations, steam distillations are usually run as fast as possible, with proper care to avoid having material splash into the receiver and to avoid having steam escape uncondensed.

Continue distillation by passing in steam until the distillate is clear and then until fluorescence appearing in the stillhead indicates that a trace of anthracene is beginning to distill. Stop the steam distillation by first opening the clamp at the bottom of the trap and then turning off the steam valve. Grasp the round-bottomed flask with a towel when disconnecting it and, using the clamp to support it, cool it under the tap. The bulk of the anthracene can be dislodged from the flask walls and collected on a small suction filter. To recover any remaining anthracene, add a little acetone to the flask, warm on the steam bath to dissolve the material, add water to precipitate it, and then collect the precipitate on the same suction filter. About 80% of the hydrocarbon in the original toluene solution should be recoverable. When dry, crystallize the material from about 1 mL of toluene, and observe that the crystals are more intensely fluorescent than the solution or the amorphous solid. The characteristic fluorescence is quenched by mere traces of impurities.

Cleaning Up   Place the toluene and the anthracene in the organic solvents container. The aqueous layer can be flushed down the drain.

MACROSCALE

Citral (mixture of isomers)
structure diagram with CH₃, H₂C, C, H(CHO), CHO(H), C—H, CH₃, CH₃

Citral (mixture of isomers)
MW 152.24
bp 229°C, den 0.888
n_D^{20} 1.4880

3. Isolation of Citral

Citral, a fragrant terpene aldehyde made up of two isoprene units, is the main component of the steam-volatile fraction of lemongrass oil and is used in a commercial synthesis of vitamin A.

Using a graduated cylinder, measure out 10 mL of lemongrass oil (*not* lemon oil) into the 250-mL boiling flask. Rinse the remaining contents of the graduated cylinder into the flask with a little *t*-butyl methyl ether. Add 100 mL of water, make connections as in Fig. 6.4, heat the flask with a small flame, and pass in steam. Distill as rapidly as the cooling facilities allow, and continue until droplets of oil no longer appear at the tip of the condenser (about 250 mL of distillate).

Pour 50 mL of *t*-butyl methyl ether into a 125-mL separatory funnel, cool the distillate if necessary, and pour a part of it into the funnel. Shake, let the layers separate, discard the lower layer, add another portion of distillate, and repeat. When the last portion of distillate has been added, rinse the flask with a little *t*-butyl methyl ether to recover adhering citral. Use the techniques described in Chapter 8 for drying, filtering, and evaporating the *t*-butyl methyl ether. Take the tare of a 1-g tincture bottle, transfer the citral to it with a Pasteur pipette, and determine the weight and the yield from the lemongrass oil. Label the bottle and store (in the dark) for later testing for the presence of functional groups.

Cleaning Up   The aqueous layer can be flushed down the drain. Any *t*-butyl methyl ether goes in the organic solvents container. Allow *t*-butyl methyl ether to evaporate from the calcium chloride pellets in the hood, then place it in the non-hazardous solid waste container.

MACROSCALE

Eugenol structure diagram with OH, OCH₃, CH₂CH=CH₂

Eugenol

4. Isolation of Eugenol from Cloves

In the fifteenth century Theophrastus Bombastus von Hohenheim, otherwise known as Paracelsus, urged all chemists to make extractives for medicinal purposes rather than try to transmute base metals into gold. He believed every extractive had a quintessence that was the most effective part in effecting cures. There was then an upsurge in the isolation of essential oils from plant materials, some of which are still used for medicinal purposes. Among these are camphor, quinine, oil of cloves, cedarwood, turpentine, cinnamon, gum benzoin, and myrrh. Oil of cloves, which consists almost entirely of eugenol and its acetate, is a food flavoring agent as well as a dental anesthetic. The Food and Drug Administration has declared clove oil to be the most effective nonprescription remedy for toothache.

The biggest market for essential oils is for perfumes, and, as might be expected, prices for these oils reflect their rarity. Recently, worldwide production of orange oil was 1500 tons and it sold for $0.75 per lb, while 400 tons of clove oil sold for $14.00 per lb, and 10 tons of jasmine oil sold for $2000 per lb. These three oils represent the most common extractive processes. Orange oil is obtained by expression (squeezing) of the peel in presses; clove oil is obtained by steam distillation, as performed in this experiment; and jasmine oil is obtained by extraction of the flower petals using ethanol.

$$O$$
$$\|$$
OCCH$_3$
OCH$_3$

CH$_2$CH=CH$_2$

Acetyleugenol

Cloves are the flower buds from a tropical tree that can grow to 50 feet in height. The buds are hand-picked in the Moluccas (the Spice Islands) of Indonesia, the East Indies, and the islands of the Indian Ocean, where a tree can yield up to 75 lb of the sun-dried buds we know as cloves.

Cloves contain between 14 and 20% by weight of essential oil, of which about half can be isolated. The principal constituent of clove oil is eugenol and its acetyl derivative. Eugenol boils at 255°C, but, being insoluble in water, it will form an azeotrope with water and steam distill at a temperature slightly below the boiling point of water. Being a phenol, eugenol will dissolve in aqueous alkali to form the phenolate anion. This forms the basis for its separation from its acetyl derivative. The relative amounts of eugenol and acetyleugenol and the effectiveness of the alkali extraction in separating them can be analyzed by thin-layer chromatography.

Procedure

Place 25 g of whole cloves in a 250-mL round-bottomed flask, add 100 mL of water, and set up an apparatus for steam distillation (Fig. 6.4) or simple distillation (see Fig. 5.5). In the latter procedure steam is generated *in situ*. Heat the flask strongly until boiling starts, then reduce the heat just enough to prevent foam from being carried over into the receiver. Instead of using a graduated cylinder as a receiver, use an Erlenmeyer flask and transfer the distillate periodically to the graduated cylinder; then, if any material does foam over, the entire distillate will not be contaminated. Collect 60 mL of distillate, remove the flame, and add 60 mL of water to the flask. Resume the distillation, and collect an additional 60 mL of distillate.

The eugenol is very sparingly soluble in water and is easily extracted from the distillate with dichloromethane. Place the 120 mL of distillate in a 250-mL separatory funnel and extract with three 15-mL portions of dichloromethane. In this extraction very gentle shaking will fail to remove all of the product, and long and vigorous shaking will produce an emulsion of the organic layer and water. The separatory funnel will appear to have three layers in it. It is better to err on the side of vigorous shaking and draw off the clear lower layer to the emulsion line for the first two extractions. For the third extraction, shake the mixture less vigorously and allow a longer period of time for the two layers to separate.

Combine the dichloromethane extracts (the aqueous layer can be discarded), and add just enough anhydrous calcium chloride pellets so that the drying agent no longer clumps together but appears to be a dry powder as it settles in the solution. This may require as little as 2 g of drying agent. Swirl the flask for a minute or two to complete the drying process, and then decant the solvents into an Erlenmeyer flask. It is quite easy to decant the dichloromethane from the drying agent, and therefore it will not be necessary to set up a filtration apparatus to make this separation.

Measure the volume of the dichloromethane in a dry graduated cylinder, place exactly one-fifth of the solution in a tared Erlenmeyer flask, add a wood boiling stick (easier to remove than a boiling chip), and evaporate the solution on a steam bath in a hood. The residue of crude clove oil will be used in the thin-layer chromatography analysis. From its weight, the total yield of crude clove oil can be calculated.

See Chapter 8 for the theory of acid/base extractions.

To separate eugenol from acetyleugenol, extract the remaining four-fifths of the dichloromethane solution (about 30 mL) with 5% aqueous sodium hydroxide solution. Carry out this extraction three times, using 10-mL portions of sodium hydroxide each time. Dry the dichloromethane layer over anhydrous calcium chloride pellets, decant the solution into a tared Erlenmeyer flask, and evaporate the solvent. The residue should consist of acetyleugenol and other steam-volatile neutral compounds from cloves.

Acidify the combined aqueous extracts to pH 1 with concentrated hydrochloric acid (use Congo red or litmus paper), and then extract the liberated eugenol with three 8-mL portions of dichloromethane. Dry the combined extracts, decant into a tared Erlenmeyer flask, and evaporate the solution on a steam bath. Calculate the weight percent yields of eugenol and acetyleugenol on the basis of the weight of cloves used.

Analyze your products by infrared spectroscopy and thin-layer chromatography. Obtain an infrared spectrum of either eugenol or acetyleugenol using the thin-film method. Compare your spectrum with the spectrum of a neighbor who has examined the other compound.

Use plastic sheets precoated with silica gel for thin-layer chromatography. One piece the size of a microscope slide should suffice. Spot crude clove oil, eugenol, and acetyleugenol (1% solutions) side by side about 5 mm from one end of the slide. The spots should be very small. Immerse the end of the slide in a dichloromethane–hexane mixture (1:2 or 2:1) about 3 mm deep in a covered beaker. After running the chromatogram observe the spots under an ultraviolet lamp or by developing in an iodine chamber.

Cleaning Up   Combine all aqueous layers, neutralize with sodium carbonate, dilute with water, and flush down the drain. Any solutions containing dichloromethane should be placed in the halogenated organic solvents container. Allow the solvent to evaporate from the calcium chloride pellets in the hood and then place the drying agent in the nonhazardous solid waste container.

Questions

1. Assign the peaks in the ^{1}H NMR spectrum of eugenol to specific protons in the molecule. See Fig. 6.5. The OH peak is at 5.1 ppm.

2. A mixture of ethyl iodide (C_2H_5I, bp 72.3°C) and water boils at 63.7°C. What weight of ethyl iodide would be carried over by 1 g of steam during steam distillation?

3. Iodobenzene (C_6H_5I, bp 188°C) steam distills at a temperature of 98.2°C. How many molecules of water are required to carry over one molecule of iodobenzene? How many grams per gram of iodobenzene?

FIG. 6.5 ^{1}H NMR spectrum of eugenol (250 MHz).

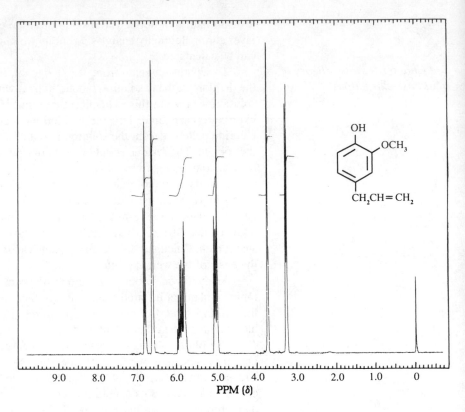

4. The condensate from a steam distillation contains 8 g of an unknown compound and 18 g of water. The mixture steam distilled at 98°C. What is the molecular weight of the unknown?

5. Bearing in mind that a sealed container filled with steam will develop a vacuum if the container is cooled, explain what will happen if a steam distillation is stopped by turning off the steam valve before opening the screw clamp on the adapter trap.

6. A mixture of toluene (bp 111°C) and water was steam distilled. Visual inspection of the distillate reveals that there is a greater volume of toluene than water present, yet water (bp 100°C) has the higher vapor pressure. Explain this observation.

7

Vacuum Distillation and Sublimation

Prelab Exercise: Compare the operation of a barometer, used to measure atmospheric pressure, and a mercury manometer, used to measure the pressure in an evacuated system.

Part 1. Vacuum Distillation

Many substances cannot be distilled satisfactorily in the ordinary way, either because they boil at such high temperatures that decomposition occurs or because they are sensitive to oxidation. In such cases purification can be accomplished by distillation at diminished pressure. A few of the many pieces of apparatus for vacuum distillation are described on the following pages. Round-bottomed Pyrex ware and thick-walled suction flasks are not liable to collapse, but even so safety glasses should be worn at all times when carrying out this type of distillation.

Vacuum Distillation Assemblies

Carry out vacuum distillation behind a safety shield. Check apparatus for scratches and cracks.

A typical macroscale vacuum distillation apparatus is illustrated in Fig. 7.1. It is constructed of a round-bottomed flask (often called the "pot") containing the material to be distilled, a Claisen distilling head fitted with a hair-fine capillary mounted through a rubber tubing sleeve, and a thermometer with the bulb extending below the side arm opening. The condenser fits into a vacuum adapter that is connected to the receiver and, via heavy-walled rubber tubing, to a mercury manometer and thence to the trap and water aspirator.

Prevention of bumping

Liquids usually bump vigorously when boiled at reduced pressure and most boiling stones lose their activity in an evacuated system; it is therefore essential to make a special provision for controlling the bumping. This is done by allowing a fine stream of air bubbles to be drawn into the boiling liquid through a glass tube drawn to a hair-fine capillary. The capillary should be so fine that even under vacuum only a few bubbles of air are drawn in each second; smooth boiling will be promoted, and the pressure will remain low. The capillary should extend to the very bottom of the flask, and it should be slender and flexible so that it will whip back and forth in the boiling liquid. Another method used to prevent bumping, when small quantities of material are being distilled, is to introduce sufficient glass wool into the flask to fill a part of the space above the liquid. Rapid magnetic stirring will also prevent bumping.

The capillary is made in three operations. First, a 6-in. length of 6-mm glass tubing is rotated and heated in a small area over a *very hot* flame to collapse the

FIG. 7.1 Vacuum distillation apparatus.

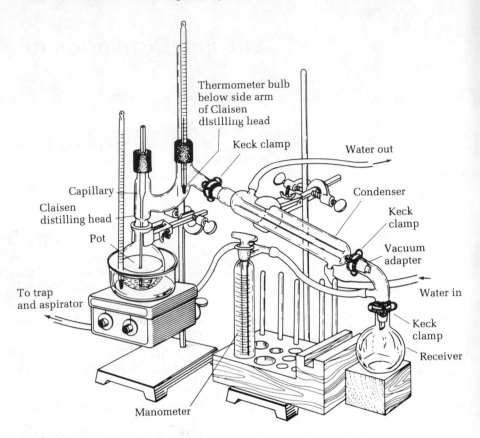

Making a hair-fine capillary

Handle flames in the organic laboratory with great care.

Heating baths

glass and thicken the side walls as seen in Fig. 7.2(a). The tube is removed from the flame, allowed to cool slightly, and then drawn into a thick-walled, coarse diameter capillary [Fig. 7.2(b)]. This coarse capillary is heated at point X over the wing top of a Bunsen burner turned 90°. When the glass is very soft, but not soft enough to collapse the tube entirely, the tubing is lifted from the flame and without hesitation drawn smoothly and rapidly into a hair-fine capillary by stretching the hands as far as they will reach (about 2 m) [Fig. 7.2(c)]. The two capillaries so produced can be snapped off to the desired length. To ascertain that there is indeed a hole in the capillary, place the end beneath a low-viscosity liquid such as acetone or *t*-butyl methyl ether, and blow in the large end. A stream of very small bubbles should be seen. If the stream of bubbles is not extremely fine, make a new capillary. The flow of air through the capillary *cannot* be controlled by attaching a rubber tube and clamp to the top of the capillary tube. Should the right-hand capillary of Fig. 7.2(c) break when in use, it can be fused to a scrap of glass (for use as a handle) and heated again at point Y [Fig. 7.2(c)]. In this way the capillary can be redrawn many times.

The pot is heated with a heating bath rather than a Thermowell to promote even boiling and make possible accurate determination of the boiling point. The bath is filled with a suitable heat transfer liquid (water, cottonseed oil, silicone oil, or molten metal) and heated to a temperature about 20°C higher than that at which

FIG. 7.2 Capillary for vacuum distillation.

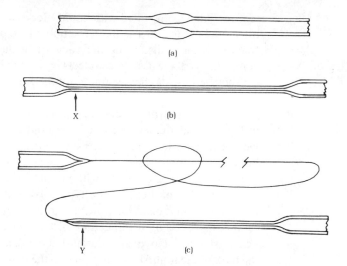

the substance in the flask distills. The bath temperature is kept constant through-out the distillation. The surface of the liquid in the flask should be below that of the heating medium, for this condition lessens the tendency to bump. Heating of the flask is begun only after the system has been evacuated to the desired pressure; otherwise the liquid might boil too suddenly on reduction of the pressure.

Changing fractions

To change fractions the following must be done in sequence: Remove the source of heat, release the vacuum, change the receiver, restore the vacuum to the former pressure, resume heating.

The pressure of the system is measured with a closed-end mercury manometer. The manometer (Fig. 7.3) is connected to the system by turning the stopcock until the V-groove in the stopcock is aligned with the side arm. To avoid contamination of the manometer, it should be connected to the system only when a read-ing is being made. The pressure, in mm Hg, is given by the height, in mm, of the

Mercury manometer

central mercury column above the reservoir of mercury and represents the difference in pressure between the nearly perfect vacuum in the center tube (closed at the top, open at the bottom) and the large volume of the manometer, which is at the pressure of the system.

Another common type of closed-end mercury manometer is shown in Fig. 7.4. At atmospheric pressure, the manometer is completely full of mercury, as shown in Fig. 7.4(a). When the open end is connected to a vacuum, the pressure is measured as the distance in millimeters between the mercury in the right and left legs. Note that the pressure in the left leg is essentially zero (mercury has a low vapor pressure). If the applied vacuum were perfect (0 mm Hg), then the two columns would have the same height.

A better vacuum distillation apparatus is shown in Fig. 7.5. The distillation neck of the Claisen adapter is longer than other adapters and has a series of indentations made from four directions, so that the points nearly meet in the center. These indentations increase the surface area over which rising vapor can come to equilibrium with descending liquid, and it then serves as a fractionating column (a Vigreux column). A column packed with a metal sponge has a great

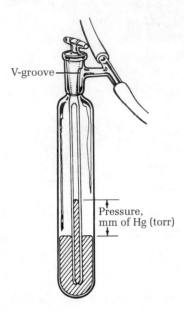

FIG. 7.3 Closed-end mercury manometer.

tendency to become filled with liquid (flood) at reduced pressure. The apparatus illustrated in Fig. 7.5 also has a fraction collector, which allows the removal of a fraction without disturbing the vacuum in the system. While the receiver is being changed, the distillate collects in the small reservoir A. The clean receiver is evacuated by another aspirator at tube B before being connected again to the system.

If only a few milliliters of a liquid are to be distilled, the apparatus shown in Fig. 7.6 has the advantage of low hold-up—that is, not much liquid is lost wetting the surface area of the apparatus. The fraction collector illustrated is known as a "cow." Rotation of the cow about the standard taper joint will allow four fractions to be collected without interrupting the vacuum.

A distillation head of the type shown in Fig. 7.7 allows fractions to be removed without disturbing the vacuum, and it also allows control of the reflux ratio (Chapter 5) by manipulation of the condenser and stopcock A. These can be adjusted to remove all material that condenses or only a small fraction, with the bulk of the liquid being returned to the distilling column to establish equilibrium between descending liquid and ascending vapor. In this way liquids with small boiling-point differences can be separated.

Microscale Vacuum Distillation Assemblies

On a truly micro scale (<10 mg), simple distillation is not practical because of mechanical losses. On a small scale (10–500 mg), distillation is still not a good method for isolating material, again because of mechanical losses, but bulb-to-bulb distillations can serve to rid a sample of all low-boiling material and non-volatile impurities. In research, preparative-scale gas chromatography (Chapters 10 and 11) and high-performance liquid chromatography (HPLC) have supplanted vacuum distillation of very small quantities of material.

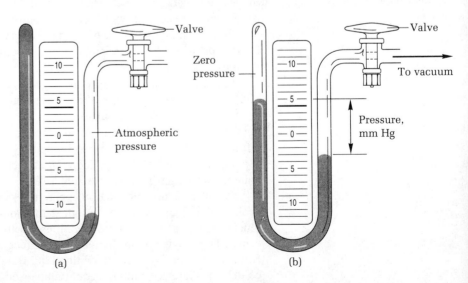

FIG. 7.4 Another type of closed-end mercury manometer.

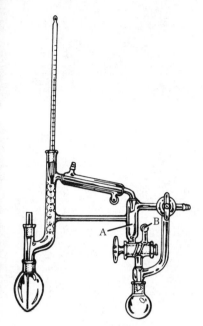

FIG. 7.5 Vacuum distillation apparatus with Vigreux column and fraction collector.

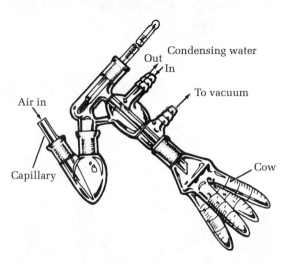

FIG. 7.6 Semimicroscale vacuum distillation apparatus.

Fractional distillation of small quantities of liquids is even more difficult. The necessity for maintaining an equilibrium between ascending vapors and descending liquid means there will inevitably be large mechanical losses. The best apparatus for the fractional distillation of about 0.5 to 20 mL under reduced pressure is the so-called micro spinning band column, which has a holdup of about 0.1 mL. Again, preparative-scale gas chromatography and HPLC are the best means for separating mixtures of liquids.

Microscale Apparatus

It is possible to assemble several pieces of apparatus that can be used for simple vacuum distillation using components from the microscale kit. One such assembly is shown in Fig. 7.8. The material to be distilled is placed in the 5-mL short-necked flask packed loosely with glass wool, which will help prevent bumping. The thermometer bulb extends entirely below the side arm, and the distillate is collected in a reaction tube, which can be cooled in a beaker of ice. The vacuum source is connected to a syringe needle, or the rubber tubing from the aspirator, if of the proper diameter, can be connected directly to the adapter, dispensing with the septum and needle.

The Water Aspirator in Vacuum Distillation

A water aspirator in good order gives a vacuum nearly corresponding to the vapor pressure of the water flowing through it. Polypropylene aspirators give good service and are not subject to corrosion, as are the brass ones. If a manometer is not available and the assembly is free of leaks and the trap and lines clean and dry, an approximate estimate of the pressure can be made by measuring the water temperature and reading the pressure from Table 7.1.

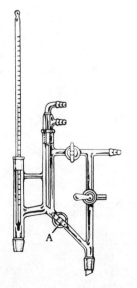

FIG. 7.7 Vacuum distillation head.

FIG. 7.8 Microscale apparatus for simple vacuum distillation.

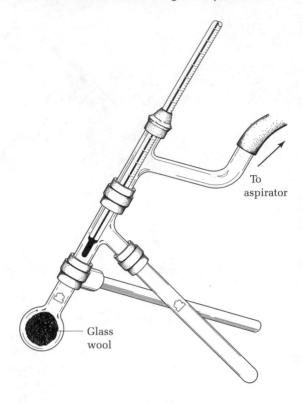

To
aspirator

Glass
wool

The Rotary Oil Pump

To obtain pressures below 10 mm Hg, a mechanical vacuum pump of the type illustrated in Fig. 7.9(a) is used. A pump of this type in good condition can give pressures as low as 0.1 mm Hg. These low pressures are measured with a tilting-type McLeod gauge [Fig. 7.9(b)]. When a reading is being made, the gauge is tilted to the vertical position shown and the pressure is read as the difference between the heights of the two columns of mercury. Between readings the gauge is rotated clockwise 90°.

Dewar should be wrapped with tape or otherwise protected from implosion.

Never use a mechanical vacuum pump before placing a mixture of dry ice and isopropyl alcohol in a Dewar flask [Fig. 7.9(c)] around the trap, and never pump corrosive vapors (e.g., HCl gas) into the pump. Should this happen, change the pump oil immediately. With care, it will give many years of good service. The

TABLE 7.1 Vapor Pressure of Water at Different Temperatures

t (°C)	p (mm Hg)	t (°C)	p (mm Hg)	t (°C)	p (mm Hg)	t (°C)	p (mm Hg)
0	4.58	20	17.41	24	22.18	28	28.10
5	6.53	21	18.50	25	23.54	29	29.78
10	9.18	22	19.66	26	24.99	30	31.55
15	12.73	23	20.88	27	26.50	35	41.85

FIG. 7.9   (a) Rotary oil pump, (b) tilting McLeod gauge, (c) vacuum system.

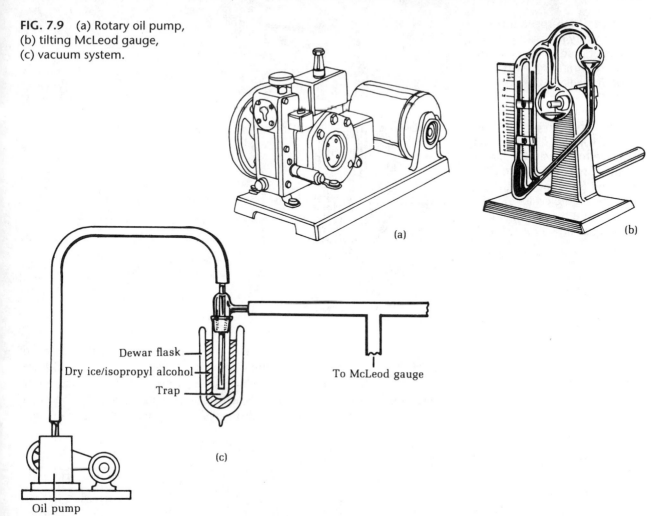

(a)

(b)

Dewar flask

Dry ice/isopropyl alcohol

Trap

To McLeod gauge

(c)

Oil pump

dry ice trap condenses organic vapors and water vapor, both of which would otherwise contaminate the vacuum pump oil and exert enough vapor pressure to destroy a good vacuum.

For an exceedingly high vacuum (5×10^{-8} mm Hg) a high-speed three-stage mercury diffusion pump is used (Fig. 7.10).

Relationship Between Boiling Point and Pressure

It is not possible to calculate the boiling point of a substance at some reduced pressure from a knowledge of the boiling temperature at 760 mm Hg, because the relationship between boiling point and pressure varies from compound to compound and is somewhat unpredictable. It is true, however, that boiling point curves for organic substances have much the same general disposition, as illustrated by the two lower curves in Fig. 7.11. These are similar and do not differ

FIG. 7.10 High-speed three-stage mercury diffusion pump capable of producing a vacuum of 5×10^{-8} mm Hg. Mercury is boiled in the flask; the vapor rises in the center tube, is deflected downward in the inverted cups, and entrains gas molecules, which diffuse in from the space to be evacuated, A. The mercury condenses to a liquid and is returned to the flask; the gas molecules are removed at B by an ordinary rotary vacuum pump.

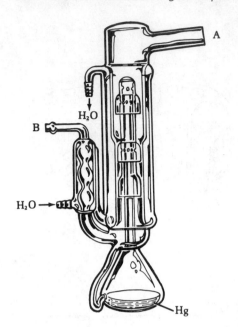

greatly from the curve for water. For substances boiling in the region of 150–250°C at 760 mm Hg, the boiling point at 20 mm Hg is 100–120°C lower than at 760 mm Hg. Benzaldehyde, which is very sensitive to air oxidation at the normal boiling point of 178°C, distills at 76°C at 20 mm Hg, and the concentration of oxygen in the rarefied atmosphere is just $\frac{20}{760}$, or 3% of that in an ordinary distillation.

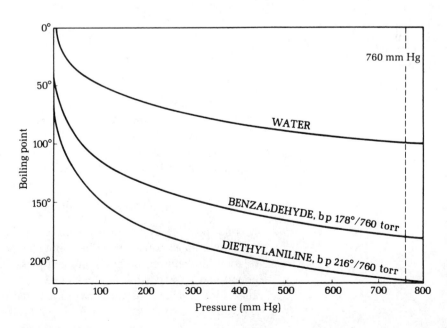

FIG. 7.11 Boiling point curves.

The curves all show a sharp upward inclination in the region of very low pressure. The lowering of the boiling point attending a reduction in pressure is much more pronounced at low than at high pressures. A drop in the atmospheric pressure of 10 mm Hg lowers the normal boiling point of an ordinary liquid by less than a degree, but a reduction of pressure from 20 to 10 mm Hg causes a drop of about 15°C in the boiling point. The effect at pressures below 1 mm is still more striking, and with development of practical forms of the highly efficient oil vapor or mercury vapor diffusion pump, distillation at a pressure of a few thousandths or ten thousandths of a millimeter has become a standard operation in many research laboratories. High-vacuum distillation, that is, at a pressure below 1 mm Hg, affords a useful means of purifying extremely sensitive or very slightly volatile substances. Table 7.2 indicates the order of magnitude of the reduction in boiling point attainable by operating in different ways and illustrates the importance of keeping vacuum pumps in good repair.

The boiling point of a substance at various pressures can be estimated from a pressure–temperature nomograph such as the one shown in Fig. 7.12. If the boiling point of a nonpolar substance at 760 mm Hg is known, for example, 145°C (column B), and the new pressure measured, for example, 28 mm Hg (column C), then a straight line connecting these values on columns B and C when extended intersects column A to give the observed boiling point of 50°C. Conversely, a substance observed to boil at 50°C (column A) at 1.0 mm Hg (column C) will boil at approximately 212°C at atmospheric pressure (column B). If the material being distilled is polar, such as a carboxylic acid, column D is used. Thus a substance that boils at 145°C at atmospheric pressure (column B) will boil at 28 mm Hg (column C) if it is nonpolar and at a pressure of 10 mm Hg (column D) if it is polar. The boiling point will be 50°C (column A).

Vacuum distillation is not confined to the purification of substances that are liquid at ordinary temperatures but often can be used to advantage for solid substances. The operation is conducted for a different purpose and by a different technique. A solid is seldom distilled to cause a separation of constituents of different degrees of volatility but rather to purify the solid. It is often possible in one vacuum distillation to remove foreign coloring matter and tar without appreciable loss of product, whereas several wasteful crystallizations might be required to attain the same purity. It is often good practice to distill a crude

TABLE 7.2 Distillation of a (Hypothetical) Substance at Various Pressures

Method		Pressure (mm Hg)	bp (°C)
Ordinary distillation		760	250
Aspirator {	summer	25	144
	winter	15	134
Rotary oil pump {	poor condition	10	124
	good condition	3	99
	excellent condition	1	89
Mercury vapor pump (high vacuum distillation)		0.01	30

FIG. 7.12 Pressure–temperature nomograph.

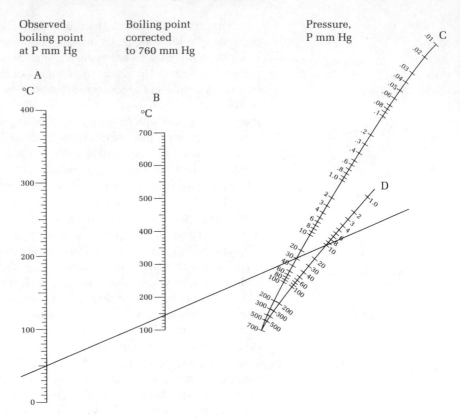

product and then to crystallize it. Time is saved in the latter operation because the hot solution usually requires neither filtration nor clarification. The solid must be dry and a test should be made to determine if it will distill without decomposition at the pressure of the pump available. That a compound lacks the required stability at high temperatures is sometimes indicated by the structure, but a high melting point should not be taken as an indication that distillation will fail. Substances melting as high as 300°C have been distilled with success at the pressure of an ordinary rotary vacuum pump.

Molecular Distillation

At 1×10^{-3} mm Hg the mean free path of nitrogen molecules is 56 mm at 25°C. This means that an average N_2 molecule can travel 56 mm before bumping into another N_2 molecule. In specially designed apparatus it is possible to distill almost any volatile molecule by operating at a very low pressure and having the condensing surface close to the material being distilled. The thermally energized molecule escapes from the liquid, moves about 10 mm without encountering any other molecules, and condenses on a cold surface. A simple apparatus for molecular distillation is illustrated in Fig. 7.13. The bottom of the apparatus is placed on a hot surface, and the distillate moves a very short distance before condens-

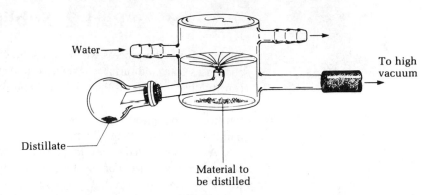

Water→

Distillate—

To high
vacuum

Material to
be distilled

FIG. 7.13 Molecular distillation apparatus.

ing. If it is not too viscous it will drip from the point on the glass condensing surface and run down to the receiver.

The efficiency of molecular distillation apparatus is less than one theoretical plate (see Chapter 5). It is used to remove very volatile substances such as solvents and to separate volatile, high-boiling substances at low temperatures from nonvolatile impurities.

In more sophisticated apparatus of quite different design, the liquid to be distilled falls down a heated surface, which is again close to the condenser. This moving film prevents a buildup of nonvolatile material on the surface of the material to be distilled, which would cause the distillation to cease. Vitamin A is distilled commercially in this manner.

Part 2. Sublimation

Sublimation is the process whereby a solid evaporates from a warm surface and condenses on a cold surface, again as a solid (Figs. 7.14 and 7.15). This technique is particularly useful for the small-scale purification of solids because there is so little loss of material in transfer. If the substance has the correct properties, sublimation is preferred over crystallization when the amount of material to be purified weighs less than 100 mg.

As demonstrated in the first experiment, sublimation can occur readily at atmospheric pressure. For substances with lower vapor pressures vacuum sublimation is used. At very low pressure the sublimation becomes very similar to molecular distillation, where the molecule leaves the warm solid and passes unobstructed to a cold condensing surface and condenses in the form of a solid.

Since sublimation occurs from the surface of the warm solid, impurities can accumulate and slow down or even stop the sublimation, in which case it is necessary to open the apparatus, grind the impure solid to a fine powder, and restart the sublimation.

Sublimation is much easier to carry out on a small scale than on a large one. It would be unusual for a research chemist to sublime more than about 10 g of material, because the sublimate tends to fall back to the bottom of the apparatus. In an apparatus such as that illustrated in Fig. 7.15(a), the solid to be sublimed is placed in the lower flask and connected via a lubricant-free rubber O-ring to the condenser, which in turn is connected to a vacuum pump. The lower flask is immersed in an oil bath at the appropriate temperature and the product sublimed and condensed onto the cool walls of the condenser. The parts of the apparatus are gently separated and the condenser inverted; the vacuum connection serves as

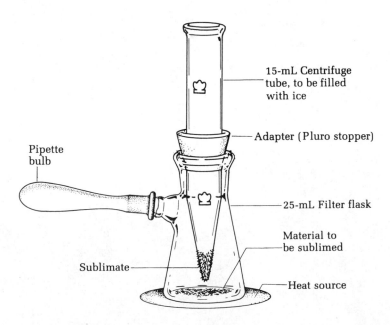

15-mL Centrifuge tube, to be filled with ice

Adapter (Pluro stopper)

Pipette bulb

25-mL Filter flask

Material to be sublimed

Sublimate

Heat source

FIG. 7.14 Small-scale vacuum sublimation apparatus.

a convenient funnel for product removal. For large-scale work the sublimator of Fig. 7.15(b) is used. The inner well is filled with a coolant (ice or dry ice). The sublimate clings to this cool surface, from which it can be removed by scraping and dissolving in an appropriate solvent.

Lyophilization

Lyophilization, also called freeze-drying, is the process of subliming a solvent, usually water, with the object of recovering the solid that remains after the solvent is removed. This technique is extensively used to recover heat- and oxygen-sensitive substances of natural origin, such as proteins, vitamins, nucleic acids, and other biochemicals from dilute aqueous solution. The aqueous solution of the substance to be lyophilized is usually frozen to prevent bumping and then subjected to a vacuum of about 1×10^{-3} mm Hg. The water sublimes and condenses as ice on the surface of a large and very cold condenser. The sample remains frozen during this entire process, without any external cooling being

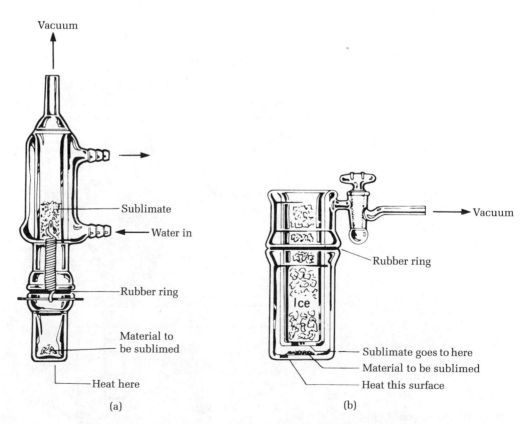

FIG. 7.15　(a) Mallory sublimator; (b) large vacuum sublimator.

supplied, because of the very high heat of vaporization of water; thus the temperature of the sample never exceeds 0°C. Freeze-drying on a large scale is employed to make "instant" coffee, tea, soup, rice, and all sorts of dehydrated foods.

The Kugelrohr

The Kugelrohr (German, meaning "bulb-tube") is a widely used piece of research apparatus that consists of a series of bulbs that can be rotated and heated under vacuum. Bulb-to-bulb distillation frees the desired compound of very low-boiling and very high-boiling or nonvolatile impurities. The crude mixture is placed in bulb A (Fig. 7.16) in the heated glass chamber. At C is a shutter mechanism that holds in the heat but allows the bulbs to be moved out one by one as distillation proceeds. The lowest boiling material collects in bulb D; then bulb C is moved out of the heated chamber, the temperature increased, and the next fraction collected in bulb C. Finally this process is repeated for bulb B. The bulbs rotate under vacuum using a mechanism such as the one used on rotary evaporators. The same apparatus is used for bulb-to-bulb fractional sublimation.

Experiment ————————————————————————————

Sublimation of an Unknown Substance

Apparatus and Technique

The apparatus consists of a 15-mL centrifuge tube thrust through an adapter (use a drop of glycerol to lubricate the adapter) fitted in a 25-mL filter flask. Once the adapter is properly positioned on the centrifuge tube it should be left there permanently. This tube is used only for sublimation.

The sample is either ground to a fine powder and uniformly distributed on

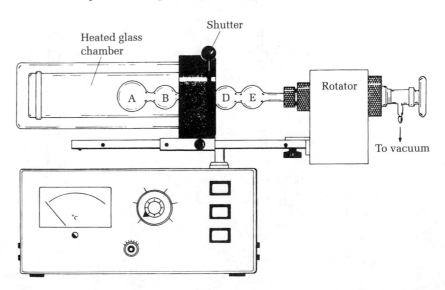

FIG. 7.16 Kugelrohr bulb-to-bulb distillation apparatus.

the bottom of the flask or is introduced into the flask as a solution and the solvent evaporated to deposit the substance on the bottom of the filter flask. If the compound sublimes easily, care must be exercised using the latter technique to ensure that the sample does not evaporate as the last of the solvent is being removed.

The 15-mL centrifuge tube and adapter are placed in the flask so that the tip of the centrifuge tube is about 3 to 8 mm above the bottom of the flask (Fig. 7.14). The flask is clamped with a three-prong clamp. Many substances sublime at atmospheric pressure, including the sublimation unknowns listed in Table 7.3. For vacuum sublimation, the side arm of the flask leads to an aspirator or vacuum pump for reduced pressure sublimation. The centrifuge tube is filled with ice and water. The ice is not added before closing the apparatus, so atmospheric moisture will not condense on the tube.

The filter flask is warmed cautiously on a hot sand bath until the product just begins to sublime. The heat is maintained at that temperature until sublimation is complete. Because some product will collect on the cool upper parts of the flask, it should be fitted with a loose cone of aluminum foil to direct heat there and cause the material to collect on the surface of the centrifuge tube. Alternatively, tilt the flask and rotate it in the hot sand, or remove the flask from the sand bath and then use a heat gun to warm the flask.

Once sublimation is judged complete, the ice water is removed from the centrifuge tube with a pipette and replaced with water at room temperature. This will prevent moisture from condensing on the product once the vacuum is turned off and the tube removed from the flask. The product is scraped from the centrifuge tube with a metal spatula onto a piece of glazed paper. It is much easier to scrape the product from a centrifuge tube than from a round-bottomed test tube. The last traces can be removed by washing off the tube with a few drops of an appropriate solvent, if that is deemed necessary.

Sublimation: Add sample to flask, close flask openings, add ice, sublime sample, remove ice, add room-temperature water, open apparatus and scrape off product.

Procedure

Into the bottom of a 25-mL filter flask, place 50 mg of an impure unknown taken from the list in Table 7.3. These substances can be sublimed at atmospheric pressure, although some will sublime more rapidly at reduced pressure. Close the flask with a rubber pipette bulb, and then place ice water in the centrifuge tube. Cautiously warm the flask until sublimation starts, and then maintain that temperature throughout the sublimation. Fit the flask with a loose cone of aluminum foil to direct heat up the sides of the flask, causing the

A very simple sublimation apparatus: an inverted stemless funnel on a watch glass

TABLE 7.3 Sublimation Unknowns

Substance	mp (°C)	Substance	mp (°C)
1,4-Dichlorobenzene	55	Benzoic acid	122
Naphthalene	82	Salicylic acid	159
1-Naphthol	96	Camphor	177
Acetanilide	114	Caffeine	235

Roll the filter flask in the sand bath or heat it with a heat gun to drive material from the upper walls of the flask.

product to collect on the tube. Once sublimation is complete, remove the ice water from the centrifuge tube and replace it with water at room temperature. Collect the product, determine its weight and the percent recovery, and from the melting point identify the unknown. Hand in the product in a neatly labeled vial.

Cleaning Up   Wash material from the apparatus with a minimum quantity of acetone, which, except for the chloro compound, can be placed in the organic solvents container. The 1,4-dichlorobenzene solution must be placed in the halogenated organic waste container.

For Further Investigation

Room deodorants, for example the Renuzit Long Last adjustable air freshener, appear to sublime. Place about 150 mg, carefully weighed, of the green solid in the filter flask and sublime it with very little heat applied to the flask. Identify the substance that sublimes. Weigh the residue. As an advanced experiment run the IR spectrum of the residue (*not* of the substance that sublimes). Comment on the economics of air fresheners.

Questions

1. Has sublimation aided in the purification of the unknown? Justify your answer.

2. From your experience and a knowledge of the process of sublimation, cite some examples of substances that sublime.

3. How might freeze-dried foods be prepared?

Extraction of Acids and Bases and the Isolation of Caffeine from Coffee, Tea, and Cola Syrup

Extraction is one of humankind's oldest chemical operations. The preparation of a cup of coffee or tea involves the extraction of flavor and odor components from dried vegetable matter with hot water. Aqueous extracts of bay leaves, stick cinnamon, peppercorns, and cloves are used as food flavorings, along with alcoholic extracts of vanilla and almond. For the last century and a half, organic chemists have been extracting, isolating, purifying, and then characterizing the myriad compounds produced by plants that have been used for centuries as drugs and perfumes—substances such as quinine from cinchona bark, morphine from the opium poppy, cocaine from coca leaves, and menthol from peppermint oil. In research a Soxhlet extractor is often used (Fig. 8.1). The organic chemist commonly employs, in addition to solid/liquid extraction, two other types of extraction: liquid/liquid extraction and acid/base extraction.

Part 1. Liquid/Liquid Extraction

After a chemical reaction has been carried out, the organic product is often separated from inorganic substances by liquid/liquid extraction. For example, in the synthesis of 1-bromobutane (Chapter 16)

$$2 \text{ CH}_3\text{CH}_2\text{CH}_2\text{CH}_2\text{OH} + 2 \text{ NaBr} + \text{H}_2\text{SO}_4 \rightarrow$$
$$2 \text{ CH}_3\text{CH}_2\text{CH}_2\text{CH}_2\text{Br} + 2 \text{ H}_2\text{O} + \text{Na}_2\text{SO}_4$$

1-butanol, also a liquid, is heated with an aqueous solution of sodium bromide and sulfuric acid to produce the product and sodium sulfate. The 1-bromobutane is isolated from the reaction mixture by extraction with *t*-butyl methyl ether, a solvent in which 1-bromobutane is soluble and in which water and sodium sulfate are insoluble. The extraction is accomplished by simply adding *t*-butyl methyl ether to the aqueous mixture and shaking it. The *t*-butyl methyl ether is

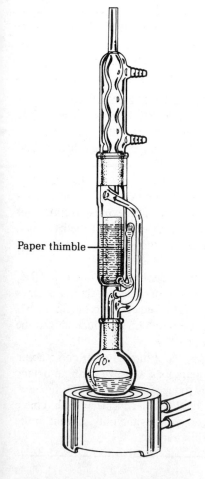

Paper thimble

FIG. 8.1 Soxhlet extractor; for extraction of solids, e.g. dried leaves or seeds. The solid is put in a filter paper thimble. Solvent vapor rises in the tube on the right, and condensate drops onto the solid in the thimble, leaches out soluble material, and, after initiating an automatic siphon, carries it to the flask where nonvolatile extracted material accumulates. Substances of very slight solubility can be extracted by prolonged operation.

less dense than water and floats on top; it is removed and evaporated to leave the bromo compound free of inorganic substances.

Properties of Extraction Solvents

The solvent used for extraction should have many properties of a satisfactory recrystallization solvent. It should readily dissolve the substance to be extracted; it should have a low boiling point so that it can readily be removed; it should not react with the solute or the other solvent; it should not be flammable or toxic; and it should be relatively inexpensive. In addition, it should not be miscible with water (the usual second phase). No solvent meets all the criteria, but several come close. Diethyl ether, usually referred to simply as ether, is probably the most common solvent used for extraction, but it is extremely flammable.

In the past, diethyl ether has been the most common solvent for extraction in the laboratory. It has high solvent power for hydrocarbons and for oxygen-containing compounds. It is highly volatile (bp 34.6°C) and therefore is easily removed from an extract. However, diethyl ether has two big disadvantages. It is highly flammable and so poses a great fire threat, and it easily forms peroxides. The reaction of diethyl ether with air is catalyzed by light. The resulting peroxides are higher boiling than the ether and are left as a residue when the ether evaporates. If the residue is heated, it will explode, since ether peroxides are treacherous high explosives. But in recent years, a new solvent has come on the scene.

$$CH_3CH_2\!-\!O\!-\!CH_2CH_3$$

Diethyl ether
"Ether"
MW 74.12, den. 0.708
bp 34.6°C, n_D^{20} 1.3530

$$H_3C\!-\!\underset{\underset{\displaystyle CH_3}{|}}{\overset{\overset{\displaystyle CH_3}{|}}{C}}\!-\!O\!-\!CH_3$$

***tert*-Butyl methyl ether**
MW 88.14, den. 0.741
bp 55.2°C, n_D^{20} 1.369

tert-Butyl Methyl Ether, the Extraction Solvent of Choice.

tert-Butyl methyl ether, called methyl *tert*-butyl ether (MTBE) in industry, has many advantages over diethyl ether as an extraction solvent. Most important, it does not easily form peroxides, so it can be stored for much longer periods than diethyl ether. And, in the United States, it is less than one-half the price of diethyl ether. It is slightly less volatile (bp 55°C), so it does not pose the same fire threat as diethyl ether, although one must be as careful in handling this solvent as in handling any other highly volatile, flammable substance. The explosion limits for *tert*-butyl methyl ether mixed with air are much narrower than for diethyl ether, the toxicity is less, the solvent power is the same, and the ignition temperature is higher (224 versus 180°C).

The weigh percent solubility of diethyl ether dissolved in water is 7.2%, while that of the butyl ether is 4.8%. The solubility of water in diethyl ether is 1.2%, while in the butyl ether it is 1.5%. Unlike diethyl ether, *t*-butyl methyl ether forms an azeotrope with water (4% water) that boils at 52.6°C, which aids in the removal of the last traces of water from an extract.

The low price and ready availability of *tert*-butyl methyl ether come about because it has replaced tetraethyl lead as the antiknock additive for high-octane gasoline and as a fuel oxygenate, which helps reduce air pollution. In this text, *tert*-butyl methyl ether is suggested wherever diethyl ether formerly would have been used in an extraction. It will not, however, work as the only solvent in the Grignard reaction, probably because of steric hindrance.

Partition Coefficient

The extraction of a compound such as 1-butanol, which is slightly soluble in water as well as very soluble in ether, is an equilibrium process governed by the solubilities of the alcohol in the two solvents. The ratio of the solubilities is known as the *distribution coefficient,* also called the *partition coefficient, k,* and is an equilibrium constant with a certain value for a given substance, pair of solvents, and temperature.

To a good approximation the *concentration* of the solute in each solvent can be correlated with the *solubility* of the solute in the pure solvent, a figure that can be found readily in tables of solubility in reference books.

$$k = \frac{\text{Concentration of C in } t\text{-butyl methyl ether}}{\text{Concentration of C in water}}$$

$$\cong \frac{\text{Solubility of C in } t\text{-butyl methyl ether (g/100 mL)}}{\text{Solubility of C in water (g/100 mL)}}$$

Consider a compound, A, that dissolves in *t*-butyl methyl ether to the extent of 12 g/100 mL and dissolves in water to the extent of 6 g/100 mL.

$$k = \frac{12 \text{ g/100 mL } t\text{-butyl methyl ether}}{6 \text{ g/100 mL water}} = 2$$

If a solution of 6 g of A in 100 mL of water is shaken with 100 mL of *t*-butyl methyl ether, then

$$k = \frac{(x \text{ grams of A/100 mL } t\text{-butyl methyl ether})}{(6 - x \text{ grams of A/100 mL water})} = 2$$

from which

$$x = 4.0 \text{ g of A in the ether layer}$$

$$6 - x = 2.0 \text{ g left in the water layer}$$

It is, however, more efficient to extract the 100 mL of aqueous solution twice with 50-mL portions of *t*-butyl methyl ether rather than once with a 100-mL portion.

$$k = \frac{(x \text{ g of A/50 mL})}{(6 - x \text{ g of A/100 mL})} = 2$$

from which

$$x = 3.0 \text{ g in the } t\text{-butyl methyl ether layer}$$

$$6 - x = 3.0 \text{ g in the water layer}$$

If this 3.0 g/100 mL of water is extracted once more with 50 mL of *t*-butyl methyl ether, we can calculate that 1.5 g will be in the ether layer, leaving 1.5 g in the water layer. So two extractions with 50-mL portions of ether will extract 3.0 g + 1.5 g = 4.5 g of A, while one extraction with a 100-mL portion of *t*-butyl methyl ether removes only 4.0 g of A. Three extractions with $33\frac{1}{3}$-mL portions of *t*-butyl methyl ether would extract 4.7 g. Obviously there is a point at which the increased amount of A extracted does not repay the effort of multiple extractions, but keep in mind that several small-scale extractions are more effective than one large-scale extraction.

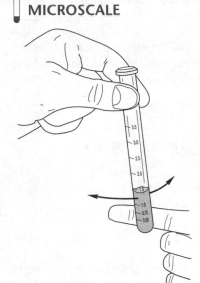

MICROSCALE

FIG. 8.2 Mixing the contents of a reaction tube by flicking it. Grasp the tube firmly at the very top, and flick it vigorously at the bottom. The contents will mix without coming out of the tube.

Always draw out the lower layer and place it in another container.

Technique of Liquid/Liquid Extraction

Place 1 to 2 mL of an aqueous solution of the compound to be extracted in a reaction tube. Add about 1 mL of dichloromethane. Note, as you add the dichloromethane, whether it is the top or bottom layer.

Mixing the Layers.   An effective way to do this is to flick the tube with a finger. Grasp the tube firmly at the very top between thumb and forefinger, and flick it vigorously at the bottom (Fig. 8.2). You will find that this violent motion mixes the contents well, but nothing comes out the top. Another good mixing technique is to pull the contents of the reaction tube into a Pasteur pipette and then expel the mixture back into the tube with force. Doing this several times will effect good mixing of the two layers. A stopper can be placed in the top of the tube and the contents mixed by shaking the tube, but the problem with this technique is that the high vapor pressure of the solvent often will force liquid out around the cork or stopper.

Separating the Layers.   After thorough mixing of the two layers, allow them to separate. Tap the tube if droplets of one layer are in the other layer or on the side of the tube. After the layers separate completely, draw up the lower dichloromethane layer into a Pasteur pipette. Leave behind any middle emulsion layer. The easiest way to do this is to attach the pipette to a pipette pump (Fig. 8.3). This allows very precise control of the liquid being removed. It takes more skill and practice to remove the lower layer cleanly with a 2-mL rubber bulb attached to the pipette, because the high vapor pressure of the solvent tends to make it dribble out of the pipette. To avoid losing any of the solution, it is best to grasp a clean, dry, empty tube in the same hand as the full tube to receive the organic layer (Fig. 8.4).

From the discussion of the distribution coefficient in the preceding section you know that several small extractions are better than one big one, so repeat the extraction process with two further 1-mL portions of dichloromethane. An experienced chemist might summarize all the preceding with the notebook entry, "Aqueous layer extracted 3 × 1-mL portions CH_2Cl_2," and in a formal report would write, "The aqueous layer was extracted three times with 1-mL portions of dichloromethane." If any water droplets have inadvertently been transferred, the easiest way to eliminate them is to again transfer the organic layer to a clean, dry reaction tube.

Larger Microscale Technique

A separatory funnel, regardless of size, should be filled only to about two-thirds of its capacity so the layers can be mixed by shaking. The microscale separatory

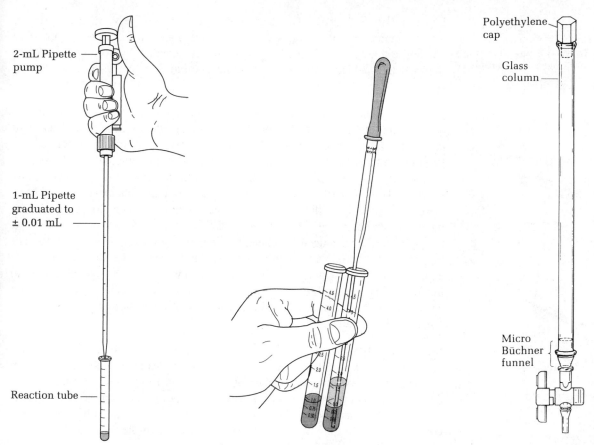

FIG. 8.3 Removal of a solvent from a reaction tube using a pipette and pipette pump.

FIG. 8.4 Grasp both reaction tubes in one hand when transferring material from one tube to another using a Pasteur pipette.

FIG. 8.5 Microscale separatory funnel. Remove the polyethylene frit from the micro Büchner funnel before using.

funnel (Fig. 8.5) has a capacity of 8.5 mL when full, so it is useful for an extraction with a total volume of about 6 mL.

Poke out the polyethylene frit from the bottom part of the separatory funnel using a wood boiling stick. Store it for later replacement. Close the valve, add up to about 5 mL of the solution to be extracted to the separatory funnel, then add the extraction solvent so that the total volume does not exceed 6 mL.

Cap the separatory funnel, and mix the contents by inverting the funnel several times. If the two layers separate fairly easily, then the contents can be shaken more thoroughly. If the layers do not separate easily, be careful not to shake the funnel too vigorously; intractable emulsions can form.

Remove the stopper from the funnel, clamp it, and then, grasping the valve with two hands, empty the bottom layer into an Erlenmeyer flask or other container. If the top layer is desired, pour it out through the top of the separatory funnel at this time—don't drain it through the valve, which may have a drop of the lower layer remaining in it.

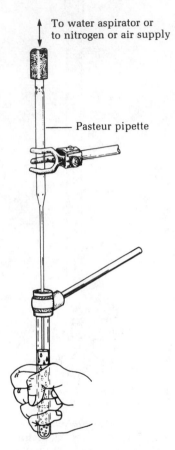

To water aspirator or
to nitrogen or air supply

Pasteur pipette

FIG. 8.6 Aspirator tube being used to remove solvent vapors.

Record the tare of the final container.

 **MICROSCALE**

Drying the Solvent.   Dichloromethane dissolves a very small quantity of water, and microscopic droplets of water are suspended in the organic layer, often making it cloudy. To remove the water, a drying agent, anhydrous calcium chloride pellets, is added to the dichloromethane solution.

How Much Drying Agent Should Be Used?   When a small quantity of the drying agent is added, the crystals or pellets become sticky with water, clump together, and fall rapidly as a lump to the bottom of the reaction tube. There will come a point when a new small quantity of drying agent no longer clumps together, but the individual particles settle slowly throughout the solution. As they say in Scandinavia, "Add drying agent until it begins to snow." The drying process takes about 10 to 15 min, during which time the tube contents should be mixed occasionally by flicking the tube. The solution should no longer be cloudy, but clear (although it may be colored).

Once drying is judged complete, the solvent is removed by forcing a Pasteur pipette to the bottom of the reaction tube and pulling the solvent in. Air is expelled from the pipette as it is being pushed through the crystals or pellets so that no drying agent will enter the pipette. It is very important to wash the drying agent left in the reaction tube with several small quantities of pure solvent to transfer all the extract.

Removing the Solvent.   If the quantity of extract is relatively small, say three mL or less, then the easiest way to remove the solvent is to blow a stream of air (or nitrogen) onto the surface of the solution from a Pasteur pipette (Fig. 8.6). Be sure that the stream of air is very gentle before inserting it into the reaction tube. The heat of vaporization of the solvent will cause the tube to get rather cold during the evaporation and, of course, slow down the process. The easiest way to add heat is to hold the tube in your hand.

Another way to remove the solvent is to attach the Pasteur pipette to an aspirator and pull air over the surface of the liquid. This is not so fast as blowing air onto the surface of the liquid and runs the danger of sucking the liquid up into the aspirator.

If the volume of liquid is more than about three mL, then put it into the 25-mL filter flask, put the plastic Hirsch funnel in place, and attach the flask to the aspirator. By placing your thumb in the Hirsch funnel, the vacuum can be controlled, and heat can be applied by holding the flask in the other hand while swirling the contents (Fig. 8.7).

The reaction tube or filter flask in which the solvent is evaporated should be tared (weighed empty), and this weight recorded in the notebook. In this way, the weight of material extracted can be determined by again weighing the container.

Experiment

Partition Coefficient of Benzoic Acid

In a reaction tube, place about 100 mg of benzoic acid (weighed to the nearest milligram), and to this add exactly equal volumes of water followed by

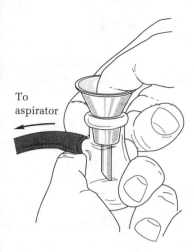

To
aspirator

FIG. 8.7 Apparatus for removal of a solvent under vacuum.

dichloromethane (about 1.6 mL each). While making this addition, note which layer is the organic layer and which the aqueous. Put a septum on the tube, and shake the contents vigorously for at least 2 min. Allow the tube to stand undisturbed until the layers separate, and then carefully draw off, using a Pasteur pipette, *all* the aqueous layer without removing any of the organic layer. It may be of help to draw out the tip of the pipette to a fine point in a flame and, using this, to tilt the reaction tube on its side to make this separation as clean as possible.

Add anhydrous calcium chloride pellets to the dichloromethane in very small quantities until it no longer clamps together. Mix the contents of the tube by flicking it, and allow it to stand for about 5 min to complete the drying process. Using a dry Pasteur pipette, transfer the dichloromethane to a tared (previously weighed) dry reaction tube or 10-mL Erlenmeyer flask containing a boiling chip. Complete the transfer by washing the drying agent with two more portions of solvent that are added to the original solution, and then evaporate the solvent. This can be done by boiling off the solvent while removing solvent vapors with an aspirator tube or by blowing a stream of air or nitrogen into the container while warming it in the hand (see Fig. 8.6). This operation should be performed in a hood.

From the weight of the benzoic acid in the dichloromethane layer, the weight in the water layer can be obtained by difference. The ratio of the weight in dichloromethane to the weight in water is the distribution coefficient, because the volumes of the two solvents were equal. Report the value of the distribution coefficient.

Cleaning Up   The aqueous layer can be flushed down the drain. Dichloromethane goes in the halogenated organic solvents container, and after allowing the solvent to evaporate from the sodium sulfate in the hood, place it in the non-hazardous solid waste container.

Part 2. Acid/Base Extraction

The third type of extraction, acid/base extraction, involves carrying out simple acid/base reactions in order to separate strong organic acids, weak organic acids, neutral organic compounds, and basic organic substances. The chemistry involved is given in the equations that follow, using benzoic acid, phenol, naphthalene, and aniline as examples of the four types of compounds. Chapter 44 is also an acid/base extraction experiment; the components of analgesic tablets are separated.

Here is the strategy: The four organic compounds are dissolved in *t*-butyl methyl ether. The ether solution is shaken with a saturated aqueous solution of sodium bicarbonate, a weak base. This will react only with the strong acid, benzoic acid (**1**) to form the ionic salt, sodium benzoate (**5**), which dissolves in the aqueous layer and is removed. The ether solution now contains just phenol (**2**), naphthalene (**4**), and aniline (**3**). A 3 *M* aqueous solution of sodium hydroxide is added and the mixture shaken. The hydroxide, a strong base, will react only with the phenol (**2**) a weak acid, to form sodium phenoxide (**6**), an ionic compound that dissolves in the aqueous layer and is removed. The ether now contains only naphthalene (**4**) and aniline (**3**). Shaking it with dilute hydrochloric acid removes the aniline, a base, as the ionic anilinium chloride (**7**). The aqueous layer is removed. Evaporation of the *t*-butyl methyl ether now leaves naphthalene (**4**), the neutral compound. The other three compounds are recovered by adding acid to the sodium benzoate (**5**) and sodium phenoxide (**6**) and base to the anilinium chloride (**7**) to regenerate the covalent compounds benzoic acid (**1**), phenol (**2**), and aniline (**3**). These operations are conveniently represented in a flow diagram (Fig. 8.8).

The ability to separate strong from weak acids depends on the acidity constants of the acids and the basicity constants of the bases as follows. In the first equation consider the ionization of benzoic acid, which has an equilibrium constant, K_a, of 6.8×10^{-5}. The conversion of benzoic acid to the benzoate anion in Eq. 4 is governed by the equilibrium constant, K (Eq. 5), obtained by combining the third and fourth equations.

The pKₐ of carboric acid, H_2CO_3, is 6.35.

$$C_6H_5COOH + H_2O \rightleftharpoons C_6H_5COO^- + H_3O^+ \qquad (1)$$

$$K_a = \frac{[C_6H_5COO^-][H_3O^+]}{[C_6H_5COOH]} = 6.8 \times 10^{-5}, pK_a = 4.17 \qquad (2)$$

$$K_w = [H_3O^+][OH^-] = 10^{-14} \qquad (3)$$

$$C_6H_5COOH + OH^- \rightleftharpoons C_6H_5COO^- + H_2O \qquad (4)$$

$$K = \frac{[C_6H_5COO^-]}{[C_6H_5COOH][OH^-]} = \frac{K_a}{K_w} = \frac{6.8 \times 10^{-5}}{10^{-14}} = 3.2 \times 10^8 \qquad (5)$$

For phenol with a K_a of 10^{-10}, the minimum hydroxide ion concentration that will produce the phenoxide anion in 99% conversion is 10^{-2} *M*. The concentration of hydroxide in 10% sodium hydroxide solution is 10^{-1} *M*, and so phenol in strong base is entirely converted to the water-soluble salt.

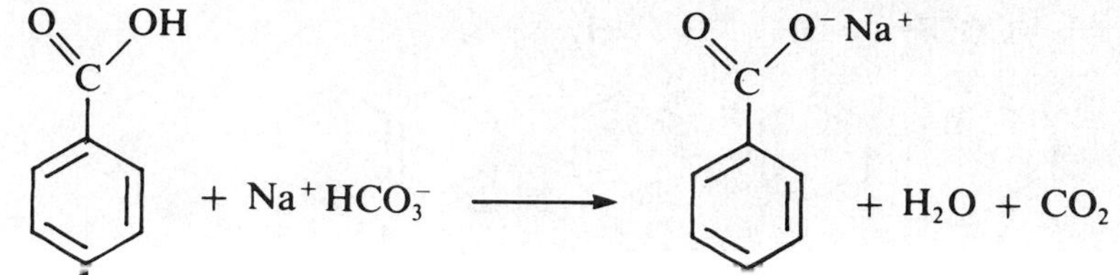

1
Benzoic acid
$pK_a = 4.17$
Covalent, sol. in org. solvents

$+ Na^+ HCO_3^-$ $\longrightarrow$

5
Sodium benzoate
Ionic, sol. in water

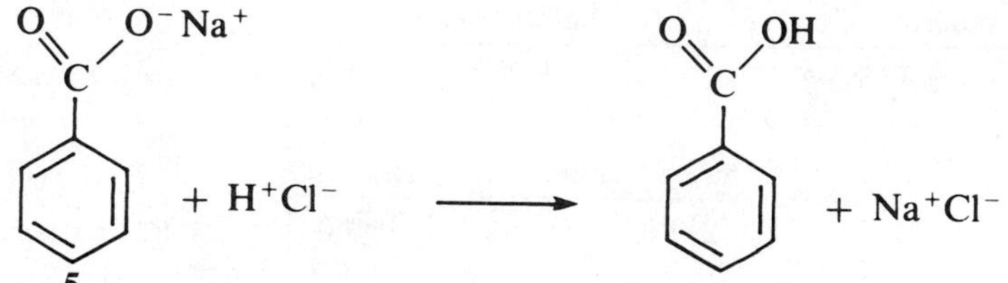

$+ H_2O + CO_2$

5

$+ H^+ Cl^-$ $\longrightarrow$

1

$+ Na^+ Cl^-$

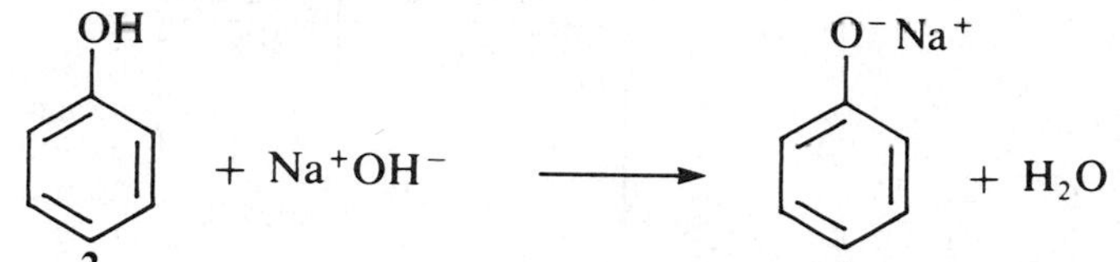

2
Phenol
$pK_a = 10$
Covalent, sol. in org. solvents

$+ Na^+ OH^-$ $\longrightarrow$

6
Sodium phenoxide
Ionic, sol. in water

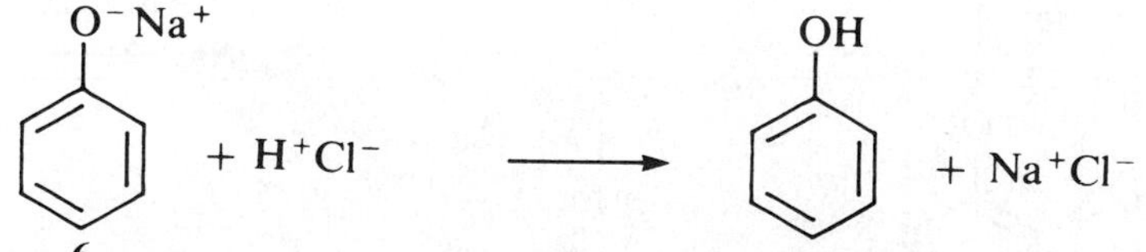

$+ H_2O$

6

$+ H^+ Cl^-$ $\longrightarrow$

2

$+ Na^+ Cl^-$

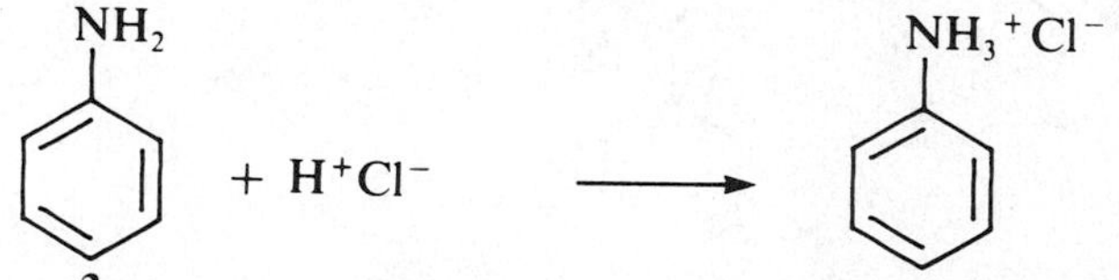

3
Aniline
$pK_b = 9.30$
Covalent, sol. in org. solvents

$+ H^+ Cl^-$ $\longrightarrow$

7
Anilinium chloride
Ionic, sol. in water

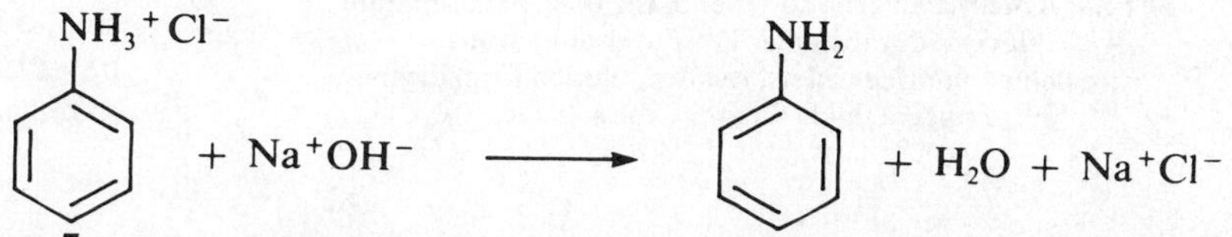

7

$+ Na^+ OH^-$ $\longrightarrow$

3

$+ H_2O + Na^+ Cl^-$

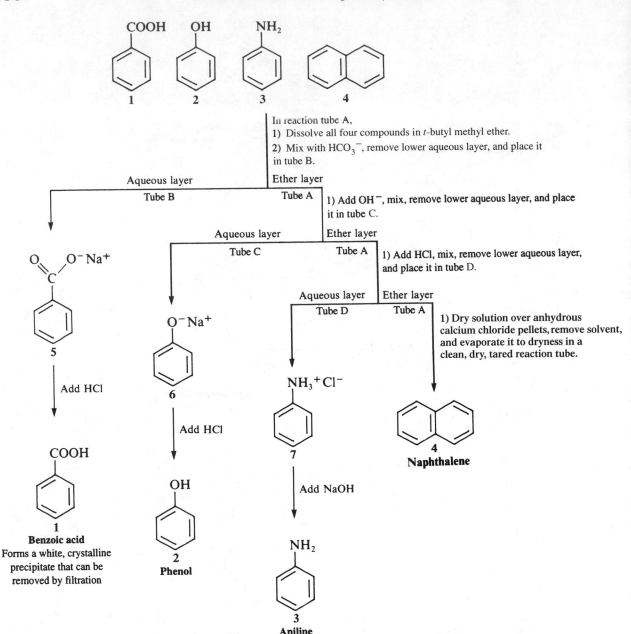

In reaction tube A,
1) Dissolve all four compounds in *t*-butyl methyl ether.
2) Mix with HCO_3^-, remove lower aqueous layer, and place it in tube B.

1) Add OH^-, mix, remove lower aqueous layer, and place it in tube C.

1) Add HCl, mix, remove lower aqueous layer, and place it in tube D.

1) Dry solution over anhydrous calcium chloride pellets, remove solvent, and evaporate it to dryness in a clean, dry, tared reaction tube.

Benzoic acid
Forms a white, crystalline precipitate that can be removed by filtration

Phenol

Naphthalene

Aniline

Phenol and aniline each form oily layers on the top of the aqueous layer. Extract each with *t*-butyl methyl ether: Add ether to the tube, mix, separate layers, dry the ether layer over anhydrous calcium chloride pellets, remove solution from drying agent, and evaporate the solvent.

FIG. 8.8 Flow diagram for the separation of a strong acid, a weak acid, a neutral compound, and a base: benzoic acid, phenol, naphthalene, and aniline.

Liquid/liquid extraction and acid/base extraction are employed in the majority of organic reactions because it is unusual to have the product crystallize from the reaction mixture or to be able to distill the reaction product directly from the reaction mixture. In the research literature, one will often see the statement "the reaction mixture was worked up in the usual way," which implies an extraction process of the type described here. Good laboratory practice dictates, however, that the details of the process be written out.

Practical Considerations

Emulsions

Imagine trying to extract a soap solution, e.g., a nonfoaming dishwasher detergent, into an organic solvent. After a few shakes with an organic solvent, you would have an absolutely intractable emulsion. An emulsion is a suspension of one liquid as droplets in another. Detergents stabilize emulsions, and so any time a detergentlike molecule happens to be in the material being extracted there is the danger that emulsions will form. Substances of this type are commonly found in nature, so one must be particularly wary of emulsion formation when making organic extracts of aqueous plant material, such as caffeine from tea. Emulsions, once formed, can be quite stable. You would be quite surprised to open your refrigerator one morning and see a layer of clarified butter floating on the top of a perfectly clear aqueous solution that had once been milk, but that is the classic example of an emulsion.

Shake gently to avoid emulsions.

Prevention is the best cure for emulsions. This means shaking the solution to be extracted *very gently* until you see that the two layers will separate readily. If a bit of emulsion forms it may break simply on standing for a sufficient length of time. Making the aqueous layer highly ionic will help. Add as much sodium chloride as will dissolve, and shake the mixture gently. Vacuum filtration sometimes works and, when the organic layer is the lower layer, filtration through silicone-impregnated filter paper is an aid. Centrifugation works very well for breaking emulsions. This is easy on a small scale, but often the equipment is not available for large-scale centrifugation of organic liquids.

Pressure Buildup

The heat of the hand or heat from acid/base reactions will cause pressure buildup in an extraction mixture that contains a very volatile solvent such as dichloromethane. The extraction container, be it a test tube or a separatory funnel, must be opened carefully to vent this pressure.

Sodium bicarbonate solution is often used to neutralize acids when carrying out acid/base extractions. The result is the formation of carbon dioxide, which can cause foaming and high pressure buildup. Whenever bicarbonate is used, add it very gradually with thorough mixing and frequent venting of the extraction device. If a large amount of acid is to be neutralized with bicarbonate, the process should be carried out in a beaker.

Removal of Water from Extraction Solvents

Saturated aqueous sodium chloride solution removes water from ether.

The organic solvents used for extraction dissolve not only the compound being extracted but also water. Evaporation of the solvent then leaves the desired compound contaminated with water. At room temperature water dissolves 4.8% of *t*-butyl methyl ether by weight, and the ether dissolves 1.5% of water. But ether is virtually insoluble in water saturated with sodium chloride (36.7 g/100 mL). If ether that contains dissolved water is shaken with a saturated aqueous solution of sodium chloride, water will be transferred from the *t*-butyl methyl ether to the aqueous layer. So, strange as it may seem, ethereal extracts routinely are dried by shaking them with aqueous saturated sodium chloride solution.

Solvents such as dichloromethane do not dissolve nearly as much water and so are dried over a chemical drying agent. Many choices of chemical drying agents are available for this purpose, and the choice of which one to use is governed by four factors: the possibility of reaction with the substance being extracted, the speed with which it removes water from the solvent, the efficiency of the process, and the ease of recovery from the drying agent.

Some very good but specialized and reactive drying agents are potassium hydroxide, anhydrous potassium carbonate, sodium metal, calcium hydride, lithium aluminum hydride, and phosphorus pentoxide. Substances that are essentially neutral and unreactive and are widely used as drying agents include anhydrous calcium sulfate (Drierite), magnesium sulfate, molecular sieves, calcium chloride, and sodium sulfate.

Drierite, $CaSO_4$

Drierite, a specially prepared form of calcium sulfate, is a fast and effective drying agent. However, it is difficult to ascertain whether enough has been used. An indicating type of Drierite is impregnated with cobalt chloride, which turns from blue to red when it is saturated with water. This works well when gases are being dried, but it should not be used for liquid extractions because the cobalt chloride dissolves in many protic solvents.

Magnesium sulfate, $MgSO_4$

Magnesium sulfate is also a fast and fairly effective drying agent, but it is so finely powdered that it always requires careful filtration for removal.

Molecular sieves, zeolites

Molecular sieves are sodium alumino-silicates (zeolites) that have well-defined pore sizes. The 4A size adsorbs water to the exclusion of almost all organic substances and is a fast and effective drying agent, but like Drierite it is impossible to ascertain by appearance whether enough has been used. Molecular sieves in the form of 1/16-in. pellets are often used to dry solvents by simply adding them to the container.

Calcium chloride ($CaCl_2$) pellets, the drying agent of choice for small-scale experiments

Calcium Chloride Pellets, the Drying Agent of Choice   *Calcium chloride,* recently available in the form of pellets[1] (4 to 80 mesh), is a very fast and effective drying agent. It has the advantage that it clumps together when excess water is present so that it is possible, by observing its behavior, to know how much to add. Unlike the older granular form, the pellets do not disintegrate

---

1. Fisher, Cat. No. 614-3.

Add anhydrous drying agent until it begins to snow.

to give a fine powder. These pellets are admirably suited to microscale experiments where the solvent is removed from the drying agent with a Pasteur pipette. It is much faster and much more effective than anhydrous sodium sulfate; after much experimentation, we have decided that this is the agent of choice for microscale experiments in particular. These pellets are used for most of the drying operations in this text. But calcium chloride reacts with some alcohols, phenols, amides, and some carbonyl-containing compounds. Advantage is sometimes taken of this property to remove not only water from a solvent but also, for example, a contaminating alcohol (see the synthesis of 1-bromobutane from 1-butanol, Chapter 16).

Sodium sulfate, Na₂SO₄

Sodium sulfate is a very poor drying agent. It has a very high capacity for water but is slow and not very efficient in the removal of water. It, like calcium chloride pellets, clumps together when wet, and solutions are easily removed from it using a Pasteur pipette. It has been used extensively in the past and should still be used for compounds that react with calcium chloride.

MICROSCALE

1. Separation of a Carboxylic Acid, a Phenol, and a Neutral Substance

A mixture of equal parts of a carboxylic acid, a phenol, and a neutral substance is to be separated by extraction from an ether solvent. Note carefully the procedure for this extraction. In the next experiment you are to work out your own extraction procedure. Your unknown will consist of either benzoic acid or 2-chlorobenzoic acid (the carboxylic acid), 4-*tert*-butyl phenol or 4-bromophenol and biphenyl or 1,4-dimethoxybenzene (the neutral substance). The object of this experiment is to identify the three substances in the mixture and to determine the percent recovery of each from the mixture.

Benzoic acid
mp 123°C, pK_a 4.17

1,4-Dimethoxybenzene
(Hydroquinone dimethyl ether)
mp 57°C

Procedure

Dissolve about 0.18 g of the mixture (record the exact weight) in 2 mL of *tert*-butyl methyl ether or diethyl ether in a reaction tube. Then add 1 mL of a saturated aqueous solution of sodium bicarbonate to the tube. Use the graduations on the side of the tube to measure the amounts, because they do not need to be exact. Mix the contents of the tube thoroughly by pulling the two layers into a Pasteur pipette and expelling them forcefully into the reaction tube. Do this for about 3 min. Allow the layers to separate completely, and then draw off the lower layer into another reaction tube (tube 2). Add another 0.15 mL of sodium bicarbonate solution to the tube, mix the contents as before, and add the lower layer to tube 2. Exactly what chemical species is in tube 2? Add 0.2 mL of ether to tube 2, mix it thoroughly, remove the ether layer, and discard it. This is called *backwashing* and serves to remove any organic material that might contaminate the contents of tube 2.

OH

Br

4-Bromophenol
mp 66°C, pK_a 10.2

Add HCl with care. CO$_2$ is released.

Biphenyl
mp 71°C

OH

H$_3$C—C—CH$_3$
       CH$_3$

4-*tert*- Butylphenol
mp 101°C, pK_a 10.17

Best way to remove the solvent: under a gentle stream of air or nitrogen.

Add 1.0 mL of 3 M aqueous sodium hydroxide to tube 1, shake the mixture thoroughly, allow the layers to separate, draw off the lower layer using a clean Pasteur pipette, and place it in tube 3. Extract tube 1 with two 0.15-mL portions of water, and add these to tube 3. Backwash the contents of tube 3 with 0.15 mL of ether, and discard the ether wash just as was done for tube 2. Exactly what chemical species is in tube 3?

To tube 1 add saturated sodium chloride solution, mix, remove the aqueous layer, and then add to the ether anhydrous calcium chloride pellets until the drying agent no longer clumps together. Wash it off with ether after the drying process is finished. Allow 5 to 10 min for drying of the ether solution.

Using the concentration information given in the inside back cover of this book, calculate exactly how much concentrated hydrochloric acid is needed to neutralize the contents of tube 2. Then, by dropwise addition of concentrated hydrochloric acid, carry out this neutralization while testing the solution with litmus paper. An excess of hydrochloric acid does no harm. This reaction must be carried out with *extreme care* because much carbon dioxide is released in the neutralization. Add a boiling stick to the tube, and very cautiously heat the tube to bring most of the solid carboxylic acid into solution. Allow the tube to cool slowly to room temperature, and then cool it in ice. Remove the solvent with a Pasteur pipette, and recrystallize the residue from boiling water. Again allow the tube to cool slowly to room temperature, and then cool it in ice. At the appropriate time stir the crystals, and collect them on the Hirsch funnel using the procedures detailed in Chapter 3. The crystals can be transferred and washed on the funnel using a small quantity of ice water. The solubility of benzoic acid in water is 1.9 g/L at 0°C and 68 g/L at 95°C. The solubility of chlorobenzoic acid is similar. Turn the crystals out onto a tared piece of paper, allow them to dry thoroughly, and determine the percent recovery of the acid. Assess the purity of the product by melting point.

In exactly the same way, neutralize the contents of tube 3 with concentrated hydrochloric acid. This time, of course, there will be no carbon dioxide evolution. Again, heat the tube to bring most of the material into solution, allow it to cool slowly, remove the solvent, and recrystallize the phenol from boiling water. At the appropriate time, after the product has cooled slowly to room temperature and then in ice, it is also collected on the Hirsch funnel, washed with a very small quantity of ice water, and allowed to dry. The percent recovery and melting point are determined.

The neutral compound is recovered using the Pasteur pipette to remove the ether from the drying agent and to transfer it to a tared reaction tube. The drying agent is washed two or three times with additional ether to ensure complete transfer of the product.

Evaporate the solvent by placing the tube in a warm water bath and directing a stream of nitrogen or air onto the surface of the ether in the hood (see Fig. 8.6). An aspirator tube also can be used for this purpose. Determine the weight of the crude product, and then recrystallize it from methanol–water if it is the low-melting compound. Reread Chapter 3 for detailed instructions on carrying out this process of crystallization from a mixed solvent. The product is dissolved in

COOH
Cl

2-Chlorobenzoic acid
mp 141°C, pK_a 2.92

about 0.5 to 1 mL of methanol, and water is added until the solution gets cloudy, indicating the solution is saturated. This process is best carried out while heating the tube in a hot water bath at 50°C. Allow the tube to cool slowly to room temperature, and then cool it thoroughly in ice. If you have the high-melting compound, recrystallize it from ethanol (8 mL/g).

The products are best isolated by collection on the Hirsch funnel using an ice-cold alcohol–water mixture to transfer and wash the compounds. Determine the percent recovery and the melting point. Turn in the products in neatly labeled 1 ½-in. × 1 ½-in. (4 cm × 4 cm) ziplock plastic bags attached to the laboratory report. If the yield on crystallization is low, concentrate the filtrate (the mother liquor) and obtain a second crop of crystals.

MICROSCALE

2. Separation of a Neutral and Basic Substance

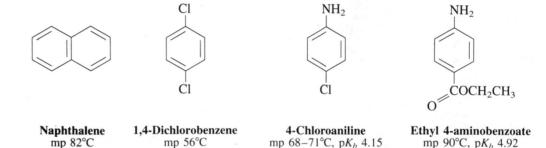

| **Naphthalene** | **1,4-Dichlorobenzene** | **4-Chloroaniline** | **Ethyl 4-aminobenzoate** |
| mp 82°C | mp 56°C | mp 68–71°C, pK_b 4.15 | mp 90°C, pK_b 4.92 |

A mixture of equal parts of a neutral substance, naphthalene or 1,4-dichlorobenzene, and a basic substance, 4-chloroaniline or ethyl 4-aminobenzoate, is to be separated by extraction from an ether solution. Naphthalene and 1,4-dichlorobenzene are completely insoluble in water. The bases will dissolve in hydrochloric acid, while the neutral compounds will remain in ether solution. The bases are insoluble in cold water but will dissolve to some extent in hot water and are soluble in ethanol. Naphthalene and 1,4-dichlorobenzene can be purified as described in Chapter 3. They sublime very easily also. Keep the samples covered.

Plan a procedure for separating 200 mg of the mixture into its components, and have the plan checked by the instructor before proceeding. A flow sheet is a convenient way to present the plan. Select the correct solvent or mixture of solvents for the recrystallization of the bases on the basis of solubility tests. Determine the weights and melting points of the isolated and purified products, and calculate the percent recovery of each. Turn in the products in neatly labeled vials or 1 ½-in. × 1 ½-in. ziplock plastic bags attached to the report.

FIG. 8.9 Correct positions for holding a separatory funnel when shaking. Point outlet away from yourself and your neighbors.

Cleaning Up   Combine all aqueous filtrates and solutions, neutralize them, and flush the resulting solution down the drain. Used ether should be placed in the organic solvents container, and the drying agent, once the solvent has evaporated from it, can be placed in the nonhazardous solid waste container. Any 4-chloroaniline or 1,4-dichlorobenzene should be placed in the halogenated waste container.

 **MACROSCALE**

Apparatus and Operations

In macroscale experiments, a frequently used method of working up a reaction mixture is to dilute the mixture with water and extract with ether in a separatory funnel (Fig. 8.9). When the stoppered funnel is shaken to distribute the components between the immiscible solvents *t*-butyl methyl ether and water, pressure always develops through volatilization of ether from the heat of the hands, and the liberation of a gas (CO_2) can increase the pressure. Consequently, the funnel is grasped so that the stopper is held in place by one hand and the stopcock by the other, as illustrated. After a brief shake or two, the funnel is held in the inverted position shown and the stopcock opened cautiously (with the funnel stem pointed away from nearby people) to release pressure. The mixture can then be shaken more vigorously and pressure released as necessary. When equilibration is judged to be complete, the slight, constant terminal pressure due to ether is released, the stopper is rinsed with a few drops of ether delivered by a Pasteur pipette, and the

Ether vapors are heavier than air and can travel along bench tops, run down drain troughs, and collect in sinks. Be extremely careful to avoid flames when working with volatile ethers.

2-Chlorobenzoic acid
mp 141°C, pK_a 2.92

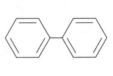

1,4-Dichlorobenzene
mp 56°C

Biphenyl
mp 71°C

4-*t*-Butylphenol
mp 101°C

Foaming occurs when bicarbonate is added to acid; use care.

layers are allowed to separate. The organic reaction product is distributed wholly or largely into the upper ether layer, whereas inorganic salts, acids, and bases pass into the water layer, which can be drawn off and discarded. If the reaction was conducted in alcohol or some other water-soluble solvent, the bulk of the solvent is removed in the water layer and the rest can be eliminated in two or three washings with 1–2 volumes of water conducted with the techniques used in the first equilibration. The separatory funnel should be supported in a ring stand as shown in Fig. 8.10 on page 145.

If acetic acid were used as the reaction solvent it would also be distributed largely into the aqueous phase, but if the reaction product is a neutral substance the residual acetic acid in the ether can be removed by one washing with excess 5% sodium bicarbonate solution. If the reaction product is a higher molecular weight acid, for example benzoic acid (C_6H_5COOH), it will stay in the ether layer, while acetic acid is being removed by repeated washing with water; the benzoic acid can then be separated from neutral byproducts by extraction with sodium bicarbonate or sodium hydroxide solution and acidification of the extract. Acids of high molecular weight are extracted only slowly by sodium bicarbonate, so sodium carbonate is used in its place; however, carbonate is more prone than bicarbonate to produce emulsions. Some-

1,4-Dimethoxybenzene
(Hydroquinone dimehtyl ether)
mp 57°C

times an emulsion in the lower layer can be settled by twirling the separatory funnel by its stem. An emulsion in the upper layer can be broken by grasping the funnel by the neck and swirling it. Because the tendency to emulsify increases with removal of electrolytes and solvents, a little sodium chloride or hydrochloric acid solution is added with each portion of wash water. If the layers are largely clear but an emulsion persists at the interface, the clear part of the water layer can be drawn off and the emulsion run into a second funnel and shaken with fresh ether.

Before adding a liquid to the separatory funnel, check the stopcock. If it is glass, see that it is properly greased, bearing in mind that too much grease will clog the hole in the stopcock and also contaminate the extract. If the stopcock is Teflon, see that it is adjusted to a tight fit in the bore. Store the separatory funnel with the Teflon stopcock loosened to prevent sticking. Because Teflon has a much larger temperature coefficient of expansion than glass, a stuck stopcock can be loosened by cooling the stopcock in ice or dry ice. Do not store liquids in the separatory funnel; they often leak or cause the stopper or stopcock to freeze. To have sufficient room for mixing the layers, fill the separatory funnel no more than three-fourths full. Withdraw the lower layer from the separatory funnel through the stopcock, and pour the upper layer out through the neck.

All too often the inexperienced chemist discards the wrong layer when using a separatory funnel. Through incomplete neutralization, a desired component may still remain in the aqueous layer or the densities of the layers may change. Cautious workers save all layers until the desired product has been isolated. The organic layer is not always the top layer. If in doubt, test the layers by adding a few drops of each to water in a test tube.

MACROSCALE

$pH = -log\,[H^+]$
$pK_a = acidity\ constant$
$pK_b = basicity\ constant$

3. Separation of Acidic and Neutral Substances

A mixture of equal proportions of benzoic acid, 2-naphthol, and 1,4-dimethoxy-benzene is to be separated by extraction from *t*-butyl methyl ether. Note the detailed directions for extraction carefully. Prepare a flowsheet (see Fig. 8.8) for this sequence of operations. In the next experiment you will work out your own extraction procedure.

Procedure

Dissolve 3 g of the mixture in 30 mL of *t*-butyl methyl ether, and transfer the mixture to a 125-mL separatory funnel (Fig. 8.10) using a little t-butyl methyl ether to complete the transfer. Add 10 mL of water, and note which layer is organic and which is aqueous. Add 10 mL of a 3 *M* aqueous solution of sodium bicarbonate to the funnel. Stopper the funnel, and cautiously mix the contents. Vent the liberated carbon dioxide, and then shake the mixture thoroughly with frequent vent-

O OH
\
C
/
(benzene ring)

Benzoic acid
mp 123°C, pK_a 4.17

OH
|
(benzene ring)
|
H₃C — C — CH₃
|
CH₃

4-*t*-Butylphenol
mp 101°C

OCH₃
|
(benzene ring)
|
OCH₃

1,4-Dimethoxybenzene
(Hydroquinone dimethyl ether)
mp 57°C

Extinguish all flames when working with t-butyl methyl ether! The best method for removing the ether is by simple distillation. Dispose of waste ether in container provided.

ing of the funnel. Allow the layers to separate completely, and then draw off the lower layer into a 50-mL Erlenmeyer flask (labeled Flask 1). What does this layer contain?

Add 10 mL of 1.5 *M* aqueous sodium hydroxide to the separatory funnel, shake the mixture thoroughly, allow the layers to separate, and draw off the lower layer into a 25-mL Erlenmeyer flask (labeled Flask 2). Add an additional 5 mL of water to the separatory funnel, shake the mixture as before, and add this to Flask 2. What does Flask 2 contain?

Add 15 mL of a saturated aqueous solution of sodium chloride to the separatory funnel, shake the mixture thoroughly, allow the layers to separate, and draw off the lower layer, which can be discarded. What is the purpose of adding saturated sodium chloride solution? Carefully pour the ether layer into a 50-mL Erlenmeyer flask (labeled Flask 3) from the top of the separatory funnel, taking great care not to allow any water droplets to be transferred. Add about 4 g of anhydrous calcium chloride pellets to the ether extract, and set it aside.

Acidify the contents of Flask 2 by dropwise addition of concentrated hydrochloric acid while testing with litmus paper. Cool the flask in an ice bath.

Cautiously add concentrated hydrochloric acid dropwise to Flask 1 until the contents are acidic to litmus, and then cool the flask in ice.

Decant (pour off) the ether from Flask 3 into a tared (previously weighed) flask, taking care to leave all of the drying agent behind. Wash the drying agent with additional ether to ensure complete transfer of the product. If decantation is difficult then remove the drying agent by gravity filtration (see Fig. 3.2). Put a boiling stick in the flask, and evaporate the ether in the hood. An aspirator tube can be used for this purpose (see Fig. 8.11). Determine the weight of the crude *p*-dimethoxybenzene, and then recrystallize it from methanol. See Chapter 3 for detailed instructions on how to carry out crystallization.

Isolate the *t*-butylphenol from Flask 2 employing vacuum filtration on a Hirsch funnel (see Fig. 3.2), and wash it on the filter with a small quantity of ice water. Determine the weight of the crude product, and then recrystallize it from ethanol. Similarly isolate, weigh, and recrystallize from boiling water the benzoic acid in Flask 1. The solubility of benzoic acid in water is 1.9 g/L at 0°C and 68 g/L at 95°C.

Dry the purified products, determine their melting points and weights, and calculate the percent recovery of each substance, bearing in mind that the original mixture contained 1 g of each compound. Hand in the three products in neatly labeled vials.

Cleaning Up   Combine all aqueous layers, washes, and filtrates. Dilute with water, neutralize using either sodium carbonate or dilute hydrochloric acid. Methanol filtrate and any ether go in the organic solvents container. Allow ether to evaporate from the calcium chloride in the hood. It can be placed in the nonhazardous solid waste container.

FIG. 8.11   Aspirator tube in use.

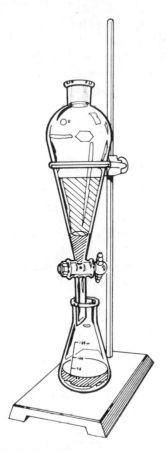

FIG. 8.10   *Separatory funnel with Teflon stopcock.*

Naphthalene
mp 80°C

4. Separation of a Neutral and Basic Substance

A mixture of equal parts of a neutral substance, naphthalene, and a basic substance, 4-chloroaniline, are to be separated by extraction from *t*-butyl methyl ether solution. The base will dissolve in hydrochloric acid while the neutral naphthalene will remain in the *t*-butyl methyl ether solution. 4-Chloroaniline is insoluble in cold water but will dissolve to some extent in hot water and is soluble in ethanol. Naphthalene can be purified as described in Chapter 3.

Plan a procedure for separating 2.0 g of the mixture into its components and have the plan checked by the instructor before proceeding. A flowsheet is a convenient way to present the plan. Using solubility tests, select the correct solvent or mixture of solvents to crystallize 4-chloroaniline. Determine the weights and melting points of the isolated and purified products, and calculate the percent recovery of each. Turn in the products in neatly labeled vials.

Handle aromatic amines with care. Most are toxic and some are carcinogenic. Avoid breathing the dust and vapor from the solid and keep the compounds off the skin, best done by wearing gloves.

NH₂

![4-Chloroaniline structure]

Cl

4-Chloroaniline
mp 68–71°C, pK_b 10.0

Cleaning Up   Combine all aqueous filtrates and solutions, neutralize them, and flush the resulting solution down the drain with a large excess of water. Used *t*-butyl methyl ether should be placed in the organic solvents container, and the calcium chloride, once the solvent has evaporated from it, can be placed in the nonhazardous solid waste container. Any 4-chloroaniline should be placed in the chlorinated organic compounds container.

5. Extraction of Caffeine from Tea

Tea and coffee have been popular beverages for centuries, primarily because they contain the stimulant caffeine. It stimulates respiration, the heart, and the central nervous system, and is a diuretic (promotes urination). It can cause nervousness and insomnia and, like many drugs, can be addictive, making it difficult to reduce the daily dose. A regular coffee drinker who consumes just four cups per day can experience headache, insomnia, and even nausea upon withdrawal from the drug. On the other hand, it helps people to pay attention and can sharpen moderately complex mental skills as well as prolong the ability to exercise.

Caffeine may be the most widely abused drug in the United States. During the course of a day an average person may unwittingly consume up to a gram of this substance. The caffeine content of some common foods and drugs is given in Table 8.1.

Caffeine belongs to a large class of compounds known as alkaloids. These are of plant origin, contain basic nitrogen, often have a bitter taste and complex structure, and usually have physiological activity. Their names usually end in "ine"; many are quite familiar by name if not chemical structure—nicotine, cocaine, morphine, strychnine.

TABLE 8.1   Caffeine Content of Common Foods and Drugs

Espresso	120 mg per 2 oz
Coffee, regular, brewed	103 mg per cup
Instant coffee	57 mg per cup
Coffee, decaffeinated	2 to 4 mg per cup
Tea	30 to 75 mg per cup
Cocoa	5 to 40 mg per cup
Milk chocolate	6 mg per oz
Baking chocolate	35 mg per oz
Coca-Cola, Classic	46 mg per 12 oz
Jolt Cola	72 mg per 12 oz
Anacin, Bromo-Seltzer, Midol	32 mg per pill
Excedrin, extra strength	65 mg per pill
Dexatrim, Dietac, Vivarin	200 mg per pill
Dristan	16 mg per pill
No-Doz	100 mg per pill

Tea leaves contain tannins, which are acidic, as well as a number of colored compounds and a small amount of undecomposed chlorophyll (soluble in dichloromethane). To ensure that the acidic substances remain water soluble and that the caffeine will be present as the free base, sodium carbonate is added to the extraction medium.

The solubility of caffeine in water is 2.2 mg/mL at 25°C, 180 mg/mL at 80°C, and 670 mg/mL at 100°C. It is quite soluble in dichloromethane, the solvent used in this experiment to extract the caffeine from water.

Caffeine can be extracted easily from tea bags. The procedure one would use to make a cup of tea—simply "steeping" the tea with very hot water for about 7 min—extracts most of the caffeine. There is no advantage to boiling the tea leaves with water for 20 min. Since caffeine is a white, slightly bitter, odorless, crystalline solid, it is obvious that water extracts more than just caffeine. When the brown aqueous solution is subsequently extracted with dichloromethane, primarily caffeine dissolves in the organic solvent. Evaporation of the solvent leaves crude caffeine, which on sublimation yields a relatively pure product. When the concentrated tea solution is extracted with dichloromethane, emulsions can form very easily. There are substances in tea that cause small droplets of the organic layer to remain suspended in the aqueous layer. This emulsion formation results from vigorous shaking. To avoid this problem, it might seem that one could boil the tea leaves with dichloromethane first and then extract the caffeine from the dichloromethane solution with water. In fact this does not work. Boiling 25 g of tea leaves with 50 mL of dichloromethane gives only 0.05 g of residue after evaporation of the solvent. Subsequent extractions give less material. Hot water causes the tea leaves to swell and is obviously a much more efficient extraction solvent. An attempt to sublime caffeine directly from tea leaves was also unsuccessful.

Microscale Procedure

In a 30-mL beaker place 15 mL of water, 2 g of sodium carbonate, and a wooden boiling stick. Bring the water to a boil on the sand bath, remove the boiling stick, and brew a very concentrated tea solution by immersing a tea bag (2.4 g tea) in the very hot water for 5 min. Squeeze as much water from the bag as possible after it cools enough to handle. Be careful not to break the bag. Again bring the water to a boil, and add a new tea bag to the hot solution. After 5 min, remove the tea bag, and squeeze out as much water as possible. This can be done easily on the Hirsch funnel. Rinse the bag with a few milliliters of very hot water, but be sure the total volume of aqueous extract does not exceed 12 mL. Pour the extract into a 15-mL centrifuge tube, and cool the solution in ice to below 40°C (the boiling point of dichloromethane).

CAUTION: Do not breathe the vapors of dichloromethane, and if possible, work with this solvent in the hood.

Balance the centrifuge tubes.

Using three 2-mL portions of dichloromethane, extract the caffeine from the tea. Cork the tube, and use a gentle rocking motion to carry out the extraction. Vigorous shaking will produce an intractable emulsion, while extremely gentle mixing will fail to extract the caffeine. If you have ready access to a centrifuge, the shaking can be very vigorous because any emulsions formed can be broken fairly well by centrifugation for about 90 sec. After each extraction, remove the lower organic layer into a reaction tube, leaving any emulsion layer behind. Dry the combined extracts over anhydrous calcium chloride pellets for 5 or 10 min in an Erlenmeyer flask. Add the drying agent in portions with shaking until it no longer clumps together. Transfer the dry solution to a tared 25-mL filter flask, wash the drying agent twice with 2-mL portions of dichloromethane, and evaporate it to dryness (see Fig. 8.7). The residue will be crude caffeine (determine its weight) that is to be purified by sublimation.

Fit the filter flask with a Pluro stopper or no. 2 neoprene adapter through which is thrust a 15-mL centrifuge tube. Put a pipette bulb on the side arm. Clamp the flask with a large three-prong clamp, fill the centrifuge tube with ice and water, and heat the flask on a hot sand bath (Fig. 8.12). Caffeine is reported to sublime at about 170°C. Tilt the filter flask, and rotate it in the hot sand bath to drive more caffeine onto the centrifuge tube. Use a heat gun to heat the upper walls of the filter flask. When sublimation ceases, remove the ice water, and allow the flask to cool somewhat before removing the centrifuge tube. Scrape the caffeine onto a tared weighing paper, weigh, and using the plastic funnel, transfer it to a small vial or a plastic bag. At the discretion of the instructor, determine the melting point using a sealed capillary. The melting point of caffeine is 238°C. Using the centrifugation technique to separate the extracts, about 30 mg of crude caffeine can be obtained. This will give you 10 to 15 mg of sublimed material, depending on the caffeine content of the particular tea being used. The isolated caffeine can be used to prepare caffeine salicylate (Experiment 7).

Cleaning Up   Discard the tea bags in the nonhazardous solid waste container. Allow the solvent to evaporate from the drying agent, and discard in the same

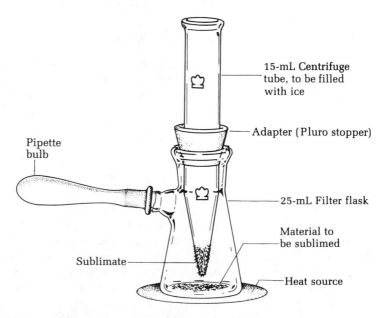

FIG. 8.12 Sublimation apparatus.

container. Place any unused and unrecovered dichloromethane in the chlorinated organic compounds container. The apparatus can be cleaned with soap and hot water. Caffeine can be flushed down the drain, since it is biodegradable.

 ### *Macroscale Procedure*

To an Erlenmeyer flask containing 25 g of tea leaves (or 10 tea bags) and 20 g of sodium carbonate, add 225 mL of vigorously boiling water. Allow the mixture to stand for 7 min, and then decant into another Erlenmeyer flask. To the hot tea leaves add another 50 mL of hot water, and then immediately decant and combine with the first extract. Very little, if any, additional caffeine is extracted by boiling the tea leaves for 20 min. Decantation works nearly as well as vacuum filtration and is much faster.

CAUTION: Carry out this work with dichloromethane in the hood.

Cool the aqueous solution to near room temperature, and extract it twice with 30-mL portions of dichloromethane. Take great care not to shake the separatory funnel so vigorously as to cause emulsion formation, bearing in mind that if it is not shaken vigorously enough the caffeine will not be extracted into the organic layer. Use a gentle rocking motion of the separatory funnel. Drain off the dichloromethane layer on the first extraction; include the emulsion layer on the second extraction. Dry the combined dichloromethane solutions and any emulsion layer with anhydrous calcium chloride pellets. Add sufficient drying agent until it no longer clumps together on the bottom of the flask. Carefully decant or filter the dichloromethane solution into a tared (previously weighed) Erlenmeyer or distilling flask. Silicone-impregnated filter paper passes dichloromethane and retains water. Wash the drying agent with a further portion of solvent, and evaporate or distill the solvent. A wooden stick is better than a boiling chip to promote smooth boiling because it is easily removed once the solvent is gone. The residue of greenish-white crystalline caffeine should weigh about 0.25 g.

Rock the separatory funnel very gently to avoid emulsions.

Dispose of used dichloromethane in the container provided.

Crystallization of Caffeine.   To recrystallize the caffeine, dissolve it in 5 mL of hot acetone, transfer it with a Pasteur pipette to a small Erlenmeyer flask, and, while it is hot, add ligroin to the solution until a faint cloudiness appears. Set the flask aside, and allow it to cool slowly to room temperature. This mixed solvent method of recrystallization depends on the fact that caffeine is much more soluble in acetone than ligroin, so a combination of the two solvents can be found where the solution is saturated in caffeine (the cloud point). Cool the solution containing the crystals, and remove them by vacuum filtration, employing the Hirsch funnel or a very small Büchner funnel. Use a few drops of ligroin to transfer the crystals and wash the crystals. If you wish to obtain a second crop of crystals, collect the filtrate in a test tube, concentrate it to the cloud point using the aspirator tube (Fig. 3.21), and repeat the crystallization process.

Cleaning Up   The filtrate can be diluted with water and washed down the drain. Any dichloromethane collected goes into the halogenated organic waste container. After the solvent is allowed to evaporate from the drying agent in the hood, it can be placed in the nonhazardous solid waste container; otherwise it goes in the hazardous waste container. The tea leaves go in the nonhazardous solid waste container.

MICROSCALE AND MACROSCALE

6. Extraction of Caffeine from Cola Syrup

Coca-Cola was originally flavored with extracts from the leaves of the coca plant and the kola nut. Coca is grown in northern South America; the Indians of Peru and Bolivia have for centuries chewed the leaves to relieve the pangs of hunger and high mountain cold. The cocaine from the leaves causes local anesthesia of the stomach. It has limited use as a local anesthetic for surgery on the eye, nose, and throat. Unfortunately it is now a widely abused illicit drug. Kola nuts contain about 3% caffeine as well as a number of other alkaloids. The kola tree is in the same family as the cacao tree from which cocoa and chocolate are obtained. Modern cola drinks do not contain cocaine; however, Coca-Cola contains 43 mg of caffeine per 12-oz serving. The acidic taste of many soft drinks comes from citric, tartaric, phosphoric, and benzoic acids.

Automatic soft drink dispensing machines mix a syrup with carbonated water. In the following experiment caffeine is extracted from concentrated cola syrup.

Caffeine
mp 238°C

Cocaine

CAUTION: Do not breathe the vapor of dichloromethane. Work with this solvent in the hood.

Microscale Procedure

Add 1 mL of concentrated ammonium hydroxide to a mixture of 5 mL commercial cola syrup and 5 mL of water in a 15-mL centrifuge tube. Add 1 mL dichloromethane, and tip the tube gently back and forth for 5 min. Do not shake the mixture as in a normal extraction, because an emulsion will form and the layers will not separate. After the layers have separated as much as possible, remove the clear lower layer, leaving the emulsion behind. Using 1.5 mL dichloromethane, repeat the extraction in the same way twice more. At the final separation, include the emulsion layer with the dichloromethane. If a centrifuge is available, the mixture can be shaken vigorously and the emulsion broken by centrifugation for 90 sec. Combine the extracts in a reaction tube, and dry the solution with anhydrous calcium chloride pellets. Add the drying agent with shaking until it no longer clumps together. After 5 to 10 min, remove the solution with a Pasteur pipette, and place it in a tared filter flask. Wash off the drying agent with more dichloromethane, and evaporate the mixture to dryness. Determine the crude weight of caffeine, and then sublime it as described in the preceding experiment.

Macroscale Procedure

Add 10 mL of concentrated ammonium hydroxide to a mixture of 50 mL of commercial cola syrup and 50 mL of water. Place the mixture in a separatory funnel, add 50 mL of dichloromethane, and swirl the mixture and tip the funnel back and forth for at least 5 min. Do not shake the solutions together as in a normal extraction, because an emulsion will form and the layers will not separate. An emulsion is made up of droplets of one phase suspended in the other (milk is an emulsion). Separate the layers. Repeat the extraction with a second 50-mL portion of dichloromethane. From your knowledge of the density of dichloromethane and water you should be able to predict which is the top layer and which is the bottom layer. If in doubt, add a few drops of each layer to water. The aqueous layer will be soluble, and the organic layer will not. Combine the dichloromethane extracts and any emulsion that has formed in a 125-mL Erlenmeyer flask, and add anhydrous calcium chloride pellets to remove water from the solution. Add the drying agent until it no longer clumps together at the bottom of the flask but swirls freely in solution. Swirl the flask with the drying agent from time to time over a 10-min period. Carefully decant (pour off) the dichloromethane or remove it by filtration through a fluted filter paper, add about 5 mL more solvent to the drying agent to wash it, and decant this also. Combine the dried dichloromethane solutions in a tared flask, and remove the dichloromethane by distillation or evaporation on the steam bath. Remember to add a wooden boiling stick to the solution to promote even boiling. Determine the weight of the crude product.

Chlorinated solvents are toxic, insoluble in water, expensive, and should never be poured down the drain.

Crystallization of Caffeine

To recrystallize the caffeine dissolve it in 5 mL of hot acetone, transfer it with a Pasteur pipette to a small Erlenmeyer flask, and, while it is hot, add ligroin to the solution until a faint cloudiness appears. Set the flask aside, and allow it to cool slowly to room temperature. This mixed solvent method of recrystallization depends on the fact that caffeine is much more soluble in acetone than ligroin, so a combination of the two solvents can be found where the solution is saturated in caffeine (the cloud point). Cool the solution containing the crystals, and remove them by vacuum filtration, employing the Hirsch funnel or a very small Büchner funnel. Use a few drops of ligroin to transfer the crystals and wash the crystals. If you wish to obtain a second crop of crystals, collect the filtrate in a test tube, concentrate it to the cloud point using the aspirator tube (Fig. 3.9), and repeat the crystallization process.

Cleaning Up   Combine all aqueous filtrates and solutions, neutralize them, and flush the resulting solution down the drain. Used dichloromethane should be placed in the halogenated waste container, and the drying agent, once the solvent has evaporated from it, can be placed in the nonhazardous solid waste container. The ligroin–acetone filtrates should be placed in the organic solvents container.

Sublimation of Caffeine

Sublimation is a fast and easy way to purify caffeine. Using the apparatus depicted in Fig. 8.12, sublime the crude caffeine at atmospheric pressure following the procedure in Part 2 of Chapter 7.

⎵ MICROSCALE

7. Isolation of Caffeine from Instant Coffee

Instant coffee, according to the manufacturer, contains between 55 and 62 mg of caffeine per 6-oz cup, and a cup is presumably made from a teaspoon of the powder, which weighs 1.3 g; so 2 g of the powder should contain 85 to 95 mg of caffeine. Unlike tea, however, coffee contains other compounds that are soluble in dichloromethane, so obtaining pure caffeine from coffee is not easy. The object of this experiment is to extract instant coffee with dichloromethane (the easy part) and then to try to devise a procedure for obtaining pure caffeine from the extract.

From the thin-layer chromatography you may deduce that certain impurities have a high R_f value in hydrocarbons (in which caffeine is insoluble). Consult reference books (see especially the *Merck Index*) to determine the solubility (and lack of solubility) of caffeine in various solvents. You might try trituration (grinding the crude solid with a solvent) to preferentially dissolve impurities. Column chromatography is another possible means of purifying the product. Or you might convert all of it to the salicylate and then regenerate the caffeine from the salicylate. Experiment! Or you can simply follow the following procedure.

Procedure

In a 10-mL Erlenmeyer flask, place 2 g of sodium carbonate and 2 g of instant coffee powder. Add to this 9 mL of boiling water, stir the mixture well, bring it to a boil again with stirring, cool it to room temperature, and then pour it into a 15-mL plastic centrifuge tube fitted with a screw cap. Add 2 mL of dichloromethane, cap the tube, shake it vigorously for 60 sec, and then centrifuge it at high speed for 90 sec. Remove the clear yellow dichloromethane layer, and place it in a 10-mL Erlenmeyer flask. Repeat this process twice more. To the combined extracts add anhydrous calcium chloride pellets until they no longer clump together, allow the solution to dry for a few minutes, and then transfer it to a tared 25-mL filter flask, and wash the drying agent with more solvent. Remove the solvent as was done in the tea extraction experiment, and determine the weight of the crude caffeine. You should obtain about 60 mg of crude product. Sublimation of this orange powder gives an impure orange sublimate that smells strongly of coffee, so sublimation is not a good way to purify this material.

Caffeine has no odor.

Dissolve a very small quantity of the product in a drop of dichloromethane, and analyze the crude material by thin-layer chromatography. Dissolve the remainder of the material in 1 mL of boiling 95% ethanol, and then dilute the mixture with 1 mL of *tert*-butyl methyl ether, heat to boiling, and allow to cool slowly to room temperature. Long, needlelike crystals should form in the orange solution. Alternatively, crystallize the product from a 1:1 mixture of ligroin [hexane(s)] and 2-propanol, using about 2 mL. Cool the mixture in ice for at least 10 min, and then collect the product on the Hirsch funnel. Complete the transfer with the filtrate, and then wash the crystals twice with cold 50/50 ethanol/*tert*-butyl methyl ether. The yield of white fluffy needles of caffeine should be more than 30 mg.

Cleaning Up   Allow the solvent to evaporate from the drying agent, and discard it in the nonhazardous waste container. Place any unused and unrecovered dichloromethane in the chlorinated organic solvents container.

MICROSCALE AND MACROSCALE

Preparation of a derivative of caffeine.

8. Caffeine Salicylate

One way to confirm the identity of an organic compound is to prepare a derivative of it. Caffeine melts and sublimes at 238°C. It is an organic base and can therefore accept a proton from an acid to form a salt. The salt formed when caffeine combines with hydrochloric acid, like many amine salts, does not have a sharp melting point; it simply decomposes when heated. But the salt formed from salicylic acid, even though ionic, has a sharp melting point and can thus be used to help characterize caffeine. See Fig. 8.14 for the ^{1}H NMR spectrum of caffeine.

Caffeine   **Salicylic acid**   **Caffeine salicylate**

Procedure

CAUTION: Petroleum ether is very flammable. Extinguish all flames.

The quantities given can be multiplied by 5 or 10 if necessary. To 10 mg of sublimed caffeine in a tared reaction tube add 7.5 mg of salicylic acid and 0.5 mL of dichloromethane. Heat the mixture to boiling, and add petroleum ether (a poor solvent for the product) dropwise until the mixture just turns cloudy, indicating the solution is saturated. If too much petroleum ether is added, then clarify it by adding a very small quantity of dichloromethane. Insulate the tube in order to allow it to cool slowly to room temperature, and then cool it in ice. The needle-like crystals are isolated by removing the solvent with a Pasteur pipette while the reaction tube is in the ice bath. Evaporate the last traces of solvent under vacuum (Fig. 8.13), and determine the weight of the derivative and its melting point. Caffeine salicylate is reported to melt at 137°C.

Crystallization from mixed solvents

Cleaning Up   Place the filtrate in the halogenated organic solvents container.

Questions

1. Suppose a reaction mixture, when diluted with water, afforded 300 mL of an aqueous solution of 30 g of the reaction product malononitrile,

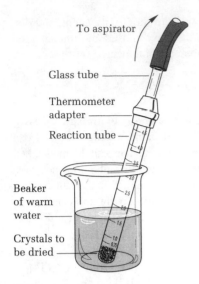

FIG. 8.13 Drying of crystals under vacuum in beaker of warm water.

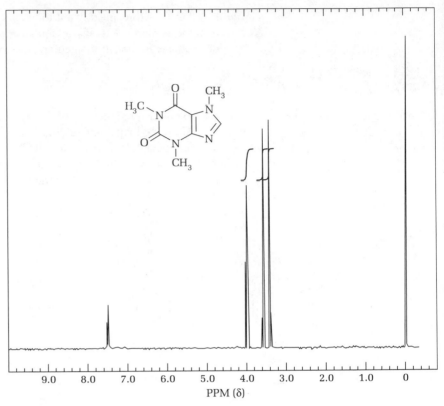

FIG. 8.14 ¹H NMR spectrum of caffeine (250 MHz).

$CH_2(CN)_2$, which is to be isolated by extraction with ether. The solubility of malononitrile in ether at room temperature is 20.0 g per 100 mL, and in water is 13.3 g per 100 mL. What weight of malononitrile would be recovered by extraction with (a) three 100-mL portions of ether; (b) one 300-mL portion of ether? *Suggestion:* For each extraction let x equal the weight extracted into the ether layer. In case (a) the concentration in the ether layer is $x/100$, and in the water layer is $(30 - x)/300$; the ratio of these quantities is equal to $k = 20/13.3$.

2. Why is it necessary to remove the stopper from a separatory funnel when liquid is being drained from it through the stopcock?

3. The pK_a of *p*-nitrophenol is 7.15. Would you expect this to dissolve in sodium bicarbonate solution? The pK_a of 2,5-dinitrophenol is 5.15. Will it dissolve in bicarbonate solution?

4. The distribution coefficient, k = (conc. in ligroin/conc. in water), between ligroin and water for solute A is 7.5. What weight of A would be removed from a solution of 10 g of A in 100 mL of water by a single extraction with 100 mL of ligroin? What weight of A would be removed by four successive

extractions with 25-mL portions of ligroin? How much ligroin would be required to remove 98.5% of A in a single extraction?

5. In Experiment 1, how many moles of benzoic acid are present? How many moles of sodium bicarbonate are contained in 1 mL of a 10% aqueous solution? (A 10% solution has 1 g of solute in 9 mL of solvent.) Is the amount of sodium bicarbonate sufficient to react with all of the benzoic acid?

6. To isolate the benzoic acid from the bicarbonate solution, it is acidified with concentrated hydrochloric acid in Experiment 1. What volume of acid is needed to neutralize the bicarbonate? The concentration of hydrochloric acid is expressed in various ways on the inside back cover of this laboratory manual.

7. How many moles of 2-naphthol are in the mixture to be separated in Experiment 1? How many moles of sodium hydroxide are contained in 1 mL of 5% sodium hydroxide solution? (Assume the density of the solution is 1.0.) What volume of concentrated hydrochloric acid is needed to neutralize this amount of sodium hydroxide solution?

8. Draw a flow diagram to show how you would separate the components of a mixture containing an acid substance, toluic acid, a basic substance, *p*-bromo-aniline, and a neutral substance, anthracene.

Surfing the Web

http://ull.chemistry.uakron.edu/organic_lab/distribution/

This University of Akron site with some 15 graphics shows the macroscale determination of the distribution coefficient (partition coefficient) of crotonic acid and benzoic acid in the water/dichloromethane system. The concentration of acid in the aqueous phase is determined by titration.

http://ull.chemistry.uakron.edu/organic_lab/cola/

The extraction of caffeine from 10 mL of cola syrup (Experiment 4) is very clearly demonstrated in 14 color photos. Preparation of caffeine salicylate (Experiment 5) is demonstrated with 8 close-up color photos.

For updated information visit:

www.mtholyoke.edu/courses/kwilliam/microscale.shtml

or

www.hmco.com/hmco/college/chemistry/Home.html

9

Thin-Layer Chromatography: Analysis of Analgesics and Isolation of Lycopene from Tomato Paste

Prelab Exercise: Compare thin-layer chromatography with column chromatography with regard to: (1) quantity of material that can be separated; (2) the speed; (3) the solvent systems; and (4) the ability to separate compounds.

TLC requires micrograms of material.

Thin-layer chromatography (TLC) is a sensitive, fast, simple, and inexpensive analytical technique that will be used repeatedly in carrying out organic experiments. It is a micro technique; as little as 10^{-9} g of material can be detected, although the usual sample size is from 1 to 100×10^{-6} g.

TLC involves spotting the sample to be analyzed near one end of a sheet of glass or plastic that is coated with a thin layer of an adsorbent. The sheet, which can be the size of a microscope slide, is placed on end in a covered jar containing a shallow layer of solvent. As the solvent rises by capillary action up through the adsorbent, differential partitioning occurs between the components of the mixture dissolved in the solvent and the stationary adsorbent phase. The more strongly a given component of the mixture is adsorbed onto the stationary phase, the less time it will spend in the mobile phase, and the more slowly it will migrate up the TLC plate.

Uses of Thin-Layer Chromatography

1. **To determine the number of components in a mixture.** TLC affords a quick and easy method for analyzing such things as a crude reaction mixture, an extract from some plant substance, or a painkiller. Knowing the number and relative amounts of the components aids in planning further analytical and separation steps.

2. **To determine the identity of two substances.** If two substances spotted on the same TLC plate give spots in identical locations, they *may* be identical. If the spot positions are not the same, the substances cannot be the same. It is possible for two closely related compounds that are not identical to have the same positions on a TLC plate.

3. **To monitor the progress of a reaction.** By sampling a reaction from time to time it is possible to watch the reactants disappear and the products appear using TLC. Thus, the optimum time to halt the reaction can be determined, and the effect of changing such variables as temperature, concentrations, and solvents can be followed without having to isolate the product.

4. **To determine the effectiveness of a purification.** The effectiveness of distillation, crystallization, extraction, and other separation and purification methods can be monitored using TLC, with the caveat that a single spot does not guarantee a single substance.

5. **To determine the appropriate conditions for a column chromatographic separation.** Thin-layer chromatography is generally unsatisfactory for purifying and isolating macroscopic quantities of material; however, the adsorbents most commonly used for TLC—silica gel and alumina—are used for column chromatography, discussed in the next chapter. Column chromatography is used to separate and purify up to about a gram of a solid mixture. The correct adsorbent and solvent used to carry out the chromatography can be determined rapidly by TLC.

6. **To monitor column chromatography.** As column chromatography is carried out, the solvent is collected in a number of small flasks. Unless the desired compound is colored, the various fractions must be analyzed in some way to determine which ones have the desired components of the mixture. TLC is a fast and effective method for doing this.

Adsorbents and Solvents

The two most common coatings for thin-layer chromatography plates are alumina, Al_2O_3, and silica gel, SiO_2. These are the same adsorbents most commonly used in column chromatography (Chapter 10) for the purification of macroscopic quantities of material. Of the two, alumina, when anhydrous, is the more active, i.e., it will adsorb substances more strongly. It is thus the adsorbent of choice when the separation involves relatively nonpolar substrates such as hydrocarbons, alkyl halides, ethers, aldehydes, and ketones. To separate the more polar substrates such as alcohols, carboxylic acids, and amines, the less active adsorbent, silica gel, is used. In an extreme situation very polar substances on alumina

R_f is the ratio of the distance the spot travels from the origin to the distance the solvent travels.

do not migrate very far from the starting point (give low R_f values), and nonpolar compounds travel with the solvent front (give high R_f values) if chromatographed on silica gel. These extremes of behavior are markedly affected, however, by the solvents used to carry out the chromatography. A polar solvent will carry along with it polar substrates, and nonpolar solvents will do the same with nonpolar compounds—another example of the generalization "like dissolves like."

Table 9.1 lists common solvents used in chromatography, both thin-layer and column. Only the environmentally safe solvents are listed; the polarities of such solvents as benzene, carbon tetrachloride, or chloroform can be matched by other less toxic solvents. In general these solvents are characterized by having

Avoid the use of benzene, carbon tetrachloride, and chloroform. Benzene is a carcinogen; the others are suspect carcinogens.

low boiling points and low viscosities that allow them to migrate rapidly. They are listed in order of increasing polarity. A solvent more polar than methanol is seldom needed. Often just two solvents are used in varying proportions; the polarity of the mixture is a weighted average of the two. Ligroin–ether mixtures are often employed in this way.

The order in which solutes migrate on thin-layer chromatography is the

TABLE 9.1 Chromatography Solvents

Solvent	bp (°C)
Petroleum ether (pentanes)	35–60
Ligroin (hexanes)	60–80
Dichloromethane	40
t-Butyl methyl ether	55
Ethyl acetate	77
Acetone	56
2-Propanol	82
Ethanol	78
Methanol	65
Water	100
Acetic acid	118

TABLE 9.2 Order of Solute Migration on Chromatography

Solute	Solute
Fastest	
Alkanes	Ketones
Alkyl halides	Aldehydes
Alkenes	Amines
Dienes	Alcohols
Aromatic hydrocarbons	Phenols
Aromatic halides	Carboxylic acids
Ethers	Sulfonic acids
Esters	*Slowest*

same as the order of solvent polarity. The largest R_f values are shown by the least polar solutes. In Table 9.2 the solutes are arranged in order of increasing polarity.

Apparatus and Procedure

A very dilute solution (1%) of the test substance is employed for TLC.

A 1% solution of the substance to be examined is spotted onto the plate about 1 cm from the bottom end, and the plate is inserted into a beaker or a 4-oz wide-mouth bottle containing 4 mL of an organic solvent. The bottle is lined with filter paper wet with solvent to saturate the atmosphere within the container. The top is put in place and the time noted. The solvent travels rapidly up in the thin layer by capillary action, and if the substance is a pure colored compound, one soon sees a spot traveling either along with the solvent front or, more usually, at some distance behind the solvent front. One can remove the slide, quickly mark the front before the solvent evaporates, and calculate the R_f value. The R_f value is the ratio of the distance the spot travels from the point of origin to the distance the solvent travels (Fig. 9.1). The best separations are achieved when the R_f value falls between 0.3 and 0.7.

Visualization of the Chromatogram

Colored compounds

And colorless ones

If the substances being chromatographed are colored, then it is possible to detect the compounds visually. Colorless substances can be detected when the solvent is allowed to evaporate and the plate allowed to stand in a stoppered 4-oz bottle containing a few crystals of iodine. Iodine vapor is adsorbed by the organic compound to form a brown spot. A spot should be outlined at once with a pencil because it will soon disappear as the iodine sublimes away; brief return to the iodine chamber will regenerate the spot. The order of elution and the elution power for solvents are the same as for column chromatography.

The use of commercial TLC sheets such as Whatman flexible plates for TLC, cat. no. 4410 222 (Fisher cat. no. 05-713-162), is strongly recommended. These poly(ethylene terephthalate) (Mylar) sheets are coated with silica gel using

FIG. 9.1 Thin-layer chromatography plate.

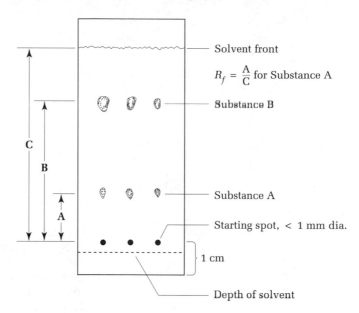

Solvent front

$$R_f = \frac{A}{C} \text{ for Substance A}$$

Substance B

C

B

Substance A

A

Starting spot, < 1 mm dia.

1 cm

Depth of solvent

TLC tests for
1. *Completeness of reaction*
2. *Purity of product*
3. *Side reactions*

polyacrylic acid as a binder. A fluorescent indicator has been added to the silica gel so that when the sheet is observed under 254-nm ultraviolet light, spots that either quench or enhance fluorescence can be seen. Iodine can also be used to visualize spots. The coating on these sheets is only 100 μm thick, so very small spots must be applied. Unlike student-prepared plates, these coated sheets (cut to 1- × 3-in. size with scissors) give very consistent results. A light pencil mark 1 cm from the end will guide spotting. A supply of these little precut sheets makes it a simple matter to examine most of the reactions in this book for completeness of reaction, purity of product, and side reactions.

A large number of specialized spray reagents have been developed that give specific colors for certain types of compounds, and there is a large amount of literature on the solvents and adsorbents to use for the separation of given types of material.[1]

Spotting Test Solutions

It is extremely important that the spots be as small as possible and that they be applied using a 1% (not 2% or more) solution of the compounds being separated. This is done with micropipettes made by drawing open-end mp capillaries in a burner flame (Fig. 9.2). The bore should be of such a size that, when the pipette is dipped deep into ligroin, the liquid flows in to form a tiny thread that, when the pipette is withdrawn, does not flow out to form a drop. To spot a test solution, let a 2–3 cm column of solution flow into the pipette, hold this vertically over a coated plate, aim it at a point on the right side of the plate and about 1 cm from the bottom, and lower the pipette until the tip just touches the adsorbent and liq-

1. Egon Stahl, *Thin-Layer Chromatography,* Springer Verlag, New York, 1969.

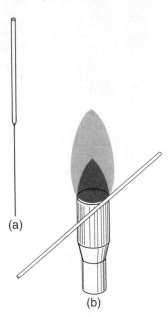

(a)

(b)

FIG. 9.2 (a) Micropipette. (b) Making a micropipette, which is pulled from a melting point capillary that is open on each end. Heat it at the base of the flame as shown.

uid flows onto the plate; withdraw when the spot is about 1 mm in diameter. Make a second 1-mm spot on the left side of the plate, let it dry, and make two more applications of the same size (1 mm) at the same place. Determine whether the large or the small spot gives the better results.

Experiments

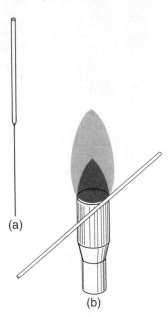 **MICROSCALE**

1. Analgesics

Analgesics are substances that relieve pain. The most common of these is aspirin, a component of more than 100 nonprescription drugs. In Chapter 32, the history and background of this most popular drug is discussed. In the present experiment, analgesic tablets will be analyzed by thin-layer chromatography to determine which analgesics they contain and whether they contain caffeine, which is often added to counteract the sedative effects of the analgesic.

In addition to aspirin and caffeine, the most common components of analgesics are, at present, acetaminophen and ibuprofen (Motrin). In addition to one or more of these substances, each tablet contains a binder, often starch, microcrystalline cellulose, or silica gel. And to counteract the acidic properties of aspirin, an inorganic buffering agent is added to some analgesics. Inspection of labels will reveal that most cold remedies and decongestants contain both aspirin and caffeine in addition to the primary ingredient.

Because of the insoluble binder, not all the unknown will dissolve.

To identify an unknown by TLC, the usual strategy is to run chromatograms of known substances (the standards) and the unknown at the same time. If the unknown has one or more spots that correspond to spots with the same R_f values as the standards, then those substances are probably present.

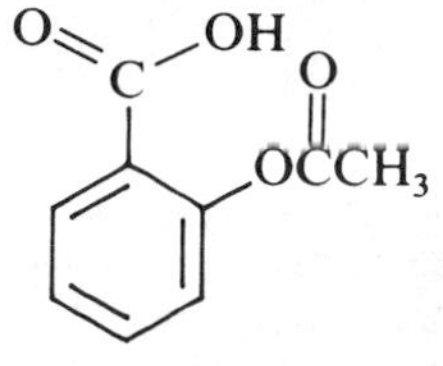

Aspirin
Acetylsalicylic acid

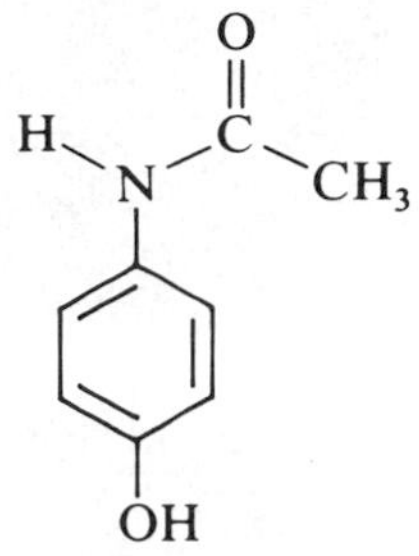

Acetaminophen
4-Acetamidophenol

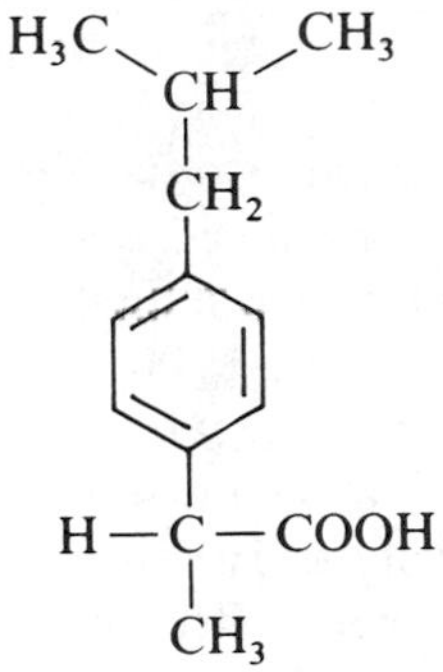

Ibuprofen
2-(4-Isobutylphenyl)propionic acid

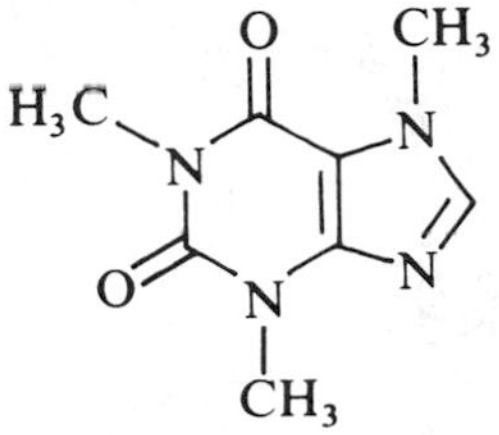

Caffeine

Proprietary drugs that contain one or more of the common analgesics and sometimes caffeine are sold under the names of Bayer Aspirin, Anacin, Datril, Advil, Excedrin, Extra Strength Excedrin, Tylenol, and Vanquish. Note that ibuprofen has a chiral carbon atom. One enantiomer is more effective than the other.

Procedure

Following the procedure outlined earlier, draw a light pencil line about 1 cm from the end of a chromatographic plate, and on this line spot aspirin, acetaminophen, ibuprofen, and caffeine, which are available as reference standards. Use a separate capillary for each standard, or rinse the capillary carefully. Make each spot as small as possible, preferably less than 0.5 mm in diameter. Examine the plate under the ultraviolet (UV) light to see that enough of each compound has been applied; if not, add more. On a separate plate run the unknown and one or more of the standards.

The unknown sample is prepared by crushing a part of a tablet, adding this powder to a test tube or small vial along with an appropriate amount of ethanol, and then mixing the suspension. Not all of the tablet will dissolve, but enough

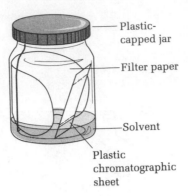

FIG. 9.3   A method for developing TLC plates.

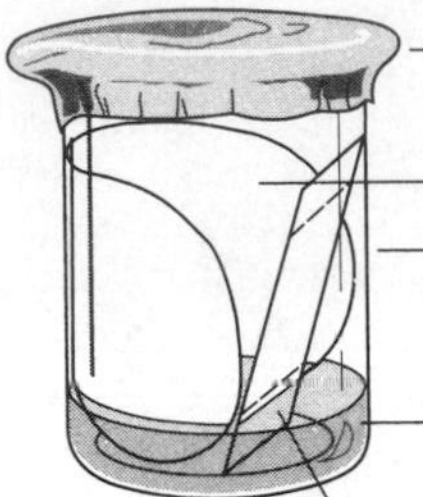

FIG. 9.4   An alternative method for developing TLC plates.

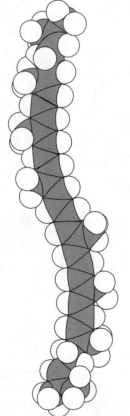

FIG. 9.5   An energy-minimized space-filling model of lycopene. The molecule is flat, but steric hindrance of the methyl groups causes the molecule to bend into the S shape.

will go into solution to spot the plate. The binder—starch or silica—will not dissolve. Try to prepare a 1% solution of the unknown.

Use as the solvent for the chromatogram a mixture of 95% ethyl acetate and 5% acetic acid (Fig. 9.3 or 9.4). After the solvent has risen to near the top of the plate, mark the solvent front with a pencil, remove the plate from the developing chamber, and allow the solvent to dry. Examine the plate under UV light to see the components as dark spots against a bright green-blue background. Outline the spots with a pencil. The spots can also be visualized by putting the plate in an iodine chamber made by placing a few crystals of iodine in the bottom of a capped 4-oz jar. Calculate the R_f values for the spots, and identify the components in the unknown.

Cleaning Up   Solvents should be placed in the organic solvents container, and dry, used chromatographic plates can be discarded in the nonhazardous solid waste container.

2. Plant Pigments

The botanist Michael Tswett discovered the technique of chromatography and applied it, as the name implies, to colored plant pigments. The leaves of plants contain, in addition to chlorophyll-a and -b, other pigments that are revealed in the fall when the leaf dies and the chlorophyll rapidly decomposes. Among the most abundant of the other pigments are the carotenoids, which include the carotenes and their oxygenated homologs, the xanthophylls. The bright orange β-carotene is the most important of these, because it is transformed in the liver to vitamin A, which is required for night vision.

Cows eat fresh, green grass that contains carotene, but they do not metabolize the carotene entirely, and so it ends up in their milk. Butter made from this milk is therefore yellow. In the winter the silage cows eat does not contain carotene because it readily undergoes air oxidation, and the butter made at that time is white. For some time an azo dye called Butter Yellow was added to winter butter to give it the accustomed color, but the dye was found to be a carcinogen. Now winter butter is colored with synthetic carotene, as is all margarine.

Lycopene (Fig. 9.5), the red pigment of the tomato, is a C_{40}-carotenoid made up of eight isoprene units. β-Carotene, the yellow pigment of the carrot, is an isomer of lycopene in which the double bonds at C_1—C_2 and C'_1—C'_2 are replaced by bonds extending from C_1 to C_6 and from C'_1 to C'_6 to form rings. The chromophore in each case is a system of 11 all-*trans* conjugated double bonds; the closing of the two rings renders β-carotene less highly pigmented than lycopene.

Fresh tomato fruit is about 96% water, and R. Willstätter and H. R. Escher isolated from this source 20 mg of lycopene per kilogram of fruit. They then found a more convenient source in commercial tomato paste, from which seeds and skin have been eliminated and the water content reduced by evaporation in vacuum to a content of 26% solids. From this they isolated 150 mg of lycopene per kilogram of paste. The expected yield in the present experiment is 0.075 mg.

A jar of strained carrots sold as baby food serves as a convenient source of β-carotene. The German investigators isolated 1 g of β-carotene per kilogram of "dried" shredded carrots of unstated water content.

The following procedure calls for dehydration of tomato or carrot paste with ethanol and extraction with dichloromethane, an efficient solvent for lipids.

Lycopene and β-carotene from tomato paste and strained carrots

As an interesting variation, try extraction of lycopene from commercial catsup.

Experimental Considerations

Carotenoids are very sensitive to light-catalyzed air oxidation. Perform this experiment as rapidly as possible; keep solutions as cool and dark as possible. This extraction gives a mixture of products that can be analyzed by both TLC and column chromatography (next chapter). If enough material for TLC only is desired, use one-tenth the quantities of starting material and solvents employed in the following procedure. This extraction also can be carried out with hexane if the ventilation is not adequate to use dichloromethane. Hexane is more prone to form emulsions than the chlorinated solvent.

CAUTION: Do not breathe the vapors of dichloromethane. Carry out the extraction in the hood. Dichloromethane is a cancer-suspect agent.

MICROSCALE AND MACROSCALE

Procedure

A 5-g sample of fresh tomato or carrot paste (baby food) is transferred to the bottom of a 25 × 150 mm test tube followed by 10 mL of acetone. The mixture is stirred and shaken before being filtered on the Hirsch funnel. Scrape as much of the material from the tube as possible, and press it as dry as possible on the funnel. Let the tube drain thoroughly. Place the filtrate in a 125-mL Erlenmeyer flask.

Return the solid residue to the test tube, shake it with a 10-mL portion of dichloromethane, and again filter the material on the Hirsch funnel. Add the filtrate to the 125-mL flask. Repeat this process twice more, and then pour the combined filtrates into a separatory funnel. Add water and sodium chloride solution (which aids in the breaking of emulsions), and shake the funnel gently. This aqueous extraction will remove the acetone and any water-soluble components from the mixture, leaving the hydrocarbon carotenoids in the dichloromethane. Dry the colored organic layer over anhydrous calcium chloride, and filter the

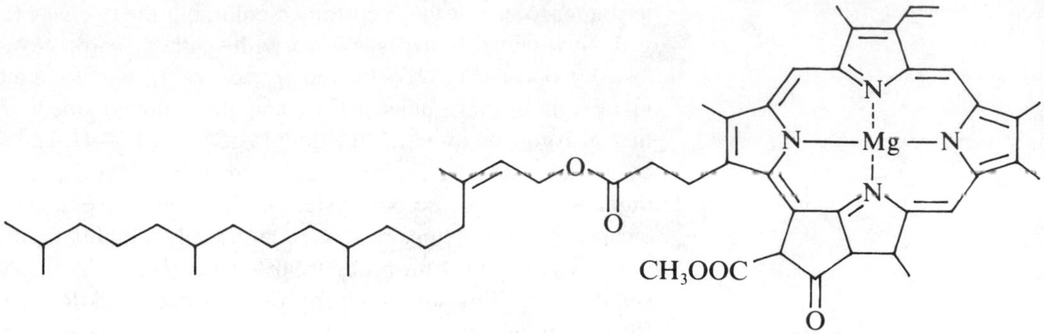

Chlorophyll-a

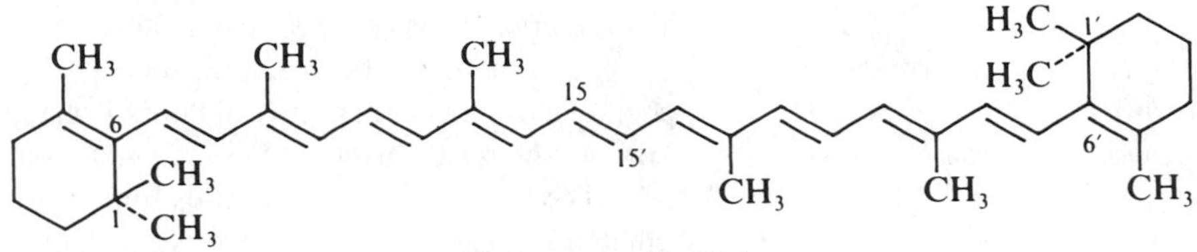

β-Carotene ($C_{40}H_{56}$)
mp 183°C, λ_{max}^{hexane} 451 nm

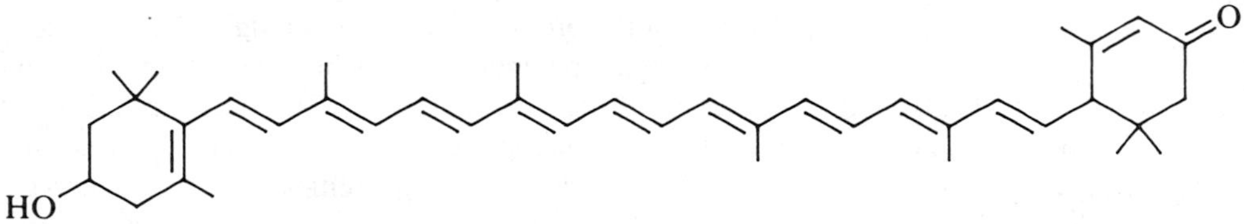

3′-Dehydrolutein (a xanthophyll)

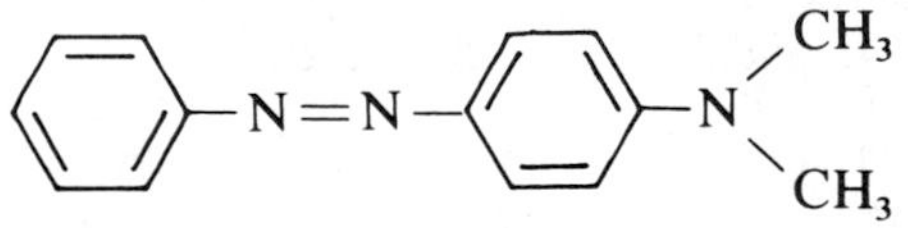

Butter Yellow

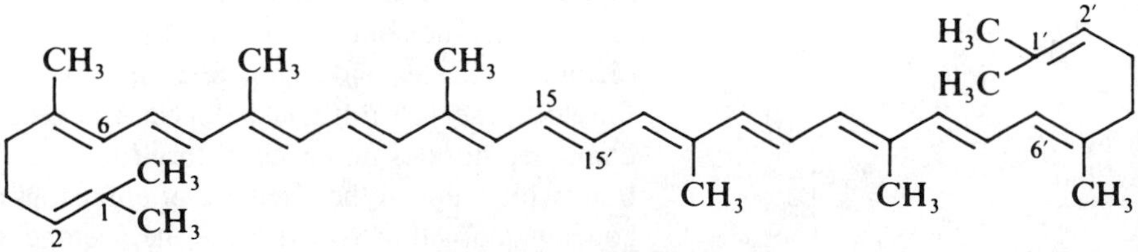

Lycopene ($C_{40}H_{56}$)
MW 536.85
mp 173°C, λ_{max}^{hexane} 475 nm

Air can be used for the evaporation, but nitrogen is better because these hydrocarbons air-oxidize with great rapidity.

solution into a dry flask. Remove about 0.5 mL of this solution and store it under nitrogen in the dark until it can be analyzed by thin-layer chromatography. Evaporate the remainder of the dichloromethane solution to dryness under a stream of nitrogen or under vacuum on a rotary evaporator. This material will be used for column chromatography (see next chapter). If it is to be stored, fill the flask with nitrogen and store it in a dark place.

Thin-Layer Chromatography

Spot the mixture on a TLC plate about 1 cm from the bottom and 8 mm from the edge. Make one spot concentrated by repeatedly touching the plate, but ensure that the spot is as small as possible, certainly less than 1 mm in diameter. The other spot can be of lower concentration. Develop the plate with 80 : 20 hexane : acetone. With other plates you could try cyclohexane and toluene as eluents and also hexane–ethanol mixtures of various compositions.

Many spots may be seen. There are two common carotene and chlorophyll isomers and four xanthophyll isomers.

The container in which the chromatography is carried out should be lined with filter paper that is wet with the solvent so that the atmosphere in the container will be saturated with solvent vapor. On completion of elution, mark the solvent front with a pencil, and outline the colored spots. Examine the plate under UV light. Also place the plate in an iodine chamber to visualize spots (see next experiment).

Cleaning Up   The aqueous saline filtrate containing acetone can be flushed down the drain. Recovered and unused dichloromethane should be placed in the halogenated organic waste container; the solvents used for TLC should be placed in the organic solvents container; and the drying agents, once free of solvents, can be placed in the nonhazardous solid waste container along with the used plant material and the TLC plates.

 **MICROSCALE AND MACROSCALE**

Procedure

In a small mortar grind 2 g of green or brightly colored fall leaves (don't use ivy or waxy leaves) with 10 mL of ethanol, pour off the ethanol, which serves to break up and dehydrate the plant cells, and grind the leaves successively with three 1-mL portions of dichloromethane that are decanted or withdrawn with a Pasteur pipette and placed in a test tube. The pigments of interest are extracted by the dichloromethane. Alternatively, place 0.5 g of carrot paste (baby food) or tomato paste in a test tube, stir and shake the paste with 3 mL of ethanol until the paste has a somewhat dry or fluffy appearance, remove the ethanol, and extract the dehydrated paste with three 1-mL portions of dichloromethane. Stir and shake the plant material with the solvent in order to extract as much of the pigments as possible.

Fill the tube containing the dichloromethane extract from leaves or vegetable paste with a saturated sodium chloride solution, and shake the mixture. Remove the aqueous layer and to the dichloromethane solution add anhydrous calcium chloride pellets until the drying agent no longer clumps together. Shake the mixture with the drying agent for about 5 min, and then withdraw the solvent with a Pasteur pipette and place it in a test tube. Add to the solvent a few pieces of Drierite to complete the drying process. Gently stir the mixture for about 5 min, transfer the solvent to a test tube, wash off the drying agent with more solvent, and then evaporate the combined dichloromethane solutions under a stream of nitrogen while warming the tube in the hand or in a beaker of warm water. Carry out this evaporation in the hood.

These hydrocarbons air-oxidize with great rapidity.

Immediately cork the tube filled with nitrogen, and then add a drop or two of dichloromethane to dissolve the pigments for TLC analysis. Carry out the analysis without delay by spotting the mixture on a TLC plate about 1 cm from the bottom and 8 mm from the edge. Make one spot concentrated by repeatedly touching the plate, but ensure that the spot is as small as possible—less than 1.0 mm in diameter. The other spot can be of lower concentration. Develop the plate with 70 : 30 hexane : acetone. With other plates try cyclohexane and toluene as eluents and also hexane–ethanol mixtures of various compositions. The container in which the chromatography is carried out should be lined with filter paper that is wet with the solvent so the atmosphere in the container will be saturated with solvent vapor. On completion of elution, mark the solvent front with a pencil, and outline the colored spots. Examine the plate under the UV light. Are any new spots seen? Report colors and R_f values for all of your spots, and identify each as lycopene, carotene, chlorophyll, or xanthophyll.

Cleaning Up   The ethanol used for dehydration of the plant material can be flushed down the drain along with the saturated sodium chloride solution. Recovered and unused dichloromethane should be placed in the halogenated organic waste container. The solvents used for TLC should be placed in the organic solvents container. The drying agents, once free of solvents, can be placed in the nonhazardous solid waste container along with the used plant material and TLC plates.

 **MICROSCALE AND MACROSCALE**

3. Colorless Compounds

You are now to apply the thin-layer technique to a group of colorless compounds. The spots can be visualized under an ultraviolet light if the plates have been coated with a fluorescent indicator, or chromatograms can be developed in a 4-oz bottle containing crystals of iodine. During development, spots appear rapidly, but remember that they also disappear rapidly. Therefore, outline each spot with a pencil immediately on withdrawal of the plate from the iodine chamber. Solvents suggested are as follows:

Cyclohexane	Toluene (3 mL)–dichloromethane (1 mL)
Toluene	9 : 1 Toluene–methanol (use 4 mL)

Make your own selections.

Compounds for trial are to be selected from the following list (all 1% solutions in toluene):

1. Anthracene*
2. Cholesterol

Except for tetraphenylthiophene, the structures for all of these molecules will be found in this book.

3. 2,7-Dimethyl-3,5-octadiyne-2,7-diol
4. Diphenylacetylene
5. *trans,trans*-1,4-Diphenyl-1,3-butadiene*
6. *p*-Di-*t*-butylbenzene
7. 1,4-Di-*t*-butyl-2,5-dimethoxybenzene
8. *trans*-Stilbene
9. 1,2,3,4-Tetraphenylnaphthalene*
10. Tetraphenylthiophene
11. *p*-Terphenyl*
12. Triphenylmethanol
13. Triptycene

*Fluorescent under UV light.

It is up to you to make selections and to plan your own experiments. Do as many as time permits. One plan would be to select a pair of compounds estimated to be separable and that have R_f values determinable with the same solvent. One can assume that a hydroxyl compound will travel less rapidly with a hydrocarbon solvent than a hydroxyl-free compound, and so you will know what to expect if the solvent contains a hydroxylic component. An aliphatic solvent should carry along an aromatic compound with aliphatic substituents better than one without such groups. However, instead of relying on assumptions, you can do brief preliminary experiments on used plates on which previous spots are visible or outlined. If you spot a pair of compounds on such a plate and let the solvent rise about 3 cm from the starting line before development, you might be able to tell if a certain solvent is appropriate for a given sample. Alternatively, make some spots on a plate (new or used) and then touch each spot with a different solvent held in a capillary. In Figure 9.6(a), the mixture did not move away from the point of origin; in Figure 9.6(b), two concentric rings are seen between the origin and the solvent front. This is how a good solvent behaves. In Figure 9.6(c), the mixture of compounds traveled with the solvent front.

Preliminary trials on used plates

Once a solvent is chosen, run a complete chromatogram on the two compounds on a fresh plate. If separation of the two seems feasible, put two spots of one compound on a plate, let the solvent evaporate, and put spots of the second compound over the first ones. Run a chromatogram and see if you can detect two spots in either lane (with colorless compounds, it is advisable not to attempt

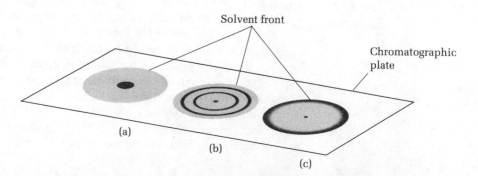

FIG. 9.6 Fast method for determining the correct solvent for TLC. See text for procedure.

a three-lane chromatogram until you have acquired considerable practice and skill).

Cleaning Up   Solvents should be placed in the organic solvents container, and dry, used chromatographic plates can be discarded in the nonhazardous solid waste container.

Discussion

If you have investigated hydroxylated compounds, you doubtless have found that it is reasonably easy to separate a hydroxylated from a nonhydroxylated compound, or a diol from a mono-ol. How, by a simple reaction followed by a thin-layer chromatogram, could you separate cholesterol from triphenylmethanol? Heating a sample of each with acetic anhydride-pyridine for 5 min on the steam bath, followed by chromatography, should do it. A first trial of a new reaction leaves questions about what has happened and how much, if any, starting material is present. A comparative chromatogram of reaction mixture with starting material may tell the story. How crude is a crude reaction product? How many components are present? The thin-layer technique may give the answers to these questions and suggest how best to process the product. A preparative column chromatogram may afford a large number of fractions of eluent (say, 1 to 30). Some fractions probably contain nothing and should be discarded, while others should be combined for evaporation and workup. How can you identify the good and the useless fractions? Take a few used plates and put numbered circles on clean places of each; spot samples of each of the fractions; and, without any chromatography, develop the plates with iodine. Negative fractions for discard will be obvious, and the pattern alone of positive fractions may allow you to infer which fractions can be combined. Thin-layer chromatograms of the first and last fractions of each suspected group would then show whether or not your inferences are correct.

Fluorescence

Do not look into a UV lamp.

Four of the compounds listed in Experiment 3 are fluorescent under UV light, and such compounds give colorless spots that can be picked up on a chromatogram by fluorescence (after removal from the UV-absorbing glass bottle). If a UV light source is available, spot the four compounds on a used plate and observe the fluorescence. Take this opportunity to examine a white shirt or handkerchief under UV light to see if it contains a brightener, that is, a fluorescent white dye or optical bleach. These substances are added to counteract the yellow color that repeated washing gives to cloth. Brighteners of the type of Calcofluor White MR, a sulfonated *trans*-stilbene derivative, are commonly used in detergent formulations for cotton; the substituted coumarin derivative formulated is typical of brighteners used for nylon, acetate, and wool. Detergents normally contain 0.1–0.2% of optical bleach. The amount of dye on a freshly laundered shirt is approximately 0.01% of the weight of the fabric.

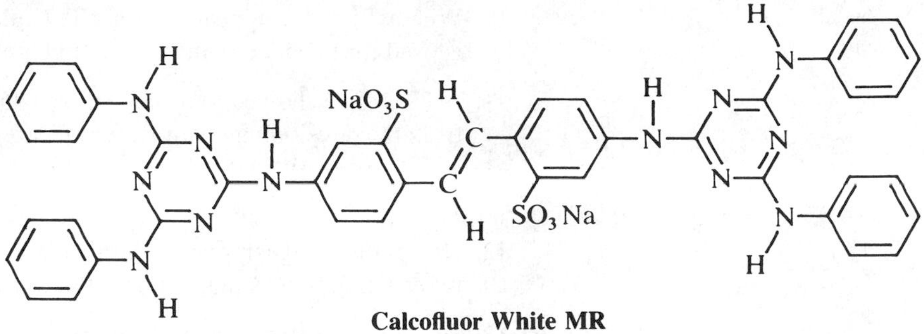

Calcofluor White MR

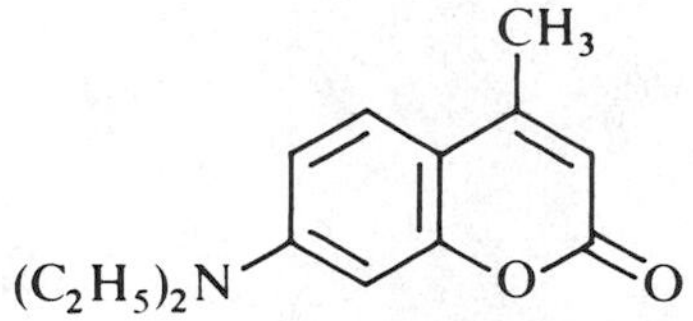

7-Diethylamino-4-methylcoumarin

Questions

1. Why might it be very difficult to visualize the separation of *cis*- and *trans*-2-butene by thin-layer chromatography?

2. What error is introduced into the determination of an R_f value if the top is left off of the developing chamber?

3. What problem will ensue if the level of the developing liquid is higher than the applied spot in a TLC analysis?

4. In what order (from top to bottom) would you expect to find naphthalene, butyric acid, and phenyl acetate on a silica gel TLC plate developed with dichloromethane?

5. In carrying out an analysis of a mixture, what do you expect to see when the TLC plate has been allowed to remain in the developing chamber too long, so that the solvent front has reached the top of the plate?

6. Arrange the following in order of increasing R_f on thin-layer chromatography: acetic acid, acetaldehyde, 2-octanone, decane, and 1-butanol.

7. Why must the spot applied to a TLC plate be above the level of the developing solvent?

8. What will be the result of applying too much compound to a TLC plate?

9. Why is it necessary to run TLC in a closed container and to have the interior vapor saturated with the solvent?

10. What will be the appearance of a TLC plate if a solvent of too low polarity is used for the development? too high polarity?

11. A TLC plate showed two spots of R_f 0.25 and 0.26. The plate was removed from the developing chamber, dried carefully, and returned to the developing chamber. What would you expect to see after the second development was complete?

12. One of the analgesics has a chiral center. Which compound is it? One of the two enantiomers is much more effective at reducing pain than the other.

13. Using a ruler to measure distances, calculate the R_f value for Substance B in Fig. 9.1.

Column Chromatography: Acetylferrocene, Cholesteryl Acetate, and Fluorenone

Column chromatography is one of the most useful methods for the separation and purification of both solids and liquids when carrying out small-scale experiments. It becomes expensive and time consuming, however, when more than about 10 g of material must be purified.

The application in the present experiment is typical: A reaction is carried out, it does not go to completion, and so column chromatography is used to separate the product from starting material, reagents, and byproducts.

The theory of column chromatography is analogous to that of thin-layer chromatography. The most common adsorbents—silica gel and alumina—are the same ones used in TLC. The sample is dissolved in a small quantity of solvent (the eluent) and applied to the top of the column. The eluent, instead of rising by capillary action up a thin layer, flows down through the column filled with the adsorbent. Just as in TLC, there is an equilibrium established between the solute adsorbed on the silica gel or alumina and the eluting solvent flowing down through the column. Under some conditions the solute may be partitioning between an adsorbed solvent and the elution solvent; the partition coefficient, just as in the extraction process, determines the efficiency of separation in chromatography. The partition coefficient is determined by the solubility of the solute in the two phases, as was discussed in the extraction experiment (Chapter 8).

Three mutual interactions must be considered in column chromatography: the polarity of the sample, the polarity of the eluting solvent, and the activity of the adsorbent.

Adsorbent

A large number of adsorbents have been used for column chromatography—cellulose, sugar, starch, inorganic carbonates—but most separations employ alumina (Al_2O_3) and silica gel (SiO_2). Alumina comes in three forms: acidic, neutral, and basic. The neutral form of Brockmann activity II or III, 150 mesh, is most commonly employed. The surface area of this alumina is about 150 m^2/g. Alumina as purchased will usually be activity I, meaning it will strongly adsorb solutes. It must be deactivated by adding water, shaking, and allowing the mixture to reach equilibrium over an hour or so. The amount of water needed to achieve certain activities is given in Table 10.1. The activity of the alumina on TLC plates is usually about III. Silica gel for column chromatography, 70–230 mesh, has a surface area of about 500 m^2/g and comes in only one activity.

TABLE 10.1 Alumina Activity

Brockmann activity	I	II	III	IV	V
Percent by weight of water	0	3	6	10	15

Solvents

The elutropic series for a number of solvents is given in Table 10.2. The solvents are arranged in increasing polarity, with *n*-pentane being the least polar. This is the order of ability of these solvents to dissolve polar organic compounds and to dislodge a polar substance adsorbed onto either silica gel or alumina, with *n*-pentane having the lowest solvent power.

As a practical matter, the following sequence of solvents is recommended in an investigation of unknown mixtures: elute first with petroleum ether; then ligroin, followed by ligroin containing 1%, 2%, 5%, 10%, 25%, and 50% *t*-butyl methyl ether; pure *t*-butyl methyl ether; *t*-butyl methyl ether and dichloromethane mixtures, followed by dichloromethane and methanol mixtures. A sudden change in solvent polarity will cause heat evolution as the alumina or silica gel adsorbs the new solvent. This will cause undesirable vapor pockets and cracks in the column.

Petroleum ether: mostly isometric pentanes; ligroin: mostly isomeric hexanes

TABLE 10.2 Elutropic Series for Solvents

n-Pentane (first)
Petroleum ether
Cyclohexane
Ligroin
Carbon disulfide
t-Butyl methyl ether
Dichloromethane
Tetrahydrofuran
Dioxane
Ethyl acetate
2-Propanol
Ethanol
Methanol
Acetic acid (last)

Solutes

The ease with which different classes of compounds elute from a column is indicated in Table 10.3. The order is similar to that of the eluting solvents—another application of "like dissolves like."

Sample and Column Size

In general the amount of alumina or silica gel used should weigh at least 30 times as much as the sample, and the column, when packed, should have a height at least 10 times the diameter. The density of silica gel is 0.4 g/mL, and the density of alumina is 0.9 g/mL, so the optimum size for any column can be calculated. A microscale column for the chromatography of about 50 mg of material is shown in Fig. 10.1.

TABLE 10.3 Elution Order for Solutes

Alkanes (first)
Alkenes
Dienes
Aromatic hydrocarbons
Ethers
Esters
Ketones
Aldehydes
Amines
Alcohols
Phenols
Acids (last)

Packing the Chromatography Column

Uniform packing of the chromatography column is critical to the success of this technique. The sample is applied as a pure liquid or, if it is a solid, as a very concentrated solution in the solvent that will dissolve it best, regardless of polarity. As elution takes place, this narrow band of sample will separate into several bands corresponding to the number of components in the mixture and their relative polarities and molecular weights. It is essential that the components move through the column as a narrow horizontal band in order to come off the column in the least volume of solvent and not overlap with other components of the mixture. Therefore, the column should be vertical, and the packing should be perfectly uniform, without voids caused by air bubbles.

The preferred method for packing silica gel and alumina columns is the slurry method, whereby a slurry of the adsorbent and the first eluting solvent is made and poured into the column. When nothing is known about the mixture being separated, the column is prepared in petroleum ether, the least polar of the eluting solvents.

Microscale Procedure

Packing the Column.   Assemble the column as depicted in Figure 10.1. To measure the adsorbent, fill the column one-half to two-thirds full, and then pour the powder into a 10-mL Erlenmeyer flask. Clamp the column in a vertical position, and close the valve. Always grasp the valve with one hand while turning it with the other. Fill the column with ligroin or hexane to the top of the glass column. Add about 8 mL hexane to the adsorbent in the flask, stir the mixture to eliminate air bubbles, and then (this is the hard part) swirl the mixture to get the adsorbent suspended in the solvent and immediately pour the entire slurry into the funnel. Open the valve, drain some solvent into the flask that had the adsorbent in it, and finish transferring the slurry to the column. Place an empty flask under the column, and allow the solvent to drain to about 5 mm above the top surface of the adsorbent. *Never allow the column to dry out;* this creates channels that will result in uneven bands and poor separation.

Adding the Sample and Eluting the Column.   Dissolve the sample completely in the very minimum volume of dichloromethane (just a few drops) in an Erlenmeyer flask. Add to this solution 300 mg of the adsorbent, stir, and evaporate the solvent completely. Heat the flask *very gently* with *constant* stirring to avoid having the material bump. Remember that dichloromethane boils at 55°C. Pour this dry powder into the funnel of the chromatography column, wash it down onto the column with a few drops of hexane, and then tap the column to remove air bubbles from this layer of adsorbent-solute mixture just added. Open the valve, and carefully add new solvent in such a manner that the top surface of the column is not disturbed. Run the solvent down near to the surface several times to apply the sample as a narrow band at the top of the column. Then fill the column with the solvent, and elute the sample from the column.

Macroscale Procedure

The column can be prepared using a 50-mL burette such as the one shown in Fig. 10.2 or using the less expensive and equally satisfactory chromatographic tube shown in Fig. 10.3, in which the flow of solvent is controlled by a screw pinchclamp. Weigh out the required amount of silica gel (12.5 g in the first experiment), close the pinchclamp on the tube, and fill about half full with 90:10

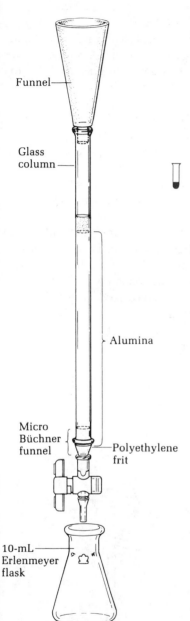

Funnel

Glass column

Alumina

Micro Büchner funnel

Polyethylene frit

10-mL Erlenmeyer flask

FIG. 10.1 Microscale chromatographic column.

Extinguish all flames; work in laboratory hood.

Dry packing

ligroin : ether. With a wooden dowel or glass rod push a small plug of glass wool through the liquid to the bottom of the tube, dust in through a funnel enough sand to form a 1-cm layer over the glass wool, and level the surface by tapping the tube. Unclamp the tube. With the right hand grasp both the top of the tube and the funnel so that the whole assembly can be shaken to dislodge silica gel that may stick to the walls, and with the left hand pour in the silica gel slowly (Fig. 10.4) while tapping the column with a rubber stopper fitted on the end of a pencil. If necessary, use a Pasteur pipette full of 90 : 10 ligroin : ether to wash down any silica gel that adheres to the walls of the column above the liquid. When the silica gel has settled, add a little sand to provide a protective layer at the top. Open the pinchclamp, let the solvent level fall until it is just a little above the upper layer of sand, and then stop the flow.

Slurry packing

Alternatively, the silica gel can be added to the column (half filled with ligroin) by slurrying the silica gel with 90 : 10 ligroin : ether in a beaker. The powder is stirred to suspend it in the solvent and immediately poured through a wide-mouth funnel into the chromatographic tube. Rap the column with a rubber stopper to cause the silica gel to settle and to remove bubbles. Add a protective layer of sand to the top. The column is now ready for use.

Prepare several Erlenmeyer flasks as receivers by taring (weighing) each one carefully and marking them with numbers on the etched circle.

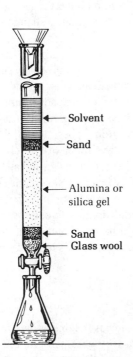

← Solvent

← Sand

← Alumina or silica gel

← Sand
← Glass wool

FIG. 10.2 Macroscale chromatographic column.

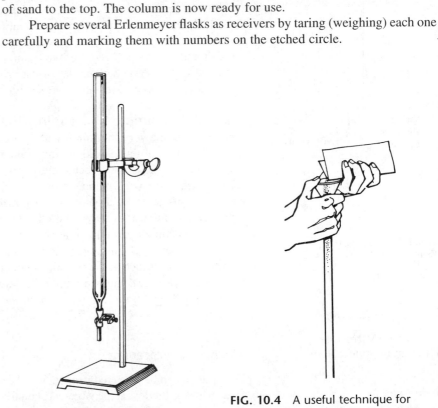

FIG. 10.3 Chromatographic tube on ring stand.

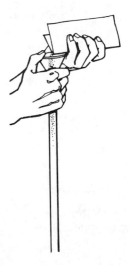

FIG. 10.4 A useful technique for filling a chromatographic tube with silica gel.

Cleaning Up   After use, the tube is conveniently emptied by pointing the open end into a beaker, opening the pinchclamp, and applying gentle air pressure to the tip. If the plug of glass wool remains in the tube after the alumina leaves, wet it with acetone and reapply air pressure. Allow the adsorbent to dry in the hood, and then dispose of it in the nonhazardous waste container.

Experiments

MICROSCALE

Ferrocene
MW 186.04
mp 172–174°C

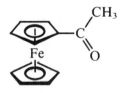

Acetylferrocene
MW 228.08, mp 85–86°C

Caution: Toxic.

1. Chromatography of a Mixture of Ferrocene and Acetylferrocene

Both these compounds are colored (see Chapter 44 for their preparation), so it is easy to follow the progress of the chromatographic separation.

Prepare the microscale alumina column exactly as described above. Then add a dry slurry of 90 mg of a 50 : 50 mixture of acetylferrocene (*Caution!* toxic) and ferrocene that has been adsorbed onto 300 mg of alumina following the above procedure for preparing and adding the sample.

Carefully add hexane to the column, open the valve (use both hands), and elute the two compounds. The first to be eluted, ferrocene, will be seen as a yellow band. Collect this in a 10-mL flask. Any crystalline material seen at the tip of the valve should be washed into the flask with a drop or two of ether. Without allowing the column to run dry, add a 50:50 mixture of hexane and *t*-butyl methyl ether, and elute the acetylferrocene, which will be seen as an orange band. Collect it in a 10-mL flask. Spot a thin-layer silica gel chromatography plate with these two solutions. Evaporate the solvents from the two flasks, and determine the weights of the residues. An easy way to evaporate the solvent is to place it in a tared 25-mL filter flask and heat the flask in the hand under vacuum while swirling the contents (Fig. 10.5).

Recrystallize the products from the minimum quantities of hot hexane or ligroin. Isolate the crystals, dry them, and determine their weights and melting points. Calculate the percent recovery of the crude and recrystallized products based on the 45 mg of each in the original mixture.

The thin-layer chromatography plate is eluted with 30 : 1 toluene : absolute ethanol. Do you detect any contamination of one compound by the other?

Cleaning Up   Empty the chromatography column onto a piece of aluminum foil in the hood. After the solvent has evaporated, place the alumina and sand in the nonhazardous waste container. Evaporate the crystallization mother liquor to dryness, and place the residue in the hazardous waste container.

MICROSCALE AND MACROSCALE

2. Acetylation of Cholesterol

Cholesterol is a solid alcohol; the average human body contains about 200 g distributed in brain, spinal cord, and nerve tissue and occasionally clogging the arteries and the gall bladder.

In the present experiment, cholesterol is dissolved in acetic acid and allowed to react with acetic anhydride to form the ester, cholesteryl acetate. The reaction does not take place rapidly and consequently does not go to completion under the conditions of this experiment. Thus, when the reaction is over, both unreacted cholesterol and the product, cholesteryl acetate, are present. Separating these by fractional crystallization would be extremely difficult; but because they differ in polarity (the hydroxyl group of the cholesterol is the more strongly adsorbed on alumina), they are easily separated by column chromatography. Both molecules are colorless and hence cannot be detected visually. Each fraction should be sampled for thin-layer chromatography. In that way not only the presence but also the purity of each fraction can be assessed. It is also possible to put a drop of each fraction on a watch glass and evaporate it to see if the fraction contains product. Solid will also appear on the tip of the column while a compound is being eluted.

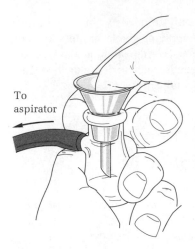

FIG. 10.5 Evaporation of a low-boiling liquid under vacuum. Heat is supplied by the hand, the contents of the flask are swirled, and the vacuum is controlled with the thumb.

Cholesterol

Acetic anhydride

Cholesteryl acetate

 Microscale Procedure

In a reaction tube, add 0.5 mL of acetic acid to 50 mg of cholesterol, which can be material isolated from human gallstones. The initial thin slurry may set into a stiff paste of the molecular complex consisting of one molecule of cholesterol and one of acetic acid. Add 0.10 mL of acetic anhydride and a boiling chip, and gently reflux the reaction mixture on a hot sand bath for no more than 30 min.

While the reaction is taking place, prepare the microscale chromatography column as described above, using silica gel as the adsorbent. Cool the mixture, add 2 mL of water, and extract the product with three 2-mL portions of *tert*-butyl methyl ether that are placed in the 15-mL centrifuge tube. Wash the ether extracts in the tube with two 2-mL portions of water and one 2.5-mL portion of 3 M sodium hydroxide (these three washes remove the acetic acid), and dry the ether by shaking it with 2.5 mL of saturated sodium chloride solution. Then complete the drying by adding enough anhydrous calcium chloride pellets to the solution so that it does not clump together.

Shake the ether solution with the drying agent for 10 min, and then transfer it in portions to a tared reaction tube and evaporate to dryness. Use a drop of this ether solution to spot a TLC plate for later analysis. If the crude material weighs more than the theoretical weight, you will know it is not dry or it contains acetic acid, which can be detected by its odor. Dissolve this crude cholesteryl acetate in the minimum quantity of *tert*-butyl methyl ether, and apply it to the top of the chromatography column.

To prevent the solution from dribbling from the pipette, use a pipette pump to make the transfer. It also could be applied as a dry powder adsorbed on silica gel as in Experiment 1. Elute the column with ligroin or hexane, and collect two 5-mL fractions in tared 10-mL Erlenmeyer flasks or other suitable containers. Add a boiling stick to each flask, and evaporate the solvent under an aspirator tube on a steam bath or sand bath. If the flask appears empty, it can be used to collect later fractions. Lower the solvent layer to the top of the sand and elute with 25 mL of 70 : 30 ligroin : *t*-butyl methyl ether, collecting the five 5-mL fractions. Evaporate the ligroin under an aspirator tube (Fig. 10.6). The last traces of ligroin can be removed using reduced pressure as shown in Fig. 10.7. Follow the 70 : 30 mixture with 20 mL of 50 : 50 ligroin : *t*-butyl methyl ether, collecting four 5-mL fractions. Save any flask that has any visible residue. Analyze the original mixture and each fraction by thin-layer chromatography on silica gel plates using 1:1 ether : ligroin to develop the plates and either ultraviolet light or iodine vapor to visualize the spots.

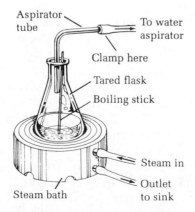

FIG. 10.6 Aspirator tube in use. A boiling stick may be necessary to promote even boiling.

FIG. 10.7 Drying a solid by reduced air pressure.

Cholesteryl acetate (mp 115°C) and cholesterol (mp 149°C) should appear, respectively, in early and late fractions with a few empty fractions (no residue) in between. If so, combine consecutive fractions of early and late material, and determine the weights and melting points. Calculate the percentage of the acetylated material compared with the total recovered, and calculate the percentage yield from cholesterol.

Cleaning Up   Acetic acid, aqueous layers, and saturated sodium chloride layers from the extraction, after neutralization, can be flushed down the drain with water. Ether, ligroin, and TLC solvents should be placed in the organic solvents container and the drying agent, and chromatography adsorbent, once free of solvent, can be placed in the nonhazardous solid water container.

Macroscale Procedure

Carry out procedure in laboratory hood.

Cover 0.5 g of cholesterol with 5 mL of acetic acid in a small Erlenmeyer flask; swirl, and note that the initially thin slurry soon sets to a stiff paste of the molecular compound $C_{27}H_{45}OH \cdot CH_3CO_2H$. Add 1 mL of acetic anhydride, and heat the mixture on the steam bath for any convenient period of time from 15 min to 1 h; record the actual heating period. While the reaction takes place, prepare the chromatographic column. Cool the reaction mixture, add 20 mL of water, and extract with two 25-mL portions of *tert*-butyl methyl ether. Wash the combined ethereal extracts twice with 15-mL portions of water and once with 25 mL of 3 M sodium hydroxide, dry by shaking the ether extracts with 25 mL saturated sodium chloride solution; then dry the ether over anhydrous calcium chloride pellets for 10 min in an Erlenmeyer flask, filter, and evaporate the ether. Save a few crystals of this material for TLC (thin-layer chromatography) analysis. Dissolve the residue in 3 to 4 mL of ether, transfer the solution with a Pasteur pipette onto a column of 12.5 g of silica gel, and rinse the flask with another small portion of *tert*-butyl methyl ether.[1]

In order to apply the ether solution to the top of the sand and avoid having it coat the interior of the column, pipette the solution down a 6-mm-diameter glass tube that is resting on the top of the sand. Label a series of 50-mL Erlenmeyer flasks as fractions 1 to 10. Open the pinchclamp, run the eluant solution into a 50-mL Erlenmeyer flask, and as soon as the solvent in the column has fallen to the level of the upper layer of sand, fill the column with a part of a measured 125 mL of 70 : 30 ligroin : *t*-butyl methyl ether. When about 25 mL of eluant has collected in the flask (fraction 1), change to a fresh flask; add a boiling stone to the first flask, and evaporate the solution to dryness on the steam bath under an aspirator tube (Fig. 10.6). Evacuation using the aspirator helps to remove last traces of ligroin (Fig. 10.7). If this fraction 1 is negative (no residue),

---

1. Ideally, the material to be adsorbed is dissolved in ligroin, the solvent of least elutant power. The present mixture is not soluble enough in ligroin and so ether is used, but the volume is kept to a minimum.

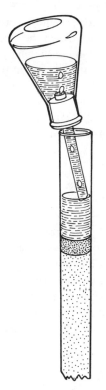

use the flask for collecting further fractions. Continue adding the ligroin–*t*-butyl methyl ether mixture until the 125-mL portion is exhausted, and then use 100 mL of a 1:1 ligroin:*t*-butyl methyl ether mixture. A convenient bubbler (Fig. 10.8) made from a 125-mL Erlenmeyer flask, a short piece of 10-mm-diameter glass tubing, and a cork will automatically add solvent. A separatory funnel with stopper and partially open stopcock serves the same purpose. Collect and evaporate successive 25-mL fractions of eluant. Save any flask that has any visible solid residue. The ideal method for removal of solvents involves the use of a rotary evaporator (Fig. 10.9). Analyze the original mixture and each fraction by TLC on silica gel plates using 1:1 ether:ligroin to develop the plates and either ultraviolet light or iodine vapor to visualize the spots.

Cholesteryl acetate (mp 115°C) and cholesterol (mp 149°C) should appear, respectively, in early and late fractions with a few empty fractions (no residue) in between. If so, combine consecutive fractions of early and of late material, and determine the weights and melting points. Calculate the percentage of the acetylated material compared with the total recovered, and compare your result with those of others in your class employing different reaction periods.

FIG. 10.8   Bubbler for adding solvent automatically.

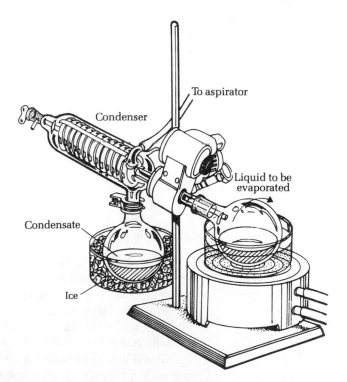

FIG. 10.9   Rotary evaporator. The rate of evaporation with this apparatus is very fast due to the thin film of liquid spread over the entire inner surface of the rotating flask, which is heated under a vacuum. Foaming and bumping are also greatly reduced.

Cleaning Up   Acetic acid, aqueous layers, and saturated sodium chloride layers from the extraction, after neutralization, can be flushed down the drain with water. Ether, ligroin, and TLC solvents should be placed in the organic solvents container, and the drying agent, once free of solvent, can be placed in the non-hazardous solid waste container. Ligroin and ether from the chromatography go into the organic solvents container. If possible, spread out the silica gel in the hood to dry. It can then be placed in the nonhazardous solid waste container. If it is wet with organic solvents, it is a hazardous solid waste and must be disposed of, at great expense, in a secure landfill.

MICROSCALE AND MACROSCALE

3. Fluorene and Fluorenone

The 9-position of fluorene is unusually reactive for a hydrocarbon. The protons on this carbon atom are acidic by virtue of being doubly benzylic, and, consequently, this carbon can be oxidized by several reagents, including elemental oxygen. In the present experiment, the very powerful and versatile oxidizing agent Cr(VI), in the form of chromium trioxide, is used to carry out this oxidation. Cr(VI) in a variety of other forms is used to carry out about a dozen oxidation reactions in this text. The *dust* of Cr(VI) salts is reported to be a carcinogen, so avoid breathing it.

$$\xrightarrow[\text{CH}_3\text{COOH}]{\text{Na}_2\text{Cr}_2\text{O}_7}$$

Fluorene
mp 114°C
MW 166.22

Fluorenone
mp 83°C
MW 180.21

 Microscale Procedure

In a reaction tube dissolve 50 mg of fluorene in a 0.25 mL of acetic acid by heating, and add this hot solution to a solution of 0.15 g of sodium dichromate dihydrate in 0.5 mL of acetic acid. Heat the reaction mixture to 80°C for 15 min in a hot water bath; then cool it, and add 1.5 mL of water. Stir the mixture for 2 min, and then filter it on the Hirsch funnel. Wash the product well with water, and press out as much water as possible. Return the product to the reaction tube, add 2 mL of *tert*-butyl methyl ether, and add anhydrous calcium chloride pellets until it no longer clumps together. Cork and shake the tube, and allow the product to dry for 5 or 10 min before evaporating the ether in another tared reaction tube. Use ether to wash off the drying agent and to complete the transfer of product. Use this ether solution to spot a TLC plate. This crude mixture of fluorene and fluorenone will be separated by column chromatography.

Column Chromatography of the Fluorene-Fluorenone Mixture.

The sample can also be applied to the column using the technique described on p. 173.

Prepare a microscale chromatographic column exactly as described above. Dissolve the crude mixture of fluorene and fluorenone in 0.2 mL of warm toluene, and add this to the surface of the alumina. Be sure to add the sample as a solution; should any sample crystallize, add a drop more toluene. Run the ligroin down to the surface of the alumina, add a few drops more ligroin, and repeat the process until the sample is seen as a narrow band at the top of the column.

Ligroin = hexane = "hexanes."

Carefully add a 3-mm layer of sand, then fill the column with ligroin, and collect 5-mL fractions in tared 10-mL Erlenmeyer flasks. Sample each flask for thin-layer chromatography (Fig. 10.10), and evaporate each to dryness. Final drying can be done under vacuum using the technique shown in Fig. 10.7. You are to decide when all the product has been eluted from the column. The thin-layer plates can be developed using 20% dichloromethane in ligroin. Combine fractions that are identical, and determine the melting points of the two substances.

$$\text{Fluorene} \xrightarrow[\text{CH}_3\text{COOH}]{\text{Na}_2\text{Cr}_2\text{O}_7} \text{Fluorenone}$$

Fluorene
mp 114°C
MW 166.22

Fluorenone
mp 83°C
MW 180.21

Macroscale Procedure

Caution: Sodium dichromate is toxic. The dust is corrosive to nasal passages and skin and is a cancer-suspect agent. Hot acetic acid is very corrosive to skin; handle with care.

One-half hour unattended heating

Handle dichromate and acetic acid in laboratory hood; wear gloves.

In a 250-mL Erlenmeyer flask dissolve 5.0 g of practical grade fluorene in 25 mL of acetic acid by heating on the steam bath with occasional swirling. In a 125-mL Erlenmeyer flask dissolve 15 g of sodium dichromate dihydrate in 50 mL of acetic acid by swirling and heating on a hot plate. Adjust the temperature of the dichromate solution to 80°C, transfer the thermometer, and adjust the fluorene–acetic acid solution to 80°C, and then, **in the hood,** pour in the dichromate solution. Note the time and the temperature of the solution, and heat on the steam bath for 30 min. Observe the maximum and final temperature, and then cool the solution and add 150 mL of water. Swirl the mixture for a full 2 min to coagulate the product and so promote rapid filtration, and collect the yellow solid in an 8.5-cm Büchner funnel (in case filtration is slow, empty the funnel and flask into a beaker and stir vigorously for a few minutes). Wash the filter cake well with water and then suck the filter cake as dry as possible. Either let the product dry overnight, or dry it quickly as follows: Put the moist solid into a 50-mL Erlenmeyer flask, add ether (20 mL) and swirl to dissolve, and add anhydrous calcium chloride (10 g) to scavenge the water. Decant the ethereal solution through a cone of anhydrous calcium chloride in a funnel into a 125-mL Erlenmeyer flask, and rinse the flask and funnel with *t*-butyl methyl ether. Evaporate on the steam bath under an aspirator, heat until the ether is all removed, and pour the hot oil into a 50-ml beaker to cool and solidify. Scrape out the yellow solid. Yield: 4.0 g.

FIG. 10.10 Spot each fraction on a chromatography plate. Examine under UV light to see which fractions contain compound.

Cleaning Up   The filtrate probably contains unreacted dichromate. To destroy it, add 3 M sulfuric acid until the pH is 1, and then complete the reduction by adding solid sodium thiosulfate until the solution becomes cloudy and blue-colored. Neutralize with sodium carbonate, and filter the flocculent precipitate of $Cr(OH)_3$ through Celite in a Büchner funnel. The filtrate can be diluted with water and flushed down the drain, while the precipitate and Celite should be placed in the heavy metal hazardous waste container.

4. Separation of Fluorene and Fluorenone

Prepare a column of 12.5 g of alumina, run out excess solvent, and pour onto the column a solution of 0.5 g of fluorene–fluorenone mixture. Elute at first with ligroin, and use tared 50-mL flasks as receivers. The yellow color of fluorenone provides one index of the course of the fractionation, and the appearance of solid around the delivery tip provides another. Wash the solid on the tip frequently into the receiver with ether. When you think that one component has been eluted completely, change to another receiver until you judge that the second component is beginning to appear. Then, when you are sure the second component is being eluted, change to a 1 : 1 ligroin : *t*-butyl methyl ether mixture, and continue until the column is exhausted. It is possible to collect practically all the two components in the two receiving flasks, with only a negligible intermediate fraction. After evaporation of solvent, evacuate each flask under vacuum (Fig. 10.7) and determine the weight and melting point of the products. A convenient method for evaporating fractions is to use a rotary evaporator (Fig. 10.9).

Cleaning Up   All organic material from this experiment can go in the organic solvents container. The alumina absorbent, if free of organic solvents, can go in the nonhazardous solid waste container. If it is wet with organic solvents, it is a hazardous solid waste. Spread it in the hood to dry.

Questions

1. Predict the order of elution of a mixture of triphenylmethanol, biphenyl, benzoic acid, and methyl benzoate from an alumina column.

2. What would be the effect of collecting larger fractions when carrying out either of the experiments described?

3. What would have been the result if a large quantity of petroleum ether alone were used as the eluent in either of the experiments described?

4. Once the chromatographic column has been prepared, why is it important to allow the level of the liquid in the column to drop to the level of the alumina before applying the solution of the compound to be separated?

5. A chemist started to carry out column chromatography on a Friday afternoon, got to the point at which the two compounds being separated were about three-fourths of the way down the column, and then returned on Monday to find that the compounds came off the column as a mixture. Speculate on the reason for this. The column had not run dry over the weekend.

CHAPTER

11

Alkenes from Alcohols: Analysis of a Mixture by Gas Chromatography

Prelab Exercise: If the dehydration of 2-methyl-2-butanol occurred on a purely statistical basis, what would be the relative proportions of 2-methyl-1-butene and 2-methyl-2-butene?

Separation of mixtures of volatile samples

Gas chromatography (GC), also called vapor phase chromatography (VPC) and gas-liquid chromatography (GLC), is a means of separating volatile mixtures, the components of which may differ in boiling points by only a few tenths of a degree. The GC process is similar to fractional distillation, but instead of a glass column 25 cm long packed with a stainless steel sponge, the GC column used is a 3–10 m long coiled metal tube (6-mm dia.), packed with ground firebrick. The firebrick serves as an inert support for a very high-boiling liquid (essentially non-volatile), such as silicone oil and low molecular weight polymers like Carbowax.

High-boiling liquid phase

These are the *liquids* of gas-*liquid* chromatography and are referred to as the stationary phase. The sample (1–25 *micro*liters) is injected through a silicone rubber septum into the column, which is being swept with a current of helium (ca. 200 mL/min). The sample first dissolves in the high-boiling liquid phase, and then the more volatile components of the sample evaporate from the liquid and pass into the gas phase. Helium, the carrier gas, carries these components along the column a short distance where they again dissolve in the liquid phase before reevaporation (Fig. 11.1).

The thermal conductivity detector

Eventually the carrier gas, which is a very good thermal conductor, and the sample reach the detector, an electrically heated tungsten wire. As long as pure helium is flowing over the detector, the temperature of the wire is relatively low and the wire has a low resistance to the flow of electric current. Organic molecules have lower thermal conductivities than helium. Hence, when a mixture of helium and an organic sample flows over the detector wire it is cooled less efficiently and heats up. When the wire is hot its electrical resistance becomes higher and offers higher resistance to the flow of current. The detector wire actually is one leg of a Wheatstone bridge, connected to a chart recorder that records, as a peak, the amount of current necessary to again balance the bridge. The record produced by the recorder is called a chromatogram.

The number and relative amounts of components in a mixture

A gas chromatogram is simply a recording of current versus time (which is equivalent to a certain volume of helium) (Fig. 11.2). In the illustration the smaller peak from component A has the shorter retention time, T_1, and so is a more volatile substance than component B (if the stationary phase is an inert liquid such as silicone oil). The areas under the two peaks are directly proportional to the molar amounts of A and B in the mixture (provided they are structurally

FIG. 11.1 Gas chromatography: Diagrammatic partition between gas and liquid phases.

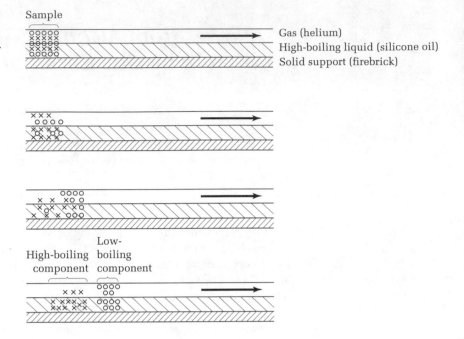

similar). The retention time of a given component is a function of the column temperature, the helium flow rate, and the nature of the stationary phase. Hundreds of stationary phases are available; picking the correct one to carry out a given analysis is somewhat of an art, but widely used stationary phases are silicone oil and silicone rubber, both of which can be used at temperatures up to 300°C and separate mixtures on the basis of boiling point differences of the mixture's components. More specialized stationary phases will, for instance, allow alkanes to pass through readily (short retention time) while holding back (long retention time) alcohols by hydrogen bonding to the liquid phase.

A diagram of a typical gas chromatograph is shown in Fig. 11.3. The carrier gas, usually helium, enters the chromatograph at ca. 60 lb/sq in. The sample (1 to 25 microliters) is injected through a rubber septum using a small hypodermic syringe (Fig. 11.4). Handle the syringe with care. The syringe needles and plungers are thin and delicate; take care not to bend either one. Be wary of the injection port. It is very hot. The sample immediately passes through the column

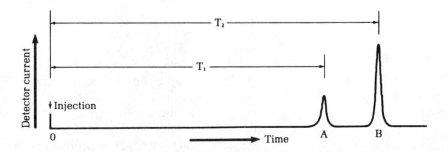

FIG. 11.2 Gas chromatogram.

FIG. 11.3   Diagram of a gas chromatograph.

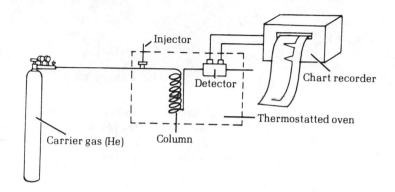

and then the detector. Injector, column, and detector are all enclosed in a thermostatted oven, which can be maintained at any temperature up to 300°C. In this way samples that would not volatilize enough at room temperature can be analyzed.

   Gas chromatography determines the number of components and their relative amounts in a very small sample. The small sample size is an advantage in many cases, but it precludes isolating the separated components. Some specialized chromatographs can separate samples as large as 0.5 mL per injection and automatically collect each fraction in a separate container. At the other extreme, gas chromatographs equipped with flame ionization detectors can detect micrograms of sample, such as traces of pesticides in food or of drugs in blood and urine. Clearly a gas chromatogram gives little information about the chemical nature of the sample being detected. However, it is sometimes possible to collect enough sample at the exit port of the chromatograph to obtain an infrared spectrum. As the peak for the compound of interest appears on the chart paper, a 2-mm-dia. glass tube, 3 in. long and packed with glass wool, is inserted into the rubber septum at the exit port. The sample, if it is not too volatile, will condense in the cold glass tube. Subsequently, the sample is washed out with a drop or two of solvent and an infrared spectrum obtained. See Fig. 11.5 for another collection device.

Collecting a sample for an infrared spectrum

FIG. 11.4   One of the commercially available chromatographs.

Dehydration

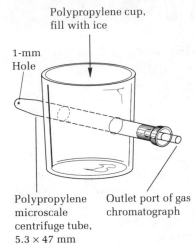

Polypropylene cup,
fill with ice

1-mm
Hole

Polypropylene
microscale
centrifuge tube,
5.3 × 47 mm

Outlet port of gas
chromatograph

FIG. 11.5 Gas chromatographic collection device. Fill container with ice or dry ice–acetone mixture and attach to outlet port of GC. Available from Kontes.

1. 2-Methyl-1-butene and 2-Methyl-2-butene[1]

The dilute sulfuric acid catalyzed dehydration of 2-methyl-2-butanol (*t*-amyl alcohol) proceeds readily to give a mixture of alkenes that can be analyzed by gas chromatography. The mechanism of this reaction involves the intermediate formation of the relatively stable tertiary carbocation followed by loss of a proton either from a primary carbon atom to give the terminal olefin, 2-methyl-1-butene, or from a secondary carbon to give 2-methyl-2-butene.

$$CH_3CH_2\overset{\overset{OH}{|}}{\underset{\underset{CH_3}{|}}{C}}CH_3 + H_2SO_4 \overset{fast}{\rightleftharpoons} CH_3CH_2\overset{\overset{\overset{+}{O}H_2}{|}}{\underset{\underset{CH_3}{|}}{C}}CH_3 + HSO_4^-$$

2-Methyl-2-butanol
bp 102°C
den 0.805
MW 88.15

$$CH_3CH_2\overset{\overset{\overset{+}{O}H_2}{|}}{\underset{\underset{CH_3}{|}}{C}}CH_3 \overset{slow}{\rightleftharpoons} CH_3CH_2\overset{+}{\underset{\underset{CH_3}{|}}{C}}CH_3 + H_2O$$

$$CH_3CH_2\overset{+}{\underset{\underset{CH_3}{|}}{C}}-CH_2 \rightleftharpoons CH_3CH_2C=CH_2$$

2-Methyl-1-butene
bp 31.16°, den 0.662
MW 70.14

$$CH_3CH-\overset{+}{\underset{\underset{CH_3}{|}}{C}}CH_3 \rightleftharpoons CH_3CH=\overset{}{\underset{\underset{CH_3}{|}}{C}}CH_3$$

2-Methyl-2-butene
bp 38.57°C
den 0.662
MW 70.14

1. C. W. Schimelpfenig *J. Chem. Ed.*, **39**, 310 (1962).

2-Methyl-2-butanol can also be dehydrated in high yield using iodine as a catalyst:

$$CH_3CH_2\underset{\underset{CH_3}{|}}{\overset{\overset{OH}{|}}{C}}CH_3 \xrightarrow{I_2 \text{ (trace)}} CH_3CH_2\underset{\underset{CH_3}{|}}{\overset{\overset{OI}{|}}{C}}CH_3 \xrightarrow{-HOI}$$

$$CH_3CH_2\underset{\underset{CH_3}{|}}{C}{=}CH_2 + CH_3CH{=}\underset{\underset{CH_3}{|}}{C}CH_3 + HI + HOI$$

$$HI + HOI \longrightarrow H_2O + I_2$$

Each step of this E_1 elimination reaction is reversible, and thus the reaction is driven to completion by removing one of the products, the alkene. In these reactions several alkenes can be produced. The Saytzeff rule states that the more substituted alkene is the more stable and thus the one formed in larger amount. And the *trans*-isomer is more stable than the *cis*-isomer. With this information it should be possible to deduce which peaks on the gas chromatogram correspond to a given alkene and to predict the ratios of the products.

In analyzing your results from this experiment, consider the fact that the carbocation can lose any of six primary hydrogen atoms but only two secondary hydrogen atoms to give the product olefins.

Microscale Dehydration Procedure

Into a 5-mL long-necked round-bottomed flask place 1 mL of water, and then add dropwise with thorough mixing 0.5 mL of concentrated sulfuric acid. Cool the hot solution in an ice bath, and add to the cold solution 1.00 mL (0.80 g) of 2-methyl-2-butanol. Mix the reactants thoroughly, add a boiling stone, and set up the apparatus for simple distillation as shown in Fig. 11.6. Hold the flask in a towel while connecting the apparatus to guard against spills or flask breakage. Cool the receiver (a vial) in ice to avoid losing the very volatile products. Warm the flask on the sand bath to start the reaction, and distill the products over the temperature range 30–45°C. After all the products have distilled, the rate of distillation will decrease markedly. Cease distillation at this point. After this, the temperature registered on the thermometer will rise rapidly as water and sulfuric acid begin to distill.

To the distillate add 0.3 mL *cold* 3 M sodium hydroxide solution to neutralize sulfurous acid, mix well, draw off the aqueous layer, and dry the product over anhydrous calcium chloride pellets, adding the drying agent in small quantities until it no longer clumps together. Keep the vial cold. While the product is drying, rinse out the distillation apparatus with water, ethanol, and then a small amount of acetone and *dry* it thoroughly, by drawing air through it using the aspirator. If this is not done carefully, the products will be contaminated with acetone. Carefully transfer the dry mixture of butenes to the distilling apparatus using a

FIG. 11.6 Apparatus for dehydration of 2-methyl-2-butanol. Beaker can be clamped with large three-prong clamp.

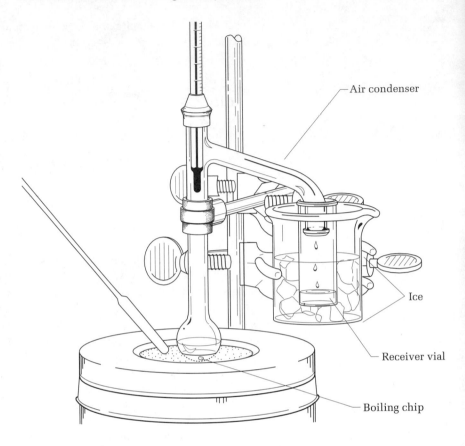

Keep the side arm down in the cold vial to avoid losing product.

Air condenser

Ice

Receiver vial

Boiling chip

Products are easily lost by evaporation. Store mixture in a spark-proof refrigerator if GC analysis must be postponed.

Pasteur pipette, add a boiling chip to the now dry products, and distill into a tared, cold receiver. Collect the portion boiling up to 43°C. The yield reported in the literature is 84% when the reaction is carried out on a much larger scale. Weigh the product, calculate the yield, and analyze it by gas chromatography. The average yield on a small scale is 50%. 2-Methyl-1-butene boils at 31.2°C, and 2-methyl-2-butene boils at 38.6°C.

Cleaning Up   The residue in the reaction tube should be diluted with water, neutralized with sodium carbonate, and then flushed down the drain.

 Macroscale Dehydration Procedure

Pour 18 mL of water into a 100-mL round-bottomed flask, and cool in an ice-water bath while slowly pouring in 9 mL of concentrated sulfuric acid. Cool this "1 : 2" acid further with swirling while slowly pouring in 18 mL (15 g) of 2-methyl-2-butanol. Shake the mixture thoroughly, and then mount the flask for fractional distillation over a flask heater, as in Fig. 5.11, with the arrangement for ice cooling of the distillate seen in Fig. 19.2. Cooling is needed because the olefin is volatile. Use a long condenser and a rapid stream of cooling water. Heat the flask slowly with an electric flask heater until distillation of the hydrocarbon is complete. If the ice-cooled test tube will not hold 15 mL, be prepared to collect half of the distillate in another ice-cooled test tube. Transfer the distillate to a separatory funnel, and shake with about 5 mL of 3 *M* sodium hydroxide solution to remove any traces of sulfurous acid. The aqueous solution sinks to the bottom and is drawn off. Dry the hydrocarbon layer by adding sufficient anhydrous calcium chloride pellets until the drying agent no longer clumps together. After about 5 min remove the drying agent by gravity filtration or careful decantation into a *dry* 25-mL round-bottomed flask, and distill the dried product through a fractionating column (see Fig. 19.2), taking the same precautions as before to avoid evaporation losses. Rinse the fractionating column with acetone, and dry it with a stream of air to remove water from the first distillation. Collect in a tared bottle the portion boiling at 30–43°C. The yield reported in the literature is 84%; the average student yield is about 50%. See Fig. 10.7 for the ^{13}C NMR spectrum of one of the products.

Cleaning Up   The pot residue from the reaction is combined with the sodium hydroxide wash and neutralized with sodium carbonate. The pot residue from the distillation of the product is combined with the acetone used to wash out the apparatus and placed in the organic solvents container. If the calcium chloride pellets are dry, they can be placed in the nonhazardous solid waste container. If they are wet with organic solvents, they must be placed in the hazardous solid waste container for solvent-contaminated drying agent.

2. Computational Chemistry

The steric energies of the two isomeric butenes produced in Experiment 1 can be calculated using a molecular mechanics program, but the results are not valid because the bonding type is not the same. It is possible to compare the steric energies of *cis*- and *trans*-isomers of alkenes, but not a 1,1-disubstituted alkene with a 1,2-disubstituted alkene.

Compare your yields of the two isomers with their heats of formation calculated using AM1 or a similar semiempirical molecular orbital program. Do the calculated heats of formation correlate with relative percentages of the isomers? What does this correlation or lack thereof tell you about the mechanism of the reaction?

U **MICROSCALE**

2. 1-Butene and *cis*- and *trans*-2-Butene[2]

2-Butanol	**1-Butene**	**cis-2-Butene**	**trans-2-Butene**
MW 74.12	MW 56.11	MW 56.11	MW 56.11
bp 98°C	bp −6.3°C	bp 3.7°C	bp 0.9°C

2. G. K. Helmkamp and H. W. Johnson, Jr. *Selected Experiments in Organic Chemistry,* 3d ed., Freeman, New York, 1983, p. 99.

FIG. 11.7 Apparatus for dehydration of 2-butanol. Butenes are collected by the downward displacement of water.

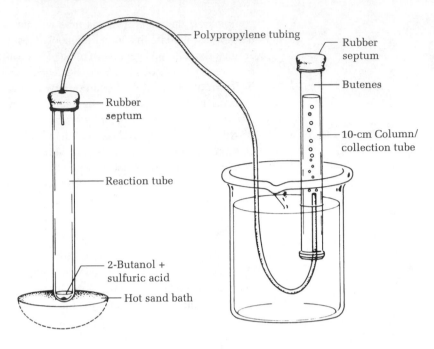

The quantities of alcohol and acid are not critical.

In a 10 × 100 mm reaction tube place 0.10 mL (81 mg) of 2-butanol and 0.05 mL (1 drop) of concentrated sulfuric acid. Mix the reactants well, and add a boiling chip. Stirring of the reaction mixture is not necessary because it is homogeneous. Insert a polypropylene tube through a septum (see Fig. 18.2), place this septum on the reaction tube, and lead the tubing into the mouth of the distilling column/collection tube, which is capped with a septum. The distilling column/collection tube is filled with water, capped with a finger, and inverted in the beaker of water and clamped for the collection of butene by the downward displacement of water (Fig. 11.7).

Bend the polypropylene tubing to shape in hot sand or steam.

Lower the reaction tube into a hot (~ 100°C) sand bath, and increase the heat slowly to complete the reaction. Collect a few milliliters of butene, and *remove the polypropylene tube from the water bath.* If the polypropylene tube is not removed from the water bath, water will be sucked back into the reaction tube when it cools. Use about 0.5 mL of the gaseous butene mixture to carry out the GC analysis.

Cleaning Up   The residue in the reaction tube should be diluted with water, neutralized with sodium carbonate, and then flushed down the drain.

Gas Chromatographic Analysis

Be sure the helium cylinder is attached to a bench or wall.

Inject a few microliters of product into a gas chromatograph maintained at room temperature and equipped with a 6-mm-dia. × 3-m column packed with 10%

Use care. The injection port is hot.

In your notebook record all information relevant to this analysis: column diameter and length, column packing, carrier gas and its flow rate, filament current, temperature of column, detector and injection port, sample size, attenuation, and chart speed.

SE-30 silicone rubber on Chromosorb-W or a similar inert packing. Mark the chart paper at the time of injection. In a few minutes two peaks should appear. After using the chromatograph, turn off the recorder, and cap the pen. Make no unauthorized adjustments on the gas chromatograph. From your knowledge of the mechanisms of dehydration of secondary alcohols, which olefin should predominate? Does this agree with the boiling points? (In general, the compound with the shorter retention time has the lower boiling point.) Measure the relative areas under the two peaks. One way to perform this integration is to cut out the peaks with scissors and weigh the two pieces of paper separately on an analytical balance. Although time consuming, this method gives very precise results. If the peaks are symmetrical, their areas can be approximated by simply multiplying the height of the peak by its width at half-height.

Questions

1. Write the structure of the three olefins produced by the dehydration of 3-methyl-3-pentanol.

2. When 2-methylpropene is bubbled into dilute sulfuric acid at room temperature, it appears to dissolve. What new substance has been formed?

3. A student wished to prepare ethylene gas by dehydration of ethanol at 140°C using sulfuric acid as the dehydrating agent. A low-boiling liquid was obtained instead of ethylene. What was the liquid, and how might the reaction conditions be changed to give ethylene?

4. What would be the effect of increasing the carrier gas flow rate on the retention time?

5. What would be the effect of raising the column temperature on the retention time?

6. What would be the effect of raising the temperature or increasing the carrier gas flow rate on the ability to resolve two closely spaced peaks?

7. If you were to try a column one-half the length of the one you actually used, how do you think the retention times of the butenes would be affected? How do you think the separation of the peaks would be affected? How do you think the width of each peak would be affected?

8. From your knowledge of the dehydration of tertiary alcohols, which olefin should predominate in the product of the dehydration of 2-methyl-2-butanol, and why?

9. What is the maximum volume (at STP) of the butene mixture that could be obtained by the dehydration of 81 mg of 2-butanol?

10. What other gases are in the collection tube besides the three butenes at the end of the second reaction?

11. Using a molecular mechanics program (see Chapter 16), calculate the steric energies or heats of formation of the three isomeric butenes produced in Experiment 2. Do these energies correlate with the product distributions found? What does a correlation or lack thereof tell you about the mechanism of the reaction?

12 *Infrared Spectroscopy*

Prelab Exercise: When an infrared spectrum is run, there is a possibility that the chart paper is not properly placed or that the spectrometer is not mechanically adjusted. Describe how you could calibrate an infrared spectrum.

The presence and also the environment of functional groups in organic molecules can be identified by infrared spectroscopy. Like nuclear magnetic resonance (NMR) and ultraviolet spectroscopy, infrared spectroscopy is nondestructive. Moreover, the small quantity of sample needed, the speed with which a spectrum can be obtained, the relatively small cost of the spectrometer, and the wide applicability of the method combine to make infrared spectroscopy one of the most useful tools available to the organic chemist.

Infrared radiation, which is electromagnetic radiation of longer wavelength than visible light, is detected not with the eyes but by a feeling of warmth on the skin. When absorbed by molecules, radiation of this wavelength (typically 2.5 to 25 μm) increases the amplitude of vibrations of the chemical bonds joining atoms.

2.5 to 25 μm equals 4000 to 400 cm^{-1}.

Wavenumbers, cm^{-1}, are proportional to frequency:

$$\bar{\nu}(cm^{-1}) = \frac{\nu}{c}$$

Infrared spectra are measured in units of frequency or wavelength. The wavelength is measured in micrometers or microns, μm (1 μm = 1×10^{-4} cm). The positions of absorption bands are measured in frequency units by wavenumbers $\bar{\nu}$, which are expressed in reciprocal centimeters, cm^{-1}, corresponding to the number of cycles of the wave in each centimeter.

$$cm^{-1} = \frac{10,000}{\mu m}$$

Examine the scale carefully.

IR spectroscopy easily detects

Hydroxyl groups —OH

Amines —NH$_2$

Nitriles —C≡N

Nitro groups —NO$_2$

Unlike ultraviolet and NMR spectra, infrared spectra are inverted and are not always presented on the same scale. Some spectrometers record the spectra on an ordinate linear in microns, but this compresses the low-wavelength region. Other spectrometers present the spectra on a scale linear in reciprocal centimeters, but linear on two different scales, one between 4000 and 2000 cm^{-1}, which spreads out the low-wavelength region, and the other a smaller one between 2000 and 667 cm^{-2}. Consequently, spectra of the same compound run on two different spectrometers will not always look the same.

To picture the molecular vibrations that interact with infrared light, imagine a molecule as being made up of balls (atoms) connected by springs (bonds). The vibration can be described by Hooke's law from classical mechanics, which says that the frequency of a stretching vibration is directly proportional to the strength of the spring (bond) and inversely proportional to the masses connected by the spring. Thus we find C—H, N—H, and O—H bond-stretching vibrations are high

IR spectroscopy is especially useful for detecting and distinguishing among all carbonyl-containing compounds:

Acids

$$\overset{\displaystyle O}{\overset{\|}{-C}}-OH$$

Amides

$$\overset{\displaystyle O}{\overset{\|}{-C}}-NH_2$$

Anhydrides

$$\overset{\displaystyle O}{\overset{\|}{-C}}-O-\overset{\displaystyle O}{\overset{\|}{C}}-$$

Aldehydes

$$\overset{\displaystyle O}{\overset{\|}{-C}}-H$$

Ketones

$$\overset{\displaystyle O}{\overset{\|}{-C}}-$$

Esters

$$\overset{\displaystyle O}{\overset{\|}{-C}}-O-$$

Lactones

$$\overset{\displaystyle O}{\overset{\|}{-C}}-O-$$

frequency (short wavelength) compared with those of C—C and C—O because of the low mass of hydrogen compared with that of carbon or oxygen. The bonds connecting carbon to bromine and iodine, atoms of large mass, vibrate so slowly that they are beyond the range of most common infrared spectrometers. A double bond can be regarded as a stiffer, stronger spring, so we find C=C and C=O vibrations at higher frequency than C—C and C—O stretching vibrations. And C≡C and C≡N stretch at even higher frequencies than C=C and C=O (but at lower frequencies than C—H, N—H, and O—H). These frequencies are in keeping with the bond strengths of single (~100 kcal/mol), double (~160 kcal/mol), and triple bonds (~220 kcal/mol).

The stretching vibrations noted above are intense and particularly easy to analyze. A nonlinear molecule of n atoms can undergo $3n - 6$ possible modes of vibration, which means cyclohexane with 18 atoms can undergo 48 possible modes of vibration. A single CH_2 group can vibrate in six different ways. Each vibrational mode produces a peak in the spectrum because it corresponds to the absorption of energy at a discrete frequency. These many modes of vibration create a complex spectrum that defies simple analysis, but even in very complex molecules, certain functional groups have characteristic frequencies that can easily be recognized. Within these functional groups are the above-mentioned atoms and bonds, C—H, N—H, O—H, C=C, C=O, C≡C, and C≡N. Their absorption frequencies are given in Table 12.1.

When the frequency of infrared light is the same as the natural vibrational frequency of an interatomic bond, light will be absorbed by the molecule, and the amplitude of the bond vibration will increase. The intensity of infrared absorption bands is proportional to the change in dipole moment that a bond undergoes when it stretches. Thus the most intense bands (peaks) in an infrared spectrum are often from C=O and C—O stretching vibrations, while the C≡C stretching band for a symmetrical acetylene is almost nonexistent because the molecule undergoes no net change of dipole moment when it stretches:

$$\overset{+}{\underset{}{\diagdown}}C=O \longleftrightarrow \overset{+}{\underset{}{\diagdown}}C=O \qquad H_3C-C\equiv C-CH_3 \longleftrightarrow H_2C-C\equiv C-CH_3$$

Change in dipole moment   No change in dipole moment

Intensity of absorption is proportional to change in dipole moment.

Unlike proton NMR spectroscopy, where the area of the peaks is strictly proportional to the number of hydrogen atoms causing the peaks, the intensities of infrared peaks are not proportional to the numbers of atoms causing them. Given the chemical shifts and coupling constants, it is possible to calculate a theoretical NMR spectrum that is an exact match to the experimental one. It is not possible, except for very simple molecules, to do this for infrared spectra. Every peak or group of peaks in an NMR spectrum can be assigned to specific hydrogens in a molecule, but the assignment of the majority of peaks in an infrared spectrum is usually not possible. Peaks to the right (longer wavelength) of 1250 cm^{-1} are the result of combinations of vibrations that are characteristic not of individual functional groups but of the molecule as a whole. This part of the spectrum is often

TABLE 12.1   Characteristic Infrared Absorption Wavenumbers

Functional Group	Wavenumber (cm^{-1})
$O-H$	3600–3400
$N-H$	3400–3200
$C-H$	3080–2760
$C\equiv N$	2260–2215
$C\equiv C$	2150–2100
$C=O$	1815–1650
$C=C$	1660–1600
$C-O$	1200–1050

referred to as the *fingerprint region*, because it is uniquely characteristic of each molecule. While two organic compounds can have the same melting points or boiling points and can have identical ultraviolet and NMR spectra, they cannot have identical infrared spectra (except, as usual, for enantiomers). Infrared spectroscopy is thus the final arbiter in deciding whether two compounds are identical.

Analysis of Infrared Spectra

Very few simple equations or rules govern infrared spectroscopy. Since it is not practical to calculate theoretical spectra, the analysis is done almost entirely by correlation with other spectra. In printed form, these comparisons take the form of lengthy discussions, so detailed analysis of a spectrum is best done with a good reference book at hand.

In a modern analytical or research laboratory, a collection of many thousands of spectra is maintained on a computer. When the spectrum of an unknown compound is run, the analyst picks out five or six of the strongest peaks and asks the computer to list all compounds that have peaks within a few reciprocal centimeters of the listed peaks. From the printout of a dozen or so compounds, it is often possible to pinpoint all the functional groups in the molecule being analyzed. There may be a perfect match of all peaks, in which case the unknown will have been identified.

For relatively simple molecules, a computer search is hardly necessary. Much information can be gained about the functional groups in a molecule from relatively few correlations.

To carry out an analysis, (1) pay most attention to the strongest absorptions, (2) pay more attention to peaks to the left (shorter wavelength) of 1250 cm^{-1}, and (3) pay as much attention to the absence of certain peaks as to the presence of others. The absence of characteristic peaks will definitely exclude certain functional groups. Be wary of weak O—H peaks because water is a common contaminant of many samples. Because potassium bromide is hygroscopic, it is often found in the spectra of KBr pellets.

Step-by-Step Analysis of Infrared Spectra

1. Is there a peak between 1820 to 1625 cm^{-1}? If not, go to Step 2.

    (a) Is there also a strong, wide O—H peak between 3200 and 2500 cm^{-1}? If so, the compound is a carboxylic acid (see Fig. 12.1, oleic acid). If not:

    (b) Is there a medium-to-weak N—H band between 3520 and 3070 cm^{-1}? If there are two peaks in this region, then it is a primary amide; if not, it is a secondary amide. If there is no peak in this region:

    (c) Are there two strong peaks, one in the region 1870 to 1800 cm^{-1} and the other in the region 1800 to 1740 cm^{-1}? If so, an acid anhydride is present. If not:

    (d) Is there a peak in the region of 2720 cm^{-1}? If so, is the carbonyl peak in the region 1715 to 1680 cm^{-1}? If so, it is a conjugated aldehyde; if not, it is an isolated aldehyde (see Fig. 12.2, benzaldehyde). However, if there is no peak near 2720 cm^{-1}:

    (e) Does the strong carbonyl peak fall in the range 1815 to 1770 cm^{-1} and the molecule give a positive Beilstein test? If so, it is an acid halide. If not:

    (f) Does the strong carbonyl peak fall in the range 1690 to 1675 cm^{-1}? If so, it is a conjugated ketone. If not:

    (g) Does the strong carbonyl peak fall in the range 1670 to 1630 cm^{-1}? If so, it is a tertiary amide. If not:

    (h) Does the molecule have a strong, wide peak in the range 1310 to 1100 cm^{-1}? If so, does the carbonyl peak fall in the range 1730 to 1715 cm^{-1}? If so, it is a conjugated ester; if not, the ester is not conjugated

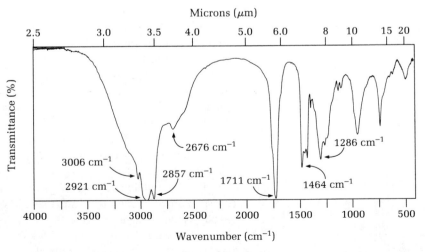

FIG. 12.1   Infrared spectrum of oleic acid (thin film).

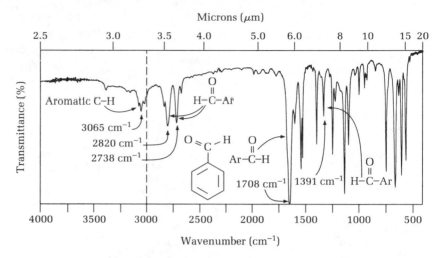

FIG. 12.2   Infrared spectrum of benzaldehyde (thin film).

(see Fig. 12.3, *n*-butyl acetate). If there is no strong, wide peak in the range 1310 to 1100 cm^{-1}, then:

(i) The molecule is an ordinary nonconjugated ketone.

2. If the molecule lacks a carbonyl peak in the range 1820 to 1625 cm^{-1}, does it have a broad band in the region 3650 to 3200 cm^{-1}? If so, does it also have a peak at about 1200 cm^{-1}, a C—H stretching peak to the left of 3000 cm^{-1}, and a peak in the region 1600 to 1470 cm^{-1}? If so, it is a phenol. If it does not meet these latter three criteria, it is an alcohol (see Fig. 12.4, cyclohexanol). However, if there is no broad band in the region 3650 to 3200 cm^{-1}, then:

(a) Is there a broad band in the region 3500 to 3300 cm^{-1}, does the mole-

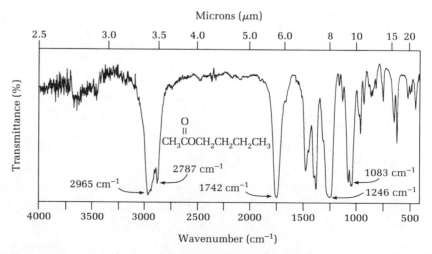

FIG. 12.3   Infrared spectrum of *n*-butyl acetate (thin film).

The effect of ring size on the carbonyl frequencies of lactones and esters

1727 cm^{-1}

1745 cm^{-1}

1740 cm^{-1}

1775 cm^{-1}

1832 cm^{-1}

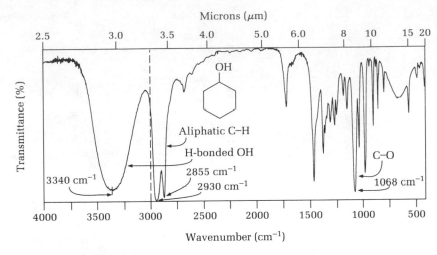

FIG. 12.4   Infrared spectrum of cyclohexanol (thin film).

cule smell like an amine, or does it contain nitrogen? If so, are there two peaks in this region? If so, it is a primary amine; if not, it is a secondary amine. However, if there is no broad band in the region 3500 to 3300 cm^{-1}, then:

(b) Is there a sharp peak of medium-to-weak intensity at 2260 to 2100 cm^{-1}? If so, is there also a peak at 3320 to 3310 cm^{-1}? If so, the molecule is a terminal acetylene. If not, then the molecule is most likely a nitrile (see Fig. 12.5, benzonitrile), although it might be an asymmetrically substituted acetylene. If there is no sharp peak of medium-to-weak intensity at 2260 to 2100 cm^{-1}, then:

(c) Are there strong peaks in the region 1600 to 1540 cm^{-1} and 1380 to 1300 cm^{-1}? If so, the molecule contains a nitro group. If not:

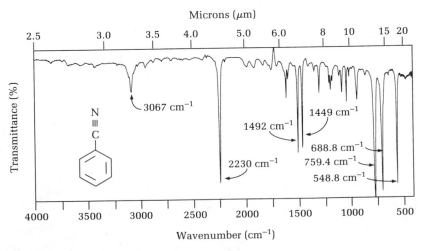

FIG. 12.5   Infrared spectrum of benzonitrile (thin film).

The effect of ring size on the carbonyl frequency

1705 cm⁻¹

1715 cm⁻¹

1715 cm⁻¹

1745 cm⁻¹

1780 cm⁻¹

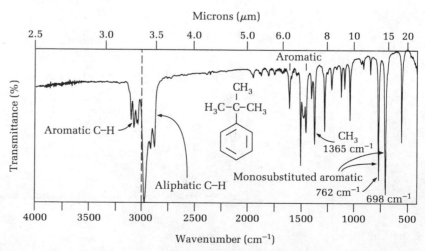

1815 cm⁻¹

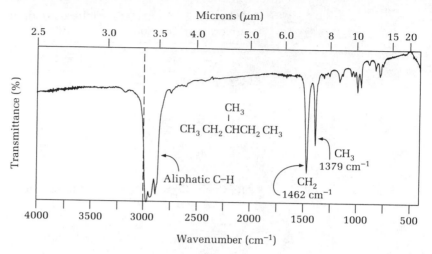

FIG. 12.6   Infrared spectrum of 3-methylpentane (thin film).

(d) Is there a strong peak in the region 1270 to 1060 cm⁻¹? If so, the molecule is an ether. If not:

(e) The molecule is either a tertiary amine (odor?), a halogenated hydrocarbon (Beilstein test?), or just an ordinary hydrocarbon (see Fig. 12.6, 3-methylpentane, and Fig. 12.7, *tert*-butylbenzene).

Many comments can be added to this bare outline. For example, dilute solutions of alcohols will show a sharp peak at about 3600 cm⁻¹ for a nonhydrogen-bonded O—H in addition to the usual broad hydrogen-bonded O—H peak.

Aromatic hydrogens give peaks just to the left of 3000 cm⁻¹, while aliphatic hydrogens appear just to the right of 3000 cm⁻¹. However, NMR spectroscopy is the best method for identifying aromatic hydrogens.

FIG. 12.7   Infrared spectrum of *tert*-butylbenzene (thin film).

TABLE 12.2   Characteristic Infrared Carbonyl Stretching Peaks (Chloroform Solutions)

	Carbonyl-Containing Compounds	Wavenumber (cm^{-1})
O‖RCR	Aliphatic ketones	1725–1705
O‖RCCl	Acid chlorides	1815–1785
R—C=C—C(O)—R	α,β-Unsaturated ketones	1685–1666
O‖ArCR	Aryl ketones	1700–1680
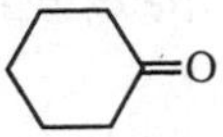 cyclohexanone =O	Cyclohexanones	1725–1705
—C(O)—CH$_2$—C(O)—	β-Diketones	1640–1540
O‖RCH	Aliphatic aldehydes	1740–1720
R—C=C—CH(O)	α,β-Unsaturated aldehydes	1705–1685
O‖ArCH	Aryl aldehydes	1715–1695
RCOOH	Aliphatic acids	1725–1700
R—C=C—COOH	α,β-Unsaturated acids	1700–1680
ArCOOH	Aryl acids	1700–1680
O‖RCOR′	Aliphatic esters	1740
R—C=C—COR′(O)	α,β-Unsaturated esters	1730–1715
O‖ArCOR	Aryl esters	1730–1715

TABLE 12.2 Characteristic Infrared Carbonyl Stretching Peaks
(Chloroform Solutions) *continued*

	Carbonyl-Containing Compounds	Wavenumber (cm^{-1})
$\overset{\overset{\displaystyle O}{\parallel}}{HCOR}$	Formate esters	1730–1715
$\underset{C_6H_5O\overset{\overset{\displaystyle O}{\parallel}}{C}CH_3}{CH_2=CHO\overset{\overset{\displaystyle O}{\parallel}}{C}CH_3}\Bigg\}$	Vinyl and phenyl acetate	1776
$\begin{array}{c} R-\overset{\overset{\displaystyle O}{\parallel}}{C}\diagdown \\ \qquad\quad O \\ R-\underset{\underset{\displaystyle O}{\parallel}}{C}\diagup \end{array}$	Acyclic anhydrides (two peaks)	$\left\{\begin{array}{l} 1840\text{–}1800 \\ \\ 1780\text{–}1740 \end{array}\right.$
$\overset{\overset{\displaystyle O}{\parallel}}{RCNH_2}$	Primary amides	1694–1650
$\overset{\overset{\displaystyle O}{\parallel}}{RCNHR'}$	Secondary amides	1700–1670
$\overset{\overset{\displaystyle O}{\parallel}}{RCNR'_2}$	Tertiary amides	1670–1630

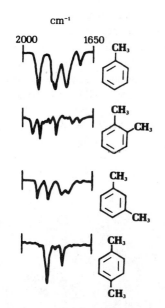

cm^{-1}

2000   1650

CH₃

CH₃
CH₃

CH₃

CH₃

CH₃

CH₃

FIG. 12.8 Band patterns of toluene and *o*-, *m*-, and *p*-xylene. These peaks are *very weak*. They are characteristic of aromatic substitution patterns in general, not just for these four molecules. Try to find these patterns in other spectra throughout this text.

The carbonyl frequencies listed earlier refer to the open chain or unstrained functional group in a nonconjugated system. If the carbonyl group is conjugated with a double bond or an aromatic ring, the peak will be displaced to the right by 30 cm^{-1} (see the margin notes). When the carbonyl group is in a ring smaller than six members or if there is oxygen substitution on the carbon adjacent to an aldehyde or ketone carbonyl, the peak will be moved to the left (see the margin notes and Table 12.2).

Methyl groups often give a peak near 1375 cm^{-1}, but NMR is a better method for detecting this group.

The pattern of substitution on an aromatic ring (mono-, *ortho*-, *meta*-, and *para*-, di-, tri-, tetra-, and penta-) can be determined from C—H out-of-plane bending vibrations in the region 670–900 cm^{-1}. Much weaker peaks between 1650 and 2000 cm^{-1} are illustrated in Fig. 12.8.

Extensive correlation tables and discussions of characteristic group frequencies can be found in the specialized references listed at the end of this chapter.

The Fourier Transform Infrared (FTIR) Spectrometer

The double-beam infrared spectrometer is rapidly being supplanted by Fourier transform instruments. Like their NMR counterparts, they are computer based, can sum a number of scans to increase the signal-to-noise ratio, and being digitally oriented, allow enormous flexibility in the ways spectra can be analyzed and displayed. In addition, they are faster and more accurate than double-beam spectrometers.

A schematic diagram of a single-beam FTIR spectrometer is given in Fig. 12.9. Infrared radiation goes to a Michelson interferometer in which half the light beam passes through a partially coated mirror to a fixed mirror and half goes to a moving mirror. The combined beams pass through the sample and are then focused on the detector. The motion of the mirror gives a signal that varies sinusoidally. Depending on the mirror position, the different frequencies of light either reinforce or cancel each other, resulting in an interferogram. The mirror, the only moving part of the spectrometer, thus gives rise to a signal that encodes the very high optical frequencies as a low-frequency signal that changes as the mirror moves back and forth. The Fourier transform converts the digitized signal into a signal that is a function of frequency, that is, a spectrum.

Like Fourier transform NMR spectroscopy, all infrared frequencies are sampled at once instead of being scanned successively, so a spectrum is obtained in seconds. Because there is no slit to sort out the wavelengths, high resolution is possible without losing signal strength. Extremely small samples or very dilute solutions can be examined because it is easy to sum hundreds of scans in the computer. It is even possible to take a spectrum of an object seen under a microscope. The spectrum of one compound can be subtracted from, say, a binary mixture spectrum to reveal the nature of the other component. Data can be smoothed, added, resized, and so on because of the digital nature of the output from the spectrometer.

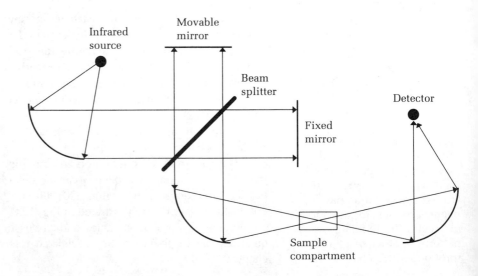

FIG. 12.9 Optical path diagram of an FTIR spectrometer.

Experimental Aspects

*Sample, solvents, and
equipment **must be dry.***

Infrared spectra can be determined on neat (undiluted) liquids, on solutions with
an appropriate solvent, and on solids as mulls and KBr pellets. Glass is opaque to
infrared radiation; therefore, the sample and reference cells used in infrared spectroscopy are sodium chloride plates. The sodium chloride plates are fragile and
can be attacked by moisture. *Handle only by the edges.*

Spectra of Neat Liquids

To run a spectrum of a neat (free of water!) liquid, remove a demountable cell
(Fig. 12.10) from the desiccator, place a drop of the liquid between the salt plates,
press the plates together to remove any air bubbles, and add the top rubber gas-
ket and metal top plate. Next, put on all four of the nuts and *gently* tighten them
to apply an even pressure to the top plate. Place the cell in the sample compart-
ment (nearest the front of the spectrometer), and run the spectrum.

  Although running a spectrum on a neat liquid is convenient and results in no
extraneous bands to interpret, it is not possible to control the path length of the
light through the liquid in a demountable cell. A low-viscosity liquid when
squeezed between the salt plates may be so thin that the short path length gives
peaks that are too weak. A viscous liquid, on the other hand, may give peaks that
are too intense. A properly run spectrum will have the most intense peak with an
absorbance of about 1.0.

  Another demountable cell is pictured in Fig. 12.11. The plates are thin
wafers of silver chloride, which is transparent to infrared radiation. This cell has
advantages over the salt cell in that the silver chloride disks are more resistant to

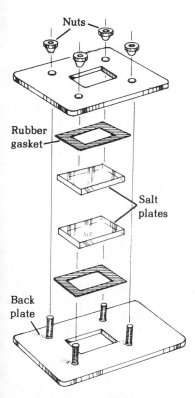

FIG. 12.10 Exploded view of a
demountable salt cell for
analyzing the infrared spectra of
neat liquids. Salt plates are
fragile and expensive. Do not
touch front surfaces. Use only
dry solvents and samples.

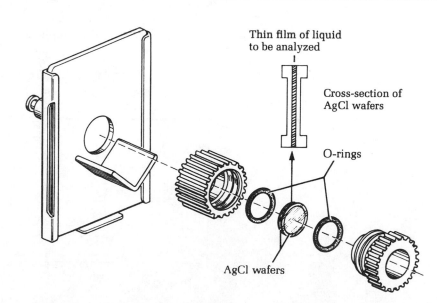

FIG. 12.11 Demountable silver chloride cell.

A fast, simple alternative: Place a drop of the compound on a round salt plate, add a top plate, and mount the two on the holder pictured in Fig. 12.11.

breakage than NaCl plates, less expensive, and not affected by water. Because silver chloride is photosensitive, the wafers must be stored in the dark to prevent them from turning black. Because one side of each wafer is recessed, the thickness of the sample can be varied according to the manner in which the cell is assembled. In general, spectra of pure liquids are run as the thinnest possible films. Most of the spectra in this chapter have been obtained in this way. The disks are cleaned by rinsing them with an organic solvent such as acetone or ethanol and wiping dry with an absorbent paper towel.

A very simple "cell" consists of two circular sodium chloride disks (1" dia. × $\frac{3}{16}$" thick, 25 mm × 4 mm). The sample, one drop of a pure liquid or solution, is applied to the center of a disk with a polyethylene pipette (to avoid scratching the sodium chloride). The other disk is added, the sample squeezed to a thin film, and the two disks placed on the V-shaped holder seen in Fig. 12.11.

Spectra of Solutions

Solvents: $CHCl_3$, CS_2, CCl_4

CAUTION: Chloroform, carbon tetrachloride, and carbon disulfide are toxic. The first two are carcinogens. Use laboratory hood. Avoid the use of these solvents if possible.

The most widely applicable method of running spectra of solutions involves dissolving an amount of the liquid or solid sample in an appropriate solvent to give a 10% solution. Just as in NMR spectroscopy, the best solvents to use are carbon disulfide and carbon tetrachloride, but because these compounds are not polar enough to dissolve many substances, chloroform is used as a compromise. Unlike NMR solvents, no solvent suitable in infrared spectroscopy is entirely free of absorption bands in the frequency range of interest (Figs. 12.12, 12.13, and 12.14). In chloroform, for instance, no light passes through the cell between 650 and 800 cm^{-1}. As can be seen from the figures, spectra obtained using carbon disulfide and chloroform cover the entire infrared frequency range. Carbon tetrachloride would appear to be a good choice because it has few interfering peaks, but it is a poor solvent for polar compounds. In practice, a base line is run with

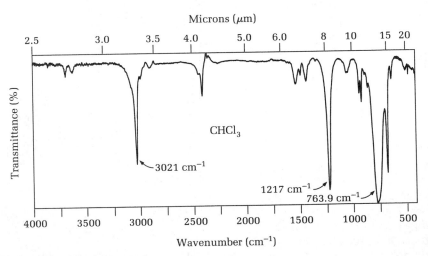

FIG. 12.12 Infrared spectrum of chloroform (thin film).

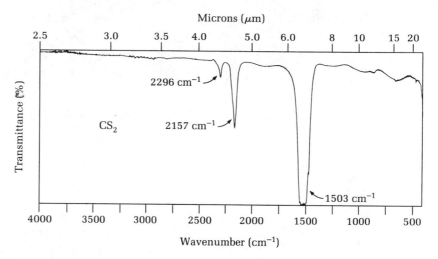

FIG. 12.13   Infrared spectrum of carbon disulfide (thin film).

the same solvent in both cells to ascertain if the cells are clean and matched. Often it is necessary to obtain only one spectrum employing one solvent, depending on which region of the spectrum you need to use.

Three drops are needed to fill the cell.

Three large drops of solution will fill the usual sealed infrared cell (Fig. 12.15). A 10% solution of a liquid sample can be approximated by dilution of one drop of the liquid sample with nine drops of the solvent. Since weights are more difficult to estimate, solid samples should be weighed to obtain a 10% solution.

Solvent and sample must be dry. Do not touch or breathe on NaCl plates. Cells are very expensive.

The infrared cell is filled by inclining it slightly and placing about three drops of the solution in the lower hypodermic port with a capillary dropper. The cell must be completely dry because the new sample will not enter if the cell contains solvent. The liquid can be seen rising between the salt plates through the window.

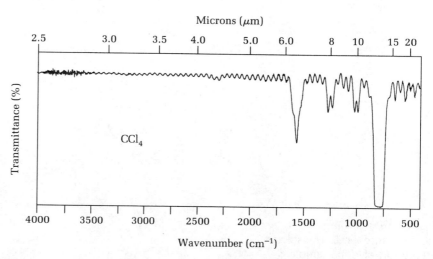

FIG. 12.14   Infrared spectrum of carbon tetrachloride (thin film).

FIG. 12.15   Sealed infrared sample cell.

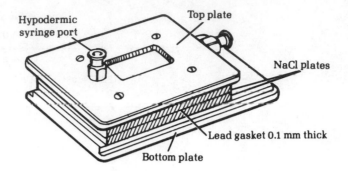

In the most common sealed cell, the salt plates are spaced 0.1 mm apart. Make sure that the cell is filled past the window and that no air bubbles are present. Then place the Teflon stopper lightly but firmly in the hypodermic port. Be particularly careful not to spill any of the sample on the outside of the cell windows.

Fill the reference cell from a clean hypodermic syringe in the same manner as the sample cell, and place both cells in the spectrometer, with the sample cell toward the front of the instrument. After running the spectrum, force clean solvent through the sample cell using a syringe attached to the top port of the cell (Fig. 12.16). Finally, with the syringe, pull the last bit of solvent from both cells or suck clean dry air through the cells to dry them, and store them in a desiccator.

Dispose of waste solvents in the container provided.

Cleaning Up   Discard halogenated liquids in the halogenated organic waste container. Other solutions should be placed in the organic solvents container.

Mulls

Solids insoluble in the usual solvents can be run as mulls. In preparing a mull, the sample is ground to a particle size less than that of the wavelength of light going

FIG. 12.16   Flushing the infrared sample cell. The solvent used to dissolve the sample is used in this process.

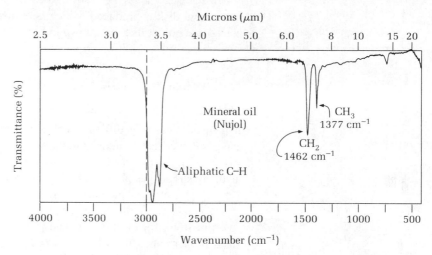

FIG. 12.17   Infrared spectrum of Nujol (paraffin oil).

Sample must be finely ground.

A fast, simple alternative: Grind a few milligrams of the solid with a drop of tetrachloro-ethylene between two round salt plates. Mount on the holder shown in Fig. 12.11.

through the sample (2.5 μm) in order to avoid scattering the light. About 15 to 20 mg of the sample is ground for 3 to 10 min in an agate mortar until it is spread over the entire inner surface of the mortar and has a caked and glassy appearance. Then, to make a mull, 1 or 2 drops of paraffin oil (Nujol) (Fig. 12.17) are added, and the sample ground 2 to 5 more minutes. The mull, which should have the consistency of thin margarine, is transferred to the bottom salt plate of a demountable cell (Fig. 12.10) using a rubber policeman, the top plate is added and twisted to distribute the sample evenly and to eliminate all air pockets, and the spectrum is run. Since the bands from Nujol obscure certain frequency regions, running another mull using Fluorolube as the mulling agent will allow the entire infrared spectral region to be covered. If the sample has not been ground sufficiently fine, there will be marked loss of transmittance at the short-wavelength end of the spectrum. After running the spectrum, the salt plates are wiped clean with a cloth saturated with an appropriate solvent.

Potassium Bromide Disk

The spectrum of a solid sample also can be run by incorporating the sample in a potassium bromide (KBr) disk. This procedure needs only one disk to cover the entire spectral range, since KBr is completely transparent to infrared radiation. Although very little sample is required, making the disk calls for special equipment and time to prepare it. Since KBr is hygroscopic, water is a problem.

Into a stainless steel capsule containing a ball bearing are weighed 1.5 to 2 mg of the compound and 200 mg of spectroscopic-grade potassium bromide (previously dried and stored in a desiccator). The capsule is shaken for 2 min on a Wig-L-Bug (the device used by dentists to mix silver amalgam). The sample is evenly distributed over the face of a 13-mm die and subjected to 14,000 to 16,000 lb/in.[2] pressure for 3 to 6 min while under vacuum in a hydraulic press. A transparent disk is produced, which is removed from the die with tweezers and placed in a holder like that shown in Fig. 12.11 prior to running the spectrum.

Gas Phase Infrared Spectroscopy: The Williamson Gas Phase IR Cell®

In the past it has been difficult to obtain the infrared spectra of gases. A commercial gas cell costs hundreds of dollars and the cell must be evacuated and the dry gas carefully introduced—not a routine process. But we have found that a uniquely simple gas phase IR cell can be assembled at virtually no cost from the 105° adapter in your microscale kit.

To make the Williamson gas phase IR cell, attach a thin piece of polyethylene film from, for example, a sandwich bag to each end of the adapter with a small rubber band, fine wire, or thread. Place a loose wad of cotton in the side arm, and mount the IR cell in the beam of the spectrometer. (Figs. 12.18 and 12.19).

The adapter can be supported on a V-block as shown in Fig. 12.19 or, temporarily, on a block of modeling clay. Be sure the beam, usually indicated by a red laser, strikes both ends of the cell. Run a background spectrum on the cell. The spectrum of polyethylene (Fig. 12.20) is very simple, so interfering bands will be few. This spectrum is automatically subtracted when you put a sample in the cell and run a spectrum.

A more permanent cell can be constructed by attaching two thin plates of silver chloride to ends of the adapter with epoxy glue. These silver chloride plates can be made in the press used for making potassium bromide discs or they can be purchased (Fisher 14-385-860). See the *Instructor's Guide*.

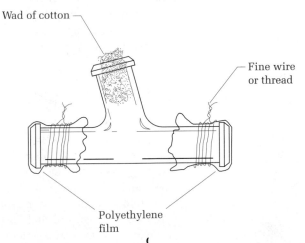

Wad of cotton

Fine wire or thread

Polyethylene film

FIG. 12.18 The Williamson Gas Phase IR Cell® utilizing the 105° adapter from the microscale kit.

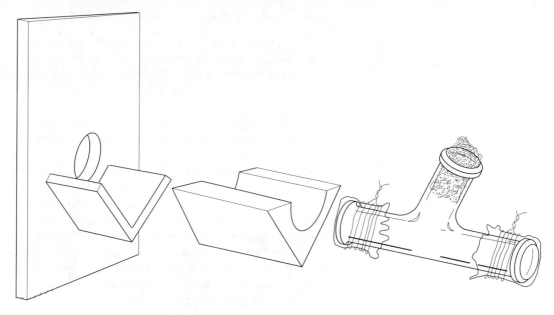

FIG. 12.19   The Williamson Gas Phase IR Cell® mounted in a cell holder.

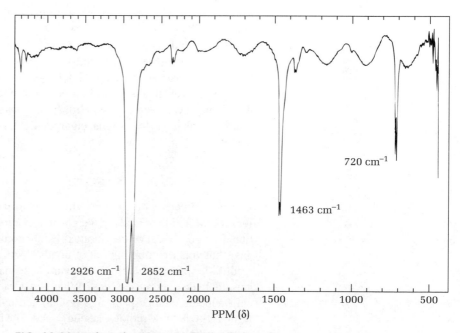

2926 cm⁻¹   2852 cm⁻¹

1463 cm⁻¹

720 cm⁻¹

PPM (δ)

FIG. 12.20   Infrared spectrum of polyethylene film.

Obtaining a Spectrum

To run a gas phase spectrum, remove the cell from the spectrometer and spray onto the cotton a very short burst of a gas from an aerosol can. The nonvolatile components of the spray will stick on the cotton, and the propellant gas will diffuse into the cell. Run the spectrum in the usual way. The concentration of the gas is about the same as a thin film of liquid run between salt plates. If the concentration of the gas is too high, remove the cell from the spectrometer, remove the cotton, and "pour out" half the invisible gas by holding the cell in a vertical position.

You may be fortunate in having in your institution's library a copy of *The Aldrich Library of FT-IR Spectra: Vapor Phase,* Volume 3 ($500). Most libraries do not have it. This 2000-page book has more than 6500 gas phase spectra that will help you analyze the IR spectra of propellants in aerosol cans and any other gases you may encounter. Otherwise you may need to compare your spectra with the few in this text and in the *Instructor's Guide* and begin to assemble your own library of gas phase spectra. It is possible to make approximate calculations of IR spectra (vibrational spectra) using computational chemistry at the AM1 level, for example with the program Spartan.

For Further Investigation

The Gas Phase IR Spectra of Aerosol Propellants

Acquire gas phase spectra from a number of aerosol cans. Some cans will be labeled with the name of the propellant gas, others will not. Analyze the spectra, bearing in mind that the gas will be a small molecule such as carbon dioxide (Fig. 12.21), methane (Fig. 12.22), nitrous oxide (Fig. 12.23), propane, butane, isobutane, or a fluorocarbon such as 1,1,1,2-tetrafluoroethane. A very old aerosol can may contain an illegal chlorofluorocarbon, now outlawed because they destroy the ozone layer. Some possibilities: hair sprays, lubricants (WD-40, Teflon), Freeze-It, inhalants for asthma, cigarette lighters, butane torches, natural gas from the lab.

Other Gas Phase Spectral Research

You can deliberately generate hydrogen chloride (or deuterium chloride) and sulfur dioxide and then analyze the gas that is evolved from a thionyl chloride reaction. Can you detect carbon monoxide in automobile exhaust? How about "marsh gas" that you see bubbling out of ponds? With perhaps dozens of scans, can you record the spectra of vanillin over vanilla beans, of eugenol vapor over cloves, of anisaldehyde over anise seeds? What is the nature of the volatile components of chili peppers, orange peel, nail polish, and nail polish remover? Send me your results (kwilliam@mtholyoke.edu), and I will post them on my Web page (www.mtholyoke.edu/courses/kwilliam/microscale.shtml).

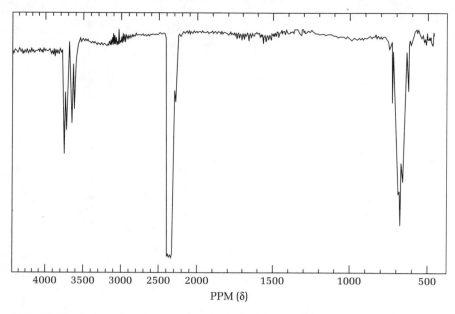

FIG. 12.21 Infrared spectrum of carbon dioxide, gas phase.

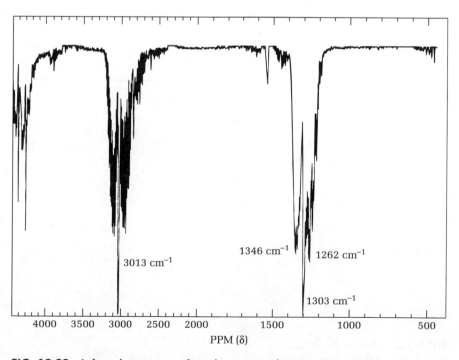

FIG. 12.22 Infrared spectrum of methane, gas phase.

FIG. 12.23 Infrared spectrum of nitrous oxide, gas phase.

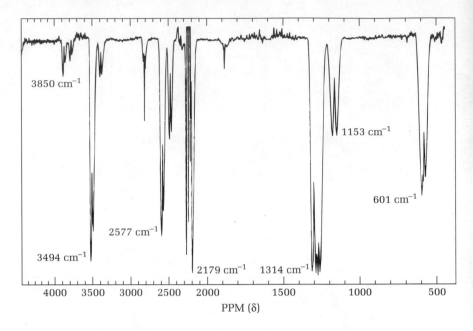

Running the Spectrum

Satisfactory spectra are easily obtained with the lower-cost analog spectrometers, even those that have only a few controls and require few adjustments. To run a spectrum on these instruments, the paper must be positioned accurately, the pen

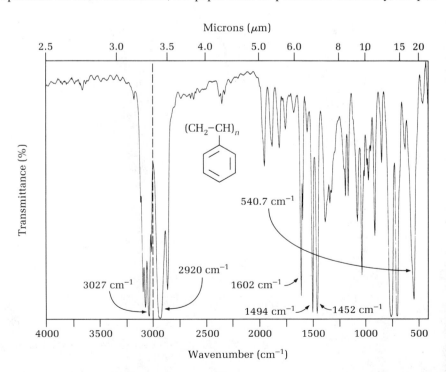

FIG. 12.24 Infrared spectrum of polystyrene film calibration standard.

Calibration with polystyrene

set between 90 and 100% transmittance with the 100% control (0.0 and 0.05 absorbance), and the speed control set for a fast scan (usually one of about 3 min). The calibration of a given spectrum can be checked by backing up the pen and superimposing a spectrum of a thin polystyrene film. This film, mounted in a cardboard holder, will be found near most spectrometers. The film is held in the sample beam, and parts of the spectrum to be calibrated are rerun. See Fig. 12.24 for the spectrum of polystyrene. The spectrometer gain (amplification) should be checked frequently and adjusted when necessary. To check the gain, put the pen on the 90% transmittance line with the 100% control. Place your finger in the sample beam so that the pen goes down to 70% *T*. Then quickly remove your finger. The pen should overshoot the 90% *T* line by 2%.

Fourier transform spectrometers are faster, more sensitive, and can print out peak locations to at least four significant figures. They do not usually require routine calibration or gain adjustments, and the size of the most intense peak can be adjusted. A background spectrum (the empty cell, of whatever nature, or the cell plus a pure solvent) is run first and automatically stored in digital form in the spectrometer. The sample is introduced (a thin film, a solution, a mull, or a gas) and the background spectrum automatically subtracted, leaving only the spectrum of the sample.

Throughout the remainder of this book, representative infrared spectra of starting materials and products are presented and the important bands in each spectrum identified.

Experiment

Unknown Carbonyl Compound

Sample, solvents, and apparatus must be dry.

Run the infrared spectrum of an unknown carbonyl compound obtained from the laboratory instructor. Be particularly careful that all apparatus and solvents are completely free of water, which will damage the sodium chloride cell plates. The spectrum can be calibrated by positioning the spectrometer pen at a wavelength of about 1610 cm^{-1} without disturbing the paper and rerunning the spectrum in the region from 1610 to 1575 cm^{-1} while holding the polystyrene calibration film in the sample beam (see Fig. 12.24). This will superimpose a sharp calibration peak at 1601 cm^{-1} and a less intense peak at 1583 cm^{-1} on the spectrum. Fourier transform spectrometers do not need frequent calibrating. Determine the frequency of the carbonyl peak in the unknown, and list the possible types of compounds that could correspond to this frequency (Table 12.2).

Surfing the Web

http://www.cc.columbia.edu/cu/chemistry/irtutor/Edison.html

An excellent demo of an infrared spectroscopy tutorial is available from this Columbia University site. The infrared spectrum of 1-hexene can, for example, be expanded and each of the 11 major peaks can be selected, giving animations of the molecular motions responsible for each peak.

For updated information visit:

www.mtholyoke.edu.courses.kwilliam.microscale.shtml

or

www.hmco.com/hmco/college/chemistry/Home.html

References

1. Noel P. G. Roeges, *A Guide to the Complete Interpretation of Infrared Spectra of Organic Structures,* John Wiley, New York, 1994.

2. Norman B. Colthup, Lawrence H. Daly, and Stephen E. Wiberley, *Introduction to Infrared and Raman Spectroscopy,* 3rd ed., Academic Press, Boston, 1990.

3. *The Handbook of Infrared and Raman Characteristic Frequencies of Organic Molecules,* Academic Press, Boston, 1991.

4. *Infrared and Raman Spectroscopy: Methods and Applications,* VCH, New York, 1995.

5. Charles J. Pouchert, *The Aldrich Library of FT-IR Spectra,* 1st ed., Aldrich Chemical Co., Milwaukee, WI, 1985.

6. J. W. Cooper, *Spectroscopic Techniques for Organic Chemists,* John Wiley, New York, 1980.

13

Nuclear Magnetic Resonance Spectroscopy

Prelab Exercise: Outline the preliminary solubility experiments you would carry out on an unknown using inexpensive solvents, before preparing a solution of a compound for nuclear magnetic resonance spectroscopy.

Proton NMR Spectroscopy

NMR: Determination of the number, kind, and relative locations of hydrogen atoms (protons) in a molecule

Nuclear magnetic resonance (NMR) spectroscopy is a means of determining the number, kind, and relative locations of certain atoms, principally hydrogen, in molecules. The number of peaks or groups of coupled peaks indicate the number of chemically and magnetically distinct hydrogen atoms in a molecule. The *integral* indicates the relative number of protons within a peak or group of peaks. The separation between the lines within a group of peaks, called the *coupling constant, J,* can give information about the locations of the protons relative to other nearby protons and, thus, the geometry of the molecule. And the location of a peak in the spectrum, called the *chemical shift, δ,* indicates what kind of proton gave rise to the peak, be it a methyl group, an alkene, or an aromatic ring.

Chemical shift, δ

Integration: The area under a peak or group of peaks

In the process of analyzing a sample, it is possible to expand portions of the spectrum so that patterns of peaks become clear, to *integrate* all peaks, which is the process of determining the relative areas of the peaks numerically or graphically and printing out the frequency of every peak. The integral indicates how many of each type of proton are present. If one or more sets of peaks are well resolved, then the coupling pattern can be analyzed to determine how many protons are adjacent to the proton or protons giving rise to that peak. And measurement of the magnitude of the coupling constant can give information about the geometry of the molecule.

Spectral Analysis: Proton Chemical Shifts

The first place to turn in analyzing a spectrum is a table of chemical shifts. Table 13.1 in this text and in similar general organic chemistry texts is just barely adequate for the task. In books devoted to NMR spectroscopy, such as the one by Friebolin (see the references at the end of this chapter), dozens of detailed tables are supplied for all types of functional groups. Many of the tables contain additivity parameters that allow the approximate calculation of chemical shifts for, say, the protons in a 1,4-disubstituted aromatic ring containing a bromine atom and a methyl group.

TABLE 13.1 Proton Chemical Shifts

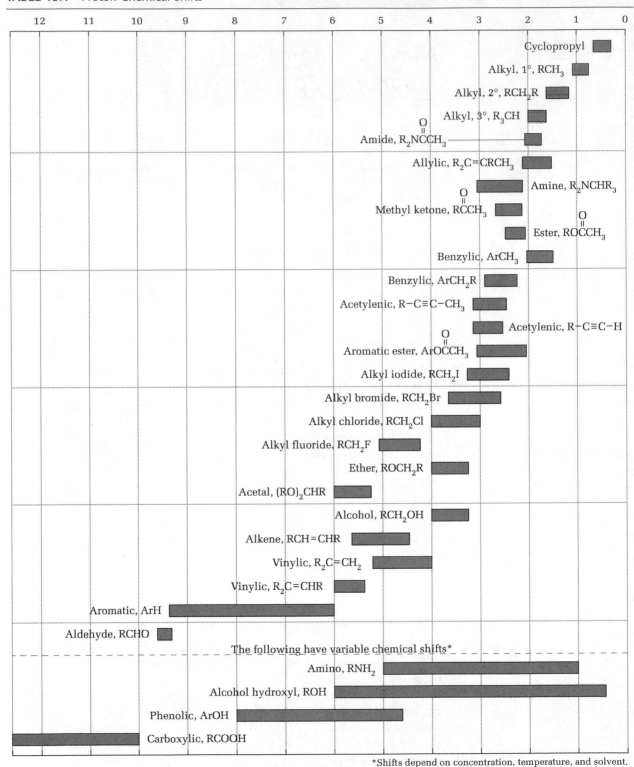

*Shifts depend on concentration, temperature, and solvent.

FIG. 13.1 Proton NMR spectrum of ethyl iodide. The staircaselike line is the integral. In the integral mode of operation, the recorder pen moves from left to right and moves vertically a distance proportional to the areas of the peaks over which it passes. Hence the relative area of the quartet of peaks at 3.20 ppm to the triplet of peaks at 1.83 ppm is given by the relative heights of the integral (4 cm is to 6 cm as 2 is to 3). The relative numbers of hydrogen atoms are proportional to the peak areas (2H and 3H) (60-MHz spectrum).

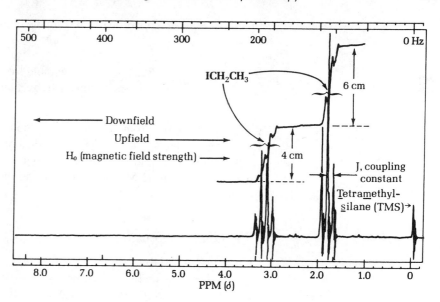

A spectrum of ethyl iodide, typical of those obtained in an older low-field spectrometer operating at 60 MHz, is seen in Fig. 13.1. Figure 13.2 shows a spectrum for the same compound obtained at 250 MHz. The information contained in these two spectra is identical. The chemical shifts reported in parts per million (δ) are identical, and expansion of the pattern of lines in Fig. 13.2 shows that the coupling is exactly the same.

Every proton and carbon spectrum contains tetramethylsilane (TMS), which serves as the zero reference point. Chemical shifts are measured in dimensionless

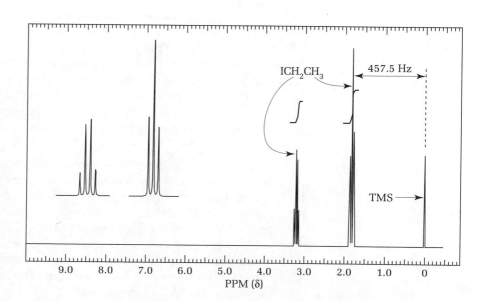

FIG. 13.2 Proton NMR spectrum of ethyl iodide (250 MHz). Compare this spectrum with Fig. 13.1, run at 60 MHz.

parts per million (ppm) downfield from the reference TMS according to the equation

$$\delta(\text{ppm}) = \frac{\text{Shift of peak downfield from TMS (measured in Hz)}}{\text{Spectrometer frequency (measured in MHz)}}$$

An NMR spectrum is characterized completely by the chemical shifts and coupling constants of the nuclei; this is the only information reported in research reports. Using just this information, one can calculate NMR spectra using programs such as gNMR (see the "Surfing the Web" section at the end of this chapter).

The Coupling Constant, J

Coupling Constants. In an open-chain molecule with free rotation about the bonds and no chiral centers, protons couple with each other over three chemical bonds to give characteristic patterns of lines. If one or more equivalent protons couple to one adjacent proton, then the coupling hydrogen(s) appears as a doublet of equal intensity lines separated by the coupling constant, *J*, measured in hertz. If the coupling proton or protons couple equally to two protons three chemical bonds away, they appear as a triplet of lines with relative intensities of 1 : 2 : 1, again separated by the coupling constant, *J*. In general, chemically equivalent protons give a pattern of lines containing one more line than the number of protons being coupled to, and the intensities of the peaks follow the binomial expansion, conveniently represented by Pascal's triangle. Thus an ethyl group, as in ethyl iodide (Figs. 13.1 and 13.2), appears as a quartet of lines at 3.20 ppm with relative intensities of 1 : 3 : 3 : 1 because the two methylene protons have coupled to the three equivalent methyl protons. The methyl peak is split into a 1 : 2 : 1 triplet centered at 1.83 ppm because the methyl protons have coupled to the two adjacent methylene protons.

Other characteristic proton coupling constants are given in Table 13.2. In alkenes *trans* coupling is larger than *cis* coupling, and both are much larger than geminal coupling. *Ortho, meta,* and *para* couplings in aromatic rings range from 9 to 0 Hz. The couplings in a rigid system of saturated bonds are strongly dependent on the dihedral angle between the coupling protons as seen in cyclohexane.

Pascal's triangle

			1				*singlet, s*
		1		1			*doublet, d*
	1		2		1		*triplet, t*
1		3		3		1	*quartet, q*
1	4		6		4	1	*pentet*
1	5	10		10	5	1	*sextet*

The Integral. The relative numbers of hydrogen atoms (protons) in the molecule of ethyl iodide are determined from the *integral,* the stair-step line over the peaks. The height of the step is proportional to the area under the NMR peak or group of peaks. In NMR spectroscopy (contrasted with infrared spectroscopy, for instance) the area of each group of peaks is directly proportional to the number of hydrogen atoms causing the peaks. Integrators are part of all NMR spectrometers, and running the integral takes little more time than running the spectrum. Most spectrometers print out a numerical value for the integral (with many more significant figures than are justified!).

TABLE 13.2 Spin–Spin Coupling Constants for Various Geometries

Fragment	J (Hz)	Fragment	J (Hz)
cis H–C=C–H	7–12	geminal =C(H)(H)	12–15
trans H–C=C–H	13–18	vicinal –C(H)–C(H)–	0–10
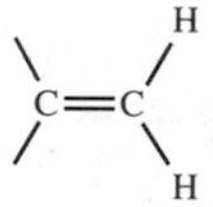 geminal C=C(H)(H)	0.5–3	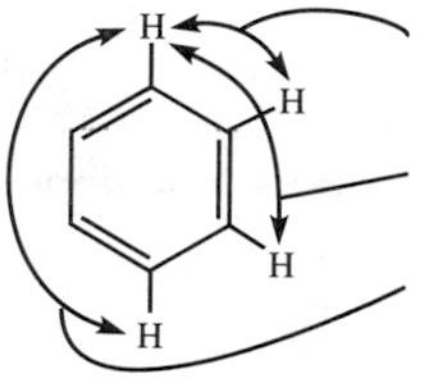 aromatic (ortho)	6–9
		aromatic (meta)	1–3
		aromatic (para)	0–1
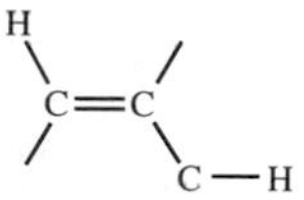 allylic C=C–C–H	0.5–2.5	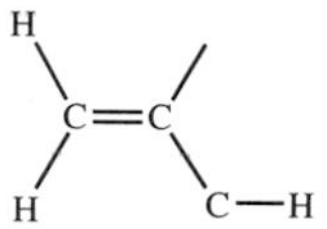 allylic (trans) C=C–C–H	0
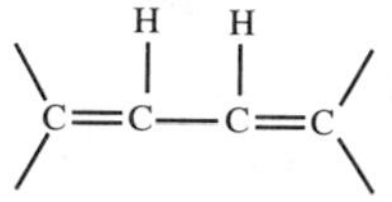 diene C=C–C=C	9–13	CH_3—CH_2—	6.5–7.5
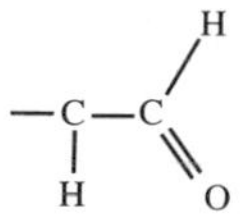 –C(H)–C(=O)H	1–3	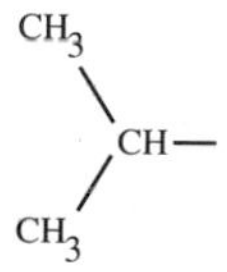 $(CH_3)_2$CH—	5.5–7
allylic C=C–C–H	4–10	cyclohexane (axial)	5–9
		cyclohexane (equatorial)	2–4

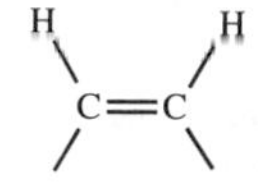

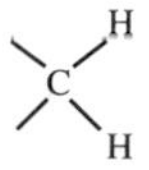

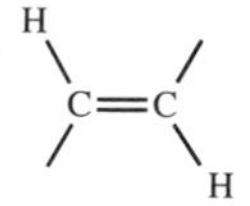

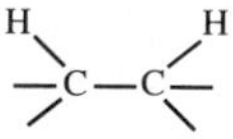

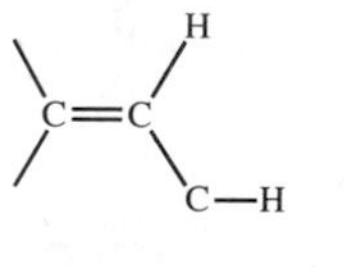

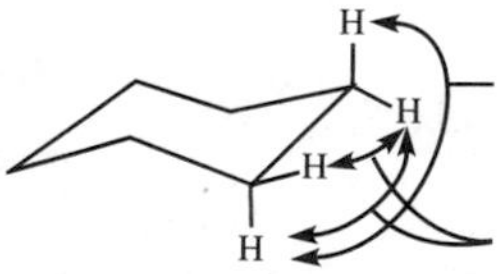

A more complex spectrum

Some NMR spectra are not as easily analyzed as the spectrum for ethyl iodide. Consider the spectra shown in Fig. 13.3. The proton spectrum of this unsaturated chloroester has been run at 500 MHz. Each chemically and magnetically nonequivalent proton is well resolved so that all of the couplings can be seen. Only the quartet of peaks at 4.1 ppm (area 2) and the triplet at 1.2 ppm (area 3) follow the simple first-order coupling rules outlined above. This quartet/triplet pattern is very characteristic of the commonly encountered ethyl group.

Because of the chiral carbon (marked with an asterisk), the protons on carbons D and E are diastereotopic, have different chemical shifts, and couple with each other and with adjacent protons to give the patterns seen in the spectrum.

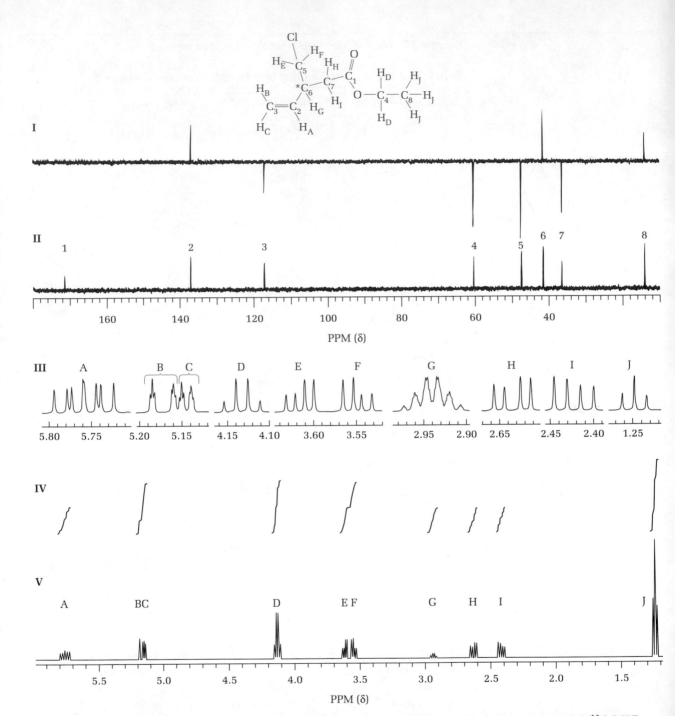

FIG. 13.3 The 500-MHz ¹H and 75-MHz ¹³C NMR spectra of ethyl (3-chloromethyl)-4-pentenoate.* *I. ¹³C DEPT spectrum.* The CH₃ and CH peaks are upright and the CH₂ peaks inverted. The quaternary carbon (the carbonyl carbon) does not appear. *II. The normal 75-MHz noise-decoupled ¹³C spectrum.* Note the small size of the carbonyl peak. *III. Expansions of each group of proton NMR peaks.* Protons E and F as well as protons H and I are not equivalent to each other. These pairs of protons are diastereotopic because they are on a carbon adjacent to a chiral carbon atom. The frequencies of all peaks are found in the *Instructor's Guide*. *IV. The integral.* The height of the integral is proportional to the number of protons under it. *V. The 500-MHz ¹H spectrum.*

*Spectra courtesy of Professor Scott Virgil.

Many of the peaks can be assigned to specific hydrogens based simply on their chemical shifts. The coupling patterns then confirm these assignments.

Carbon-13 Spectroscopy

The element carbon consists of 98.9% carbon with mass 12 and spin 0 (NMR inactive) and only 1.1% ^{13}C with spin 1/2 (NMR active). Carbon, with such a low concentration of spin 1/2 nuclei, gives a very small signal when run under the same conditions as a proton spectrum. Carbon resonates at 75 MHz in a spectrometer where protons resonate at 300 MHz. Because only 1 in 100 carbon atoms has mass 13, the chances of a molecule having two ^{13}C atoms adjacent to one another are small. Consequently, coupling of one carbon with another is not observed.

^{13}C spectra: Broadband noise decoupling gives a single line for each carbon.

Each ^{13}C atom couples to hydrogen atoms over one, two, and three bonds. The coupling constants are large, so there is much overlap of peaks. To simplify the spectra as well as to increase the signal-to-noise ratio, a special technique is routinely used in obtaining ^{13}C spectra: *broadband noise decoupling*. Decoupling has the effect of collapsing all multiplets (quartets, triplets, etc.) into a single peak. Furthermore, the energy put into decoupling the protons can be looked on as appearing in the carbon spectrum in the form of an enhanced peak. This nuclear Overhauser enhancement (NOE) effect makes the peak appear three times larger than it would otherwise be. The result of decoupling is that every chemically and magnetically distinct carbon atom will appear as a single sharp line in the spectrum. Because the NOE effect is somewhat variable and does not affect carbons bearing no protons, one cannot obtain reliable information about peak areas by integration of the carbon spectrum.

Carbon Chemical Shifts

The range of carbon chemical shifts is 200 ppm compared with the 10-ppm range for protons. It is not common to have accidental overlap of carbon peaks. In the unsaturated chloroester seen in Fig. 13.3 there are eight sharp peaks corresponding to the eight carbon atoms in the molecule.

The generalities governing carbon chemical shifts are very similar to those governing proton shifts, as seen by comparing Table 13.1 with Table 13.3. Most of the downfield peaks are due to those carbon atoms near electron-withdrawing groups. In Fig. 13.3 the most downfield peak is that from the carbonyl carbon of the ester. The attached electronegative oxygen makes the peak appear at about 172 ppm. The peak is smaller than any other in the spectrum because it does not have an attached proton and thus does not benefit from a full NOE effect.

The ^{13}C spectrum of sucrose seen in Fig. 13.4 displays a single line for each carbon atom, and in Fig. 20.3 a single line is seen for each of the 27 carbon atoms in cholesterol.

TABLE 13.3 Carbon Chemical Shifts

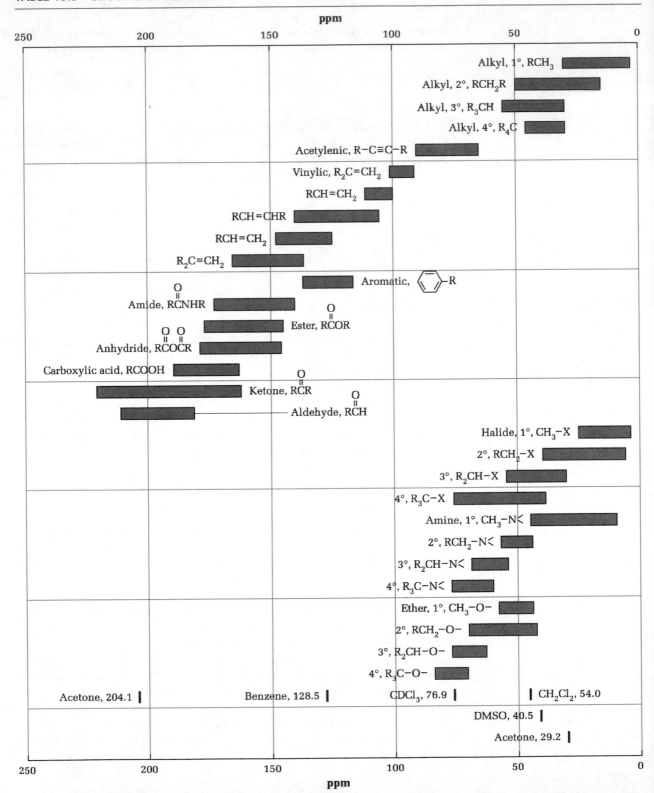

FIG. 13.4 ^{13}C NMR spectrum of sucrose (22.6 MHz). Not all lines have been assigned to individual carbon atoms.

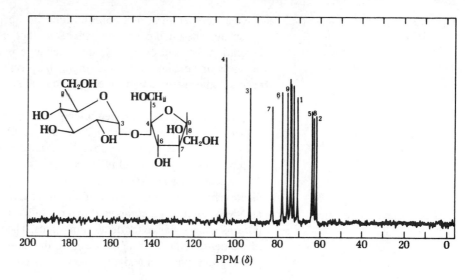

Fourier Transform Spectroscopy

The Fourier transform spectrometer

Even with broadband decoupling, the signal from ^{13}C spectra can get lost in the random noise produced by the spectrometer. Similarly, dilute solutions of samples give very noisy proton spectra. To increase the signal-to-noise ratio, the data for a number of spectra are averaged in the spectrometer computer. Because noise is random and the signal coherent, the signal will increase in size and the noise decrease as many spectra are averaged together. The improvement in the signal-to-noise ratio is proportional to the square root of the number of scans.

Spectra are accumulated rapidly by applying a very short pulse of radio-frequency energy to the sample and then storing the resulting free induction decay (FID) signal in digital form in the computer. The FID signal, which takes seconds to acquire, contains frequency information about all the signals in the spectrum. In a few minutes several hundred FID signals can be obtained and averaged in the computer. The FID signal is converted to a spectrum of conventional appearance by carrying out a Fourier transform computation on the signal using the spectrometer's computer.

Because the spectral information is in digital form, a large number of operations can be performed on it. The spectrum can be smoothed, at the expense of resolution, or the resolution can be enhanced, at the expense of noise. As indicated earlier, the spectrum can be numerically integrated and the position of each peak, in hertz or parts per million, printed above the peak. The spectra can be expanded both vertically and horizontally to clarify complex couplings, and all the parameters used to acquire the spectrum can be printed as well.

Protons with Variable Chemical Shifts

Protons attached to oxygen and nitrogen form hydrogen bonds to each other and with protonic solvents. As a result the chemical shifts of these protons depend on

Labile Protons	δ *(ppm)*
ROH	*0.5–6*
ArOH	*4.5–8*
RCOOH	*10–12*
Enols	*10–17*
RNII$_2$	*1–5*
Amides	*5–6.5*

Deuterium exchange

Chiral shift reagents

solvent, temperature, and concentration and can appear almost anywhere in a spectrum. They will exchange with each other, a process catalyzed by small concentrations of acid or base. If this exchange is rapid (a common occurrence in alcohols, for example), only a single sharp line for the hydroxyl proton will be seen. At intermediate rates of exchange the line becomes broad, and at slow rates of exchange it couples to other protons within three chemical bonds.

Protons bound to nitrogen often give a broad line because of the nonuniform distribution of charge on the nucleus of ^{14}N. Obviously NMR spectroscopy is a poor means for detecting hydroxylic, amine, and amide protons; infrared spectroscopy on the other hand is an excellent means.

Because labile protons can undergo rapid exchange with each other they can also exchange with deuterium. If a drop of D_2O is added to a $CDCl_3$ solution of an NMR sample of a compound that contains such protons and gently mixed, the protons will exchange for deuterium atoms. The water will, of course, float on top of the denser chloroform. The peak for the labile proton will disappear, thus simplifying the spectrum and allowing for assignment of the peak.

Shift Reagents

Some spectra can be simplified by addition of shift reagents to the sample. The resulting spectrum is similar to that one might obtain in a high-field spectrometer.

Addition of a few milligrams of a hexacoordinate complex of europium {tris(dipivaloylmethanato)europium(III), [Eu(dpm)$_3$]} to an NMR sample that contains a Lewis base center (an amine or basic oxygen, such as a hydroxyl group) has a dramatic effect on the spectrum. This lanthanide (soluble in $CDCl_3$) causes large shifts in the positions of peaks arising from the protons near the metal atom in this molecule and is therefore referred to as a *shift reagent*. It produces the shifts by complexing with the unshared electrons of the hydroxyl oxygen, the amine nitrogen, or other Lewis base center.

With no shift reagent present, the NMR spectrum of 2-methyl-3-pentanol [Fig. 13.5(a)] is not readily analyzed. Addition of 10-mg portions of the shift reagent to the sample causes very large downfield shifts of peaks owing to protons near the coordination site [Figs. 13.5(b)–(g)].[1] The two protons on C-4 and the two methyls on C-2 are diastereotopic and thus magnetically nonequivalent because they are adjacent to a chiral center C-3; each gives a separate set of peaks. When enough shift reagent is added, this spectrum can be analyzed by inspection. (See Fig. 13.6.)

Quantitative information about molecular geometry can be obtained from shifted spectra. The shift induced by the shift reagent is related to the distance and the angle that the proton bears to the europium atom.

Chiral shift reagents (derivatives of camphor) will cause differential shifts of the protons or carbon atoms in enantiomers. The separated peaks can be

1. K. L. Williamson, D. R. Clutter, R. Emch, M. Alexander, A. E. Burroughs, C. Chua, and M. E. Bogel, *J. Am. Chem. Soc.*, **96,** 1471 (1974).

FIG. 13.5 The 60-MHz ^{1}H NMR spectrum of 2-methyl-3-pentanol (0.4 M in CS_2) with various amounts of shift reagent present. (A) No shift reagent present. All methyl peaks are superimposed. The peak for the proton adjacent to the hydroxyl group is downfield from the others because it is adjacent to the electronegative oxygen atom. (B) 2-Methyl-3-pentanol + Eu(dpm)$_3$. Mole ratio of Eu(dpm)$_3$ to alcohol = 0.05. The hydroxyl proton peak at 1.6 ppm in spectrum A appears at 6.2 ppm in spectrum B because it is closest to the Eu in the complex formed between Eu(dpm)$_3$ and the alcohol. The next closest proton, the one on the hydroxyl-bearing carbon atom, gives a peak at 4.4 ppm. Peaks due to the three different methyl groups at 1.1 to 1.5 ppm begin to differentiate. (C) Mole ratio of Eu(dpm)$_3$ to alcohol = 0.1. The hydroxyl proton does not appear in this spectrum because its chemical shift is greater than 8.6 ppm with this much shift reagent present. (D) Mole ratio of Eu(dpm)$_3$ to alcohol = 0.25. Further differentiation of methyl peaks (2.3 to 3.0 ppm) is evident. (E) Mole ratio of Eu(dpm)$_3$ to alcohol = 0.5. Separate groups of peaks begin to appear in the region 4.2 to 6.2 ppm. (F) Mole ratio of Eu(dpm)$_3$ to alcohol = 0.7. Three groups of peaks (at 6.8, 7.7, and 8.1 ppm) due to the protons on C-2 and C-4 are evident, and three different methyls are now apparent. (G) Mole ratio of Eu(dpm)$_3$ to alcohol = 0.9. Only the methyl peaks appear on the spectrum. The two doublets come from the methyls attached to C-2, and the triplet comes from the terminal methyl at C-5.

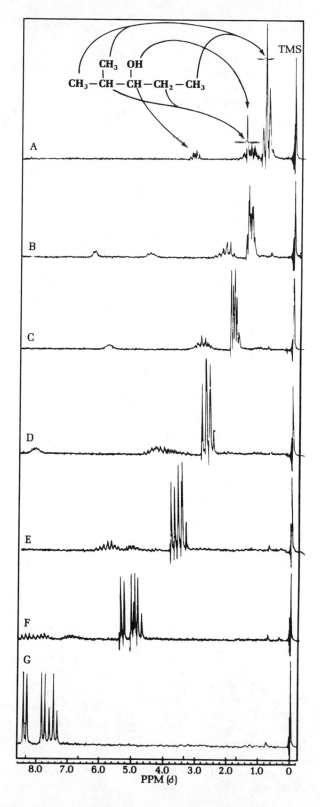

Tris(dipivaloylmethanato)europium(III), Eu(dpm)$_3$
MW 701.78, mp 188–189°C

FIG. 13.6 Proton NMR spectrum of 2-methyl-3-pentanol containing Eu(dpm)$_3$ (60 MHz). Mole ratio of Eu(dpm)$_3$ to alcohol = 1.0. Compare this spectrum with those shown in Fig. 13.5. Protons nearest the hydroxyl group are shifted most. Methyl groups are recorded at reduced spectrum amplitude. Note the large chemical shift difference between the two diastereotopic protons on C-4.

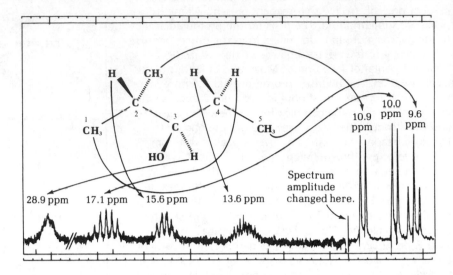

integrated to determine enantiomeric purity. A chiral shift reagent is used in Chapter 66 to determine the enantiomeric purity of a chiral alcohol made by enzymatic reduction of a ketone.

Experiments

1. Running an NMR Spectrum

Sample Preparation

A typical ^{1}H NMR sample consists of about 10–50 mg of sample dissolved in about 0.9 mL of deuterochloroform, CDCl$_3$, that contains a very small percentage of TMS, the reference compound. Because the tiniest particles of dust, especially metal particles, can adversely affect resolution, it is best to filter every sample routinely through a wad of cotton or glass wool in a Pasteur pipette (Fig. 13.7).

If very high resolution spectra (all lines very sharp) are desired, oxygen, a paramagnetic impurity, must be removed by bubbling a fine stream of pure nitrogen through the sample for 60 sec. Routine samples do not require this treatment.

It is most convenient if all tubes in a laboratory are filled to exactly the same height with the CDCl$_3$ solution; this will greatly facilitate tuning of the spectrometer. High-quality NMR tubes give the best spectra, free of spinning side bands. A good tube should roll at a slow, even rate down a very slightly inclined piece of plate glass.

Chemical shifts of protons and carbon are measured relative to the sharp peak in the TMS (taken as 0.0 ppm). Because the deuterochloroform is not 100% pure there will always be a very small peak at 7.27 ppm in the proton spectrum that arises from the tiny amount of residual ordinary chloroform, CHCl$_3$, in the

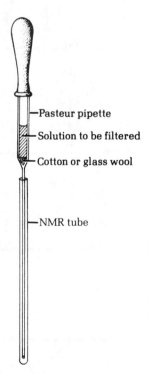

FIG. 13.7 Microfilter for NMR samples. Solution to be filtered is placed in the top of the Pasteur pipette, the rubber bulb is put in place, and pressure is applied to force the sample through the cotton or glass wool into an NMR sample tube.

— Pasteur pipette
— Solution to be filtered
— Cotton or glass wool
— NMR tube

sample. In the carbon spectrum three lines of equal intensity appear at 77 ppm due to coupling of the deuterium atom with the carbon in $CDCl_3$.

If the sample does not dissolve in deuterochloroform, a number of other deuterated solvents are available including deuteroacetone (CD_3COCD_3), deuterodimethylsulfoxide (DMSO-d_6), and deuterobenzene (C_6D_6). All of these solvents are expensive, so the solubility of the sample should be checked in non-deuterated solvents first. For highly polar samples, a mixture of the more expensive deuterodimethylsulfoxide with the less expensive deuterochloroform is often satisfactory. If it is necessary to use D_2O as the solvent, then a special water-soluble reference, sodium 2,2-dimethyl-2-silapentane-5-sulfonate [$(CH_3)_3$ $Si(CH_2)_3SO_3^-Na^+$ (DSS)], must be used. The protons on the three methyl groups bound to the silicon in this salt absorb at 0.0 ppm.

The usual NMR sample has a volume of about 0.9 mL, even though the volume sensed by the spectrometer receiver coils (referred to as the *active volume*) is much smaller. To average the magnetic fields produced by the spectrometer within the sample, the tube is spun by an air turbine at about 30 revolutions per second while taking the spectrum. Too rapid spinning or an insufficient amount of solution will cause the vortex produced by the spinning to penetrate the active volume, giving erratic nonreproducible spectra (Fig. 13.8).

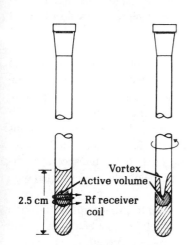

FIG. 13.8 Effect of too rapid spinning or insufficient sample. The active volume is the only part of the sample detected by the spectrometer.

Cleaning Up   Place halogenated solvents and compounds in the halogenated organic waste container. All others go into the organic solvents container.

Adjusting the Spectrometer

To be certain the spectrometer is correctly adjusted and working properly, record the spectrum of the standard sample of chloroform and TMS usually found with the instrument.

Small peaks symmetrically placed on each side of a principal peak are artifacts called *spinning side bands*. They are recognized as such by changing the spin speed, which causes the spinning side bands to change positions. They can be minimized by proper adjustment of the homogeneity controls.

Since Fourier transform (FT) spectrometers lock on the resonance of deuterium to achieve field/frequency stabilization, all samples must be dissolved in a solvent containing deuterium. The fact that lock is obtained is registered on a meter or oscilloscope. The Z and Z^2 controls are used to maximize the field homogeneity and achieve the highest resolution. This adjustment is needed for virtually every sample, but, as noted earlier, if all sample tubes in a laboratory are filled to exactly the same height, these adjustments will be minimized.

Two-Dimensional NMR Spectroscopy

The fast computers associated with FT spectrometers allow for a series of precisely timed pulses and data accumulations to give a large data matrix that can be subjected to Fourier transformation in two dimensions to produce a two-dimensional (2D) NMR spectrum.

One of the most common and useful of these is the COSY (<u>co</u>rrelated <u>s</u>pec-troscop<u>y</u>) spectrum (Fig. 13.9). In this spectrum, two ordinary one-dimensional (1D) spectra are correlated with each other through spin–spin coupling. The 2D spectrum is a topographic representation, where spots represent peaks. The 1D spectra at the top and side are projections of these peaks. Along the diagonal of the 2D spectrum is a spot for each group of peaks in the molecule.

Figure 13.9 shows the 2D spectrum of citronellol. (See Fig. 13.10 for the 1D spectrum of this compound.) Consider spot A (Fig. 13.9) on the diagonal of the 2D spectrum. From a table of chemical shifts we know that this is a vinyl proton, the single proton on the double bond. From the structure of citronellol, we can expect this proton to have a small coupling, over four bonds, to the methyl groups and a stronger coupling to the protons on carbon-6. In the absence of a 2D spectrum, it is not obvious which group of peaks belongs to C-6, but the spot at B correlates with spot C on the diagonal, which is directly below the peak labeled 6 on the spectrum.

The diagonal spot C, which we have just assigned to C-6, correlates through the off-diagonal spot D with the diagonal spot E, which lies just below the group of peaks labeled 5. Spot A on the diagonal also correlates through spots F and G

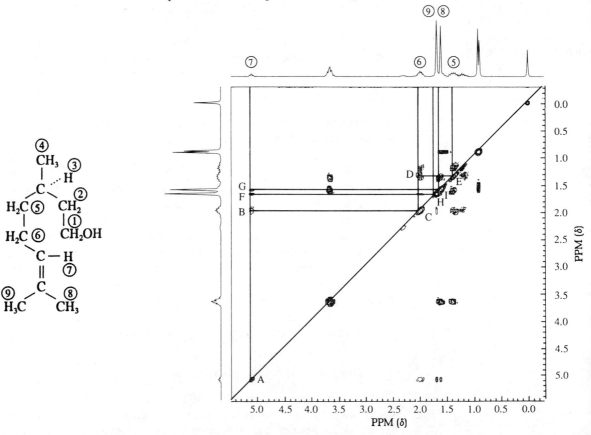

FIG. 13.9   The 2D COSY NMR spectrum of citronellol.

FIG. 13.10 ¹H NMR spectrum of citronellol (250 MHz).

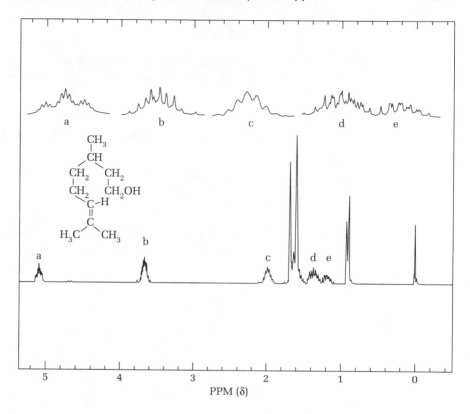

with spots H and I on the diagonal, which lie directly below methyl peaks 8 and 9. In this way we can determine the complete connectivity pattern of the molecule, seeing which protons are coupled to other protons.

Other 2D experiments allow proton spectra on one axis to be correlated with carbon spectra on the other axis. NOESY (<u>n</u>uclear <u>O</u>verhauser <u>e</u>ffect <u>s</u>pectroscop<u>y</u>) experiments give cross-peaks for protons that are near to each other in space but not spin coupled to each other.

2. Identification of Unknown Alcohol or Amine by ¹H NMR

Prepare a solution of the unknown in deuterochloroform, and run a spectrum in the normal way. To the unknown solution add about 5 mg of $Eu(dpm)_3$ (the shift reagent), shake thoroughly to dissolve, and run another spectrum. Continue adding $Eu(dpm)_3$ in 5- to 10-mg portions until the spectrum is shifted enough for easy analysis. Integrate peaks and groups of peaks if in doubt about their relative areas. To protect the $Eu(dpm)_3$ from moisture, store it in a desiccator.

Shift reagents are expensive.

Cleaning Up   All samples containing shift reagents go into a hazardous waste container for heavy metals.

Questions

1. Propose a structure or structures consistent with the proton NMR spectrum of Fig. 13.11. Numbers adjacent to groups of peaks refer to relative peak areas. Account for missing lines.

2. Propose a structure or structures consistent with the proton NMR spectrum of Fig. 13.12. Numbers adjacent to groups of peaks refer to relative peak areas.

3. Propose a structure or structures consistent with the proton NMR spectrum of Fig. 13.13.

4. Propose structures for (a), (b), and (c) consistent with the ^{13}C NMR spectra of Figs. 13.14, 13.15, and 13.16. These are isomeric alcohols with the empirical formula $C_4H_{10}O$.

Surfing the Web

The digital nature of modern NMR spectroscopy naturally lends itself to the graphical and computer-intensive material available on the Internet. The demonstration programs available for prospective buyers (which can be downloaded from the Internet) are often very useful, the complete program even more so. From this wealth of material the following sites have been chosen:

http://www.cherwell.com/ProdHome/gnmrhome.html

gNMR is a demonstration program that allows you to calculate spectra given the chemical shifts and coupling constants for the molecule (for PCs and Macs). You cannot save files, and the word DEMO will be superimposed on any printout, but otherwise it is very worthwhile. Ten nonequivalent nuclei can be handled if your computer's memory is large enough.

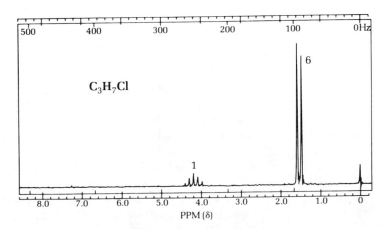

FIG. 13.11 Proton NMR spectrum (60 MHz), Question 1.

FIG. 13.12   Proton NMR spectrum (60 MHz), Question 2.

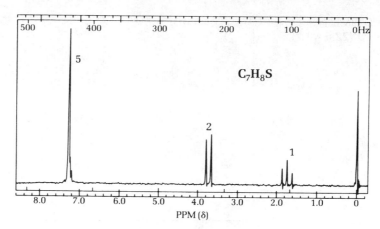

FIG. 13.13   Proton NMR spectrum (60 MHz), Question 3.

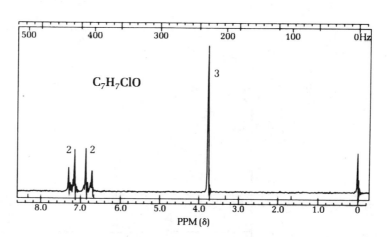

FIG. 13.14   ¹H NMR spectrum $C_4H_{10}O$ (90 MHz), Question 4(a).

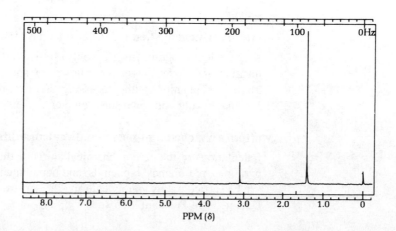

FIG. 13.15 ^{13}C NMR spectrum $C_4H_{10}O$ (22.6 MHz), Question 4(b).

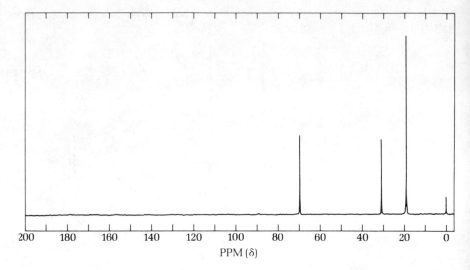

FIG. 13.16 ^{13}C NMR spectrum $C_4H_{10}O$ (22.6 MHz), Question 4(c).

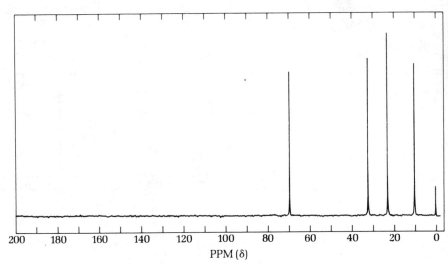

http://www.chem.vt.edu/simulation/VTNMR.html

An excellent program (for PCs only) by Harold Bell from Virginia Polytechnic Institute and State University. Allows for spectral simulation as well as spectrometer simulation, with processing of FIDs, data enhancement, phasing, etc. Just like having your own spectrometer.

http://www.chem.uni-potsdam.de/calcnmr.html

Calculations of the carbon chemical shifts of the aromatic carbons in benzenes, pyridines, pyridizines, biphenyls, and naphthalenes with 32 different substituents can be carried out by simply entering the substituents in the table provided at this University of Potsdam Web site.

http://www.chem.uni-potsdam.de/cgi-bin/perl_prog/lineform.pg

Dynamic NMR line shape simulator. This program allows for the very easy simulation of spectra resulting from exchange in a two-spin system.

http://bmrl.med.uiuc.edu:8080/EduNMRSoft13.2.html

This site contains a complete compilation of software programs dealing with all aspects of NMR spectroscopy. It lists the above sites as well as 100 others, including shareware, commercial programs, demos, and freeware. A commercial program, for instance, will list the carbon chemical shifts for molecules whose structures are entered into the program.

For updated informatin visit:

www.mtholyoke.edu/courses/kwilliam/microscale.shtml

or

www.hmco.com/hmco/college/chemistry/Home.html

References

1. R. J. Abraham, J. Fisher, and P. Loftus, *Introduction to NMR Spectroscopy,* John Wiley & Sons, Chichester, 1988.

2. J. Hore, *Nuclear Magnetic Resonance,* Oxford University Press, Oxford, 1995.

3. J. K. M. Sanders and B. K. Hunter, *Modern NMR Spectroscopy,* 2nd ed., Oxford University Press, Oxford, 1993.

4. H. Friebolin, *Basic One- and Two-Dimensional NMR Spectroscopy,* 2nd ed., VCH, New York, 1993.

5. A. E. Derome, *Modern NMR Techniques for Chemistry Research,* Pergamon Press, Oxford, 1987.

6. D. Canet, *Nuclear Magnetic Resonance. Concepts and Methods,* John Wiley and Sons, West Sussex, 1996.

7. D. G. Gadian, *NMR and Its Applications to Living Systems,* Oxford University Press, Oxford, 1996.

8. D. M. Grant and R. K. Harris (eds.), *Encyclopedia of NMR,* 8-vol. set, John Wiley and Sons, New York, 1996.

9. W. R. Croasmun and R. M. K. Carlson, *Two-Dimensional NMR Spectroscopy. Applications for Chemists and Biochemists,* 2nd ed., VCH Publishers, New York, 1994.

10. J. W. Akitt, *NMR and Chemistry,* 2nd ed., Chapman and Hall, London, 1983.

11. Charles J. Pouchert, *The Aldrich Library of C and H FT NMR Spectra,* 3 vols., Aldrich Chemical Co., Milwaukee, WI, 1993.

12. E. L. Eliel and S. H. Wilen, *Stereochemistry of Organic Compounds,* John Wiley and Sons, New York, 1994.

CHAPTER

14

Ultraviolet Spectroscopy

Prelab Exercise: Predict the appearance of the ultraviolet spectrum of 1-butyl-amine in acid, of methoxybenzene in acid and base, and of benzoic acid in acid and base.

UV: Electronic transitions within molecules

Ultraviolet (UV) spectroscopy gives information about electronic transitions within molecules. Whereas absorption of low-energy infrared radiation causes bonds in a molecule to stretch and bend, the absorption of short-wavelength, high-energy ultraviolet radiation causes electrons to move from one energy level to another with energies that are often capable of breaking chemical bonds.

We shall be most concerned with transitions of π-electrons in conjugated and aromatic ring systems. These transitions occur in the wavelength region of 200 to 800 nm (nanometers, 10^{-9} meters, formerly known as mμ, millimicrons). Most common ultraviolet spectrometers cover the region of 200 to 400 nm as well as the visible spectral region of 400 to 800 nm. Below 200 nm air (oxygen) absorbs UV radiation; spectra in that region must therefore be obtained in a vacuum or in an atmosphere of pure nitrogen.

Consider ethylene, even though it absorbs UV radiation in the normally inaccessible region at 163 nm. The double bond in ethylene has two *s* electrons in a σ-molecular orbital and two, less tightly held, *p* electrons in a π-molecular orbital. Two unoccupied, high-energy-level, antibonding orbitals are associated with these orbitals. When ethylene absorbs UV radiation, one electron moves up from the bonding π-molecular orbital to the antibonding π^*-molecular orbital (Fig. 14.1). As the diagram indicates, this change requires less energy than the excitation of an electron from the σ to the σ^* orbital.

By comparison with infrared spectra and NMR spectra, UV spectra are fairly featureless (Fig. 14.2). This condition results as molecules in a number of different vibrational states undergo the same electronic transition, to produce a band spectrum instead of a line spectrum.

Band spectra

Unlike IR spectroscopy, ultraviolet spectroscopy lends itself to precise quantitative analysis of substances. The intensity of an absorption band is usually given by the molar extinction coefficient, ε, which, according to the Beer–Lambert law, is equal to the absorbance, A, divided by the product of the molar concentration, c, and the path length, l, in centimeters.

Beer–Lambert law

$$\varepsilon = \frac{A}{cl}$$

FIG. 14.1 Electronic energy levels of ethylene.

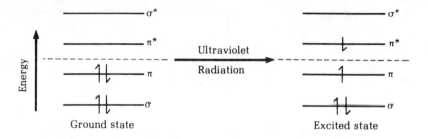

Ground state   Excited state

λ_{max}, wavelength of maximum absorption

The wavelength of maximum absorption (the tip of the peak) is given by λ_{max}. Because UV spectra are so featureless, it is common practice to describe a spectrum like that of cholesta-3,5-diene (Fig. 14.2) as λ_{max} 234 nm (ε = 20,000) and not bother to reproduce the actual spectrum.

The extinction coefficients of conjugated dienes and enones are in the range of 10,000–20,000, so only very dilute solutions are needed for spectra. In the example of Fig. 14.2 the absorbance at the tip of the peak, *A*, is 1.2, and the path length is the usual 1 cm; so the molar concentration needed for this spectrum is 6×10^{-5} mole per liter,

ε, extinction coefficient

$$c = \frac{A}{l\varepsilon} = \frac{1.2}{20,000} = 6 \times 10^{-5} \text{ mole per liter}$$

which is 0.221 mg per 10 mL of solvent.

Spectro grade solvents

The usual solvents for UV spectroscopy are 95% ethanol, methanol, water, and saturated hydrocarbons such as hexane, trimethylpentane, and isooctane; the three hydrocarbons are often specially purified to remove impurities that absorb in the UV region. Any transparent solvent can be used for spectra in the visible region.

Sample cells for spectra in the visible region are made of glass, but UV cells must be of the more expensive fused quartz, since glass absorbs UV radiation. The cells and solvents must be clean and pure, since very little of a substance produces a UV spectrum. A single fingerprint will give a spectrum!

UV cells are expensive; handle with care.

Ethylene has λ_{max} 163 nm (ε = 15,000), and butadiene has λ_{max} 217 nm (ε = 20,900). As the conjugated system is extended, the wavelength of maximum

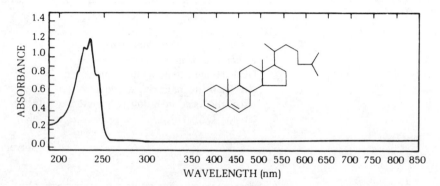

FIG. 14.2 The ultraviolet spectrum of cholesta-3,5-diene in ethanol.

FIG. 14.3 The ultraviolet–visible spectrum of lycopene in isooctane.

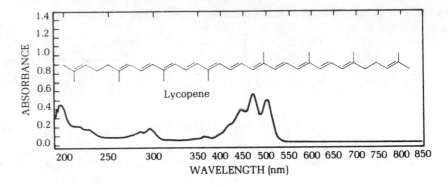

absorption moves to longer wavelengths (toward the visible region). For example, lycopene with 11 conjugated double bonds has λ_{max} 470 nm ($\varepsilon = 185{,}000$; Fig. 14.3). Because lycopene absorbs blue visible light at 470 nm, the substance appears bright red. It is responsible for the color of tomatoes; its isolation is described in Chapter 9.

Woodward and Fieser rules for dienes and dienones

The wavelengths of maximum absorption of conjugated dienes and polyenes and conjugated enones and dienones are given by the Woodward and Fieser rules, Tables 14.1 and 14.2. The application of the rules in these tables is demonstrated by the spectra of pulegone (1) and carvone (2) in Fig. 14.4. The solvent correction is given in Table 14.3. The calculations are given in Tables 14.4 and 14.5.

These rules will be applied in a later experiment in which cholesterol is converted into an α,β-unsaturated ketone.

No simple rules exist for calculation of aromatic ring spectra, but several generalizations can be made. From Fig. 14.5 it is obvious that as polynuclear aromatic rings are extended linearly, λ_{max} shifts to longer wavelengths.

As alkyl groups are added to benzene, λ_{max} shifts from 255 nm for benzene to 261 nm for toluene to 272 nm for hexamethylbenzene. Substituents bearing nonbonding electrons also cause shifts of λ_{max} to longer wavelengths, e.g., from 255 nm for benzene to 257 nm for chlorobenzene, 270 nm for phenol, and 280

TABLE 14.1 Rules for the Prediction of λ_{max} for Conjugated Dienes and Polyenes

	Increment (nm)
Parent acyclic diene (butadiene)	217
Parent heteroannular diene	214
Double bond extending the conjugation	30
Alkyl substituent or ring residue	5
Exocyclic location of double bond to any ring	5
Groups: OAc, OR	0
	0
Solvent correction, see Table 14.3 ( )	
λ_{max}^{EtOH} = Total	

TABLE 14.2   Rules for the Prediction of λ_{max} for Conjugated Enones and Dienones

$$
\begin{array}{ccc}
\beta & \alpha & R \\
| & | & | \\
\end{array}
$$

$$\beta-C{=}C-C{=}O \quad \text{and} \quad \delta-C{=}C-C{=}C-C{=}O$$

	Increment (nm)
Parent α,β-unsaturated system	215
Double bond extending the conjugation	30
R (alkyl or ring residue), OR, OCOCH$_3$ α	10
β	12
λ, δ, and higher	18
α-Hydroxyl, enolic	35
α-Cl	15
α-Br	23
exo-Location of double bond to any ring	5
Homoannular diene component	39

Solvent correction, see Table 14.3 ( )

λ_{max}^{EtOH} = Total

TABLE 14.3   Solvent Correction

Solvent	Factor for Correction to Ethanol
Hexane	+11
Ether	+7
Dioxane	+5
Chloroform	+1
Methanol	0
Ethanol	0
Water	−8

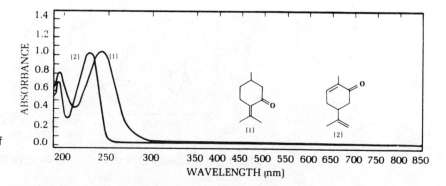

FIG. 14.4   Ultraviolet spectra of (1) pulegone and (2) carvone in hexane.

TABLE 14.4 Calculation of λ_{max} for Pulegone (See Fig. 14.4)

Parent α,β-unsaturated system	215 nm
α-Ring residue, R	10
β-Alkyl group (two methyls)	24
Exocyclic double bond	5
Solvent correction (hexane)	−11

Calcd λ_{max} 245 nm; found 244 nm

TABLE 14.5 Calculation of λ_{max} for Carvone (See Fig. 14.4)

Parent α,β-unsaturated system	215 nm
α-Alkyl group (methyl)	10
β-Ring residue	12
Solvent correction (hexane)	−11

Calcd λ_{max} 226 nm; found 229 nm

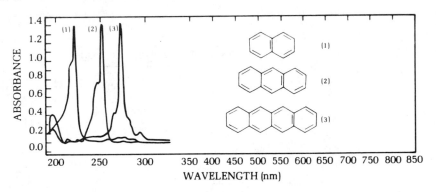

FIG. 14.5 The ultraviolet spectra of (1) naphthalene, (2) anthracene, and (3) tetracene.

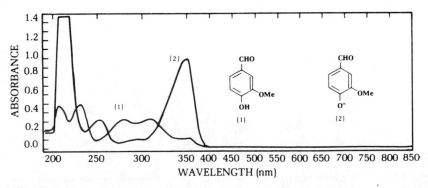

FIG. 14.6 Ultraviolet spectra of (1) neutral vanillin and (2) the anion of vanillin.

FIG. 14.7 Ultraviolet spectra of (1) aniline and (2) aniline hydrochloride.

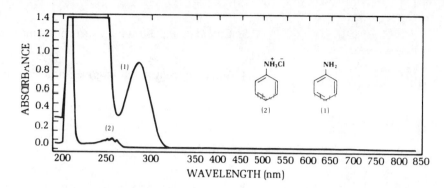

Effect of acid and base on λ_{max}

nm for aniline ($\varepsilon = 6200–8600$). That these effects are the result of interaction of the π-electron system with the nonbonded electrons is seen dramatically in the spectra of vanillin and the derived anion (Fig. 14.6). Addition of two more nonbonding electrons in the anion causes λ_{max} to shift from 279 to 351 nm and ε to increase. Removing the electrons from the nitrogen of aniline by making the anilinium cation causes λ_{max} to decrease from 280 to 254 nm (Fig. 14.7). These changes of λ_{max} as a function of pH have obvious analytical applications.

Intense bands result from π–π conjugation of double bonds and carbonyl groups with the aromatic ring. Styrene, for example, has λ_{max} 244 nm ($\varepsilon = 12,000$) and benzaldehyde λ_{max} 244 nm ($\varepsilon = 15,000$).

Experiment

Ultraviolet Spectrum of Unknown Acid, Base, or Neutral Compound

Determine whether an unknown compound obtained from the instructor is acidic, basic, or neutral from the ultraviolet spectra in the presence of acid and base as well as in neutral media.

Cleaning Up   Because UV samples are extremely dilute solutions in ethanol, they can normally be flushed down the drain.

Questions

1. Calculate the ultraviolet absorption maximum for 2-cyclohexene-1-one.

2. Calculate the ultraviolet absorption maximum for 3,4,4-trimethyl-2-cyclo-hexene-1-one.

3. Calculate the ultraviolet absorption maximum for 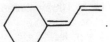 .

4. What concentration, in g/mL, of a substance with MW 200 should be prepared in order to give an absorbance value A equal to 0.8 if the substance has $\varepsilon = 16{,}000$ and a cell with a path length of 1 cm is employed?

References

1. A. E. Gillam and E. S. Stern, *An Introduction to Electronic Absorption Spectroscopy in Organic Chemistry,* Edward Arnold, London, 1967.

2. C. N. R. Rao, *Ultraviolet and Visible Spectroscopy,* 2nd ed., Butterworths, London, 1967.

3. H. H. Jaffe and M. Orchin, *Theory and Application of Ultraviolet Spectroscopy,* John Wiley, New York, 1962.

4. R. M. Silverstein, C. G. Bassler, and T. C. Morrill, *Spectrometric Identification of Organic Compounds,* 4th ed., John Wiley, New York, 1981. (Includes IR, UV, and NMR.)

CHAPTER

15

Molecular Mechanics and Computational Chemistry

No, molecular mechanics are not little people with wrenches who run around fixing molecules. *Molecular mechanics* is the term given to a method for calculating the structures of molecules based not on a complete solution of the Schrödinger equation, but on a mechanical model for molecules, a model that regards molecules as being masses (atoms) connected by forces somewhat like springs (bonds) at certain preferred lengths and angles.

With a computer it is possible to draw an approximate structure of a molecule, after which a molecular mechanics program will automatically, through a series of iterative calculations, adjust every bond angle, bond length, dihedral angle, and van der Waals interaction to produce a new structure having the lowest possible energy.

This calculational approach to chemistry is being used extensively in drug design, protein mutagenesis, biomimetics, catalysis, the study of DNA/protein interactions, and the determination of structures of molecules using NMR spectroscopy, to name just a few applications. In this course we use molecular mechanics calculations to predict the structures of the products of reactions.

Every organic chemist routinely uses a set of models to examine the three-dimensional structures of molecules, because it is difficult to visualize these from a two-dimensional drawing. Molecular mechanics programs allow one to determine quickly the "best" (lowest energy) structure for a molecule. Experienced chemists have been doing this intuitively for years, but simple plastic or metal models can lead to incorrect structures in many cases and, of course, one obtains no quantitative data regarding the stability of one structure compared to another.

Unlike quantum mechanical calculations, molecular mechanics calculations are empirical in nature; that is, the parameters used in the equations have been derived from experimental work and not from first principles. These calculations do not take the electronic structure of molecules into account. Molecular mechanics assumes that all bond properties are the same in all molecules. For exceptional molecules such as cyclopropane and cyclobutane, it simply substitutes new parameters for these unusual rings. Experimental observations of this nature are taken into account in constructing the parameter tables for molecular mechanics calculations. The experimental data on which many of the parameters for molecular mechanics rest come from x-ray diffraction studies of crystals, electron diffraction studies of gases, and microwave spectroscopy.

A number of molecular mechanics programs are available. The simpler ones can now be run on a personal computer; the more complex, faster programs capable of handling hundreds of atoms in, say, a protein, run on workstations. The authors of these computer programs (Norman Allinger, the author of MM2-MM4, is a pioneer in this area) have considerable leeway in constructing their parameter sets. They make judicious decisions regarding, for example, how hard or soft to make the carbon and hydrogen atoms in their programs, because atoms do not behave like hard billiard balls.

Molecular mechanics calculations can find minima but cannot evaluate the energy of a transition state and thus the activation energy from which reaction rates can be derived, nor can the heat of formation or of combustion be calculated. This is the province of the next higher level: semiempirical calculations. All that molecular mechanics calculations provide is an arbitrarily defined *strain energy* that differs from one program to another. It is truly, as stated, a mechanical approach to molecular structure.

If a molecule has a single minimum energy conformation and that conformation has been found, then the shape of the final molecule is used to predict or confirm approximate vicinal (three-bond) proton–proton NMR coupling constants, which are dependent on the dihedral angle between the protons. (Couplings are also dependent to a smaller degree on electronegativity of substituents, bond angles, etc.) And optical rotatory dispersion curves can be predicted knowing molecular conformations. From slightly higher-level calculations, it is also possible to calculate an approximate infrared spectrum once an accurate geometry for a molecule has been calculated.

In a typical empirical force field, the steric energy of a molecule can be represented as the sum of five energies given by the following equation:

$$E_{steric} = E_{str} + E_{bnd} + E_{tor} + E_{oop} + E_{vdW}$$

where E_{str} is the energy needed to stretch or compress a bond. Plastic student molecular models are realistic in that there is no way to adjust this value; in molecular mechanics calculations one finds there is a big force constant (more correctly "potential constant") for this parameter. Most sets of molecular models are constructed so that modest bending of bond angles is possible, and similarly one finds that the force constant for angle bending, E_{bnd}, has an intermediate value. The force constant for torsional motion, E_{tor}, is very small as seen in the free rotation about bonds in most mechanical models. The out-of-plane bending term, E_{oop}, arises in molecules such as methylene cyclobutene where, without this term, the methylene group would want to be bent out of the plane to relieve angle strain. The energy due to van der Waals interactions, E_{vdW}, is very important. It takes two forms, either attractive or repulsive, depending on internuclear distance.

Molecular mechanics allows the calculation of the equilibrium conformation of a molecule (see Fig. 15.1). The individual mathematical equations describing the mechanics of a molecule are not terribly complex, as we shall see, but a fast computer is needed to carry out the millions of calculations and summations needed to describe the pairwise interactions between atoms described by these equations.

Methylene cyclobutene

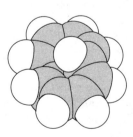

FIG. 15.1 Cyclodecapentaene. From this energy-minimized structure, it is easy to see that the molecule cannot be flat and thus cannot benefit from aromatic stabilization.

Let us examine in detail the five most important factors entering into molecular mechanics calculations. The bond between two atoms can be regarded just like a spring. It can be stretched, compressed, or bent, and, provided these distortions are small, will follow Hooke's law from classical physics.

$$C \text{\textemdash} C$$

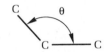

For bond stretching the energy is given by

$$E_{str} = 1/2k_r(r - r_0)^2$$

where E_{str} is the bond stretching energy in kJ/mol, k_r is the bond stretching force constant in kJ/mol·Å², r is the bond length, and r_0 is the equilibrium bond length.

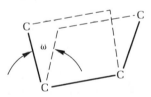

All of these energies are empirical in nature, not calculated from first principles.

Similarly, the energy for bond angle bending (in-plane bending) is given by

$$E_{bnd} = 1/2k_\theta(\theta - \theta_0)^2$$

where E_{bnd} is the bond bending energy in kJ/mol, k_θ is the angle bending force constant in kJ/mol·deg², θ is the angle between two adjacent bonds in degrees, and θ_0 is the equilibrium value for the angle between the two bonds in degrees.

The energy for twisting a bond, E_{tor}, is given by

$$E_{tor} = 1/2k_\omega[1 - \cos(j\omega)]$$

where E_{tor} is the torsional energy in kJ/mol, k_ω is the torsional force constant in kJ/mol·deg, ω is the torsion angle in degrees, and j is usually 2 or 3 depending on the symmetry of the bond (e.g., twofold in ethylene or threefold in ethane).

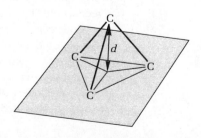

The energy term for out-of-plane bending, E_{oop}, is not used very often, but it is a necessary term. It arises in cases such as cyclobutanone, where the CCO bond angle is not the sp^2 bond angle of 120° but 133°. To relieve this bond angle strain, the oxygen could tip up out of the plane, but this would require a large amount of energy because of distortion of the π bonding of the carbonyl. So the force constant for out-of-plane bending is different from that for in-plane bending (bond angle bending), E_{bnd}, discussed earlier.

$$E_{oop} = 1/2 k_\delta (d^2)$$

where E_{oop} is the out-of-plane bending energy in kJ/mol, k_δ is the out-of-plane bending constant in kJ/mol·Å^2, and d is the height of the central atom in angstroms above the plane of its substituents.

Very important is the van der Waals energy term, E_{vdW}, given by

$$E_{vdW} = \varepsilon[(r_0/r)^{12} - 2(r_0/r)^6]$$

where E_{vdW} is the van der Waals energy in kJ/mol, ε is the van der Waals potential of the two interacting atoms in kJ/mol, r_0 is the equilibrium distance between the two atoms, and r is the interatomic distance in the molecule in angstroms.

This important function is best understood by reference to Fig. 15.2, which shows a curve that describes the energy of two atoms as a function of their distance, r, from each other. At very short distances, the two atoms repel each other very strongly. This interaction raises the energy of the system as the 12th power of the interatomic distance. At a slightly longer distance, a weak attraction is described by an r^{-6} term, and then at large separation the interaction energy goes asymptotically to zero. The value of ε determines the depth of the potential well. At distance r_0, when the attractive forces overwhelm the repulsive forces, the system has minimum energy. It is r_0 that we call the *bond length*. The molecular mechanics van der Waals term given above is just an approximation of the ideal curve. It comes close enough and is easy to compute.

Carrying out molecular mechanics calculations on a personal computer or a small workstation is not difficult. One first draws the structure of the molecule on the computer screen in the same way or even using the same program that is used to draw organic structures for publication in reports and papers. This results in a "wire model" such as the hexagon we associate with cyclohexane with the proper x,y,z coordinates for each carbon. The program can be directed to add hydrogens at all the unfilled valences. Then the molecular mechanics program is told to "minimize" the structure, and in a few seconds or a few minutes a minimum energy structure is calculated. Usually, the molecule appears to move on the screen as the "best" conformation is calculated.

Consider, for example, the molecule butane. When you initially draw the molecule on the screen you will undoubtedly not draw a perfectly correct structure in terms of bond lengths, bond angles, and dihedral angles. The computer will calculate the energy of the structure you draw and then make a slight change in the coordinates of the atoms and recalculate the energy of the molecule. If the newly calculated energy is lower than the previously calculated one, it will

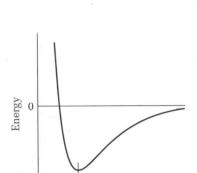

FIG. 15.2 Van der Waals energy function.

continue to change the coordinates until a minimum energy is reached. This iterative process is continued until it is found that no change in any bond angle, bond length, torsion angle, or van der Waals interaction can be made without raising the energy of the molecule. At this point the calculation stops. These millions of calculations can be made in a few seconds or a few minutes on a modern computer.

But there is one serious pitfall in this process: The computer program usually has no way of knowing whether it has calculated the structure with the lowest possible energy for the molecule (the global minimum) or just a local minimum. A geographical analogy may make this statement more clear. It is as if you were instructing the computer to roll a ball to the lowest place in California (Death Valley). The computer might find the lowest place in the Sacramento Valley, but the algorithm on which it operates has no way of knowing whether that is the global minimum or any way to climb the mountains that separate the local from the global minimum. Or consider once more the molecule butane. The global minimum is the *anti*-conformer (Fig. 15.3), but the calculation can easily fall into one of the two local minima, the *gauche*-conformations, which are not the lowest energy conformations.

There are approaches to a solution to this problem, but none is completely satisfactory. If one has enough experience and intuition, it is often possible to spot the fact that the minimum-energy conformation calculated by the computer is not the global minimum. A better solution, however, is to ask the computer to calculate energies for an arbitrary set of dihedral angles about one of the bonds.

In butane, one could ask for the energies to be calculated at dihedral angles of 0°, 30°, 60°, 90°, 120°, 150°, and 180° using a part of the program called a *dihedral driver*. In this way you would be sure to catch the lowest energy form.

But some molecular mechanics programs do not have this feature, in which case it is up to the chemist to look at the calculated structure and try to decide whether the very lowest energy form has been calculated.

A simple example of this problem is butane. If the first trial conformation (A) put into the computer has a torsion angle (dihedral angle) of between 0° and just a bit less than 120°, the molecular mechanics program will calculate the *gauche* conformation (B) [60° torsion angle (dihedral angle)] as the lowest energy form. In fact, the lowest energy form has a torsion angle of 180° (C) (see Fig. 15.3).

A   B   C

FIG. 15.3 The potential energy of butane as a function of the dihedral angle between the two central carbons.

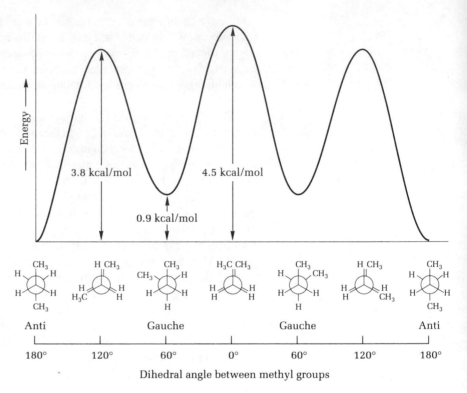

In the plot of potential energy as a function of torsion angle, we see that the *gauche*-conformations are local minima, while the anti-conformation is the global minimum.

Consider a larger molecule. If one were to search for the best conformation at angles of 0°, 30°, 60°, 90°, 120°, 150°, and 180° around each carbon–carbon bond, it would be necessary to make 7^n calculations, where n is the number of carbon–carbon bonds. In 2-bromodecane, for example, this would be 78 = 5,764,801 calculations. The problem gets out of hand very rapidly.

It is not possible to use a dihedral driver within a ring, so the lowest energy form is found by other means (ring flip and flap, molecular dynamics, and Monte Carlo methods).

Experiments

With a molecular mechanics program, calculate the energy of butane at dihedral angles of 0°, 60°, 120°, and 180°. You should be able to reproduce the *form* of the curve shown in Fig. 15.3, but you usually will not obtain the experimentally derived energies shown.

Remember that the calculated steric energy has no physical significance. Differences in steric energies can be useful, however, if the molecules being compared have the same bonding pattern. The molecule with the lower energy will be the more stable isomer (geometric, conformational, or stereo). Other types of comparisons cannot be made. A molecular mechanics calculation cannot be used to compare the stabilities of unrelated molecules.

If the program you are using breaks down the steric energy into the five terms $E_{\text{steric}} = E_{\text{str}} + E_{\text{bnd}} + E_{\text{tor}} + E_{\text{oop}} + E_{\text{vdW}}$, do not give the individual terms much credence. Different parameter sets in other programs may weigh these differently. You should pay the most attention to the total steric energy.

Molecular mechanics calculations, like IR and NMR spectroscopy, are another tool available to the chemist. As such, they are presented in this text as adjuncts to experimental work in the laboratory, not as an end in themselves. In most of the computational problems presented in the following chapters, you are asked to carry out computations of the steric energies or, better, of the heats of formations of isomers.

One must be careful when using molecular mechanics computations to draw conclusions. The computation can be made on any molecule, but the resulting "steric energy" cannot always be compared with the steric energy of another molecule, even though they are isomers. For example, *cis*- and *trans*-butene and other similar *cis*- and *trans*-isomers can be compared with each other because they are both 1,2-disubstituted alkenes, but a valid comparison cannot be made with 1,1-disubstituted alkenes.

Although not intending to mislead you deliberately, we have included computations on reactions and products that for one reason or another are not expected to give valid comparisons or answers. It is for you to decide whether the computations are in agreement with your experiment.

Higher levels of computation, the semiempirical molecular orbital methods, and the even higher *ab initio* methods will give heats of formation that can be compared with one another. The problem with these higher levels of computation is that they require more time and more computer power. But often an AM1 semiempirical molecular orbital calculation on a relatively small molecule can be done in just minutes. The cost (in computer time) of *ab initio* calculations goes up more than the fourth power of the number of atoms, whereas molecular mechanics calculations go up as the square of the number of atoms in the molecule.

The increase in time needed for a calculation on a molecule having n *atoms:*

Molecular mechanics	n^2
Semiempirical	n^3
Ab initio	$>n^4$

The radical chlorination of 1-chlorobutane gives four isomeric dichlorobutanes. In Chapter 11 these butanes are synthesized and their relative amounts measured by gas chromatography with the object of determining the relative reactivity of the various hydrogens in the starting material. Using a molecular mechanics program, you can calculate the energies of the four products to see whether or not these energies correlate with the product distributions. A correlation or lack thereof may give information regarding the mechanism of the reaction.

Similarly, the dehydration of 2-butanol (Chapter 11) gives three isomeric butenes that are separated and analyzed by gas chromatography. Calculation of their heats of formation or steric energies may or may not correlate with the percentage of each isomer formed in the reaction. Again, a correlation or lack thereof may give information about the mechanism of the reaction or about the applicability of molecular mechanics to problems of this type.

Treatment of *cis*-norbornene-5,6-endo-dicarboxylic acid with concentrated sulfuric acid gives the isomeric compound X (Chapter 49). Using molecular mechanics, the energies of any proposed structures for X can be calculated and compared with that of the starting material. If the reaction conditions (hot concentrated sulfuric acid) are regarded as favoring an equilibrium between product and starting material, then X would be expected to have the lower energy if the correct structure is proposed and molecular mechanics calculations are applicable. If molecular mechanics calculations are not applicable, then a higher level of calculation will be necessary.

In Chapter 51 you are asked to calculate the most stable conformation of the product *p*-terphenyl. Hexaphenylethane is a molecule that does not exist, but this does not preclude calculation of its steric energy and bond lengths (Chapter 31). One can calculate the energy of the isomeric dimer that does form. This calculation gives a clear picture of why the triphenylmethyl radical does not simply dimerize, even though the steric energy may not be correct.

The oxidation of citronellol can give four possible isomeric isopulegols (Chapter 23). Molecular mechanics calculations can indicate which of these is the most stable. Similarly, oxidation of these alcohols can give two possible isopulegones that theoretically can equilibrate. Molecular mechanics calculations can again indicate which is the more stable. These rearrange to pulegone. Again, calculation may disclose whether this product is more stable than the starting material, throwing light on the mechanism of the isomerization reaction.

The Wittig reaction usually gives a mixture of *cis*- and *trans*-isomers. In Chapter 39 two Wittig reactions are carried out in which one isomer predominates in each reaction. A molecular mechanics calculation not only discloses what these isomers look like but also their energies.

A very easy reaction to carry out is the aldol condensation of benzaldehyde with acetone to give in high yield dibenzalacetone (mp 110–111°C). It is interesting to explore the products of this reaction with molecular mechanics because three geometric isomers can be formed and seven single-bond *cis*- or *trans*-isomers of the geometric isomers lie at energy minima. The question is, which of these 10 isomers is formed in the reaction? Calculational aspects of this question are dealt with at length in Chapter 37.

In all these calculations, not only will you calculate the energies of the molecules, but you will be able to see what the lowest energy conformation of the molecule looks like. In this regard it is interesting to look at the calculated low-energy conformation of benzophenone (Chapter 38), tetraphenylcyclopentadienone (Chapter 52), and pseudopellitierene (Chapter 68).

Z- and *E*-stilbene are synthesized in Chapters 59 and 61, and in Chapter 60 the Perkin reaction is used to make a mixture of *Z*- and *E*-α-phenylcinnamic acids. Steric energy calculations, as well as a picture of the molecular conformations, help rationalize the relative amounts of the isomers formed in these synthetic reactions as well as their UV spectra.

Surfing the Web

http://cmm.info.nih.gov/modeling/web_chemistry/web_chemistry_modeling.html

This is the address for the molecular modeling page of the National Institutes of Health with many links to modeling throughout the world.

For updated information visit:

www.mtholyoke.edu/courses/kwilliam/microscale.shtml

or

www.hmco.com/hmco/college/chemistry/Home.html

References

1. U. Burkert and N. L. Allinger, *Molecular Mechanics,* ACS Monograph 177, American Chemical Society, Washington, D.C., 1982.

2. Tim Clark, *A Handbook of Computational Chemistry,* John Wiley and Sons, New York, 1985.

3. D. B. Boyd and K. B. Lipkowitz, *J. Chem. Educ.,* **59,** 269–274 (1982).

4. K. B. Lipkowitz, *J. Chem. Educ.,* **66,** 275–277 (1989).

5. R. J. Jarret and N. Sin, *J. Chem. Educ.,* **67,** 153–155 (1990).

6. K. B. Lipkowitz, *J. Chem. Educ.,* **72,** 1070–1075 (1995).

7. *Journal of Chemical Education.* Since 1989, a large number of articles on molecular mechanics calculations and computational chemistry in general have been published.

8. W. J. Hehre, L. D. Burke, A. J. Shusterman, and W. J. Pietro, *Experiments in Computational Organic Chemistry,* Wavefunction, Inc., 18401 Von Karman, Suite 370, Irvine, CA 92715.

9. K. B. Lipkowitz and D. B. Boyd (eds.), *Reviews in Computational Chemistry,* VCH, New York. A continuing series.

10. G. H. Grant and W. G. Richards, *Computational Chemistry,* Oxford University Press, Oxford, 1995.

The S_N2 Reaction: 1-Bromobutane

Prelab Exercise: Prepare a detailed flow sheet for the isolation and purification of *n*-butyl bromide. Indicate how each reaction byproduct is removed and which layer is expected to contain the product in each separation step.

$$CH_3CH_2CH_2CH_2OH \xrightarrow{\text{NaBr, H}_2\text{SO}_4} CH_3CH_2CH_2CH_2Br + NaHSO_4 + H_2O$$

1-Butanol
bp 118°C
den. 0.810
MW 74.12
n_D^{20} 1.399

1-Bromobutane
bp 101.6°C
den. 1.275
MW 137.03
n_D^{20} 1.439

In this experiment 1-butanol is converted to 1-bromobutane by an S_N2 reaction. In general, a primary alkyl bromide can be prepared by heating the corresponding alcohol with (1) constant-boiling hydrobromic acid (47% HBr); (2) an aqueous solution of sodium bromide and excess sulfuric acid, which is an equilibrium mixture containing hydrobromic acid; or (3) with a solution of hydrobromic acid produced by bubbling sulfur dioxide into a suspension of bromine in water. Reagents (2) and (3) contain sulfuric acid at a concentration high enough to dehydrate secondary and tertiary alcohols to undesirable byproducts (alkenes and ethers), and hence the HBr method (1) is preferred for preparation of halides of the types R_2CHBr and R_3CBr. Primary alcohols are more resistant to dehydration and can be converted efficiently to the bromides by the more economical methods (2) and (3), unless they are of such high molecular weight as to lack adequate solubility in the aqueous mixtures. The $NaBr$-H_2SO_4 method is preferred to the Br_2-SO_2 method because of the unpleasant, choking property of sulfur dioxide. The overall equation is given above, along with key properties of the starting material and principal product.

Choice of reagents

The procedure that follows specifies a certain proportion of 1-butanol, sodium bromide, sulfuric acid, and water; defines the reaction temperature and time; and describes operations to be performed in working up the reaction mixture. The prescription of quantities is based on considerations of stoichiometry as modified by the results of experimentation. Before undertaking a preparative experiment you should analyze the procedure and calculate the molecular proportions of the reagents. Construction of tables (see following example) of properties of starting material, reagents, products, and byproducts provides guidance

TABLE 16.1   Properties of Reagents, Products, and Byproducts

Reagents

				Wt used (g)	Moles	
Reagent	MW	Den	Bp (°C)		*Theory*	*Used*
n-C$_4$H$_9$OH	74.12	0.810	118	8.0	0.108	0.108
NaBr	102.91	—	—	13.3	0.108	0.129
H$_2$SO$_4$	98.08	1.84	—	20.	0.108	0.200

Product and Byproducts

			Bp (°C)		Theoretical Yield		Found	
Compound	MW	Den	*Given*	*Found*	*Moles*	*g*	*g*	*%*
n-C$_4$H$_9$Br	137.03	1.275	101.6	—	0.11	14.8 (100%)	_g	_%
CH$_3$CH$_2$CH=CH$_2$			−6.3	—				
C$_4$H$_9$OC$_4$H$_9$			141	—				

The laboratory notebook

in regulation of temperature and in separation and purification of the product and should be entered in the laboratory notebook. Construct a flowchart of the operations to be performed.

$$\text{NaBr} + \text{H}_2\text{SO}_4 \rightleftharpoons \text{HBr} + \text{NaHSO}_4$$

One mole of 1-butanol theoretically requires one mole each of sodium bromide and sulfuric acid, but the procedure calls for use of a slight excess of bromide and twice the theoretical amount of acid. Excess acid is used to shift the equilibrium in favor of a high concentration of hydrobromic acid. The amount of sodium bromide taken, arbitrarily set at 1.2 times the theory as an insurance measure, is calculated as follows:

$$\frac{8 \text{ (g of C}_4\text{H}_9\text{OH)}}{74.12 \text{ (MW of C}_4\text{H}_9\text{OH)}} \times 102.91 \text{ (MW of NaBr)} \times 1.2 = 13.3 \text{ g of NaBr}$$

Theoretical Yield Calculation

The theoretical yield is 0.11 mole of product, corresponding to the 0.11 mole of butyl alcohol taken; the maximal weight of product is calculated thus:

0.11 (mole of alcohol) $\times$ 137.03 (MW of product)
$$= 14.8 \text{ g 1-bromobutane (theoretical yield of product)}$$

The probable byproducts are 1-butene, dibutyl ether, and the starting alcohol. The alkene is easily separable by distillation, but the other substances are in the same boiling point range as the product. However, all three possible byproducts can be eliminated by extraction with concentrated sulfuric acid.

Experiments

1. Synthesis of 1-Bromobutane

$$CH_3CH_2CH_2CH_2OH \xrightarrow{\text{NaBr, H}_2\text{SO}_4} CH_3CH_2CH_2CH_2Br + NaHSO_4 + H_2O$$

1-Butanol
bp 118°C
den 0.810
MW 74.12
n_D^{20} 1.399

1-Bromobutane
bp 101.6°C
den 1.275
MW 137.03
n_D^{20} 1.439

In a 5-mL round-bottomed long-necked flask, dissolve 1.33 g of sodium bromide in 1.5 mL of water and 0.80 g of 1-butanol. Cautiously, with constant swirling, add 1.1 mL (2.0 g) of concentrated sulfuric acid dropwise to the solution. The NaBr will dissolve during heating. Fit the flask with a distillation head and ice-cooled receiver (Fig. 16.1), and reflux the reaction mixture on the sand bath for 45 min, taking care that none of the reactants distills during the reaction period. Wrap the upper end of the apparatus with a damp pipe cleaner if escaping vapor is a problem. The upper layer that soon separates in the reaction flask is the alkyl bromide, because the aqueous solution of inorganic salts has the greater density.

Control heating by piling sand around flask or scraping it away.

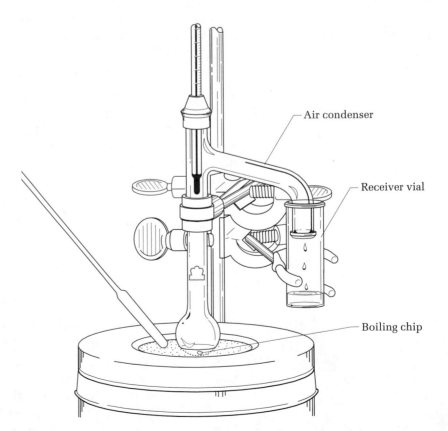

- Air condenser
- Receiver vial
- Boiling chip

FIG. 16.1 Apparatus for the refluxing and distillation of a reaction mixture, shown in the distillation mode.

Remove the pipe cleaner, insulate the flask with cotton or glass wool, and distill the product into a collection vial until no more water-insoluble droplets come over, by which time the temperature of the distillate should have reached 115°C. If in doubt about whether all the product has distilled, collect some of the distillate in a small tube and examine it carefully. The sample collected in the receiver is an azeotrope of 1-bromobutane and water containing some sulfuric acid, 1-butene, unreacted 1-butanol, and di-*n*-butyl ether. To ease cleanup, wash the round-bottomed flask immediately. Rinse the distillation head with acetone so that it will be dry for use later in the experiment.

Transfer the distillate to a reaction tube, rinsing the vial with about 1 mL of water, which is then mixed with the sample in the reaction tube. Note that the 1-bromobutane now forms the lower layer. Remove the 1-bromobutane with a Pasteur pipette, and place it in a dry reaction tube. Add 1 mL of concentrated sulfuric acid, and mix the contents well by flicking the tube. The acid removes any unreacted starting material as well as any alkenyl or ethereal byproducts. Allow the two layers to separate completely, and then remove the sulfuric acid layer. The relative densities given above will help identify the two layers. An empirical method of distinguishing the layers is to remove a drop of the lower layer into a test tube of water to see whether the material is soluble (H$_2$SO$_4$) or not (1-bromobutane). Separate the layers, and wash the 1-bromobutane layer with 1 mL of 3 M sodium hydroxide solution (den. 1.11) to remove traces of acid, separate, and be careful to save the proper layer. In experiments of this type, it is good practice to save all layers until the product is in hand.

Identifying the organic layer.

Dry the cloudy 1-bromobutane by adding anhydrous calcium chloride pellets and mixing until the liquid clears and the calcium chloride no longer clumps together. After 5 min, decant the dried liquid into the dry 5-mL round-bottomed flask. This flask is conveniently dried by rinsing it with a milliliter of ethanol followed by a milliliter of acetone and then drawing air through it using a water aspirator. It is very important when drying an apparatus of this type to remove all the wash solvent, acetone in this case; otherwise, it will contaminate the final product. Rinse the drying agent with two 1-mL portions of *p*-xylene (bp 137–138°C), which is then transferred to the round-bottomed flask. The high-boiling xylene is a "chaser"; it chases all the bromobutane from the distilling flask. Otherwise, about 0.3 mL would remain behind.

CH$_3$

CH$_3$

p-Xylene
MW 106.17
bp 137–138°C

Add a boiling stone, pack the neck of the flask with a stainless steel sponge for fractional distillation, and fit it with a dry distilling head and thermometer. Use the Viton (black) connector between the flask and the distilling head. Wrap the column with cotton or glass wool and distill, collecting material boiling in the range 99 to 103°C. Stop collecting the moment the temperature begins to rise above 103°C. Most of the product will boil at 102°C. A typical yield is in the range of 1 to 1.2 g.

Viton is very resistant to aromatic solvents.

Put the sample in a vial of appropriate size, and make a neatly printed label giving the name and formula of the product and your name.

Cleaning Up   Carefully dilute all nonorganic material with water (the pot residue, the sulfuric acid wash, the sodium hydroxide wash), combine, and neutralize with sodium carbonate before flushing down the drain. All organic material, because it is contaminated with small quantities of product, must be placed in the halogenated organic waste container. The drying agent, wet with *p*-xylene, goes in the hazardous waste container for solvent-contaminated drying agent.

MACROSCALE

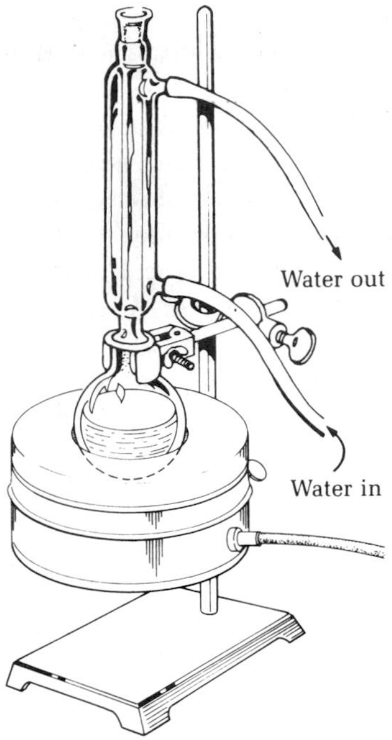

FIG. 16.2 Refluxing a reaction mixture.

Calcium chloride removes both water and alcohol from a solution. Not as efficient a drying agent as anhydrous sodium sulfate.

Synthesis of 1-Bromobutane

Put 13.3 g of sodium bromide, 15 mL of water, and 10 mL of *n*-butyl alcohol in a 100-mL round-bottomed flask, cool the mixture in an ice-water bath, and slowly add 11.5 mL of concentrated sulfuric acid with swirling and cooling. Place the flask in an electric flask heater, clamp it securely, and fit it with a short condenser for reflux condensation (Fig. 16.2). Heat to the boiling point, note the time, and adjust the heat for brisk, steady refluxing. The upper layer that soon separates is the alkyl bromide, since the aqueous solution of inorganic salts has a greater density. Reflux for 45 min, remove the heat, and let the condenser drain for a few minutes (extension of the reaction period to 1 h increases the yield by only 1–2%). Remove the condenser, mount a stillhead in the flask, and set the condenser for downward distillation (see Fig. 5.10) through a bent or vacuum adapter into a 50-mL Erlenmeyer. Distill the mixture, make frequent readings of the temperature, and distill until no more water-insoluble droplets come over, by which time the temperature should have reached 115°C (collect a few drops of distillate in a test tube, and see if it is water soluble). The increasing boiling point is due to azeotropic distillation of *n*-butyl bromide with water containing increasing amounts of sulfuric acid, which raises the boiling point.

Pour the distillate into a separatory funnel, shake with about 10 mL of water, and note that *n*-butyl bromide (1-bromobutane) now forms the lower layer. A pink coloration in this layer due to a trace of bromine can be discharged by adding a pinch of sodium bisulfite and shaking again. Drain the lower layer of 1-bromobutane into a clean flask, clean and dry the separatory funnel, and return the 1-bromobutane to it. Then cool 10 mL of concentrated sulfuric acid thoroughly in an ice bath, add the acid to the funnel, shake well, and allow 5 min for separation of the layers. Use care in handling concentrated sulfuric acid. Check to see that the stopcock and stopper don't leak. The relative densities given in the tables presented in the introduction of this experiment identify the two layers; an empirical method of telling the layers apart is to draw off a few drops of the lower layer into a test tube and see whether the material is soluble in water (H_2SO_4) or insoluble in water (bromobutane). Separate the layers, allow 5 min for further drainage, and separate again. Then wash the 1-bromobutane with 10 mL of 3 *M* sodium hydroxide (den. 1.11) solution to remove traces of acid, separate, and be careful to save the proper layer.

Dry the cloudy 1-bromobutane by adding 1 g of anhydrous calcium chloride pellets with swirling until the liquid clears. After 5 min decant the dried liquid into a 25-mL flask, or filter it through a fluted filter paper, add a boiling stone, distill, and collect material boiling in the range 99–103°C. A typical student yield is in the range 10–12 g. Note the approximate volumes of forerun and residue.

A proper label is important.

Check purity by TLC and IR. The refractive index can also be checked.

Put the sample in a narrow-mouth bottle of appropriate size; make a neatly printed label giving the name and formula of the product and your name. Press the label onto the bottle under a piece of filter paper, and make sure that it is secure. After all the time spent on the preparation, the final product should be worthy of a carefully executed and secured label.

The ^{13}C NMR spectra of 1-bromobutane (Fig. 16.3) and 1-butanol (Fig. 16.4) as well as the ^{1}H NMR spectrum of 1-butanol (Fig. 16.5) are found at the end of the chapter.

Cleaning Up   Carefully dilute all nonorganic material with water (the reaction pot residue, the sulfuric acid wash, and the sodium hydroxide wash), and combine and neutralize with sodium carbonate before flushing down the drain with excess water. The residue from the distillation of 1-bromobutane goes in the container for halogenated organic solvents. The drying agent, after the solvent is allowed to evaporate from it in the hood, goes in the nonhazardous solid waste container.

Questions

1. What experimental method would you recommend for the preparation of 1-bromooctane? Of *t*-butyl bromide?

2. Explain why the crude product is apt to contain certain definite organic impurities.

3. How does each of these impurities react with sulfuric acid when the crude 1-bromobutane is shaken with this reagent?

4. How should the reaction conditions in the present experiment be changed to try to produce 1-chlorobutane?

5. Write a balanced equation for the reaction of sodium bisulfite with bromine.

6. What is the purpose of refluxing the reaction mixture for 30 min? Why not simply boil the mixture in an Erlenmeyer flask?

7. Why is it necessary to remove all water from 1-bromobutane before distilling it?

8. Write reaction mechanisms showing how 1-butene and di-*n*-butyl ether are formed.

9. Why is the resonance of the bromine-bearing carbon atom and the hydroxyl-bearing carbon atom farthest downfield in Figs. 16.3 and 16.4?

FIG. 16.3 ¹³C NMR spectrum
of *n*-butyl bromide (22.6 MHz).

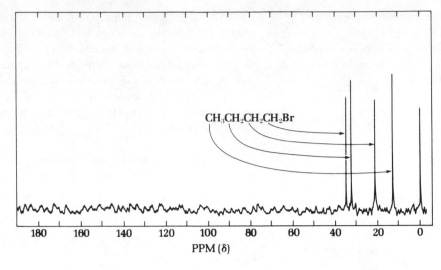

FIG. 16.4 ¹³C NMR spectrum
of 1-butanol (22.6 MHz).

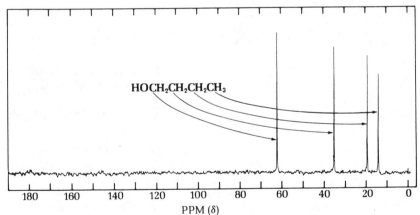

Surfing the Web ───────────────────

http://odin.chemistry.uakron.edu/organic_lab/butyl/butyl01.html

The small-scale synthesis (one-third the scale in this experiment) of 1-bromo-butane is illustrated in this University of Akron Web site with 12 very clear color photos.

For updated information visit:

www.mtholyoke.edu/courses/kwilliam/microscale.shtml

or

www.hmco.com/hmco/college/chemistry/Home.html

FIG. 16.5 ^{1}H NMR spectrum of 1-butanol (250 MHz).

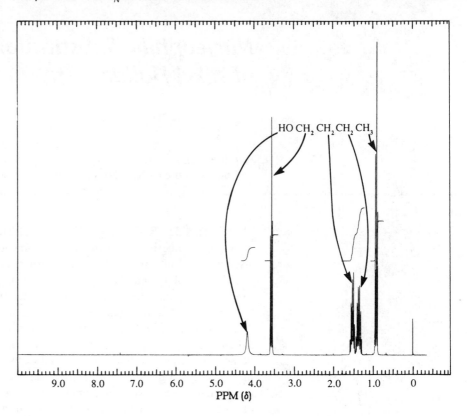

Nucleophilic Substitution Reactions of Alkyl Halides

Prelab Exercise: Predict the outcomes of the two sets of experiments to be carried out with the eight halides used in the present experiment.

The alkyl halides, R—X, where X = Cl, Br, I, and sometimes F, play a central role in organic synthesis. They can easily be prepared from, among others, alcohols, alkenes, and, industrially, alkanes. In turn, they are the starting materials for the synthesis of a large number of new functional groups. These syntheses are often carried out by nucleophilic substitution reactions in which the halide is replaced by some nucleophile such as cyano, hydroxyl, ether, ester, alkyl—the list is long. As a consequence of the importance of this substitution reaction, it has been studied carefully by employing reactions such as the two used in this experiment. Some of the questions that can be asked include these: How does the structure of the alkyl part of the alkyl halide affect the reaction? What is the effect of changing the nature of the halide, the nature of the solvent, the relative concentrations of the reactants, the temperature of the reaction, or the nature of the nucleophile? In this experiment we explore the answers to a few of these questions.

In free radical reactions the covalent bond undergoes homolysis when it breaks

$$R{:}\ddot{C}\kern-0.3em\ddot{}\,l{:} \longrightarrow R{\cdot} + {\cdot}\ddot{C}\kern-0.3em\ddot{}\,l{:}$$

whereas in ionic reactions it undergoes heterolysis

$$R{:}\ddot{C}\kern-0.3em\ddot{}\,l{:} \longrightarrow R^+ + {:}\ddot{C}\kern-0.3em\ddot{}\,l{:}^-$$

A carbocation is often formed as a reactive intermediate in these reactions. This carbocation is sp^2 hybridized and trigonal-planar in structure with a vacant *p*-orbital. Much experimental evidence of the type obtained in the present experiment indicates that the order of stability of carbocations is

$$
\begin{array}{cccc}
\text{R} & \text{R} & \text{H} & \text{H} \\
| & | & | & | \\
\text{R}-\text{C}^+ > \text{R}-\text{C}^+ > \text{R}-\text{C}^+ > \text{H}-\text{C}^+ \\
| & | & | & | \\
\text{R} & \text{H} & \text{H} & \text{H}
\end{array}
$$

The alkyl (R) groups stabilize the positive charge of the carbocation by displacing or releasing electrons toward the positive charge. Delocalization of the charge over several atoms stabilizes the charge.

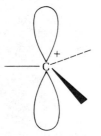

Order of carbocation stability

Nucleophilic substitution

Many organic reactions occur when a nucleophile (a species with an unshared pair of electrons) reacts with an alkyl halide to replace the halogen with the nucleophile.

$$\text{Nu:}^- + \text{R:}\ddot{\text{X}}\text{:} \longrightarrow \text{Nu:R} + \text{:}\ddot{\text{X}}\text{:}^-$$

This substitution reaction can occur in one smooth step

One step

$$\text{Nu:}^- + \text{R:}\ddot{\text{X}}\text{:} \longrightarrow \left[\overset{\delta-}{\text{Nu}} \cdots \text{R} \cdots \overset{\delta-}{\ddot{\text{X}}\text{:}} \right] \longrightarrow \text{Nu:R} + \text{:}\ddot{\text{X}}\text{:}^-$$

or it can occur in two discrete steps

Two steps

$$\text{R:}\ddot{\text{X}}\text{:} \longrightarrow \text{R}^+ + \text{:}\ddot{\text{X}}\text{:}^-$$

$$\text{Nu:}^- + \text{R}^+ \longrightarrow \text{Nu:R}$$

depending primarily on the structure of the R group. The nucleophile (Nu:$^-$) can be a substance with a full negative charge, such as $\text{:}\ddot{\text{I}}\text{:}^-$ or $\text{H:}\ddot{\text{O}}\text{:}^-$, or an uncharged molecule with an unshared pair of electrons such as exists on the oxygen atom in water, $\text{H}—\ddot{\text{O}}—\text{H}$. Not all of the halides, $\text{:}\ddot{\text{X}}\text{:}^-$, depart with equal ease in nucleophilic substitution reactions. In this experiment we investigate the ease with which the different halogens leave in one of the substitution reactions.

Reaction kinetics

To distinguish between the reaction that occurs as one smooth step and the reaction that occurs as two discrete steps, it is necessary to study the kinetics of the reaction. If the reaction were carried out with several different concentrations of $\text{R:}\ddot{\text{X}}\text{:}$ and Nu:$^-$, we could determine if the reaction is bimolecular or unimolecular. In the case of the smooth, one-step reaction, the nucleophile must collide with the alkyl halide. The kinetics of the reaction

$$\text{Nu:}^- + \text{R:}\ddot{\text{X}}\text{:} \longrightarrow \left[\overset{\delta-}{\text{Nu}} \cdots \text{R} \cdots \overset{\delta-}{\ddot{\text{X}}\text{:}} \right] \longrightarrow \text{Nu:R} + \text{:}\ddot{\text{X}}\text{:}^-$$

$$\text{Rate} = k\left[\text{Nu:}^- \right]\left[\text{R}\ \ddot{\text{X}}\text{:} \right]$$

are found to depend on the concentration of both the nucleophile and the halide. Such a reaction is said to be a *bimolecular nucleophilic substitution reaction*, S_N2. If the reaction occurs as a two-step process

$$\text{R:}\ddot{\text{X}}\text{:} \xrightarrow{\text{slow}} \text{R}^+ + \text{:}\ddot{\text{X}}\text{:}^-$$

$$\text{Nu:}^- + \text{R}^+ \xrightarrow{\text{fast}} \text{Nu:R}$$

$$\text{Rate} = k\left[\text{R:}\ddot{\text{X}}\text{:} \right]$$

the rate of the first step, the slow step, depends only on the concentration of the halide, and it is said to be a *unimolecular nucleophilic substitution reaction*, S_N1.

The S_N1 reaction proceeds through a planar carbocation. Even if the starting material were chiral, the product would be a $50:50$ mixture of enantiomers, because the intermediate is planar.

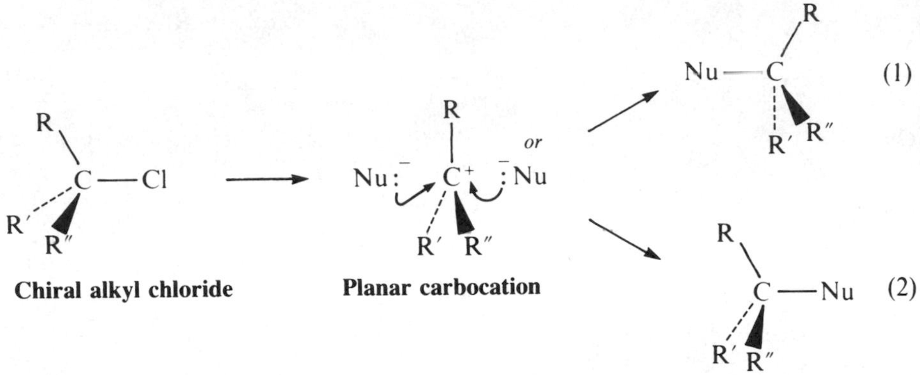

Chiral alkyl chloride   **Planar carbocation**   **Enantiomers**

(1)

(2)

The S_N2 reaction occurs with inversion of configuration to give a product of the opposite chirality from the starting material.

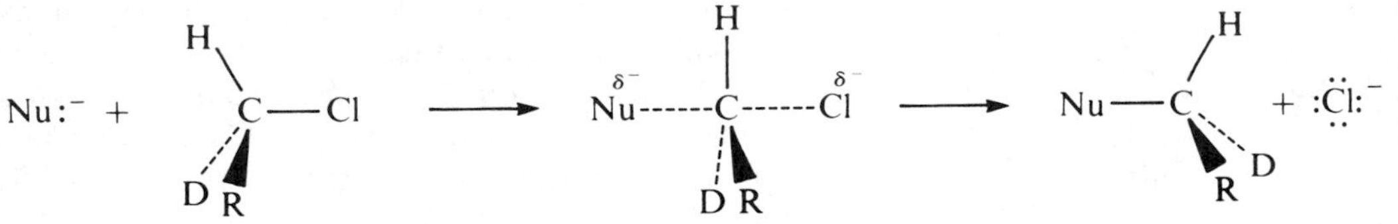

The order of reactivity for *simple* alkyl halides in the S_N2 reaction is

$$CH_3-X > R-CH_2-X > R-\underset{\underset{R}{|}}{CH}-X > (R-\underset{\underset{R}{|}}{\overset{\overset{R}{|}}{C}}-X)$$

The tertiary halide is in parentheses because it usually does not react by an S_N2 mechanism. The primary factor in this order of reactivity is steric hindrance, that is, the ease with which the nucleophile can come within bonding distance of the alkyl halide;. 2,2-dimethyl-1-bromopropane

$$CH_3-\underset{\underset{CH_3}{|}}{\overset{\overset{CH_3}{|}}{C}}-CH_2-Br$$

even though it is a primary halide, reacts 100,000 times slower than ethyl bromide, CH_3CH_2Br, because of steric hindrance to attack on the bromine atom in the dimethyl compound.

The primary factor in S_N1 reactivity is the relative stability of the carbocation that is formed. For simple alkyl halides, this means that only tertiary halides

The allyl carbocation

react by this mechanism. The tertiary halide must be able to form a planar carbocation. Only slightly less reactive are the allyl carbocations, which derive their great stability from the delocalization of the charge on the carbon by resonance.

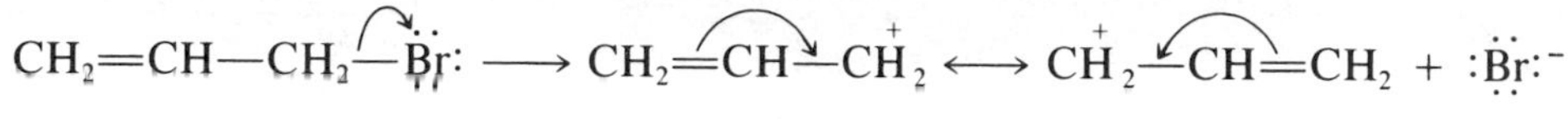

$$CH_2{=}CH{-}CH_2{-}\ddot{B}r: \longrightarrow CH_2{=}CH{-}\overset{+}{C}H_2 \longleftrightarrow \overset{+}{C}H_2{-}CH{=}CH_2 + :\ddot{B}r:^-$$

Solvent effects

The nature of the solvent has a large effect on the rates of S_N2 reactions. In a solvent with a hydrogen atom attached to an electronegative atom such as oxygen, the protic solvent forms hydrogen bonds to the nucleophile.

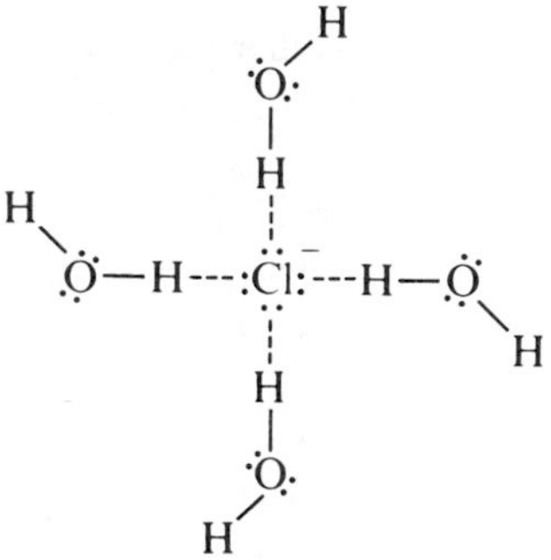

Aprotic: no ionizable protons

These solvent molecules get in the way during an S_N2 reaction. If the solvent is polar and aprotic, solvation of the nucleophile cannot occur, and the S_N2 reaction can occur up to a million times faster. Some common polar, aprotic solvents are

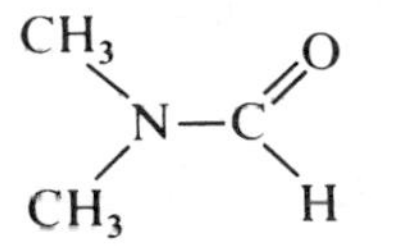

N,N-Dimethylformamide 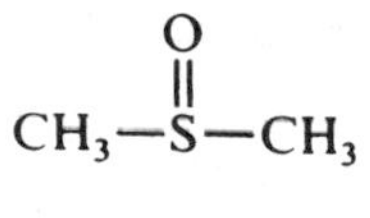   **Dimethylsulfoxide**

In the S_N1 reaction a polar protic solvent such as water stabilizes the transition state more than it does the reactants, lowering the energy of activation for the reaction and thus increasing the rate, relative to the rate in a nonpolar solvent. Acetic acid, ethanol, and acetone are relatively nonpolar solvents and have lower dielectric constants than the polar solvents water, dimethylsulfoxide, and N,N-dimethylformamide.

The leaving group

The rate of S_N1 and S_N2 reactions depends on the nature of the leaving group, the best leaving groups being the ones that form stable ions. Among the halogens we find that iodide ion, I^-, is the best leaving group as well as the best nucleophile in the S_N2 reaction.

Vinylic and aryl halides

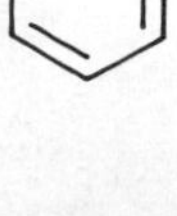

Table 17.1   Summary of S_N1 and S_N2 Reactions

	Unimolecular Nucleophilic Substitution (S_N1)	Second-Order Nucleophilic Substitution (S_N2)
Kinetics	First order	Second order
Mechanism	Two-step that is unimolecular in the rate-determining step via a carbocation	One-step, bimolecular
Stereochemistry	Racemization predominates	Inversion of configuration
Reactivity of reaction	$3° > 2° > 1° > CH_3 > $ vinyl	$3° < 2° < 1° < CH_3 < $ vinyl
Rearrangements	May occur	No rearrangements because no carbocation intermediate
Effect of leaving group	$-I > -Br > -Cl \gg -F$	$-I > -Br > -Cl \gg -F$
Effect of nucleophile	Not important because it is not in the rate-determining step	$I^- > Br^- > Cl^- > F^-$
Concentration of nucleophile	S_N1 is favored by low concentration	S_N2 is favored by high concentration
Solvent polarity	High, favors S_N1	Low, favors S_N2

do not normally react by S_N1 or S_N2 reactions because the resulting carbocations

are relatively unstable. The electrons in the nearby double bonds repel the nucleophile, which is either an ion or a polarized neutral species.

Temperature dependence

   The rates of both S_N1 and S_N2 reactions depend on the temperature of the reaction. As the temperature increases the kinetic energy of the molecules increases, leading to a greater rate of reaction. The rate of many organic reactions will approximately double when the temperature increases about 10°C.

   This information is summarized in Table 17.1.

Experiments

In the experiments that follow, eight representative alkyl halides are treated with sodium iodide in acetone and with an ethanolic solution of silver nitrate.

Sodium Iodide in Acetone

Acetone, with a dielectric constant of 21, is a relatively nonpolar solvent that will readily dissolve sodium iodide. The iodide ion is an excellent nucleophile, and

Organic halides that can react by an S_N2 mechanism give a precipitate of NaX with sodium iodide in acetone.

the nonpolar solvent, acetone, favors the S_N2 reaction; it does not favor ionization of the alkyl halide. The extent of reaction can be observed because sodium bromide and sodium chloride are not soluble in acetone and precipitate from solution if reaction occurs.

$$Na^+I^- + R\!-\!Cl \longrightarrow R\!-\!I + NaCl\downarrow$$

$$Na^+I^- + R\!-\!Br \longrightarrow R\!-\!I + NaBr\downarrow$$

Ethanolic Silver Nitrate Solution

Organic halides that can react by an S_N1 mechanism give a precipitate of AgX with ethanolic silver nitrate solution

When an alkyl halide is treated with an ethanolic solution of silver nitrate, the silver ion coordinates with an electron pair of the halogen. This weakens the carbon–halogen bond as a molecule of insoluble silver halide is formed, thus promoting an S_N1 reaction of the alkyl halide. The solvent, ethanol, favors ionization of the halide, and the nitrate ion is a very poor nucleophile, so alkyl nitrates do not form by an S_N2 reaction.

$$R\!-\!\overset{..}{\underset{..}{X}}\!: \underset{Ag^+}{\overset{Ag^+}{\rightleftharpoons}} \overset{\delta+}{R}\overset{..}{\underset{..}{X}}\overset{\delta+}{Ag} \longrightarrow R^+ + AgX\downarrow$$

On the basis of the foregoing discussion tertiary halides would be expected to react with silver nitrate most rapidly and primary halides least rapidly.

Microscale Procedure

Label eleven small containers (reaction tubes, 3-mL centrifuge tubes, 10 × 75-mm test tubes, or 1-mL vials), and place 0.1 mL or 100 mg of each of the following halides in the tubes.

$CH_3CH_2CH_2CH_2Cl$

1-Chlorobutane
bp 77–78°C

$CH_3CH_2CH_2CH_2Br$

1-Bromobutane
bp 100–104°C

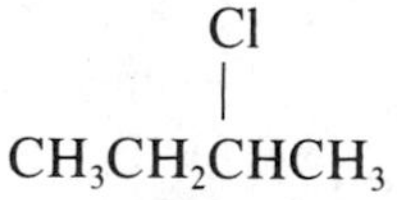

$$CH_3CH_2\overset{\overset{\displaystyle Cl}{|}}{C}HCH_3$$

2-Chlorobutane
bp 68–70°C

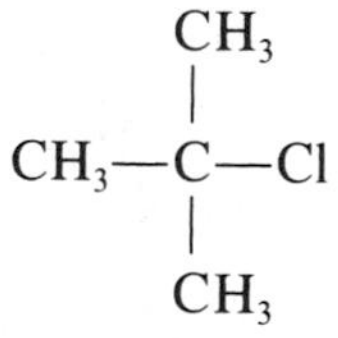

$$CH_3\!-\!\overset{\overset{\displaystyle CH_3}{|}}{\underset{\underset{\displaystyle CH_3}{|}}{C}}\!-\!Cl$$

2-Chloro-2-methylpropane
bp 51–52°C

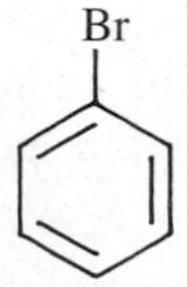

2-Bromobenzene
bp 156° C

$CH_3CH\!=\!CHCH_2Cl$

**1-Chloro-2-butene mixture
of *cis* and *trans* isomers**
bp 63.5°C (*cis*)
68°C (*trans*)

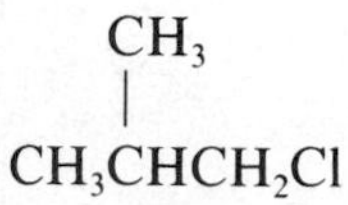

$$CH_3\overset{\overset{\displaystyle CH_3}{|}}{C}HCH_2Cl$$

1-Chloro-2-methylpropane
bp 68–69°C

$$CH_3CH_2\overset{\overset{\displaystyle Br}{\displaystyle |}}{C}HCH_3 \quad CH_2CH{=}C\,(Cl)\,CH_3 \quad CH_3CH_2I$$

2-Bromobutane   **2-Chloro-2-butene**   **Iodoethane**   **1-Chloroadamantane**
bp 91°C   **(mixture of isomers)**   bp 72°C   mp 165–166°C
                        bp 62–67°C

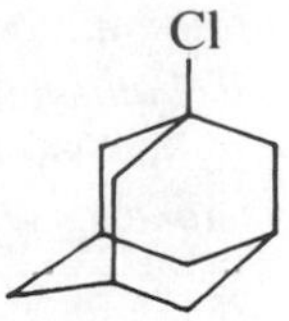

To each tube then rapidly add 1 mL of an 18% solution of sodium iodide in acetone, stopper each tube, mix the contents thoroughly, and note the time. Note the time of first appearance of any precipitate. If no reaction occurs within about 5 min, place those tubes in a 50°C water bath and watch for any reaction over the next 5 or 6 min.

Empty the tubes, rinse them with ethanol, place the same amount of each of the alkyl halides in each tube as in the first part of the experiment, add 1 mL of 1% ethanolic silver nitrate solution to each tube, mix the contents well, and note the time of addition as well as the time of appearance of the first traces of any precipitate. If a precipitate does not appear in 5 min, heat those tubes in a 50°C water bath for 5 to 6 min and watch for any reaction.

To test the effect of solvent on the rate of S_N1 reactivity, compare the time needed for a precipitate to appear when 2-chlorobutane is treated with 1% ethanolic silver nitrate solution (above) and when treated with 1% silver nitrate in a mixture of 50% ethanol and 50% water.

In your analysis of the results from these experiments consider the following for both S_N1 and S_N2 conditions: the nature of the leaving group (Cl vs. Br) in the 1-halobutanes; the effect of structure, that is, compare simple primary, secondary, and tertiary halides, unhindered primary vs. hindered primary halides, a simple tertiary halide vs. a complex tertiary halide, and an allylic halide vs. a tertiary halide; the effect of solvent polarity on the S_N1 reaction; and the effect of temperature on the reaction.

Cleaning Up   Since all of the test solutions contain halogenated material, all test solutions and washes as well as unused starting materials should be placed in the halogenated organic waste container.

Questions

1. What would be the effect of carrying out the sodium iodide in acetone reaction with the alkyl halides using an iodide solution half as concentrated?

2. The addition of sodium or potassium iodide catalyzes many S_N2 reactions of alkyl chlorides or bromides. Explain.

18

Alkanes and Alkenes: Radical Initiated Chlorination of 1-Chlorobutane; Reactions of Alkanes and Alkenes

Most of the alkanes from petroleum are used to produce energy by combustion, but a few percent are converted to industrially useful compounds by controlled reaction with oxygen or chlorine. The alkanes are inert to attack by most chemical reagents and will react with oxygen and halogens only under the special conditions of radical initiated reactions.

At room temperature an alkane such as butane will not react with chlorine. In order for a reaction to occur

$$CH_3CH_2CH_2CH_3 + Cl_2 \longrightarrow CH_3CH_2CH_2CH_2Cl + HCl$$

a precisely oriented four-center collision of the Cl_2 molecule with the butane molecule must occur with two bonds broken and two bonds formed simultaneously:

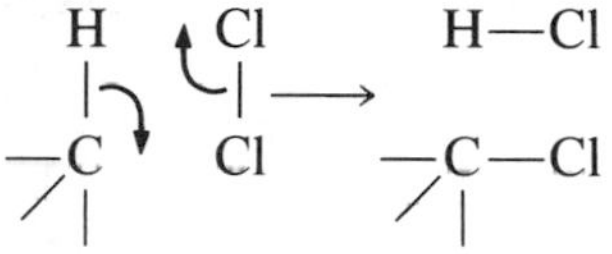

It is unlikely that the necessarily precise orientation of the two reacting molecules will be found. They must, of course, be within bonding distance of each other as well, so steric factors play a part.

If a concerted four-center reaction won't work, then an alternative possibility is a stepwise mechanism:

$$:\ddot{C}l:\ddot{C}l: \xrightarrow{\text{slow}} 2\ :\ddot{C}l\cdot \tag{1}$$

$$CH_3CH_2CH_2CH_3 \xrightarrow{\text{slow}} CH_3CH_2CH_2CH_2\cdot + H\cdot \tag{2}$$

$$:\ddot{C}l\cdot + CH_3CH_2CH_2CH_2\cdot \xrightarrow{\text{fast}} CH_3CH_2CH_2CH_2Cl \tag{3}$$

$$:\ddot{C}l\cdot + H\cdot \xrightarrow{\text{fast}} H:\ddot{C}l: \tag{4}$$

For this series of reactions to occur, the chlorine molecule must dissociate into two chlorine atoms. Because chlorine has a bond energy of 58 kcal mole^{-1}, we would not expect any significant numbers of molecules to dissociate at room temperature. Thermal motion at 25°C can only break bonds having energies less than 30–35 kcal mole^{-1}. Although thermal dissociation would require a high temperature, the dissociation of chlorine into atoms (chlorine radicals) can be caused by violet and ultraviolet light:

$$Cl_2 \xrightarrow{h\nu} 2 \; :\ddot{\underset{..}{C}l}\cdot$$

The photon energy of red light is 48 kcal mole^{-1} while light of 300 nm (ultraviolet) has a photon energy of 96 kcal mole^{-1}.

The chlorine radical can react with butane by abstraction of a hydrogen atom:

$$CH_3CH_2CH_2CH_3 \; + \; :\ddot{\underset{..}{C}l}\cdot \rightleftharpoons CH_3CH_2CH_2CH_2\cdot \; + \; HCl$$

a reaction that is very slightly exothermic (and thus written as a reversible reaction). The butyl radical can react with chlorine:

$$CH_3CH_2CH_2CH_2\cdot \; + \; Cl_2 \longrightarrow CH_3CH_2CH_2CH_2Cl \; + \; :\ddot{\underset{..}{C}l}\cdot$$

a reaction that evolves 26 kcal mole^{-1} of energy. The net result of these two reactions is a reaction of Cl_2 with butane "catalyzed" by $:\ddot{\underset{..}{C}l}\cdot$, the chlorine radical. The whole process can be terminated by the reaction of radicals with each other:

$$CH_3CH_2CH_2CH_2\cdot \; + \; :\ddot{\underset{..}{C}l}\cdot \longrightarrow CH_3CH_2CH_2CH_2Cl$$

$$2 \; CH_3CH_2CH_2CH_2\cdot \longrightarrow CH_3(CH_2)_6CH_3$$

$$2 \; :\ddot{\underset{..}{C}l}\cdot \longrightarrow Cl_2$$

To summarize, the process of light-induced radical chlorination involves three steps: chain initiation in which chlorine radicals are produced, chain propagation that involves no net consumption of chlorine radicals, and chain termination that destroys radicals.

$$Cl_2 \longrightarrow 2 \; :\ddot{\underset{..}{C}l}\cdot \qquad \textit{Chain initiation}$$

$$\left.\begin{array}{l} CH_3CH_2CH_2CH_3 \; + \; :\ddot{\underset{..}{C}l}\cdot \rightleftharpoons CH_3CH_2CH_2CH_2\cdot \; + \; HCl \\ CH_3CH_2CH_2CH_2\cdot \; + \; Cl_2 \longrightarrow CH_3CH_2CH_2CH_2Cl \; + \; :\ddot{\underset{..}{C}l}\cdot \end{array}\right\} \; \begin{array}{l}\textit{Chain}\\ \textit{propagation}\end{array}$$

$$\left.\begin{array}{l} CH_3CH_2CH_2CH_2\cdot \; + \; :\ddot{\underset{..}{C}l}\cdot \longrightarrow CH_3CH_2CH_2CH_2Cl \\ 2 \; CH_3CH_2CH_2CH_2\cdot \longrightarrow CH_3(CH_2)_6CH_3 \\ 2 \; :\ddot{\underset{..}{C}l}\cdot \longrightarrow Cl_2 \end{array}\right\} \; \begin{array}{l}\textit{Chain}\\ \textit{termination}\end{array}$$

Relative Reactivities of Hydrogen

In the preceding example the product is shown to be 1-chlorobutane, but in fact the reaction produces a mixture of 1-chlorobutane and 2-chlorobutane. If the reaction were to occur purely by chance, we would expect the ratio of products to be 6:4, because there are six primary hydrogens and four secondary hydrogens on butane. But because a secondary C—H bond is weaker than a primary C—H bond (95 vs. 98 kcal mole^{-1}), we might expect more 2-chlorobutane than chance would dictate. The chlorination of 2-methylbutane is summarized in Table 18.1.

As can be seen from the table, the relatively weak tertiary C—H bond (92 kcal mole^{-1}) gives rise to 22% of product in contrast to the 8% expected on the basis of a random attack of $:\overset{..}{\text{Cl}}\cdot$ on the starting material.

The relative reactivities of the various hydrogens of 2-methylbutane on a per-hydrogen basis (referred to the primary hydrogens of C-4 as 1.0) can be calculated. The molecule has nine primary hydrogens, chlorination of which accounts for 48% of the products. This means that each primary hydrogen is responsible for 48%/9 = 5.3% of the product. Similarly, the two secondary hydrogens account for 33% of the product, so each secondary hydrogen is responsible for 16.5% of the product. Finally, the one tertiary hydrogen is responsible for 22% of the product. These results mean that tertiary hydrogens are 22/5.3 = 4.2 times as reactive as primary hydrogens toward chlorination, and secondary hydrogens are 16.5/5.3 = 3.1 times as reactive toward chlorination as primary hydrogens. Results very similar to these have been found for a large number of hydrocarbons.

Table 18.1   Product Distribution for Chlorination of 2-Methylbutane

	Found	Statistical Expectation	
		Ratio	*Percent*
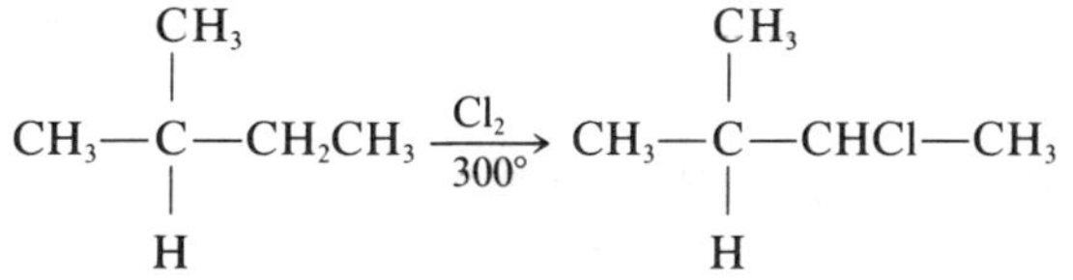 $\text{CH}_3-\overset{\overset{\displaystyle CH_3}{\displaystyle \vert}}{\underset{\underset{\displaystyle H}{\displaystyle \vert}}{C}}-\text{CH}_2\text{CH}_3 \xrightarrow[300°]{\text{Cl}_2} \text{CH}_3-\overset{\overset{\displaystyle CH_3}{\displaystyle \vert}}{\underset{\underset{\displaystyle H}{\displaystyle \vert}}{C}}-\text{CHCl}-\text{CH}_3$	33%	2/12	17
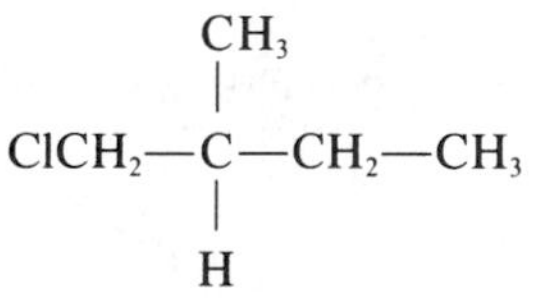 $\text{ClCH}_2-\overset{\overset{\displaystyle CH_3}{\displaystyle \vert}}{\underset{\underset{\displaystyle H}{\displaystyle \vert}}{C}}-\text{CH}_2-\text{CH}_3$	30%	6/12	50
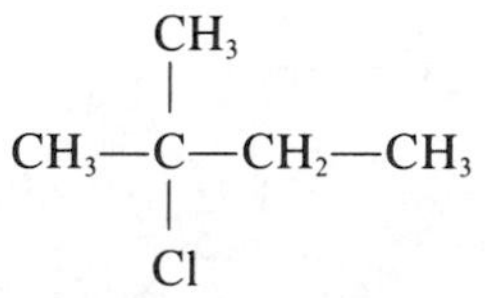 $\text{CH}_3-\overset{\overset{\displaystyle CH_3}{\displaystyle \vert}}{\underset{\underset{\displaystyle Cl}{\displaystyle \vert}}{C}}-\text{CH}_2-\text{CH}_3$	22%	1/12	8
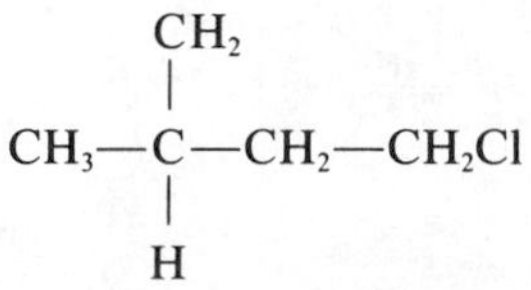 $\text{CH}_3-\overset{\overset{\displaystyle CH_2}{\displaystyle \vert}}{\underset{\underset{\displaystyle H}{\displaystyle \vert}}{C}}-\text{CH}_2-\text{CH}_2\text{Cl}$	15%	3/12	25

Relative reactivity toward chlorination:

$$R_3CH > R_2CH_2 > RCH_3$$
$$\quad 4.2 \quad\quad 3.1 \quad\quad 1$$

A reaction of this type is of little use unless you happen to need the four products in the ratios found and can manage to separate them (their boiling points are very similar). Industrially, however, the radical chlorination of methane and ethane is important, and the products can be separated easily:

$$CH_4 \xrightarrow[\Delta]{Cl_2} \quad CH_3Cl \quad + \quad CH_2Cl_2 \quad + \quad CHCl_3 \quad + \quad CCl_4$$

	Chloro-methane	**Dichloro-methane**	**Trichloro-methane**	**Tetrachloro-methane**
Common names:	Methyl chloride	Methylene chloride	Chloroform	Carbon tetrachloride
Boiling points:	−24°C	40°C	62°C	77°C

In the first experiment we will chlorinate 1-chlorobutane because it is easier to handle in the laboratory than gaseous butane, and we will use sulfuryl chloride as our source of chlorine radicals because it is easier to handle than gaseous chlorine. Instead of using light to initiate the reaction, we will use a chemical initiator, 2,2′-azobis-(2-methylpropionitrile). This azo compound (R—N=N—R) decomposes at moderate temperatures (80–100°C) to give two relatively stable radicals and nitrogen gas:

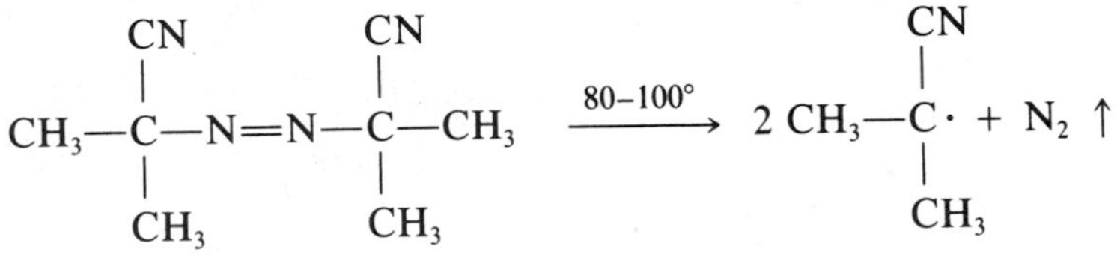

2,2′-Azobis-(2-methylpropionitrile)
MW 164.21, mp 102–103° (dec.)

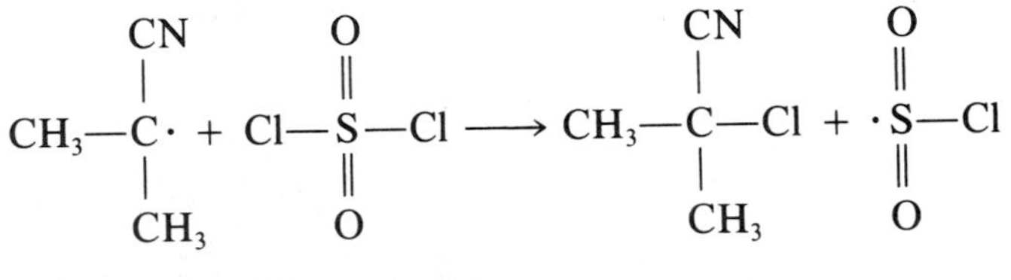

The common name for this catalyst is AIBN, which stands for azoisobutylnitrile.

Sulfuryl chloride

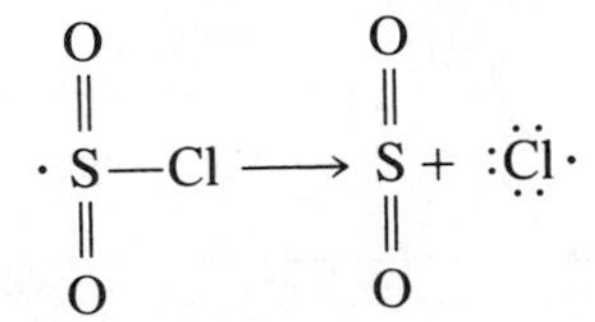

Before conducting the experiment, try to predict the ratio of products.

Will these ratios correlate with the calculated heats of formation of the products?

The radical monochlorination of 1-chlorobutane can give four products: 1,1-, 1,2-, 1,3-, and 1,4-dichlorobutane. If the reaction occurred completely at random we would expect products in the ratios of the number of hydrogen atoms on each carbon, i.e., $2:2:2:3$, respectively (22%, 22%, 22%, 33%). The object of the present experiment is to carry out the radical chlorination of 1-chlorobutane and then to determine the ratio of products using gas-liquid chromatography. From these ratios the relative reactivities of the hydrogens can be calculated.

$$CH_3CH_2CH_2CH_2Cl + SO_2Cl_2 \xrightarrow{R\cdot} CH_3CH_2CH_2CHCl_2 + CH_3CH_2CHClCH_2Cl$$

| | | bp 114°C | bp 124°C |

1-Chlorobutane

MW 92.57
den. 0.886
bp 77–78°C

Sulfuryl chloride
MW 134.97
den. 1.67
bp 69°C

$$+ CH_3CHClCH_2CH_2Cl + CH_2ClCH_2CH_2CH_2Cl$$

bp 134°C   bp 162°C

$$+ SO_2 + HCl$$

Experiments

MICROSCALE

1. Radical Chlorination of 1-Chlorobutane

To a 10×100 mm reaction tube add 1-chlorobutane (0.50 mL, 0.432 g, 4.6 mmol), sulfuryl chloride (0.16 mL, 0.27 g, 2.0 mmol), 2,2′-azobis-(2-methyl-propionitrile) (4 mg, 0.025 mmol), and a boiling chip. Use a 1-mL graduated pipette to measure the 1-chlorobutane, and use a dispenser or a 0.5-mL syringe to measure the sulfuryl chloride (in the hood). Rinse the syringe immediately after using, and leave it disassembled. Weigh the azo compound on the balance. In this experiment none of the reagents need be measured with great care; the 1-chlorobutane is in large excess, and the azo compound is present in catalytic amounts.

Fit the reaction tube with a rubber septum and a piece of polyethylene tubing that leads down into another reaction tube, the mouth of which has a piece of damp cotton placed in it (Fig. 18.1). During this reaction, sulfur dioxide and hydrogen chloride gas are evolved, and the damp cotton will absorb the gas. The technique for threading a polyethylene tube through a septum is shown in Fig. 18.2. The amount of HCl evolved in this experiment (2 mmol) is equal to 0.15 mL of concentrated hydrochloric acid. Be sure that the end of the polyethylene tube does not touch any water because it would be sucked back into the reaction tube. Clamp the reaction tube with the reactants in a beaker of hot water maintained at 80°C so that just the tip of the tube is immersed in the water. This will cause the contents to boil gently and the vapors to condense on the cool upper walls of the reaction tube.

At the end of the reaction period, remove the tube from the beaker of water, allow it to cool, and then carefully add 0.5 mL of water dropwise to the tube from a Pasteur pipette. Note which layer is the aqueous one. Mix the contents thoroughly, and then draw off and discard the water layer. Wash the organic phase in the same way with a 0.5-mL portion of 5% sodium bicarbonate solution and once with 0.5 mL water. Carefully remove all the water with a Pasteur pipette, and then

FIG. 18.1 Microscale apparatus for chlorination of 1-chlorobutane with HCl and SO$_2$ trap. Heat the polyethylene tubing carefully in a steam bath to bend it permanently. To be clamped appropriately.

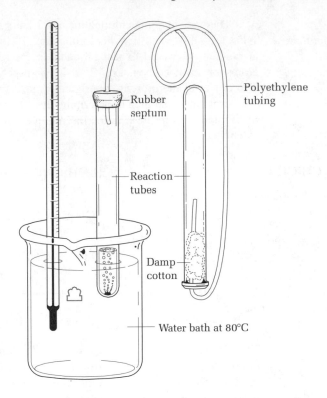

Rubber septum

Polyethylene tubing

Reaction tubes

Damp cotton

Water bath at 80°C

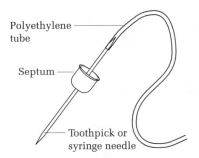

Polyethylene tube

Septum

Toothpick or syringe needle

FIG. 18.2 To thread a polyethylene tube through a septum, make a hole through the septum with a needle, and then push a toothpick through the hole. Push the polyethylene tube firmly onto the toothpick, and then pull and push on the toothpick. The tube will slide through the septum. Finally, pull the tube from the toothpick. A blunt syringe needle can be used instead of a toothpick.

add anhydrous calcium chloride pellets to dry the product. Transfer the dry product (it should be perfectly clear, not cloudy) to a small tared screw-capped sample vial or corked reaction tube. Do *not* store the product in a polyethylene or rubber-capped container, because it will dissolve in the cap material. It is not a good idea to store the product before analysis because the composition will change depending on which components evaporate or are absorbed by the cap on the container. Determine the weight of the product, and then analyze it by gas-liquid chromatography.

Cleaning Up   Rinse the damp cotton with water, and combine the rinse with all the aqueous layers and washes. Neutralize the aqueous solution with sodium carbonate, and flush the solution down the drain. The drying agent will be coated with chlorinated product and therefore must be disposed of in the hazardous waste container. Any unused starting material and product must be placed in the halogenated organic waste container.

2. Free-Radical Chlorination of 1-Chlorobutane

To a 25-mL round-bottomed flask add 1-chlorobutane (5 mL, 4.32 g, 0.046 mole), sulfuryl chloride (1.6 mL, 2.7 g, 0.02 mole), 2,2'-azobis-(2-methylpropionitrile)

Conduct this experiment in a laboratory hood.

(0.03 g), and a boiling chip. Equip the flask with a condenser and gas trap as seen in Fig. 18.3. Heat the mixture to gentle reflux on the steam bath for 20 min. Remove the flask from the steam bath, allow it to cool somewhat, and *quickly,* to minimize the escape of sulfur dioxide and hydrochloric acid, lift the condenser from the flask and add a second 0.03-g portion of the initiator. Heat the reaction mixture for an additional 10 min, remove the flask and condenser from the steam bath, and cool the flask in a beaker of water. Pour the contents of the flask through a funnel into about 10 mL of water in a small separatory funnel, shake the mixture, and separate the two phases. Wash the organic phase with two 4-mL portions of 0.5 *M* sodium bicarbonate solution, once with a 4-mL portion of water, and then dry the organic layer over anhydrous calcium chloride pellets (about 1 g) in a dry Erlenmeyer flask. The mixture can be analyzed by gas chromatography at this point, or the unreacted 1-chlorobutane can be removed by fractional distillation (up to bp 85°C) and the pot residue analyzed by gas chromatography.

Cleaning Up   Empty the gas trap, and combine the contents with all the aqueous layers and washes. Neutralize the aqueous solution with sodium carbonate, and flush the solution down the drain with a large excess of water. The drying agent will be coated with chlorinated product and therefore must be

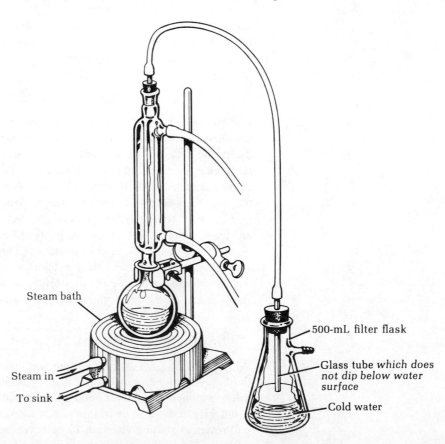

Steam bath

500-mL filter flask

Glass tube *which does not dip below water surface*

Steam in

To sink

Cold water

FIG. 18.3 Apparatus for chlorination of 1-chlorobutane with SO₂ and HCl gas trap.

disposed of in the hazardous waste container for solvent-contaminated drying agent. Any unused starting material and product must be placed in the halogenated organic waste container.

Gas Chromatography

See Chapter 11 for information about gas chromatography. A Carbowax column works best, although any other nonpolar phase such as silicone rubber should work as well. With a nonpolar column packing, the products are expected to come out in the order of their boiling points. A typical set of operating conditions would be column temperature 100°C, He flow rate 35 mL/min, column size 5-mm dia × 2 m, sample size 5 μL, attenuation 16.

The molar amounts of each compound present in the reaction mixture are proportional to the areas under the peaks in the chromatogram. Because 1-chlorobutane is present in large excess, let this peak run off the paper, but be sure to keep the four product peaks on the paper. To determine relative peak areas, simply cut out the peaks with a pair of scissors and weigh them (if the chromatograph is not equipped with an integrator). This method of peak integration works very well and depends on the uniform thickness of paper, which results in its weight being proportional to its area. Calculate the relative percent of each product molecule and the partial rate factors relative to the primary hydrogens on carbon-4. Compare your results with those for the chlorination of 2-methylbutane, the data for which were given above.

3. Computational Chemistry

Construct the four dichlorobutanes and minimize their energies using a molecular mechanics program. Compare steric energies for the different conformers to determine if the global minimum for each one has been found. Record the four steric energies, then submit each energy-minimized molecule to an AM1 semiempirical molecular orbital calculation to calculate the heat of formation. Only the heats of formation can be used to compare the stabilities of the different isomers, so one would not expect a correlation to exist between the steric energies and the heats of formation. Do the heats of formation correlate with the product distributions of the four dichlorobutanes? Does this correlation or lack thereof give information regarding the mechanism of the reaction? Remember that the isomers would need to be in equilibrium with each other in order for the most stable to also be the one formed in highest yield.

Distinguishing Between Alkanes and Alkenes

When an olefin such as cyclohexene is allowed to react with bromine at a very low concentration, substitution instead of addition occurs. Although addition of a halogen atom to the double bond occurs readily, the intermediate free radical (**1**) is not very stable and, if it does not encounter another halogen molecule soon, will revert to starting material. If, however, an allylic free radical (**2**) is formed initially, it is much more stable and will survive long enough to react with a halogen molecule, even when the halogen is at low concentration. Thus, low halogen concentration favors allylic substitution, whereas high halogen concentration favors addition.

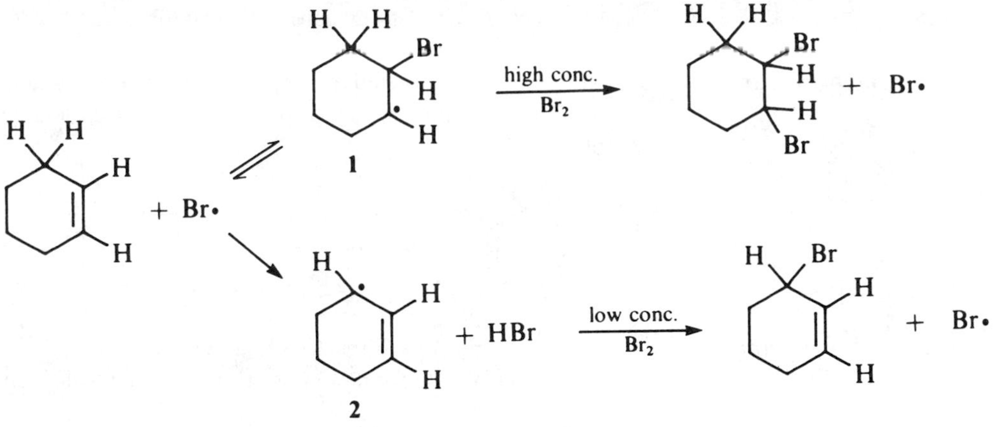

N-Bromosuccinimide (NBS) is a reagent that will continuously generate bromine molecules at a low concentration. This occurs when the HBr molecule from an allylic substitution reacts with NBS:

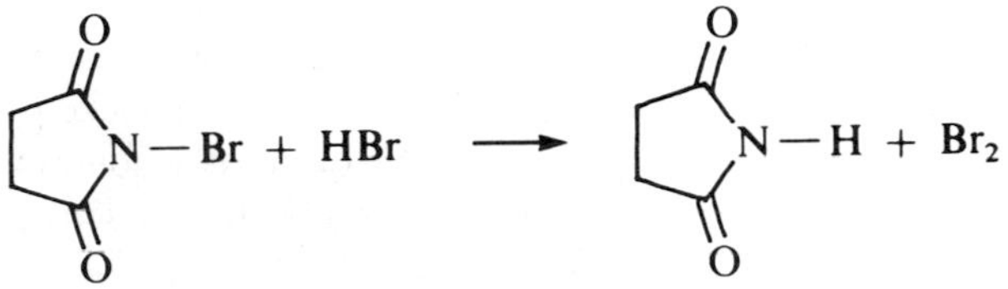

N-Bromosuccinimide

Alkenes react with NBS.   The net reaction is one of allylic bromination:

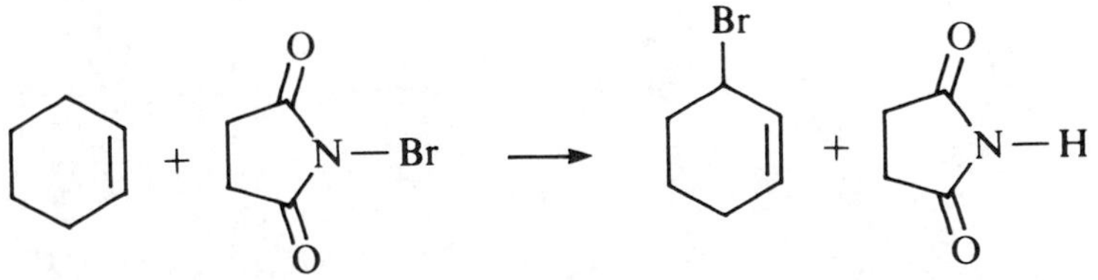

In aqueous solution NBS will react with cyclohexene to form the bromohydrin, a reaction that may or may not involve the intermediate formation of HOBr:

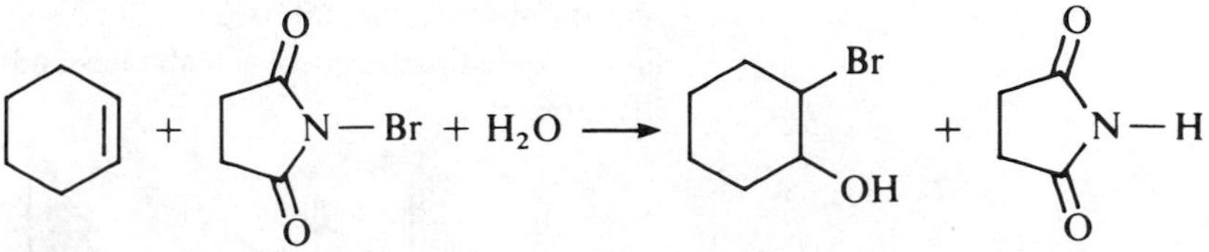

As seen in the preceding experiment, free radical halogenation of alkanes proceeds by substitution to give the chloroalkane and hydrogen chloride gas. Halogenation of an alkene, on the other hand, proceeds by addition across the double bond. These two different modes of reaction can be used to distinguish alkenes from alkanes. Bromine is used as the test reagent. The disappearance of the red bromine color indicates a reaction has taken place. Cyclohexene and dibromocyclohexane are both colorless. If hydrogen bromide has evolved, it can be detected by breathing across the test tube. The hydrogen bromide dissolves in the moist air to give a cloud of hydrobromic acid.

Alkenes decolorize bromine.

Bromine at a relatively high concentration in a nonaqueous solution such as dichloromethane can add to an alkene such as cyclohexene through a free radical process:

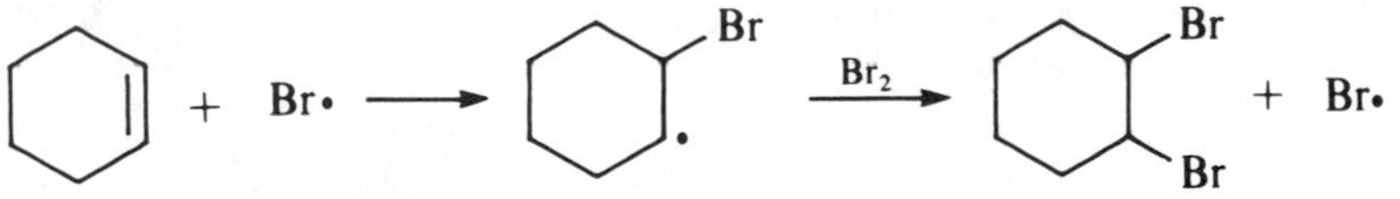

Alkenes react with bromine and evolve HBr

Similarly, it will react with an alkane and will evolve a molecule of hydrogen bromide. The formation of bromine radicals is promoted by light:

$$Br_2 \xrightarrow{h\nu} 2 \cdot \ddot{B}\ddot{r}:$$

$$CH_3CH_2CH_3 + \cdot \ddot{B}\ddot{r}: \longrightarrow HBr + CH_3\dot{C}HCH_3 \xrightarrow{Br_2} CH_3\overset{\underset{|}{Br}}{C}HCH_3 + \cdot \ddot{B}\ddot{r}:$$

In aqueous solution a bromonium ion is formed as the intermediate when bromine reacts with an alkene. This intermediate bromonium ion can react with a molecule of bromine to form a dibromide, and it can react with water to form a bromohydrin. Both reactions give a mixture of products:

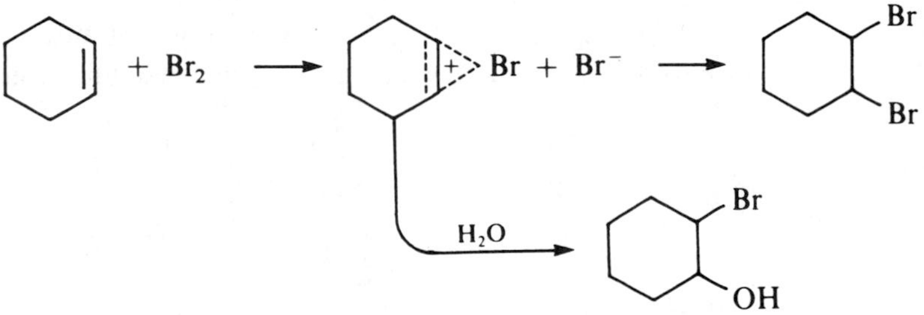

Alkenes decolorize permanganate.

A 10% solution of potassium permanganate is bright purple. When this reagent reacts with an alkene, the purple color disappears and a fine brown suspension of manganese dioxide may be seen. Although we will use permanganate simply to distinguish between alkenes and alkanes, it is also an important preparative reagent:

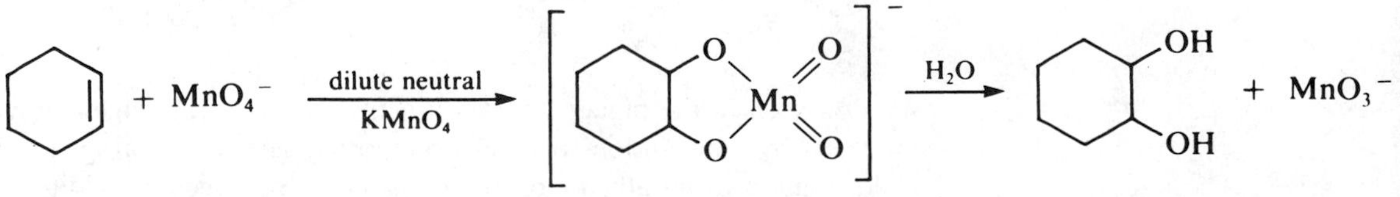

$$\text{(cyclohexane-1,2-diol)} + MnO_4^- + H^+ \xrightarrow[\text{KMnO}_4]{\text{acidic}} \text{(cyclohexane-1,2-dicarboxylic acid)} + HMnO_4^{2-}$$

$$3HMnO_4^{2-} + H_2O \rightarrow 2MnO_2 + MnO_4^- + 5OH^-$$

Alkenes dissolve in sulfuric acid.

Alkenes will react with sulfuric acid to form alkyl hydrogen sulfates. In the process the alkene will appear to dissolve in the concentrated sulfuric acid. Alkanes are completely unreactive toward sulfuric acid.

$$\text{(cyclohexene)} + H_2SO_4 \longrightarrow \text{(cyclohexyl hydrogen sulfate)} \quad OSO_3^- H^+$$

MICROSCALE AND MACROSCALE

4. Tests for Alkanes and Alkenes

The following tests demonstrate properties characteristic of alkanes and alkenes, provide means of distinguishing between compounds of the two types, and distinguish between pure and impure alkanes. Use your own preparation of cyclohexene (dried over anhydrous calcium chloride pellets) as a typical alkene, purified 66–75°C ligroin (Eastman Organic Chemicals No. 513) as a typical alkane mixture (the boiling point of hexane is 69°C), and unpurified ligroin (see *Instructor's Guide*) as an impure alkane. In the following experiments, distinguish clearly, as you take notes, between *observations* and *conclusions* based on those observations. Write equations for all positive tests.

Ligroin = "hexanes"
≈ hexane

(A) Bromine in Nonaqueous Solution. Treat 0.5-mL samples of purified ligroin and cyclohexene with 2 to 3 drops of a 3% solution of bromine in dichloromethane. In case decolorization occurs, breathe across the mouth of the tube to see if hydrogen bromide can be detected. If the bromine color persists, illuminate the solution, and if a reaction occurs, test as before for hydrogen bromide.

These tests can be carried out on a much smaller scale.

(B) Bromine Water. Measure 1 mL of a 3% aqueous solution of bromine into each of three reaction tubes, and then add 0.3-mL portions of purified ligroin to two of the tubes and 0.3 mL of cyclohexene to the third. Shake each tube, and record the initial results. Put one of the ligroin-containing tubes in the desk out of the light, and expose the other to bright sunlight or hold it close to a light bulb. When a change is noted, compare the appearance with that of the mixture kept in the dark.

Bromine (3%) in dichloromethane should be freshly prepared and stored in a brown bottle; test the reagent for decomposition by breathing across the bottle.

(C) Acid Permanganate Test. To 0.3-mL portions of purified ligroin and cyclohexene, add a drop of an aqueous solution containing 1% potassium permanganate and 10% sulfuric acid and shake. If the initial portion of reagent is decolorized, add further portions.

CAUTION: Student-prepared cyclohexene may be wet. Dry over anhydrous calcium chloride pellets before use.

All these tests can be carried out on a much smaller scale.

**Pyridinium hydrobromide perbromide
MW 319.84**

**Dioxane
bp 101.5°C, den. 1.04
(dissolves organic
compounds, miscible
with water)**

CAUTION: Carcinogen

Note for the instructor

(D) Sulfuric Acid.

Cool 0.3-mL portions of purified ligroin and cyclohexene in ice, treat each with 1 mL of concentrated sulfuric acid, and shake. Observe and interpret the results. Is any reaction apparent? Any warming? If the mixture separates into two layers, identify them.

(E) Bromination with Pyridinium Hydrobromide Perbromide ($C_5H_5\overset{+}{N}HBr_3^-$).[1]

This substance is a crystalline, nonvolatile, odorless complex of high molecular weight (319.84), which, in the presence of a bromine acceptor such as an alkene, dissociates to liberate 1 mol of bromine. For small-scale experiments it is much more convenient and agreeable to measure and use than free bromine.

Add 80 mg of the reagent to a reaction tube, and add 0.5 mL of acetic acid. Swirl the mixture, and note that the solid is sparingly soluble. Add 20 mg (0.025 mL) of cyclohexene to the suspension of reagent. Swirl, crush any remaining crystals with a flattened stirring rod, and if after a time the amount of cyclohexene appears insufficient to exhaust the reagent, add a little more. When the solid is all dissolved, dilute with water, and note the character of the product. By what property can you be sure that it is the reaction product and not starting material?

(F) Formation of a Bromohydrin.

N-Bromosuccinimide in an aqueous solution will react with an olefin to form a bromohydrin:

Weigh 90 mg of *N*-bromosuccinimide, put it into reaction tube, and add 0.25 mL of dioxane and 0.5 mmol (40 mg) of cyclohexene. In another tube chill 0.1 mL water, and add to it 0.5 mmol concentrated sulfuric acid. Transfer the cold dilute solution to the first tube with a Pasteur pipette. Note the result and the nature of the product that separates on dilution with water.

(G) Tests for Unsaturation.

Determine which of the following hydrocarbons are saturated and which are unsaturated or contain unsaturated material. Use any of the preceding tests that seem appropriate.

Camphene
Pinene, the principal constituent of turpentine oil
Paraffin oil, a purified petroleum product

---

1. Crystalline material suitable for small-scale experiments is supplied by Aldrich Chemical Co. Massive crystals commercially available should be recrystallized from acetic acid (4 mL/g). Preparation: Mix 15 mL of pyridine with 30 mL of 48% hydrobromic acid and cool; add 25 g bromine gradually with swirling, cool, and collect the product with use of acetic acid for rinsing and washing. Without drying the solid, crystallize it from 100 mL of acetic acid. Yield of orange needles, 33 g (69%).

Gasoline produced by cracking

Cyclohexane

Rubber (The adhesive Grippit and other rubber cements are solutions of unvulcanized rubber. Squeeze a drop of one of these onto a stirring rod, and dissolve it in toluene. For tests with permanganate or with bromine in dichloromethane, use only a drop of the former and just enough of the latter to produce coloration.)

Cleaning Up   Place the entire test solutions from tests 3(A), 3(B), 3(E), 3(F), and 3(G) in the halogenated organic solvents container. Dilute the solutions from tests 3(C) and 3(D) with water, place the organic layer in the organic solvents container, and after neutralization with sodium carbonate, flush the aqueous layer down the drain.

Questions

1. Draw the structure of the compound formed when cyclohexene dissolves in concentrated sulfuric acid.

2. The reaction of cyclohexene with cold dilute aqueous potassium permanganate gives a compound having the empirical formula $C_6H_{12}O_2$. What is the structure of this compound? What is the stereochemistry of the compound?

3. Cyclohexene reacts with a high concentration of bromine to give what compound? What is its stereochemistry?

4. 1-Hexene is brominated with pyridinium hydrobromide perbromide as in test 3(E). The reaction mixture is diluted with water. By what physical property can you be sure that a reaction product has been produced and not starting material? What is its density?

19

Alkenes from Alcohols: Cyclohexene from Cyclohexanol

Prelab Exercise: Prepare a detailed flow sheet for the preparation of cyclohexene, indicating at each step which layer contains the desired product.

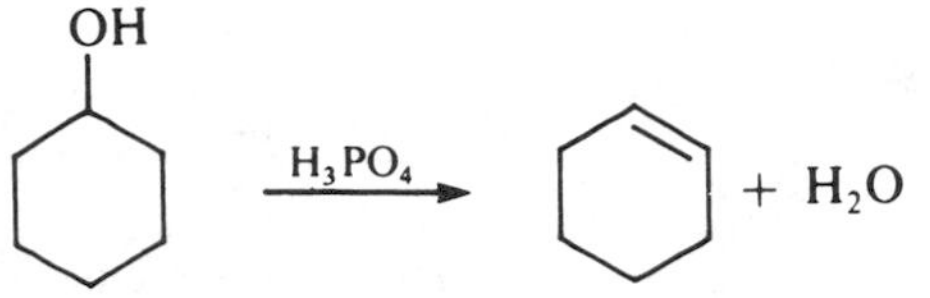

OH

$$\xrightarrow{\text{H}_3\text{PO}_4}$$

$+ \text{ H}_2\text{O}$

Cyclohexanol
mp 25°C, bp 161°C
den. 0.96, MW 100.16

Cyclohexene
bp 83°C
den. 0.81, MW 82.14

Dehydration of cyclohexanol to cyclohexene can be accomplished by pyrolysis of the cyclic secondary alcohol with an acid catalyst at a moderate temperature or by distillation over alumina or silica gel. The procedure selected for this experiment involves catalysis by phosphoric acid; sulfuric acid is no more efficient, causes charring, and gives rise to sulfur dioxide. When a mixture of cyclohexanol and phosphoric acid is heated in a flask equipped with a fractionating column, the formation of water is soon evident. On further heating, the water and the cyclohexene that form distill together by the principle of steam distillation, and any high-boiling cyclohexanol that may volatilize is returned to the flask. However, after dehydration is complete and the bulk of the product has distilled, the column remains saturated with water–cyclohexene that merely refluxes and does not distill. Hence, for recovery of otherwise lost reaction product, a chaser solvent is added and distillation is continued. A suitable chaser solvent is the water-immiscible, aromatic solvent toluene (bp 110°C); as it steam-distills it carries over the more volatile cyclohexene. When the total water-insoluble layer is separated, dried, and redistilled through the dried column, the chaser again drives the cyclohexene from the column; the difference in boiling points is such that a sharp separation is possible. The holdup in the metal sponge-packed column is so great (about 1.0 mL) that if a chaser solvent is not used in the procedure the yield will be much lower.

The mechanism of this reaction involves initial rapid protonation of the hydroxyl group by the phosphoric acid:

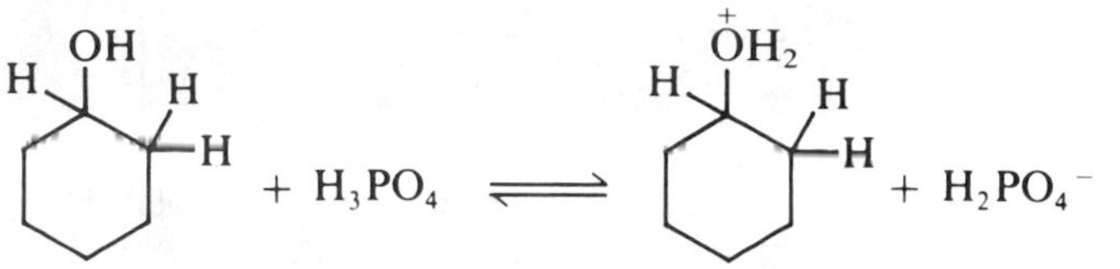

This is followed by loss of water to give the unstable secondary carbocation, which quickly loses a proton to water or the conjugate acid to give the alkene:

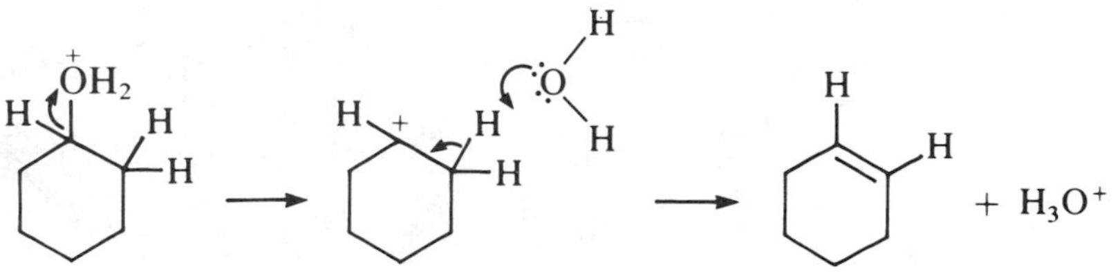

Experiments

MICROSCALE

1. Preparation of Cyclohexene

Handle sulfuric acid with care because it is corrosive to tissue.

Use Viton connector.

Use of a chaser

NOTE: Cyclohexene is extremely flammable: keep away from open flame.

Introduce 2.0 g of cyclohexanol followed by 0.5 mL of 85% phosphoric acid and a boiling chip into a 5-mL round-bottomed long-necked flask. Add a copper sponge to the neck of the flask, and shake to mix the layers. Heat will be evolved. Use the arrangement for fractional distillation as shown in Fig. 19.1. Note that the bulb of the thermometer must be *completely* below the side arm of the distilling head. Wrap the fractionating column and distilling head with glass wool or cotton. (See the experiment on fractional distillation for details of this technique.)

Heat the mixture gently on the sand bath, and then distill until the residue in the flask has a volume of about 0.5 to 1.0 mL and very little distillate is being formed; note the temperature range. Let the assembly cool a little after removing it from the sand, remove the thermometer briefly, and add 2 mL of toluene (the chaser solvent) into the top of the column using a Pasteur pipette. Note the amount of the upper layer in the boiling flask, and distill again until the volume of this layer has been reduced by about half. Transfer the contents of the vial into a reaction tube, and rinse with a little toluene. Use toluene for rinsing in subsequent operations. Wash the mixture with an equal volume of saturated sodium chloride solution, remove the aqueous layer, and then add sufficient anhydrous calcium chloride to the reaction tube so that it does not clump together. Shake the solution with the drying agent, and let it dry for at least 5 min. While this is taking place, clean the distilling apparatus first with water, then ethanol, and finally a little acetone. It is absolutely essential that the apparatus be completely dry; otherwise, the product will be contaminated with whatever solvent is left in the apparatus. Transfer the dry cyclohexene solution to the distilling flask, add a boiling chip, record the atmospheric pressure, and distill the product. At the moment the temperature starts to rise above the plateau at which the product distills

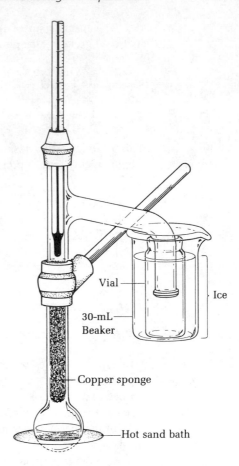

Vial

30-mL
Beaker

Ice

Copper sponge

Hot sand bath

FIG. 19.1   Apparatus for the synthesis of cyclohexene from cyclohexanol. The column is packed with a copper sponge.

(83°C), stop the distillation to avoid contamination of the cyclohexene with toluene. A typical yield of this volatile alkene is about 1 g. Report your yield in grams and your percentage yield. Run the infrared spectrum, and interpret it for purity. Look for peaks due to starting material and toluene. Gas chromatography is especially useful in the analysis of this compound because the expected impurities differ markedly in boiling point from the product. Carry out the tests for alkenes described in Chapter 18, Part 3, on the cyclohexene and, for comparison, on cyclohexane.

Cleaning Up   The aqueous solutions (pot residues and washes) should be diluted with water and neutralized before flushing down the drain. The ethanol wash and sodium chloride solution also can be flushed down the drain, while the acetone wash and all toluene-containing solutions should be placed in the organic solvents container. Once free of solvent, the calcium chloride can be placed in the nonhazardous solid waste container.

MACROSCALE

Handle phosphoric acid with care because it is corrosive to tissue.

Use of a chaser

2. Preparation of Cyclohexene

Introduce 20.0 g of cyclohexanol (technical grade), 5 mL of 85% phosphoric acid, and a boiling stone into a 100-mL round-bottomed flask and shake to mix the layers. Note the evolution of heat. Use the arrangement for fractional distillation shown in Fig. 5.11 but modified by use of a bent adapter delivering into an ice-cooled test tube in a 125-mL Erlenmeyer receiver, as shown in Fig. 19.2.

Note the initial effect of heating the mixture, and then distill until the residue

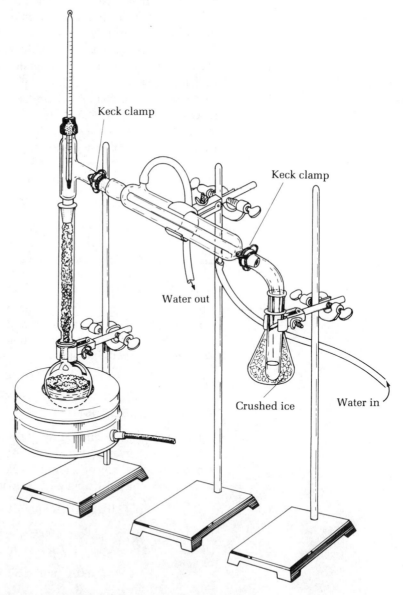

Keck clamp

Keck clamp

Water out

Crushed ice

Water in

FIG. 19.2 Fractionation into an ice-cooled receiver.

in the flask has a volume of 5–10 mL and very little distillate is being formed; note the temperature range. Then let the assembly cool a little, remove the thermometer briefly, and pour 20 mL of toluene (the chaser solvent) into the top of the column through a long-stemmed funnel. Note the amount of the upper layer in the boiling flask, and distill again until the volume of the layer has been reduced by about half. Pour the contents of the test tube into a small separatory funnel and rinse with a little chaser solvent; use this solvent for rinsing in subsequent operations. Wash the mixture with an equal volume of saturated sodium chloride solution, separate the water layer, run the upper layer into a clean flask, and add 5 g of anhydrous calcium chloride (10×75-mm test tube-full) to dry it. Before the final distillation note the barometric pressure, apply any necessary thermometer corrections, and determine the reading expected for a boiling point of 83°C. Dry the boiling flask, column, and condenser, decant the dried liquid into the flask through a stemless funnel plugged with a bit of cotton, and fractionally distill, with all precautions against evaporation losses. The ring of condensate should rise very slowly as it approaches the top of the fractionating column so that the thermometer can record the true boiling point soon after distillation starts. Record both the corrected boiling point of the bulk of the cyclohexene fraction and the temperature range, which should not be more than 2°C. If the cyclohexene has been dried thoroughly it will be clear and colorless; if wet it will be cloudy. A typical student yield is 13.2 g. Report your yield in grams and your percent yield.

CAUTION: Cyclohexene is extremely flammable; keep away from open flame.

Cleaning Up   The aqueous solutions (pot residues and washes) should be diluted with water and neutralized before flushing down the drain with a large excess of water. The ethanol wash and sodium chloride solution can also be flushed down the drain; the acetone wash and all toluene-containing solutions should be placed in the organic solvents container. Once free of solvent, the calcium chloride can be placed in the nonhazardous solid waste container. Allow it to dry on a tray in the hood.

Yield Calculations

Rarely do organic reactions give 100% yields of one pure product; an important objective of every experiment is a high yield of the desired product. The present experiment uses 2 g of starting cyclohexanol. This corresponds to 0.02 mol because the molecular weight of cyclohexanol is 100:

$$(2.0 \text{ g})/(100.16 \text{ g/mol}) = 0.020 \text{ mol} = 20 \text{ mmol}$$

If the reaction gave a 100% yield of cyclohexene, then 0.02 mol of the alkene would be produced. The weight of 0.020 mol of cyclohexene is

$$0.020 \text{ mol} \times 82.14 \text{ g/mol} = 1.64 \text{ g}$$

We call the 1.64 g the *theoretical yield;* it could be obtained if the reaction proceeded perfectly. If the actual yield is only 0.82 g, then the reaction would be said to give a 50% yield:

$$(0.82 \text{ g})/(1.64 \text{ g}) \times 100 = 50\%$$

Typical student yields are included throughout this text. These are *not* theoretical yields; they suggest what an average or above-average student can expect to obtain for the experiment.

Questions

1. Assign the peaks in the ¹H NMR spectrum of cyclohexene (Fig. 19.3) to specific groups of protons on the molecule.

2. What product(s) would be obtained by the dehydration of 2-heptanol? Of 2-methyl-1-cyclohexanol?

3. Mixing cyclohexanol with phosphoric acid is an exothermic process, whereas the production of cyclohexene is endothermic. Referring to the two chemical reactions on p. 279, construct an energy diagram showing the course of this reaction. Label the diagram with the starting alcohol, the oxonium ion (the protonated alcohol), the carbocation, and the product.

4. Is it reasonable that the ¹H spectrum of cyclohexene (Fig. 19.3) should closely resemble the ¹³C spectrum (Fig. 19.4)? Compare these spectra with the spectrum of cyclohexanol (Fig. 19.5).

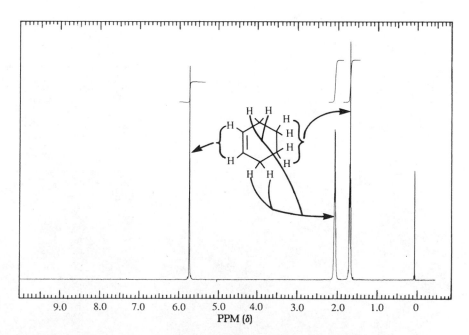

FIG. 19.3 ¹H NMR spectrum of cyclohexene (250 MHz).

FIG. 19.4 ^{13}C NMR spectrum of cyclohexene.

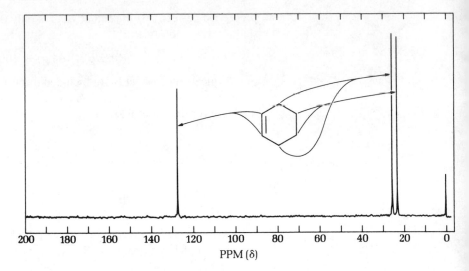

FIG. 19.5 ^{13}C NMR spectrum of cyclohexanol.

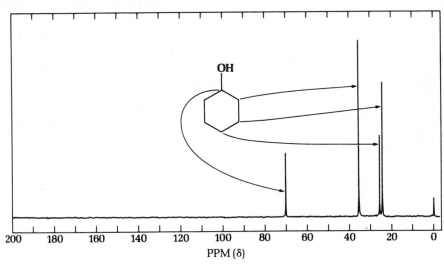

Bromination and Debromination: Purification of Cholesterol

Prelab Exercise: Make a molecular model of 5α,6β-dibromocholestan-3β-ol, and convince yourself that the bromine atoms in this molecule are *trans* and *diaxial.*

The bromination of a double bond is an important and well-understood organic reaction. In the present experiment it is employed for the very practical purpose of purifying crude cholesterol through the process of bromination, crystallization, and then zinc dust debromination.

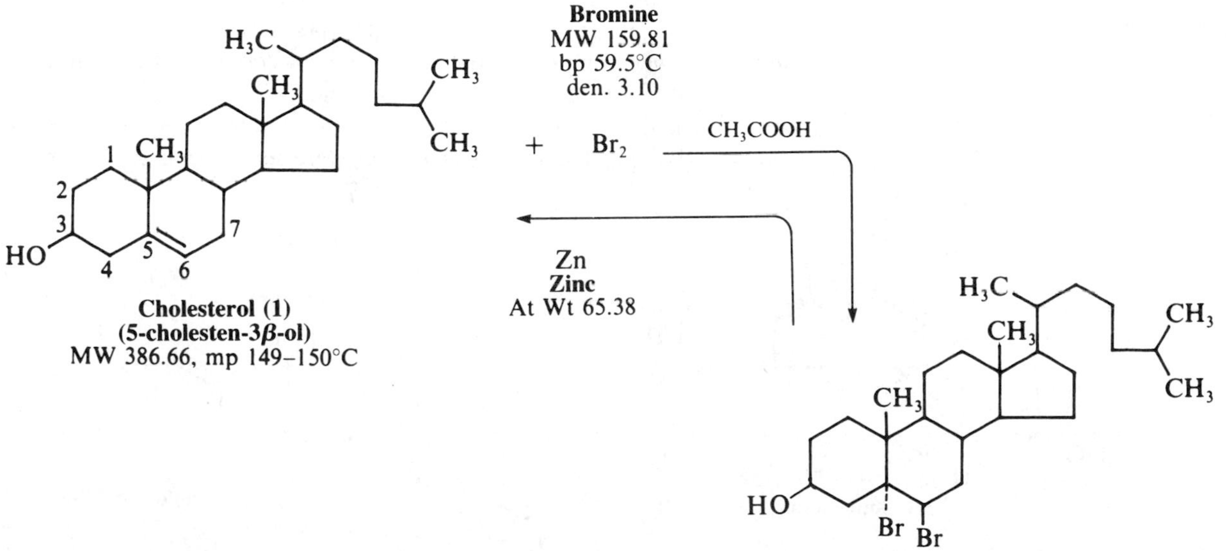

Cholesterol (1)
(5-cholesten-3β-ol)
MW 386.66, mp 149–150°C

Bromine
MW 159.81
bp 59.5°C
den. 3.10

+ Br₂

Zn
Zinc
At Wt 65.38

5α,6β-Dibromocholestan-3β-ol (3)

The reaction involves nucleophilic attack by the alkene on bromine with the formation of a tertiary carbocation that probably has some bromonium ion character resulting from sharing of the nonbonding electrons on bromine with the electron-deficient C-5 carbon. This ion is attacked from the backside by bromide ion to form dibromocholesterol with the bromine atoms in the *trans* and diaxial configuration, the usual result when brominating a cyclohexene.

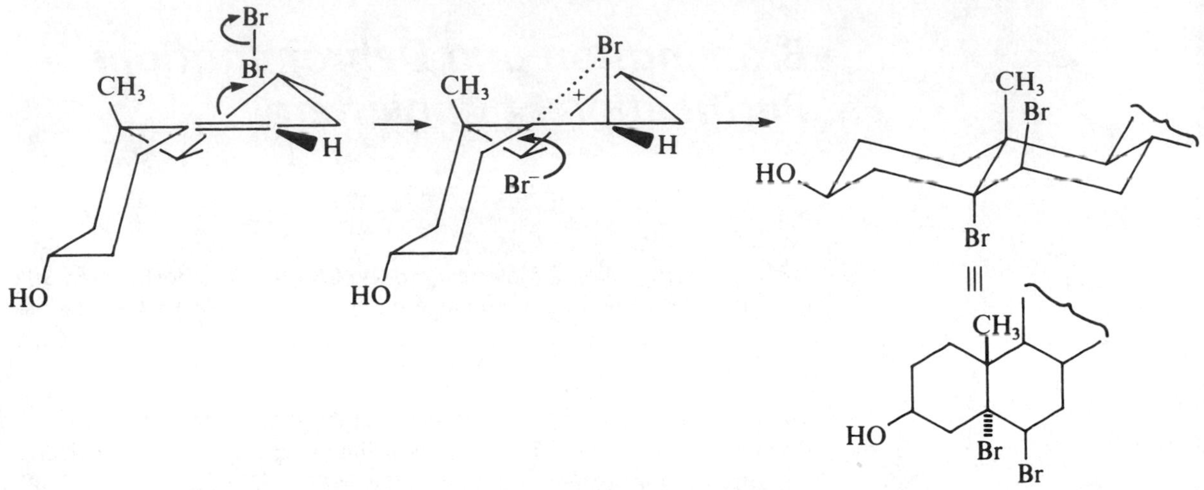

Cholesterol isolated from natural sources contains small amounts (0.1–3%) of 3β-cholestanol, 7-cholesten-3β-ol, and 5,7-cholestadien-3β-ol.[1] These are so very similar to cholesterol in solubility that their removal by crystallization is not feasible. However, complete purification can be accomplished through the sparingly soluble dibromo derivative 5α,6β-dibromocholestan-3β-ol. 3β-Cholestanol is saturated and does not react with bromine; thus, it remains in the mother liquor. 7-Cholesten-3β-ol and 5,7-cholestadien-3β-ol are dehydrogenated by bromine to

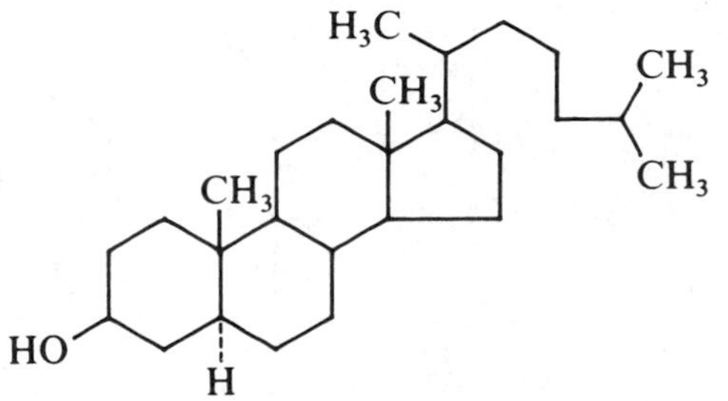

3β-Cholestanol

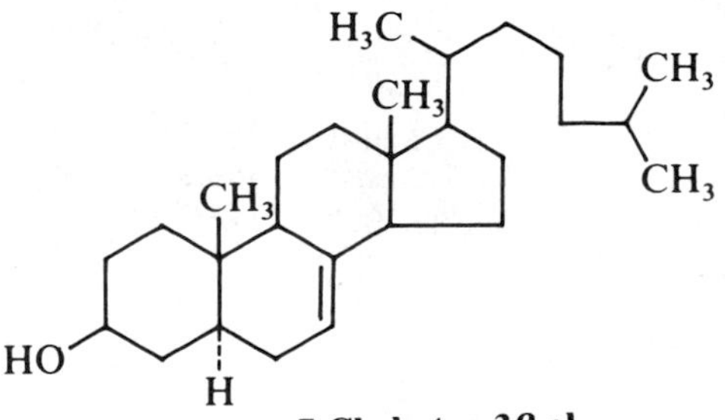

7-Cholesten-3β-ol

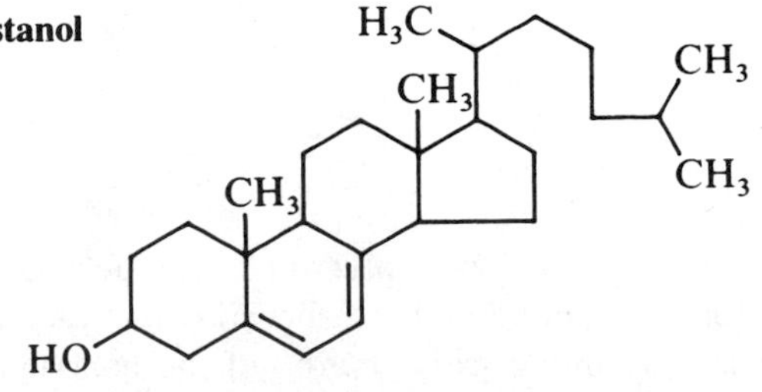

5,7-Cholestadien-3β-ol

---

1. A fourth companion, cerebrosterol, or 25-hydroxycholesterol, is easily eliminated by crystallization from alcohol.

dienes and trienes, respectively, that likewise remain in the mother liquor and are eliminated along with colored byproducts.

The cholesterol dibromide that crystallizes from the reaction solution is collected, washed free of the impurities or their dehydrogenation products, and debrominated with zinc dust, with regeneration of cholesterol in pure form. Specific color tests can differentiate between pure cholesterol and tissue cholesterol purified by ordinary methods.

Experiments

1. Microscale Bromination of Cholesterol

CAUTION: Be extremely careful to avoid getting the solution of bromine and acetic acid on the skin.

In a 10 × 75 mm test tube, dissolve 100 mg of gallstone cholesterol or of commercial cholesterol (content of 7-cholesten-3β-ol about 0.6%) in 0.7 mL of *tert*-butyl methyl ether by gentle warming, and then add 0.5 mL of a solution of bromine and sodium acetate in acetic acid.[2] Cholesterol dibromide begins to crystallize in a minute or two. Cool in an ice bath, and stir the crystalline paste with a stirring rod for about 10 min to ensure complete crystallization; at the same time, cool a mixture of 0.3 mL of ether and 0.7 mL of acetic acid in ice. Then collect the crystals on the Hirsch funnel, and wash with the iced ether–acetic acid solution to remove the yellow mother liquor. The short test tube is used to facilitate this transfer of the crystalline paste. This material is not easy to filter on a very small scale because the crystals are very small. Alternatively, carry out the bromination in a reaction tube and isolate the product with the Wilfilter (Fig. 20.1). Finally, wash the crystals on the filter with a few drops of ethanol, continuing to apply suction, and transfer the white solid without drying it (dry weight 120 mg) to a reaction tube.

Cleaning Up   The filtrate from this reaction contains halogenated material and so must be placed in the halogenated organic waste container.

2. Bromination of Cholesterol

In a 25-mL Erlenmeyer flask dissolve 1 g of commercial cholesterol (content of 7-cholesten-3β-ol about 0.6%) in 7 mL of *t*-butyl methyl ether by gentle warming, and, with a pipette fitted with a pipetter or with a plastic syringe, or from a burette in the hood, add 5 mL of a solution of bromine and sodium acetate in acetic acid.[2] Cholesterol dibromide begins to crystallize in a minute or two. Cool in an ice bath and stir the crystalline paste with a stirring rod for about 10 min to ensure complete crystallization, and at the same time cool a mixture of 3 mL of *t*-butyl methyl ether and 7 mL of acetic acid in ice. Then collect the crystals on a small suction funnel, and wash with the iced ether–acetic acid solution to remove

2. Weigh a 125-mL Erlenmeyer flask on a balance placed in the hood, add 4.5 g bromine by a capillary dropping tube (avoid breathing the vapor), and add 50 mL acetic acid and 0.4 g sodium acetate (anhydrous).

FIG. 20.1 Wilfilter filtration apparatus.

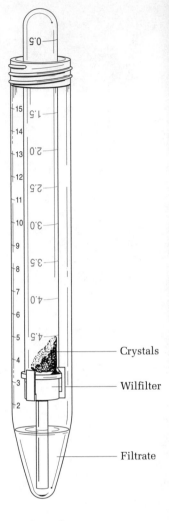

Crystals

Wilfilter

Filtrate

CAUTION: Be extremely careful to avoid getting the solution of bromine and acetic acid on the skin. Carry out the reaction in the hood and wear disposable gloves for maximum safety.

the yellow mother liquor. Finally, wash with a little methanol, continuing to apply suction, and transfer the white solid without drying it (dry weight 1.2 g) to a 50-mL Erlenmeyer flask.

Cleaning Up   The filtrate from this reaction contains halogenated material and so must be placed in the halogenated organic waste container.

3. Microscale Zinc Dust Debromination

Add 2 mL of *tert*-butyl methyl ether, 0.5 mL of acetic acid, and 40 mg of Zn dust[3] to the dibromocholesterol from Experiment 1, and mix the contents. In about 3

3. If the reaction is slow, add more zinc dust or place the flask in an ultrasonic cleaning bath, which will activate the zinc. The amount specified is adequate if material is taken from a freshly opened bottle, but zinc dust deteriorates on exposure to air.

min, the dibromide dissolves; after 5 to 10 min of mixing (a magnetic stirrer could be used), zinc acetate usually separates to form a white precipitate (the dilution sometimes is such that no separation occurs). Stir for 5 min more, and then add water (no more than one drop) until any solid present (zinc acetate) dissolves to make a clear solution. Decant the solution from the zinc into a reaction tube, and wash the ethereal solution twice with 2 mL of water and then with 1 mL of 3 M sodium hydroxide and 1 mL of saturated sodium chloride solution. Then shake the ether solution with anhydrous calcium chloride pellets, pipette off the dry solution, wash the drying agent with a little ether, add 1 mL methanol (and a boiling stick), and evaporate the solution to the point where most of the *tert*-butyl methyl ether is removed and the purified cholesterol begins to crystallize. Remove the solution from the heat, let crystallization proceed at room temperature, and then, in an ice bath, collect the crystals on the Hirsch funnel or with the aid of the Wilfilter (Fig. 20.1) and wash them with cool methanol: you should obtain 60 to 70 mg (mp 149–150°C).

Collect the product on a Wilfilter.

Cleaning Up   The unreacted zinc dust can be discarded in the nonhazardous solid waste container after it has been allowed to dry and then sit exposed to the air for about one-half hour on a watch glass. Sometimes the zinc dust at the end of this reaction will get quite hot as it air oxidizes. The aqueous and acetic acid solutions after neutralization can be flushed down the drain, and the organic filtrates containing ether and methanol should be placed in the organic solvents container. After the solvent has evaporated from the drying agent, it can be placed in the nonhazardous solid waste container.

4. Zinc Dust Debromination

To the flask containing dibromocholesterol add 20 mL of *t*-butyl methyl ether, 5 mL of acetic acid, and 0.2 g of zinc dust and swirl.[4] In about 3 min the dibromide dissolves; after 5–10 min swirling, zinc acetate usually separates to form a white precipitate (the dilution sometimes is such that no separation occurs). Stir for 5 min more, and then add water by drops (no more than 0.5 mL) until any solid present (zinc acetate) dissolves to make a clear solution. Decant the solution from the zinc into a separatory funnel, and wash the ethereal solution twice with water and then with 10% sodium hydroxide (to remove traces of acetic acid). Then shake the *t*-butyl methyl ether solution with an equal volume of saturated sodium chloride solution to reduce the water content, dry the ether with anhydrous calcium chloride, remove the drying agent, add 10 mL of methanol (and a boiling stone), and evaporate the solution on the steam bath to the point where most of the ether is removed and the purified cholesterol begins to crystallize. Remove the solution from the steam bath, let crystallization proceed at room temperature and then in an ice bath, collect the crystals, and wash them with cold methanol; you should obtain 0.6–0.7 g, mp 149–150°C. Proton (Fig. 20.2) and carbon (Fig. 20.3) NMR spectra of cholesterol are found at the end of the chapter.

Aspirator tube

---

4. If the reaction is slow, add more zinc dust or place the flask in an ultrasonic cleaning bath, which will activate the zinc. The amount specified is adequate if material is taken from a freshly opened bottle, but zinc dust deteriorates on exposure to air.

Cleaning Up   The unreacted zinc dust can be discarded in the nonhazardous solid waste container after it has been allowed to dry and then sit exposed to the air for about one-half hour on a watch glass. Sometimes the zinc dust at the end of this reaction will get quite hot as it air oxidizes. The aqueous and acetic acid solutions after neutralization can be flushed down the drain, and the organic filtrates containing *t*-butyl methyl ether and methanol should be placed in the organic solvents container. After the solvent has evaporated from the calcium chloride it can be placed in the nonhazardous solid waste container.

Questions

1. What is the purpose of the acetic acid in this reaction?

2. Why might old zinc dust not react in the debromination reaction?

3. Why doesn't acetate ion attack the intermediate bromonium ion to give the 5-acetoxy-6-bromo compound in the bromination of cholesterol?

4. Assign the 3400 cm^{-1} peak in the infrared spectrum of cholesterol (Fig. 20.4).

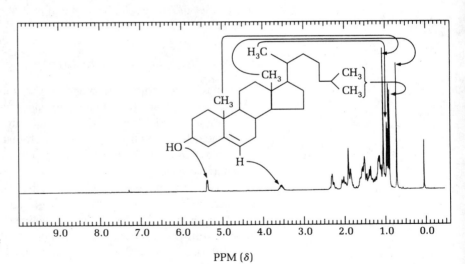

FIG. 20.2 ^{1}H NMR spectrum of cholesterol (60 MHz).

PPM (δ)

FIG. 20.3 ^{13}C NMR spectrum of cholesterol (22.6 MHz).

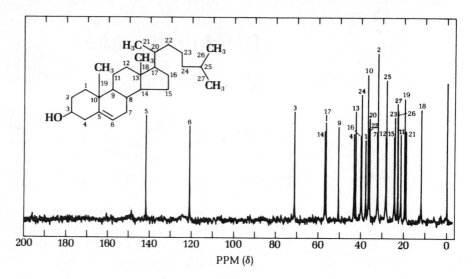

FIG. 20.4   Infrared spectrum of cholesterol (KBr disk).

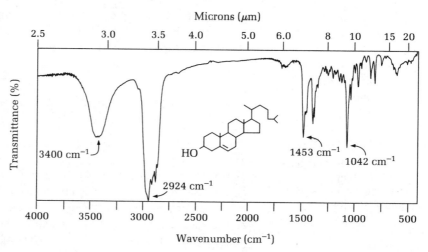

21

Dichlorocarbene

Prelab Exercise: Propose a synthesis of

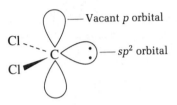

using the principles of phase transfer catalysis. Why is phase transfer catalysis particularly suited to this reaction?

Dichlorocarbene

— Vacant p orbital

Cl

C — sp^2 orbital

Cl

Like a carbanion—electrons in sp^2 orbital. Like a carbocation—vacant p orbital.

$CH_3OCH_2CH_2OCH_2CH_2OCH_3$

2-Methoxyethyl ether
Diglyme
MW 134.17, bp 161°C
miscible with water

Test the diglyme for peroxides.

Dichlorocarbene is a highly reactive intermediate of bivalent carbon with only six valence electrons around the carbon. It is electrically neutral and a powerful electrophile. As such it reacts with alkenes, forming cyclopropane derivatives by *cis*-addition to the double bond.

There are a dozen or so ways by which dichlorocarbene can be generated. In this experiment thermal decomposition of anhydrous sodium trichloroacetate in an aprotic solvent (diglyme) in the presence of *cis, cis*-1,5-cyclooctadiene generates dichlorocarbene to give 5,5,10,10-tetrachlorotricyclo[7.1.0.0^{4,6}]decane (**1**).

Thermal Decomposition of Sodium Trichloroacetate; Reaction of Dichlorocarbene with 1,5-Cyclooctadiene

The thermal decomposition of sodium trichloroacetate initially gives the trichloromethyl anion (Eq. 1). In the presence of a proton-donating solvent (or moisture) this anion gives chloroform; in the absence of these reagents the anion decomposes by loss of chloride ion to give dichlorocarbene (Eq. 2).

The conventional method for carrying out the reaction is to add the salt portionwise, during 1–2 h, to a magnetically stirred solution of the olefin in diethylene glycol dimethyl ether, 2-methoxyethyl ether ("diglyme"), at a temperature maintained in a bath at 120°C. Under these conditions the reaction mixture becomes almost black, isolation of a pure product is tedious, and the yield is low.

Tetrachloroethylene, a nonflammable solvent widely used in the dry-cleaning industry, boils at 121°C and is relatively inert toward electrophilic dichlorocarbene. On generation of dichlorocarbene from either chloroform or sodium trichloroacetate in the presence of tetrachloroethylene, the yield of hexachlorocyclopropane (mp 104°C) is only 0.2–10%. The first idea for simplifying the procedure[1] was to use tetrachloroethylene to control the temperature to the desired range, but sodium trichloroacetate is insoluble in this solvent and no reaction occurs. Diglyme, or an equivalent, is required to provide some solubil-

---

1. L. F. Fieser and David H. Sachs, *J. Org. Chem.*, **29**, 1113 (1964).

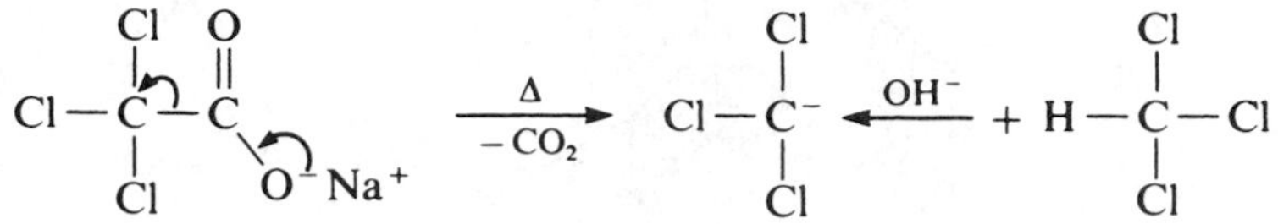

Sodium trichloroacetate   Trichloromethyl anion   Chloroform
MW 185.39   MW 119.39, bp 61°C

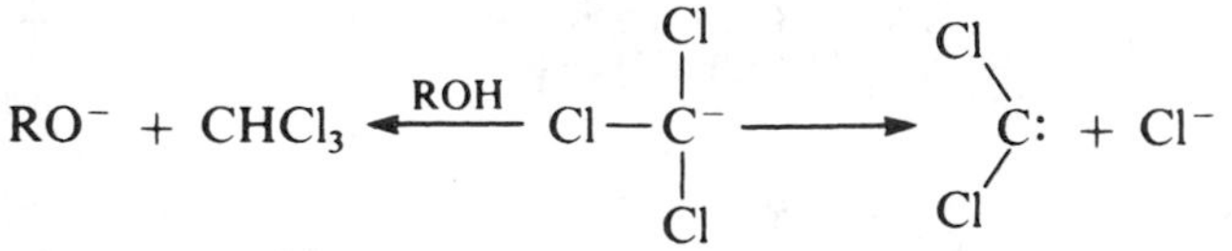

$$ (1) $$

Dichlorocarbene

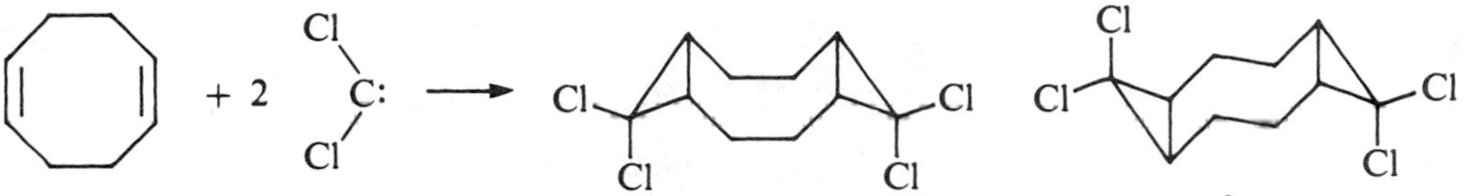

$$ (2) $$

cis,cis-**1,5-Cyclo-**   **1**   **2**
octadiene   *cis*, mp 176°C   *trans*, mp 230°C dec
MW 108.14
bp 149–150°C   **5,5,10,10-Tetrachlorotricyclo[7.1.0.0^{4,6}]decane**
                                        MW 274.03

$$ (3) $$

Rationale of the procedure

ity. The reaction proceeds better in a 7:10 mixture of diglyme to tetra-chloroethylene than in diglyme alone, but the salt dissolves rapidly in this mixture and has to be added in several small portions and the reaction mixture becomes very dark. The situation is vastly improved by the simple expedient of decreasing the amount of diglyme to a 2.5:10 ratio. The salt is so sparingly soluble in this mixture that it can be added at the start of the experiment, and it dissolves slowly as the reaction proceeds. The boiling and evolution of carbon dioxide provide adequate stirring, the mixture can be left unattended, and the little color that develops is eliminated by washing the crude product with methanol.

The main reaction product crystallizes from ethyl acetate in beautiful prismatic needles, mp 174–175°C, and this was the sole product encountered in runs made in the solvent mixture recommended. In earlier runs made in diglyme with manual control of temperature, the ethyl acetate mother liquor material on repeated crystallization from toluene afforded small amounts of a second isomer, mp 230°C, dec. Analyses checked for a pair of *cis-trans* isomers, and both gave negative permanganate tests. To distinguish between them, the junior author of the paper undertook an X-ray analysis, which showed that the lower melting isomer is *cis* and the higher melting isomer is *trans*.

X-ray analysis

4
5,5,10,10-Tetrabromotricyclo-
[7.1.0.0⁴,⁶]decane

5
Cyclodeca-1,2,6,7-tetracene

A striking reaction of a bis dihalocarbene adduct

The bis adduct (**4**) of *cis,cis*-1,5-cyclooctadiene with dibromocarbene is described as melting at 174–180°C and may be a *cis-trans* mixture. Treatment of the substance with methyl lithium at 240°C gave a small amount of the ring-expanded bisallene (**5**).

1. Sodium Trichloroacetate

Use commercial sodium trichloroacetate (dry), or prepare it as follows. Place 6.4 g of trichloroacetic acid in a 125-mL Erlenmeyer flask, dissolve 1.6 g of sodium hydroxide pellets in 6 mL of water in a 25-mL Erlenmeyer flask, cool the solution thoroughly in an ice bath, and swirl the flask containing the acid in the ice bath while slowly dropping in about nine-tenths of the alkali solution. Then add a drop of 0.04% Bromocresol green solution to produce a faint yellow color, visible when the flask is dried and placed on white paper. With a Pasteur pipette, titrate the solution to an end point where a single drop produces a change from yellow to blue. If the end point is overshot, add a few crystals of acid and titrate more carefully. Close the flask with a rubber stopper, connect to an aspirator, place the flask within the rings of the steam bath, and wrap a towel around both for maximum heat. Turn on the water at full force for maximum suction. The evaporation requires no further attention and should be complete in 15–20 min. When you have an apparently dry white solid, scrape it out with a spatula and break up the large lumps. If you see any evidence of moisture, or if the weight exceeds the theory, dry it to constant weight under vacuum.

Use caution in handling the corrosive trichloroacetic acid. Carry out procedure in hood.

Sodium trichloroacetate will decompose completely to sodium chloride in a drying oven at 150°C.

Make sure that the salt is completely dry.

MICROSCALE AND MACROSCALE

MICROSCALE

Tetrachloroethylene
bp 121°C, den 1.623

Note for the instructor

2. Dichlorocarbene Reaction

Place 0.91 g of *dry* sodium trichloroacetate in a 5-mL round-bottomed short-necked flask mounted over a hot sand bath and a magnetic stirrer, and add 1 mL of tetrachloroethylene, 0.25 mL of diglyme, 0.25 mL (220 mg) of *cis,cis*-1,5-cyclooctadiene and a magnetic stirring bar.[2] Attach the empty distilling column

2. Store diglyme, tetrachloroethylene, and 1,5-cyclooctadiene over Linde 5A molecular sieves or anhydrous magnesium sulfate to guarantee that the reagents are dry. The reaction will not work if any reagent is wet. For safety reasons, dispense tetrachloroethylene in the hood. Do the same with the diene; it has a bad odor. Anhydrous diglyme is available from Aldrich in septum-capped bottles.

as an air condenser, and cap it with a septum bearing polyethylene tubing, which in turn dips below the surface of 0.5 mL of diglyme contained in a reaction tube (Fig. 21.1). This bubbler will show when the evolution of carbon dioxide ceases. The solvent should be the same as that of the reaction mixture, should there be a suckback. Heat the reaction mixture to boiling, note the time, and reflux gently until the reaction is complete. You will notice foaming, due to liberated carbon dioxide, and separation of finely divided sodium chloride. Inspection of the bottom of the flask will show lumps of undissolved sodium trichloroacetate, which gradually disappear.

When the reaction is complete, add 3.5 mL of water to the hot mixture, connect an adapter and a distilling head to the flask (Fig. 21.2), add a boiling chip, heat the mixture to boiling, and steam distill until the tetrachloroethylene is eliminated and the product separates in the flask as an oil or semisolid. If necessary, inject more water through the septum to complete the steam distillation. Cool the flask to room temperature, transfer the contents to a reaction tube using dichloromethane, and remove the aqueous layer. Dry the solution over anhydrous calcium chloride pellets, remove the solvent, wash the drying agent with more dichloromethane, and evaporate the solvent in a reaction tube. The weight of the crude product should be about 400 mg.

Cover the crude product with methanol, break up the cake with a flattened stirring rod, and crush the lumps. Cool in ice, collect, and wash the product with methanol. The yield of colorless, or nearly colorless, 5,5,10,10-tetrachlorotricyclo[7.1.0.0^{4,6}]decane is about 300 mg. This material (mp 174–175°C) consists almost entirely of the *cis* isomer. Dissolve it in ethyl acetate (1.5–2 mL), and let the solution stand undisturbed for crystallization at room temperature. The pure *cis* isomer separates in large, prismatic needles (mp 175–176°C).

Cleaning Up   The aqueous layer should be neutralized with hydrochloric acid, diluted with water, and flushed down the drain. Organic layers that contain halogenated material must be placed in the halogenated organic solvents container. After the solvent has been allowed to evaporate from the drying agent in the hood, it can be placed in the nonhazardous solid waste container. Methanol and ethyl acetate used in crystallization can be placed in the organic solvents container.

 **MACROSCALE**

3. Dichlorocarbene Reaction

Place 7.1 *g* of *dry* sodium trichloroacetate in a 100-mL round-bottomed flask mounted over a flask heater, and add 10 mL of tetrachloroethylene, 2.5 mL of diglyme, and 2.5 mL (2.2 g) of *cis,cis*-1,5-cyclooctadiene.[3] Attach a reflux condenser, and, in the top opening of the condenser, insert a rubber stopper carrying a glass tube connected by rubber tubing to a short section of glass tube, which can

Note for the instructor

3. Store diglyme, tetrachloroethylene, and 1,5-cyclooctadiene over Linde 5A molecular sieves or anhydrous magnesium sulfate to guarantee that the reagents are dry. The reaction will not work if any reagent is wet.

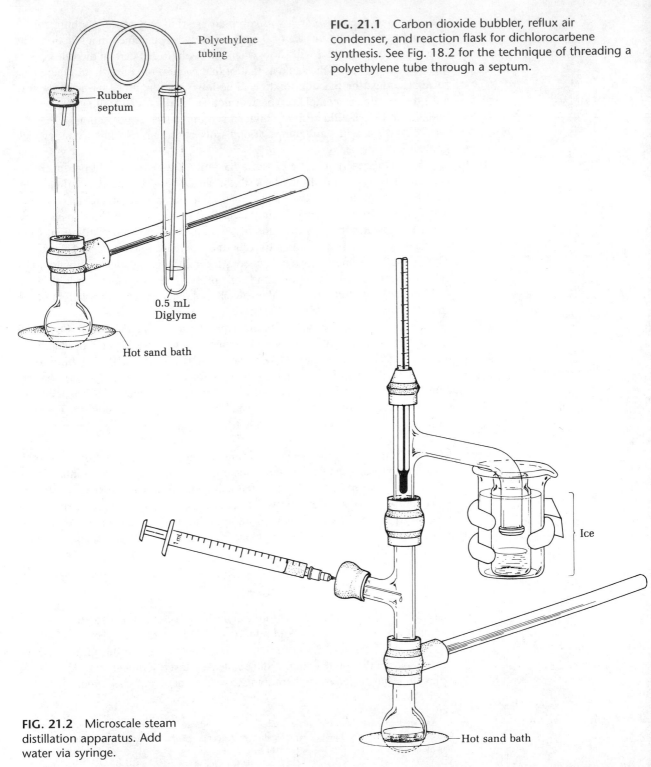

FIG. 21.1 Carbon dioxide bubbler, reflux air condenser, and reaction flask for dichlorocarbene synthesis. See Fig. 18.2 for the technique of threading a polyethylene tube through a septum.

Polyethylene tubing

Rubber septum

0.5 mL Diglyme

Hot sand bath

FIG. 21.2 Microscale steam distillation apparatus. Add water via syringe.

1 mL

Ice

Hot sand bath

Reflux time: about 75 min

be inserted below the surface of 2–3 mL of tetrachloroethylene in a 20 × 150-mm test tube mounted at a suitable level. This bubbler will show when the evolution of carbon dioxide ceases; the solvent should be the same as that of the reaction mixture, should there be a suckback. Heat to boiling, note the time, and reflux gently until the reaction is complete. You will notice foaming, due to liberated carbon dioxide, and separation of finely divided sodium chloride. Inspection of the bottom of the flask will show lumps of undissolved sodium trichloroacetate, which gradually disappear. A large flask is specified because it will serve later for removal of tetrachloroethylene by steam distillation. Make advance preparation for this operation.

Easy workup

When the reaction is complete, add 40 mL of water to the hot mixture, heat over a hot electric flask heater, and steam distill until the tetrachloroethylene is eliminated and the product separates as an oil or semisolid. Cool the flask to room temperature, decant the supernatant liquid into a separatory funnel, and extract with dichloromethane. Run the extract into the reaction flask, and use enough more dichloromethane to dissolve the product; use a Pasteur pipette to rinse down material adhering to the adapter. Run the dichloromethane solution of the product into an Erlenmeyer flask through a cone of anhydrous calcium chloride pellets in a funnel, and evaporate the solvent (bp 41°C). The residue is a tan or brown solid (14 g). Cover it with methanol, break up the cake with a flattened stirring rod, and crush the lumps. Cool in ice, collect, and wash the product with methanol. The yield of colorless, or nearly colorless, 5,5,10,10-tetrachlorotricyclo[7.1.0.0^{4,6}]decane is 3.3 g. This material, mp 174–175°C, consists almost entirely of the *cis* isomer. Dissolve it in ethyl acetate (15–20 mL), and let the solution stand undisturbed for crystallization at room temperature. The pure *cis* isomer separates in large, prismatic needles, mp 175–176°C. The proton NMR spectrum (Fig. 21.3) is found at the end of the chapter.

Cleaning Up   The aqueous layer can be diluted with water and flushed down the drain. Organic layers, which contain halogenated material, must be placed in the halogenated organic solvents container. After the solvent has been allowed to evaporate from the drying agent in the hood, it can be placed in the nonhazardous solid waste container. Methanol and ethyl acetate used in crystallization can be placed in the organic solvents container.

4. Phase Transfer Catalysis—Reaction of Dichlorocarbene with Cyclohexene

$$CCl_3^- + H_2O \longrightarrow CHCl_3 + OH^-$$
3

$$:CCl_2 + 2\,H_2O$$
4 $\downarrow$
$$CH_2Cl_2 + 2\,OH^-$$

Ordinarily water must be scrupulously excluded from carbene-generating reactions. Both the intermediate trichloromethyl anion (**3**) and dichlorocarbene (**4**) react with water. But in the presence of a phase transfer catalyst (a quaternary ammonium salt such as benzyltriethylammonium chloride) it is possible to carry out the reaction in aqueous medium. Chloroform, 50% aqueous sodium hydroxide, and the olefin, in the presence of a catalytic amount of the quaternary ammonium salt, are stirred for a few minutes to produce an emulsion; and after the exothermic reaction is complete (about 30 min) the product is isolated.

Both benzyltriethylammonium chloride (**1**) and benzyltriethylammonium hydroxide (**2**) partition between the aqueous and organic phases. In the aqueous phase, the quaternary ammonium chloride (**1**) reacts with concentrated hydroxide to give the quaternary ammonium hydroxide (**2**). In the chloroform phase, **2** reacts with chloroform to give the trichloromethyl anion (**3**), which eliminates chloride ion to give dichlorocarbene, :CCl2 (**4**), and **1**. The carbene (**4**) reacts with the alkene to give product (**5**).

Aqueous phase

$$\overset{Cl^-}{\underset{1}{(C_2H_5)_3\overset{+}{N}CH_2C_6H_5}} + OH^- \rightleftharpoons \overset{OH^-}{\underset{2}{(C_2H_5)_3\overset{+}{N}CH_2C_6H_5}} + Cl^-$$

CHCl₃ phase

$$\underset{\underset{4}{:CCl_2}}{\overset{Cl^-}{(C_2H_5)_3\overset{+}{N}CH_2C_6H_5}} \longleftarrow CCl_3^- \longleftarrow \underset{\underset{3}{CHCl_3}}{\overset{OH^-}{(C_2H_5)_3\overset{+}{N}CH_2C_6H_5}} +$$

$$+ :CCl_2 \longrightarrow \overset{Cl}{\underset{Cl}{\diagdown}}$$

$$\quad\quad\quad\quad 4 \quad\quad\quad\quad\quad\quad 5$$

Macroscale Procedure

CAUTION: Chloroform is a carcinogen and should be handled in the hood. Avoid all contact with the skin.

50% sodium hydroxide solution is very corrosive. Handle with care.

To a mixture of 8.2 g of cyclohexene, 12.0 g of chloroform (*Caution!*), and 20 mL of 50% aqueous sodium hydroxide in a 125-mL Erlenmeyer flask containing a thermometer, add 0.2 g of benzyltriethylammonium chloride.[4] Swirl the mixture to produce a thick emulsion. The temperature of the reaction will rise gradually at first and then markedly accelerate. As it approaches 60°C, prepare to immerse the flask in an ice bath. With the flask alternately in and out of the ice bath, stir the thick paste and maintain the temperature between 50 and 60°C. After the exothermic reaction is complete (about 10 min) allow the mixture to cool spontaneously to 35°C, and then dilute with 50 mL of water. Separate the layers (test to determine which is the organic layer), extract the aqueous layer once with 10 mL of *t*-butyl methyl ether, and wash the combined organic layer and ether extract once with 20 mL of water. Dry the cloudy organic layer by shaking it with anhydrous calcium chloride pellets until the liquid is clear. Decant into a 50-mL round-bottomed flask and distill the mixture (boiling chip), first from a steam bath to remove the *t*-butyl methyl ether and some unreacted cyclohexene and chloroform, then over a very hot sand bath or a free flame. Collect 7,7-dichloro-

Note for the instructor

---

4. Many other quaternary ammonium salts, commercially available, work equally well (e.g., cetyltrimethylammonium bromide).

bicyclo[4.1.0]heptane over the range 195–200°C. Yield, about 9 g. A purer product will result if the distillation is carried out under reduced pressure.

Cleaning Up   Organic layers go in the organic solvents container. The aqueous layer should be diluted with water and flushed down the drain. After the *t*-butyl methyl ether evaporates from the calcium chloride in the hood, it can be placed in the nonhazardous solid waste container. The pot residue from the distillation goes in the halogenated organic solvents container.

The success of this reaction is critically dependent on forming a thick emulsion, for only then will the surface between the two phases be large enough for complete reaction to occur. James Wilbur at Southwest Missouri State University found that when freshly distilled, pure reagents were used, no exothermic reaction occurred, and very little or no dichloronorcarane was formed. We have subsequently found that old, impure cyclohexene gives very good yields of product. Cyclohexene is notorious for forming peroxides at the allylic position. Under the reaction conditions this peroxide will be converted to the ketone:

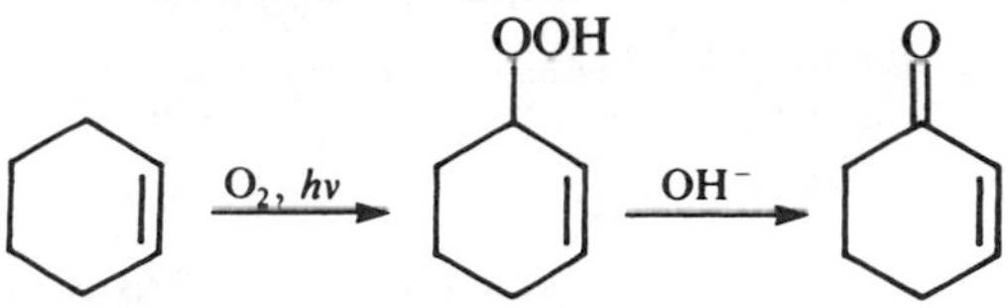

Apparently, small amounts of this ketone are responsible for the formation of stable emulsions. This appears to be one of the rare cases when impure reagents lead to successful completion of a reaction. If only pure reagents are available, then the reaction mixture must be stirred vigorously overnight, using a magnetic stirrer. Alternatively, a small quantity of cyclohexene hydroperoxide can be prepared by pulling air through the alkene for a few hours, employing an aspirator, and using this impure reagent, or—as Wilbur found—by adding 1 mL of cyclohexanol to the reaction mixture. Detergents such as sodium lauryl sulfate do not increase the rate of reaction or emulsion formation.

Computational Chemistry

Construct models of the *cis*- and *trans*-dichlorocarbene adducts of cyclooctadiene, **1** and **2,** using a molecular mechanics program. Carry out energy minimization and note the steric energies of each isomer. Then carry out AM1 semiempirical molecular orbital calculations on each isomer to determine their heats of formation. Do these calculations confirm that the predominant product, the *cis*-isomer, is also the most stable? Would you have predicted that the *cis*-isomer would be the most stable by simple inspection of the energy-minimized conformers?

Questions ———————————————————————

1. If water is not rigorously excluded from sodium trichloroacetate dichlorocarbene synthesis, what side reaction occurs?

2. Why is vigorous stirring or emulsion formation necessary in a phase transfer reaction?

3. Assign the three groups of peaks in the NMR spectrum of 7,7-dichlorobicyclo[4.1.0]heptane (Fig. 21.4).

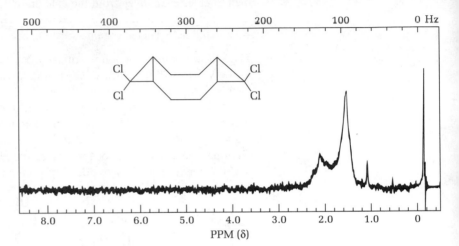

FIG. 21.3 ¹H NMR spectrum of 5,5,10,10-tetrachlorotricyclo-[7.1.0.0⁴,⁶]decane (60 MHz).

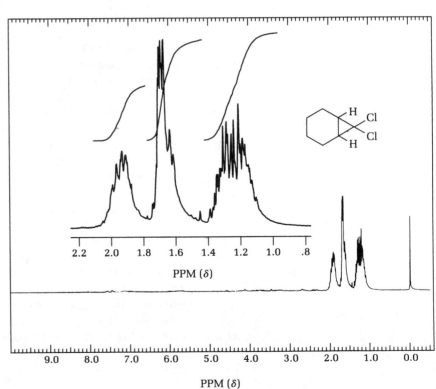

FIG. 21.4 ¹H NMR spectrum of 7,7-dichlorobicyclo-[4.1.0]heptane (250 MHz).

Oxidation: Cyclohexanol to Cyclohexanone; Cyclohexanone to Adipic Acid

Prelab Exercise: Write balanced equations for the dichromate and hypochlorite oxidations of cyclohexanol to cyclohexanone and for the permanganate oxidation of cyclohexanone to adipic acid.

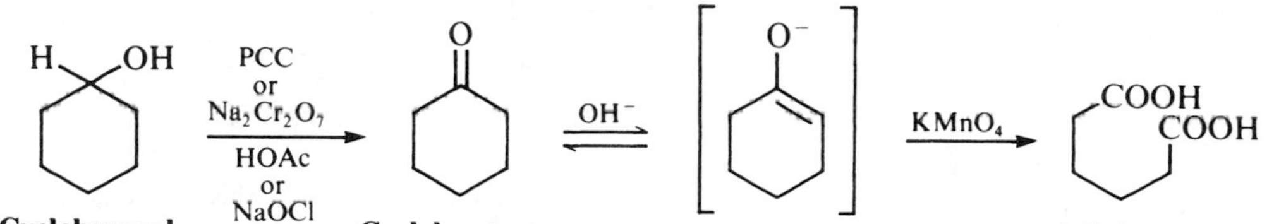

Cyclohexanol
bp 161.5°C, den. 0.96
MW 100.16

Cyclohexanone
bp 157°C, den. 0.95
MW 98.14
solubility 1.5 g/100 mL $H_2O^{10°}$

Adipic acid
mp 153°C, MW 146.14
solubility 1.4 g/100 g $H_2O^{15°}$

The oxidation of a secondary alcohol to a ketone is done by a very large number of oxidizing agents, including sodium dichromate, pyridinium chlorochromate, and sodium hypochlorite (household bleach). The ketone can be oxidized further to the dicarboxylic acid giving adipic acid. Both of these oxidations can be carried out by permanganate ion to give the diacid. Nitric acid is a powerful oxidizing agent that can oxidize cyclohexane, cyclohexene, cyclohexanol, or cyclohexanone to adipic acid.

Dichromate oxidation   The mechanism of oxidation of an alcohol to a ketone by dichromate appears to be the following:

$$H_2O + Cr_2O_7^{2-} \rightleftharpoons 2\ HCrO_4^-$$

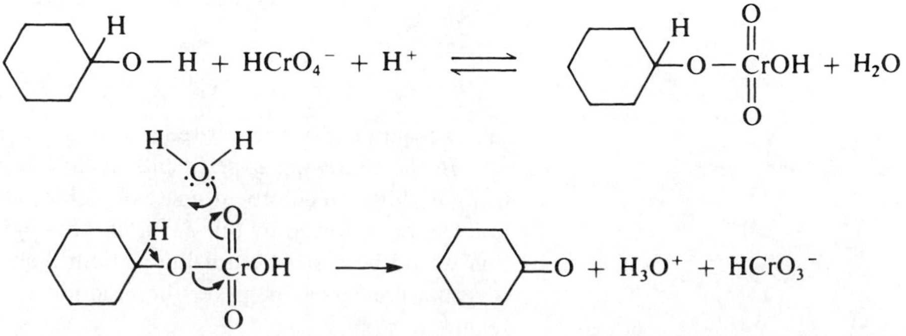

A number of intermediate valence states of chromium are involved in this reaction as orange Cr^{6+} ultimately is reduced to green Cr^{3+}. The course of the oxidation can be followed by these color changes.

Permanganate oxidation

The oxidation of a ketone by permanganate to the dicarboxylic acid takes place through the enol form of the ketone. The reaction can be followed as the bright purple permanganate solution reacts to give a brown precipitate of manganese dioxide. A possible mechanism for this reaction is the following:

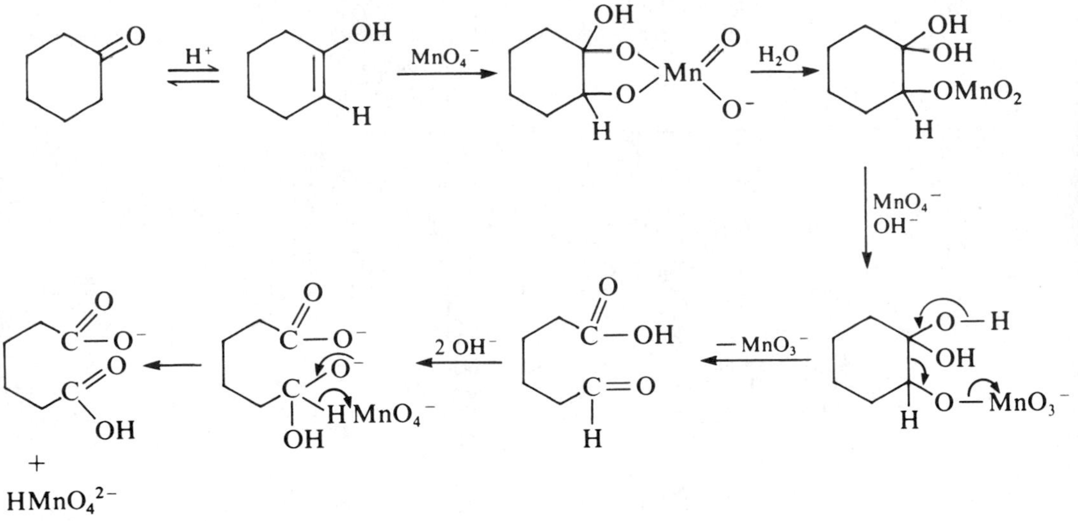

$$3 \; HMnO_4^{2-} + H_2O \longrightarrow 2 \; MnO_2 + MnO_4^- + 5 \; OH^-$$

Nitric acid oxidation

The balanced equation for the nitric acid oxidation of cyclohexanone to adipic acid is:

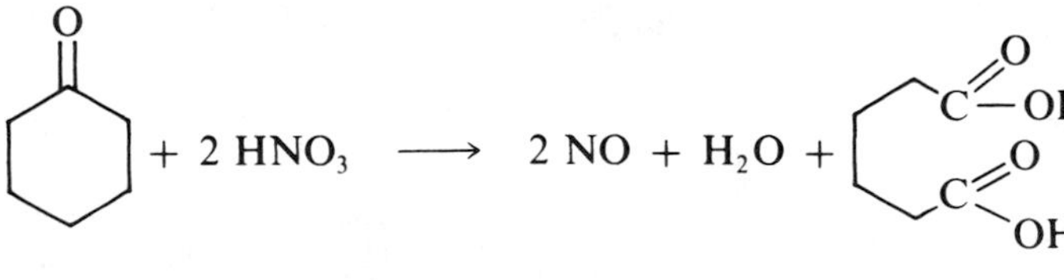

In this reaction nitric acid is reduced to nitric oxide.

In the following experiments cyclohexanol is oxidized to cyclohexanone using pyridinium chlorochromate in dichloromethane. The progress of the reaction can be followed by thin-layer chromatography. On a larger scale this reaction would be carried out using sodium dichromate in acetic acid because the reagents are less expensive, the reaction is faster, and much less solvent is required.

Cr(VI) is probably the most widely used and versatile laboratory oxidizing agent and is used in a number of different forms to carry out selective oxidations in this text. But from an environmental standpoint, it is far from ideal. Inhalation of the dust from insoluble Cr(VI) compounds can lead to cancer of the respiratory system. The product of the reaction (Cr(III)) should not be flushed down the drain, because it is toxic to aquatic life at extremely low concentrations, so, as stated in the "Cleaning Up" section of the dichromate oxidation experiment, the Cr(III) must be precipitated as insoluble $Cr(OH)_3$, and this material dealt with as a hazardous waste.

Sodium hypochlorite oxidation

There is an alternative oxidant for secondary alcohols that is just as efficient and much safer from an environmental standpoint: 5.25% (0.75 M) sodium hypochlorite solution available in the grocery store as household bleach.[1] The mechanism of the reaction is not clear. It is not a free radical reaction; the reaction is much faster in acid than in base; elemental chlorine is presumably the oxidant; and hypochlorous acid must be present. It may form an intermediate alkyl hypochlorite ester, which, by an E_2 elimination, gives the ketone and chloride ion.

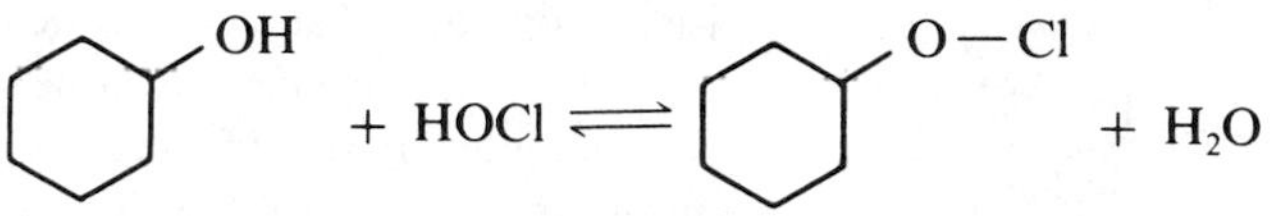

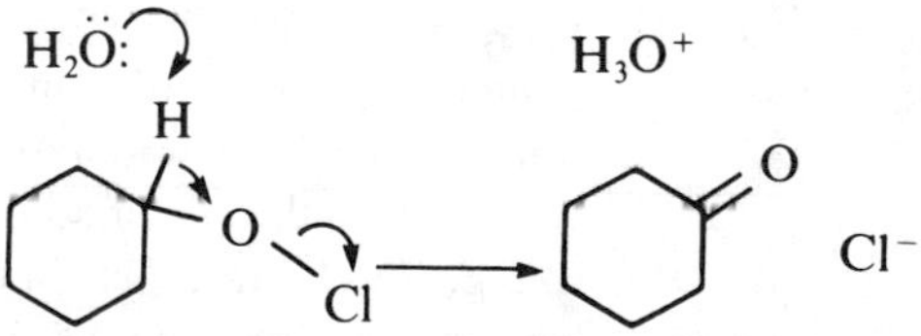

Excess hypochlorite is easily destroyed with bisulfite; the final product is chloride ion, much less toxic to the environment than Cr(III).

Experiments

1. Cyclohexanone from Cyclohexanol

Microscale Procedure

To a 10-mL Erlenmeyer flask add 0.62 g of finely powdered pyridinium chlorochromate (see Chapter 30 for preparation), 4 mL of dichloromethane, and

---

1. J. Mohrig, D. M. Nienhuis, C. F. Linck, C. Van Zoeren, and B. G. Fox, *J. Chem. Ed.,* **62,** 519 (1985).

$$\text{Cyclohexanol} \xrightarrow[\substack{\text{HOAc} \\ \text{or} \\ \text{NaOCl}}]{\substack{\text{PCC} \\ \text{or} \\ \text{Na}_2\text{Cr}_2\text{O}_7}} \text{Cyclohexanone}$$

Cyclohexanol
bp 161.5°C, den. 0.96
MW 100.16

Cyclohexanone
bp 157°C, den. 0.95
MW 98.14
solubility 1.5 g/100 mL $H_2O^{10°}$

250 mg of cyclohexanol. The mixture is stirred magnetically or shaken over a period of days (a week does no harm) until thin-layer chromatography on silica gel indicates the reaction is complete. Elute the TLC plates with dichloromethane. Handle open containers of dichloromethane in the hood; it is a cancer-suspect agent.

At the end of the reaction period, remove the chromium salts by filtration on the Hirsch funnel or, if the salts are coarse enough, by removal of the solvent using a Pasteur pipette. Wash the chromium salts with a few drops of dichloromethane, and evaporate the solvent under a gentle stream of nitrogen or air to leave the product, cyclohexanone. Confirm the structure of this product by obtaining an IR spectrum of the pure liquid between sodium chloride or silver chloride plates. Look for a band at 3600 to 3400 cm^{-1} indicative of unoxidized alcohol (or water if you are not careful).

Cleaning Up   Place used dichloromethane in the halogenated organic waste container. The chromium salts should be dissolved in hydrochloric acid and diluted to <5%. The solution, which should have pH < 3, should then be treated with a 50% excess of sodium bisulfite to reduce the orange dichromate ion to the green chromic ion. Make the solution basic with ammonium hydroxide to precipitate chromium as the hydroxide. This precipitate is collected on a filter paper, which is placed in the hazardous solid waste container for heavy metals. The filtrate from this latter treatment can go down the drain.

The use of Celite will speed the filtration.

2. Cyclohexanone from Cyclohexanol

Macroscale Procedure

CAUTION: The dust from Cr⁶⁺ salts is toxic and a cancer suspect agent; weigh in hood.

In a 50-mL Erlenmeyer flask, dissolve 7.5 g of sodium dichromate dihydrate in 12.5 mL of acetic acid by swirling the mixture on the hot plate, and then cool the solution with ice to 15°C. In a second Erlenmeyer flask chill a mixture of 7.5 g of cyclohexanol and 5 mL of acetic acid in ice. After the first solution is cooled to 15°C, transfer the thermometer and adjust the temperature in the second flask to 15°C. Wipe the flask containing the dichromate solution, pour the solution into the cyclohexanol–acetic acid mixture, rinse the flask with a little solvent (acetic acid), note the time, and take the initially light orange solution

Reaction time: 45 min

from the ice bath, but keep the ice bath ready for use when required.[2] The exothermic reaction that is soon evident can get out of hand unless controlled. When the temperature rises to 60°C, cool in ice just enough to prevent a further rise, and then, by intermittent brief cooling, keep the temperature close to 60°C for 15 min. No further cooling is needed, but the flask should be swirled occasionally and the temperature watched. The usual maximal temperature is 65°C (25–30 min). When the temperature begins to drop and the solution becomes pure green, the reaction is over. Allow 5–10 min more reaction time, and then pour the green solution into a 100-mL round-bottomed flask, rinse the Erlenmeyer flask with 50 mL of water, and add the solution to the flask for steam distillation (Chapter 6) of the product (Fig. 22.1). Distill as long as any oil passes over with the water and, because cyclohexanone is appreciably soluble in water, continue somewhat beyond this point (about 40 mL will be collected).

Alternatively, instead of setting up the apparatus for steam distillation, simply add a boiling chip to the 100-mL flask and distill 20 mL of liquid, cool the flask slightly, add 20 mL of water to it, and distill 20 mL more. Note the temperature during the distillation. This is a steam distillation in which steam is generated *in situ* rather than from an outside source.

Isolation of Cyclohexanone from Steam Distillate

Cyclohexanone is fairly soluble in water. Dissolving inorganic salts such as potassium carbonate or sodium chloride in the aqueous layer will decrease the solubility of cyclohexanone such that it can be completely extracted with *t*-butyl methyl ether. This process is known as "salting out."

To salt out the cyclohexanone, add to the distillate 0.2 g of sodium chloride per milliliter of water present and swirl to dissolve the salt. Then pour the mixture into a separatory funnel, rinse the flask with *t*-butyl methyl ether, add more *t*-butyl methyl ether to a total volume of 10–15 mL, shake, and draw off the water layer. Then wash the ether layer with 10 mL of 3 *M* sodium hydroxide solution to remove acetic acid, test a drop of the wash liquor to make sure it contains excess alkali, and draw off the aqueous layer.

To dry the *t*-butyl methyl ether, which contains dissolved water, shake the ether layer with an equal volume of saturated aqueous sodium chloride solution. Draw off the aqueous layer, pour the ether out of the neck of the separatory funnel into an Erlenmeyer flask, add about 2.5 g of anhydrous calcium chloride pellets, and complete final drying of the ether solution by occasional swirling of

---

2. When the acetic acid solutions of cyclohexanol and dichromate were mixed at 25°C rather than at 15°C the yield of crude cyclohexanone was only 3.5 g. A clue to the evident importance of the initial temperature is suggested by an experiment in which the cyclohexanol was dissolved in 6 mL of benzene instead of 5 mL of acetic acid and the two solutions were mixed at 15°C. Within a few minutes orange-yellow crystals separated and soon filled the flask; the substance probably is the chromate ester, $(C_6H_{11}O)_2CrO_2$. When the crystal magma was let stand at room temperature, the crystals soon dissolved, exothermic oxidation proceeded, and cyclohexanone was formed in high yield. Perhaps a low initial temperature ensures complete conversion of the alcohol into the chromate ester before side reactions set in.

FIG. 22.1   Steam distillation apparatus.

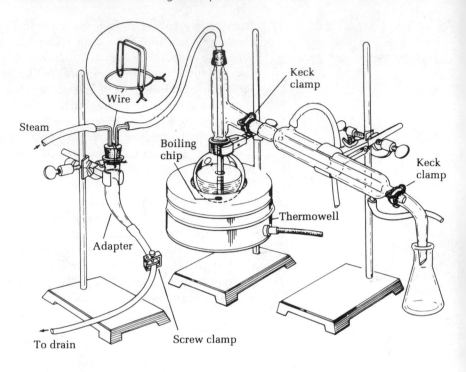

the solution over a 5-min period. Remove the drying agent by decantation or gravity filtration into a tared Erlenmeyer flask, and rinse the flask that contained the drying agent, the calcium chloride, and the funnel with ether. Add a boiling chip to the ether solution, and evaporate the *t*-butyl methyl ether on the steam bath under an aspirator tube (Fig. 8.5). Cool the contents of the flask to room temperature, evacuate the crude cyclohexanone under aspirator vacuum to remove final traces of ether (Fig. 10.6), and weigh the product. Yield is 15–16 g.

The crude cyclohexanone can be purified by simple distillation or used directly to make adipic acid (Experiment 6).

Cleaning Up   To the residue from the steam distillation add sodium bisulfite to destroy any excess dichromate ion, neutralize the solution, dilute it with a large quantity of water, and pour it down the drain if local laws allow. If not, collect the precipitate of chromium hydroxide that forms when the solution is *The use of Celite will speed up*   just slightly basic, and dispose of the filter paper and precipitate in the hazardous *the filtration.*   waste container designated for heavy metals. The filtrate can be flushed down the drain. Combine water layers from the extraction process, neutralize with dilute hydrochloric acid, and dispose of the solution down the drain. Any *t*-butyl methyl ether should be placed in the organic solvents container, and the drying agent, if completely dry, can be placed in the nonhazardous waste container. If it still has organic solvent on it, then it must be placed in the hazardous waste container. It is *much* more expensive to dispose of hazardous waste.

3. Cyclohexanone from Cyclohexanol by Hypochlorite Oxidation

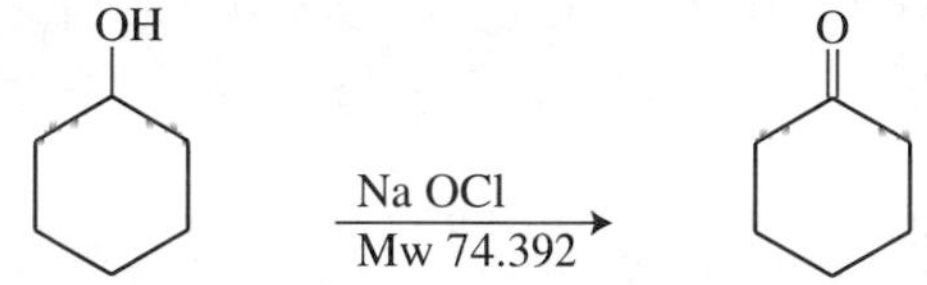

$$\text{Cyclohexanol} \xrightarrow[\text{Mw 74.392}]{\text{Na OCl}} \text{Cyclohexanone}$$

Cyclohexanol
bp 161°C, den. 0.96
MW 100.16

Cyclohexanone
bp 157°C, den. 0.97
MW 98.14
solubility 1.5g /100 mL $H_2O^{10°}$

To a 5-mL long-necked, round-bottomed flask fitted with a connector with support rod, add 150 mg (1.5 mmol) of cyclohexanol followed by a mixture of 80 mg of acetic acid and 2.3 mL of a 5.25% solution of sodium hypochlorite (household bleach, e.g., Clorox). This mixture should be acidic. If necessary add more acetic acid. Add a thermometer. The temperature will rise to 45°C. Swirl the contents of the flask, and as soon as the temperature begins to drop, remove the thermometer, cap the flask with a septum, and shake it vigorously for about 3 min. Then place the flask in a beaker of water at 45°C for 15 min to complete the reaction.

Destroy excess oxidizing agent by adding saturated sodium bisulfite solution (about 1 or 2 drops). At this point the reaction mixture no longer gives a positive test with starch–iodide paper (a positive test is a blue-purple color from the starch–triiodide complex).

Add a drop of thymol blue indicator to the solution, and then neutralize it with 0.3 or 0.4 mL 6 M sodium hydroxide solution. Add a boiling chip to the solution, attach to the flask the distilling head, thermometer adapter, and thermometer, and then distill about 0.8 mL of a mixture of water and cyclohexanone (Fig. 22.2). This is a steam distillation in which the steam is generated *in situ* (see Chapter 6).

Isolation of Cyclohexanone from the Steam Distillate

Cyclohexanone is fairly soluble in water. Dissolving inorganic salts such as potassium carbonate or sodium chloride in the aqueous layer will decrease the solubility of cyclohexanone such that it can be completely extracted with ether. This process is known as *salting out*.

To salt out the cyclohexanone, transfer the distillate to a reaction tube, add 0.1 g of sodium chloride to the distillate, stir, and shake the tube until most of it dissolves. Then extract the solution with three 0.4-mL portions of *tert*-butyl methyl ether. As a reminder, this process consists of adding 0.4-mL of the ether to the reaction tube, shaking the tube, allowing the layers to separate, drawing off the top layer with a Pasteur pipette, and placing it in a clean, dry reaction tube.

FIG. 22.2 Microscale simple distillation apparatus. The temperature is regulated by piling up or removing sand from near the flask.

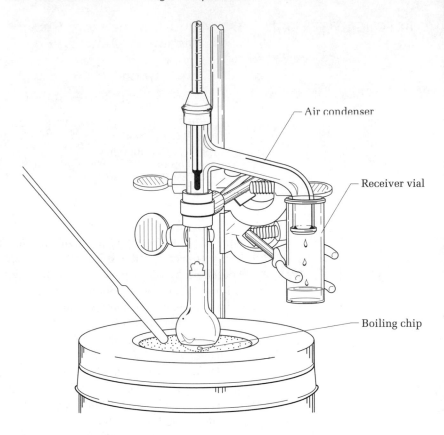

Air condenser

Receiver vial

Boiling chip

Since *tert*-butyl methyl ether dissolves water, the ether extract is wet. A convenient way to dry it is to pass it through a short column of drying agent and alumina (a powerful drying agent). To the chromatography column add about an inch-high column of alumina and on top of this an inch-high column of anhydrous sodium sulfate. Transfer the ether solution of cyclohexanone to the column, let it sit for a minute or two, and then collect the dry ether in a tared 5-mL round-bottomed flask. Wash out the column with about 2 mL of ether, which is collected in the same flask.

Add a boiling chip to the dry ether solution, and fit the flask for simple distillation. Distill all the ether (bp 55°C) using the apparatus shown in Fig. 22.3. The perfectly water-clear residue is cyclohexanone, which can be analyzed by IR and NMR spectroscopy. If it is cloudy, it is wet. Determine the weight of the product, and calculate the percentage yield.

Cleaning Up　　The aqueous reaction mixture after extraction of the product contains primarily sodium, chloride, and acetate ions and can therefore be disposed of down the drain. Any unused ether goes in the organic solvents container, and the drying agent and alumina, after they are dry, can go in the nonhazardous waste container.

FIG. 22.3 Microscale simple distillation apparatus.

Boiling chip

4. Cyclohexanone from Cyclohexanol by Hypochlorite Oxidation

Macroscale Procedure

Into a 250-mL Erlenmeyer flask in the hood place 8 mL (0.075 mol) of cyclohexanol. Introduce a thermometer, and slowly add to the flask with swirling a mixture of 4 mL of acetic acid and 115 mL of a commercial household bleach such as Clorox (usually 5.25% by weight NaOCl, which is 0.75 molar). This can be added from a separatory funnel clamped to a ring stand or from another Erlenmeyer flask. Take care not to come in contact with the reagent. It should be acidic. Test it with indicator paper and if necessary add more acetic acid. During the addition, keep the temperature in the range of 40–50°C. Have an ice bath available in case the temperature goes above 50°C, but do not allow the temperature to go below 40°C because oxidation will be incomplete. The addition should take about 15–20 min. Swirl the reaction mixture periodically for the next 20 min to complete the reaction.

Because the exact concentration of the hypochlorite in the bleach depends on its age, it is necessary to analyze the reaction mixture for unreacted hypochlorite and to reduce excess to chloride ion with bisulfite. Add a drop of the reaction mixture to a piece of starch-iodide paper. Any unreacted hypochlorite will cause the appearance of the blue starch–triiodide complex. Add 1-mL portions of saturated sodium bisulfite solution to the reaction mixture until the starch-iodide test is negative.

Add a few drops of thymol blue indicator solution to the mixture, and then slowly add with swirling 15–20 mL of 6 *M* sodium hydroxide solution until the mixture is neutral as indicated by the blue color.

Note the temperature of the vapor during this distillation.

Transfer the reaction mixture to a 250-mL round-bottomed flask, add a boiling chip, and set up the apparatus for simple distillation (Fig. 22.2). Distillation under these conditions is a steam distillation (see Chapter 6) in which the steam is generated *in situ*. Continue the distillation until no more cyclohexanone comes over with the water (40 mL of distillate should be collected).

See the section titled "Isolation of Cyclohexanone from Steam Distillate" in Experiment 2 to complete the reaction sequence.

Cleaning Up   The residue from the steam distillation contains only chloride, sodium, and acetate ions and can therefore be flushed down the drain.

5. Adipic Acid from Cyclohexanone

Cyclohexanone $\xrightarrow[\text{KMnO}_4]{\text{HNO}_3 \atop \text{or}}$ **Adipic acid**
mp 153°C, MW 146.14
solubility 1.4 g/100 g H_2O

Microscale Procedure

Because of the highly exothermic nature of this reaction with nitric acid, it is not advisable to scale up this procedure. In a 10 × 100 mm reaction tube place 1.0 mL of concentrated nitric acid and a boiling chip. Clamp the tube and also a Pasteur pipette, which dips into the tube and is connected to a water aspirator pulling a gentle vacuum to remove nitrogen oxides. Add to the tube one small drop of cyclohexanone from a vial containing 150 mg of the ketone. Warm the tube on the sand bath until brown oxides of nitrogen are seen emanating from the nitric acid and an exothermic reaction begins. Do not add more cyclohexanone until it is quite clear (evolution of brown oxides of nitrogen, generation of heat) that the reaction has started. Remove the tube from the heat. Add the remainder of the cyclohexanone to the hot nitric acid over a period of about 3 min. The reaction is extremely exothermic; no external heating will be necessary, but add the ketone rapidly enough to keep the reaction going. After the completion of cyclohexanone addition, heat the flask to boiling for 1 min.

As the tube cools to room temperature, fine crystals of adipic acid should appear. If they do not, scratch the inside of the tube at the liquid–air interface

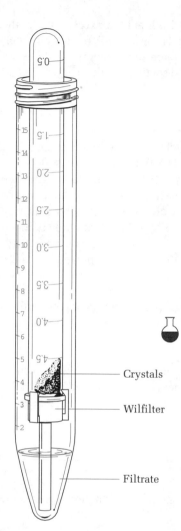

FIG. 22.4 Wilfilter filtration apparatus.

Choice of controlling the temperature for 30 min or running an unattended overnight reaction

with a small glass rod to initiate crystallization. Cool the tube in a mixture of ice and water for at least 3 min, then stir the crystals with the tip of a glass Pasteur pipette and remove the solvent by forcing the pipette to the bottom of the tube. Rap the tube on a hard surface, and remove more solvent from the crystals, taking care to keep the tube cold. Add 0.2 mL of ice water to the crystals, cool thoroughly in ice, and remove as much of the wash liquid as possible using the Pasteur pipette, bearing in mind that the product is *very* soluble in water. Collect the damp crystals on a Wilfilter, by centrifugation (Fig. 22.4) or dry the crystals under vacuum in the reaction tube while heating it on a steam bath. Determine the melting point, the IR spectrum, and the homogeneity of the product by thin-layer chromatography. For TLC, use as the eluent three parts 95% ethanol and one part ammonium hydroxide.

Cleaning Up   The nitric acid filtrate and aqueous wash should be neutralized with sodium carbonate before flushing down the drain. The TLC solvent should be neutralized with dilute hydrochloric acid and then flushed down the drain.

6. Adipic Acid from Cyclohexanone

Macroscale Procedure

The reaction to prepare adipic acid is conducted with 10.0 g of cyclohexanone, 30.5 g of potassium permanganate, and amounts of water and alkali that can be adjusted to provide an attended reaction period of 1/2 h, or an unattended overnight reaction.

Short-Term Procedure

For the short-term reaction, mix the cyclohexanone and permanganate with 125 mL of water in a 250-mL Erlenmeyer flask, adjust the temperature to 30°C, note that there is no spontaneous temperature rise, and then add 1 mL of 3 *M* sodium hydroxide solution. A temperature rise is soon registered by the thermometer. It may be of interest to determine the temperature at which you can just detect warmth, by holding the flask in the palm of the hand and by touching the flask to your cheek. When the temperature reaches 45°C (15 min) slow the oxidation process by brief ice-cooling, and keep the temperature at 45°C for 20 min. Wait for a slight further rise (47°C) and an eventual drop in temperature (25 min), and then heat the mixture by swirling it on a hot plate to complete the oxidation and to coagulate the precipitated manganese dioxide. Make a spot test by withdrawing reaction mixture on the tip of a stirring rod and touching it to a filter paper; permanganate, if present, will appear in a ring around the spot of manganese dioxide. If permanganate is still present, add small amounts of sodium bisulfite until the spot test is negative. Then filter the mixture by suction on an 11-cm Büchner funnel, wash the brown precipitate well with water, add a boiling chip, and evaporate the filtrate over a flame from a large beaker to a volume of 35 mL. If the solution is not clear and colorless, clarify it with decolorizing charcoal and evaporate again to 35 mL. Acidify the hot solution with concentrated hydrochloric acid to pH 1–2, add 5 mL acid in excess, and let the solution stand to crystallize. Collect the crystals on a small Büchner funnel, wash them with a very small quantity of *cold* water, press the crystals between sheets of filter paper to remove excess water, and set them aside to dry. A typical yield of adipic acid, mp 152–153°C, is 3.5 g.

Long-Term Procedure

In the alternative procedure, the weights of cyclohexanone and permanganate are the same, but the amount of water is doubled (250 mL) to moderate the reaction and make temperature control unnecessary. All the permanganate must be dissolved before the reaction is begun. Heat the flask, and swirl the contents vigorously. Test for undissolved permanganate with a glass rod. After adjusting the temperature of the solution to 30°C, 5 mL of 3 M sodium hydroxide is added and the mixture is swirled briefly and let stand overnight (maximum temperature 45–46°C). The workup is the same as in the short-term procedure, and a typical yield of adipic acid is 4.1 g.

A longer oxidizing time does no harm.

Cleaning Up   Place the manganese dioxide precipitate in the hazardous waste container for heavy metals. Neutralize the aqueous solution with sodium carbonate and flush it down the drain.

The proton and carbon-13 NMR spectra of cyclohexanone are given in Figs. 22.5 and 22.6. The IR spectrum of cyclohexanol is presented in Fig. 22.7 and of cyclohexanone in Fig. 22.8. The carbon spectrum of adipic acid is given in Fig. 22.9.

Questions _____

1. In the oxidation of cyclohexanol to cyclohexanone, what purpose does the acetic acid serve?

2. Explain the order of the chemical shifts of the carbon atoms in the ^{13}C spectra of cyclohexanone (Fig. 22.6) and adipic acid (Fig. 22.9).

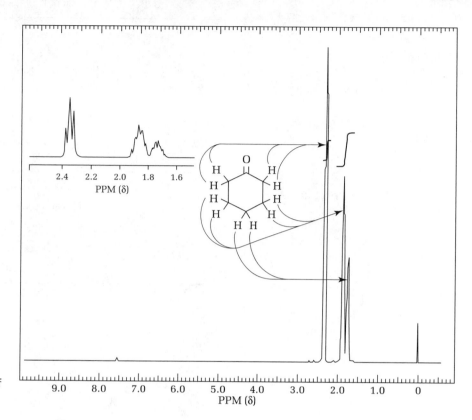

FIG. 22.5   ¹H NMR spectrum of cyclohexanone (250 MHz).

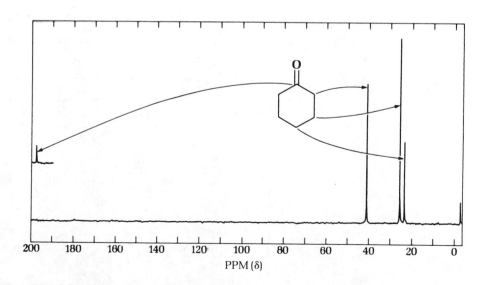

FIG. 22.6   ¹³C NMR spectrum of cyclohexanone (22.6 MHz).

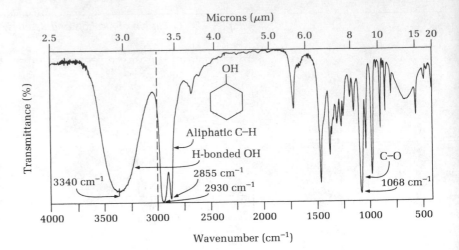

FIG. 22.7 Infrared spectrum of cyclohexanol (thin film).

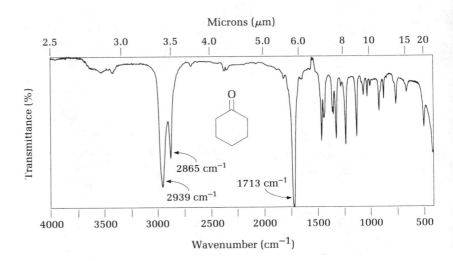

FIG. 22.8 Infrared spectrum of cyclohexanone (thin film).

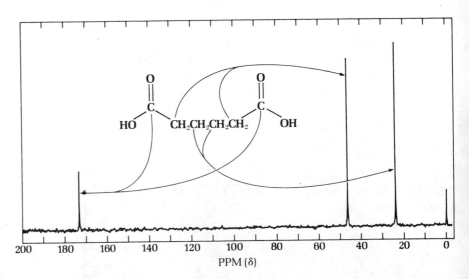

FIG. 22.9 ^{13}C NMR spectrum of adipic acid (22.6 MHz).

Pulegone from Citronellol: Oxidation with Pyridinium Chlorochromate

Prelab Exercise: Write a balanced equation for the oxidation of citronellol to citronellal.

Oxidizing agents:
$Cr_2O_7^{2-}$ in HOAc,
$KMnO_4$,
CrO_3 in aq H_2SO_4, Jones reagent,
CrO_3 + pyridine, Collins reagent,
Corey reagents, PCC

E. J. Corey of Harvard, winner of the 1991 Nobel prize in chemistry

A variety of reagents and conditions can be employed to oxidize alcohols to carbonyl compounds. The choice of which reagents and set of conditions to use depends on such factors as the scale of the reaction, the speed of the reaction, anticipated yield, and ease of isolation of the products. Cyclohexanol can be oxidized to cyclohexanone by dichromate in acetic acid. This is much faster than permanganate oxidation of the same alcohol. Cholesterol can be oxidized to 5-cholesten-3-one by either the Jones reagent (chromium trioxide in sulfuric acid and water) or the Collins reagent, $(C_5H_5N)_2CrO_3$, prepared from chromium trioxide in anhydrous pyridine.

E. J. Corey has added two new Cr^{6+} reagents to the repertory. Although they might seem similar to the other dichromate reagents, they each have unique advantages. While not nearly as fast as the Collins or Jones reagents, the Corey reagents are characterized by extraordinary ease of product isolation, because one merely removes the reagent by filtration and evaporates the solvent to obtain the product. Both reagents are easily prepared or can be purchased commercially (Aldrich). The reagent used in this experiment, pyridinium chlorochromate, $C_5H_5NH^+$ $ClCrO_3^-$, will oxidize primary alcohols to aldehydes in high yield without oxidizing the aldehyde further to a carboxylic acid.[1] It will oxidize secondary alcohols to ketones, but, being somewhat acidic, it will cause rearrangement of unconjugated to conjugated ketones.

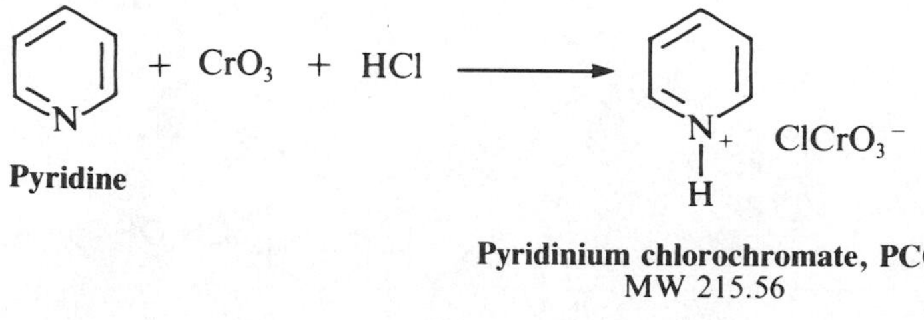

Pyridine

Pyridinium chlorochromate, PCC
MW 215.56

1. E. J. Corey and J. W. Suggs, *Tetrahedron Letters*, 2647 (1975).

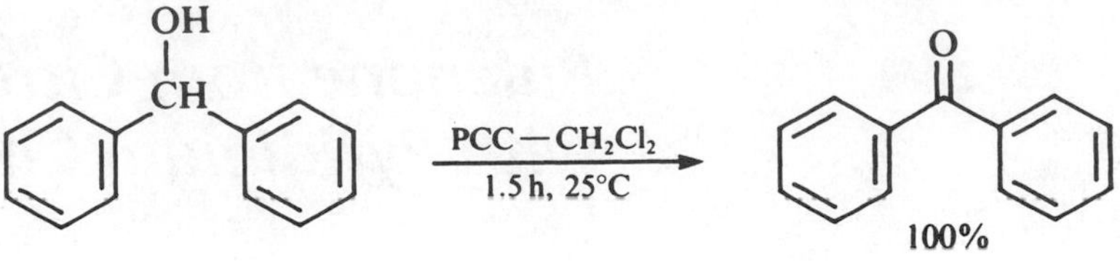

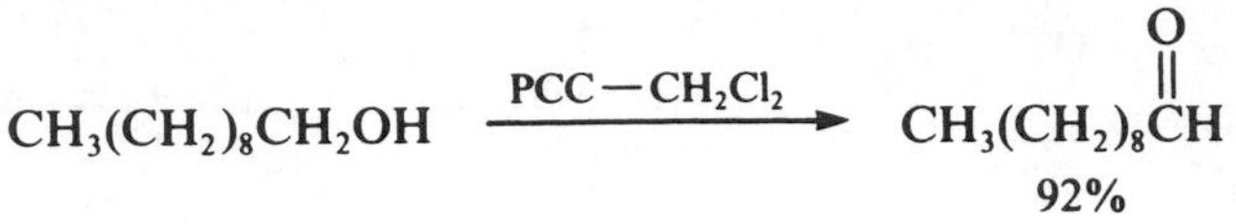

$$CH_3(CH_2)_8CH_2OH \xrightarrow[]{PCC-CH_2Cl_2} CH_3(CH_2)_8\overset{\overset{\displaystyle O}{\|}}{C}H$$

92%

The other Corey reagent, pyridinium dichromate (PDC), $(C_5H_5NH^+)_2Cr_2O_7^{2-}$, dissolved in dimethylformamide (DMF) will oxidize allylic alcohols to α,β-unsaturated aldehydes without oxidizing the aldehyde to the carboxylic acid.[2]

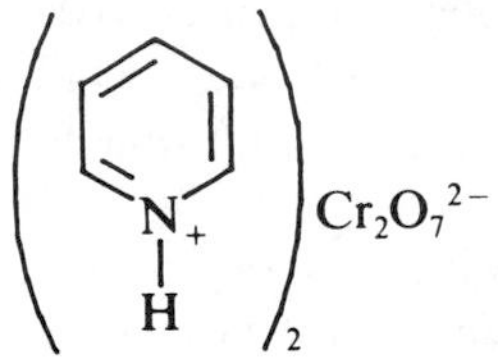

Pyridinium dichromate, PDC

$$C_6H_5CH=CH-CH_2OH \xrightarrow[4-5\,h,\ 0°C]{PDC-DMF} C_6H_5CH=CH-\overset{\overset{\displaystyle O}{\|}}{C}H$$

97%

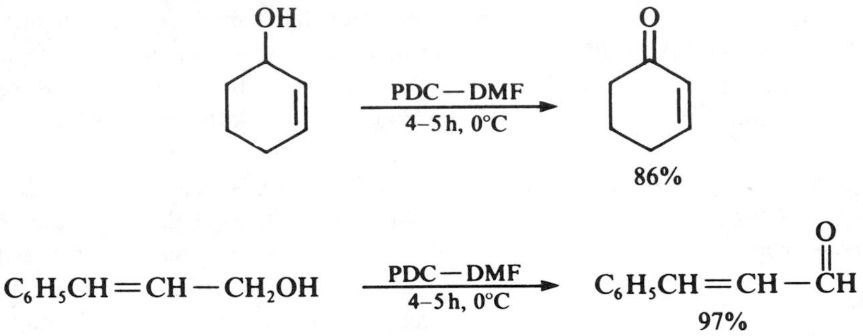

If the primary alcohol is not conjugated, then it is oxidized to the carboxylic acid:

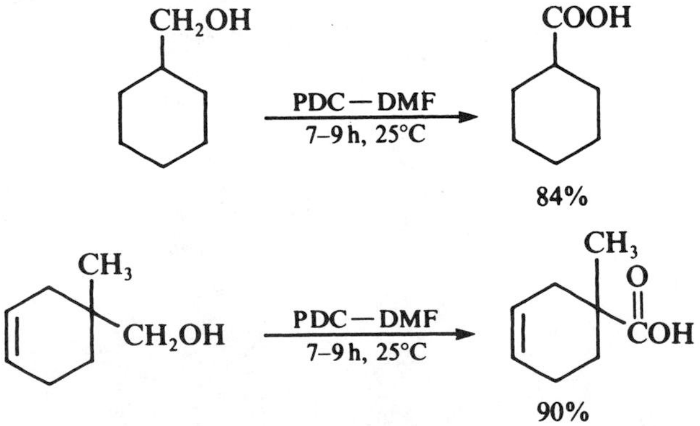

84%

90%

---

2. E. J. Corey and G. Schmidt, *Tetrahedron Letters,* 399 (1979).

When suspended in CH_2Cl_2, pyridinium dichromate will oxidize primary alcohols to the corresponding aldehydes and no further:

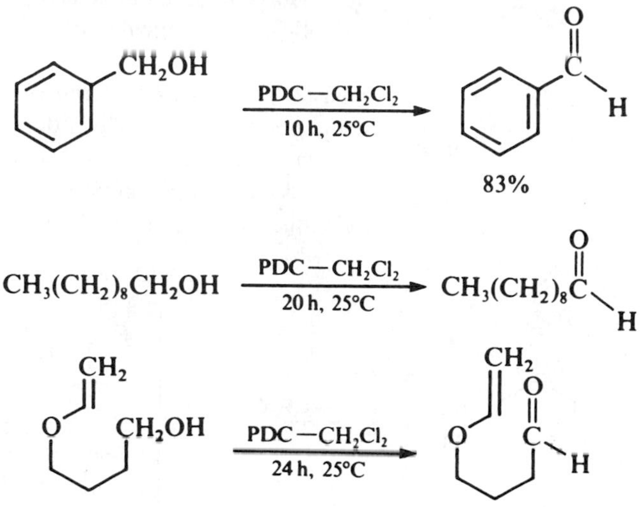

$$CH_3(CH_2)_8CH_2OH \xrightarrow[\text{20 h, 25°C}]{PDC-CH_2Cl_2} CH_3(CH_2)_8C\overset{O}{\underset{H}{\parallel}}$$

Citronellol-A Terpene

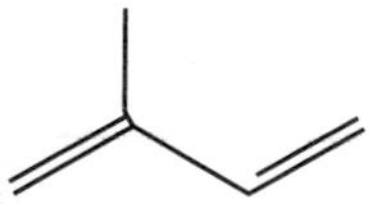

Citronellol is a member of the large class of natural products known as terpenes. Biosynthesized from isopentenyl pyrophosphate, these molecules contain 10, 15, 20, etc., atoms, are made up of repeating isoprene units, and have pleasant odors. Many of the ten-carbon molecules are very familiar to you by odor, if not by name, for example, pinene from turpentine, limonene from citrus oil, camphor, and menthol.

Geraniol is a sex pheromone released by female bees to attract males. Certain orchids mimic the female bee by also releasing geraniol. The male bee is attracted to the orchid and will even try to copulate with it, in the course of which, as it wanders from orchid to orchid, it collects pollen and pollinates the flowers.

Isoprene

Experiments

PCC on alumina

$RCH_2OH \longrightarrow RCHO$

In the present experiment pyridinium chlorochromate (PCC) is used to oxidize citronellol, a constituent of rose and geranium oil, to the corresponding aldehyde. If PDC in CH_2Cl_2 were used, the reaction would stop at this point and citronellal could be isolated in 92% yield. Or if sodium acetate (a buffer) were added to the PCC reaction, the oxidation would also stop at the aldehyde stage, in 82% yield. It has been found that PCC deposited on alumina will effect the same oxidation in less time and even higher yield.[3]

---

3. Y.-S. Cheng, W.-L. Liu, and S. Chen, *Synthesis,* 223 (1980).

If the buffer is omitted, the pyridinium chlorochromate is acidic enough to cause the intermediate citronellal to cyclize to isopulegols (four possible isomers). These secondary alcohols are oxidized to two isomeric isopulegones. On treatment with base, the double bond will migrate to give pulegone, as shown in the following chemical structures.

The course of the reaction can easily be followed by thin-layer chromatography.[4] The isolation of the oxidation products, citronellal and isopulegone, could not be simpler: The reduced reagent is removed by filtration and the solvent is evaporated. In the case of the second experiment, the reaction can be stopped once isopulegone is formed, or the isopulegone can be isomerized to pulegone as a separate reaction.

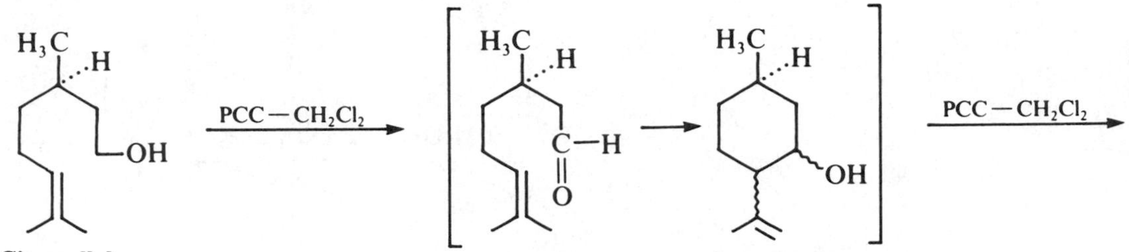

Citronellol
MW 156.27
bp 222°C

Citronellal
MW 154.24

Isopulegols
(four possible isomers)

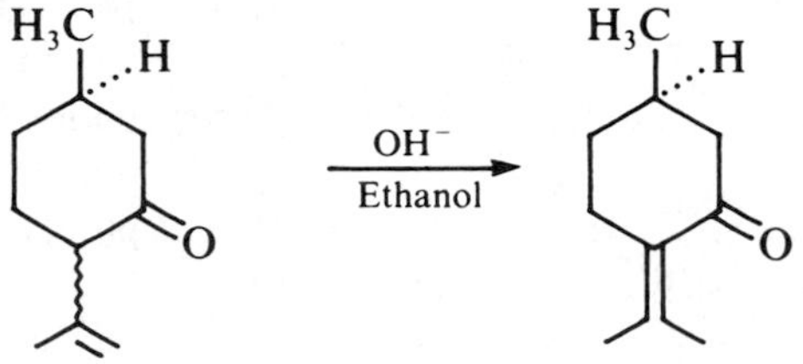

Isopulegone
(two possible isomers)

Pulegone
MW 152.23
bp 224°C

MICROSCALE AND MACROSCALE   1. Preparation of Pyridinium Chlorochromate, PCC

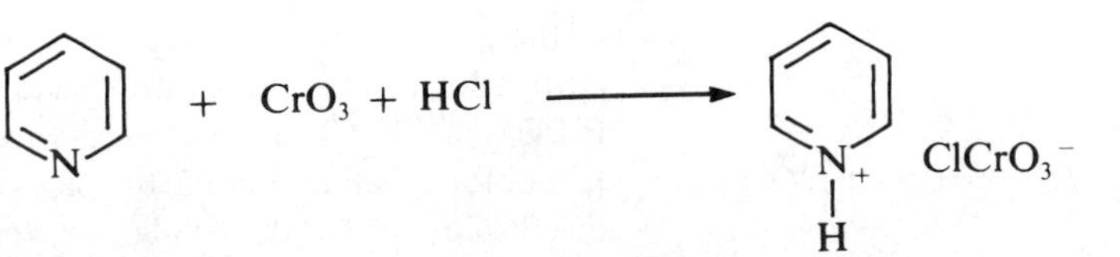

Pyridine

Pyridinium chlorochromate, PCC
MW 215.56

---

4. E. J. Corey and J. W. Suggs, *J. Org. Chem.,* **41,** 380 (1976). TLC R_f values: citronellol, 0.17; citronellal, 0.65; isopulegols, 0.27 and 0.35; isopulegone, 0.41; pulegone, 0.36, CH_2Cl_2 developer.

Dissolve 12 g of chromium trioxide in 22 mL of 6 *N* hydrochloric acid, and add 9.5 g of pyridine during a 10-min period while maintaining the temperature at 45°C. Cool the mixture to 0°C, and collect the crystalline yellow-orange pyridinium chlorochromate on a sintered glass funnel and dry it for 1 h *in vacuo*. The compound is not hygroscopic and is stable at room temperature.

Cleaning Up   The filtrate should have pH < 3 and can then be treated with a 50% excess of sodium bisulfite to reduce the orange dichromate ion to the green chromic ion. This solution should then be made basic with ammonium hydroxide to precipitate chromium as the hydroxide. This precipitate is collected on a filter paper, which is placed in the hazardous solid waste container for heavy metals. The filtrate from this latter treatment can go down the drain.

MICROSCALE AND MACROSCALE

2. Preparation of PCC-Alumina Reagent

To prepare PCC on alumina, cool the reaction mixture described above to 10°C until the product crystallizes; then reheat the solution to 40°C to dissolve the solid. To the resulting solution add 100 g of alumina with stirring at 40°C. The solvent is removed on a rotary evaporator, and the orange solid is dried in vacuum for 2 h at room temperature. If stored under vacuum in the dark, the reagent will retain its activity for several weeks.

CAUTION: *The dust (but not the solutions) of Cr^{6+} is a cancer-suspect agent.*

Cleaning Up   The used alumina, since it contains adsorbed hexavalent chromium ion, should be placed in the hazardous solid waste container for Cr^{6+} contaminated alumina. Place dichloromethane in the halogenated waste container.

3. Citronellal from Citronellol Using Pyridinium Chlorochromate on Alumina

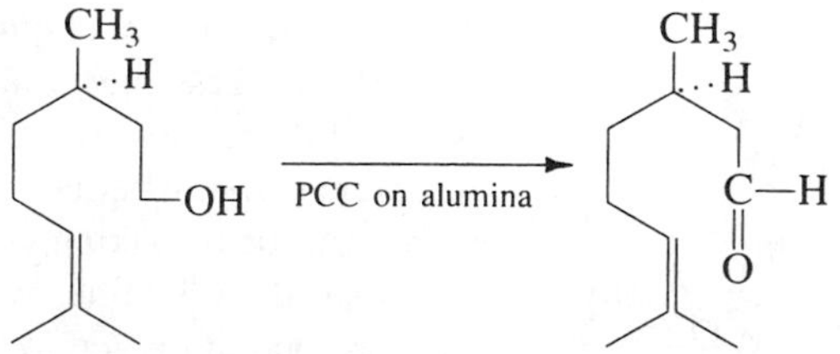

Citronellol
MW 156.27
bp 222°C
n_D^{20} 1.4560

Citronellal
MW 145.25
bp 207°C
n_D^{20} 1.4480

Microscale Procedure

To a 10-mL Erlenmeyer flask, add 1.5 g (1.22 mmol) of PCC on alumina, 120 mg of citronellol, and 2 mL of hexane. Stir the mixture using a magnetic stirrer or by

Pyridinium chlorochromate
MW 215.56

shaking for up to 3 h (follow the course of the reaction by TLC), remove the stirring bar, and then remove the solid by filtration on the Hirsch funnel under vacuum. Wash the alumina on the filter with three 2-mL portions of ether, and remove the solvents from the filtrate by evaporation under a gentle stream of air while warming the flask in a beaker of water. The last trace of solvent can be removed under vacuum. The residue should be pure citronellal, bp 90°C at 14 mm Hg. Check its purity by TLC on silica gel using dichloromethane as eluent. Using IR spectroscopy as a thin film between sodium chloride plates, look for the presence of unoxidized hydroxyl at 3600 to 3400 cm^{-1}. See next experiment for "Cleaning Up" procedure.

Macroscale Procedure

PCC on alumina (7.5 g, 6.1 mmol) is added to a flask containing a solution of citronellol (0.60 g, 3.8 mmol) in 10 mL of *n*-hexane. After stirring or shaking for up to 3 h (follow the course of the reaction by TLC), remove the solid by filtration, wash it with three 10-mL portions of ether, and remove the solvents from the filtrate by distillation or evaporation. The last trace of solvent can be removed under vacuum (see Fig. 3.22). The residue should be pure citronellal, bp 90°C at 14 mm Hg. Check its purity by TLC and IR spectroscopy. A large number of other primary and secondary alcohols can be oxidized to aldehydes and ketones using this same procedure.

Cleaning Up   The used alumina, since it contains adsorbed hexavalent chromium ion, should be placed in the hazardous solid waste container. Organic material goes in the organic solvents container.

4. Isopulegone from Citronellol Using Pyridinium Chlorochromate in CH$_2$Cl$_2$

Microscale Procedure

To 1 g of pyridinium chlorochromate suspended in 6.25 mL of dichloromethane in a 10-mL Erlenmeyer flask, add 0.25 g of citronellol. The best yields are obtained if the mixture is stirred at room temperature for 36 h or more using a magnetic stirrer. Alternatively, let the mixture sit at room temperature for a week with occasional shaking. The chromium salts are removed by vacuum filtration and washed on the filter with 1 mL of dichloromethane. Remove the solvent by evaporation with a gentle stream of nitrogen or air to give isopulegone. Follow the course of the reaction, and analyze the product by TLC on silica gel plastic sheets, developing with dichloromethane. Save 1 mg of the product for the UV spectrum (see the next experiment). See next experiment for "Cleaning Up" procedure.

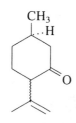

Isopulegone
(Two possible isomers)

Macroscale Procedure

To 16 g of pyridinium chlorochromate suspended in 100 mL of dichloromethane, add 4.0 g citronellol. The best yields are obtained if the mixture is stirred at room temperature for 36 h or more using a magnetic stirrer. Alternatively, let the mixture sit at room temperature for a week with occasional shaking. The chromium salts are removed by vacuum filtration and washed on the filter with 15 mL of dichloromethane. Remove the solvent by evaporation to give isopulegone. Follow the course of the reaction, and analyze the product by TLC on silica gel plastic sheets, developing with dichloromethane.

Cleaning Up   Dichloromethane should be placed in the halogenated organic waste container. The chromium salts should be dissolved in hydrochloric acid, diluted to < 5%, reduced, and precipitated as described above under the preparation of pyridinium chlorochromate.

5. Pulegone from Isopulegone

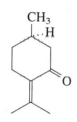

Pulegone

Ultraviolet spectra α,β-unsaturated ketone

Microscale Procedure

Add 250 mg of crude isopulegone isomers to a reaction tube followed by 2 mL of a 2% solution of sodium hydroxide in ethanol. Warm the mixture on a hot water bath or on a steam bath for 1 h. Evaporate the ethanol under a stream of air or nitrogen, and then dissolve the residue in 1 mL of dichloromethane. Prepare a chromatographic column from 2.5 g of silica gel in dichloromethane, and pour the solution of pulegone through the column followed by 2 mL of dichloromethane. This very crude chromatography is to remove the very small quantity of sodium hydroxide present in the pulegone. Evaporate the solvent to leave pure pulegone.

The pulegone also can be isolated from the reaction mixture by evaporating the ethanol, adding 1 mL of water, and extracting the product with two 1-mL portions of ether. The ether extract is washed with 0.5 mL of water, followed by 0.5 mL of saturated sodium chloride solution, and then is dried over calcium chloride pellets. The ether is removed from the drying agent with a Pasteur pipette, the drying agent is washed with more ether, and the ether is evaporated under a stream of air or nitrogen to leave pure pulegone.

Pulegone occurs naturally in pennyroyal oil and has a pleasant odor between those of peppermint and camphor. It is an α,β-unsaturated ketone and therefore should have λ$_{max}$ 254 nm as calculated by application of the Fieser and Woodward rules (see Chapter 21). The actual value is 253.3 nm. The starting material is not a conjugated ketone and so should not have intense UV absorption. Compare the two compounds by dissolving 1 mg of each in 50 mL ethanol and determining their UV spectra.

Cleaning Up   Dichloromethane should be placed in the halogenated organic waste container, ether and ethanol in the organic solvents container, and the alumina and drying agent, once completely dry, in the nonhazardous waste container.

Macroscale Procedure

Crush and then dissolve one sodium hydroxide pellet under 30 mL of ethanol in a 50-mL Erlenmeyer flask, add 4 g of isopulegone, and warm the mixture on the steam bath for 1 h. Evaporate most of the ethanol on the steam bath; then add 10 mL of cold water and extract the product with 20 mL of ether. Wash the ether extract with about 5 mL of water, followed by 5 mL of saturated sodium chloride solution, and then dry the ether over anhydrous calcium chloride. Decant the ether into an Erlenmeyer flask, wash the calcium chloride with a few milliliters more ether, and evaporate the ether to leave crude pulegone. Pulegone can be distilled at atmospheric pressure (bp 224°C) but is best distilled at reduced pressure (bp 103°C at 17 mm Hg). Pulegone occurs naturally in pennyroyal oil and has a pleasant odor somewhere between those of peppermint and camphor. If the final product is not purified by vacuum distillation, this entire experiment can be scaled down by a factor of four or more.

Cleaning Up   Combine the aqueous layers, neutralize with dilute hydrochloric acid, and after dilution flush down the drain. Dry the calcium chloride in the hood, and then place it in the nonhazardous waste container. Combine all organic material, and place it in the organic solvents container.

Computational Chemistry

Calculate, using a molecular mechanics program, the steric energies of the two isomeric isopulegone isomers to determine which is more stable. Similarly calculate the steric energies of the four isomeric isopulegols. In each of these two cases, valid comparisons can be made on the basis of steric energies because the bonding types are identical. Note that finding the global minimum, especially for the isopulegols, is not easy. You have considerable latitude when picking just how the isopropyl and hydroxyl groups are arranged. Compare your results with others to determine who has the lowest steric energy and why.

Isopulegone isomerizes to pulegone. To compare these, molecular heats of formation must be calculated using AM1 calculations. Is pulegone more stable than each of the isopulegone isomers?

Questions

1. Draw the chair conformations of the four possible isopulegols.

2. Assign as many peaks as possible to specific protons in the ^{1}H NMR spectra of citronellol (Fig. 23.1) and citronellal (Fig. 23.2).

3. Give the mechanism for the base-catalyzed isomerization of isopulegone to pulegone.

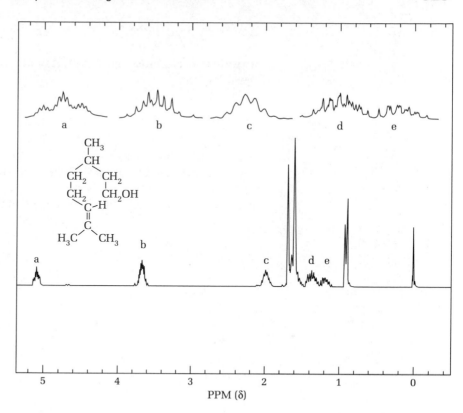

FIG. 23.1 ¹H NMR spectrum of citronellol (250 MHz).

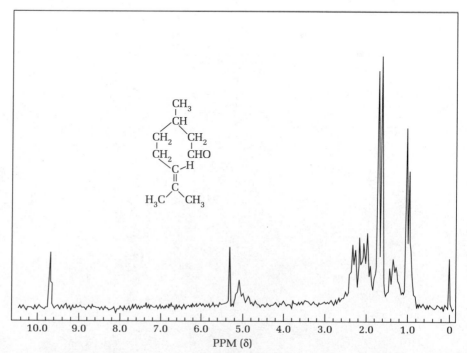

FIG. 23.2 ¹H NMR spectrum of citronellal (90 MHz).

Surfing the Web

http://www.colostate.edu/Depts/Entomology/courses/en570/papers_1996/bernklau.html

"Chemical Mimicry in Pollination" reports on the use of terpenes such as citronellol, geraniol, nerol, farnesol, etc., by flowers, principally orchids, to attract insects.

For updated information visit:
www.mtholyoke.edu/courses/kwilliam/microscale.shtml
or
www.hmco.com/hmco/college/chemistry/home.html

Oxidative Coupling of Alkynes:
2,7-Dimethyl-3,5-octadiyn-2,7-diol

Prelab Exercise: Write a balanced equation for the coupling of 2-methyl-3-butyn-2-ol. What role does cuprous chloride play in this reaction?

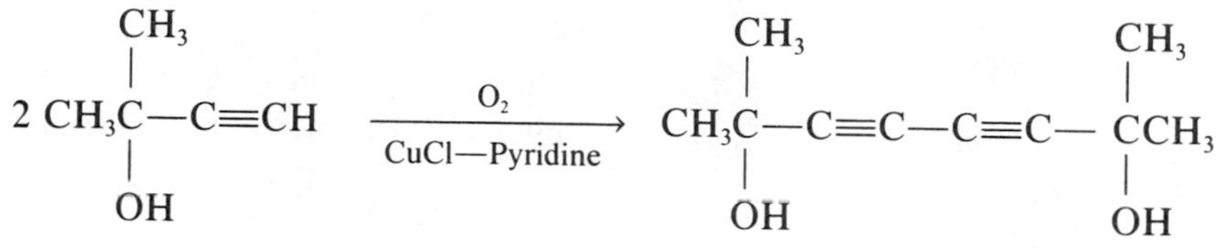

$$2\ CH_3\underset{\underset{OH}{|}}{\overset{\overset{CH_3}{|}}{C}}-C{\equiv}CH \xrightarrow[\text{CuCl—Pyridine}]{O_2} CH_3\underset{\underset{OH}{|}}{\overset{\overset{CH_3}{|}}{C}}-C{\equiv}C-C{\equiv}C-\underset{\underset{OH}{|}}{\overset{\overset{CH_3}{|}}{C}}CH_3$$

2-Methyl-3-butyn-2-ol
MW 84.11, den. 0.868, bp 103°C

2,7-Dimethyl-3,5-octadiyn-2,7-diol
MW 166.21, mp 130°C

The starting material, 2-methyl-3-butyn-2-ol, is made commercially from acetone and acetylene and is convertible into isoprene. This experiment illustrates the oxidative coupling of a terminal acetylene to produce a diacetylene, the Glaser reaction.

Although Glaser reported the coupling of acetylenes using cuprous ion in 1869, the details of the mechanism of this reaction are still obscure. It appears to involve both Cu^+ and Cu^{2+} ions; as noted by Glaser, the cuprous acetylide is oxidized by oxygen,

$$C_6H_5C{\equiv}CH \xrightarrow[\text{NH}_4\text{OH}]{\text{CuCl}} C_6H_5C{\equiv}CCu \xrightarrow{\text{air}} C_6H_5C{\equiv}C-C{\equiv}CC_6H_5$$

but Cu^{2+} is the actual oxidizing agent. The amine seems to be needed to keep the acetylide in solution.

The reaction is very useful synthetically in the synthesis of polyenes, vitamins, fatty acids, and the annulenes. Baeyer used the reaction in his historic synthesis of indigo as long ago as 1882. The reaction allowed the unequivocal establishment of the carbon skeleton of this dye:

In the present experiment the reaction time is shortened by use of excess

The reaction scheme is shown with structures:

ArC≡C—COOH → (CuCl, NH₄OH, −CO₂) → ArC≡C—Cu → (K₃Fe(CN)₆, oxidizing agent) → ArC≡C—C≡CAr → (H₂SO₄, (NH₄)₂S) → **Indigo**

catalyst and by supplying oxygen under the pressure of a balloon; the state of the balloon provides an index of the course of the reaction.

Experiment

2,7-Dimethyl-3,5-octadiyn-2,7-diol

Microscale Procedure

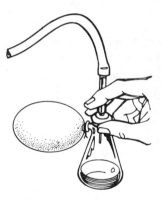

Measure pyridine in hood.

FIG. 24.1 Balloon technique of oxygenation. About 10 lb of oxygen pressure is needed to inflate the pipette bulbs.

The reaction vessel is a 25-mL filter flask equipped with a magnetic stirrer and with a rubber bulb secured to the side arm with copper wire, evident from Fig. 24.1. Add 10 mL of 2-methyl-3-butyn-2-ol, 1.0 mL of methanol, 0.3 mL of pyridine, and 0.1 g of copper(I) chloride. Before going to the oxygen station,[1] practice capping the flask with a large serum stopper until you can do this quickly. You are to flush out the flask with oxygen and cap it before air can diffuse in; the reaction is about twice as fast in an atmosphere of oxygen as in air. Insert the oxygen delivery syringe into the flask with the needle under the surface of the liquid, open the valve, and let oxygen bubble through the solution in a brisk stream for 2 min. Close the valve, and quickly cap the flask and wire the cap. Next, thrust the needle of the delivery syringe through the center of the serum stopper, open the valve, and run in oxygen until you have produced a sizable inflated bulb (see Fig. 24.1). Close the valve and withdraw the needle (the hole is self-sealing), note the time, and start swirling the reaction mixture. The rate of oxygen uptake depends on the efficiency of mixing of the liquid and gas phases. Swirl or stir the reaction mixture continuously for 20 to 25 min. The reaction mixture warms up and becomes deep green. Introduce a second charge of oxygen of the same size as the first, note the time, and swirl. In 25 to 30 min, the balloon reaches a constant volume and

1. The valves of a cylinder of oxygen should be set to deliver gas at a pressure of 10 lb/in.² when the terminal valve is opened. The barrel of a 2.5-mL plastic syringe is cut off and thrust into the end of a ¼ × 3/16-in. rubber delivery tube. Read about the handling of compressed gas cylinders in Chapter 2. Be sure the cylinder is secured to a bench or wall.

the reaction is complete. A pair of calipers is helpful in recognizing the constant size of the balloon and thus the end of the reaction.[2]

Open the reaction flask, cool if warm, add 0.5 mL of concentrated hydrochloric acid to neutralize the pyridine and keep copper compounds in solution, and cool again; the color changes from green to yellow. Use a spatula to dislodge copper salts adhering to the walls, leaving them in the flask. Then add 2.5 mL of saturated sodium chloride solution to precipitate the diol, and stir and cool the thick paste that results. Scrape out the paste onto the Hirsch funnel with the spatula; press down the material to an even cake. Rinse out the flask with water, and wash the filter cake with enough more water to remove all the color from the cake and leave a colorless product. Since the moist product dries slowly, drying is accomplished in ether solution, and the operation is combined with recovery of diol retained in the mother liquor. Transfer the moist product to a 10-mL Erlenmeyer flask, extract the mother liquor in the filter flask with one 1.5-mL portion of *tert*-butyl methyl ether, wash the extract once with water, and add the ethereal solution to the flask containing the solid product. Add enough more ether to dissolve the product, transfer the solution to a tube, remove the water layer, wash with saturated sodium chloride solution, and dry over anhydrous sodium sulfate. The solvent is evaporated on the steam bath, and the solid residue is heated and evacuated until the weight is constant. Crystallization from toluene then gives colorless needles of the diol (mp 129–130°C). The yield of crude product is about 0.5 g. In the crystallization, the recovery is practically quantitative.

Analyze the product by thin-layer chromatography and IR spectroscopy. Compare the diyn with the starting material to determine whether the product is pure.

Cleaning Up   Combine all aqueous and organic layers from the reaction, add sodium carbonate to make the aqueous layer neutral, and shake the mixture gently to extract pyridine into the organic layer. Place the organic layer in the organic solvents container. Remove any solid (copper salts) from the aqueous layer by filtration. The solid can be placed in the nonhazardous solid waste container and the aqueous layer diluted with water and flushed down the drain.

Macroscale Procedure

The reaction vessel is a 125- or 50-mL, filter flask or Erlenmeyer flask with side tube with a rubber bulb secured to the side arm with copper wire, evident from Figs. 24.1 or 24.2. Add 5 mL of 2-methyl-3-butyn-2-ol, 5 mL of methanol, 1.5 mL of pyridine, and 0.25 g of copper(I) chloride. Before going to the oxygen station,[3] practice capping the flask with a large serum stopper until you can do this quickly. You are to flush out the flask with oxygen and cap it before air can

---

2. A magnetic stirrer does not materially shorten the reaction time.

3. The valves of a cylinder of oxygen should be set to deliver gas at a pressure of 10 lb/sq in. when the terminal valve is opened. The barrel of a 2.5-mL plastic syringe is cut off and thrust into the end of a $\frac{1}{4} \times \frac{3}{16}$-in. rubber delivery tube. Read about the handling of compressed gas cylinders in Chapter 2. Be sure the cylinder is secured to a bench or wall.

diffuse in; the reaction is about twice as fast in an atmosphere of oxygen as in air. Insert the oxygen delivery syringe into the flask with the needle under the surface of the liquid, open the valve, and let oxygen bubble through the solution in a brisk stream for 2 min. Close the valve and quickly cap the flask and wire the cap. Next thrust the needle of the delivery syringe through the center of the serum stopper, open the valve, and run in oxygen until you have produced a sizable inflated bulb (see Fig. 24.1). Close the valve and withdraw the needle (the hole is self-sealing), note the time, and start swirling the reaction mixture. The rate of oxygen uptake depends on the efficiency of mixing of the liquid and gas phases. By vigorous and continuous swirling it is possible to effect deflation of the balloon to a constant size (see Fig. 24.2) in 20–25 min. The reaction mixture warms up and becomes deep green. Introduce a second charge of oxygen of the same size as the first, note the time, and swirl. In 25–30 min the balloon reaches a constant volume (e.g., a 5-cm sphere), and the reaction is complete. A pair of calipers is helpful in recognizing the constant size of the balloon and thus the end of the reaction.[4]

Measure pyridine in hood.

FIG. 24.2 Appearance of pipette bulb after oxygenation.

Open the reaction flask, cool if warm, add 2.5 mL of concentrated hydrochloric acid to neutralize the pyridine and keep copper compounds in solution, and cool again; the color changes from green to yellow. Use a spatula to dislodge copper salts adhering to the walls, leaving it in the flask. Then add 13 mL of saturated sodium chloride solution to precipitate the diol, and stir and cool the thick paste that results. Scrape the paste onto a small Büchner funnel with the spatula; press down the material to an even cake. Rinse out the flask with water, and wash the filter cake with enough more water to remove all the color from the cake and leave a colorless product. Since the moist product dries slowly, drying is accomplished in ether solution, and the operation is combined with recovery of diol retained in the mother liquor. Transfer the moist product to a 50-mL Erlenmeyer flask, extract the mother liquor in the filter flask with one portion of *t*-butyl methyl ether, wash the extract once with water, and run the ethereal solution into the flask containing the solid product. Add enough more ether to dissolve the product, transfer the solution to a separatory funnel, drain off the water layer, wash with saturated sodium chloride solution, and filter through anhydrous calcium chloride pellets into a tared flask. The solvent is evaporated on the steam bath, and the solid residue is heated and evacuated until the weight is constant. Crystallization from toluene then gives colorless needles of the diol, mp 129–130°C. The yield of crude product is 3–4 g. In the crystallization, the recovery is practically quantitative.

Cleaning Up   Combine all aqueous and organic layers from the reaction, add sodium carbonate to make the aqueous layer neutral, and shake the mixture gently to extract pyridine into the organic layer. Place the organic layer in the organic solvents container. Remove any solid (copper salts) from the aqueous layer by filtration. The solid can be placed in the nonhazardous solid waste container and the aqueous layer diluted with water and flushed down the drain.

---

4. A magnetic stirrer does not materially shorten the reaction time.

Questions

1. Write a balanced equation for this reaction.

2. What volume of oxygen at standard temperature and pressure is consumed in this reaction?

Surfing the Web

http://ull.chemistry.uakron.edu/organic_lab/oxidative/index.html

This experiment, on a mini scale, is illustrated with 16 close-up color photos from the University of Akron. An ordinary balloon and slightly different apparatus are used in these photos; the various dramatic color changes are clearly seen.

For updated information visit:
www.mtholyoke.edu/courses/kwilliam/microscale.shtml
or
www.hmco.com/hmco/college/chemistry/Home.html

CHAPTER

25

Catalytic Hydrogenation

Prelab Exercise: Calculate the volume of hydrogen gas generated when 3 mL of 1 *M* sodium borohydride reacts with concentrated hydrochloric acid. Write a balanced equation for the reaction of sodium borohydride with platinum chloride. Calculate the volume of hydrogen that can be liberated by reacting 1 g of zinc with acid.

Catalytic reduction is a very important and widely used industrial process; usually no harmful wastes are produced in the process. Catalytic hydrogenation and dehydrogenation are carried out on an enormous scale, for example, in the catalytic cracking and reforming of crude oil to make gasoline.

Nitrobenzene can be reduced catalytically to aniline with water as the only byproduct:

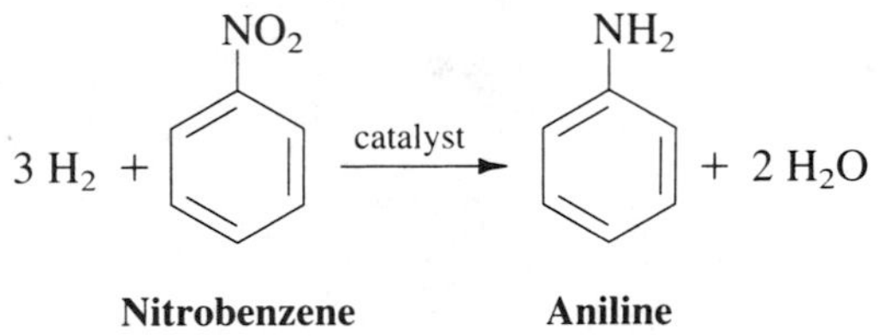

$$3\ H_2 + \underset{\textbf{Nitrobenzene}}{\underset{NO_2}{\bigcirc}} \xrightarrow{\text{catalyst}} \underset{\textbf{Aniline}}{\underset{NH_2}{\bigcirc}} + 2\ H_2O$$

Styrene is made by the catalytic dehydrogenation of ethylbenzene at very high temperature, but it also can be hydrogenated back to ethylbenzene very easily. Palladium, as a catalyst, lowers the energy barrier for the reaction in both directions.

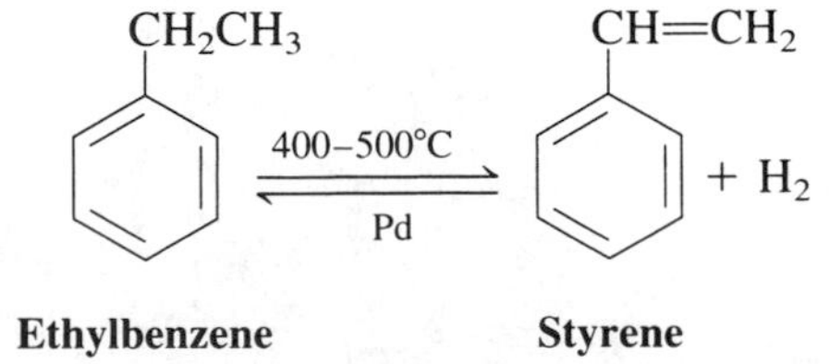

$$\underset{\textbf{Ethylbenzene}}{\underset{CH_2CH_3}{\bigcirc}} \underset{\text{Pd}}{\overset{\text{400–500°C}}{\rightleftharpoons}} \underset{\textbf{Styrene}}{\underset{CH=CH_2}{\bigcirc}} + H_2$$

The addition of hydrogen to alkenes is one of the most common reactions. The alkene is more reactive toward this process than is the aromatic ring or functional groups such as esters or ketones.

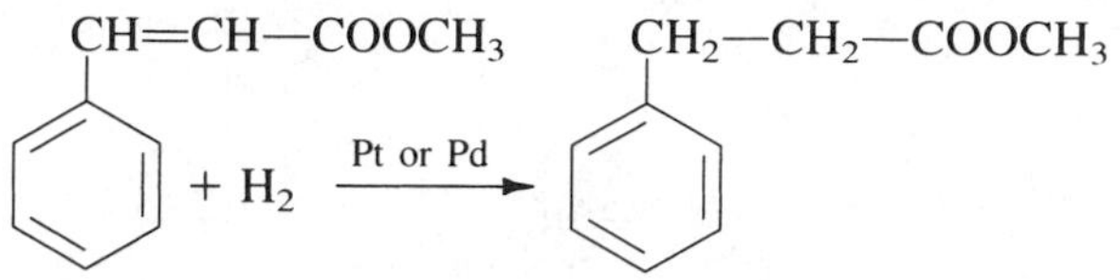

Hydrogenation is stereospecific, so alkynes are reduced to *cis*-alkenes. The metal is usually supported on a high-surface-area material such as charcoal. The alkene and the hydrogen probably are both adsorbed onto the surface of the catalyst before the transfer occurs. This heterogeneous reaction, a reaction that involves reactants in the liquid or gas phase and a catalyst in the solid phase, is difficult to study.

In the present experiments, catalyic hydrogenation is carried out in several different ways. The first experiment is a puzzle for you to solve. In the second experiment, hydrogen gas from an external supply is used to hydrogenate a long-chain unsaturated alcohol to the corresponding saturated alcohol. In the third experiment, the hydrogen is generated *in situ* using the Brown hydrogenation technique. The fourth experiment utilizes a process called *transfer hydrogenation* to produce a saturated fat from an unsaturated one (olive oil).

Experiments

1. A Small Research Project

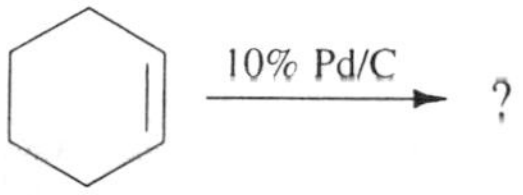

The purpose of this experiment is to determine whether a chemical reaction has occurred when cyclohexene is boiled for a short period of time with a small quantity of a catalyst, 10% palladium on carbon, and if so, to determine the structure of the product(s).

Microscale Procedure

In a reaction tube place about 50 mg of 10% palladium on carbon. To this add about 0.8 mL of cyclohexene and a boiling chip. Reflux this mixture on a warm sand bath (low setting of controller) for about 10 to 15 min. Be careful not to boil the alkene out of the reaction tube. If necessary, place a coil of damp pipe cleaner around the top of the tube. This will keep the top of the tube cool so that the cyclohexene will not escape. Further insurance against boiling away the reactant is afforded by adding the empty distilling column to the reaction tube, although with careful adjustment of the heat input (depth of tube in sand) these measures will not be necessary (Fig. 25.1).

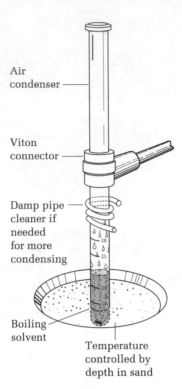

Air
condenser

Viton
connector

Damp pipe
cleaner if
needed
for more
condensing

Boiling
solvent

Temperature
controlled by
depth in sand

FIG. 25.1 Apparatus for
refluxing a reaction mixture
with maximum cooling.

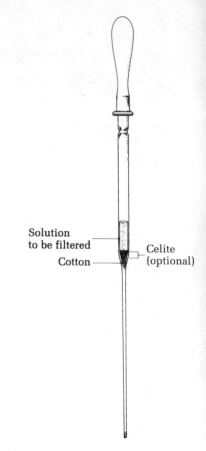

Solution
to be filtered

Cotton

Celite
(optional)

FIG. 25.2 Filtration of a
solution in a Pasteur pipette.
Usually used to remove
charcoal.

At the end of the reflux period, cool the tube to room temperature and filter
the solution to remove the catalyst. This is done by stuffing a small ball of cotton
into a Pasteur pipette and using another Pasteur pipette to transfer the reaction
mixture containing charcoal. Filter the solution into a clean, dry reaction tube. If
one filtration does not remove all the charcoal, recycle the filtrate through the filter (Fig. 25.2).

Macroscale Procedure

In a 15-mL tapered 14/20 standard taper flask, place about 100 mg of 10% palladium on carbon. To this add about 2 mL of cyclohexene and a boiling chip.
Reflux this mixture on a warm sand bath (low setting of controller) for at least
15 min (Fig. 25.3).

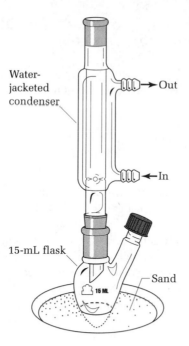

FIG. 25.3 Apparatus for refluxing a small volume of a liquid.

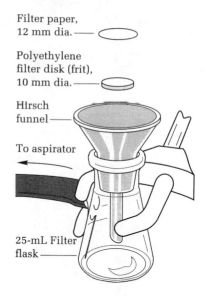

FIG. 25.4 Filtration of a reaction mixture in a Hirsch funnel to remove charcoal. Use fine porosity filter paper.

At the end of the reflux period, cool the flask to room temperature and filter the solution to remove the catalyst. This is done by stuffing a small ball of cotton into a Pasteur pipette and using another Pasteur pipette to transfer the reaction mixture containing palladium on charcoal. Filter the solution into a clean, dry reaction tube or culture tube. Alternatively the reaction mixture can be filtered in a Hirsch funnel (Fig. 25.4). Use fine porosity filter paper in the funnel.

Analysis of the Product

Consult Chapter 18, Part 3, for tests for alkanes and alkenes. Try one or more of these tests on both your starting material, cyclohexene, and your product. The decolorization of bromine is probably the best test to run. Count the number of drops of bromine solution that are decolorized by 0.10 mL of cyclohexene and by 0.10 mL of your filtered reaction mixture.

Inject about 4 μL of your product into the gas chromatograph. Read Chapter 11 for the theory of the gas chromatography and the procedure for analysis.

Run and interpret NMR, IR, and UV spectra of the starting material and product. The most useful of these is the NMR spectrum, followed by the UV spectrum. On the basis of these data, interpret the results of this experiment.

2. Catalytic Hydrogenation of Oleyl Alcohol Using Palladium on Charcoal

$$CH_3(CH_2)_7CH\!\!=\!\!CH(CH_2)_8OH + H_2 \xrightarrow{\text{10\% Pd/C}} CH_3(CH_2)_{17}OH$$

Oleyl alcohol
MW 268.49, den 0.849
bp 207°C/13 mm
n_D^{20} 1.4600

Octadecanol
MW 270.50
mp 60°C

In this experiment the 18-carbon unsaturated alcohol oleyl alcohol is reduced to the corresponding saturated alcohol using one of the most widely employed techniques for catalytic hydrogenation. The catalyst is 10% palladium on carbon, which currently sells for $4.00 per gram; since this reaction is being carried out on a microscale, we will use only a few cents worth.

The hydrogen could come from a compressed gas cylinder, but it is just as convenient to prepare a small quantity by the action of acid on zinc.

The apparatus for this experiment can be modified in a number of ways. If you merely want to hydrogenate a double bond and are not concerned about the quantitative aspects of the reaction, fit the apparatus with a rubber balloon that holds the hydrogen. As soon as the balloon has reached a constant size, hydrogen uptake is over and the product can be isolated.

In the modification of the apparatus employed in the present experiment, the hydrogen uptake can be followed as water rises in the hydrogen reservoir (a graduated cylinder). With this apparatus, the volume of hydrogen absorbed as a function of time can be measured.

In research-grade apparatus, the hydrogen is stored in a gas burette over mercury. The volume of hydrogen can be determined very accurately so that quantitative estimates of the numbers of double bonds in unknowns can be made or so that selective hydrogenation of the more reactive of several double bonds can be carried out. The reaction is stopped when the theoretical amount of hydrogen has been absorbed.

Experimental Considerations

The apparatus in Fig. 25.5 will be assembled. The graduated cylinder is filled with hydrogen, and the reaction flask containing the solvent, catalyst, and a magnetic stirring bar is flushed with hydrogen. The stirrer is started, and as soon as the water level in the reservoir reaches a constant level, the compound to be hydrogenated is injected through the septum of the reaction flask. It can either be a pure liquid or a solid dissolved in alcohol.

A millimole of hydrogen at standard temperature and pressure has a volume of 22.4 mL. This will hydrogenate a millimole of a compound that has one double bond. The apparatus of Fig. 25.5 holds about 9 mL of gas (0.4 mmol).

For highly accurate measurement of the hydrogen uptake, we would take into account the temperature, the atmospheric pressure, and the vapor pressure of the

FIG. 25.5 Catalytic hydrogenation apparatus.

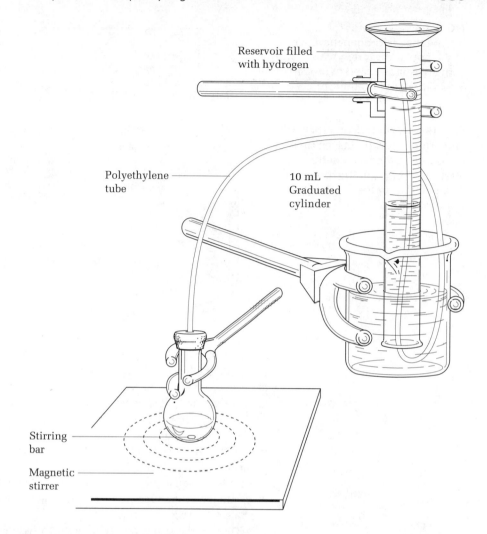

Reservoir filled with hydrogen

Polyethylene tube

10 mL Graduated cylinder

Stirring bar

Magnetic stirrer

methanol in the reaction flask. We would be careful to equalize the internal and external pressures so that precise volume measurements could be made (Fig. 25.6).

The hydrogen to fill the apparatus can come from a variety of sources. A tank of hydrogen is most convenient for an individual, but perhaps not for a large group of users. The hydrogen could be generated by reacting acid with borohydride as in the following experiment, but it is cheaper to use zinc and sulfuric acid.

The apparatus for hydrogen generation is seen in Fig. 25.7. The 5-mL long-necked flask contains 1 g of zinc. As hydrogen is needed, sulfuric acid is injected through the septum.

Generation and Storage of Hydrogen

See Fig. 18.2 for threading a polyethylene tube through a septum.

Assemble the apparatus shown in Fig. 25.5. In a 5-mL short-necked round-bottomed flask place 20 mg of 10% palladium on carbon, a small magnetic

FIG. 25.6 Apparatus for quantitative hydrogenation. Air in the apparatus is removed with an aspirator, and hydrogen is admitted and brought to atmospheric pressure by raising or lowering the leveling bulb. The stirrer is started, and the amount of hydrogen taken up is measured in the gas burette after again bringing the pressure inside to that of the atmosphere.

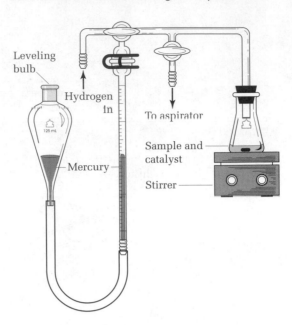

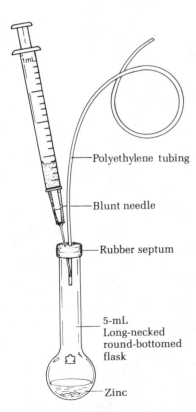

FIG. 25.7 Hydrogen generator.

stirring bar, and 1.5 mL of methanol. Fill a 250-mL beaker with water. Fill a 10- or 25-mL graduated cylinder with water, and clamp it in an inverted position over the beaker of water. In the 5-mL long-necked flask place 1 g of granulated or mossy zinc, and close it with a septum bearing a polyethylene tube (see Fig. 25.7). Inject a few drops of 6 M (33%) sulfuric acid to generate hydrogen, and displace the air from the 5-mL short-necked flask, which should then be stoppered. Thread the polyethylene tubing into the bottom of the graduated cylinder clamped over a beaker of water (see Fig. 25.5). Inject more acid onto the zinc, and fill the graduated cylinder reservoir with about 9 mL of hydrogen. Remove the septum and the tube from the gas-generating flask, and connect them to the short-necked flask as in Fig. 25.5.

Turn on the stirrer briefly in order to saturate the methanol with hydrogen. Note the level of the water in the cylinder. If it comes to the same level as the water on the outside when the cylinder is moved up and down, there is a leak in the system.

Hydrogenation

Once the pressure in the system has stabilized with the stirrer off, inject 0.4 mmol of a compound that has one olefinic double bond to be hydrogenated. In this particular experiment, this will be 107 mg of oleyl alcohol dissolved in 0.6 mL of methanol. If a larger reservoir of hydrogen has been used, for example, a 25-mL graduated cylinder, a proportionately larger quantity of alkene can be hydrogenated. Turn on the stirrer, and note the height of the hydrogen in the reservoir as a function of time. Be sure the end of the tube is always above the water level in the reservoir. When the water level no longer changes, the reaction is complete.

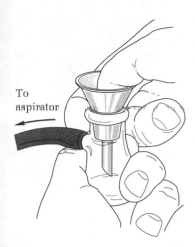

FIG. 25.8 Evaporation of solvent under reduced pressure. Heat and swirling motion is supplied by one hand; vacuum is controlled by thumb of other.

 **MACROSCALE**

Handle platinum chloride carefully to prevent any waste.

Filter the reaction mixture into a tared 25-mL filter flask using a Pasteur pipette; push a piece of cotton firmly into a Pasteur pipette followed by a 5-mm layer of Celite filter aid. With a second pipette, transfer the reaction mixture to the filter pipette and allow the solution to filter through the cotton to remove the charcoal. Force the solution through the cotton with a rubber bulb (see Fig. 25.2). Rinse the reaction flask, stirring bar, and transfer pipette with a few drops of methanol, and filter this solution also. If the filtrate is gray, refilter the solution in the same pipette without changing the filter.

Remove the methanol from the filter flask under aspirator vacuum (Fig. 25.8). Heat the filter flask on the steam bath or a sand bath under vacuum to remove the last traces of methanol. The residue should be pure 1-octadecanol (mp 60°C), although the melting point will depend on the quality of the starting alcohol. Test a small portion of the product for unsaturation using bromine (see Chapter 18, Part 3).

Cleaning Up   Decant the sulfuric acid from the zinc, dilute the solution, neutralize with sodium carbonate, and remove zinc hydroxide by filtration. The filtrate can be flushed down the drain. Zinc and zinc hydroxide are placed in the nonhazardous solid waste container. Wash out the syringe and needle carefully.

3. Brown Hydrogenation[1]

$$\text{cis-Norbornene-5,6-}endo\text{-dicarboxylic acid} \xrightarrow[\text{PtCl}_4]{\text{NaBH}_4 - \text{HCl}} \text{cis-Norbornane-5,6-}endo\text{-dicarboxylic acid}$$

cis-Norbornene-5,6-endo-dicarboxylic acid
mp 180–190°C, dec
MW 182.17

cis-Norbornane-5,6-endo-dicarboxylic acid
mp 170–175°C, dec
MW 184.19

See Fig. 25.6 for a quantitative hydrogenation apparatus. However, in this experiment we will measure the uptake of hydrogen in a semiquantitative manner. The reaction vessel is a 125-mL filter flask with a white rubber pipette bulb wired onto the side arm (Fig. 25.8). Introduce 10 mL of water, 1 mL of platinum(IV) chloride solution,[2] and 0.5 g of decolorizing charcoal, and swirl during addition of 3 mL of stabilized 1 M sodium borohydride solution.[3] While

1. H. C. Brown and C. A. Brown, *J. Am. Chem. Soc.*, **84**, 1495 (1962).

2. A solution of 200 mg of PtCl$_4$ in 4 mL of water.

3. Dissolve 1.6 g of sodium borohydride and 0.30 g of sodium hyroxide (stabilizer) in 40 mL of water. When not in use, the solution should be stored in a refrigerator. If left for some time at room temperature in a tightly stoppered container, gas pressure may develop sufficient to break the vessel.

allowing 5 min for formation of the catalyst, dissolve 1 g of *cis*-norbornene-5,6-*endo*-dicarboxylic acid in 10 mL of hot water. Pour 4 mL of concentrated hydrochloric acid into the reaction flask, followed by the hot solution of the unsaturated acid. Cap the flask with a large serum stopper, and wire it on. Draw 1.5 mL of the stabilized sodium borohydride solution into the barrel of a plastic syringe, thrust the needle through the center of the stopper, and add the solution dropwise with swirling. The initial uptake of hydrogen is so rapid that the balloon may not inflate until you start injecting a second 1.5 mL of borohydride solution through the stopper. When the addition is complete and the reaction appears to be reaching an end point (about 5 min), heat the flask on the steam bath with swirling and try to estimate the time at which the balloon is deflated to a constant size (about 5 min). When balloon size is constant, heat and swirl for 5 min more and then release the pressure by injecting the needle of an open syringe through the stopper.

Reaction time: about 15 min

Filter the hot solution by suction on a Hirsch funnel, and place the catalyst in a jar marked "Catalyst Recovery." Cool the filtrate, and extract it with three 15-mL portions of *t*-butyl methyl ether. The combined extracts are to be washed with saturated sodium chloride solution and dried over anhydrous calcium chloride pellets. Evaporation of the ether gives about 0.8–0.9 g of white solid.

Common ion effect

The only solvent of promise for crystallization of the saturated *cis*-diacid is water, and the diacid is very soluble in water and crystallizes extremely slowly and with poor recovery. However, the situation is materially improved by addition of a little hydrochloric acid to decrease the solubility of the diacid.

Scrape out the bulk of the solid product, and transfer it to a 25-mL Erlenmeyer flask. Add 1–2 mL of water to the 125-mL flask, heat to boiling to dissolve residual solid, and pour the solution into the 25-mL flask. Bring the material into solution at the boiling point with a total of not more than 3 mL of water (as a guide, measure 3 mL of water into a second 25-mL Erlenmeyer). With a Pasteur pipette add 3 drops of concentrated hydrochloric acid, and let the solution stand for crystallization. Clusters of heavy prismatic needles soon separate; the recovery is about 90%. The product should give a negative test for unsaturation with acidified permanganate solution.

Observe what happens when a sample of the product is heated in a melting point capillary to about 170°C. Account for the result. You may be able to confirm your inference by letting the oil bath cool until the sample solidifies and then noting the mp temperature and behavior on remelting.

Cleaning Up   The catalyst removed by filtration may be pyrophoric (spontaneously flammable in air). Immediately remove it from the Hirsch funnel, wet it with water, and place it either in the hazardous waste container or in the catalyst recovery container. It should be kept wet with water at all times. The combined aqueous filtrates, after neutralization, are diluted with water and flushed down the drain. After the *t*-butyl methyl ether is allowed to evaporate from the calcium chloride pellets in the hood, it can be placed in the nonhazardous solid waste container.

4. Transfer Hydrogenation of Olive Oil[4]

Olive oil is an ester made from an alcohol and a long-chain carboxylic, or fatty, acid. The alcohol is the trihydric alcohol, glycerol, and the carboxylic acid is mostly the 18-carbon monounsaturated compound, oleic acid. We can therefore regard olive oil as being primarily glycerol trioleate. This is a poly- (more than one) unsaturated fat. Three places in this molecule are carbon–carbon double bonds, which are not saturated with hydrogen.

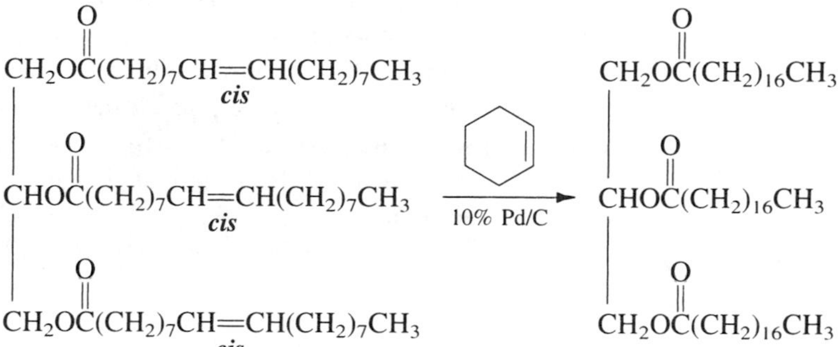

Glycerol trioleate
mp −5.5°C
MW 885.47

Glycerol tristearate
mp 69.9°C
MW 891.52

As you know, this fat is a liquid at room temperature and, when consumed in food, it does not contribute as much to the formation of cholesterol as saturated fat does.

To manufacture oleomargarine, a vegetable oil such as olive oil or, more commonly, corn or soybean oil is partially hydrogenated. This gives a mixture of liquid and solid fats that is the consistency of butter. When oils such as these are completely hydrogenated, the resulting fat is harder. Many saturated fats are found in animals. The hard fat (tallow) on a raw steak is made up of saturated fats.

In this experiment, we use 100% extra-virgin olive oil from Italy. The label says that of each 14 g of oil, 11 are monounsaturated (mostly from oleic acid), 2 are polyunsaturated (mostly linoleic acid, the 18-carbon acid with two double bonds), and 1 is saturated (this is a mixture of predominantly 18-carbon saturated stearic acid and a lesser amount of 16-carbon palmitic acid).

From this rough analysis, it can be concluded that complete hydrogenation of all of the double bonds in olive oil will give a product containing about 90 to 95% stearic acid (C-18) and 5 to 10% palmitic acid (C-18) esterified to glycerol. Although the product will not be pure glycerol tristearate, it should be a solid with a melting point above 50°C.

---

4. Discussions with Gottfried Brieger regarding transfer hydrogenation are gratefully acknowledged.

The technique of transfer hydrogenation is employed. Hydrogen is transferred from cyclohexene to the double bonds of olive oil through the intervention of the catalyst. What other product is formed? Although it cannot be used in all cases, this technique of hydrogenation is quite convenient, since it consists of simply refluxing the substance to be hydrogenated with cyclohexene and the catalyst for a few minutes.

Microscale Procedure

In a reaction tube, place 400 mg of olive oil, 1 mL of cyclohexene, 50 mg of 10% palladium-on-carbon catalyst, and a boiling chip. Reflux the mixture for 15 min. Be careful not to boil the alkene out of the reaction tube. If necessary, place a coil of damp pipe cleaner around the top of the tube. This will keep the top of the tube cool so the cyclohexene will not escape. The empty distilling column could also be mounted on top of the reaction tube to provide even more condensing area (see Fig. 25.1), but with careful adjustment of the heat input (depth of the tube in the sand), this will not be necessary.

Macroscale Procedure

In a standard taper 14/20 flask, place 0.80 g of olive oil, 2 mL of cyclohexene, 100 mg of 10% palladium-on-carbon catalyst, and a boiling chip (see Fig. 25.3). Reflux the mixture for at least 15 min.

**MICROSCALE
AND MACROSCALE**

Isolation of Hydrogenated Fat

Prepare a Pasteur pipette as a micro filter by forcing a small piece of cotton firmly down to the constriction (see Fig. 25.2). Add the reaction mixture to the filter pipette, and then force the solution through the cotton into a tared, 25-mL filter flask. Rinse the reaction tube, filter with a few drops of hexane, and then evaporate the solution to dryness (see Fig. 25.9). To remove the last traces of volatile liquid from the product, heat the filter flask in a hot sand bath under vacuum. Upon cooling on ice, the product should solidify to a hard, white fat. Analyze a portion of the product, but save most of it for the synthesis of soap (Chapter 40).

Vacuum evaporation while heating is necessary to ensure the fat will solidify when cooled on ice.

Analysis of Hydrogenated Fat

Determine the weight of the product and its physical properties. You may wish to carry out one or more of the tests for alkenes and alkanes; for example, reaction with bromine in carbon tetrachloride or with permanganate (see Chapter 18) on both the product and on olive oil. Count the number of drops of bromine in carbon tetrachloride (or dichloromethane) that can be decolorized by equivalent quantities of the product and of olive oil. Write a balanced equation for the for-

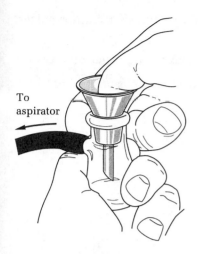

To
aspirator

FIG. 25.9 Apparatus for
removal of a solvent under
vacuum.

mation of the product of the hydrogenation reaction, assuming it is mostly glycerol trioleate. What is the driving force for this reaction?

The infrared spectrum of olive oil is presented in Fig. 25.10. The peak at
723 cm^{-1} has been assigned to the *cis-* double bond. Is it missing in the product?
Analysis of the NMR spectra of the starting material and product will also disclose whether the double bonds have been hydrogenated.

Computational Chemistry

Calculate, using a semiempirical molecular orbital program such as AM1, the
heats of formation of cyclohexene, cyclohexane, and benzene. Use this information to rationalize the results of this experiment. Molecular mechanics steric
energies for these molecules do not allow valid comparisons to be made among
the three.

Questions

1. How would you convert the reduced bicyclic diacid product to the corresponding bicyclic anhydride?

2. Why does the addition of hydrochloric acid in Experiment 2 cause the solubility of the bicyclic diacid in water to decrease?

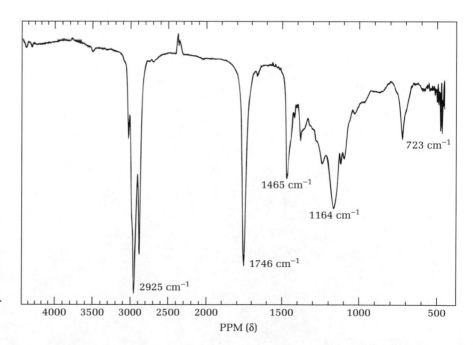

FIG. 25.10 Infrared spectrum
of olive oil (Bertoli, extra virgin).
Primarily glycerol trioleate. Run
as thin film.

3. Why is decolorizing charcoal added to the hydrogenation reaction mixture in the third experiment?

4. What is the source of hydrogen in the transfer hydrogenation of olive oil?

5. What is the driving force for the transfer hydrogenation of olive oil?

6. Assign the peak at 1746 cm^{-1} in the IR spectrum of olive oil.

Sodium Borohydride Reduction of 2-Methylcyclohexanone: A Problem in Conformational Analysis

Prelab Exercise: Using carefully prepared drawings, try to predict whether *cis*- or *trans*-2-methylcyclohexanol will predominate in this reduction. Then use molecular models to check your conclusion. Finally, if you have access to a workstation and the appropriate software, see if semiempirical molecular orbital calculations will change your conclusions.

The object of this experiment is to determine the structures and relative percentage yields of the products formed when 2-methylcyclohexanone is reduced with sodium borohydride. We also compare these results to predictions made using molecular mechanics and semiempirical molecular orbital calculations.

In the reduction of 2-methylcyclohexanone, both the *cis* and *trans* isomers of 2-methylcyclohexanol can be formed:

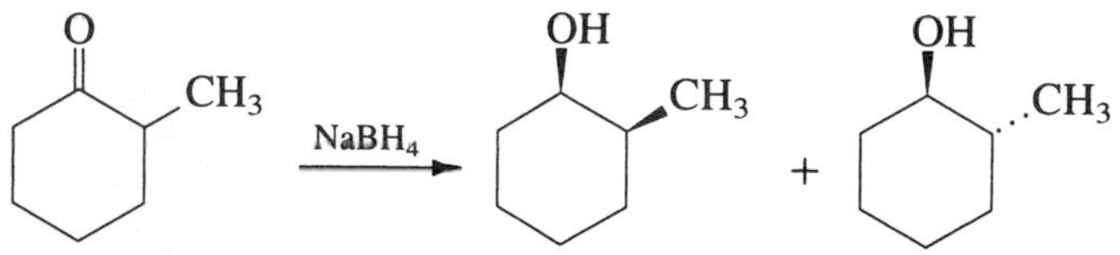

cis   trans

2-Methylcyclohexanone   **2-Methylcyclohexanol**

Examination of models reveals that the two cyclohexanols can, in principle, exist in four chair conformations:

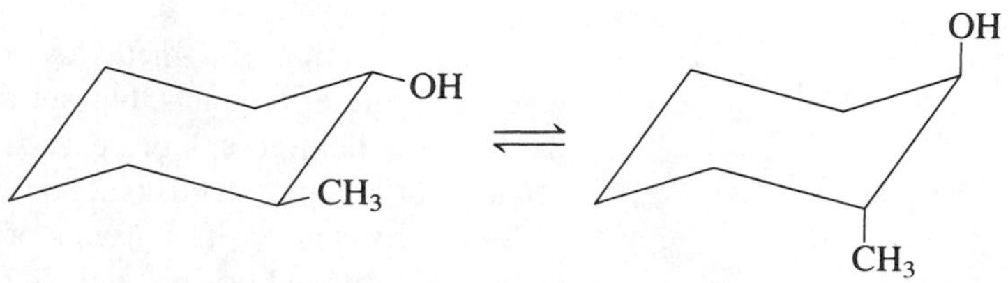

trans, **diequatorial**   *trans*, **diaxial**

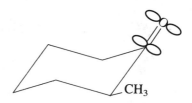

cis, **equatorial methyl,**
axial hydroxyl

cis, **axial methyl,**
equatorial hydroxyl

A molecular mechanics program can be used to calculate the relative energies of these four isomers and allow you to predict the most stable *trans* and *cis* conformations. Even without calculation, you should be able to predict which of the two *trans* conformations is the more stable.

By carrying out semiempirical molecular orbital calculations—for example, calculations using modified neglect of diatomic overlap (MNDO)—on the starting material (see Chapter 15), you may be able to decide, based on the shape and location of the lowest unoccupied molecular orbital (LUMO), whether the borohydride anion will attack the carbonyl group from the top, to give predominately the *trans* isomer, or from the bottom, to give mostly the *cis* isomer. You may also be able to make this prediction by studying a molecular model or even a drawing of 2-methylcyclohexanone:

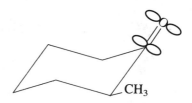

You can determine the structures and relative percentage yields of the products in this reaction using NMR spectroscopy. The 250-MHz ^{1}H NMR spectrum of a 50/50 mixture of methylcyclohexanols is seen in Fig. 26.1. The two lowest field groups of peaks are from the hydrogen atom on the hydroxyl-bearing carbon atom.

The reduction of 2-methylcyclohexanone with sodium borohydride will give a mixture of products, but not necessarily a 50/50 mixture. Integration of the two low-field groups of peaks will allow determination of the actual percentages of products in this reaction. But first, the peaks must be assigned to the *cis* and *trans* isomers. The groups of peaks have been expanded and numbered on the spectrum in Fig. 26.1 and the frequencies of the 11 peaks are listed in the caption.

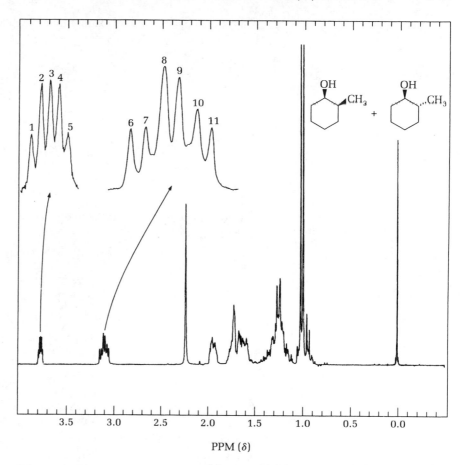

FIG. 26.1   The 250-MHz NMR spectrum of a mixture of *cis*- and *trans*-2-methylcyclohexanol. The frequencies of peaks 1 to 5 are 950.3, 947.5, 944.8, 942.2, and 939.5 Hz, respectively. The frequencies of peaks 6 to 11 are 791.2, 786.9, 781.6, 777.1, 771.6, and 767.4 Hz, respectively.

The NMR coupling constant of the proton on the hydroxyl-bearing carbon is a function of the dihedral angle between that proton and an adjacent vicinal proton. For example, the dihedral angle between H_A and H_B is 60°:

This is seen most easily by examining molecular models. A Newman projection may also make this clear:

The coupling constant for these two equatorial hydrogens, H_A and H_B, is usually in the range of 2 to 5 Hz.

The coupling of H_A and H_C (equatorial-axial) is usually in the range of 4 to 7 Hz, and the H_D to H_E coupling (diaxial) is in the range of 9 to 12 Hz. From this information, it should be possible to assign the groups of peaks at 3.1 and 3.8 ppm to either *cis-* or *trans-*2-methylcyclohexanol. From the areas of the two sets of peaks, the relative percentages of the isomers can be determined.

Experiment _____

MICROSCALE AND MACROSCALE

Borohydride Reduction of 2-Methylcyclohexanone

This reaction can be run on a scale four times larger in a 25-mL Erlenmeyer flask.

To a reaction tube, add 1.25 mL of methanol and 300 mg of 2-methylcyclohexanone. Cool this solution in an ice bath contained in a small beaker. While the reaction tube is in the ice bath, carefully add 50 mg of sodium borohydride to the solution. After the vigorous reaction has ceased, remove the tube from the ice bath, and allow it to stand at room temperature for 10 min, at which time the reaction should appear to be finished. To decompose the borate ester, add 1.25 mL of 3 M sodium hydroxide solution. To the resulting cloudy solution, add 1 mL of water. The product will separate as a small, clear upper layer. Remove as much of this as possible, place it in a reaction tube, and then extract the remainder of the product from the reaction mixture with two 0.5-mL portions of dichloromethane.

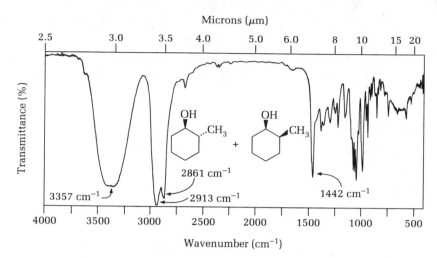

FIG. 26.2 Infrared spectrum of a mixture of *cis-* and *trans-*2-methylcyclohexanol (thin film).

Add these dichloromethane extracts to the small product layer, and dry the combined extracts over anhydrous sodium sulfate (not calcium chloride). After a few minutes, transfer the solution to a dry reaction tube containing a boiling chip. In the hood, boil off the dichloromethane (and any accompanying methanol), and use the residue to run an NMR spectrum in the usual way.

Integrate the two low-field sets of peaks at 3.1 and 3.8 ppm, analyze the coupling constant patterns to assign the two sets of peaks, and report the percentage distribution of *cis-* and *trans-*2-methylcyclohexanol formed in this reaction. The relative amounts of the products, of course, can also be determined by gas chromatography. How do these results compare with your predictions and calculations? Which *cis* and which *trans* conformation is the more stable?

Run an infrared spectrum of the product as a thin film to determine whether the reaction of the starting material has been complete (see Fig. 26.2).

Cleaning Up   The reaction mixture is neutralized with acetic acid (to react with sodium borohydride) and flushed down the drain with water.

Computational Chemistry

Using a molecular mechanics program, calculate the steric energies of *cis-* and *trans-*2-methylcyclohexanol. Each of these isomers has two principal conformations. Does reduction of 2-methylcyclohexanone give the more stable isomer? Explain.

Questions _____

1. Draw the NMR spectrum expected when H_A couples with H_B. (See the formulas above.)

2. Draw the NMR spectrum expected when H_A couples with both H_B and H_C. Remember that the coupling constants are not equal.

3. What is the approximate frequency of the most important peak in the starting material that should be absent in the IR spectrum of the product?

CHAPTER

27

Epoxidation of Cholesterol

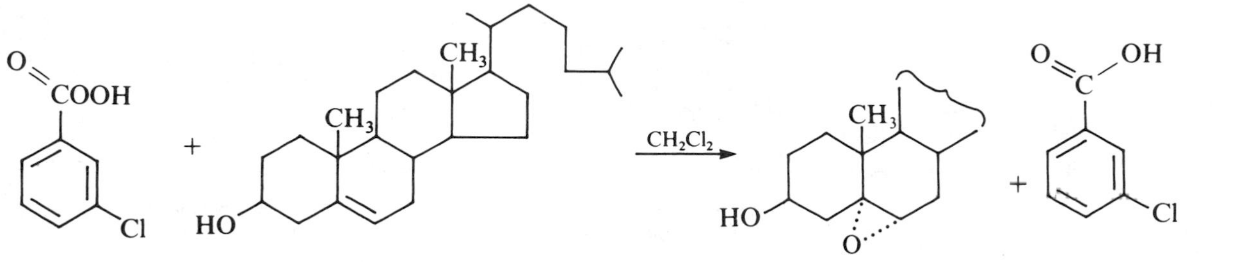

3-Chloroperoxybenzoic acid
MW 172.57
mp 92–94°C (dec)

Cholesterol
MW 386.66
mp 149°C

5α,6α-Epoxycholestan-3β-ol
MW 402.66

3-Chlorobenzoic acid
MW 156.57
mp 157°C

This experiment carries out an epoxidation reaction on cholesterol, which is a representative of a very important group of molecules, the steroids. The rigid cholesterol molecule gives products of well-defined stereochemistry. The epoxidation reaction is stereospecific, and the product can be used to carry out further stereospecific reactions.

Cholesterol itself is the principal constituent of gallstones and can be readily isolated from them. The average person contains about 200 g of cholesterol, primarily in brain and nerve tissue. The closing of arteries by cholesterol leads to the disease arteriosclerosis (hardening of the arteries).

Certain naturally occurring and synthetic steroids have powerful physiological effects. Progesterone and estrone are the female sex hormones, and testosterone is the male sex hormone; they are responsible for the development of secondary sex characteristics. The closely related synthetic steroid, norethisterone, is an oral contraceptive, and addition of four hydrogen atoms (reduction of the ethynyl group to the ethyl group) and a methyl group gives an anabolic steroid, ethyltestosterone. This muscle-building steroid is now outlawed for use by Olympic athletes. RU-486, mifepristone, is the recently approved abortifacient. Fluorocortisone is used to treat inflammations such as arthritis, and ergosterol on irradiation with ultraviolet light is converted to vitamin D_2.

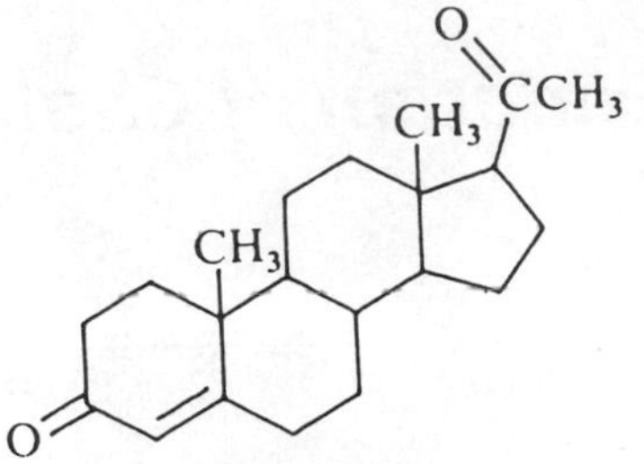

Progesterone
Female sex hormone

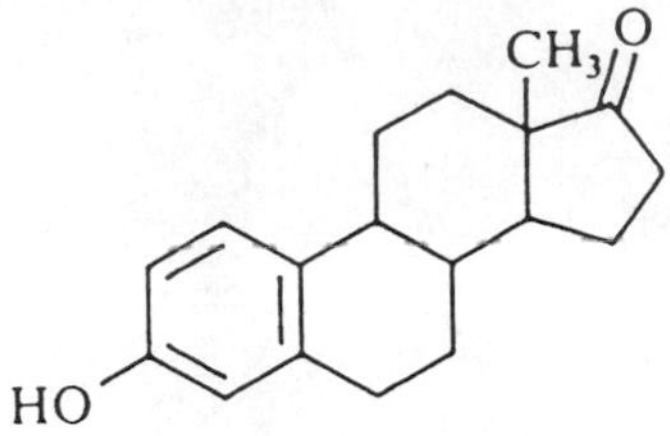

Estrone
Female sex hormone

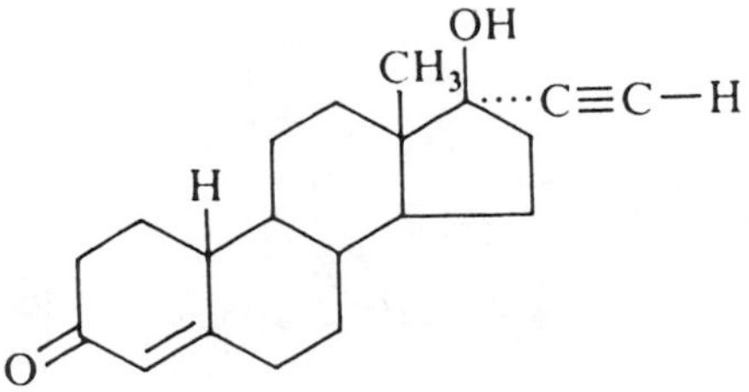

Norethisterone
Oral contraceptive

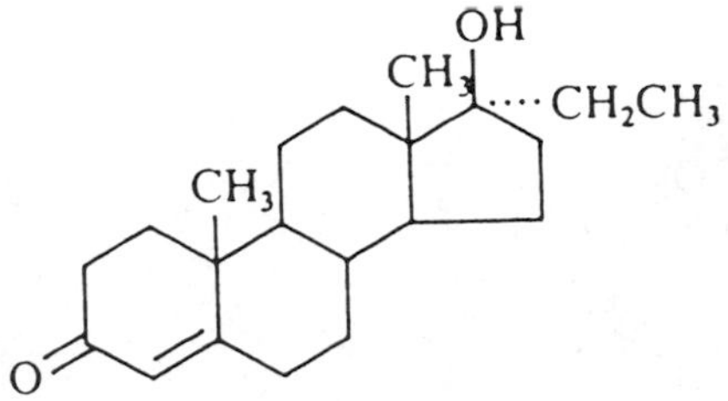

Ethyltestosterone
Anabolic steroid

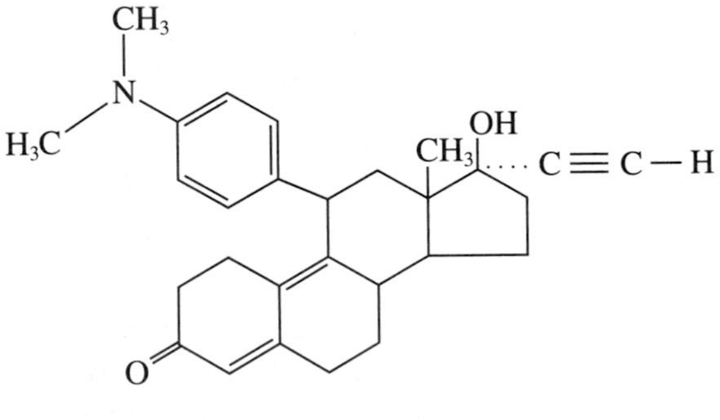

RU-486
Mifepristone
Abortifacient

Much of our present knowledge about the stereochemistry of reactions was developed from steroid chemistry. In this experiment the double bond of cholesterol is stereospecifically converted to the $5\alpha,6\alpha$ epoxide. The α designation indicates the epoxide is on the back side of the molecule as usually printed. A substituent on the top side (above the plane of the paper) is designated β. Study of molecular models reveals that the angular methyl group prevents topside attack on the double bond by the perbenzoic acid; hence the epoxide forms exclusively on the back or α side of the molecule.

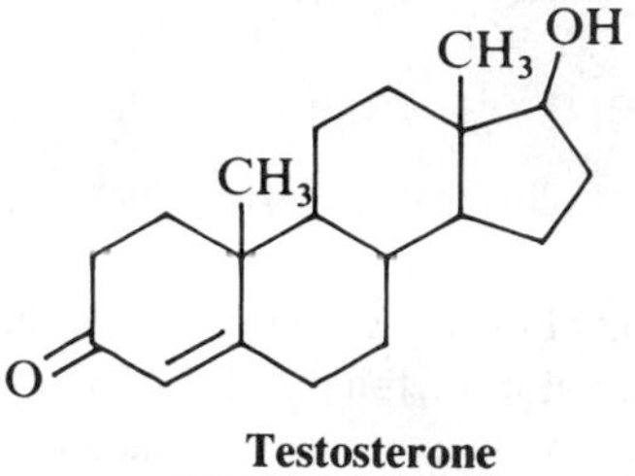
CH3
OH
O
Testosterone
Male sex hormone

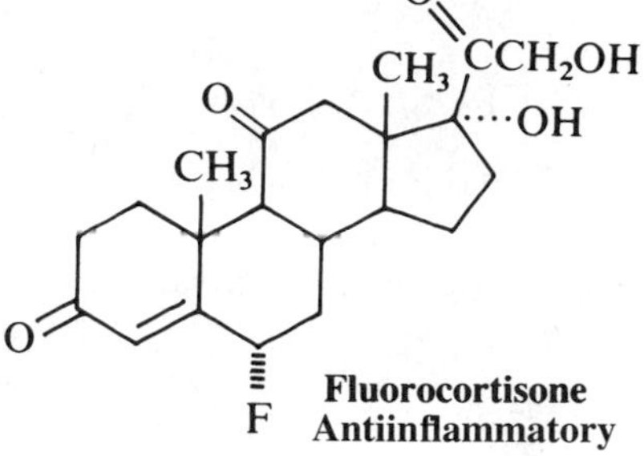
CH3
CCH2OH
OH
O
F
Fluorocortisone
Antiinflammatory

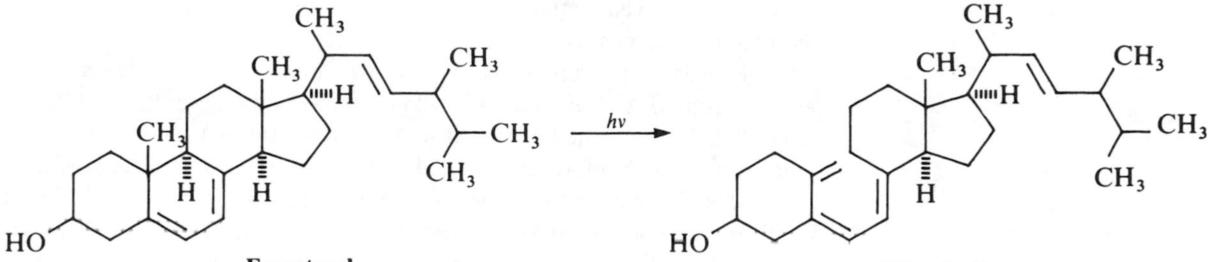
CH3
H
HO
Ergosterol
hν
Vitamin D2

Epoxides are most commonly formed by reaction of a peroxycarboxylic acid with an olefin at room temperature. It is a one-step cycloaddition reaction:

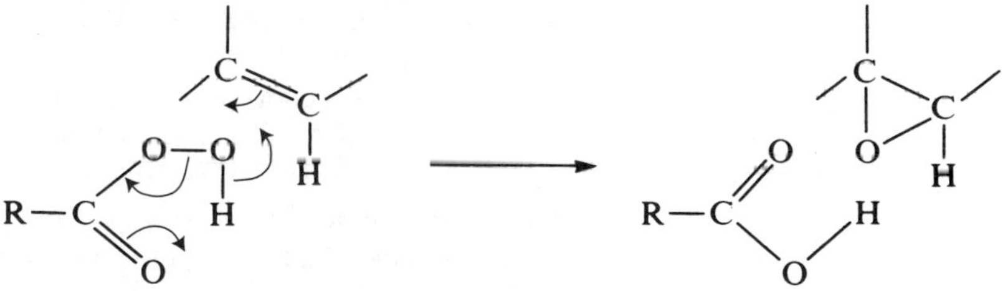
R
C
O
H

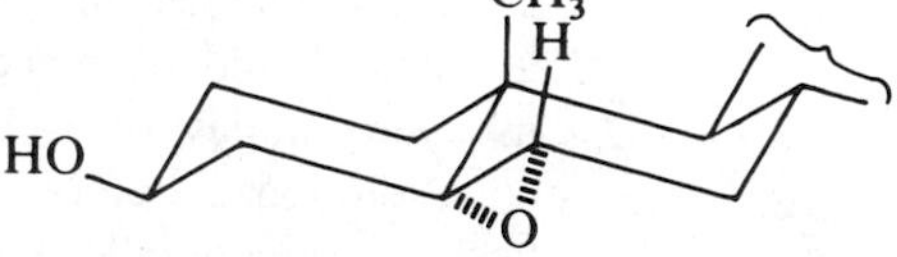
CH3
H
HO
O

Some peroxyacids are explosive; the reagent used in the present experiment is a particularly stable and convenient peroxycarboxylic acid.

The reaction is carried out in an inert solvent, dichloromethane, and the product is isolated by chromatography. No great care is required in the chromatography to collect fractions because the 3-chlorobenzoic acid, being polar, is adsorbed strongly onto the alumina while the relatively nonpolar product is eluted easily by *t*-butyl methyl ether. After removal of *t*-butyl methyl ether the product is easily recrystallized from a mixture of acetone and water.

*Experiments*_____

1. Cholesterol Epoxide

To 194 mg of cholesterol dissolved by gentle warming in 0.8 mL of dichloromethane in a 10 × 100 mm reaction tube is added a solution obtained by gently warming 117 mg of 80% 3-chloroperoxybenzoic acid (or 187 mg of 50% material) in 0.8 mL of dichloromethane. The two solutions must be cool before mixing because the reaction is exothermic. Place the stoppered reaction tube in a beaker of water at 40°C for 30 min to complete the reaction. The progress of the reaction can be followed by thin-layer chromatography on silica gel plates using *tert*-butyl methyl ether as the eluent.

    The reaction mixture is pipetted onto a chromatography column (Fig. 27.1) prepared from 3 g of alumina following the procedure described in Chapter 11, except that *tert*-butyl methyl ether is used to fill the column and to prepare the alumina slurry. The 3-chlorobenzoic acid will be adsorbed strongly by the alumina. The product is eluted with 30 mL of *tert*-butyl methyl ether collected in a tared 50-mL round-bottomed flask. The ether can flow through the column by gravity or can be forced out using pressure from a rubber bulb (flash chromatography).

    Most of the *tert*-butyl methyl ether is removed on the rotary evaporator, and the last traces are removed using the apparatus depicted in Fig. 10.7 for drying a solid under reduced pressure or that depicted in Fig. 10.5. The residue should weigh more than 150 mg. If it does not, pass more ether through the column, and collect the product as before. Dissolve the product in 1.5 mL of warm acetone, and, using a Pasteur pipette, transfer it to a reaction tube. Add 0.2 mL of water to the solution, warm the mixture to bring the solid into solution, and then let the tube and contents cool slowly to room temperature. Cool the mixture in ice, and collect the product on a Hirsch funnel. Press the solid down on the filter to squeeze solvent from the crystals, and then wash the product with 0.25 mL of ice-cold 90% acetone. Spread the product out on a watch glass to dry. Determine the weight and melting point of the product, and calculate the percentage yield.

Cleaning Up   Place dichloromethane solutions in the halogenated organic solvents container and organic solvents in the organic solvents container. The alumina should be placed in the hazardous waste container. If it should be necessary to destroy 3-chloroperoxybenzoic acid, add it to an excess of an ice-cold solution of saturated sodium bisulfite in the hood. A peracid will give a positive starch/iodide test (blue-purple color).

2. Cholesterol Epoxide

To 1 g of cholesterol dissolved by gentle warming in 4 mL of dichloromethane in a 25-mL Erlenmeyer flask is added a solution obtained by gently warming 0.6 g of 3-chloroperoxybenzoic acid in 4 mL of dichloromethane. The two solutions must be cool before mixing because the reaction is exothermic. Clamp the flask in a beaker of water at 40°C for 30 min to complete the reaction. The

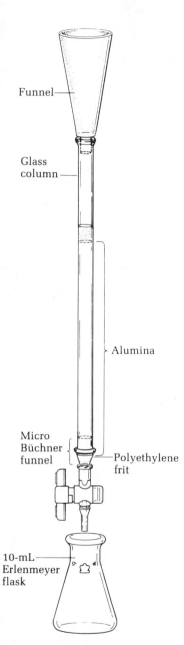

Funnel

Glass
column

Alumina

Micro
Büchner
funnel

Polyethylene
frit

10-mL
Erlenmeyer
flask

FIG. 27.1   Chromatography
column.

progress of the reaction can be followed by thin-layer chromatography on silica gel plates using *t*-butyl methyl ether as the eluent.

The reaction mixture is pipetted onto a chromatography column prepared from 15 g of alumina following the procedure described in Chapter 10, except that *t*-butyl methyl ether is used to fill the column and to prepare the alumina slurry. The 3-chlorobenzoic acid will be adsorbed strongly by the alumina. The product is eluted with 150 mL of *t*-butyl methyl ether collected in a tared 250-mL round-bottomed flask.

Most of the *t*-butyl methyl ether is removed on the rotary evaporator and the last traces are removed using the apparatus depicted in Fig. 10.7 for drying a solid under reduced pressure. The residue should weigh more than 0.75 g. If it does not, pass more *t*-butyl methyl ether through the column, and collect the product as before. Dissolve the product in 7.5 mL of warm acetone, and, using a Pasteur pipette, transfer it to a 25-mL Erlenmeyer flask. Add 1 mL of water to the solution, warm the mixture to bring the solid into solution, and then let the flask and contents cool slowly to room temperature. Cool the mixture in ice, and collect the product on a Hirsch funnel. Press the solid down on the filter to squeeze solvent from the crystals, and then wash the product with 1 mL of ice-cold 90% acetone. Spread the product out on a watch glass to dry. Determine the weight and mp of the product, and calculate the percent yield.

Cleaning Up   Place dichloromethane solutions in the halogenated organic solvents container and organic solvents in the organic solvents container. The alumina should be placed in the alumina hazardous waste container. If it should be necessary to destroy 3-chloroperoxybenzoic acid, add it to an excess of an ice-cold solution of saturated sodium bisulfite in the hood. A peracid will give a positive starch/iodide test (blue-purple color).

Steroid Biosynthesis Through an Epoxide Intermediate

The biosynthesis of cholesterol and the other naturally occurring steroids shown proceeds through an epoxide intermediate. The C_{30} triterpene hydrocarbon squalene (*L. squalus,* whale) is epoxidized to give squalene oxide. This epoxide is protonated and then cyclizes to form the characteristic four-ring steroid skeleton and, after several steps, cholesterol, in a series of very carefully studied enzyme-catalyzed reactions.

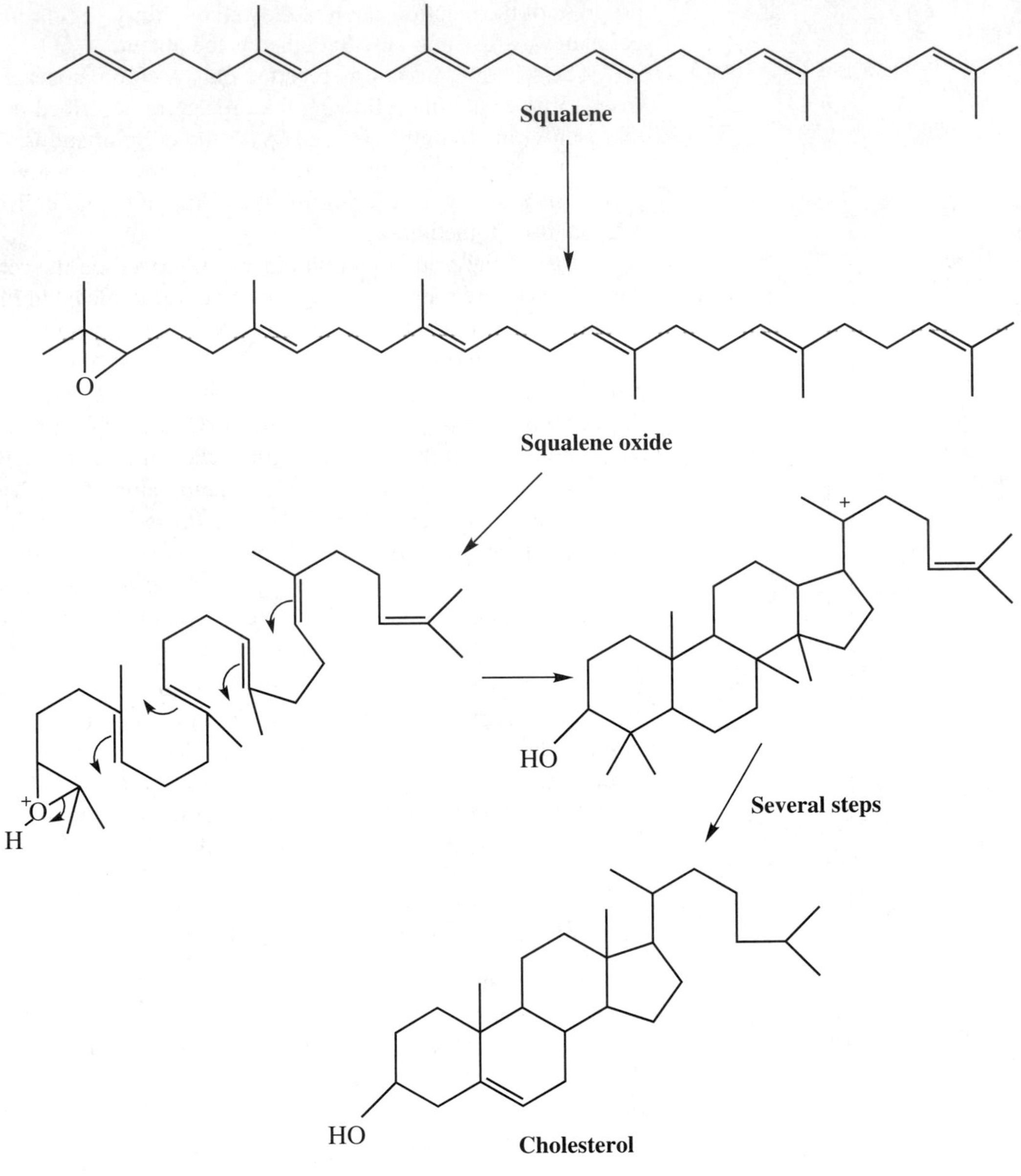

Squalene

Squalene oxide

HO

Several steps

HO   **Cholesterol**

Questions

1. What are the numbers of moles of the two reactants used in this experiment? Assume the 3-chloroperoxybenzene acid is 80% pure.

2. What simple test could you perform to show that 3-chlorobenzoic acid is not eluted from the chromatography column?

Nitration of Methyl Benzoate

Prelab Exercise: Draw the complete mechanism for the nitration of chloroben-zene. Chlorine is an *ortho-para* director and deactivator of the benzene ring.

The nitration of methyl benzoate is a typical electrophilic aromatic substitution reaction. The electrophile is the nitronium ion generated by the interaction of concentrated nitric and sulfuric acids:

$$HNO_3 + 2H_2SO_4 \rightleftharpoons NO_2^+ + 2HSO_4^- + H_3O^+$$

Nitronium ion

The solvent sulfuric acid protonates the methyl benzoate:

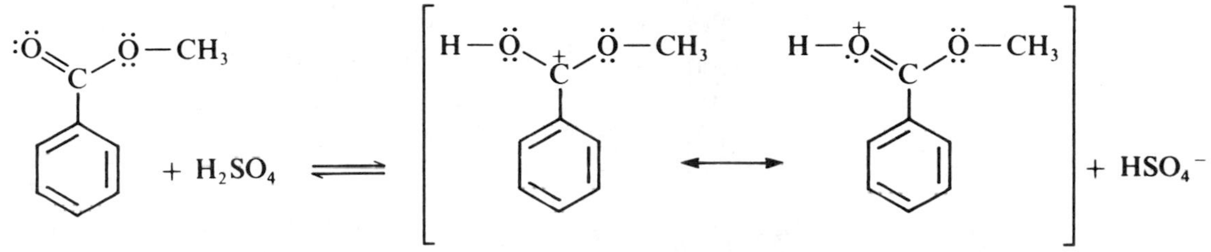

The nitronium ion then reacts with this protonated intermediate at the *meta* position, where the electron density is highest, i.e., where there is *not* a positively charged resonance form:

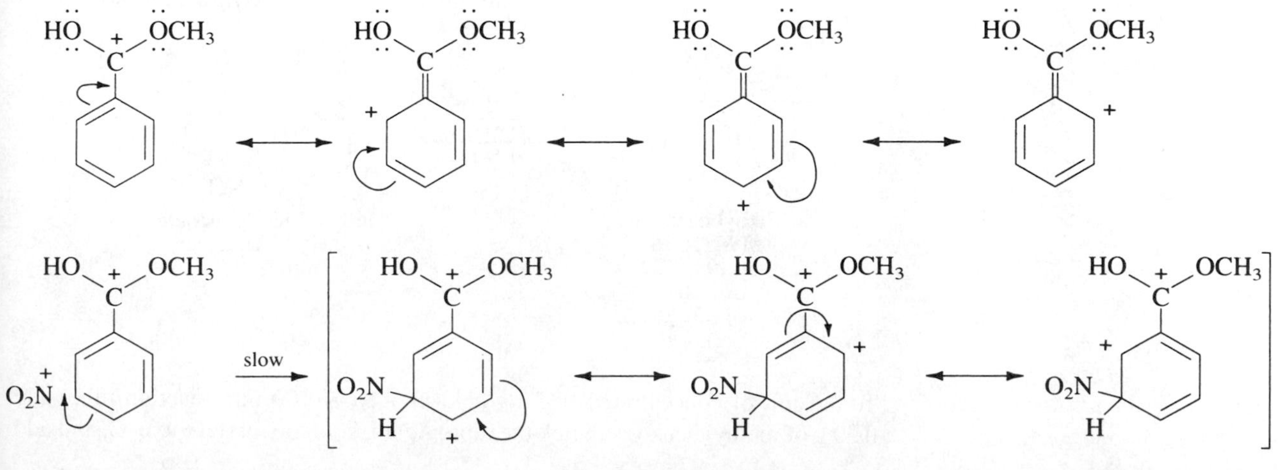

The arenium ion intermediate then transfers a proton to the basic bisulfate ion to give methyl 3-nitrobenzoate:

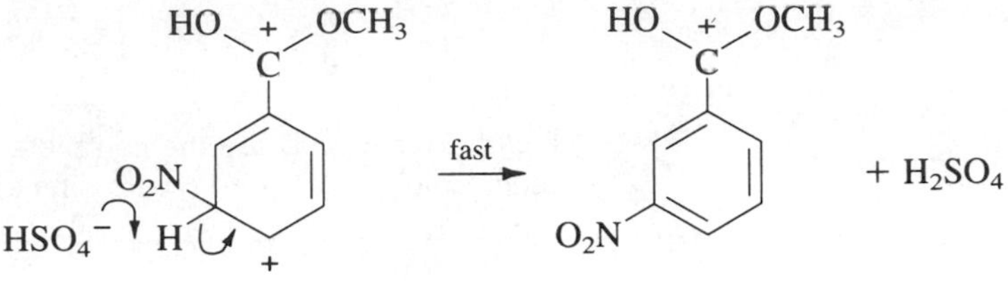

The ester group is a *meta* director and a deactivator of the benzene ring. It is much easier to nitrate a molecule such as phenol, where the hydroxyl group is an *ortho-para* director and an activator of the benzene ring, as the following resonance structures indicate:

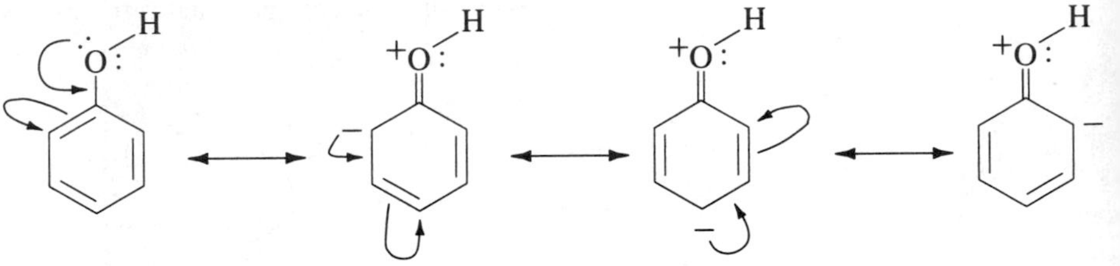

Experiments

1. Microscale Nitration of Methyl Benzoate

$$HNO_3 + 2H_2SO_4 \rightleftharpoons NO_2^+ + 2HSO_4^- + H_3O^+$$

Nitronium ion   **Hydronium ion**

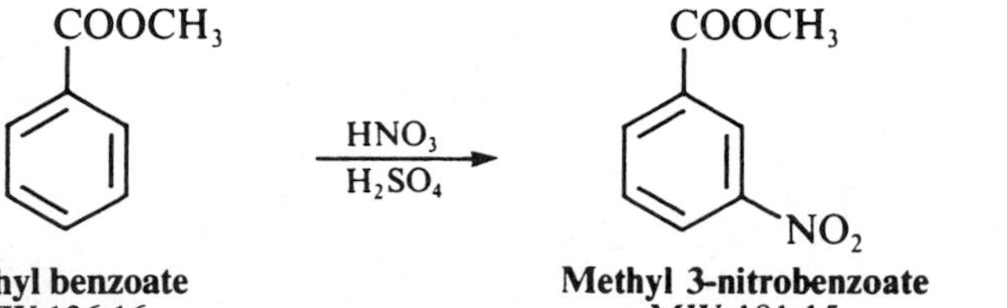

Methyl benzoate
MW 136.16
bp 199.6°C
den 1.09
n_D^{20} 1.5170

Methyl 3-nitrobenzoate
MW 181.15
mp 78°C

Do not use the plastic syringe and needle to measure acid.

To 0.6 mL of concentrated sulfuric acid in a 10 × 100 mm reaction tube, add 0.30 g of methyl benzoate. Flick the tube. Again, cool the mixture to 0°C, and add

Use care in handling concentrated sulfuric and nitric acids.

dropwise, using a Pasteur pipette, a mixture of 0.2 mL of concentrated sulfuric acid and 0.2 mL of concentrated nitric acid. Keep the reaction mixture in ice. Using a stirring rod, keep the reaction well mixed during the addition of the acids, and do not allow the temperature of the mixture to rise above about 15°C as judged by touching the reaction tube.

After all the nitric acid has been added, warm the mixture to room temperature, and, after 15 min, pour it onto 2.5 g of ice in a small beaker. Isolate the solid product by suction filtration using the Hirsch funnel and a 25-mL filter flask. Wash the product well with water and then with one 0.2-mL portion of ice-cold methanol. If the methanol is not ice-cold, product can be lost in this washing step. Save a small sample for melting-point determination and analysis by thin-layer chromatography and IR spectroscopy.

The remainder is weighed and crystallized from an equal weight of methanol in a reaction tube. Alternatively, the sample can be dissolved in a *slightly* larger quantity of methanol and water added dropwise to make the hot solution saturated with the product. Slow cooling should produce large crystals, with a melting point of 78°C. The crude material can be obtained in about 80% yield with a melting point of 74 to 76°C. If the yield is not as large as expected, concentrate the filtrate and collect a second crop of product.

Cleaning Up   Dilute the filtrate from the reaction with water, neutralize with sodium carbonate, and flush down the drain. The methanol from the crystallization should be placed in the organic solvents container.

 ## 2. Macroscale Nitration of Methyl Benzoate

Use care in handling concentrated sulfuric and nitric acids.

In a 125-mL Erlenmeyer flask, cool 12 mL of concentrated sulfuric acid to 0°C and then add 6.1 g of methyl benzoate. Again, cool the mixture to 0 to 10°C. Now add dropwise, using a Pasteur pipette, a cooled mixture of 4 mL of concentrated sulfuric acid and 4 mL of concentrated nitric acid. During the addition of the acids, swirl the mixture frequently and maintain the temperature of the reaction mixture in the range of 5 to 15°C.

When all the nitric acid has been added, warm the mixture to room temperature, and after 15 min pour it on 50 g of cracked ice in a 250-mL beaker. Isolate the solid product by suction filtration using a small Büchner funnel, and wash well with water, then with two 10-mL portions of ice-cold methanol. A small sample is saved for a melting-point determination. The remainder is weighed and crystallized from an equal weight of methanol. The crude product should be obtained in about 80% yield and with a mp of 74–76°C. The recrystallized product should have a mp of 78°C. The carbon and proton NMR spectra of the product are presented at the end of the chapter (Figs. 28.2 and 28.3).

Cleaning Up   Dilute the filtrate from the reaction with water, neutralize with sodium carbonate, and flush down the drain. The methanol from the crystallization should be placed in the organic solvents container.

Questions

1. Why does methyl benzoate dissolve in concentrated sulfuric acid? Write an equation showing the ions that are produced.

2. What would you expect the structure of the dinitro ester to be? Consider the directing effects of the ester and the first nitro group on the addition of the second nitro group.

3. Draw resonance structures to show in which position nitrobenzene will nitrate to form dinitrobenzene.

4. Assign the peaks at 3101, 1709, and 1390 cm^{-1} in the IR spectrum of methyl 3-nitrobenzoate (Fig. 28.1).

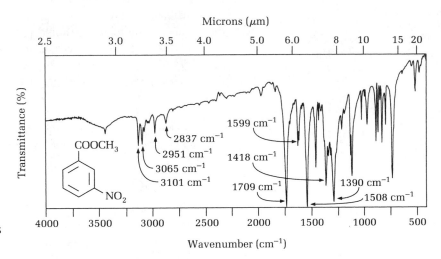

FIG. 28.1 IR spectrum of methyl 3-nitrobenzoate. The broad peak at 3400 cm^{-1} comes from water in the KBr disk.

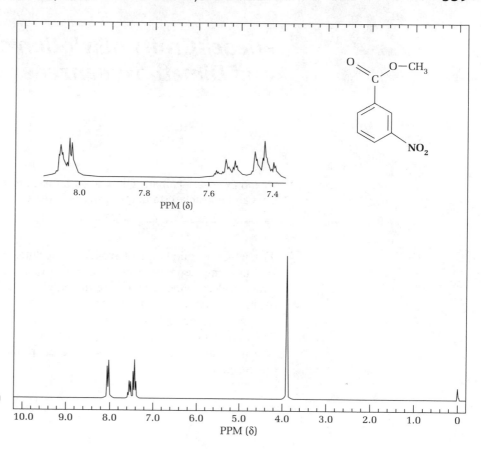

FIG. 28.2 ¹H NMR spectrum of methyl 3-nitrobenzoate (250 MHz).

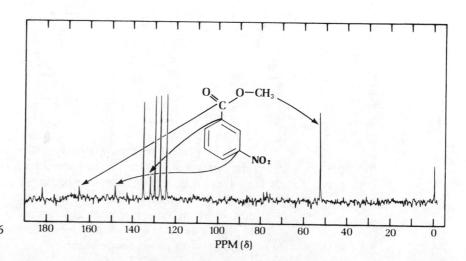

FIG. 28.3 ¹³C NMR spectrum of methyl 3-nitrobenzoate (22.6 MHz).

Friedel-Crafts Alkylation of Benzene and Dimethoxybenzene; Host-Guest Chemistry

Prelab Exercise: Prepare a flow sheet for each reaction, indicating how the catalysts and unreacted starting materials are removed from the reaction mixture.

Friedel-Crafts alkylation of aromatic rings most often employs an alkyl halide and a strong Lewis acid catalyst. Some of the catalysts that can be used, in order of decreasing activity, are the halides of Al, Sb, Fe, Ti, Sn, Bi, and Zn. Although useful, the reaction has several limitations. The aromatic ring must be unsubstituted or bear activating groups, and because the product, an alkylated aromatic molecule, is more reactive than the starting material, multiple substitution usually occurs. Furthermore, primary halides will rearrange under the reaction conditions:

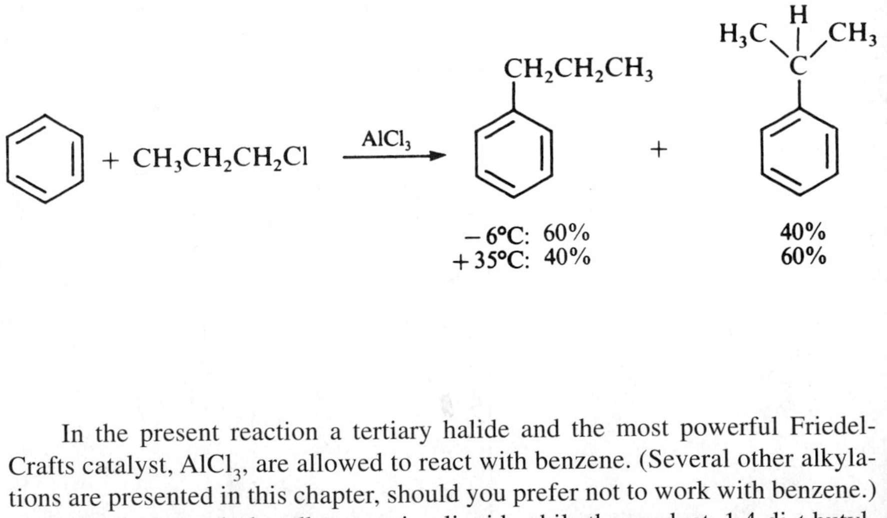

$$
-6°C: \ 60\% \qquad 40\%
$$
$$
+35°C: \ 40\% \qquad 60\%
$$

In the present reaction a tertiary halide and the most powerful Friedel-Crafts catalyst, $AlCl_3$, are allowed to react with benzene. (Several other alkylations are presented in this chapter, should you prefer not to work with benzene.) The initially formed *t*-butylbenzene is a liquid, while the product, 1,4-di-*t*-butylbenzene, which has a symmetrical structure, is a beautifully crystalline solid. The alkylation reaction probably proceeds through the carbocation under the conditions of the present experiment:

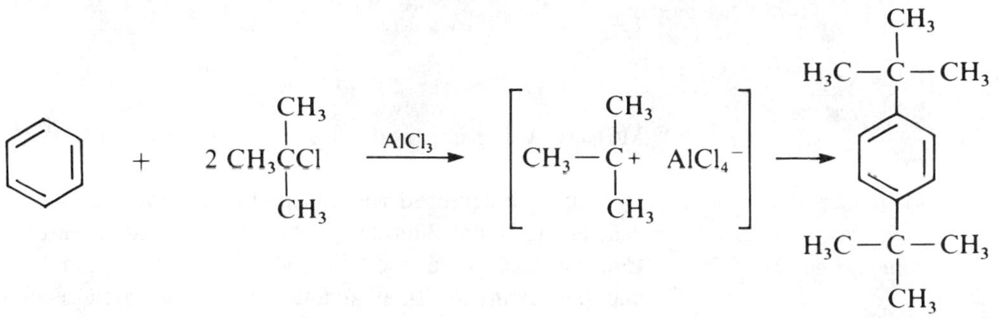

Benzene
MW 78.11, den. 0.88
bp 80°C

2-Chloro-2-methylpropane
(*t*-Butyl chloride)
MW 92.57, den. 0.85
bp 51°C

1,4-Di-*t*-butylbenzene
MW 190.32, mp 77–79°C
bp 167°C

The reaction is reversible. If 1,4-di-*t*-butylbenzene is allowed to react with *t*-butyl chloride and aluminum chloride (1.3 mol) at 0–5°C, 1,3-di-*t*-butylbenzene, 1,3,5-tri-*t*-butylbenzene, and unchanged starting material are found in the reaction mixture. Thus, the mother liquor from crystallization of 1,4-di-*t*-butylbenzene in the present experiment probably contains *t*-butylbenzene, the desired 1,4-di-product, the 1,3-di-isomer, and 1,3,5-tri-*t*-butylbenzene.

Inclusion Complexes: Host-Guest Chemistry

Although the mother liquor probably contains a mixture of several components, the 1,4-di-*t*-butylbenzene can be isolated easily as an inclusion complex. Inclusion complexes are examples of host-guest chemistry. The molecule thiourea, NH_2CSNH_2, the host, has the interesting property of crystallizing in a helical crystal lattice that has a cylindrical hole in it. The guest molecule can reside in this hole if it is the correct size. It is not bound to the host; NMR studies indicate the guest molecule can rotate longitudinally within the helical crystal lattice. There are often nonintegral numbers (on the average) of host molecules per guest. The inclusion complex of thiourea and 1,4-di-*t*-butylbenzene crystallizes very nicely from a mixture of the other hydrocarbons, and thus more of the product can be obtained. Because thiourea is very soluble in water, the product is recovered from the complex by shaking it with a mixture of *t*-butyl methyl ether and water. The complex immediately decomposes, and the product dissolves in the ether layer, from which it can be recovered.

Compare the length of the 1,4-di-*t*-butylbenzene molecule with the length of various *n*-alkanes, and predict the host-guest ratio for a given alkane. You can then check your prediction experimentally. *n*-Hexane can be isolated from the solvent mixture sold as "hexanes."

Experiments

1. 1,4-Di-*t*-butylbenzene

Measure, using a 1.0-mL plastic syringe, 0.40 mL of dry 2-chloro-2-methylpropane (*t*-butyl chloride) and 0.20 mL of dry benzene into a dry 10 × 100 mm reaction tube equipped with a septum and tubing as seen in Fig. 29.1. The benzene and the alkyl chloride will usually be found in septum-stoppered containers. Cool the tube in ice, and then add to it 20 mg of aluminum chloride. Weighing and transferring this small quantity are difficult because aluminum chloride reacts with great rapidity with moist air. Keep the reagent bottle closed as much of the time as possible while weighing the reagent into a very small, dry, capped vial. Since the aluminum chloride is a catalyst, the amount need not be exactly 20 mg.

Mix the contents of the reaction tube by flicking the tube with the finger. After an induction period of about 2 min, a vigorous reaction sets in, with bubbling and liberation of hydrogen chloride. The hydrogen chloride is trapped using the apparatus depicted in Fig. 29.1. The wet cotton in the empty reaction tube will dissolve the hydrogen chloride. See Fig. 29.2 for threading a polyethylene tube through a septum. Near the end of the reaction, the product separates as a white solid. When this occurs, remove the tube from the ice, and let it stand at room temperature for 5 min.

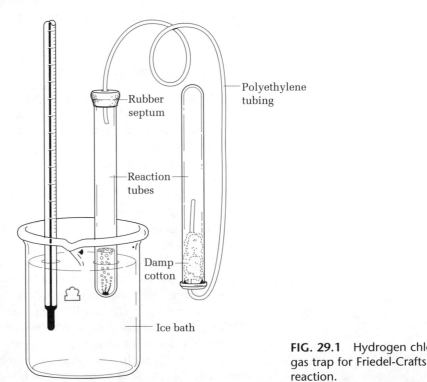

FIG. 29.1 Hydrogen chloride gas trap for Friedel-Crafts reaction.

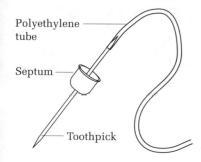

Polyethylene tube

Septum

Toothpick

Add anhydrous calcium chloride until it no longer clumps together.

Spontaneous crystallization gives beautiful needles or plates.

FIG. 29.2 To thread a polyethylene tube through a septum, make a hole through the septum with a needle, and then push a toothpick through the hole. Push the tube firmly onto the toothpick, and then pull and push on the toothpick. The tube will slide through the septum. Finally, pull the tube from the toothpick. A blunt syringe needle can be used in place of a toothpick.

Add about 1.0 mL of ice water to the reaction mixture, mix the contents thoroughly, and extract the product with three 0.8-mL portions of *tert*-butyl methyl ether. Wash the combined ether extracts with about 1.5 mL of saturated sodium chloride solution, and dry the ether over anhydrous calcium chloride pellets. Add sufficient drying agent so that it does not clump together. After 5 min, transfer the ether solution to a dry, tared reaction tube, using more ether to wash the drying agent, and evaporate the ether under a stream of air in the hood. Remove the last traces of ether under water aspirator vacuum (Fig. 29.3). The oily product should solidify on cooling and weigh about 300 mg.

For crystallization, dissolve the product in 0.40 mL of methanol, and let the solution come to room temperature without disturbance. After thorough cooling at 0°C, remove the methanol with a Pasteur pipette, and rinse the crystals with a drop of ice-cold methanol while keeping the reaction tube in ice. Save this methanol solution for analysis by thin-layer chromatography. The yield of recrystallized material after drying under aspirator vacuum should be about 160 mg. Remove a sample of crystals for analysis by IR spectroscopy, thin-layer chromatography, and melting-point determination (Figs. 29.4 and 29.5). Using thin-layer chromatography, compare the pure crystalline product with the residue left after evaporation of the methanol.

Cleaning Up   Place any unused *t*-butyl chloride in the halogenated organic waste container and any unused benzene in the hazardous waste container for benzene. Any unused aluminum chloride should be mixed thoroughly with a

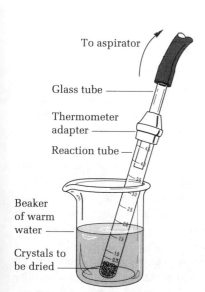

To aspirator

Glass tube

Thermometer adapter

Reaction tube

Beaker of warm water

Crystals to be dried

FIG. 29.3 Drying of crystals under reduced pressure.

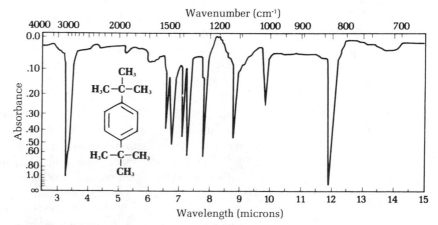

FIG. 29.4 IR spectrum of 1,4-di-*t*-butylbenzene.

FIG. 29.5 ^{1}H NMR spectrum of 1,4-di-t-butylbenzene (90 MHz).

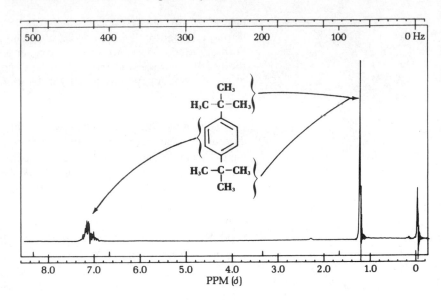

large excess of sodium carbonate and the solid mixture added to a large volume of water before being flushed down the drain. The combined aqueous layers from the reaction should be neutralized with sodium carbonate and then flushed down the drain. Methanol from the crystrallization is to be placed in the organic solvents container.

▯ MICROSCALE

CAUTION: Thiourea is a mild carcinogen. Handle the solid in a hood. Do not breathe dust.

$$\overset{S}{\underset{\parallel}{H_2NCNH_2}}$$

Thiourea
MW 76.12

Inclusion complex starts to crystallize in 10 min.

2. Preparation of Thiourea Inclusion Complex

In a tared reaction tube dissolve 200 mg of thiourea (*Caution! See margin note*) and 120 mg of 1,4-di-t-butylbenzene in 2.0 mL of methanol at room temperature; then cool the mixture in ice, at which time the inclusion complex will crystallize. Using a Pasteur pipette, remove the solvent and wash the product twice with just enough methanol to cover the crystals while keeping the tube on ice. Connect the reaction tube to a water aspirator, and, using the heat of the hand, evaporate the remaining methanol under reduced pressure until the weight of the tube is constant. The yield should be about 200 mg.

Remove a small sample, determine carefully the weight of the remaining complex, and then add about 1.2 mL of water and 1.2 mL of *tert*-butyl methyl ether to the tube. Shake the mixture until the crystals disappear. This causes the breakup of the complex, with the thiourea remaining in the aqueous layer and the 1,4-di-t-butylbenzene passing into the ether layer. Draw off the aqueous layer, and dry the ether layer with anhydrous calcium chloride pellets. Add sufficient drying agent so that it does not clump together. More ether can be added if necessary. Transfer the ether to a tared reaction tube, and wash the drying agent twice with fresh portions of ether. The object is to make a quantitative transfer of the butylbenzene. Evaporate the ether, and remove the last traces under aspirator vacuum as before. After the weight of the tube is constant, record the weight of the hydrocarbon. Calculate the number of molecules of thiourea per molecule of hydrocarbon (probably *not* an integral number).

Cleaning Up   Place any unused thiourea and the 1.2 mL of the aqueous solution containing thiourea in the hazardous waste container for thiourea. Alternatively, treat the thiourea with excess aqueous 5.25% sodium hypochlorite solution (household bleach), dilute the mixture with a large amount of water, and flush it down the drain. Allow the ether to evaporate from the calcium chloride, and then place it in the nonhazardous solid waste container.

3. 1,4-Di-*t*-butylbenzene

🏺 **MACROSCALE**

CAUTION: Benzene is a mild carcinogen. Handle in the hood; do not breathe vapors or allow liquid to come in contact with the skin.

Measure in the hood 20 mL of 2-chloro-2-methylpropane (*t*-butyl chloride) and 10 mL of benzene in a 125-mL filter flask equipped with a one-holed rubber stopper fitted with a thermometer. Place the flask in an ice-water bath to cool. Weigh 1 g of fresh aluminum chloride onto a creased paper, and scrape it with a small spatula into a 10 × 75-mm test tube; close the tube at once with a cork.[1] Connect the side arm of the flask to the aspirator (preferably one made of plastic), and operate it at a rate sufficient to carry away hydrogen chloride formed in the reaction, or make a trap for the hydrogen chloride similar to that shown in Fig. 18.3.

Reaction time: about 15 min

Aluminum chloride dust is extremely hygroscopic and irritating. It hydrolyzes to hydrogen chloride on contact with moisture. Clean up spilled material immediately.

Cool the liquid to 0–3°C, add about one-quarter of the aluminum chloride, replace the thermometer, and swirl the flask vigorously in the ice bath. After an induction period of about 2 min a vigorous reaction sets in, with bubbling and liberation of hydrogen chloride. Add the remainder of the catalyst in three portions at intervals of about 2 min. Toward the end, the reaction product begins to separate as a white solid. When this occurs, remove the flask from the bath and let stand at room temperature for 5 min. Add ice and water to the reaction mixture, allow most of the ice to melt, and then add ether for extraction of the product, stirring with a rod or spatula to help bring the solid into solution. Transfer the solution to a separatory funnel and shake; draw off the lower layer and wash the upper ether layer with water and then with a saturated sodium chloride solution. Dry the ether solution over anhydrous calcium chloride pellets for 5 min, filter the solution to remove the drying agent, remove the ether by evaporation on the steam bath, and evacuate the flask using the aspirator to remove traces of solvent until the weight is constant; yield of crude product should be 15 g.

Add anhydrous calcium chloride pellets until they no longer clump together.

Spontaneous crystallization gives beautiful needles or plates.

The oily product should solidify on cooling. For crystallization, dissolve the product in 20 mL of hot methanol and let the solution come to room temperature without disturbance. If you are in a hurry, lift the flask without swirling; place it in an ice-water bath and observe the result. After thorough cooling at 0°C, collect the product and rinse the flask and product with a little ice-cold methanol. The yield of 1,4-di-*t*-butylbenzene from the first crop is 8.2–8.6 g of satisfactory material. Save the product for the next step as well as the mother liquor, in case you later wish to work it up for a second crop.

---

1. Alternative scheme: Put a wax pencil mark on the test tube 37 mm from the bottom, and fill the tube with aluminum chloride to this mark.

Inclusion Complex Formation

$$S$$
$$\parallel$$
$$H_2NCNH_2$$

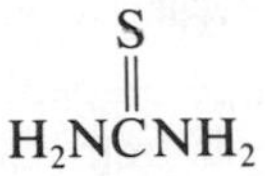

Thiourea
MW 76.12

Inclusion complex starts to crystallize in 10 min.

Workup of mother liquor

In a 25-mL Erlenmeyer flask dissolve 5 g of thiourea (see margin note) and 3 g of 1,4-di-*t*-butylbenzene in 50 mL of warm methanol (break up lumps with a flattened stirring rod), and let the solution stand for crystallization of the complex, which occurs with ice cooling. Collect the crystals, rinse with a little methanol, and dry to constant weight; yield is 5.8 g. Bottle a small sample, determine carefully the weight of the remaining complex, and place the material in a separatory funnel along with about 25 mL each of water and ether. Shake until the crystals disappear, draw off the aqueous layer containing thiourea, wash the ether layer with saturated sodium chloride, and dry the ether layer over anhydrous calcium chloride pellets. Remove the drying agent by filtration; collect the filtrate in a tared 125-mL Erlenmeyer. Evaporate and evacuate as before, making sure the weight of hydrocarbon is constant before you record it. Calculate the number of molecules of thiourea per molecule of hydrocarbon (probably *not* an integral number).

To work up the mother liquor from the crystallization from methanol, first evaporate the solvent. Note that the residual oil does not solidify on ice cooling. Next, dissolve the oil, together with 5 g of thiourea, in 50 mL of methanol, collect the inclusion complex that crystallizes (3.2 g), and recover 1,4-di-*t*-butylbenzene from the complex as before (0.8 g before crystallization). The IR and ^{1}H NMR spectra of the product are seen in Figs. 29.4 and 29.5.

Cleaning Up　Place any unused *t*-butyl chloride in the halogenated organic waste container and any unused benzene in the hazardous waste container for benzene. Any unused aluminum chloride should be mixed thoroughly with a large excess of sodium carbonate and the solid mixture added to a large volume of water before being flushed down the drain. The combined aqueous layers from the reaction should be neutralized with sodium carbonate and then flushed down the drain. Methanol from the crystallization is to be placed in the organic solvents container.

4. 1,4-Di-*t*-butyl-2,5-dimethoxybenzene

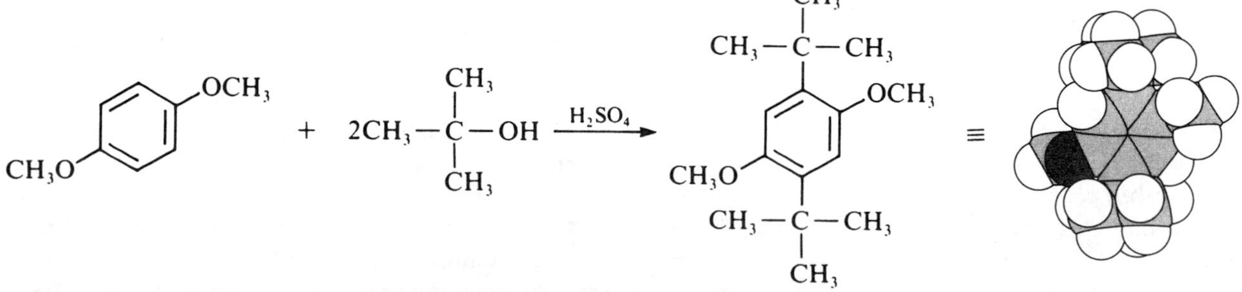

1,4-Dimethoxybenzene
(Hydroquinone dimethyl ether)
MW 138.16, mp 57°C

2-Methyl-2-propanol
(*t*-Butyl alcohol)
MW 74.12, den. 0.79
mp 25.5°C, bp 82.8°C
n_D^{20} 1.3820

1,4-Di-*t*-butyl-2,5-dimethoxybenzene
MW 250.37, mp 104–105°C

$$CH_3 - \overset{\overset{\displaystyle CH_3}{|}}{\underset{\underset{\displaystyle CH_3}{|}}{C^+}}$$

Trimethylcarbocation

This experiment illustrates the Friedel-Crafts alkylation of an activated benzene molecule with a tertiary alcohol in the presence of sulfuric acid as the Lewis acid catalyst. As in the reaction of benzene and *t*-butyl chloride, the substitution involves attack by the electrophilic trimethylcarbocation.

Microscale Procedure

In a 10 × 100 mm reaction tube dissolve 120 mg of 1,4-dimethoxybenzene (hydroquinone dimethyl ether) in 0.4 mL of acetic acid with gentle warming, and add to it 0.2 mL of *t*-butyl alcohol (it may be necessary to melt this alcohol). Cool the mixture in ice, and then add to it dropwise from a Pasteur pipette 0.4 mL of concentrated sulfuric acid. After each drop of acid is added, mix the solution thoroughly. At the end of this addition, considerable solid reaction product should have separated. Stir the mixture thoroughly with a glass stirring rod, then remove the reaction tube from the ice, and allow it to warm to room temperature and remain at 20 to 25°C for at least 10 min to complete the reaction. Then cool the mixture in ice to cause crystallization to occur. *Very carefully* add a drop of water to the mixture, stir it with the glass rod, and continue to add water dropwise with cooling and mixing until 2.5 mL have been added. Remove the solvent from the cold solution with a Pasteur pipette, and wash the crystals *thoroughly* with water. Recrystallize the product from methanol. Remove the solvent using a Pasteur pipette after allowing the mixture to cool to room temperature and then to 0°C in ice. Remove the last traces of methanol under aspirator vacuum while warming the tube in the hand or in a beaker of warm water (see Fig. 29.3). The yield of large plates of 1,4-di-*t*-butyl-2,5-dimethoxybenzene should be about 80 to 100 mg. Analyze the product by IR spectroscopy and thin-layer chromatography using ligroin as the eluent. Determine the melting point and the percentage yield.

*Stir **thoroughly** after each drop of water is added.*

Cleaning Up   Combine the aqueous layer, methanol washes, and crystallization mother liquor, dilute with water, neutralize with sodium carbonate, and flush down the drain. Any spilled sulfuric acid should be covered with a large excess of solid sodium carbonate and the mixture added to water before being flushed down the drain.

Macroscale Procedure

Place 3 g of 1,4-dimethoxybenzene (hydroquinone dimethyl ether) in a 50-mL Erlenmeyer flask, add 5 mL of *t*-butyl alcohol and 10 mL of acetic acid, and put the flask in an ice-water bath to cool. Measure 5 mL of concentrated sulfuric acid into a 25-mL Erlenmeyer flask, add 15 mL of concentrated sulfuric acid, and put the flask, properly supported, in the ice bath to cool. For good thermal contact the ice bath should be an ice–water mixture. Put a thermometer in the larger flask, swirl in the ice bath until the temperature is in the range 0–3°C, and remove the thermometer (solid, if present, will dissolve later). Don't use the thermometer as a stirring rod. Clamp a small separatory funnel

*Caution: $H_2SO_4 + SO_3$ is **fuming sulfuric acid.** Highly corrosive to human tissue. Reacts violently with water. Measure these reagents in the hood.*

Reaction time: about 12 min

in a position to deliver into the 50-mL Erlenmeyer flask so that the flask can remain in the ice-water bath, wipe the smaller flask dry, and pour the chilled sulfuric acid solution into the funnel. While swirling the 50-mL flask in the ice bath, run in the chilled sulfuric acid by rapid drops during the course of 4–7 min.

By this time considerable solid reaction product should have separated, and insertion of a thermometer should show that the temperature is in the range 15–20°C. Swirl the mixture while maintaining the temperature at about 20–25°C for 5 min more, and then cool in ice. Add ice to the mixture to dilute the sulfuric acid, then add water to nearly fill the flask, cool, and collect the product on a Büchner funnel with suction. It is good practice to clamp the filter flask so it does not tip over. Apply only very gentle suction at first to avoid breaking the filter paper, which is weakened by the strong sulfuric acid solution. Wash liberally with water, and then turn on the suction to full force. Press down the filter cake with a spatula, and let drain well. Meanwhile, cool a 15-mL portion of methanol for washing to remove a little oil and a yellow impurity. Release the suction, cover the filter cake with a third of the chilled methanol, and then apply suction. Repeat the washing a second and a third time.

Since air-drying of the crude reaction product takes time, the following short procedure is suggested: Place the moist material in a 50-mL Erlenmeyer flask, add a little dichloromethane (5–8 mL) to dissolve the organic material, and note the appearance of aqueous droplets. Add enough anhydrous calcium chloride pellets to the flask so that the drying agent no longer clumps together, let drying proceed for 10 min, and then remove the drying agent by gravity filtration or careful decantation into another 50-mL Erlenmeyer flask. Add 15 mL of methanol (bp 65°C) to the solution, and start evaporation (on the steam bath in the hood) to eliminate the dichloromethane (bp 41°C). When the volume is estimated to be about 15 mL, let the solution stand for crystallization. When crystallization is complete, cool in ice and collect.

From an environmental standpoint it would be better to eliminate the solvents by simple distillation using the apparatus depicted in Fig. 5.10. Leave 15 mL in the flask, and allow the mixture to cool slowly. Large crystals will form. Collect the product on a small Büchner funnel. The yield of large plates of pure 1,4-di-*t*-butyl-2,5-dimethoxybenzene is 2–2.5 g.

Antics of Growing Crystals

R. D. Stolow of Tufts University reported[2] that growing crystals of the di-*t*-butyldimethoxy compound change shape in a dramatic manner: Thin plates curl and roll up and then uncurl so suddenly that they propel themselves for a distance of several centimeters. If you do not observe this phenomenon during crystallization of a small sample, you may be interested in consulting the papers cited and pooling your sample with others for trial on a large scale. The solvent

---

2. R. D. Stolow and J. W. Larsen, *Chemistry and Industry,* 449 (1963). See also J. M. Blatchly and N. H. Hartshorne, *Transactions of the Faraday Society,* **62,** 513 (1966).

mixture recommended by the Tufts workers for observation of the phenomenon is 9.7 mL of acetic acid and 1.4 mL of water per gram of product.

Figures 29.6 and 29.7 present the infrared and NMR spectra of the starting hydroquinone dimethyl ether. Can you predict the appearance of the NMR spectrum of the product?

Cleaning Up   Combine the aqueous layer and methanol washes and crystallization mother liquor, dilute with water, neutralize with sodium carbonate, and flush down the drain. Any spilled sulfuric acid should be covered with a large excess of solid sodium carbonate and the mixture added to water before being

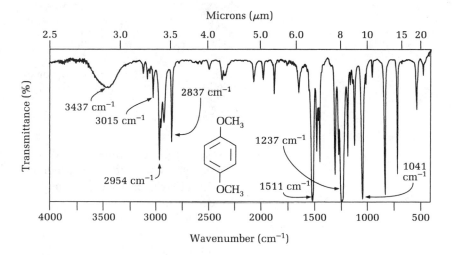

FIG. 29.6   Infrared spectrum of 1,4-dimethoxybenzene (KBr disk). Note water contaminant at 3437 cm⁻¹.

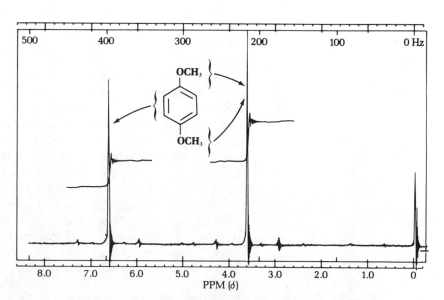

FIG. 29.7   ¹H NMR spectrum of 1,4-dimethoxybenzene (60 MHz).

flushed down the drain. Dichloromethane mother liquor from the crystallization is placed in the halogenated organic waste container. After the solvent evaporates from the drying agent it is placed in the nonhazardous solid waste
container.

For Further Investigation

Alkylation of *m*-Xylene

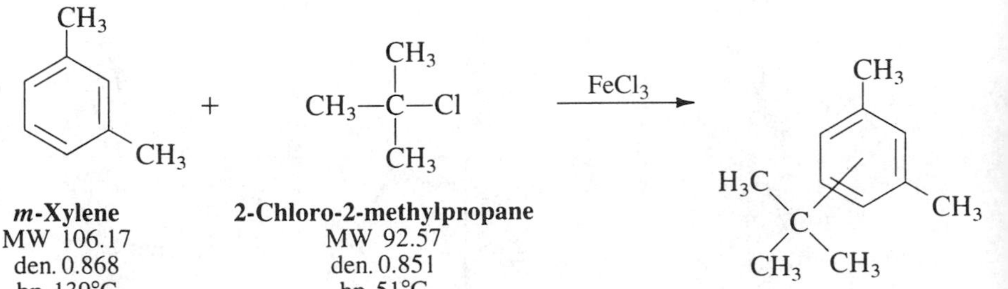

m-Xylene	2-Chloro-2-methylpropane
MW 106.17	MW 92.57
den. 0.868	den. 0.851
bp 139°C	bp 51°C

The object of this experiment is to determine the structure of the product formed
when *m*-xylene (not benzene, as in Experiment 1) is alkylated with *t*-butyl
chloride. A Lewis acid catalyst is employed, either iron(III) chloride or aluminum
chloride. The methyl groups in *m*-xylene (1,3-dimethylbenzene) are *ortho-, para-*
directors and activators of the benzene ring, but other factors may intervene in
this particular experiment. Read about the factors affecting alkylations in the
first part of this chapter.

Aluminum chloride is the catalyst most often used for the Friedel-Crafts
reaction, but it is difficult to store and to weigh out because it reacts very
rapidly with moisture in the air. In Experiment 4, iron(III) chloride is used as
the catalyst.

An excess of *m*-xylene is used in this reaction to ensure that the product
will be monoalkylated by the *t*-butyl cation. Because the boiling points of the
reactants and the product are quite different, a crude but very rapid and efficient
distillation is done to remove unreacted xylene and *t*-butyl chloride, leaving only
the product. This instant microscale distillation is carried out by boiling the
mixture in a reaction tube and then pulling the hot vapors into a Pasteur pipette,
where they condense. After about half the material has been distilled in this
way, the high-boiling residue will consist of almost pure product.

Infrared spectroscopy can be used to determine unequivocally the structure
of the product. It has been found that the hydrogen atoms on a benzene ring
give rise to one or two intense, characteristic peaks in the region of 730 to
885 cm^{-1} regardless of the nature of the substituents on the benzene ring. For
instance, monosubstituted benzenes have two peaks, one in the range of 770 to

TABLE 29.1   C—H Out-of-Plane Bending Vibrations of Substituted Benzenes

Substituted Benzene	Peak 1 (cm^{-1})	Peak 2 (cm^{-1})
Benzene	671	—
Monosubstituted benzenes	770–730	710–690
1,2-Disubstituted	770–735	—
1,3-Disubstituted	810–750	710–690
1,4-Disubstituted	835–810	—
1,2,3-Trisubstituted	780–760	745–705
1,2,4-Trisubstituted	825–805	885–870
1,3,5-Trisubstituted	865–810	730–675
1,2,3,4-Tetrasubstituted	810–800	—
1,2,3,5-Tetrasubstituted	850–840	—
1,2,4,5-Tetrasubstituted	870–855	—
Pentasubstituted	870	—

730 cm^{-1} and the other in the range of 710 to 690 cm^{-1} (Table 29.1). NMR spectroscopy can also be used to determine the structure of the product.

Molecular mechanics calculations on the three possible monosubstitution products from this reaction may help you to predict the outcome or help confirm conclusions drawn from experimental evidence.

MICROSCALE

5. Iron(III) Chloride Catalyzed Reaction

Place 0.6 mL of *m*-xylene (1,3-dimethylbenzene) and 0.5 mL of *t*-butyl chloride (2-chloro-2-methylpropane) in a dry reaction tube equipped with a septum and a polyethylene tube that leads to a tube containing a small wad of damp cotton. See Fig. 29.2 for threading a polyethylene tube through a septum. The cotton, dampened with 2 or 3 drops of water, serves to trap the hydrogen chloride evolved during the reaction. Cool the mixture in an ice bath (see Fig. 29.1), add 30 mg of iron(III) chloride (purple, free-flowing crystals when pure and dry), and replace the septum and polyethylene tube. After a short induction period, the reaction will begin a vigorous evolution of hydrogen chloride.

After this vigorous part of the reaction is over, remove the ice bath and allow the reaction mixture to warm to room temperature. When bubbles of hydrogen chloride cease to be evolved or after 15 min, add 1 mL of water to the reaction mixture, mix well, and then remove the water layer with a Pasteur pipette. Repeat this process using about 1 mL of saturated aqueous sodium bicarbonate solution followed by 1 mL of saturated sodium chloride solution. Transfer the organic layer to another reaction tube, and dry it with anhydrous calcium chloride pellets.

Transfer the organic layer to a dry reaction tube once more, add a boiling chip, and heat the mixture to boiling. Allow the refluxing vapors to rise about

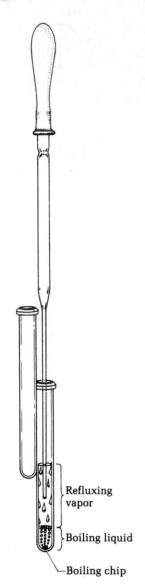

Refluxing
vapor

Boiling liquid

Boiling chip

FIG. 29.8 Apparatus for instant microscale distillation.

3 cm in the tube, and then draw them into a Pasteur pipette. Squirt the condensate into another reaction tube held in the same hand (Fig. 29.8). Repeat this until about half the mixture has been distilled.

Analyze the residue by infrared spectroscopy, which is most easily done as a thin film between sodium chloride or silver chloride plates. The IR spectrum of *m*-xylene is shown in Fig. 29.9. Refer to Table 29.1 for assignments. Note the two strong peaks in Fig. 29.6 for *m*-xylene and the frequencies expected for 1,3-disubstitution from Table 29.1. From this correlation chart for the aromatic C—H out-of-plane bending modes, deduce the structure of your product. In your report, write a mechanism for this reaction that illustrates all the details of your conclusions. Discuss the structure of the product in terms of directive effects of substituents on the benzene ring, steric effects of substituents, and thermodynamic versus kinetic control of the reaction.

Carry out molecular mechanics calculations on the three possible products to determine their relative steric energies or heats of formation. Do these calculations contribute to your conclusions?

Cleaning Up   Place any unused *t*-butyl chloride in the halogenated organic waste container and any unused xylene in the organic solvents container. Unused iron(III) chloride and wash solutions should be combined and neutralized with sodium bicarbonate solution, diluted with water, and flushed down the drain.

Questions

1. Explain why the reaction of 1,4-di-*t*-butylbenzene with *t*-butyl chloride and aluminum chloride gives 1,3,5-tri-*t*-butylbenzene.

2. Why must aluminum chloride be protected from exposure to the air?

3. Would you expect to find two strong peaks at ~690 and ~770 cm^{-1} (see Fig. 29.8) in your product from the alkylation of *m*-xylene? Why or why not?

4. Draw a detailed mechanism for the formation of *t*-butyl-2,5-dimethoxybenzene.

5. Why is the 1,4 isomer, 1,4-di-*t*-butyl-2,5-dimethoxybenzene, the major product in the alkylation of dimethoxybenzene? Would you expect either of the following compounds to be formed as side products: 1,3-di-*t*-butyl-2,5-dimethoxybenzene or 1,4-dimethoxy-2,3-di-*t*-butylbenzene? Why or why not?

6. Suggest two other compounds that might be used in place of *t*-butyl alcohol to form 1,4-di-*t*-butyl-2,5-dimethoxybenzene.

7. Can you locate the two peaks in Fig. 28.1 that show methyl-3-nitrobenzoate is indeed 1,3-disubstituted? (See Experiment 5 and Table 29.1).

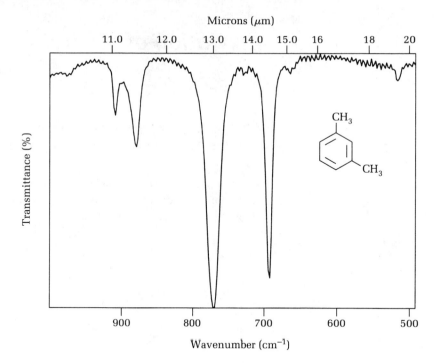

FIG. 29.9 IR spectrum of *m*-xylene (thin film) from 500 to 1000 cm^{-1} to show C—H out-of-plane bending vibrations.

30 *Alkylation of Mesitylene*

Prelab Exercise: How much formic acid is consumed in this reaction?

The reaction of an alkyl halide with an aromatic molecule in the presence of aluminum chloride, the Friedel-Crafts reaction, proceeds through the formation of an intermediate carbocation:

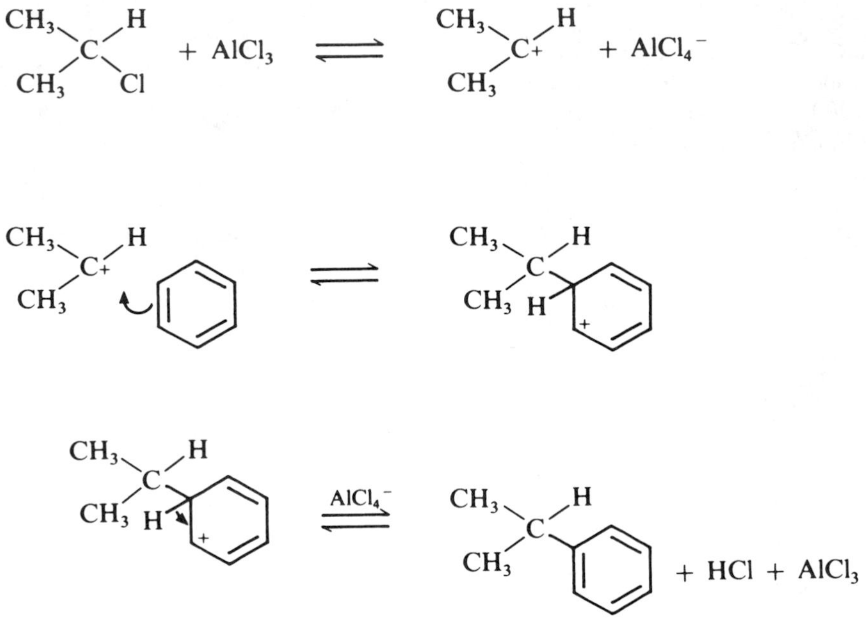

Friedel and Crafts (who later became the president of MIT) discovered this reaction in 1879, but seven years before, Baeyer and his colleagues carried out very similar reactions using aldehydes as the alkylating agent and strong acids as catalysts. These reactions, like the Friedel-Crafts reaction, proceed through carbocation intermediates. Consider the synthesis of DDT (1,1,1-trichloro-2,2-di(*p*-chlorophenyl)ethane):

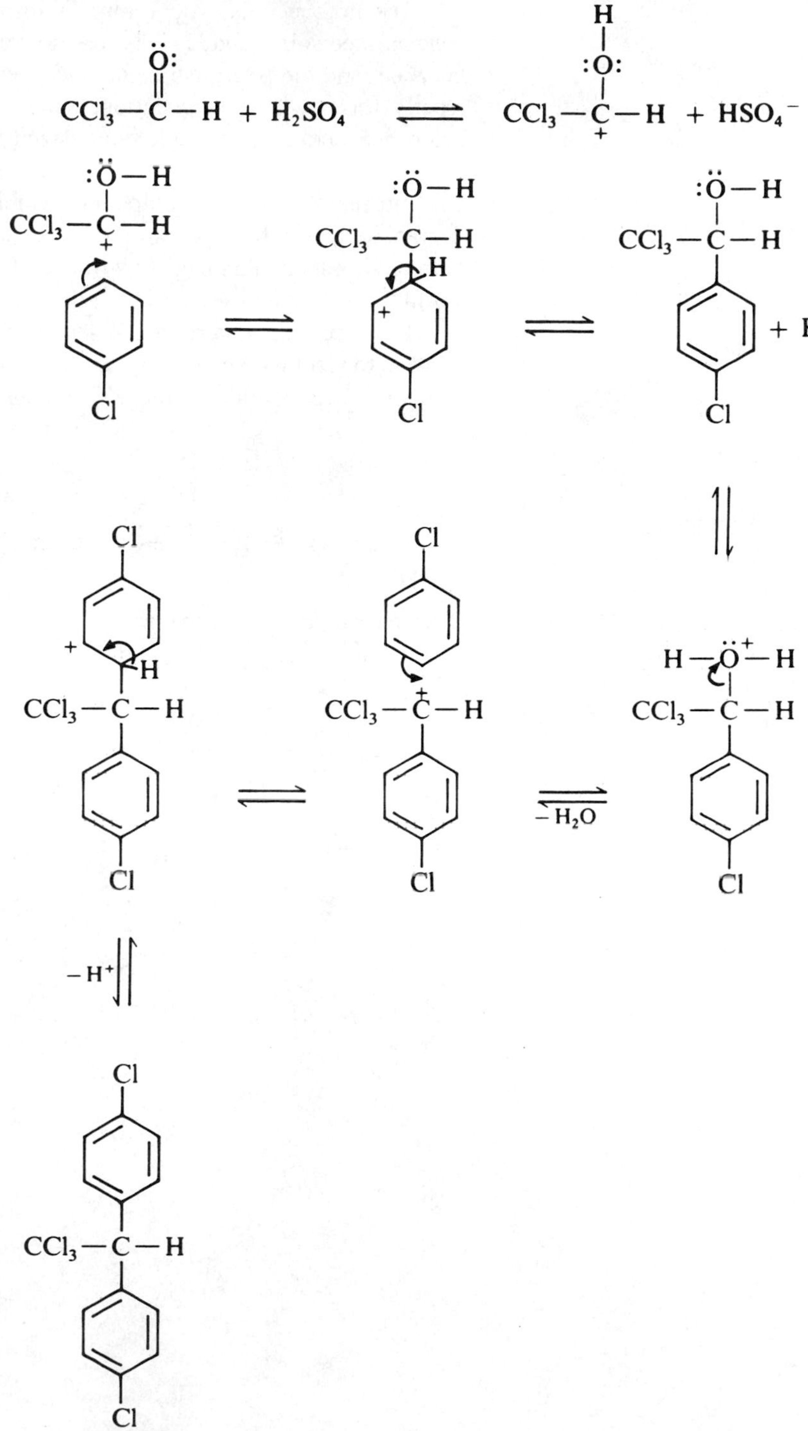

$$CCl_3-\overset{\overset{\displaystyle \ddot{O}:}{\|}}{C}-H \;+\; H_2SO_4 \;\rightleftharpoons\; CCl_3-\overset{\overset{\displaystyle :\ddot{O}:}{|}}{\underset{+}{C}}-H \;+\; HSO_4^{-}$$

$$CCl_3-\overset{\overset{\displaystyle :\ddot{O}-H}{|}}{\underset{+}{C}}-H$$

$$\rightleftharpoons$$

$$\rightleftharpoons \quad + H^+$$

DDT

Trichloroacetaldehyde (chloral) forms a carbocation on reaction with concentrated sulfuric acid. This reacts primarily at the *para*-position of chlorobenzene; and the intermediate alcohol, being benzylic, in the presence of acid readily forms a new carbocation, which in turn attacks another molecule of chlorobenzene. Even though synthesized in 1872, the remarkable insecticidal properties of DDT were not recognized until about 1940. It took another 25 years to realize that this compound, which is resistant to normal biochemical degradation, was building up in rivers, lakes, and streams and causing long-term environmental damage to wildlife. It is now outlawed in many parts of the world.

In the present experiment, discovered by Baeyer in 1872, formaldehyde is allowed to react with mesitylene in the presence of formic acid. The sequence of reactions is very similar to those that form DDT:

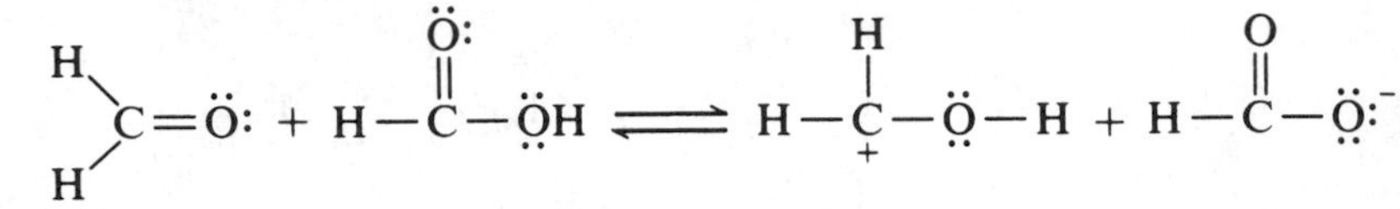

Formaldehyde   **Formic acid**

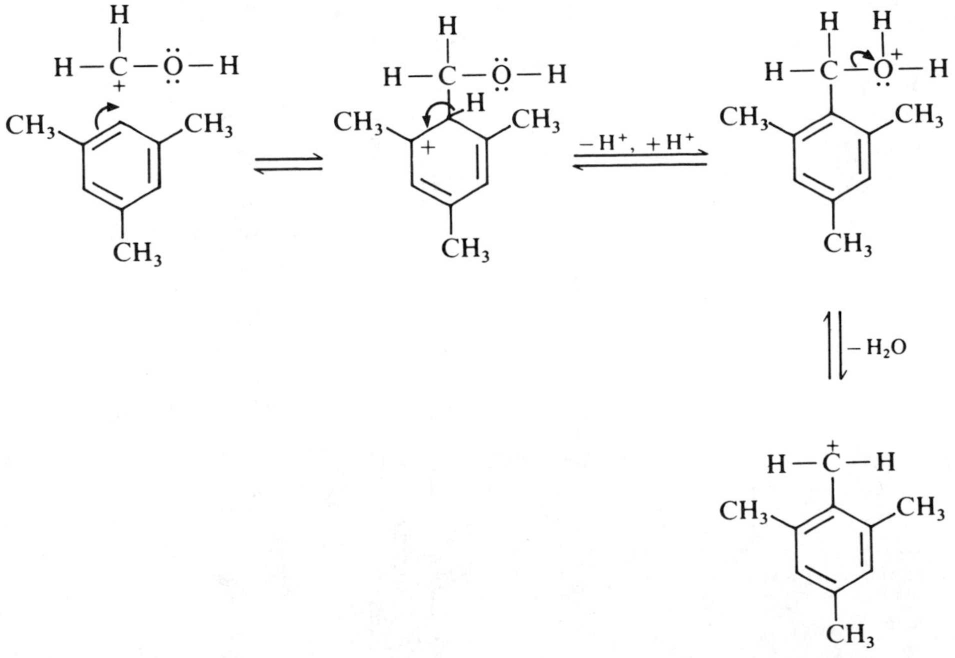

Benzylic carbocation

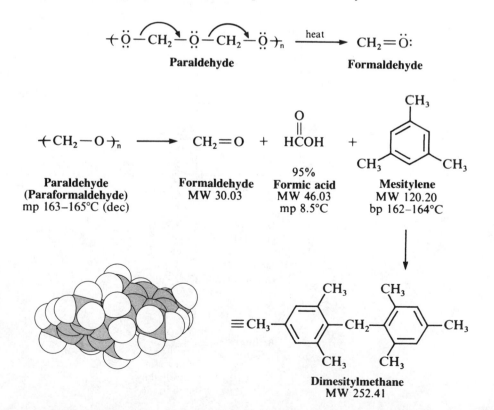

Dimesitylmethane

When aluminum chloride, a much more powerful catalyst, is used in this reaction, the methyl groups on the mesitylene rearrange and disproportionate to form a number of products, including polymeric material. The strongly activated ring of phenol reacts with formaldehyde at the *ortho-* and *para-*positions to form a polymer, Bakelite (see Chapter 69, Polymers).

A convenient form of formaldehyde to use in a reaction of this type is paraldehyde, a polymer that readily decomposes to formaldehyde:

$$+\ddot{O}-CH_2-\ddot{O}-CH_2-\ddot{O}+_n \xrightarrow{\text{heat}} CH_2=\ddot{O}:$$

Paraldehyde   **Formaldehyde**

$+CH_2-O+_n$	$CH_2=O$	$HCOH$	
Paraldehyde (**Paraformaldehyde**) mp 163–165°C (dec)	**Formaldehyde** MW 30.03	95% **Formic acid** MW 46.03 mp 8.5°C	**Mesitylene** MW 120.20 bp 162–164°C

Dimesitylmethane
MW 252.41

Experiment

1. Dimesitylmethane

Microscale Procedure

To 10 mg of paraformaldehyde in a 10 × 100 mm reaction tube add, in the hood, 0.06 mL (73 mg) of 95% formic acid and a carborundum (black) boiling chip. Dissolve the paraformaldehyde in the formic acid by boiling on a hot sand bath, and then add 0.133 mL (0.115 g) of mesitylene and another carborundum boiling chip. Add the distilling column as an air condenser (Fig. 30.1), reflux the reaction mixture for 2 h, and cool the mixture to room temperature and then in ice. Remove the excess formic acid using a Pasteur pipette, then wash the crystals with water, aqueous sodium carbonate solution, and again with water, and then scrape them out onto a piece of filter paper. Squeeze the crystals between sheets of filter paper to dry.

Determine the weight of the crude product, and save a few crystals for a melting-point determination. Recrystallize the product by dissolving it in the minimum quantity of boiling ligroin. Allow the solution to cool to room temperature, add a seed crystal if necessary, and then cool the mixture for at least 15 min in ice before removing the solvent using a Pasteur pipette. Another solvent for crystallization is a mixture of 0.75 mL of toluene and 0.1 mL of methanol, which is adequate for 0.5 g of product. Obtain the IR and NMR spectra of the pure mate-

CAUTION: Formic acid is corrosive to tissue. Avoid all contact with it or its vapors. Should any come in contact with your skin, wash it off immediately with a large quantity of water.

A carborundum boiling chip is specified because most white boiling chips will react with the formic acid and stop the reaction.

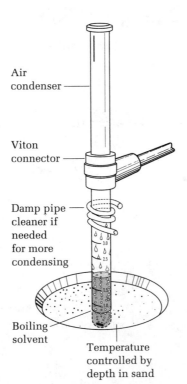

Air condenser

Viton connector

Damp pipe cleaner if needed for more condensing

Boiling solvent

Temperature controlled by depth in sand

FIG. 30.1 Apparatus for refluxing a reaction mixture with maximum cooling.

rial and the melting points of the crude and recrystallized product. Turn in the pure product along with a card giving relevant data on the substance, including the number of moles of each of the starting materials. From the spectra and any tests you may wish to run, confirm the structure of the product.

Cleaning Up At the end of this reaction there should be no formaldehyde remaining in the reaction mixture. Should it be necessary to destroy formaldehyde, it should be diluted with water and 7 mL of household bleach added to oxidize 100 mg of paraformaldehyde. After 20 min it can be flushed down the drain. The formic acid solvent from the reaction and aqueous washings should be combined, neutralized with sodium carbonate, and flushed down the drain. Mother liquor from recrystallization is placed in the organic solvents container.

 Macroscale Procedure

CAUTION: Formic acid is corrosive to tissue. Avoid all contact with it or its vapors. Should any come in contact with your skin, wash it off immediately with a large quantity of water. Carrying out this reaction in the hood protects not only against formic acid but also against formaldehyde, a suspected carcinogen formed transitorily in the reaction.

To 0.75 g of paraformaldehyde in a 50-mL round-bottomed flask add, in the hood, 4.5 mL of 95% formic acid. Fit the flask with a reflux condenser (see Fig. 16.2), and bring the mixture to a boil on the electrically heated sand bath. After about 5 min most of the paraformaldehyde will have dissolved. Add through the condenser 10.0 mL of mesitylene. Reflux the reaction mixture for 2 h, cool the mixture to room temperature, and then cool it in ice. Remove the product by vacuum filtration in the hood, taking great care not to come into contact with the residual formic acid. Wash the crystals with water, aqueous sodium carbonate solution, and again with water, and squeeze them between sheets of filter paper to dry. Determine the weight of the crude product, and save a few crystals for a melting point determination. Recrystallize the product. A suggested solvent is a mixture of 7.5 mL of toluene and 1 mL of methanol, which is adequate for 5 g of product. Obtain the melting points of the crude and recrystallized product. Calculate the percent yield, and turn in the pure product to your instructor. IR and NMR specta of the product are found in Figs. 30.2, 30.3, and 30.4.

Cleaning Up At the end of this reaction there should be no formaldehyde remaining in the reaction mixture. Should it be necessary to destroy formaldehyde it should be diluted with water and 7 mL of household bleach added to oxidize 100 mg of paraformaldehyde. After 20 min it can be flushed down the drain. The formic acid solvent from the reaction and aqueous washings should be combined, neutralized with sodium carbonate, and flushed down the drain. Mother liquor from recrystallization is placed in the organic solvents container.

Questions

1. The intermediate benzylic carbocation is stabilized by resonance. Draw the contributing resonance structures.

2. Predict, with peak intensities and approximate chemical shifts, the appearance of the ^1H NMR spectrum of dimesitylmethane. The two different types of methyls differ from each other by 0.15 ppm.

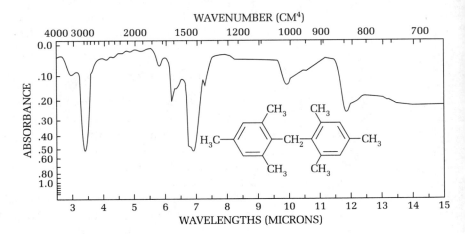

FIG. 30.2 Infrared spectrum of dimesitylmethane.

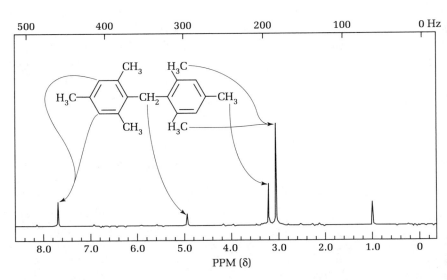

FIG. 30.3 ^1H NMR spectrum of dimesitylmethane (90 MHz).

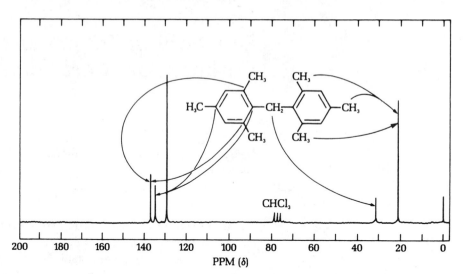

FIG. 30.4   ^{13}C NMR spectrum of dimesitylmethane (22.6 MHz).

Reactions of Triphenylmethyl Carbocation, Carbanion, and Radical

Prelab Exercise: Write all of the resonance structures of the triphenylmethyl carbocation.

Stable carbocation, carbanion, and radical

Triphenylmethanol, prepared in a later experiment (Chapter 38), has played an interesting part in the history of organic chemistry. It was converted to the first stable carbocation and the first stable free radical. In this experiment triphenylmethanol is easily converted to the triphenylmethyl (trityl) carbocation, carbanion, and radical. Each of these is stabilized by ten contributing resonance forms and consequently is unusually stable. Because of their long conjugated systems, these forms absorb radiation in the visible region of the spectrum and thus can be detected visually.

Experiments

The Triphenylmethyl Carbocation, the Trityl Carbocation

The reactions of triphenylmethanol are dominated by the ease with which it dissociates to form the relatively stable triphenylmethanol carbocation. When colorless triphenylmethanol is dissolved in concentrated sulfuric acid, an orange-yellow solution results that gives a fourfold depression of the melting point of sulfuric acid, meaning that 4 mols of ions are produced. If the triphenylmethanol were simply protonated, only 2 mols of ions would result.

$$(C_6H_5)_3COH + 2 H_2SO_4 \rightleftharpoons (C_6H_5)_3C^+ + H_3O^+ + 2 HSO_4^-$$

$$(C_6H_5)_3COH + H_2SO_4 \not\rightleftharpoons (C_6H_5)_3COH_2^+ + HSO_4^-$$

The central carbon atom in the carbocation is sp^2 hybridized, and thus the three carbons attached to it are coplanar and disposed at angles of 120°:

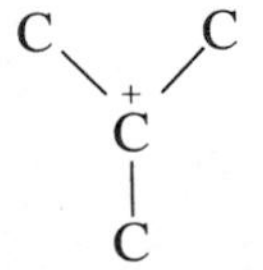

However, the three phenyl groups, because of steric hindrance, cannot lie on one plane. Therefore, the carbocation is propeller-shaped:

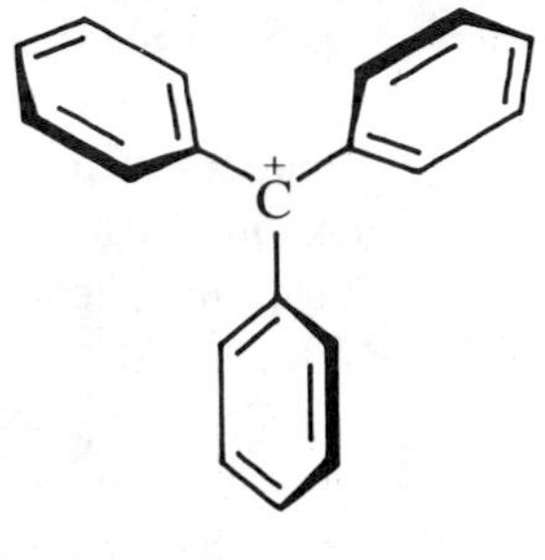

Triphenylmethanol is a tertiary alcohol and undergoes, as expected, S_N1 reactions. The intermediate cation, however, is stable enough to be seen in sulfuric acid solution as a red-brown to yellow solution. Upon dissolution in concentrated sulfuric acid, the hydroxyl is protonated; then the OH_2^+ portion is lost as H_2O (which is itself protonated) leaving the carbocation. The bisulfate ion is a very weak nucleophile and does not compete with methanol in the formation of the product, trityl methyl ether.

1. Trityl Methyl Ether

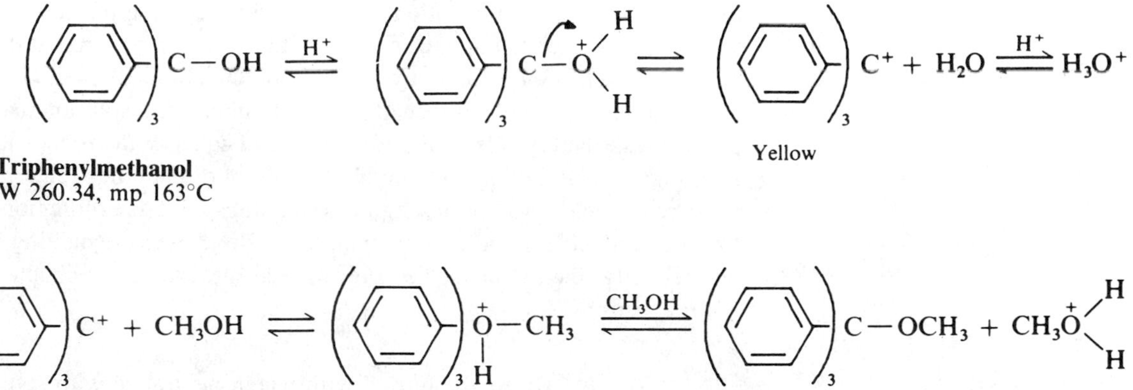

Triphenylmethanol
MW 260.34, mp 163°C

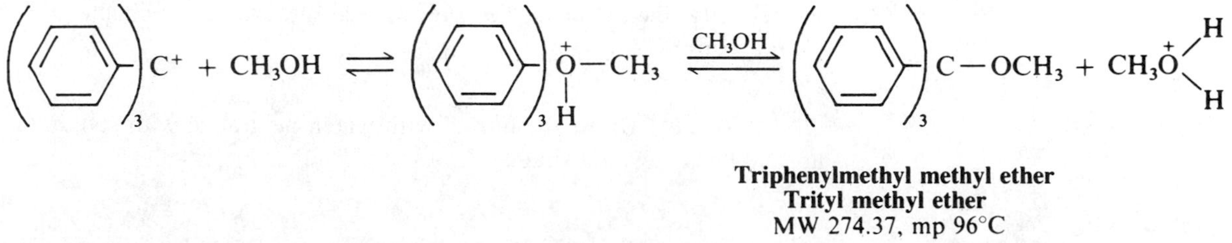

Yellow

Triphenylmethyl methyl ether
Trityl methyl ether
MW 274.37, mp 96°C

Microscale Procedure

In a 10 × 100 mm reaction tube, place 50 mg of triphenylmethanol, grind the crystals to a fine powder with a glass stirring rod, and add about 0.4 to 0.5 mL of concentrated sulfuric acid (use the tube calibration to judge the amount). Continue to stir and grind the solid to dissolve all the alcohol. Note and explain the color. Using a Pasteur pipette, transfer the sulfuric acid *solution* to 3 mL of ice-cold methanol in another reaction tube. Use some of the cold methanol to rinse out the first tube. Induce crystallization, if necessary, by cooling the solution, adding a seed crystal, or scratching the tube with a glass stirring rod at the liquid-air interface. Collect the product by filtration on the Hirsch funnel. Do not use filter paper on top of the polyethylene frit. This strongly acid solution may attack the paper. Wash the crystals well with water, and squeeze them between sheets of filter paper to aid drying. Determine the weight of the crude material, calculate the percentage yield, and save a sample for melting-point determination. Recrystallize the product from boiling methanol, and determine the melting point of the purified trityl methyl ether.

Use the Pasteur pipette technique to isolate the crystals. Dry them under aspirator vacuum in the reaction tube.

Cleaning Up   Dilute the filtrate with water, neutralize with sodium carbonate, and flush down the drain.

Macroscale Procedure

In a 10-mL Erlenmeyer flask, place 0.5 g of triphenylmethanol, grind the crystals to a fine powder with a glass stirring rod, and add 5 mL of concentrated sulfuric acid. Continue to stir the mixture to dissolve all of the alcohol. Using a Pasteur pipette, transfer the sulfuric acid solution to 30 mL of ice-cold methanol in a 50-mL Erlenmeyer flask. Use some of the cold methanol to rinse out the first tube. Induce crystallization if necessary by cooling the solution, adding a seed crystal, or scratching the solution with a glass stirring rod at the liquid–air interface. Collect the product by filtration on the Hirsch funnel. Don't use filter paper on top of the polyethylene frit. This strongly acid solution may attack the paper. Wash the crystals well with water, and squeeze them between sheets of filter paper to aid drying. Determine the weight of the crude material, calculate the percent yield, and save a sample for melting-point determination. Recrystallize the product from boiling methanol, and determine the melting point of the purified trityl methyl ether. See Fig. 31.2 at the end of the chapter for the ^{1}H NMR spectrum.

Cleaning Up   Dilute the filtrate with water, neutralize with sodium carbonate, and flush down the drain.

2. Triphenylmethyl Bromide, Trityl Bromide

In this experiment the triphenylmethanol is dissolved in a good ionizing solvent, acetic acid, and allowed to react with a strong acid and nucleophile, hydrobromic acid. The intermediate carbocation reacts immediately with bromide ion. Acetate ion, to the extent it is present, is not a good nucleophile.

Reactions that proceed through the carbocation

Experiments 2 to 9 can be run on five times the quantities specified here.

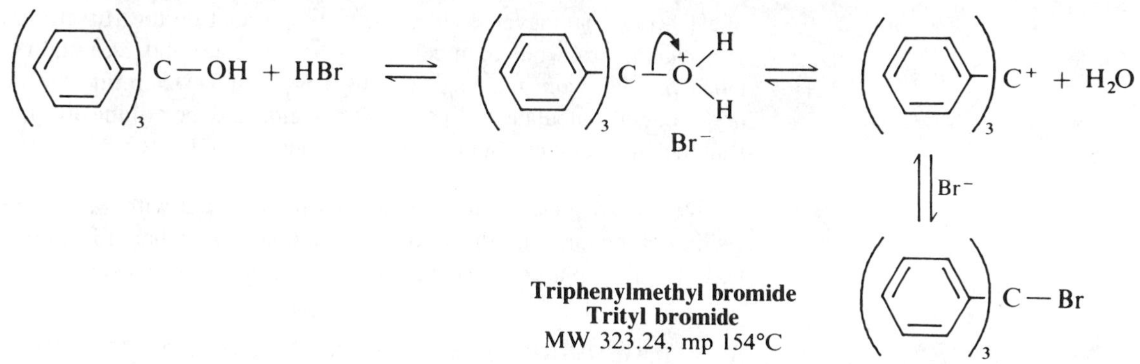

Triphenylmethyl bromide
Trityl bromide
MW 323.24, mp 154°C

**MICROSCALE
AND MACROSCALE**

Dissolve 100 mg of triphenylmethanol in 2 mL of warm acetic acid on a steam or hot water bath, add 0.2 mL of 47% hydrobromic acid, and heat the mixture for 5 min on the steam bath or in a beaker of boiling water. Cool it in ice, collect the product on the Hirsch funnel, wash the product with water and ligroin (hexane), allow it to dry, determine the weight, and calculate the percentage yield. Recrystallize the product from ligroin (hexane), and compare the melting points of the recrystallized and crude materials. The compound crystallizes slowly; allow adequate time for crystals to form. Use the Beilstein test (Chapter 69) to test for halogen in both the product and starting materials.

Measure the acid in a 1-mL graduated pipette fitted with a pipette pump.

Cleaning Up   Dilute the filtrate from the reaction with water, neutralize with sodium carbonate, and flush down the drain. Ligroin mother liquor from the crystallization goes in the organic solvents container.

For Further Investigation

MICROSCALE

3. Triphenylmethyl Iodide, Trityl Iodide?

In a reaction very similar to the preparation of the bromide, the iodide might be prepared. Bisulfite is added to react with any iodine formed.

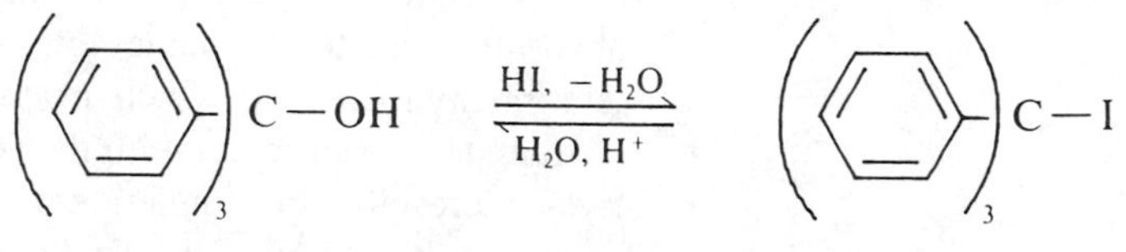

$$\left(\!\left(\bigcirc\!\right)\!\right)_3 C\!-\!OH \quad \xrightarrow[\text{H}_2\text{O, H}^+]{\text{HI, }-\text{H}_2\text{O}} \quad \left(\!\left(\bigcirc\!\right)\!\right)_3 C\!-\!I$$

Triphenylmethyl iodide
Trityl iodide
MW 370.22, mp 183°C

Dissolve 100 mg of triphenylmethanol in 2 mL of warm acetic acid, add 0.2 mL of 47% hydriodic acid, heat the mixture for 1 h on the steam bath, cool it, and add it to a solution of 0.1 g of sodium bisulfite dissolved in 2 mL water in a 10-mL Erlenmeyer flask. Collect the product on the Hirsch funnel, wash it with water, press out as much water as possible, and recrystallize the crude, moist product from methanol (about 3 or 4 mL). Determine the weight of the dry product, calculate the percentage yield, and determine the melting point. Run the Beilstein test. What has been produced and why?

Cleaning Up   Dilute the filtrate from the reaction with water, neutralize with sodium carbonate, and flush down the drain. The methanol from the crystallization should be placed in the halogenated organic solvents container.

MICROSCALE AND MACROSCALE   4. Triphenylmethyl fluoborate and the Tropilium Ion

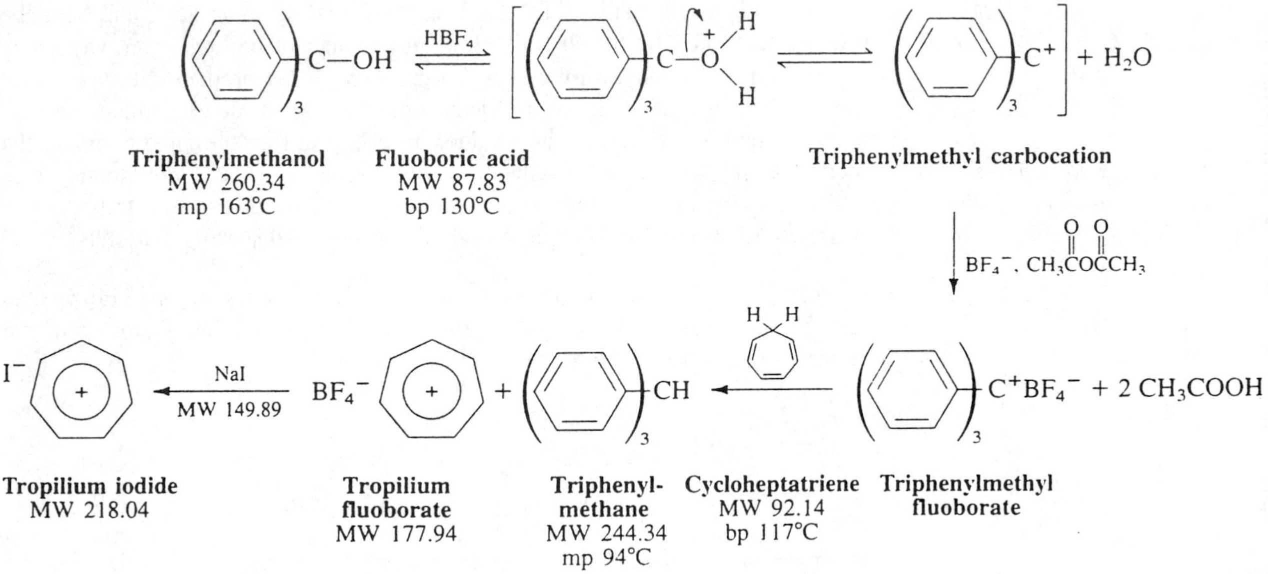

Triphenylmethanol
MW 260.34
mp 163°C

Fluoboric acid
MW 87.83
bp 130°C

Triphenylmethyl carbocation

Tropilium iodide
MW 218.04

Tropilium fluoborate
MW 177.94

Triphenyl-methane
MW 244.34
mp 94°C

Cycloheptatriene
MW 92.14
bp 117°C

Triphenylmethyl fluoborate

Tropilium ion—an aromatic ion

Reaction of triphenylmethanol with fluoboric acid in the presence of acetic anhydride generates the stable salt, trityl fluoborate. The fluoboric acid protonates the triphenylmethanol, which loses the elements of water in an equilibrium reaction. The water reacts with the acetic anhydride to form acetic acid and thus drives the reaction to completion.

The salt so formed is the fluoborate of the triphenyl carbocation; it is a powerful base. It can be isolated but in this case will be used *in situ* to react with the hydrocarbon cycloheptatriene.

This hydrocarbon is more basic than most hydrocarbons and will lose a proton to the triphenylmethyl carbocation to give triphenylmethane and the cycloheptatrienide carbocation, the tropilium ion. To demonstrate that hydride ion transfer has taken place, isolate triphenylmethane.

The Hückel Rule: 4n + 2 π electrons

The tropilium ion has a planar structure, each carbon bears a single proton, and the ion contains 6 π electrons—it is aromatic, a characteristic that can be confirmed by NMR spectroscopy.

The fluoborate group can be displaced by the iodide ion to prepare tropilium iodide. You can see that the iodide is ionic by watching the reaction of the aqueous solution of this ion with silver ion.

Procedure

In a 10 × 100 mm reaction tube place 1.75 mL of acetic anhydride, cool the tube, and add 88 mg of fluoboric acid. Add 195 mg of triphenylmethanol with thorough stirring. Warm the mixture to give a homogeneous dark solution of the triphenylmethyl fluoborate, and then add 78 mg cycloheptatriene. The color of the trityl cation should fade during this reaction, and the tropilium fluoborate should begin to precipitate. Add 2 mL of anhydrous *tert*-butyl methyl ether to the reaction tube, stir the contents well while cooling on ice, and collect the product by filtration on the Hirsch funnel. Wash the product with 2 mL of dry ether, and then dry the product between sheets of filter paper.

CAUTION: Handle fluoboric acid and acetic anhydride with great care. These reagents are toxic and corrosive. Avoid breathing the vapors or any contact with the skin. In case of skin contact, rinse the affected part under running water for at least 10 min. Carry out this experiment in the hood.

To the filtrate add 3 M sodium hydroxide solution, and shake the flask to allow all the acetic anhydride and fluoboric acid to react with the base. Test the aqueous layer with indicator paper to ascertain that neutralization is complete, and then transfer the mixture to a reaction tube and draw off the aqueous layer. Wash the ether once with 2 mL of water, and dry the ether over calcium chloride pellets, adding the drying agent until it no longer clumps together.

Transfer the ether to a tared reaction tube, and evaporate the solvent to leave crude triphenylmethane. Remove a sample for melting-point determination, and recrystallize the residue from an appropriate solvent, determined by experimentation. Prove to yourself that the compound isolated is indeed triphenylmethane. Obtain an IR spectrum and an NMR spectrum.

Structure proof: NMR

To determine the NMR spectrum of the tropilium fluoborate collected on the Hirsch funnel, dissolve about 50 mg of the product in 0.3 mL of deuterated dimethyl sulfoxide that contains 1% tetramethylsilane as a reference. Compare the NMR spectrum obtained with that of the starting material, cycloheptatriene. The spectrum of the latter can be obtained in deuterochloroform, again using tetramethylsilane as the reference compound.

The tropilium fluoborate can be converted to tropilium iodide by dissolving the fluoborate in the minimum amount of hot, but not boiling, water (a few drops) and adding to this solution 0.25 mL of a saturated solution of sodium iodide. Cool the mixture in ice, remove the solvent with a pipette, and wash the crystals with 0.5 mL of ice-cold methanol. Scrape most of the crystals onto a piece of filter paper, and allow them to dry before determining the weight. To the crystalline residue in the reaction tube add a few drops of water, warm the tube if necessary to dissolve the tropilium iodide, add a drop of 2% aqueous silver nitrate solution, and note the result.

Cleaning Up   Solutions that contain the fluoborate ion (the neutralized filtrate, the NMR sample, the filtrate from the iodide preparation) should be treated with aqueous $CaCl_2$ to precipitate insoluble CaF_2, which is removed by filtration and placed in the nonhazardous solid waste container. The filtrate can be flushed down the drain. Allow the ether to evaporate from the calcium chloride, and then place it in the nonhazardous solid waste container. Recrystallization solvent goes in the organic solvents container.

The Trityl Carbanion

The triphenylmethyl carbanion, the trityl anion, can be generated by the reaction of triphenylmethane with the very powerful base, *n*-butyllithium. The reaction generates the blood-red lithium triphenylmethide and butane. The triphenylmethyl anion reacts much as a Grignard reagent does. In the present experiment it reacts with carbon dioxide to give triphenylacetic acid after acidification. Avoid an excess of *n*-butyllithium; on reaction with carbon dioxide, it gives the vile-smelling pentanoic acid.

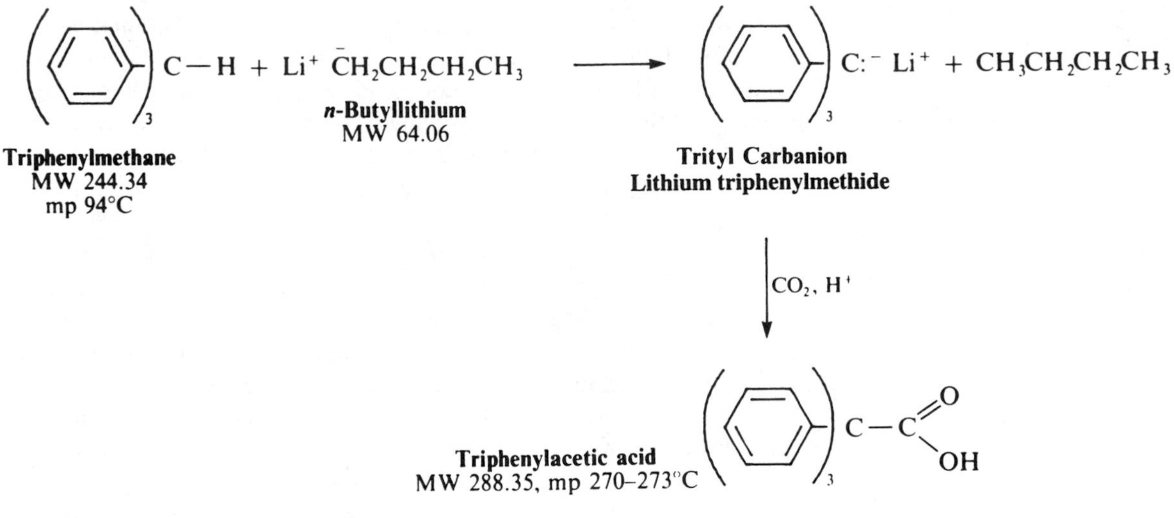

Triphenylmethane
MW 244.34
mp 94°C

n-**Butyllithium**
MW 64.06

Trityl Carbanion
Lithium triphenylmethide

CO_2, H^+

Triphenylacetic acid
MW 288.35, mp 270–273°C

5. Synthesis of the Trityl Carbanion and Triphenylacetic Acid

MICROSCALE AND MACROSCALE

To a reaction tube (Fig. 31.1) that has been dried for at least 30 min in 110°C oven and capped with a rubber septum, add 100 mg of triphenylmethane and 3 mL of anhydrous diethyl or *tert*-butyl methyl ether. Add an empty syringe needle to the septum; flush out the air in the tube by passing a slow current of nitrogen into the tube for about 30 s. Remove the nitrogen inlet needle, and, with thorough mixing of the solution, add to the reaction mixture 0.4 mL of a 1.0 M solution of *n*-butyllithium in hexane (a commercial product). Use great care in handling *n*-butyllithium. It reacts avidly with air and violently with water.

Or use 0.25 mL of a 1.6 M solution of n-butyllithium.

Reaction with dry ice

The sodium hydroxide converts the acid to its water-soluble anion.

In a 30-mL beaker place one or two small pieces of dry ice (solid carbon dioxide) that have been wiped free of any adhering frost. Handle the dry ice with a towel or gloves. Contact with the skin can cause frostbite, because the sublimation temperature of dry ice is $-78.5°C$. Calculate how much carbon dioxide is needed to react with the anion prepared from 100 mg of triphenylmethane; use at least a tenfold excess. Using a Pasteur pipette transfer the solution of the anion to the dry ice; reaction is immediate. Allow the unreacted dry ice to sublime; then add 3 mL of 1.5 M aqueous sodium hydroxide solution, dissolve as much solid as possible, and transfer the liquid to a reaction tube. Shake the mixture, draw off the ether layer, and wash the aqueous layer with two 2.5-mL portions of *tert*-butyl methyl ether, which are also discarded. Using 3 M hydrochloric acid, acidify the aqueous layer to pH < 4. Collect the precipitate, wash it with water, and allow it to dry in the air. Take the melting point. When pure, triphenylacetic acid melts at 267°C. The purity and identity of the product can be assessed using thin-layer chromatography and IR spectroscopy (using a potassium bromide disk).

Cleaning Up   There should be no excess *n*-butyllithium, but a few drops of the solution in hexane can be destroyed by reaction with *t*-butyl alcohol. Place ether

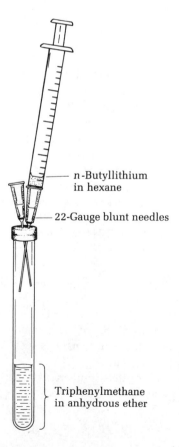

n-Butyllithium
in hexane

22-Gauge blunt needles

Triphenylmethane
in anhydrous ether

FIG. 31.1 Apparatus for the synthesis of the trityl carbanion.

in the organic solvents container. Dilute the aqueous layer with water, and then neutralize with sodium carbonate and flush down the drain.

Triphenylmethyl, a Stable Free Radical

Gomberg: the first free radical

In the early part of the nineteenth century many attempts were made to prepare methyl, ethyl, and similar radicals in a free state, as sodium had been prepared from sodium chloride. Many well-known chemists tried: Gay-Lussac's CN turned out to be cyanogen, $(CN)_2$; Bunsen's cacodyl from $(CH_3)_2AsCl$ proved to be $(CH_3)_2As—As(CH_3)_2$; and the Kolbe electrolysis gave $CH_3CH_2CH_2CH_3$ instead of CH_3CH_2. Moses Gomberg at the University of Michigan prepared the first free radical in 1900. He had prepared triphenylmethane and chlorinated it to give triphenylmethyl chloride, which he hoped to couple to form hexaphenylethane:

$$2\left(\left\langle\!\!\!\!\bigcirc\!\!\!\!\right\rangle\right)_3 C—Cl + 2Ag \;\not\!\longrightarrow\; \left(\left\langle\!\!\!\!\bigcirc\!\!\!\!\right\rangle\right)_3 C—C\left(\left\langle\!\!\!\!\bigcirc\!\!\!\!\right\rangle\right)_3 + 2AgCl$$

Triphenylmethyl chloride **Hexaphenylethane**

The solid that he obtained instead was a high-melting, sparingly soluble, white solid, which turned out on analysis to be not a hydrocarbon but instead an oxygen-containing compound $(C_{38}H_{30}O_2)$. Repeating the experiment in the absence of air, he obtained a yellow solution that, on evaporation in the absence of air, deposited crystals of a colorless hydrocarbon that was remarkably reactive. It readily reacted with oxygen, bromine, chlorine, and iodine. On dissolution the white hydrocarbon gave the yellow solution in the absence of air; the hydrocarbon was deposited when he evaporated the solution once more. Gomberg interpreted these results as follows: "The experimental evidence . . . forces me to the conclusion that we have to deal here with a free radical, triphenylmethyl, $(C_6H_5)_3C\cdot$. The action of zinc results, as it seems to me, in a mere abstraction of the halogen:

$$2(C_6H_5)_3C—Cl + Zn \longrightarrow 2(C_6H_5)_3C\cdot + ZnCl_2$$

Now as a result of the removal of the halogen atom from triphenylchloromethane, the fourth valence of the methane is bound either to take up the complicated group $(C_6H_5)_3C\cdot$ or remain as such, with carbon as trivalent. Apparently the latter is what happens."

For a long time after Gomberg first carried out this reaction in 1900, it was assumed the radical dimerized to form hexaphenylethane:

$$2(C_6H_5)_3C\cdot \longrightarrow (C_6H_5)_3C—C(C_6H_5)_3$$

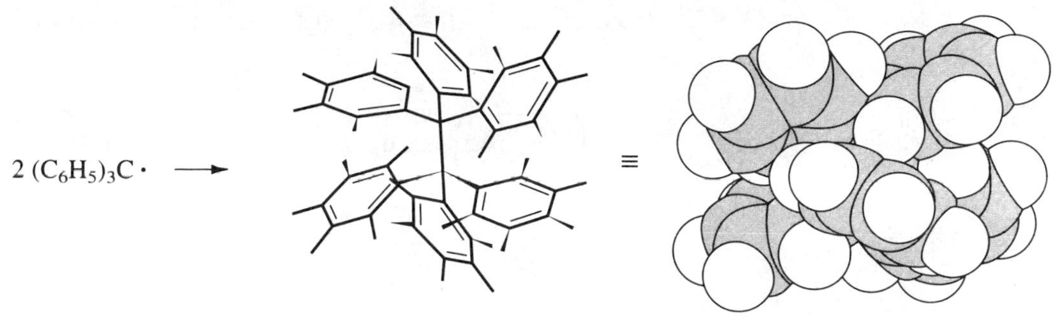

But in 1968 NMR and UV evidence showed that the radical is in equilibrium with a different substance:

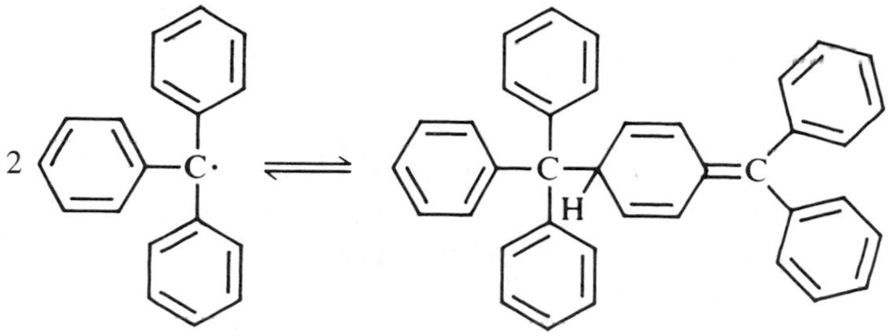

Hexaphenylethane has not yet been synthesized, presumably because of steric hindrance.

In the following experiment the trityl radical is prepared in much the same fashion as Gomberg used, by the reaction of trityl bromide with zinc in the absence of oxygen. The yellow solution is then deliberately exposed to air to give the peroxide.

6. Computational Chemistry

With a molecular mechanics program, construct a model of hexaphenylethane, and then carry out an energy minimization of the molecule. Note the steric energy. Construct a model of the compound (above) actually formed when the trityl radical dimerizes. Note the steric energy. What is the length of the central C—C bond in your model of hexaphenylethane? What is the C—C bond length in 1,2-diphenylethane? These numbers are probably not "correct" because valid parameters do not exist for severely sterically hindered molecules of this type, but the space-filling models are worth studying. Do these calculations give any insight into why hexaphenylethane has not yet been synthesized?

7. Synthesis of Triphenylmethyl Peroxide

$$\left(\left\langle\bigcirc\right\rangle\right)_3 C-Br + Zn \longrightarrow \left(\left\langle\bigcirc\right\rangle\right)_3 C\cdot + ZnBr_2$$

Trityl radical

$$\Big\downarrow O_2$$

$$\left(\left\langle\bigcirc\right\rangle\right)_3 C-O-O-C\left(\left\langle\bigcirc\right\rangle\right)_3$$

Triphenylmethyl peroxide
MW 518.67, mp 186°C

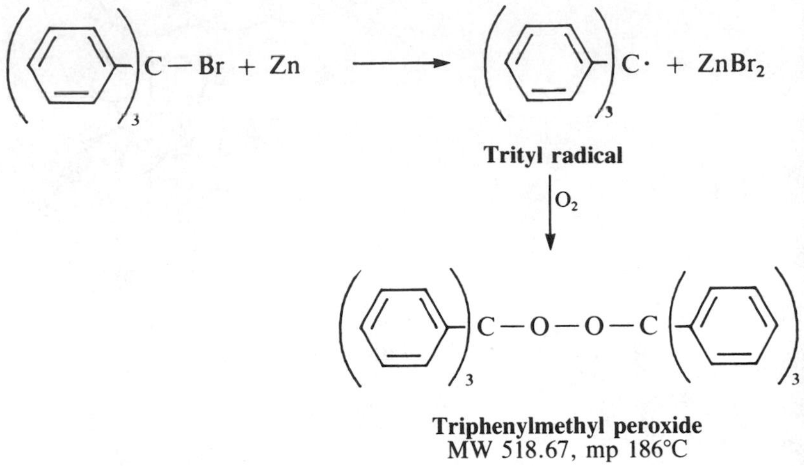

The trityl radical

In a small test tube dissolve 100 mg of trityl bromide or chloride in 0.5 mL of toluene. Material prepared in Experiment 2 may be used; it need not be recrystallized. Cap the tube with a septum, insert an empty needle, and flush the tube with nitrogen while shaking the contents. Add 0.2 g of fresh zinc dust as quickly as possible, then flush the tube with nitrogen once more. Shake the tube vigorously for about 10 min, and note the appearance of the reaction mixture. Using a Pasteur pipette transfer the solution, but not the zinc, to a 30-mL beaker. Roll the solution around the inside of the beaker to give it maximum exposure to the air, and after a few minutes collect the peroxide on the Hirsch funnel. Wash the solid with a little cold toluene, allow it to dry in the air, and determine the melting point. It is reported to melt at 186°C.

Cleaning Up   Transfer the mixture of zinc and zinc bromide to the hazardous waste container.

For Further Investigation

Carry out both of the following reactions. Compare the products formed. Run necessary solubility and simple qualitative tests. Interpret your results, and propose structures for the compounds produced in each of the reactions.

MICROSCALE AND MACROSCALE

8. Synthesis of Compound 1

HOOCCH$_2$COOH
Malonic acid

In a reaction tube, place 100 mg of triphenylmethanol and 200 mg of malonic acid. Grind the two solids together with a glass stirring rod, and then heat the reaction tube at 150°C for 7 min. Use an aspirator tube mounted at the mouth of the tube to carry away undesirable fumes. Allow the tube to cool, and then dissolve the contents in about 0.2 mL toluene. Dilute the solution with 1 mL of 60 to 80°C ligroin, and after cooling the tube in ice, isolate the crystals, wash them with a little cold ligroin, and once they are dry determine their weight and melting point.

Cleaning Up   Place the toluene/ligroin mother liquor in the organic solvents container.

MICROSCALE AND MACROSCALE

9. Synthesis of Compound 2

CICH$_2$COOH
Chloroacetic acid

In a reaction tube dissolve 100 mg of triphenylmethanol in 2 mL of acetic acid, and then add to this solution 0.4 mL of an acetic acid solution containing 5% of chloroacetic acid and 1% of sulfuric acid. Heat the mixture for 5 min, add about 1 mL of water to produce a hot saturated solution, and let the tube cool slowly to room temperature. Collect the product by filtration, and wash the crystals with 1:1 methanol-water. Once they are dry, determine the weight and melting point, and then perform tests on compounds 1 and 2 to deduce their identities. Use 1:10 *t*-butyl methyl ether/hexane for TLC.

Cleaning Up   Dilute the filtrate with water, neutralize with sodium carbonate, and flush down the drain.

Questions ⎯⎯⎯⎯⎯⎯⎯⎯⎯⎯⎯⎯⎯⎯⎯⎯⎯⎯⎯⎯⎯

1. Give the mechanism for the free radical chlorination of triphenylmethane.

2. Is the propeller-shaped triphenylmethyl carbocation a chiral species?

3. Without carrying out the experiments, speculate on the structures of Compounds 1 and 2 made in Experiments 6 and 7.

4. Is the production of triphenylmethyl peroxide a chain reaction?

5. What product would you expect from the reaction of carbon tetrachloride, benzene, and aluminum chloride?

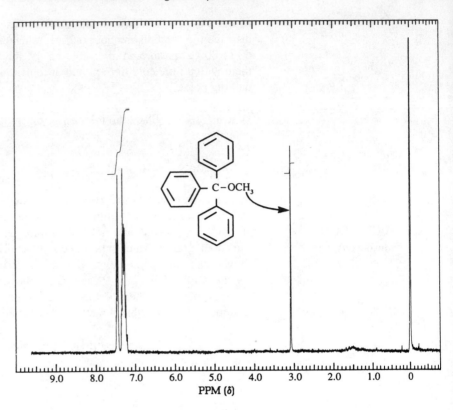

FIG. 31.2 ¹H NMR spectrum of triphenylmethyl methyl ether (trityl methyl ether) (250 MHz).

The Friedel-Crafts Reaction: Anthraquinone and Anthracene

Prelab Exercise: Draw the mechanism for the cyclization of 2-benzoylbenzoic acid to anthraquinone using concentrated sulfuric acid.

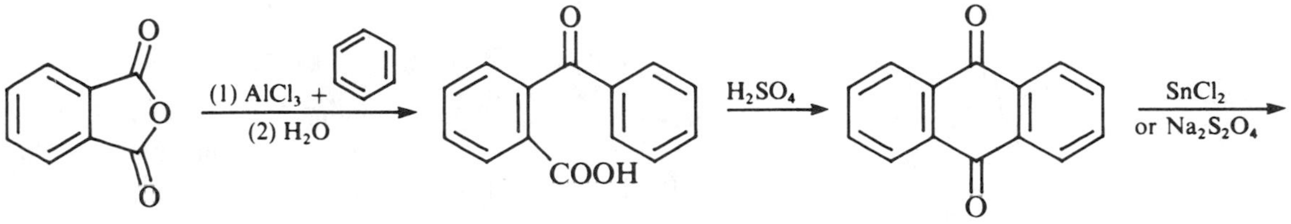

Phthalic anhydride
MW 148.11, mp 132°C

2-Benzoylbenzoic acid
MW 226.22, mp 127°C

Anthraquinone
MW 208.20, mp 286°C

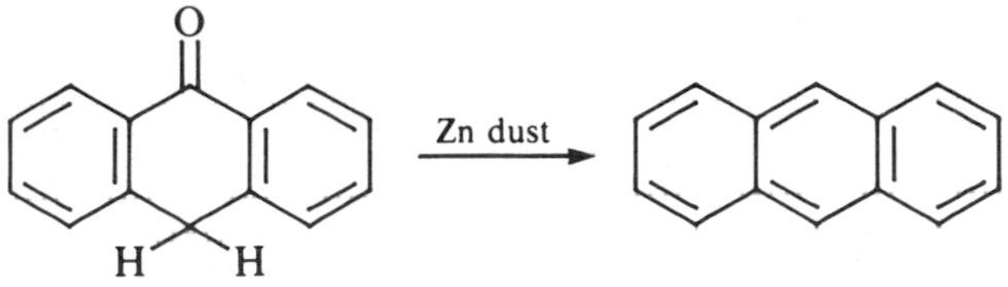

Anthrone
MW 194.22, mp 156°C

Anthracene
MW 178.22, mp 216°C

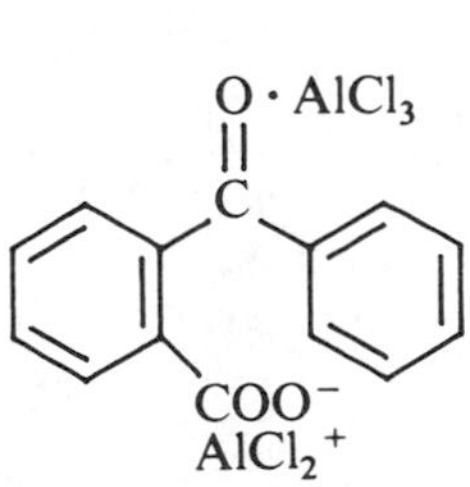

Complex salt

The Friedel-Crafts reaction of phthalic anhydride with excess benzene as solvent and two equivalents of aluminum chloride proceeds rapidly and gives a complex salt of 2-benzoylbenzoic acid in which one mole of aluminum chloride has reacted with the acid function to form the salt RCO_2^- $AlCl_2^+$ and a second mole is bound to the carbonyl group. On addition of ice and hydrochloric acid the complex is decomposed and basic aluminum salts are brought into solution (see Experiment 1).

Treatment of 2-benzoylbenzoic acid with concentrated sulfuric acid effects cyclodehydration to anthraquinone, a pale-yellow, high-melting compound of great stability. Because anthraquinone can be sulfonated only under forcing conditions, a high temperature can be used to shorten the reaction time without loss in yield of product; the conditions are so adjusted that anthraquinone separates from the hot solution in crystalline form favoring rapid drying (see Experiment 2).

Reduction of anthraquinone to anthrone (see Experiment 3) can be accomplished rapidly on a small scale with tin(II) chloride in acetic acid solution. A

$Na_2S_2O_4$
Sodium hydrosulfite

second method, which involves refluxing anthraquinone with an aqueous solution of sodium hydroxide and sodium hydrosulfite, is interesting to observe because of the sequence of color changes: Anthraquinone is reduced first to a deep red solution containing anthrahydroquinone diradical dianion; the red color then gives way to a yellow color characteristic of anthranol radical anion; as the alkali is neutralized by the conversion of $Na_2S_2O_4$ to 2 $NaHSO_3$, anthranol ketonizes to the more stable anthrone. The second method is preferred in industry, because sodium hydrosulfite costs less than half as much as tin(II) chloride and because water is cheaper than acetic acid and no solvent recovery problem is involved.

Reduction of anthrone to anthracene (see Experiment 5) is accomplished by refluxing in aqueous sodium hydroxide solution with activated zinc dust. The method has the merit of affording pure, beautifully fluorescent anthracene.

Fieser's Solution, an Oxygen Scavenger

A solution of 2 g of sodium anthraquinone-2-sulfonate and 15 g of sodium hydrosulfite in 100 mL of a 20% aqueous solution of potassium hydroxide affords a blood-red solution of the diradical dianion:

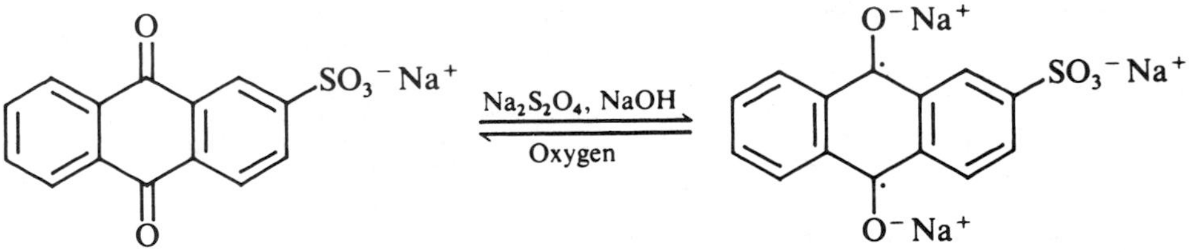

This solution has a remarkable affinity for oxygen. It is used to remove traces of oxygen from gases such as nitrogen or argon when it is desirable to render them absolutely oxygen-free. This solution has a capacity of about 800 mL of oxygen. The color fades, and the solution turns brown when it is exhausted.

Experiments

MICROSCALE

1. 2-Benzoylbenzoic Acid

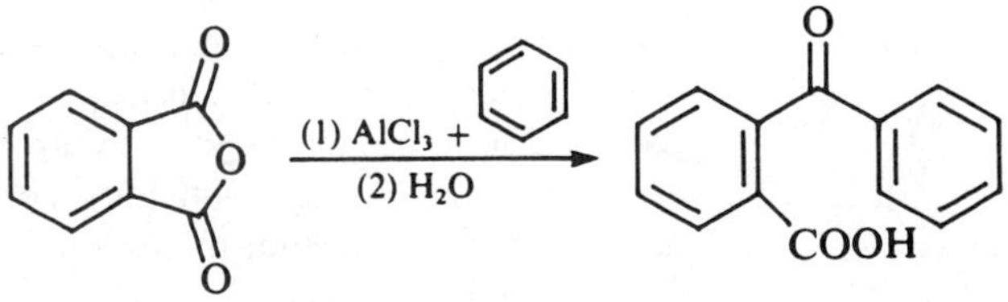

Phthalic anhydride
MW 148.11, mp 132°C

2-Benzoylbenzoic acid
MW 226.22, mp 127°C

FIG. 32.1 Hydrogen chloride trap for Friedel-Crafts reaction. The tubing can be bent permanently by heating it in a steam bath.

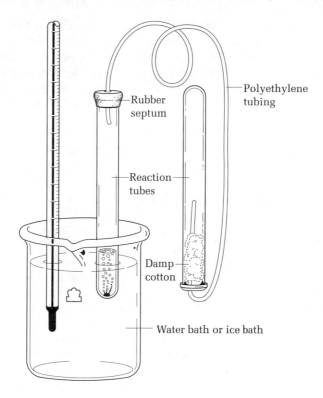

Rubber septum

Polyethylene tubing

Reaction tubes

Damp cotton

Water bath or ice bath

CAUTION: Benzene is a carcinogen. Carry out this experiment in a hood. Avoid contact of the benzene with the skin or wear gloves. Do not carry out this reaction if adequate hood facilities are unavailable.

Into a dry 10 × 100 mm reaction tube place 150 mg of phthalic anhydride and 0.75 mL of dry benzene. Cool the mixture in an ice bath, and then add 300 mg of anhydrous aluminum chloride.[1] Cap the tube with a septum connected to polyethylene tubing leading to another reaction tube, which contains a piece of damp cotton to act as a trap for the hydrogen chloride liberated in the reaction (Fig. 32.1). See Fig. 32.2 for threading a polyethylene tube through a septum. Mix the contents of the tube thoroughly by flicking the tube. Warm the tube by the heat of your hand. If the reaction does not start, warm the tube *very gently* in a beaker of hot water, or hold it over the sand bath or steam bath for a few seconds. At the first sign of vigorous boiling or evolution of hydrogen chloride, hold the tube over the ice bath in readiness to cool it if the reaction becomes too vigorous. Continue this gentle, cautious heating until the reaction proceeds smoothly enough to reflux it on the hot sand bath. This will take about 5 min.

Heat the reaction mixture on the sand bath until evolution of hydrogen chloride almost ceases, then cool it in ice, and add 1 g of ice in small pieces. Allow each little piece to react before adding the next. Mix the reaction mixture

---

1. Aluminum chloride should be weighed in a small, dry, stoppered vial or reaction tube. The aluminum chloride should be from a freshly opened bottle and should be weighed and transferred in the hood. It is very hygroscopic and releases hydrogen chloride upon reaction with water. Weigh the reagent rapidly to avoid exposure of the compound to air. The quality of the aluminum chloride determines the success of this experiment.

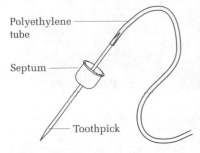

FIG. 32.2 To thread a polyethylene tube through a septum, make a hole through the septum with a needle, and then push a toothpick through the hole. Push the polyethylene tube firmly onto the toothpick, and then pull and push on the toothpick. The tube will slide through the septum. Finally pull the tube from the toothpick.

Celite may speed up the filtration.

CAUTION: Benzene is a carcinogen. Carry out this experiment in a hood. Avoid contact of the benzene with the skin or wear gloves. Do not carry out this reaction if adequate hood facilities are unavailable.

well during this hydrolysis using a glass rod. After the reaction subsides, add 0.3 mL of concentrated hydrochloric acid and 0.5 mL of water. Mix the contents of the tube thoroughly, and make sure the mixture is at room temperature. Add 0.5 mL more water, mix the solution well, ascertain that it is at room temperature, and then add 1.5 mL of *tert*-butyl methyl ether and break up any lumps in the tube with a stirring rod. Stopper the tube and shake it vigorously to complete the hydrolysis and extraction. Allow the layers to separate, and then remove the aqueous layer with a Pasteur pipette. Add 0.1 mL of concentrated hydrochloric acid and 0.25 mL of water. Shake the mixture vigorously again, and remove the aqueous layers. Transfer the organic layers to a small test tube containing calcium chloride pellets. Allow the solution to dry for 5 min, and then transfer it back to the clean dry reaction tube, rinsing the drying agent with a small quantity of ether. Add a boiling chip, and boil off the solvents until the volume in the tube is 0.5 mL; then add ligroin until the solution is slightly turbid, indicating that the product is beginning to crystallize. Allow the product to crystallize at room temperature and then at 0°C. Collect the crystals on the Hirsch funnel or on the Wilfilter by centrifugation. This product may be the monohydrate. Heat it at 100°C to drive off any water. Pure 2-benzoylbenzoic acid melts at 127 to 128°C; the yield should be about 0.2 g.

Cleaning Up   Combine all aqueous layers, neutralize with sodium carbonate, remove the aluminum hydroxide by filtration, and flush the filtrate down the drain. The aluminum hydroxide and calcium chloride pellets, after being freed of organic solvent, can be placed in the nonhazardous solid waste container. Benzene, ether, and ligroin go in the organic solvents waste container.

2. 2-Benzoylbenzoic Acid

This Friedel-Crafts reaction is conducted in a 100-mL round-bottomed flask equipped with a short condenser. A trap for collecting liberated hydrogen chloride is connected to the top of the condenser by rubber tubing of sufficient length to make it possible either to heat the flask on the steam bath or to plunge it into an ice bath. The trap is a suction flask half-filled with water and fitted with a delivery tube inserted to within 1 cm of the surface of the water (see Fig. 18.3).

Three grams of phthalic anhydride and 15 mL of reagent grade benzene (see margin note) are placed in the flask, and this solution is cooled in an ice bath until the benzene begins to crystallize. Ice cooling serves to moderate the vigorous reaction, which otherwise might be difficult to control. Six grams of anhydrous aluminum chloride[2] are added, the condenser and trap are connected, and the flask is shaken well and warmed for a few minutes by the heat of your hand. If the reaction does not start, the flask is warmed *very gently* by holding it for a few seconds over the steam bath. At the first sign of vigorous boiling, or

---

2. This is best weighed in a stoppered test tube. The chloride should be from a freshly opened bottle and should be weighed and transferred in the hood. It is very hygroscopic; work rapidly to avoid exposure of the compound to air. It releases hydrogen chloride on exposure to water. The quality of the aluminum chloride determines the success of this experiment.

evolution of hydrogen chloride, the flask is held over the ice bath in readiness to cool it if the reaction becomes too vigorous. This gentle, cautious heating is continued until the reaction is proceeding smoothly enough to be refluxed on the steam bath. This point is reached in about 5 min.

Short reaction period, high yield

Continue the heating on the steam bath, swirl the mixture, and watch it carefully for sudden separation of the addition compound, since the heat of crystallization is such that it may be necessary to plunge the flask into the ice bath to moderate the process. Once the addition compound has separated as a thick paste, heat the mixture for 10 min more on the steam bath, remove the condenser, and swirl the flask in an ice bath until cold. (Should no complex separate, heat for 10 min more, and then proceed as directed.) Take the flask and ice bath to the hood, weigh out 20 g of ice, add a few small pieces of ice to the mixture, swirl and cool as necessary, and wait until the ice has reacted before adding more. After the 20 g of ice have been added and the reaction of decomposition has subsided, add 4 mL of concentrated hydrochloric acid and 20 mL of water, swirl vigorously, and make sure that the mixture is at room temperature. Then add 10 mL of water, swirl vigorously, and again make sure the mixture is at room temperature. Add 10 mL of *t*-butyl methyl ether and, with a flattened stirring rod, dislodge solid from the neck and walls of the flask and break up lumps at the bottom. To further promote hydrolysis of the addition compound, extraction of the organic product, and solution of basic aluminum halides, stopper the flask with a cork and shake vigorously for several minutes.

When most of the solid has disappeared, pour the mixture through a funnel into a separatory funnel until the separatory funnel is nearly filled. Discard the lower aqueous layer. Pour the rest of the mixture into the separatory funnel, rinse the reaction flask with fresh *t*-butyl methyl ether, and again drain off the aqueous layer. To reduce the fluffy, dirty precipitate that appears at the interface, add 2 mL of concentrated hydrochloric acid and 5 mL of water, shake vigorously for 2–3 min, and drain off the aqueous layer. If some interfacial dirty emulsion still persists, decant the benzene-ether solution through the mouth of the funnel into a filter paper for gravity filtration, and use fresh ether to rinse the funnel. Clean the funnel, and pour in the filtered benzene–ether solution. Shake the solution with a portion of dilute hydrochloric acid, and then isolate the reaction product by either of the following procedures:

Alternative procedures

Discard benzene-ether filtrates in the container provided.

1. Add 10 mL of 3 *M* sodium hydroxide solution, shake thoroughly, and separate the aqueous layer.[3] Extract with a further 5-mL portion of aqueous alkali, and combine the extracts. Wash with 2 mL of water, and add this aqueous solution to the 15 mL of aqueous extract already collected. Discard the benzene–ether solution. Acidify the 17 mL of combined alkaline extract with concentrated hydrochloric acid to pH 1–2 and, if the *o*-benzoylbenzoic acid separates as an oil, cool in ice and rub the walls of the flask with a stirring rod to induce crystallization of the hydrate; collect the product and wash it well with water. This material is the monohydrate

---

3. The nature of a yellow pigment that appears in the first alkaline extract is unknown; the impurity is apparently transient, for the final product dissolves in alkali to give a colorless solution.

Drying time: about 1 h

$C_6H_5COC_6H_4CO_2H \cdot H_2O$. To convert it into anhydrous *o*-benzoylbenzoic acid, put it in a tared, 50-mL round-bottomed flask, evacuate the flask at the full force of the aspirator, and heat it in the open rings of a steam bath covering the flask with a towel. Check the weight of the flask and contents for constancy after 45 min, 1 h, and 1.25 h. The yield is usually about 4 g, mp 126–127°C.

Extinguish flames.

2. Filter the benzene-ether solution through anhydrous calcium chloride pellets for superficial drying, put it into a 50-mL round-bottomed flask, and distill over the steam bath through a condenser into an ice-cooled receiver until the volume in the distilling flask is reduced to about 11 mL. Add ligroin slowly until the solution is slightly turbid, and let the product crystallize at 25°C and then at 5°C. The yield of anhydrous, colorless, well-formed crystals, mp 127–128°C, is about 4 g.

MICROSCALE

3. Anthraquinone

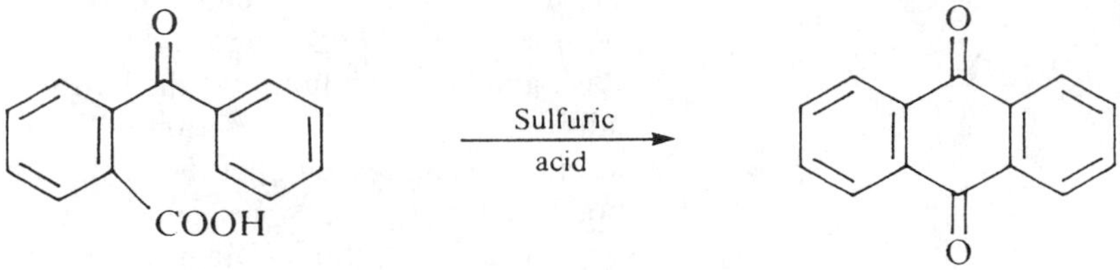

2-Benzoylbenzoic acid
MW 226.22
mp 127°C

Sulfuric
acid

Anthraquinone
MW 208.20
mp 286 C

Anthraquinone can be purified by sublimation. Vapor pressure of anthraquinone:

t(°C)	Pressure (mm Hg)
200	1.8
220	4.4
240	12.6
260	33.0

In a reaction tube dissolve 100 mg of 2-benzoylbenzoic acid[4] in 0.5 mL of concentrated sulfuric acid by gently heating and stirring. Immerse a thermometer in the reaction mixture, and heat it at 150 to 155°C for 45 min. Allow the tube to cool to below 100°C, and then, using extreme caution, add a very small drop of water to the mixture. Mix the contents of the tube, and continue adding water in minute drops to the mixture. This will cause the product to crystallize, and if done slowly enough, the crystals will be large enough to collect easily by filtration. Fill the tube with water, and collect the product by filtration on the Hirsch funnel or Wilfilter (Fig 32.3). Return the damp product to the reaction tube, and boil the product with 0.5 mL of concentrated ammonium hydroxide to remove unreacted starting material. Filter, then wash the product well with water and then with acetone, dry, determine the weight, and then calculate the percentage yield. Determine the melting point.

Cleaning Up   The aqueous filtrate, after neutralization with sodium carbonate, is diluted with water and flushed down the drain.

---

4. Available from the Aldrich Chemical Co.

4. Anthraquinone

Place 4.0 g of 2-benzoylbenzoic acid[5] (anhydrous) in a 125-mL round-bottomed flask, add 20 mL of concentrated sulfuric acid, and heat on the steam bath with swirling until the solid is dissolved. Then clamp the flask over a microburner, insert a thermometer, raise the temperature to 150°C, and heat to maintain a temperature of 150–155°C for 5 min. Let the solution cool to 100°C, remove the thermometer after letting it drain, and, with a Pasteur pipette, add 4 mL of water by drops with swirling to keep the precipitated material dissolved as long as possible so that it will separate as small, easily filtered crystals. Let the mixture cool further, dilute with water until the flask is full, again let cool, collect the product by suction filtration, and wash well with water. Then remove the filtrate, wash the filter flask, return the funnel to the filter flask without applying suction, and test the filter cake for unreacted starting material as follows: Dilute 8 mL of concentrated ammonia solution with 40 mL of water, pour the solution onto the filter, and loosen the cake so that it is well leached. Then apply suction, wash the cake with water, and acidify a few milliliters of the filtrate. If there is no precipitate upon acidifying, the yield of anthraquinone should be close to the theoretical because it is insoluble in water. Dry the product to constant weight, but do not take the melting point since it is so high (mp 286°C).

Cleaning Up   The aqueous filtrate, after neutralization with sodium carbonate, is diluted with water and flushed down the drain.

5. Anthrone

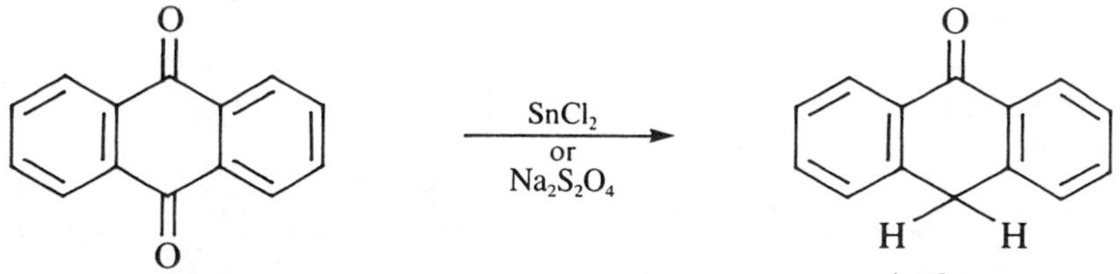

Anthraquinone
MW 208.20
mp 286°C

SnCl$_2$
or
Na$_2$S$_2$O$_4$

Anthrone
MW 194.22
mp 156°C

Tin(II) Chloride Reduction

In a reaction tube put 50 mg of crude anthraquinone from the previous experiment, 0.40 mL of acetic acid, and a solution made by warming 0.13 g of tin(II) chloride dihydrate with 0.13 mL of concentrated hydrochloric acid. Add a boiling stone, note the time, and reflux gently until crystals of anthraquinone have

---

5. Available from the Aldrich Chemical Co.

completely disappeared (8–10 min); reflux 15 min longer, and record the total time. Then add water (about 0.12 mL) dropwise until the solution is saturated. Let the solution stand for crystallization. Collect the product using the Wilfilter (Fig. 32.3), dry it, and take the melting point (156°C reported). The yield of pale-yellow crystals is about 40 mg.

Cleaning Up   Dilute the filtrate with a large volume of water, and flush the solution down the drain.

Hydrosulfite Reduction

Procedure 2

$Na_2S_2O_4$
Sodium hydrosulfite

Decomposes on storage.

In a 5-mL long-necked round-bottomed flask put 50 mg of anthraquinone, 60 mg of sodium hydroxide, 150 mg of fresh sodium hydrosulfite, and 1.3 mL of water. Heat over a hot sand bath, and swirl for a few minutes to convert the anthraquinone into the deep-red anthrahydroquinone anion. Note that particles of different appearance begin to separate even before the anthraquinone has all dissolved. Add an empty distilling column to the flask, and reflux for 45 min; cool, filter the product using the Hirsch funnel or the Wilfilter (Fig. 32.3), and wash it well and let dry. Determine the weight and melting point of the crude material, and crystallize it from 95% ethanol. Record the approximate volume of solvent used, and if the first crop of crystals recovered is not satisfactory, concentrate the mother liquor and secure a second crop. The usual yield is about 40 mg.

Cleaning Up   The aqueous filtrate is treated with household bleach (sodium hypochlorite) until reaction is complete. Remove the solid by filtration. It may be placed in the nonhazardous solid waste container. The filtrate can be diluted with water and flushed down the drain.

🏺 MACROSCALE

6. Anthrone

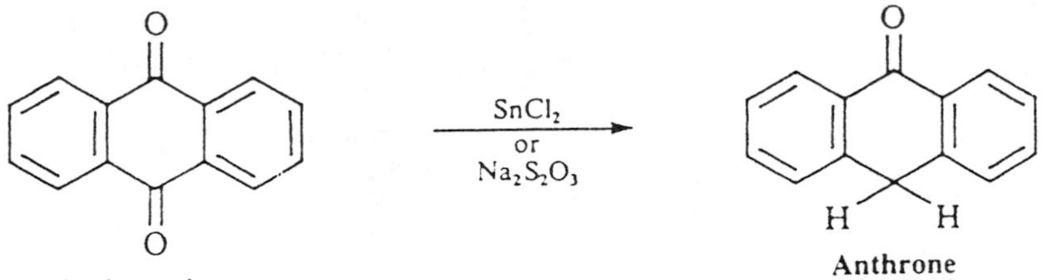

Anthraquinone
MW 208.20
mp 286°C

$SnCl_2$
or
$Na_2S_2O_3$

Anthrone
MW 194.22
mp 156°C

Tin(II) Chloride Reduction

Procedure 1

$SnCl_2 \cdot 2H_2O$
Tin(II) chloride dihydrate

In a 50-mL round-bottomed flask provided with a reflux condenser put 2.0 g of anthraquinone, 16 mL of acetic acid, and a solution made by warming 5.3 g of tin(II) chloride dihydrate with 5.3 mL of concentrated hydrochloric acid. Add a boiling stone to the reaction mixture, note the time, and reflux gently until crystals of anthraquinone have completely disappeared (8–10 min); then reflux 15 min longer and record the total time. Disconnect the flask, heat it on the steam bath, and add water (about 5 mL) in 0.5-mL portions until the solution is saturated. Let the solution stand for crystallization. Collect and dry the product, and take the melting point (156°C given). Dispose of the filtrate in the container provided. The yield of pale-yellow crystals is about 1.7 g.

Cleaning Up   Dilute the filtrate with a large volume of water, and flush the solution down the drain.

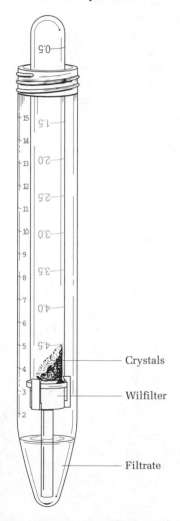

Procedure 2

$Na_2S_2O_4$
Sodium hydrosulfite

Crystals

Wilfilter

Filtrate

FIG. 32.3   Wilfilter filtration apparatus. See page 52 in Chapter 3.

Hydrosulfite Reduction

In a 250-mL round-bottomed flask, which can be heated under reflux, put 2.0 g of anthraquinone, 2.4 g of sodium hydroxide, 6 g of fresh sodium hydrosulfite, and 50 mL of water. Heat over a free flame, and swirl for a few minutes to convert the anthraquinone into the deep-red anthrahydroquinone anion. Note that particles of different appearance begin to separate even before all of the anthraquinone has dissolved. Arrange for refluxing and reflux for 45 min; cool, filter the product, and wash it well and let dry. Note the weight of the product and melting point of the crude material, and then crystallize it from 95% ethanol; the solution may require filtration to remove insoluble impurities. Record the approximate volume of solvent used and, if the first crop of crystals recovered is not satisfactory, concentrate the mother liquor and secure a second crop.

Cleaning Up   The aqueous filtrate is treated with household bleach (sodium hypochlorite) until reaction is complete. Remove the solid by filtration. It may be placed in the nonhazardous solid waste container. The filtrate can be diluted with water and flushed down the drain.

7. Comparison of Results

Compare your results with those obtained by neighbors using the alternative procedure with respect to yield, quality of product, and working time. Which is the better laboratory procedure? Then consider the cost of the three solvents concerned, the cost of the two reducing agents (current prices can be looked up in a catalog), the relative ease of recovery of the organic solvents, and the prudent disposal of byproducts, and decide which method would be preferred as a manufacturing process.

8. Fluorescent Anthracene

Anthrone
MW 194.22, mp 156°C

$\xrightarrow[\text{(Cu)}]{\text{Zn dust}}$

Anthracene
MW 178.22, mp 216°C

Microscale Procedure

Put 100 mg of zinc dust into a 5-mL round-bottomed flask, and activate the dust by adding 0.6 mL of water and 0.1 mL of a 5% aqueous copper(II) sulfate solution and swirling for a minute or two (or use a magnetic stirring bar). Add 40 mg of anthrone, 100 mg of sodium hydroxide, and 1 mL of water, heat to boiling on the sand bath, note the time, and start refluxing the mixture (Fig. 32.4). Anthrone at first dissolves as the yellow anion of anthranol, but anthracene soon begins to separate as a white precipitate.

In about 15 min, the yellow color initially observed on the walls disappears, but refluxing should be continued for a full 30 min. Then remove the heat, rinse down any anthracene in the condenser with a few drops of water, and cool the flask in ice. Remove the solvent with a Pasteur pipette, and add very carefully and dropwise 0.2 mL of concentrated hydrochloric acid to dissolve the remaining zinc. Again, cool the mixture and add water to wash down any unreacted zinc; after all has reacted, remove the acid and wash the product liberally with water. Collect the product on the Hirsch funnel, and then wash once with ice-cold methanol to remove as much of the water as possible from the product.

The product need not be dried before crystallization from about 0.5 mL of boiling toluene. Slow cooling of the toluene solution will give thin, colorless, beautifully fluorescent plates, yield about 20 mg. Collect the product on the Wilfilter (Fig. 32.3). In washing the equipment with acetone, you should be able to observe the striking fluorescence of very dilute solutions. The fluorescence is quenched by a bare trace of impurity.

Hydrogen liberated

Methanol removes water without dissolving much anthracene.

Anthracene can also be recrystallized from ethanol. See Chapter 3, Experiment 2, p. 59.

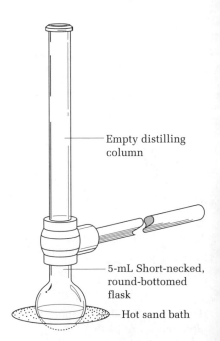

Empty distilling column

5-mL Short-necked, round-bottomed flask

Hot sand bath

FIG. 32.4 Apparatus for refluxing the reaction mixture in a reaction flask.

Cleaning Up   The aqueous solution is diluted with water, neutralized with sodium carbonate, and filtered under vacuum to remove zinc hydroxide; the filtrate is diluted with water and flushed down the drain. The solid can be placed in the nonhazardous solid waste container. Toluene and acetone are placed in the organic solvents container.

Anthrone
MW 194.22, mp 156°C

Anthracene
MW 178.22, mp 216°C

Macroscale Procedure

Reflux time: 30 min

Put 3.8 g of zinc dust into a 125-mL round-bottomed flask, and activate the dust by adding 23 mL of water and 0.4 mL of copper(II) sulfate solution (Fehling's solution I) and swirling for a minute or two. Add 1.5 g of anthrone, 3.8 g of sodium hydroxide, and 40 mL of water; attach a reflux condenser, heat to boiling, note the time, and start refluxing the mixture. Anthrone at first dissolves as the yellow anion of anthranol, but anthracene soon begins to separate as a white precipitate. In about 15 min the yellow color initially observed on the walls disappears, but refluxing should be continued for a full 30 min. Then stop the heating, use a water wash bottle to rinse down anthracene that has lodged in the condenser, and filter the still-hot mixture on a large Büchner funnel. It usually is possible to decant from, and so remove, a mass of zinc.

Extinguish flames. Hydrogen liberated.

Methanol removes water without dissolving much anthracene.

After liberal washing with water, blow or shake out the gray cake into a 250-mL beaker, and rinse the funnel and paper with water. To remove most of the zinc metal and zinc oxide, add 8 mL of concentrated hydrochloric acid and heat on the steam bath with stirring for 20–25 min, when initial frothing due to liberated hydrogen should have ceased. Collect the now nearly white precipitate on a Büchner funnel and, after liberal washing with water, release the suction, rinse the walls of the funnel with methanol, use enough methanol to cover the cake, and then apply suction. Wash again with enough methanol to cover the cake, and then remove solvent thoroughly by suction.

CAUTION: Highly flammable solvent

The product need not be dried before crystallization from toluene. Transfer the methanol-moist material to a 50-mL Erlenmeyer and add 22 mL of toluene; a liberal excess of solvent is used to avoid crystallization in the funnel. Make sure there are no flames nearby, and heat the mixture on the hot plate to bring the anthracene into solution with a small residue of zinc remaining. Filter by gravity by the usual technique to remove zinc. From the filtrate anthracene is obtained as thin, colorless, beautifully fluorescent plates; yield about 1 g. In washing the equipment with acetone, you should be able to observe the striking fluorescence of very dilute solutions. The fluorescence is quenched by a bare trace of impurity.

Cleaning Up   The aqueous solution is diluted with water, neutralized with sodium carbonate, and filtered under vacuum to remove zinc hydroxide, and the filtrate diluted with water and flushed down the drain. The solid waste can be placed in the nonhazardous solid waste container. Toluene and acetone are placed in the organic solvents container.

Questions

1. Calculate the number of moles of hydrogen chloride liberated in the synthesis of 2-(4-benzoyl)benzoic acid. If this gaseous acid were dissolved in water, hydrochloric acid would be formed. How many milliliters of concentrated hydrochloric acid would be formed in this reaction? The concentrated acid is 12 molar in HCl.

2. Write a mechanism for the formation of anthraquinone from 2-benzoylbenzoic acid, indicating clearly the role of sulfuric acid. What is the name commonly given to this type of reaction?

Friedel-Crafts Acylation of Ferrocene: Acetylferrocene

Prelab Exercise: How many possible isomers could exist for diacetylferrocene?

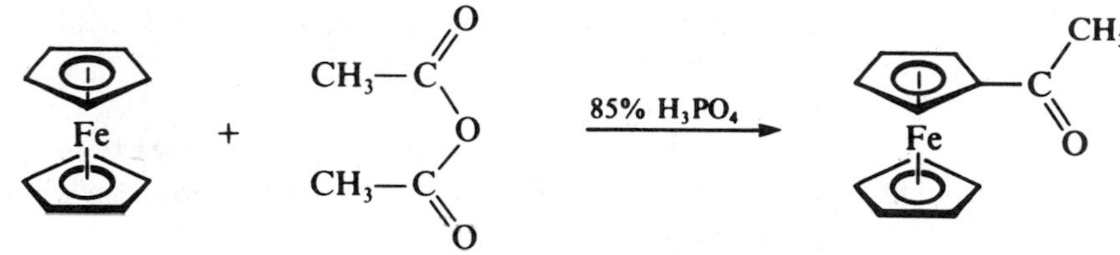

| Ferrocene [Bis(cyclopentadienyl)iron] MW 186.04 | Acetic Anhydride bp 139.5°C, den. 1.08 MW 102.09 | Acetylferrocene (Acetylcyclopentadienyl)cyclopentadienyliron mp 85–86°C MW 228.08 |

The Friedel-Crafts acylation of benzene requires aluminum chloride as the catalyst, but ferrocene, which has been referred to as a "superaromatic" compound, can be acylated under much milder conditions with phosphoric acid as catalyst. Since the acetyl group is a deactivating substituent, the addition of a second acetyl group, which requires more vigorous conditions, will occur in the nonacetylated cyclopentadienyl ring to give 1,1′-diacetylferrocene. Because ferrocene gives just one monoacetyl derivative and just one diacetyl derivative, it was assigned an unusual sandwich structure.

Acetylferrocene and ferrocene (both highly colored) are easily separated by column chromatography (see Chapter 10).

Experiment

Acetylferrocene

Microscale Procedure

Handle acetic anhydride and phosphoric acid with care. Wipe up spills immediately.

To 93 mg of dry sublimed ferrocene (material from Chapter 50) in a 10 × 100 mm reaction tube, add 0.35 mL (0.38 g) of acetic anhydride followed by 0.1 mL (170 mg) of 85% phosphoric acid. Cap the tube with a septum bearing an empty syringe needle, and warm it on a steam bath or in a beaker of boiling water while agitating the mixture to dissolve the ferrocene. Heat the mixture for 10 min more, and then cool the tube thoroughly in ice. Carefully add to the solution 0.5 mL of ice water dropwise with thorough mixing, followed by the dropwise addition of 3 M aqueous sodium hydroxide solution until the mixture is neutral (test with indicator paper, and avoid an excess of base). Collect the product on the Hirsch funnel, wash it thoroughly with water, and press it as dry as possible between sheets of filter paper. Save a sample of this material for melting-point and thin-layer chromatographic analyses, and purify the remainder by column chromatography.

Column Chromatography

Follow exactly the procedure given on page 173 of Chapter 10 for packing a microscale chromatography column and adding the sample (about 90 mg of a ferrocene/acetylferrocene mixture). Use Brockman activity III alumina as the adsorbent.

Elution.   Carefully add ligroin or hexane to the column, and begin to elute the product from the column. Unreacted ferrocene will come down the column as an orange-yellow band. Collect this in a 10-mL flask. Wash any crystalline material that collects on the tip of the valve into the flask with a few drops of ether. Then elute the column with a 50:50 mixture of ligroin or hexane and *tert*-butyl methyl ether. This will bring down the acetylferrocene as an orange-red band. Collect this in a separate 10-mL Erlenmeyer flask. Spot a thin-layer plate with these two solutions as well as the crude acetylferrocene. Any diacetylferrocene will be seen as a dark band at the top of the column. Evaporate the solvents under reduced pressure, and determine the weights of the residues (Fig. 33.1).

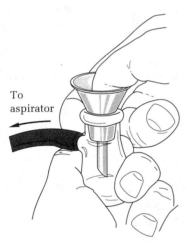

To
aspirator

FIG. 33.1 Evaporation of solvent under reduced pressure. Heat and swirling motion is supplied by one hand; vacuum is controlled by thumb of other.

Flash Chromatography

Equilibrium between the solutes adsorbed on the alumina and the eluting solvent is established rapidly, so it does no harm to increase the flow rate of the solvent through the column. This can be done by applying pressure with a rubber bulb above the solvent. There is, however, the possibility that in removing the bulb air will be sucked into the bottom of the column, which will destroy the uniform packing of the column. Practice this technique on a column that has been eluted in the usual way using gravity flow of the solvent. On a macroscale, special fittings are available for applying air pressure to larger columns.

Recrystallize the acetylferrocene from hot ligroin or hexane. Dissolve the residue in the filter flask in the minimum quantity of boiling solvent (about 1 mL), and transfer this hot solution with a Pasteur pipette to a reaction tube. Allow the tube to cool slowly to room temperature, and then cool it in ice for at least 10 min. Acetylferrocene crystallizes as dark-red rosettes of needles. Remove the solvent with a Pasteur pipette, and then, under vacuum, scrape out the product onto filter paper; after it is dry, determine the weight, calculate the percentage yield, and determine the melting point.

Analyze your product by thin-layer chromatography. Dissolve very small samples of pure ferrocene, the crude reaction mixture, and recrystallized acetylferrocene, each in a few drops of toluene; spot the three solutions with micro-capillaries on silica-gel plates; and develop the chromatogram with 30 : 1 tol-uene–absolute ethanol. Visualize the spots under a UV lamp if the silica gel has a fluorescent indicator or by adsorption of iodine vapor. Do you detect unreacted ferrocene in the reaction mixture and/or a spot that might be attributed to diacetylferrocene?

Cleaning Up The reaction mixture filtrate can be flushed down the drain. Unused chromatography, recrystallization, and thin-layer chromatography solvents should be placed in the organic solvents container. The alumina from the column should be placed in the hood to allow the ligroin and ether to evaporate from it. Once free of organic solvents, it can be placed in the nonhazardous solid waste container.

Macroscale Procedure

Both acetic anhydride and sulfuric acid are corrosive to tissue: handle both with care and wipe up any spills immediately.

Hexane and ligroin are flammable. No flames!

In a 25-mL round-bottomed flask place 3.0 g of ferrocene, 10.0 mL of acetic anhydride, and 2.0 mL of 85% phosphoric acid. Equip the flask with a reflux condenser and a calcium chloride drying tube. Warm the flask gently on the steam bath with swirling to dissolve the ferrocene, then heat strongly for 10 min more. Pour the reaction mixture onto 50 g of crushed ice in a 400-mL beaker, and rinse the flask with 10 mL of ice water. Stir the mixture for a few minutes with a glass rod, add 75 mL of 3 *M* sodium hydroxide solution (the solution should still be acidic), then add solid sodium bicarbonate (be careful of foaming) until the remaining acid has been neutralized. Stir and crush all lumps, allow the mixture to stand for 20 min, then collect the product by suction filtration. Press the crude material as dry as possible between sheets of filter paper, save a few crystals for thin-layer chromatography, transfer the remainder to an Erlenmeyer flask, and add 40 mL of hexane or ligroin to the flask. Boil the solvent on the steam bath for a few minutes, then decant the dark-orange solution into another Erlenmeyer flask, leaving a gummy residue of polymeric material. Treat the solution with decolorizing charcoal, and filter it, through a fluted filter paper placed in a warm stemless funnel, into an appropriately sized Erlenmeyer flask. Evaporate the solvent (use an aspirator tube; see Fig. 8.11) until the volume is about 20 mL. Set the flask aside to cool slowly to room temperature. Beautiful rosettes of dark orange-red needles of acetylferrocene will form. After the product has been cooled in ice, collect it on a Büchner funnel and wash the crystals with a small quantity of cold solvent. Pure acetylferrocene has mp 84–85°C. Your yield should be about 1.8 g.

Analyze your product by thin-layer chromatography. Dissolve very small samples of pure ferrocene, the crude reaction mixture, and recrystallized acetylferrocene, each in a few drops of toluene; spot the three solutions with micro-capillaries on silica gel plates; and develop the chromatogram with 3 : 1 toluene–absolute ethanol. Visualize the spots under a UV lamp, if the silica gel has a fluorescent indicator, or by adsorption of iodine vapor. Do you detect unreacted ferrocene in the reaction mixture and/or a spot that might be attributed to diacetylferrocene?

Cleaning Up   The reaction mixture filtrate can be flushed down the drain. Unused recrystallization and thin-layer chromatography solvents should be placed in the organic solvents container. The decolorizing charcoal, once free of solvent, can be placed in the nonhazardous solid waste container.

Questions

1. What is the structure of the intermediate species that attacks ferrocene to form acetylferrocene? What other organic molecule is formed?

2. Why does the second acetyl group enter the unoccupied ring to form diacetylferrocene?

34

1,2,3,4-Tetraphenylnaphthalene via Benzyne

Prelab Exercise: Speculate on the reasons for the stability of the iodonium ion. What type of strain exists in the benzyne molecule?

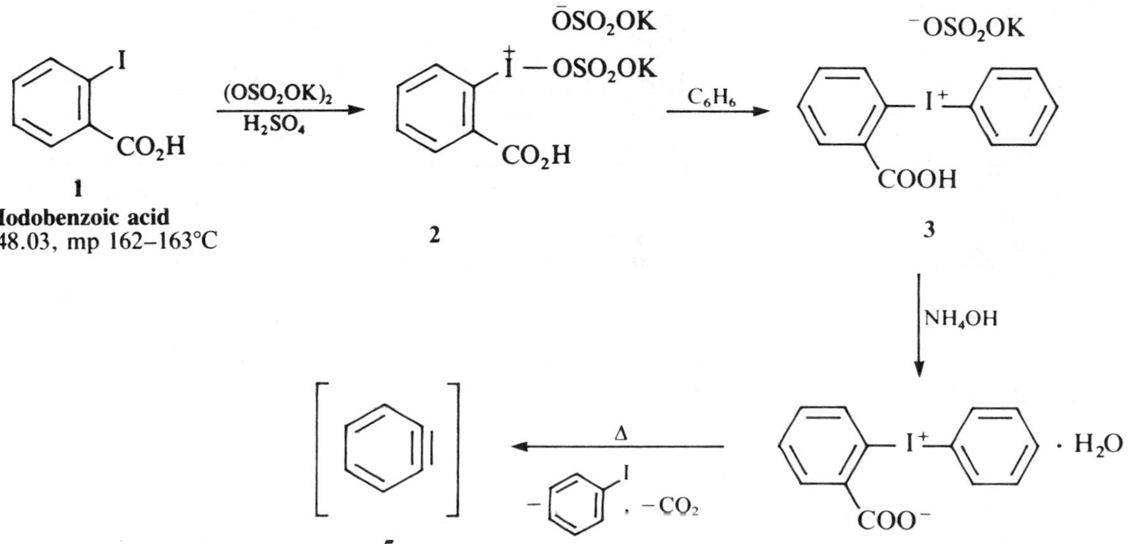

1
2-Iodobenzoic acid
MW 248.03, mp 162–163°C

2

3

4
Diphenyliodonium-2-carboxylate monohydrate
MW 341.13

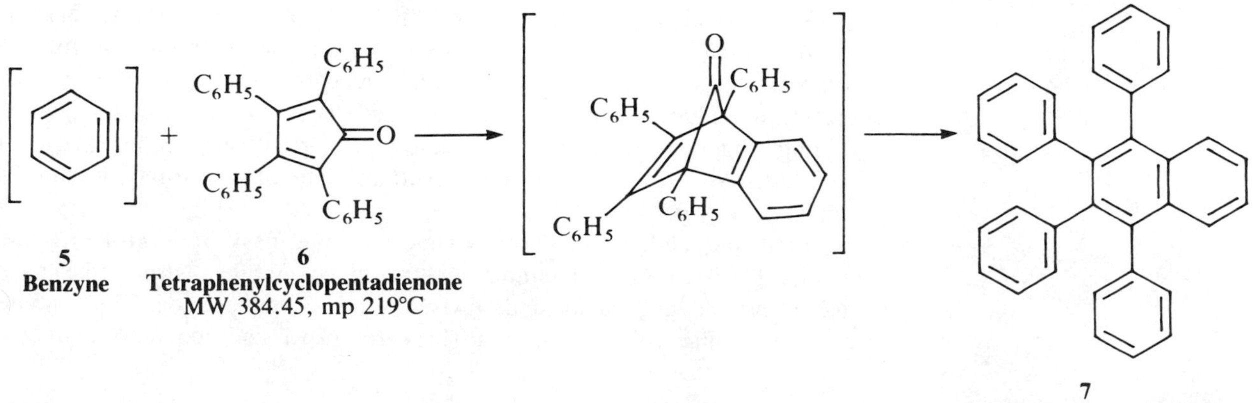

5
Benzyne

5
Benzyne

6
Tetraphenylcyclopentadienone
MW 384.45, mp 219°C

7
1,2,3,4-Tetraphenylnaphthalene
MW 432.53, mp 219–220°C

Synthetic use of an intermediate not known as such

This synthesis of 1,2,3,4-tetraphenylnaphthalene (**7**) demonstrates the transient existence of benzyne (**5**), a hydrocarbon that has not been isolated as such. The precursor, diphenyliodonium-2-carboxylate (**4**), is heated in an inert solvent to a temperature at which it decomposes to benzyne, iodobenzene, and carbon dioxide in the presence of tetraphenylcyclopentadienone (**6**) as trapping agent. The preparation of the precursor (**4**) illustrates oxidation of a derivative of iodobenzene to an iodonium salt (**2**) and the Friedel-Crafts-like reaction of the substance with benzene to form the diphenyliodonium salt (**3**). Neutralization with ammonium hydroxide then liberates the precursor, inner salt (**4**), which, when obtained by crystallization from water, is the monohydrate.

Experiments

MICROSCALE

$K_2S_2O_8$

Potassium persulfate (Potassium peroxydisulfate)
MW 270.33

Strong oxidant!

1st step: about 20 min

CAUTION: *Benzene is a mild carcinogen. Carry out this experiment in the hood.*

2nd step: about 30 min

1. Diphenyliodonium-2-carboxylate Monohydrate (4)

Measure 0.85 mL of concentrated sulfuric acid into a reaction tube, and place it in an ice bath to cool. In a 10-mL Erlenmeyer flask that is clamped in an ice bath, place 200 mg of 2-iodobenzoic acid, prepared in Chapter 45, and 260 mg of potassium persulfate.[1] Remove the tube containing the cold sulfuric acid from the ice bath, wipe it dry, and add the acid to the two solids in the flask. Stir the mixture in the ice bath for 4–5 min to produce an even suspension, and then remove it and note the time. The reaction mixture foams somewhat and acquires a succession of colors. After it has stood at room temperature for 20 min, swirl the flask vigorously in an ice bath for 2–3 min, add 0.2 mL of benzene (*caution!*), and swirl and cool until the benzene freezes. Then remove and wipe the flask, and note the time at which the benzene melts. Warm the flask in the palm of your hand, and swirl frequently at room temperature for 20 min to promote completion of reaction in the two-phase mixture. If available, magnetic stirring of the reaction mixture is convenient but not required.

While the reaction is going to completion, place three reaction tubes in an ice bath to chill: one containing 1.9 mL of distilled water, another 2.3 mL of concentrated ammonium hydroxide solution, and another 4 mL of dichloromethane. At the end of the 20-min reaction period, chill the reaction mixture thoroughly in an ice bath and add to it dropwise, with vigorous swirling and mixing, the 1.9 mL of ice-cold water that has been chilling. The solid that separates is the potassium bisulfate salt of diphenyliodonium-2-carboxylic acid, **3**.

Pour the chilled dichloromethane into the flask for extracting the product liberated when the ammonia is added. While swirling the flask vigorously, add the ammonia dropwise over a 10-min period. The mixture must be alkaline (pH 9). Test with indicator paper and add more ammonia

Note for the instructor

1. The fine granular material supplied by Fisher Scientific Co. is satisfactory. Persulfate in the form of large prisms should be finely ground prior to use.

if necessary. Using a Pasteur pipette, remove the dichloromethane layer, transfer the aqueous layer to a 4-in. test tube, and extract it with two 1.5-mL portions of dichloromethane, which are combined with the original dichloromethane extract. Dry the dichloromethane extracts over anhydrous calcium chloride pellets, and decant into a tared 10-mL flask. Rinse the drying agent with more solvent, and then evaporate the dichloromethane under a stream of air in the hood. Connect the flask to an aspirator, and heat it on the steam bath until the weight is constant. The yield is about 240 mg.

For crystallization, dislodge the solid with a spatula, and dissolve it in 2.8 mL of boiling water. Use a wood boiling stick to prevent bumping of the solution. If desired, the tan solution may be decolorized by the addition of about 15 pieces of pelletized Norit. On cooling the solution after removal from the Norit, diphenyliodonium-2-carboxylate monohydrate (**4**), the benzyne precursor, separates in colorless, rectangular prisms, mp 219–220°C. The yield should be about 200 mg.

Cleaning Up   Make the aqueous layer neutral with dilute hydrochloric acid. It should then be diluted with water and flushed down the drain. Allow the dichloromethane to evaporate from the drying agent, and then place it in the nonhazardous solid waste container along with the Norit.

2. Diphenyliodonium-2-carboxylate Monohydrate (4)

MACROSCALE

$K_2S_2O_8$

**Potassium persulfate
(Potassium peroxydisulfate)**
MW 270.33

Strong oxidant!

1st step: about 20 min

CAUTION: *Benzene is a mild carcinogen. Carry out this experiment in the hood.*

2nd step: about 30 min

Note for the instructor

Measure 8 mL of concentrated sulfuric acid into a 25-mL Erlenmeyer flask, and place the flask in an ice bath to cool. Place 2.0 g of *o*-iodobenzoic acid (prepared in Chapter 45) and 2.6 g of potassium persulfate[2] in a 125-mL Erlenmeyer flask, swirl the flask containing sulfuric acid vigorously in the ice bath for 2–3 min, and then remove it and wipe it dry. Place the larger flask in the ice bath, and pour the chilled acid down the walls to dislodge any particles of solid. Swirl the flask in the ice bath for 4–5 min to produce an even suspension, and then remove it and note the time. The reaction mixture foams somewhat and acquires a succession of colors. After it has stood at room temperature for 20 min, swirl the flask vigorously in an ice bath for 3–4 min, add 2 mL of benzene (*caution!*), and swirl and cool until the benzene freezes. Then remove and wipe the flask, and note the time at which the benzene melts. Warm the flask in the palm of your hand, and swirl frequently at room temperature for 20 min to promote completion of reaction in the two-phase mixture.

While the reaction is going to completion, place three 50-mL Erlenmeyer flasks in an ice bath to chill: one containing 19 mL of distilled water, another 23 mL of 29% ammonium hydroxide solution, and another 40 mL of dichloromethane (bp 40.8°C). At the end of the 20-min reaction period, chill the reaction mixture thoroughly in an ice bath, mount a separatory funnel to deliver into the flask containing the reaction mixture in benzene, and place in the funnel the

---

2. The fine granular material supplied by Fisher Scientific Co. is satisfactory. Persulfate in the form of large prisms should be finely ground prior to use.

chilled 19 mL of water. Swirl the reaction flask vigorously while running in the water slowly. The solid that separates is **3,** the potassium bisulfate salt of diphenyliodonium-2-carboxylic acid. Pour the chilled ammonia solution into the funnel, and pour the chilled dichloromethane into the reaction flask so that it will be available for efficient extraction of reaction product **4** as it is liberated from **3** on neutralization. While swirling the flask vigorously in the ice bath, run in the chilled ammonia solution during the course of about 10 min. The mixture must be alkaline (pH 9). If not, add more ammonia solution. Pour the mixture into a separatory funnel, and rinse the flask with a little fresh dichloromethane. Let the layers separate, and run the lower layer into a tared 125-mL Erlenmeyer flask, through a funnel fitted with a paper containing anhydrous calcium chloride pellets. Extract the aqueous solution with two 10-mL portions of dichloromethane, and run the extracts through the drying agent into the tared flask. Evaporate the dried extracts to dryness on the steam bath (use an aspirator tube), and remove the solvent from the residual cake of solid by connecting the flask, with a rubber stopper, to the aspirator and heating the flask on the steam bath until the weight is constant; yield is about 2.4 g.

For crystallization, dislodge the bulk of the solid with a spatula and transfer it onto weighing paper and then into a 50-mL Erlenmeyer flask. Measure 28 mL of distilled water into a flask, and use part of the water to dissolve the residual material in the tared 125-mL flask by heating the mixture to the boiling point over a free flame. Pour this solution into the 50-mL flask. Add the remainder of the 28 mL of water to the 50-mL flask, and bring the solid into solution at the boiling point. Add a small portion of charcoal for decolorization of the pale tan solution, swirl, and filter at the boiling point through a funnel, preheated on the steam bath and fitted with moistened filter paper. Diphenyliodonium-2-carboxylate monohydrate (**4**), the benzyne precursor, separates in colorless, rectangular prisms, mp 219–220°C (with decomposition); yield is about 2.1 g.

Cleaning Up   Make the aqueous layer neutral with dilute hydrochloric acid. It should then be diluted with water and flushed down the drain. Allow the dichloromethane to evaporate from the drying agent, and then place it in the nonhazardous solid waste container along with the decolorizing charcoal.

3. Preparation of 1,2,3,4-Tetraphenylnaphthalene (7)

Microscale Procedure

Diels-Alder reaction

Place 100 mg of the diphenyliodonium-2-carboxylate monohydrate just prepared and 100 mg of tetraphenylcyclopentadienone, prepared in Chapter 40, in a reaction tube. Add 0.6 mL of triethylene glycol dimethyl ether (bp 222°C) so that the solvent rinses the walls of the tube. Support the tube vertically, insert a thermometer, and heat it over a hot sand bath. When the temperature reaches 200°C, remove the tube from the heat and note the time. Then keep the mixture at 200–205°C by intermittent heating until the purple color is discharged, the evolution of gas (CO_2 and CO) subsides, and a pale-yellow solution is obtained. In case a purple or red color persists after 3 min at

200–205°C, add additional small amounts of the benzyne precursor and continue to heat until all the solid is dissolved.

Cool the hot solution to below 100°C, and add to it 0.6 mL of ethanol. Bring the solution to boiling, using a wood stick to prevent bumping. If shiny prisms do not separate at once, add water dropwise at the boiling point of the ethanol until crystals begin to separate. Let crystallization proceed at room temperature and then at 0°C. Collect the product using the Pasteur pipette method or on the Hirsch funnel. Wash the crystals with cold methanol. The yield of dry product should be about 80 mg. The pure hydrocarbon, **7**, exists in two crystalline forms (allotropes) and has a double melting point, the first of which is in the range 196–199°C. Let the melting point bath cool to about 195°C, remove the sample, and let the sample solidify. Then determine the second melting point, which, for the pure hydrocarbon, is 203–204°C.

Cleaning Up The filtrate contains ethanol, iodobenzene, triglyme, and probably small amounts of the reactants and the product. Place the mixture in the halogenated organic solvents container.

Macroscale Procedure

Diels-Alder reaction

Use hood; CO is evolved.

Place 1.0 g of the diphenyliodonium-2-carboxylate monohydrate just prepared and 1.0 g of tetraphenylcyclopentadienone in a 25 × 150-mm test tube. Add 6 mL of triethylene glycol dimethyl ether (bp 222°C) in a way that the solvent will rinse the walls of the tube. Support the test tube vertically, insert a thermometer, and heat with a Thermowell. When the temperature reaches 200°C, remove from the heat and note the time. Then keep the mixture at 200–205°C by intermittent heating until the purple color is discharged, the evolution of gas (CO_2 + CO) subsides, and a pale-yellow solution is obtained. In case a purple or red color persists after 3 min at 200–205°C, add additional small amounts of the benzyne precursor and continue to heat until all the solid is dissolved. Let the yellow solution cool to 90°C while heating 6 mL of 95% ethanol to the boiling point on the steam bath. Pour the yellow solution into a 25-mL Erlenmeyer flask, and use portions of the hot ethanol, drawn into a capillary dropping tube, to rinse the test tube. Add the remainder of the ethanol to the yellow solution, and heat at the boiling point. If shiny prisms do not separate at once, add a few drops of water by drops at the boiling point of the ethanol until prisms begin to separate. Let crystallization proceed at room temperature, and then at 0°C. Collect the product and wash it with methanol. The yield of colorless prisms is 0.8–0.9 g. The pure hydrocarbon, **7**, exists in two crystalline forms (allotropes) and has a double melting point, the first of which is in the range of 196–199°C. Let the bath cool to about 195°C, remove the thermometer, and let the sample solidify. Then determine the second melting point, which, for the pure hydrocarbon, is 203–204°C.

Cleaning Up   The filtrate contains ethanol, iodobenzene, triglyme, and probably small amounts of the reactants and the product. Place the mixture in the halogenated organic solvents container.

Questions

1. To what general class of compounds does potassium persulfate belong?

2. Calculate the volume of carbon monoxide, at standard temperature and pressure, released during the reaction.

Triptycene via Benzyne

Prelab Exercise: Outline the preparation of triptycene, writing balanced equations for each reaction, including the reactions that are used to remove excess anthracene.

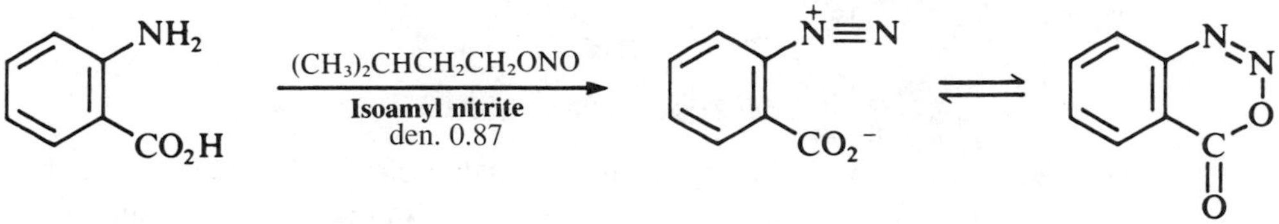

1
Anthranilic acid
MW 137.14

2a

2b

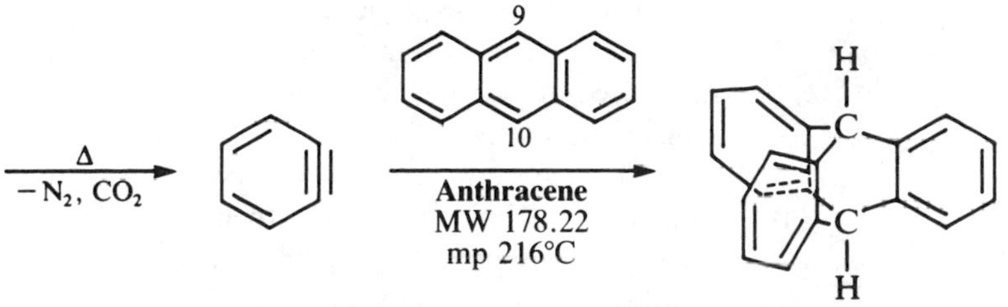

3
Benzyne

Anthracene
MW 178.22
mp 216°C

4
Triptycene
MW 254.31, mp 225°C

This interesting cage-ring hydrocarbon results from 9,10-addition of benzyne to anthracene. In one procedure presented in the literature, benzyne is generated under nitrogen from *o*-fluorobromobenzene and magnesium in the presence of anthracene, but the workup is tedious and the yield only 24%. Diazotization of anthranilic acid to benzenediazonium-2-carboxylate (**2a**) or to the covalent form (**2b**) gives another benzyne precursor, but the isolated substance can be kept only at a low temperature and is sometimes explosive. However, isolation of the precursor is not necessary. On slow addition of anthranilic acid to a solution of anthracene and isoamyl nitrite in an aprotic solvent, the precursor **2** reacts with anthracene as fast as the precursor is formed. If the anthranilic acid is all present at the start, a side reaction of this substance with benzyne drastically reduces the yield. A low-boiling solvent (CH_2Cl_2, bp 41°C) is used, in which the desired reaction goes slowly, and a solution of anthranilic acid is added dropwise over a period of 4 h. To bring the reaction time into the limits of a

Aprotic means "no protons"; HOH and ROH are protic.

laboratory period, the higher-boiling solvent 1,2-dimethoxyethane (bp 83°C, water soluble) is specified in this procedure and a large excess of anthranilic acid and isoamyl nitrite is used. Treatment of the dark reaction mixture with alkali removes acidic byproducts and most of the color, but the crude product inevitably contains anthracene. However, brief heating with maleic anhydride at a suitable temperature leaves the triptycene untouched and converts the anthracene into its maleic anhydride adduct. Treatment of the reaction mixture with alkali converts the adduct into a water-soluble salt and affords colorless, pure triptycene.

Removal of anthracene

Experiment

Triptycene

Microscale Procedure

Test dimethoxyethane for peroxides.

Place 100 mg of anthracene,[1] 0.1 mL of isoamyl nitrite, 1 mL of dimethoxyethane, and a boiling chip in a 10 × 100 mm reaction tube. In a small filter paper, place 130 mg of anthranilic acid, and push this about halfway down the reaction tube. Bring the mixture in the tube to a gentle boil, and note that the anthracene does not all dissolve. Add to the filter paper 0.5 mL of dimethoxyethane dropwise to leach the acid into the solution below over a period of not less than 20 min. Remove the filter paper, add to the solution 0.1 mL more isoamyl nitrite, add 130 mg more anthranilic acid to the filter paper, replace the paper in the reaction tube, and again leach out the anthranilic acid with 0.5 mL of dimethoxyethane added dropwise over a 20-min period. Reflux for 10 min more, and then add 0.5 mL of ethanol and a solution of 0.15 g of sodium hydroxide in 2 mL of water to produce a suspension of solid in a brown alkaline solvent. Cool the mixture thoroughly in ice, and also cool a 4 : 1 methanol-water mixture for rinsing.

Carry out reaction in hood.

Technique for slow addition of a solid

Reaction time: 1 h

Collect the solid on the Hirsch funnel by vacuum filtration, and wash it with the chilled solvent to remove brown mother liquor. Transfer the moist, nearly colorless solid to a tared, clean reaction tube, and evacuate on the steam bath until the weight is constant; the anthracene-triptycene mixture (mp about 190–230°C) weighs about 100 mg.

Add 50 mg of maleic anhydride and 1 mL of triethylene glycol dimethyl ether ("triglyme," bp 222°C), and reflux the mixture for 5 min. Cool to about 100°C, add 0.5 mL of ethanol and a solution of 150 mg of sodium hydroxide in 2 mL of water; then cool in ice, along with 1.25 mL of 4 : 1 methanol–water for rinsing. Triptycene separates as nearly white crystals from the slightly brown alkaline solution. The washed and dried product weighs about 75 mg and melts at 255°C. Recrystallize it from methylcyclohexane (23 mL/g). If an insoluble black impurity is seen in the hot solution, transfer the clear solution to a clean tube using a Pasteur pipette, leaving the impurity behind. Slow

---

1. Practical grade anthracene and anthranilic acid are satisfactory.

cooling will produce flat, rectangular, laminated prisms that can be isolated by removing the solvent with a Pasteur pipette.

Cleaning Up   Dilute the alkaline filtrate from the reaction with water, and flush it down the drain. Methylcyclohexane mother liquor from the crystallization goes in the organic solvents container.

Macroscale Procedure

CH$_3$OCH$_2$CH$_2$OCH$_3$
1,2-Dimethoxyethane
bp 83°C

Test dimethoxyethane for peroxides. Carry out reaction in hood.

Technique for slow addition of a solid

Reaction time: 1 h

CH$_3$OCH$_2$CH$_2$OCH$_2$
|
CH$_3$OCH$_2$CH$_2$OCH$_2$

**Triethylene glycol
dimethyl ether (triglyme)**
bp 222°C
Miscible with water

Place 2 g of anthracene,[2] 2 mL of isoamyl nitrite, and 20 mL of 1,2-dimethoxyethane in a 125-mL round-bottomed flask mounted in a hot sand bath and fitted with a short reflux condenser. Insert a filter paper into a 55-mm short stem funnel, moisten the paper with 1,2-dimethoxyethane, and rest the funnel in the top of the condenser. Weigh 2.6 g of anthranilic acid on a folded paper, scrape the acid into the funnel with a spatula, and pack it down. Bring the mixture in the reaction flask to a gentle boil, and note that the anthracene does not all dissolve. Measure 20 mL of 1,2-dimethoxyethane into a graduate, and use a capillary dropping tube to add small portions of the solvent to the anthranilic acid in the funnel, to slowly leach the acid into the reaction flask. If you make sure that the condenser is exactly vertical and the top of the funnel is centered, it should be possible to arrange for each drop to fall free into the flask and not touch the condenser wall. Once dripping from the funnel has started, add fresh batches of solvent to the acid, but only 2–3 drops at a time. Plan to complete leaching the first charge of anthranilic acid in a period of not less than 20 min, using about 10 mL of solvent. Then add a second 2.6-g portion of anthranilic acid to the funnel, remove from heat, and, by lifting the funnel up a little, run in 2 mL of isoamyl nitrite through the condenser. Replace the funnel, resume heating, and leach the anthranilic acid as before in about 20 min time. Reflux for 10 min more, and then add 10 mL of 95% ethanol and a solution of 3 g of sodium hydroxide in 40 mL of water to produce a suspension of solid in a brown alkaline liquor. Cool thoroughly in ice, and also cool a 4:1 methanol–water mixture for rinsing.

Collect the solid on a small Büchner funnel, and wash it with the chilled solvent to remove brown mother liquor. Transfer the moist, nearly colorless solid to a tared 125-mL round-bottomed flask, and evacuate on the steam bath until the weight is constant; the anthracene-triptycene mixture (mp about 190–230°C) weighs 2.1 g. Add 1 g of maleic anhydride and 20 mL of triethylene glycol dimethyl ether ("triglyme," bp 222°C), heat the mixture to the bp under reflux (see Fig. 16.2 for apparatus), and reflux for 5 min. Cool to about 100°C, add 10 mL of 95% ethanol and a solution of 3 g of sodium hydroxide in 40 mL of water; then cool in ice, along with 25 mL of 4:1 methanol–water for rinsing. Triptycene separates as nearly white crystals from the slightly brown alkaline liquor. The washed and dried product weighs 1.5 g and melts at 255°C. It will appear to be colorless, but it contains a trace of black insoluble material.

---

2. Practical grade anthracene and anthranilic acid are satisfactory.

Dissolve the product in methylcyclohexane (23 mL/g), decant from the specks of black material, and allow the solution to cool slowly. The product crystallizes as flat, rectangular, laminated prisms.

Cleaning Up   Dilute the alkaline filtrate from the reaction with water, and flush it down the drain. Methylcyclohexane mother liquor from the crystallization goes in the organic solvents container.

For Further Investigation

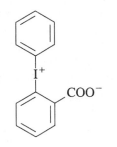

Diphenyliodonium-2-carboxylate

Diphenyliodonium-*o*-carboxylate is utilized as a benzyne precursor in the synthesis of 1,2,3,4-tetraphenylnaphthalene from tetraphenylcyclopentadienone. For reasons unknown, the reaction of anthracene with benzyne generated in this way proceeds very poorly and gives only a trace of triptycene. What about the converse proposition? Would benzyne generated by diazotization of anthranilic acid in the presence of tetraphenylcyclopentadienone afford 1,2,3,4-tetraphenylnaphthalene in satisfactory yield? If you are interested in exploring this possibility, plan and execute a procedure and see what you can discover.

If benzyne is generated from anthranilic acid in the absence of anthracene, as in this experiment, then dibenzocyclobutadiene is formed.[3]

Devise a procedure for the isolation of this interesting hydrocarbon.

Computational Chemistry

Using the AM1 semiempirical molecular orbital program, calculate the heats of formation of the isomeric compounds **2a** and **2b** in order to determine which might be formed in this reaction. Bear in mind that these will be gas phase heats of formation. Repeat the calculation in water, a polar solvent. Does this change the relative stabilities of the two isomers?

It is instructive to examine the molecular orbitals of benzyne and dibenzocyclobutadiene. Benzyne is made by joining two *ortho* hydrogens into a new bond. Does the central cyclobutadiene ring, which is antiaromatic, have any effect on the appearance of the orbitals in comparison with, say, naphthalene? Compare the bond lengths of benzyne with those of benzene.

3. (a) F. M. Logullo, A. H. Seitz, and L. Friedman, *Org. Syn.* **48,** 12 (1968). (b) L. Friedman and F. M. Logullo, *J. Org. Chem.,* **34,** 3089 (1969).

Questions

1. Write equations showing how benzyne might be generated from *o*-fluoro-bromobenzene and magnesium.

2. Study the mechanism of diazonium salt formation in Chapter 45, and then propose a mechanism for diazotization using isoamyl nitrite.

3. The ¹H NMR spectrum of anthracene is given in Fig. 35.1. Predict the appearance of the NMR spectrum of tryptycene. Which peak in Fig. 35.1 will be missing?

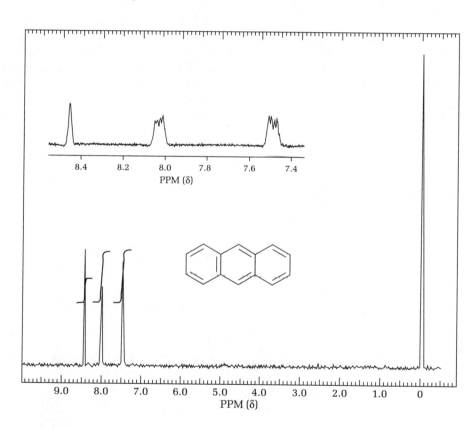

FIG. 35.1   ¹H NMR spectrum of anthracene (250 MHz).

CHAPTER

36

Aldehydes and Ketones

Prelab Exercise: Outline a logical series of experiments designed to identify an unknown aldehyde or ketone with the least effort. Consider the time required to complete each identification reaction.

The carbonyl group occupies a central place in organic chemistry. Aldehydes and ketones—compounds such as formaldehyde, acetaldehyde, acetone, and 2-butanone—are very important industrial chemicals used by themselves and as starting materials for a host of other substances. For example, in 1991 10 *billion* lb of formaldehyde-containing plastics were produced in the United States.

The carbonyl carbon is sp^2 hybridized, the bond angles between adjacent groups are 120°, and the four atoms R, R′, C, and O lie in one plane:

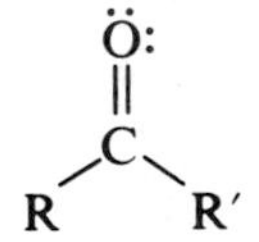

$$\ddot{O}:$$
$$\|$$
$$C$$
$$R \qquad R'$$

The electronegative oxygen polarizes the carbon–oxygen bond, rendering the carbon electron deficient and hence subject to nucleophilic substitution.

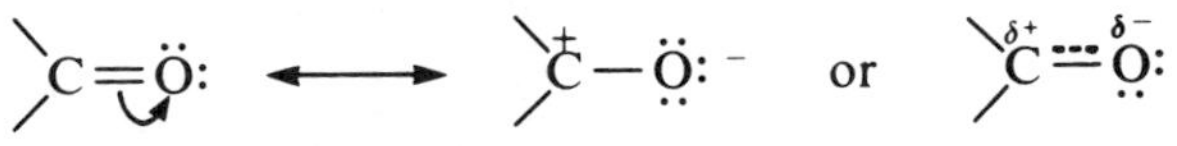

$$\sideset{}{}{\mathop{C}}=\ddot{O}: \longleftrightarrow \overset{+}{C}-\ddot{O}:^- \quad \text{or} \quad \overset{\delta+}{C}\cdots\overset{\delta-}{\ddot{O}}:$$

Geometry of the carbonyl group

Attack on the sp^2 hybridized carbon occurs via the π-electron cloud above and below the plane of the carbonyl group:

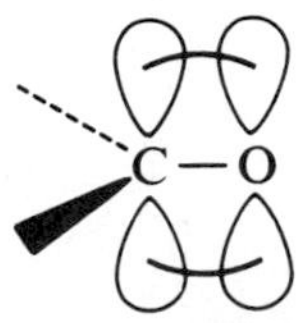

Reactions of the Carbonyl Group

Many reactions of carbonyl groups are acid catalyzed. The acid attacks the electronegative oxygen, which bears a partial negative charge, to create a carbocation that subsequently reacts with the nucleophile:

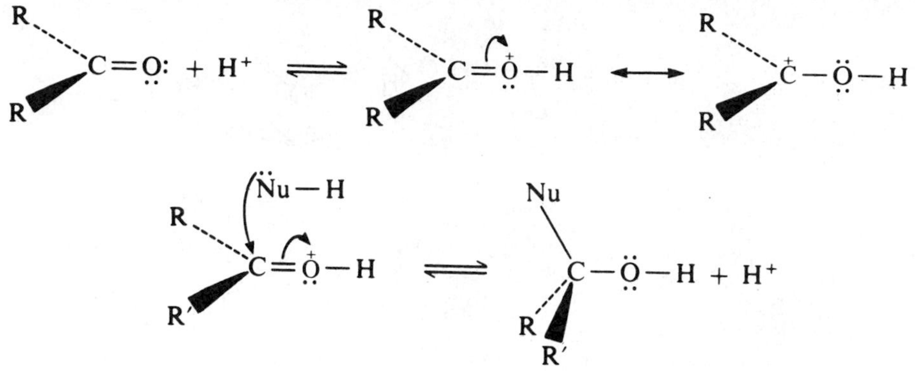

The strength of the nucleophile and the structure of the carbonyl compound determine whether the equilibrium lies on the side of the carbonyl compound or the tetrahedral adduct. Water, a weak nucleophile, does not usually add to the carbonyl group to form a stable compound:

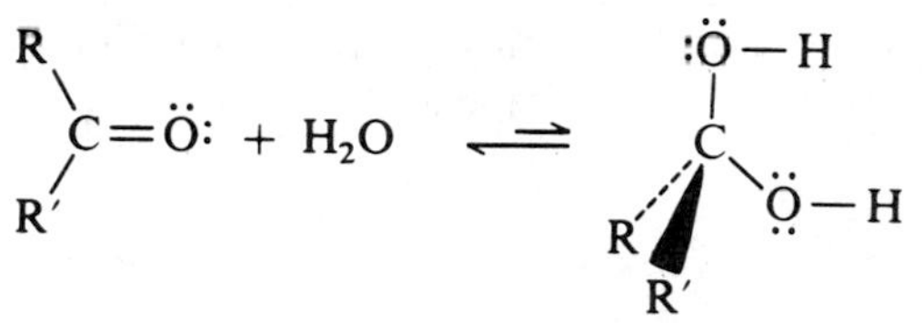

A stable hydrate: chloral hydrate

but in the special case of trichloroacetaldehyde, the electron-withdrawing trichloromethyl group allows a stable hydrate to form:

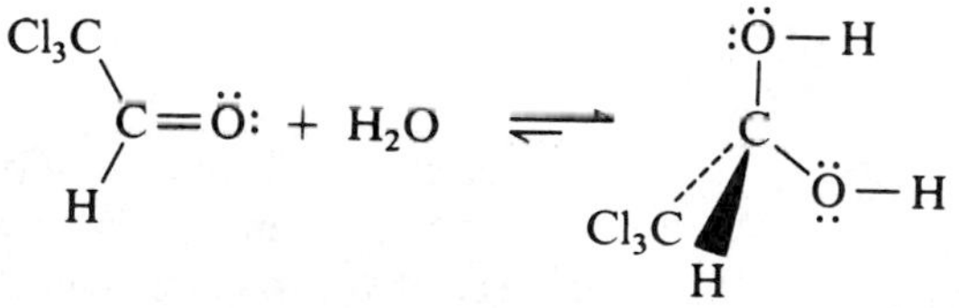

The compound so formed, chloral hydrate, was discovered by Liebig in 1832 and was introduced as one of the first sedatives and hypnotics (sleep-inducing substances) in 1869. It is now most commonly encountered in detective fiction as a "Mickey Finn" or "knockout drops."

Reaction with an alcohol; hemiacetals

In an analogous manner, an aldehyde or ketone can react with an alcohol. The product, a hemiacetal or hemiketal, is usually not stable, but in the case of certain cyclic hemiacetals the product can be isolated. Glucose is an example of a stable hemiacetal.

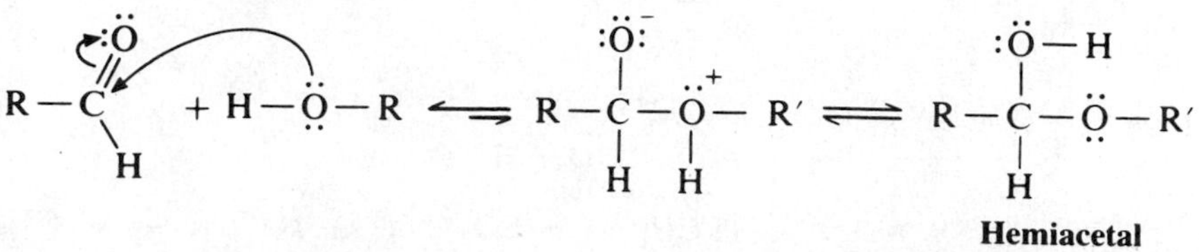

Hemiacetal
usually not stable

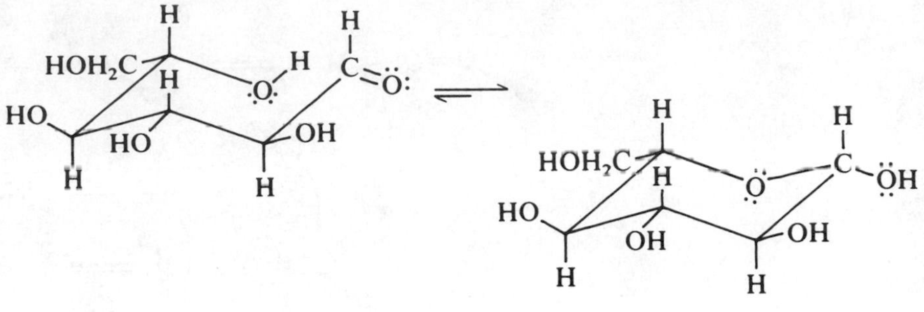

Glucose
a stable cyclic hemiacetal

Bisulfite Addition

The bisulfite ion is a strong nucleophile but a weak acid. It will attack the unhindered carbonyl group of an aldehyde or methyl ketone to form an addition product:

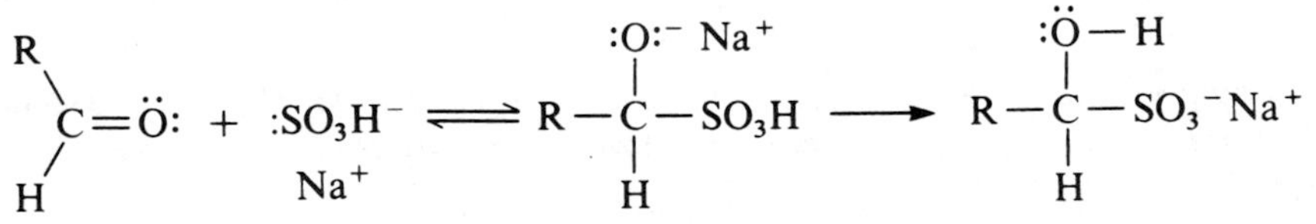

Since these bisulfite addition compounds are ionic water-soluble compounds and can be formed in up to 90% yield, they serve as a useful means of separating aldehydes and methyl ketones from mixtures of organic compounds. At high sodium bisulfite concentrations these adducts crystallize and can be isolated by filtration. The aldehyde or ketone can be regenerated by adding either a strong acid or base:

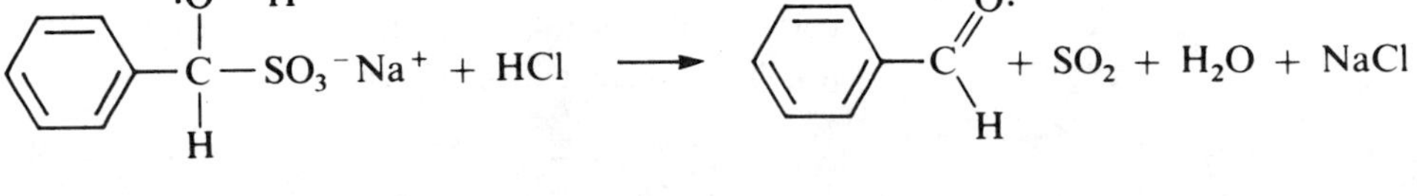

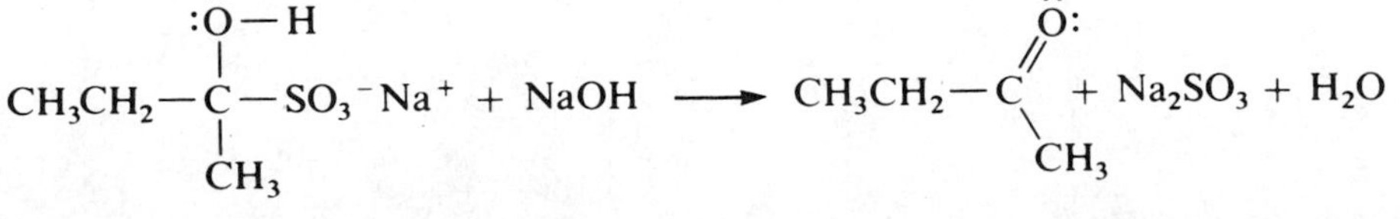

Cyanide Addition

A similar reaction occurs between aldehydes and ketones and hydrogen cyanide, which, like bisulfite, is a weak acid but a strong nucleophile. The reaction is hazardous to carry out because of the toxicity of cyanide, but the cyanohydrins are useful synthetic intermediates:

$$
\underset{\substack{\ddot{O}: \\ \|}}{CH_3CH_2-C-CH_3} + HCN \longrightarrow \underset{\substack{\ddot{O}-H \\ | \\ CN}}{CH_3CH_2\underset{|}{C}CH_3} \xrightarrow{H_2SO_4} \underset{\substack{| \\ COOH}}{CH_3CH=CCH_3}
$$

Cyanohydrin formation and reactions

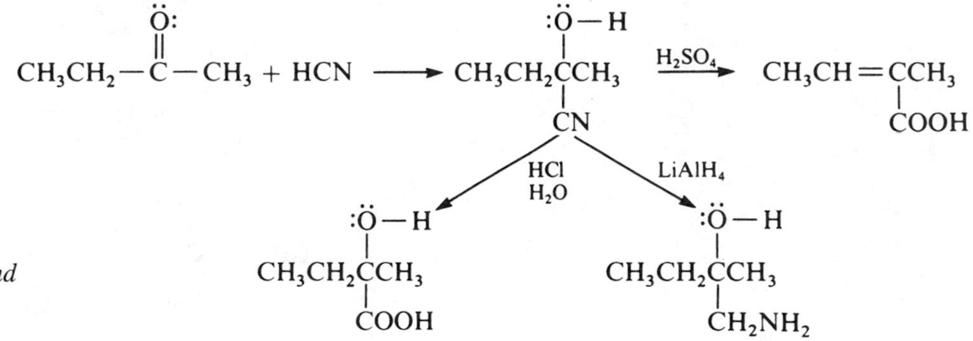

$$
\text{(HCl, } H_2O): \quad \underset{\substack{:\ddot{O}-H \\ | \\ COOH}}{CH_3CH_2\underset{|}{C}CH_3}
$$

$$
\text{(LiAlH}_4): \quad \underset{\substack{:\ddot{O}-H \\ | \\ CH_2NH_2}}{CH_3CH_2\underset{|}{C}CH_3}
$$

Amines are good nucleophiles and readily add to the carbonyl group:

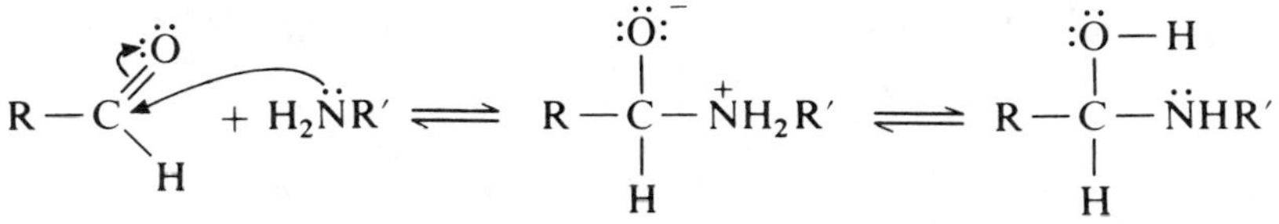

$$
\underset{\substack{\overset{\displaystyle \ddot{O}}{\diagup} \\ R-C \\ \diagdown \\ H}}{} + H_2\ddot{N}R' \rightleftharpoons \underset{\substack{:\ddot{O}:^- \\ | \\ R-C-\overset{+}{N}H_2R' \\ | \\ H}}{} \rightleftharpoons \underset{\substack{:\ddot{O}-H \\ | \\ R-C-\ddot{N}HR' \\ | \\ H}}{}
$$

The reaction is strongly dependent on the pH. In acid the amine is protonated ($R\overset{+}{N}H_3$) and is no longer a nucleophile. In strong base there are no protons available to catalyze the reaction. But in weak acid solution (pH 4–6) the equilibrium between acid and base (**a**) is such that protons are available to protonate the carbonyl (**b**), and yet there is free amine present to react with the protonated carbonyl (**c**):

(a) $\qquad CH_3\ddot{N}H_2 + HCl \rightleftharpoons CH_3\overset{+}{N}H_3 + Cl^-$

(b) $\quad \underset{\substack{:\ddot{O} \\ \diagup\diagup \\ CH_3-C \\ \diagdown \\ H}}{} + HCl \rightleftharpoons \left[\underset{\substack{:\overset{+}{\ddot{O}}-H \\ \diagup\diagup \\ CH_3-C \\ \diagdown \\ H}}{} \longleftrightarrow \underset{\substack{:\ddot{O}-H \\ | \\ CH_3-\overset{+}{C} \\ \diagdown \\ H}}{} \right] + Cl^-$

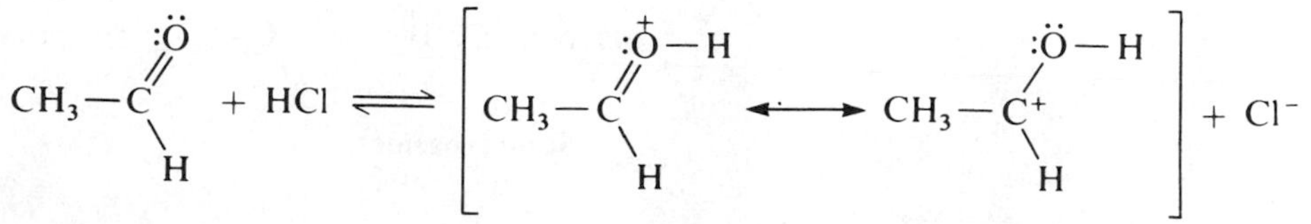

(c) $\quad \underset{\substack{\overset{+}{:\ddot{O}}-H \\ \diagup\diagup \\ CH_3-C \\ \diagdown \\ H}}{} + CH_3\ddot{N}H_2 \rightleftharpoons \underset{\substack{:\ddot{O}-H \\ | \\ CH_3-C-\overset{+}{N}H_2CH_3 \\ | \\ H}}{} \xrightarrow{-H^+} \underset{\substack{:\ddot{O}-H \\ | \\ CH_3-C-\ddot{N}-CH_3 \\ | \quad | \\ H \quad H}}{}$

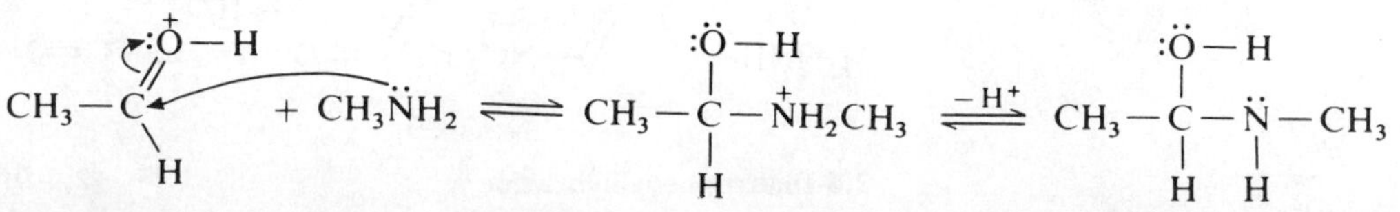

Schiff Bases

The intermediate hydroxyamino form of the adduct is not stable and spontaneously dehydrates under the mildly acidic conditions of the reaction to give a imine, commonly referred to as a *Schiff base:*

$$H^+ + CH_3-\overset{\overset{\displaystyle :\ddot{O}-H}{|}}{\underset{\underset{\displaystyle H}{|}}{C}}-\overset{\overset{\displaystyle H}{|}}{\underset{\underset{\displaystyle H}{|}}{\ddot{N}}}-CH_3 \;\rightleftharpoons\; CH_3-\overset{\overset{\displaystyle :\ddot{O}-H}{|}}{\underset{\underset{\displaystyle H}{|}}{C}}-\overset{+}{N}H_2CH_3 \;\rightleftharpoons\; CH_3-\overset{\overset{\displaystyle \overset{\displaystyle H}{|}}{\overset{\displaystyle :\overset{+}{O}-H}{|}}}{\underset{\underset{\displaystyle H}{|}}{C}}-\overset{\overset{\displaystyle H}{|}}{\underset{\underset{\displaystyle H}{|}}{\ddot{N}}}-CH_3$$

$$H^+ + \quad \overset{\overset{\displaystyle CH_3}{|}}{\underset{\underset{\displaystyle H}{|}}{C}}=\overset{\overset{\displaystyle CH_3}{\diagup}}{\ddot{N}} \quad \rightleftharpoons\; CH_3-\overset{\overset{\displaystyle CH_3}{||}}{C}=\overset{+}{\underset{\underset{\displaystyle H}{|}}{N}}-CH_3 + H_2O$$

Schiff base

Imine or Schiff base formation   The biosynthesis of most amino acids proceeds through Schiff base intermediates.

Oximes, Semicarbazones, and 2,4-Dinitrophenylhydrazones

Three rather special amines form useful stable imines:

$$H_2\ddot{N}\ddot{O}H + \overset{\displaystyle R}{\underset{\displaystyle R'}{\diagup}}C=\ddot{O}: \quad\overset{-H_2O}{\rightleftharpoons}\quad \overset{\displaystyle R}{\underset{\displaystyle R'}{\diagup}}C=\overset{\overset{\displaystyle \ddot{O}-H}{\diagdown}}{\ddot{N}}$$

Hydroxylamine   **Oxime**

$$H_2NNH\overset{\overset{\displaystyle O}{||}}{C}NH_2 + \overset{\displaystyle R}{\underset{\displaystyle R'}{\diagup}}C=O \quad\overset{-H_2O}{\rightleftharpoons}\quad \overset{\displaystyle R}{\underset{\displaystyle R'}{\diagup}}C=N-\overset{\overset{\displaystyle }{\underset{\underset{\displaystyle H}{|}}{N}}}{}-\overset{\overset{\displaystyle O}{||}}{C}-NH_2$$

Semicarbazide   **Semicarbazone**

$$H_2NNH-\!\!\!\!\bigcirc\!\!\!\!-NO_2 + \overset{\displaystyle R}{\underset{\displaystyle R'}{\diagup}}C=O \quad\overset{-H_2O}{\rightleftharpoons}\quad \overset{\displaystyle R}{\underset{\displaystyle R'}{\diagup}}C=N-\underset{\underset{\displaystyle H}{|}}{N}-\!\!\!\!\bigcirc\!\!\!\!-NO_2$$

(with NO_2 substituents on the rings)

2,4-Dinitrophenylhydrazine   **2,4-Dinitrophenylhydrazone**

These imines are solids and are useful for the characterization of alde-hydes and ketones. For example, infrared and NMR spectroscopy may indicate that a certain unknown is acetaldehyde. It is difficult to determine the boiling point of a few milligrams of a liquid, but if it can be converted to a solid deriv-ative the mp *can* be determined with that amount. The 2,4-dinitrophenylhydra-zones are usually the derivatives of choice because they are nicely crystalline compounds with well-defined melting or decomposition points and they increase the molecular weight by 180. Ten milligrams of acetaldehyde will give 51 mg of the 2,4-DNP:

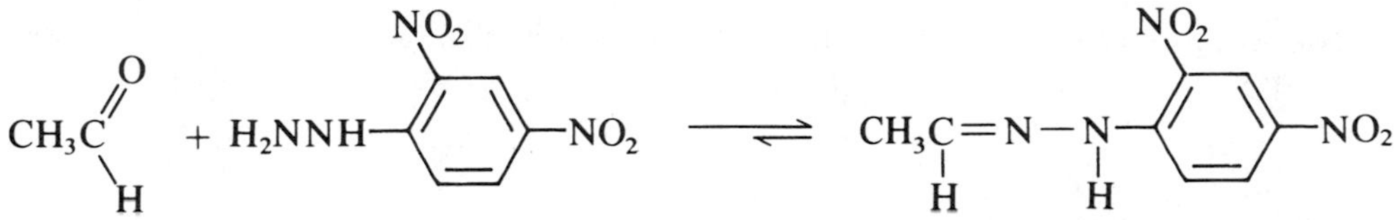

Acetaldehyde
MW 44.05
bp 20.8°C

2,4-Dinitrophenylhydrazine
MW 198.14
mp 196°C

Acetaldehyde 2,4-dinitrophenylhydrazone
MW 224.19
mp 168.5°C

Tollens' Reagent

Before the advent of NMR and IR spectroscopy the chemist was often called on to identify aldehydes and ketones by purely chemical means. Aldehydes can be distinguished chemically from ketones by their ease of oxidation to carboxylic acids. The oxidizing agent, an ammoniacal solution of silver nitrate, Tollens' reagent, is reduced to metallic silver, which is deposited on the inside of a test tube as a silver mirror.

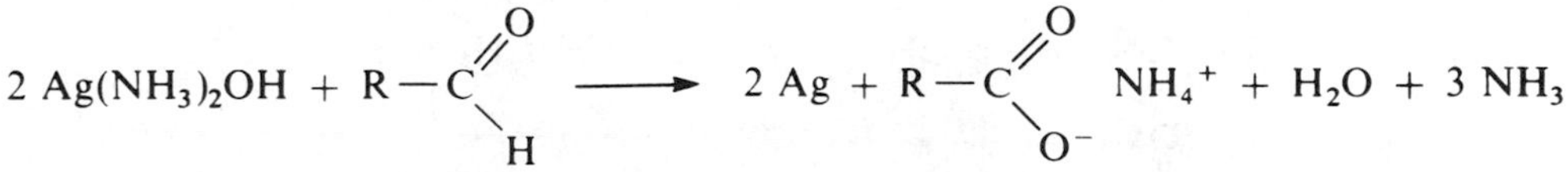

Schiff's Reagent

Another way to distinguish aldehydes from ketones is to use Schiff's reagent. This is a solution of the red dye Basic Fuchsin, which is rendered colorless on treatment with sulfur dioxide. In the presence of an aldehyde the colorless solu-tion turns magenta.

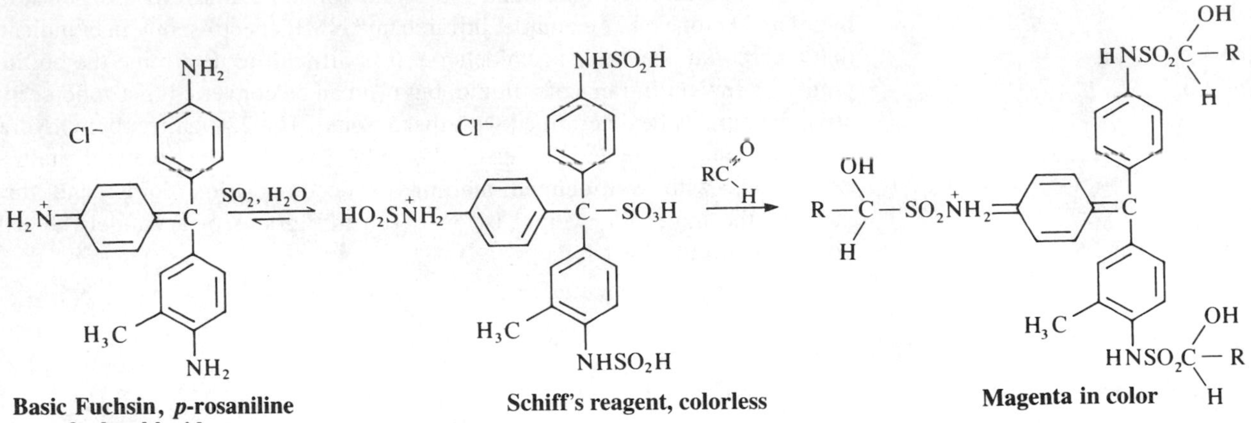

Basic Fuchsin, *p*-rosaniline hydrochloride   **Schiff's reagent, colorless**   **Magenta in color**

A test for methyl ketones

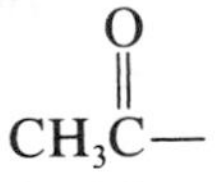

$$\underset{\text{Methyl ketone}}{CH_3\overset{\displaystyle O}{\overset{\|}{C}}-}$$

Iodoform Test

Methyl ketones can be distinguished from other ketones by the iodoform test. The methyl ketone is treated with iodine in a basic solution. Introduction of the first iodine atom increases the acidity of the remaining methyl protons, so halogenation stops only when the triiodo compound has been produced. The base then allows the relatively stable triiodomethyl carbanion to leave, and a subsequent proton transfer gives iodoform, a yellow crystalline solid of mp 119–123°C. The test is also positive for fragments easily oxidized to methyl ketones, such as CH_3CHOH- and ethanol. Acetaldehyde also gives a positive test because it is both a methyl ketone and an aldehyde.

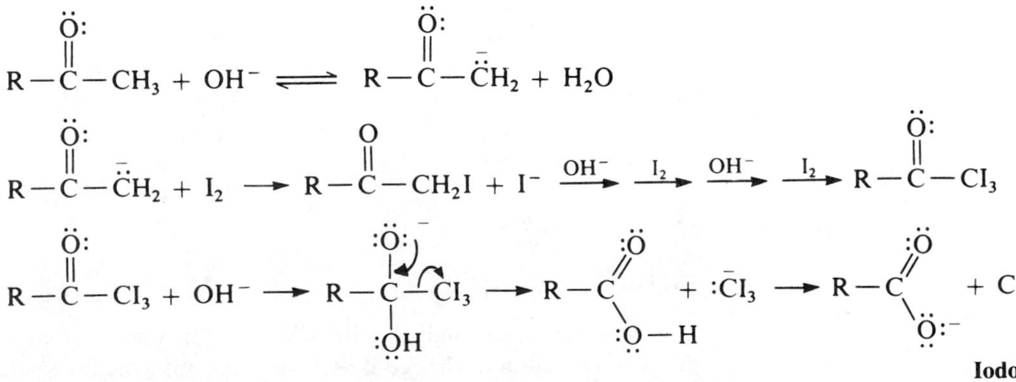

Iodoform
mp 123°C

**MICROSCALE
AND MACROSCALE**

Experiments _____

1. Unknowns

You will be given an unknown that may be any of the aldehydes or ketones listed in Table 36.1. At least one derivative of the unknown is to be submitted to the instructor; but if you first do the bisulfite and iodoform characterizing tests, the results may suggest derivatives whose melting points will be particularly revealing.

You can further characterize the unknown by determining the boiling point, best done with a digital thermometer and a reaction tube (Fig. 4.7). The boiling points of the unknowns are given in Tables 70.3 and 70.14.

Carry out three tests:
  Known positive
  Known negative
  Unknown

In conducting the following tests you should perform three tests simultaneously: on a compound known to give a positive test, on a compound known to give a negative test, and on the unknown. In this way you will be able to determine whether the reagents are working as they should as well as interpret a positive or a negative test.

MICROSCALE

2. 2,4-Dinitrophenylhydrazones

$$\underset{R'}{\overset{R}{\diagdown}}C=\ddot{O}: \; + \; H_2\ddot{N}-\underset{\underset{H}{|}}{\ddot{N}}-\text{benzene ring with } NO_2, NO_2 \xrightarrow{\;H^+\;} \underset{R'}{\overset{R}{\diagdown}}C=\ddot{N}-\underset{\underset{H}{|}}{\ddot{N}}-\text{benzene ring with } NO_2, NO_2$$

An easily prepared derivative of aldehydes and ketones

To 5 mL of the stock solution[1] of 2,4-dinitrophenylhydrazine in phosphoric acid add about 0.05 g of the compound to be tested. Five milliliters of the 0.1 *M* solution contains 0.5 millimole (0.0005 mol) of the reagent. If the compound to be tested has a molecular weight of 100, then 0.05 g is 0.5 millimole. Warm the reaction mixture for a few minutes in a water bath, and then let crystallization proceed. Collect the product by suction filtration (Fig. 36.1), wash the crystals with a large amount of water to remove all phosphoric acid, press a piece of moist litmus paper onto the crystals, and if they are acidic wash them with more water. Press the product between sheets of filter paper until it is as dry as possible, and recrystallize from ethanol. Occasionally a high molecular weight derivative won't dissolve in a reasonable quantity (10 mL) of ethanol. In that case cool the hot suspension, and isolate the crystals by suction filtration. The boiling ethanol treatment removes impurities so that an accurate melting point can be obtained on the isolated material.

An alternative procedure is applicable when the 2,4-dinitrophenylhydrazone is known to be sparingly soluble in ethanol. Measure 0.5 millimole (0.1 g) of crystalline 2,4-dinitrophenylhydrazine into a 50-mL Erlenmeyer flask, add

_____

1. Dissolve 2.0 g of 2,4-dinitrophenylhydrazine in 50 mL of 85% phosphoric acid by heating, cool, add 50 mL of 95% ethanol, cool again, and clarify by suction filtration from a trace of residue.

TABLE 36.1   Melting Points of Derivatives of Some Aldehydes and Ketones

Compound*	Formula	n_D^{20}	MW	Water Solubility	Phenyl-hydrazone	2,4-DNP	Semi-carbazone
Acetone	CH_3COCH_3	1.3590	58.08		42	126	187
n-Butanal	$CH_3CH_2CH_2CHO$	1.3790	72.10	4 g/100 g	Oil	123	95(106)†
3-Pentanone (diethyl ketone)	$CH_3CH_2COCH_2CH_3$	1.3920	86.13	4.7 g/100 g	Oil	156	138
2-Furaldehyde (furfural)	$C_4H_3O \cdot CHO$	1.5260	96.08	9 g/100 g	97	212 (230)†	202
Benzaldehyde	C_6H_5CHO	1.5450	106.12	Insol.	158	237	222
Hexane-2,5-dione	$CH_3COCH_2CH_2COCH_3$	1.4260	114.14	∞	120‡	257‡	224‡
2-Heptanone	$CH_3(CH_2)_4COCH_3$	1.4080	114.18	Insol.	Oil	89	123
3-Heptanone	$CH_3(CH_2)_3COCH_2CH_3$	1.4080	114.18	Insol.	Oil	81	101
n-Heptanal	$n\text{-}C_6H_{13}CHO$	1.4125	114.18	Insol.	Oil	108	109
Acetophenone	$C_6H_5COCH_3$	1.5325	120.66	Insol.	105	238	198
2-Octanone	$CH_3(CH_2)_5COCH_3$	1.4150	128.21	Insol.	Oil	58	122
Cinnamaldehyde	$C_6H_5CH{=}CHCHO$	1.6220	132.15	Insol.	168	255	215
Propiophenone	$C_6H_5COCH_2CH_3$	1.5260	134.17	Insol.	About 48	191	182

*See Tables 70.3 and 70.4 for additional aldehydes and ketones.
†Both melting points have been found, depending on crystalline form of derivative.
‡Monoderivative or diderivative.

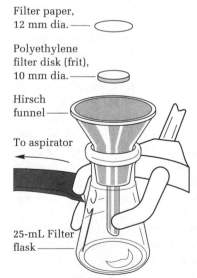

Filter paper, 12 mm dia. —

Polyethylene filter disk (frit), 10 mm dia. —

Hirsch funnel —

To aspirator

25-mL Filter flask —

FIG. 36.1   Hirsch funnel filtration apparatus.

15 mL of 95% ethanol, digest on the steam bath until all particles of solid are dissolved, and then add 0.5 millimole of the compound to be tested and continue warming. If there is no immediate change, add, from a Pasteur pipette, 3–4 drops of concentrated hydrochloric acid as catalyst and note the result. Warm for a few minutes, then cool and collect the product. This procedure would be used for an aldehyde like cinnamaldehyde ($C_6H_5CH{=}CHCHO$).

The alternative procedure strikingly demonstrates the catalytic effect of hydrochloric acid, but it is not applicable to a substance like diethyl ketone, whose 2,4-dinitrophenylhydrazone is much too soluble to crystallize from the large volume of ethanol. The first procedure is obviously the one to use for an unknown.

Cleaning Up   The filtrate from the preparation of the 2,4-dinitrophenylhydrazone should have very little 2,4-dinitrophenylhydrazine in it, so after dilution with water and neutralization with sodium carbonate it can be flushed down the drain. Similarly the mother liquor from crystallization of the phenylhydrazone should have very little product in it and so should be diluted and flushed down the drain. If solid material is detected, it should be collected by suction filtration, the filtrate flushed down the drain, and the filter paper placed in the solid hazardous waste container, because hydrazines are toxic.

MICROSCALE

3. Semicarbazones

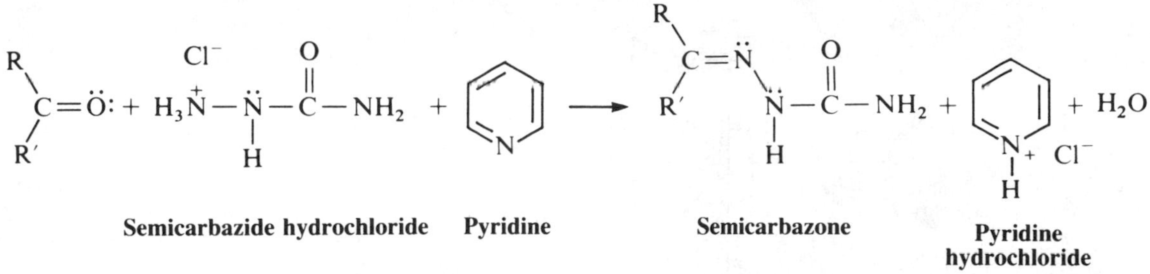

Semicarbazide hydrochloride   **Pyridine**   **Semicarbazone**   **Pyridine hydrochloride**

Semicarbazide (mp 96°C) is not very stable in the free form and is used as the crystalline hydrochloride (mp 173°C). Since this salt is insoluble in methanol or ethanol and does not react readily with typical carbonyl compounds in alcohol-water mixtures, a basic reagent, pyridine, is added to liberate free semicarbazide.

To 0.5 mL of the stock solution[2] of semicarbazide hydrochloride, which contains 1 millimole of the reagent, add 1 millimole of the compound to be tested and enough methanol (1 mL) to produce a clear solution; then add 10 drops of pyridine (a twofold excess) and warm the solution gently on the steam bath for a few minutes. Cool the solution slowly to room temperature. It may be necessary to scratch the inside of the test tube in order to induce crystallization. Cool the tube in ice, collect the product by suction filtration, and wash it with water followed by a small amount of cold methanol. Recrystallize the product from methanol, ethanol, or ethanol/water. The product can easily be collected on the Wilfilter (Fig. 36.2).

Cleaning Up   Combine the filtrate from the reaction and the mother liquor from the crystallization, dilute with water, make very slightly acidic with dilute hydrochloric acid, and flush the mixture down the drain.

MICROSCALE

4. Tollens' Test

Test for aldehydes

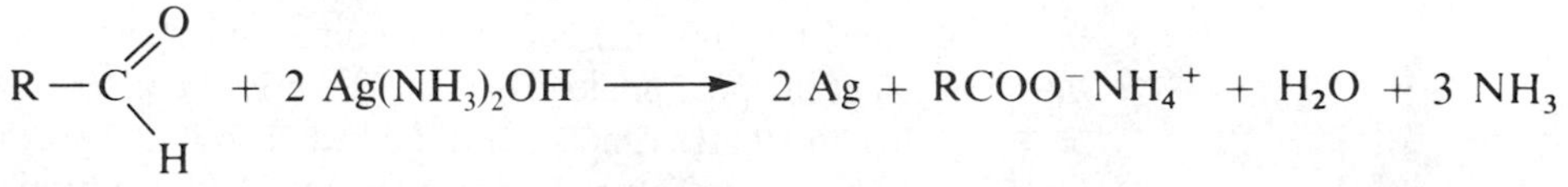

$$R-C\begin{matrix}O\\\\H\end{matrix} + 2\ Ag(NH_3)_2OH \longrightarrow 2\ Ag + RCOO^-NH_4^+ + H_2O + 3\ NH_3$$

Clean four or five test tubes by adding a few milliliters of 3 M sodium hydroxide solution to each and heating them in a water bath while preparing Tollens' reagent.

---

2. Prepare a stock solution by dissolving 1.11 g of semicarbazide hydrochloride in 5 mL of water; 0.5 mL of this solution contains 1 millimole of reagent.

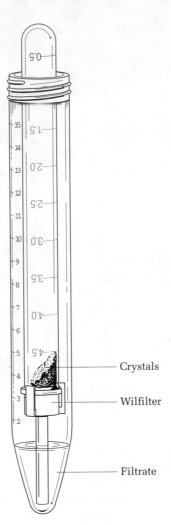

Crystals

Wilfilter

Filtrate

FIG. 36.2 Wilfilter filtration apparatus. See page 52 in Chapter 3.

To 2.0 mL of 0.03 *M* silver nitrate solution, add 1.0 mL of 3 *M* sodium hydroxide in a test tube. To the gray precipitate of silver oxide, Ag_2O, add 0.5 mL of a 2.8% ammonia solution (10 mL of concentrated ammonium hydroxide diluted to 100 mL). Stopper the tube, and shake it. Repeat the process until *almost* all of the precipitate dissolves (3.0 mL of ammonia at most); then dilute the solution to 10 mL. Empty the test tubes of sodium hydroxide solution, rinse them, and add 1 mL of Tollens' reagent to each. Add one drop (no more) of the substance to be tested by allowing it to run down the inside of the inclined test tube. Set the tubes aside for a few minutes without agitating the contents. If no reaction occurs, warm the mixture briefly on a water bath. As a known aldehyde try one drop of a 0.1 *M* solution of glucose. A more typical aldehyde to test is benzaldehyde.

At the end of the reaction promptly destroy excess Tollens' reagent with nitric acid: It can form an explosive fulminate on standing. Nitric acid can also be used to remove silver mirrors from the test tubes.

Cleaning Up   Place all solutions used in this experiment in a beaker (unused ammonium hydroxide, sodium hydroxide solution used to clean out the tubes, Tollens' reagent from all tubes). Remove any silver mirrors from reaction tubes with a few drops of nitric acid, which is added to the beaker. Make the mixture acidic with nitric acid to destroy unreacted Tollens' reagent, then neutralize the solution with sodium carbonate and add some sodium chloride solution to precipitate silver chloride (about 40 mg). The whole mixture can be flushed down the drain or the silver chloride collected by suction filtration and the filtrate flushed down the drain. The silver chloride would go in the nonhazardous solid waste container.

⊍ MICROSCALE

Very sensitive test for aldehydes

5. Schiff's Test

Add three drops of the unknown to 2 mL of Schiff's reagent.[3] A magenta color will appear within 10 min with aldehydes. As in all of these tests, compare the colors produced by a known aldehyde, a known ketone, and the unknown compound.

Cleaning Up   Neutralize the solution with sodium carbonate, and flush it down the drain. The amount of *p*-rosaniline in this mixture is negligible (1 mg).

⊍ MICROSCALE

6. Iodoform Test

$$R - \overset{\overset{\ddot{O}}{\|}}{C} - CH_3 + 3\ I_2 + 4\ OH^- \longrightarrow R - C\overset{\ddot{O}}{\underset{\ddot{O}:^-}{\diagdown}} + CHI_3 + 3\ \ddot{\underset{..}{I}}:^- + 3\ H_2O$$

A methyl ketone   **Iodoform**
                                                         mp 119–123°C

Test for methyl ketones

      The reagent contains iodine in potassium iodide solution[4] at a concentration such that 2 mL of solution, on reaction with excess methyl ketone, will yield 174 mg of iodoform. If the substance to be tested is water soluble, dissolve four drops of a liquid or an estimated 50 mg of a solid in 2 mL of water in a 20 × 150-mm test tube; add 2 mL of 3 *M* sodium hydroxide, and then slowly add 3 mL of the iodine solution. In a positive test the brown color of the reagent disappears, and yellow iodoform separates. If the substance to be tested is insoluble in water, dissolve it in 2 mL of 1,2-dimethoxyethane, proceed as above, and at the end dilute with 10 mL of water.

---

3. Schiff's reagent is prepared by dissolving 0.1 g Basic Fuchsin (*p*-rosaniline hydrochloride) in 100 mL of water and then adding 4 mL of a saturated aqueous solution of sodium bisulfite. After 1 h, add 2 mL of concentrated hydrochloric acid.

4. Dissolve 25 g of iodine in a solution of 50 g of potassium iodide in 200 mL of water.

Suggested test substances are hexane-2,5-dione (water soluble), *n*-butyraldehyde (water soluble), and acetophenone (water insoluble).

Iodoform can be recognized by its odor and yellow color and, more securely, from the melting point (119–123°C). The substance can be isolated by suction filtration of the test suspension or by adding 0.5 mL of dichloromethane, shaking the stoppered test tube to extract the iodoform into the small lower layer, withdrawing the clear part of this layer with a Pasteur pipette, and evaporating it in a small tube on the steam bath. The crude solid is crystallized from methanol-water. It can be collected on the Wilfilter (Fig. 36.2).

Cleaning Up   Combine all reaction mixtures in a beaker, add a few drops of acetone to destroy any unreacted iodine in potassium iodide reagent, remove the iodoform by suction filtration, and place it in the halogenated organic waste container. The filtrate can be flushed down the drain after neutralization (if necessary).

MICROSCALE

7. Bisulfite Test

Forms with unhindered carbonyls

$$\underset{H}{\overset{R}{\diagdown}}C=\overset{..}{\underset{..}{O}}: \; + \; Na^+SO_3H^- \; \rightleftharpoons \; R-\underset{H}{\overset{:\overset{..}{O}H}{\underset{|}{\overset{|}{C}}}}-SO_3^-Na^+$$

Put 1 mL of the stock solution[5] into a 13 × 100-mm test tube, and add five drops of the substance to be tested. Shake each tube during the next 10 min, and note the results. A positive test will result from aldehydes, unhindered cyclic ketones such as cyclohexanone, and unhindered methyl ketones.

If the bisulfite test is applied to a liquid or solid that is very sparingly soluble in water, formation of the addition product is facilitated by adding a small amount of methanol before the addition to the bisulfite solution.

Cleaning Up   Dilute the bisulfite solution or any bisulfite addition products (they will dissociate) with a large volume of water, and flush the mixture down the drain. The amount of organic material being discarded is negligible.

MICROSCALE

8. IR and NMR Spectroscopy

Infrared spectroscopy is extremely useful in analyzing all carbonyl-containing compounds, including aldehydes and ketones (see Figs. 36.5 and 36.12). See the extensive discussion in Chapter 12. In the modern laboratory spectroscopy has almost completely supplanted the qualitative tests described in this chapter. Figures 36.3 through 36.12 present IR and NMR spectra of typical aldehydes and

---

5. Prepare a stock solution from 50 g sodium bisulfite dissolved in 200 mL of water with brief swirling.

ketones. Compare these spectra with those of your unknown. Also compare the IR and ^{1}H NMR spectra of the hydrocarbon fluorene (Figs. 36.3 and 36.4) with those of the ketone derivative, fluorenone (Figs. 36.5 and 36.6).

A peak at 9.6–10 ppm in the ^{1}H NMR spectrum is highly characteristic of aldehydes, because almost no other peaks appear in this region (see Figs. 36.8 and 36.11). On some spectrometers an offset will be needed to detect this region. Similarly a sharp singlet at 2.2 ppm is very characteristic of methyl ketones; but beware of contamination of the sample by acetone, which is often used to clean glassware.

Questions

1. What is the purpose of making derivatives of unknowns?

2. Why are 2,4-dinitrophenylhydrazones better derivatives than phenylhydrazones?

3. Using chemical tests, how would you distinguish among 2-pentanone, 3-pentanone, and pentanal?

4. Draw the structure of a compound, C_5H_8O, that gives a positive iodoform test and does not decolorize permanganate.

5. Draw the structure of a compound, C_5H_8O, that gives a positive Tollens' test and does not react with bromine in dichloromethane.

6. Draw the structure of a compound, C_5H_8O, that reacts with phenylhydrazine, decolorizes bromine in dichloromethane, and does not give a positive iodoform test.

7. Draw the structure of two geometric isomers, C_5H_8O, that give a positive iodoform test.

8. What vibrations cause the peaks at about 3.4 μm (3000 cm^{-1}) in the IR spectrum of fluorene (Fig. 36.3)?

9. Locate the carbonyl peak in Fig. 36.5.

10. Assign the various peaks in the ^{1}H NMR spectrum of 2-butanone to specific protons in the molecule (Fig. 36.7).

11. Assign the various peaks in the ^{1}H NMR spectrum of crotonaldehyde to specific protons in the molecule (Fig. 36.8).

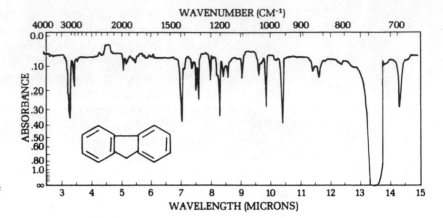

FIG. 36.3 Infrared spectrum of fluorene in CS_2.

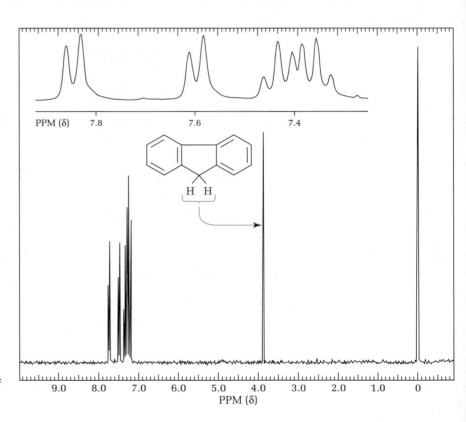

FIG. 36.4 ^{1}H NMR spectrum of fluorene (250 MHz).

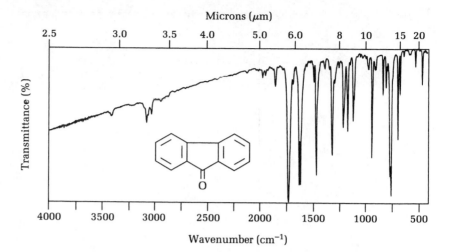

FIG. 36.5   IR spectrum of fluorenone (KBr disk).

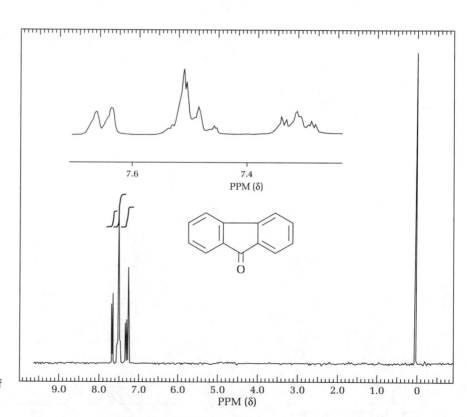

FIG. 36.6   ¹H NMR spectrum of fluorenone (250 MHz).

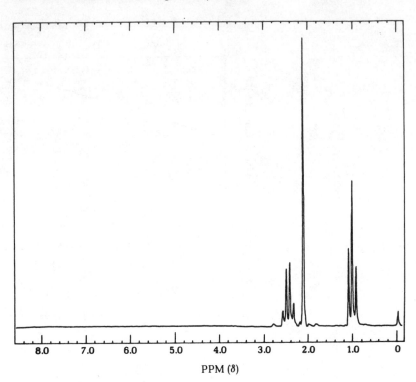

FIG. 36.7 ¹H NMR spectrum of 2-butanone, CH₃COCH₂CH₃ (90 MHz).

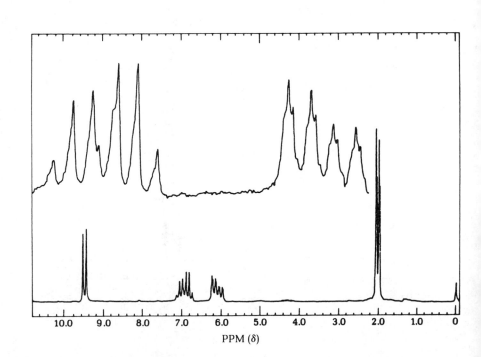

FIG. 36.8 ¹H NMR spectrum of crotonaldehyde, CH₃CH=CHCHO (90 MHz).

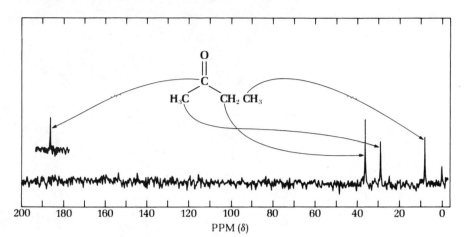

FIG. 36.9 ^{13}C NMR spectrum of 2-butanone, $CH_3COCH_2CH_3$ (22.6 MHz).

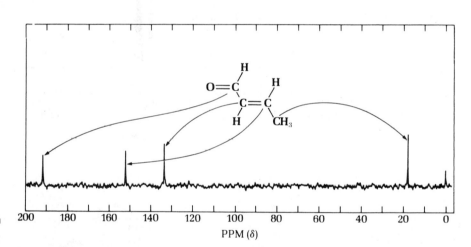

FIG. 36.10 ^{13}C NMR spectrum of crotonaldehyde (22.6 MHz).

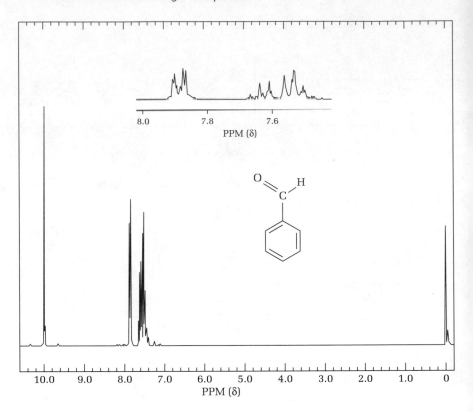

FIG. 36.11 ¹H NMR spectrum of benzaldehyde (250 MHz).

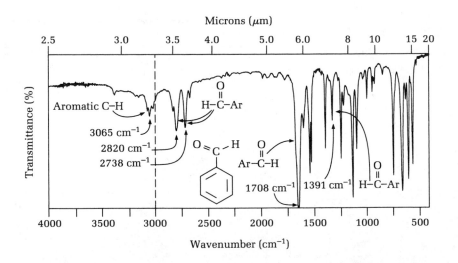

FIG. 36.12 IR spectrum of benzaldehyde (thin film).

Dibenzalacetone by the Aldol Condensation

Prelab Exercise: Calculate the volumes of benzaldehyde and acetone needed for this reaction, taking into account the densities of the liquids and the number of moles of each required.

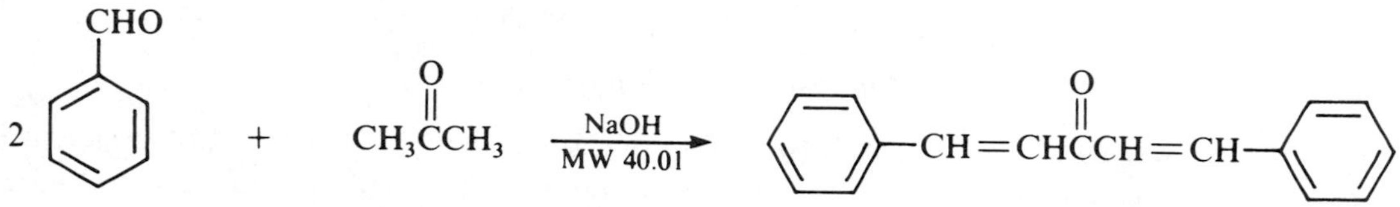

Benzaldehyde
MW 106.13, bp 178°C
den. 1.04

Acetone
(2-Propanone)
MW 58.08, bp 56°C
den. 0.790

Dibenzalacetone
(1,5-Diphenyl-1,4-pentadien-3-one)
MW 234.30

The reaction of an aldehyde with a ketone employing sodium hydroxide as the base is an example of a mixed aldol condensation reaction, the Claisen-Schmidt reaction. Dibenzalacetone is readily prepared by condensation of acetone with two equivalents of benzaldehyde. The aldehyde carbonyl is more reactive than that of the ketone and therefore reacts rapidly with the anion of the ketone to give a β-hydroxyketone, which easily undergoes base-catalyzed dehydration. Depending on the relative quantities of the reactants, the reaction can give either mono- or dibenzalacetone.

Dibenzalacetone is a fairly innocuous substance; its spectral properties (UV absorbance) indicate why it is used in sun-protection preparations. In the present experiment sufficient ethanol is present as solvent to readily dissolve the starting material, benzaldehyde, and also the intermediate, benzalacetone. The benzalacetone, once formed, can then easily react with another mole of benzaldehyde to give the product, dibenzalacetone. The detailed mechanism for the formation of benzalacetone is:

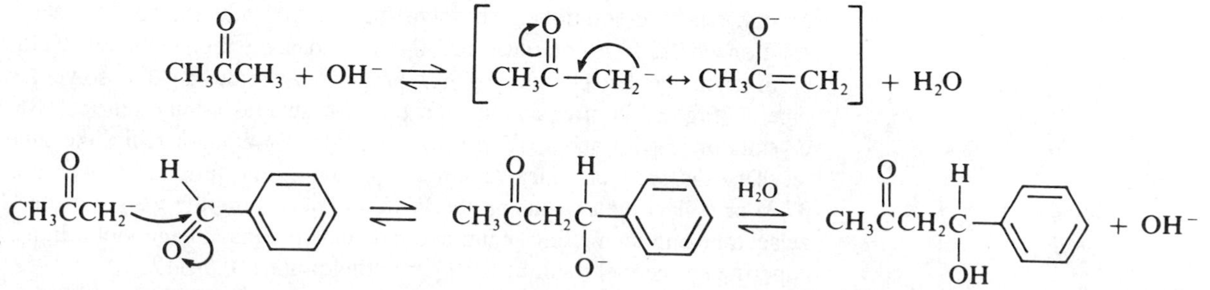

Benzalacetone
4-Phenyl-3-butene-2-one
mp 42°C

Experiment

Synthesis of Dibenzalacetone

Microscale Procedure

If sodium hydroxide gets on the skin, wash until the skin no longer has a soapy feeling.

Into a 10 × 100 mm reaction tube, place 2 mL of 3 M sodium hydroxide solution. To this solution add 1.6 mL of 95% ethanol and 0.212 g of benzaldehyde. Then add 0.058 g of acetone to the reaction mixture. Alternatively, the instructor may provide a solution that contains 58 mg of acetone in each 1.6 mL of ethanol. Cap the tube immediately, and shake the mixture vigorously. The benzaldehyde, initially insoluble, goes into solution, and a water-clear, pale-yellow solution results. After a minute or so it suddenly becomes cloudy, and a yellow precipitate of the product forms. Continue to shake the tube from time to time for the next 30 min. If the product fails to crystallize, open the tube and scratch the inside of the tube with a glass rod. Remove the liquid from the tube using a Pasteur pipette by squeezing the bulb of the pipette, pressing the tip against the bottom of the tube, and bringing the liquid into the pipette, leaving the crystals in the tube (Fig. 37.1). Add 3 mL of water, cap, and shake the tube vigorously. Remove the wash liquid as before, and wash the crystals twice more with 3-mL portions of water.

After the final washing, add 3 mL of water to the tube and collect the crystals on a Hirsch funnel using vacuum filtration. Use the filtrate to complete transfer of the crystals. Squeeze the product between sheets of filter paper to dry it, and then recrystallize the crude dibenzalacetone from 70 : 30 ethanol–water. Insert a wooden boiling stick to promote even boiling. Remove the tube from the hot sand bath, and place it in an insulated container to cool slowly to room temperature. Should the product separate as an oil, try to obtain a seed crystal, heat the solution to dissolve the oil, and add the seed crystal as the solution cools. If it continues to oil out, add a bit more ethanol. Collect the product on the Hirsch funnel (Fig. 37.2) or by removing the solvent with a pipette (Fig. 37.1) after cooling the tube for several minutes in ice. Wash the crystals once with about 0.5 mL of ice-cold 70% ethanol while the tube is in ice. Dry the product under vacuum by attaching the tube to an aspirator (Fig. 37.3) or collect the crystals on the Wilfilter. Determine the weight of the dibenzalacetone and its melting point, and calculate the percentage yield. In a typical experiment, the yield will be 0.10 g, melting point 110.5–112°C.

Cleaning Up Dilute the filtrate from the reaction mixture with water, and neutralize it with dilute hydrochloric acid before flushing down the drain. The ethanol filtrate from the crystallization should be placed in the organic solvents container.

Macroscale Procedure

If sodium hydroxide gets on the skin, wash until the skin no longer has a "soapy" feeling.

Mix 0.05 mol of benzaldehyde with the theoretical quantity of acetone, and add one-half the mixture to a solution of 5 g of sodium hydroxide dissolved in 50 mL of water and 40 mL of ethanol at room temperature (<25°C). After 15 min add the remainder of the aldehyde-ketone mixture, and rinse the container with a little ethanol to complete the transfer. After one-half hour, during which time the mixture is swirled frequently, collect the product by suction filtration on a Büchner funnel. Break the suction, and carefully pour 100 mL of water on the product. Reapply the vacuum. Repeat this process three times in order to remove all traces of sodium hydroxide. Finally, press the product as dry as possible on the filter using a cork, then press it between sheets of filter paper to remove as much water as possible. Save a small sample for melting point determination, and then recrystallize the product from ethanol using about 10 mL of ethanol for each 4 g of dibenzalacetone. The yield after recrystallization should be about 4 g.

Cleaning Up Dilute the filtrate from the reaction mixture with water and neutralize it with dilute hydrochloric acid before flushing down the drain. The ethanol filtrate from the crystallization should be placed in the organic solvents container.

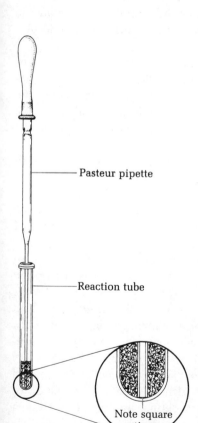

FIG. 37.1 Pasteur pipette filtration technique. Solvent is removed between pipette tip and bottom of tube, leaving crystals in reaction tube.

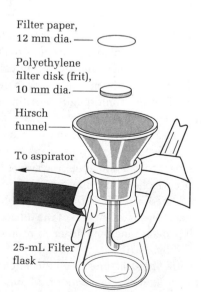

FIG. 37.2 Hirsch funnel filtration apparatus.

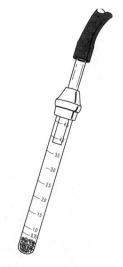

FIG. 37.3 Drying of crystals under vacuum.

Molecular Modeling

The name *dibenzalacetone* does not completely characterize the molecule made in this experiment. There are actually three isomeric dibenzalacetones, one melting at 110 to 111°C, λ_{max} 330 nm, ϵ 34,300; another melting at 60°C, λ_{max} 295 nm, ϵ 20,000; and a third, a liquid with λ_{max} 287 nm, ϵ 11,000.

Both the melting points and the UV spectral data give some hints regarding the structures of these molecules. The first one is very symmetrical and can pack well into a crystal lattice. The long wavelength of the ultraviolet light absorption maximum and the high value of the molar absorbance ϵ indicate a long, planar conjugated system (see Chapter 14). The other two molecules are increasingly less able to pack nicely into a crystal lattice or to have a planar conjugated system. In the last step of the aldol condensation, loss of water from the β-hydroxyketone can form molecules in which the alkene hydrogen atoms are either *cis* or *trans* to each other. Write the structures of the three geometric isomers of dibenzalacetone, and assign each one to the three molecules described above.

Enter the structures of these three isomers into a molecular modeling program, and carry out an energy minimization to calculate the relative steric energies or heats of formation of each molecule. Note, once the calculation is complete, that the lowest energy conformation of each isomer will be as planar as possible in order that there can be maximum overlap of the p orbitals on each sp^2 hybridized carbon. To test this idea, calculate the steric energy or heat of formation of benzalacetone [Fig. 37.4(a)] using the usual energy minimization procedure. The result should be an almost planar molecule. Then deliberately hold the dihedral angle defined by atoms 1, 2, 3, and 4 at 90° [Fig. 37.4(b)], and again calculate the energy of the molecule. In the latter conformation, the p orbitals of the carbonyl group are orthogonal to the p orbitals of the alkene.

As you may have discovered in calculating the energies of the three geometric isomers of dibenzalacetone, there is still another form of isomerization entering into the conformations of these molecules: single-bond *cis* and *trans* isomers, exemplified by isomers (a) and (c) of benzalacetone (Fig. 37.4). Both these conformers are planar in order to achieve maximum overlap of p orbitals, but in (a) the carbonyl group is *cis* to the alkene bond, while in (c) it is *trans*. The barrier to rotation about the single bond is not very high, so these isomers cannot be isolated at room temperature. If your molecular modeling program has a "dihedral driver" routine, you can calculate the heats of formation of benzalacetone as a function of the dihedral angle defined by atoms 1, 2, 3, and 4 and thus determine the barrier to rotation around this bond in kilocalories per mole.

Draw the structures of all the single-bond *cis* and *trans* isomers for each of the three geometric isomers of dibenzalacetone. There are a total of ten such isomers. Pick out the one you regard as the most stable and calculate its steric energy. Which three are represented by the solid (mp 110–111°C), the solid (mp 60°C), and the liquid?

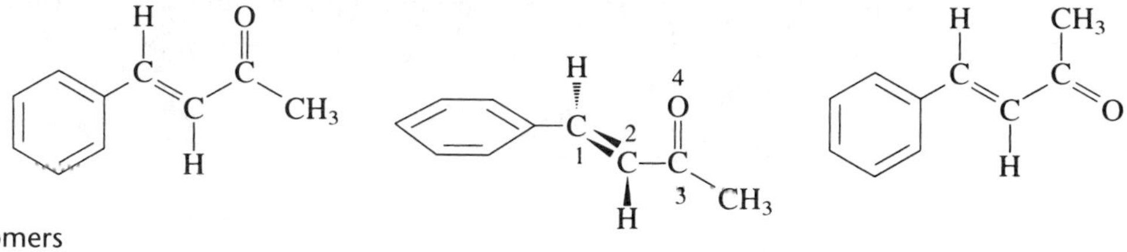

FIG. 37.4 Single-bond isomers of benzalacetone.

(a)   (b)   (c)

Questions

1. Why is it important to maintain equivalent proportions of reagents in this reaction?

2. What side products do you expect in this reaction? How are they removed?

3. What evidence do you have that your product consists of a single geometric isomer or a mixture of isomers? Does the melting point give such information?

4. From the ^{1}H NMR spectrum of dibenzalacetone (Fig. 37.5), can you deduce what geometric isomer(s) is (are) formed? See Table 13.1.

5. How would you change the procedures in this chapter if you wished to synthesize benzalacetone, $C_6H_5CH=CHCOCH_3$? Benzalacetophenone, $C_6H_5CH=CHCOC_6H_5$?

6. Assign the peak at 1639 cm^{-1} in Fig. 37.6.

7. Write the structures of the three geometric isomers of dibenzalacetone, and assign each one to the three molecules described. Disregard s-*cis* and s-*trans* isomerism.

8. Draw the structures of all the single-bond *cis* and *trans* isomers for each of the three geometric isomers of dibenzalacetone. There are a total of ten such isomers.

9. Carry out an energy minimization to calculate the relative steric energies or heats of formation of three of the ten possible isomeric dibenzalacetones.

10. Calculate the steric energy or heat of formation for one single-bond isomer of *trans*-benzalacetone (a) using the usual energy minimization procedure. The result should be a planar molecule. Then deliberately hold the dihedral angle defined by atoms 1, 2, 3, and 4 at 0°, 90°, and 180° (b) and again calculate the energies of the molecule. Approximately what is the barrier to rotation about the single bond?

11. Pick out the isomer you regard as the most stable from Question 7, and calculate its steric energy (if this was not done in Question 8).

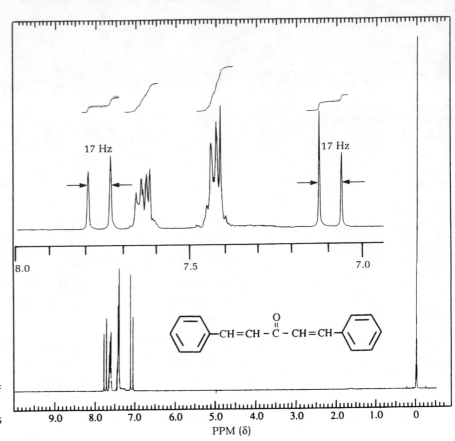

FIG. 37.5 ¹H NMR spectrum of dibenzalacetone (250 MHz). (An expanded spectrum appears above the normal spectrum.)

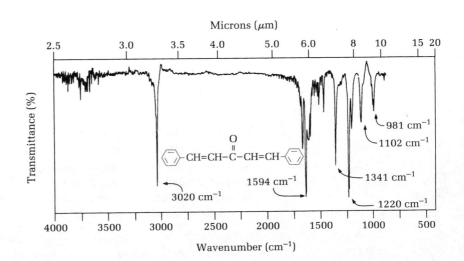

FIG. 37.6 IR spectrum of dibenzalacetone (KBr disk).

Grignard Synthesis of Triphenylmethanol and Benzoic Acid

Prelab Exercise: Prepare a flow sheet for the preparation of triphenylmethanol. Through a knowledge of the physical properties of the solvents, reactants, and products, show how the products are purified. Indicate which layer in separations should contain the product.

In 1912 Victor Grignard received the Nobel prize in chemistry for his work on the reaction that bears his name, a carbon–carbon bond-forming reaction by which almost any alcohol may be formed from appropriate alkyl halides and carbonyl compounds. The Grignard reagent is easily formed by reaction of an alkyl halide, in particular a bromide, with magnesium metal in anhydrous diethyl ether. Although the reaction can be written and thought of as simply

$$R—Br + Mg \longrightarrow R—Mg—Br$$

it appears that the structure of the material in solution is rather more complex. There is evidence that dialkylmagnesium is present

$$2 R—Mg—Br \rightleftharpoons R—Mg—R + MgBr_2$$

Structure of the Grignard reagent

and that the magnesium atoms, which have the capacity to accept two electron pairs from donor molecules to achieve a four-coordinated state, are solvated by the unshared pairs of electrons on diethyl ether:

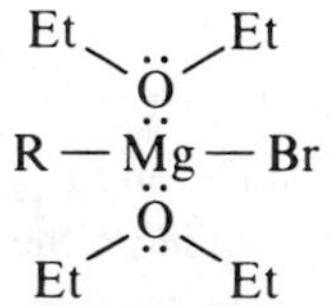

$$
\begin{array}{c}
Et \diagdown \quad \diagup Et \\
\overset{..}{\underset{..}{O}} \\
R — \overset{\overset{\displaystyle |}{}}{Mg} — Br \\
\overset{..}{\underset{..}{O}} \\
Et \diagup \quad \diagdown Et
\end{array}
$$

A strong base and strong nucleophile

The Grignard reagent is a strong base and a strong nucleophile. As a base it will react with all protons that are more acidic than those found on alkenes and alkanes. Thus, Grignard reagents react readily with water, alcohols, amines, thiols, etc., to regenerate the alkane:

$$R\text{—}Mg\text{—}Br + H_2O \longrightarrow R\text{—}H + MgBrOH$$

$$R\text{—}Mg\text{—}Br + R'OH \longrightarrow R\text{—}H + MgBrOR'$$

$$R\text{—}Mg\text{—}Br + R'NH_2 \longrightarrow R\text{—}H + MgBrNHR'$$

$$R\text{—}Mg\text{—}Br + R'C \equiv C\text{—}H \longrightarrow R\text{—}H + R'C \equiv CMgBr$$

(an acetylenic Grignard reagent)

The starting material for preparing the Grignard reagent can contain no acidic protons. The reactants and apparatus must all be completely dry; otherwise the reaction will not start. If proper precautions are taken, however, the reaction proceeds smoothly.

The magnesium metal, in the form of a coarse powder, has a coat of oxide on the outside. A fresh surface can be exposed by crushing the powder under the absolutely dry ether in the presence of the organic halide. Reaction will begin at exposed surfaces, as evidenced by a slight turbidity in the solution and evolution of bubbles. Once the exothermic reaction starts it proceeds easily, the magnesium dissolves, and a solution of the Grignard reagent is formed. The solution is often turbid and gray due to impurities in the magnesium. The reagent is not isolated but reacted immediately with, most often, an appropriate carbonyl compound

$$R\text{—}Mg\text{—}Br + R'\text{—}\overset{\overset{\displaystyle \ddot{O}:}{\|}}{C}\text{—}R'' \longrightarrow R'\text{—}\underset{\underset{\displaystyle R}{|}}{\overset{\overset{\displaystyle :\ddot{O}:^- MgBr^+}{|}}{C}}\text{—}R''$$

to give, in another exothermic reaction, the magnesium alkoxide, a salt insoluble in ether. In a simple acid–base reaction this alkoxide is reacted with acidified ice water to give the covalent, ether-soluble alcohol and the ionic water-soluble magnesium salt:

$$R'\text{—}\underset{\underset{\displaystyle R}{|}}{\overset{\overset{\displaystyle :\ddot{O}:^- MgBr^+}{|}}{C}}\text{—}R'' + H^+Cl^- \longrightarrow R'\text{—}\underset{\underset{\displaystyle R}{|}}{\overset{\overset{\displaystyle :\ddot{O}\text{—}H}{|}}{C}}\text{—}R'' + Mg^{2+}Br^-Cl^-$$

The great versatility of this reaction lies in the wide range of reactants that undergo the reaction. Thirteen representative reactions are shown on the following page.

A versatile reagent

In every case except reaction 1 the intermediate alkoxide must be hydrolyzed to give the product. The reaction with oxygen (reaction 2) is usually not a problem because the ether vapor over the reagent protects it from attack by oxygen, but this reaction is one reason why the reagent cannot usually be stored without special precautions. Reactions 6 and 7 with ketones and aldehydes giving, respectively, tertiary and secondary alcohols are among the most common. Reactions 8–13 are not nearly so common.

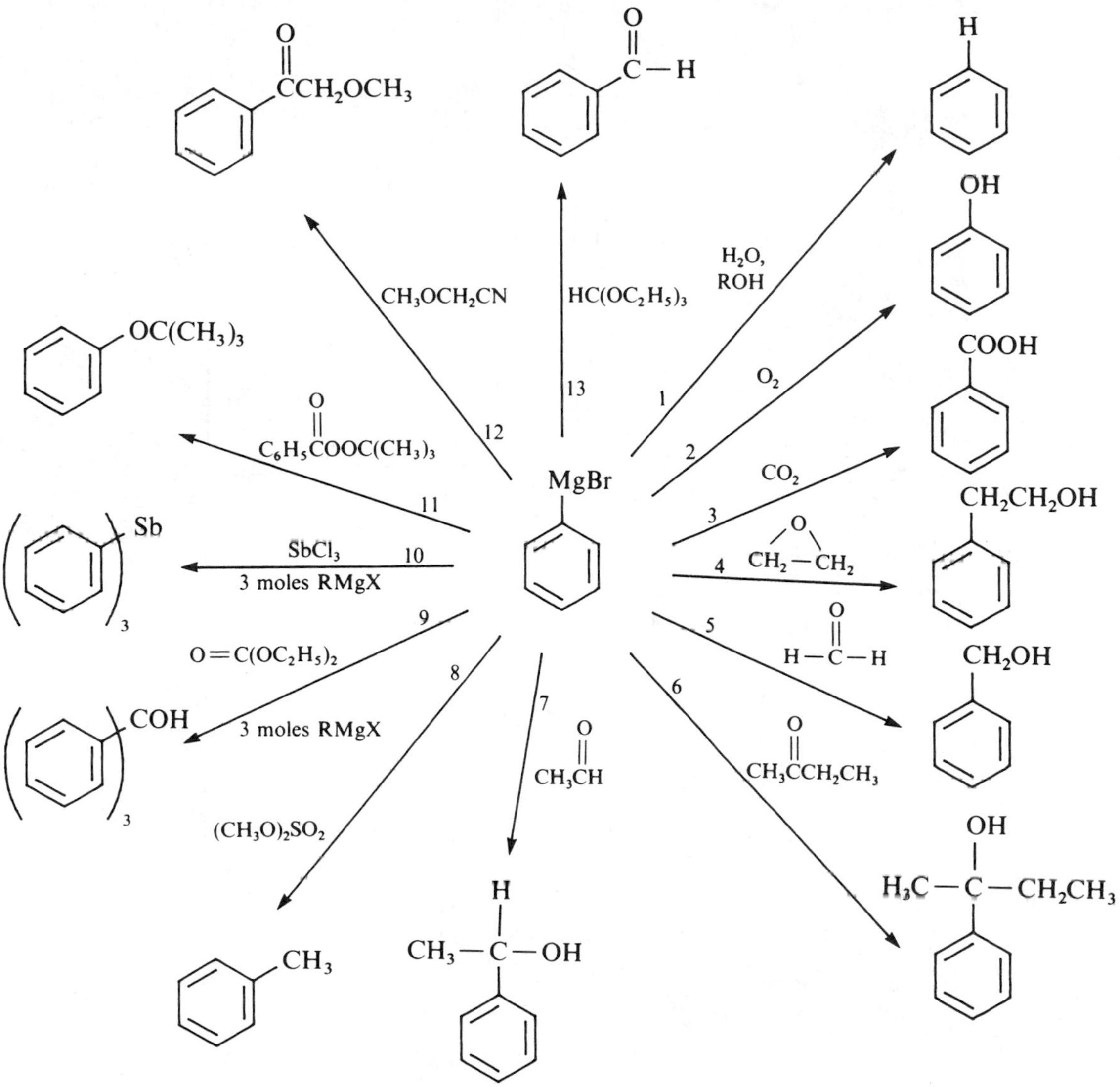

In the present experiment we shall carry out another common type of Grignard reaction, the formation of a tertiary alcohol from 2 mol of the reagent and one of an ester. The ester employed is the methyl benzoate synthesized in Chapter 40. The initially formed product is unstable and decomposes to a ketone, which, being more reactive than an ester, immediately reacts with more Grignard reagent:

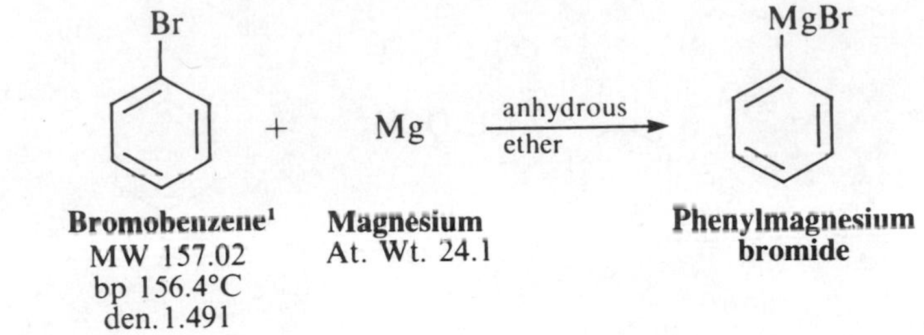
Br
+
Mg
anhydrous
ether
MgBr
Bromobenzene[1]
MW 157.02
bp 156.4°C
den. 1.491
Magnesium
At. Wt. 24.1
Phenylmagnesium
bromide

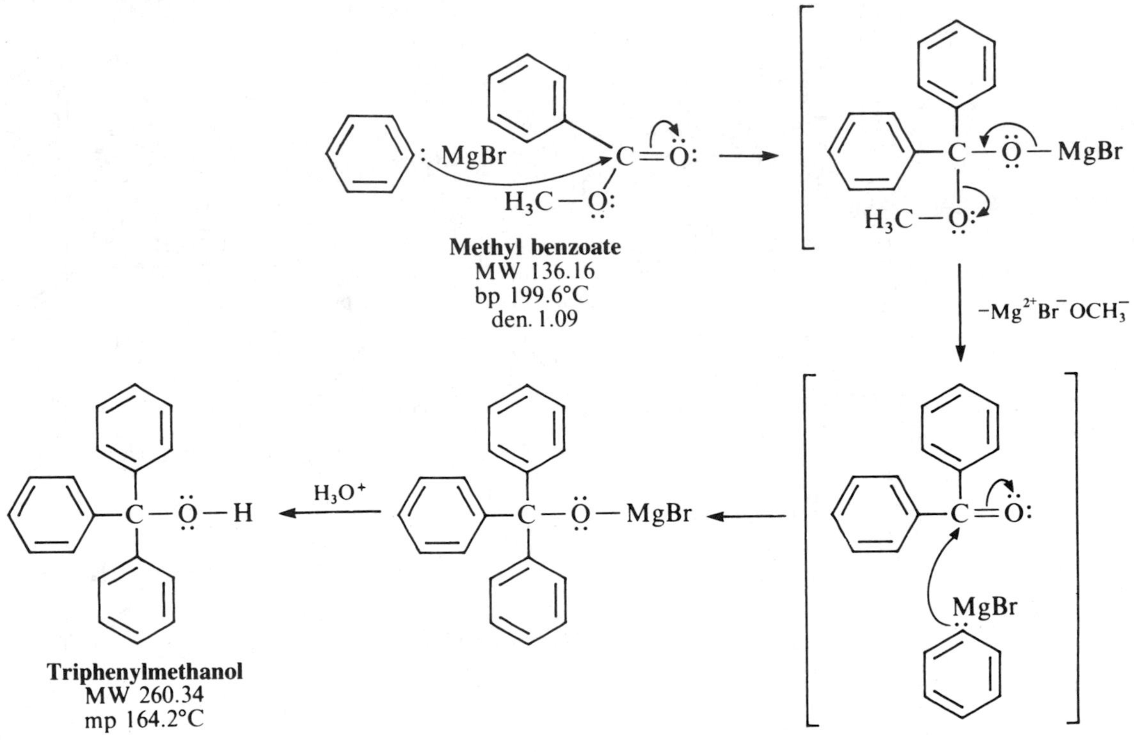
: MgBr
C=Ö:
$H_3C—\ddot{O}:$
Methyl benzoate
MW 136.16
bp 199.6°C
den. 1.09
C—Ö—MgBr
$H_3C—\ddot{O}:$
$-Mg^{2+}Br^{-}OCH_3^{-}$
C=Ö:
MgBr
C—Ö—MgBr
H_3O^+
C—Ö—H
Triphenylmethanol
MW 260.34
mp 164.2°C

Biphenyl has a characteristic odor. Triphenylmethanol is odorless.

The primary impurity in the present experiment is biphenyl, formed by the reaction of phenylmagnesium bromide with unreacted bromobenzene. The most effective way to lessen this side reaction is to add the bromobenzene slowly to the reaction mixture so it will react with the magnesium and not be present in high concentration to react with previously formed Grignard reagent. The impurity is easily eliminated because it is much more soluble in hydrocarbon solvents than triphenylmethanol.

1. See Fig. 38.7 at the end of the chapter for the ^{13}C NMR spectrum.

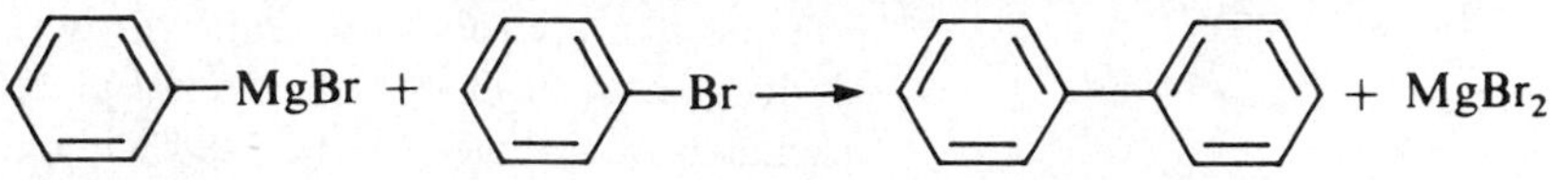

Biphenyl
mp 72°C

Triphenylmethanol can also be prepared from benzophenone.

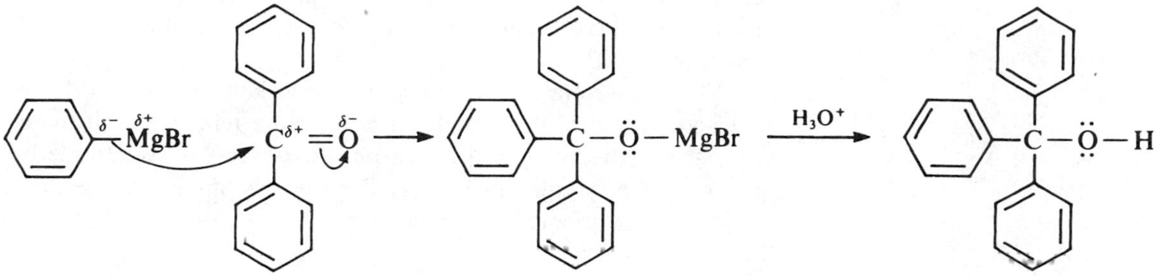

| **Phenylmagnesium bromide** | **Benzophenone** MW 182.22 mp 48°C | | **Triphenylmethanol** MW 260.34 mp 164.2°C |

Experiments

1. Phenylmagnesium Bromide (Phenyl Grignard Reagent)

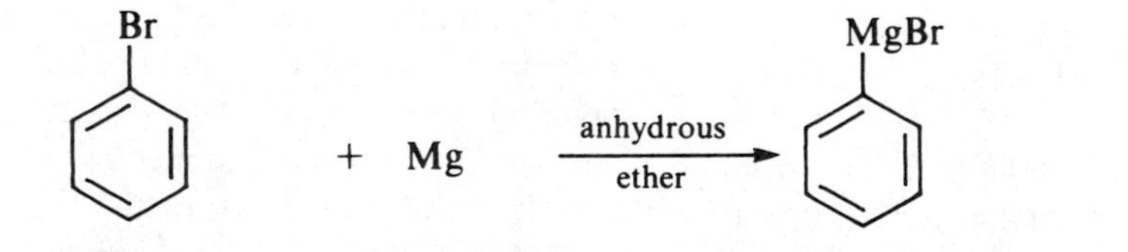

| **Bromobenzene** MW 157.02 bp 156°C den. 1.491 | **Magnesium** At. Wt. 24.31 | **Phenylmagnesium bromide** not isolated, used *in situ* |

Advance Preparation

It is imperative that all equipment and reagents be absolutely dry. The glassware to be used—two reaction tubes, two 1-dram vials, and a stirring rod—can be dried in a 110°C oven for at least 30 min, along with the magnesium. Alternatively, if the glassware, syringe, septa, and magnesium appear to be perfectly dry, they can be used without special drying. The plastic and rubber ware should be rinsed with acetone if it appears to be either dirty or wet with water and then placed in a desiccator for at least 12 h. Do not place plastic ware in the oven. New, sealed packages of syringes can be used without prior drying. The ether used throughout this reaction must be absolutely dry (absolute ether). To prepare the Grignard reagent, absolute diethyl ether must be used; elsewhere, *tert*-butyl methyl ether can be used. Ether extractions of aqueous solutions do not need to be carried out with dry ether.

A very convenient container for absolute diethyl ether is a 50-mL septum-capped bottle. This method of dispensing the solvent has three advantages: The ether is kept anhydrous, the exposure to oxygen is minimized, and there is little possibility of its catching fire. Ether is extremely flammable; do not work with this solvent near flames.

To remove ether from a septum-capped bottle, inject a volume of air into the bottle equal to the amount of ether being removed. Pull more ether than needed into the syringe, and then push the excess back into the bottle before removing the syringe. In this way there will be no air bubbles in the syringe, and it will not dribble (Fig. 38.1).

Procedure

Remove a reaction tube from the oven, and immediately cap it with a septum. In the operations that follow, keep the tube capped except when it is necessary to open it. After it cools to room temperature, add about 2 mmol (about 50 mg) of magnesium powder. Record the weight of magnesium used to the nearest milligram. We will make it the limiting reagent by using a 5% molar excess of bromobenzene (about 2.1 mmol). (See below).

Diethyl ether can be made and kept anhydrous by storing over Linde 5A molecular sieves. Discard diethyl ether within 90 days because of peroxide formation. tert-*Butyl methyl ether does not have this problem.*

Using a dry syringe, add to the magnesium by injection through the septum 0.5 mL of anhydrous diethyl ether. Your laboratory instructor will demonstrate transfer from the storage container used in your laboratory.

Into an oven-dried vial weigh about 2.1 mmol (about 330 mg) of dry (stored over molecular sieves) bromobenzene. Using a syringe, add to this vial 0.7 mL anhydrous diethyl ether, and *immediately,* with the same syringe, remove all the solution from the vial. This can be done virtually quantitatively so you do not need to rinse the vial. Immediately cap the empty vial to keep it dry for later use. Inject about 0.1 mL of the bromobenzene–ether mixture into the reaction tube, and mix the contents by flicking the tube. Pierce the septum with another syringe needle for pressure relief (Fig. 38.2).

It takes much force to crush the magnesium. Place the tube on a hard surface and bear down with the stirring rod while twisting the reaction tube. Do not pound the magnesium.

The reaction will not ordinarily start at this point, so remove the septum, syringe, and empty syringe needle and crush the magnesium with a dry stirring rod. You can do this easily in the confines of the 10-mm-diameter reaction tube while it is placed on a hard surface. There is little danger of poking the stirring rod through the bottom of the tube (Fig. 38.3). Immediately replace the septum, syringe, and empty syringe needle (for pressure relief). The reaction should start within seconds. The formerly clear solution becomes cloudy and soon begins to boil as the magnesium metal reacts with the bromobenzene to form the Grignard reagent, phenylmagnesium bromide.

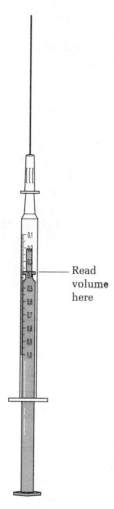

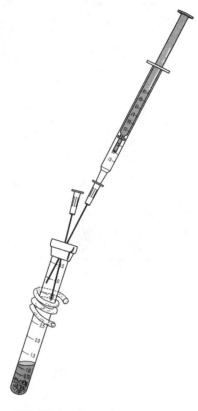

Read
volume
here

FIG. 38.1   Polypropylene syringe (1 mL, with 0.01-mL graduations). The needle is blunt. When there are no air bubbles in the syringe, it will not dribble ether.

FIG. 38.2   Once the reaction has started, bromobenzene in ether is added slowly from the syringe. The empty needle is for pressure relief, but if condensation is complete (aided by the damp pipe cleaner), it will not be needed. Once the reaction slows down, stir it with the magnetic stirring bar.

   If the reaction does not start within 1 min, begin again with completely different, dry equipment (syringe, syringe needle, reaction tube, etc.). Once the Grignard reaction starts, it will continue. To prevent the ether from boiling away, wrap a pipe cleaner around the top part of the reaction tube. Dampen this with water or, if the room temperature is very hot, with alcohol.
   To the refluxing mixture add slowly and dropwise over a period of several

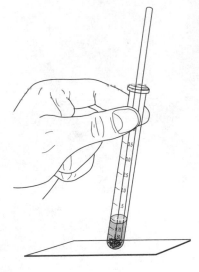

FIG. 38.3 To start the Grignard reaction, remove the septum and apply pressure to the stirring rod while rotating the reaction tube on a hard surface that will not scratch the tube, such as a book.

minutes the remainder of the solution of bromobenzene in ether at such a rate that the reaction remains under control at all times. After all the bromobenzene solution is added, spontaneous boiling of the diluted mixture may be slow or become slow. At this point, add a magnetic stirring bar to the reaction tube and stir the reaction mixture with a magnetic stirrer. If the rate of reaction is too fast, slow down the stirrer. The reaction is complete when none or a very small quantity of the metal remains. Check to see that the volume of ether has not decreased. If it has, add more anhydrous diethyl ether. Since the solution of the Grignard reagent deteriorates on standing, the next step should be started at once. The phenylmagnesium bromide can be converted to triphenylmethanol or to benzoic acid.

MICROSCALE   ## 2. Triphenylmethanol

| Phenylmagnesium bromide | Benzophenone MW 182.22 mp 48°C | | Triphenylmethanol MW 260.34 mp 164.2°C |

Make t-butyl methyl ether anhydrous by storing it over molecular sieves.

In an oven-dried vial dissolve 2.0 mmol (0.364 g) of benzophenone in 1.0 mL of anhydrous *tert*-butyl methyl ether by capping the vial and mixing the contents thoroughly. With a dry syringe, remove all the solution from the vial and add it dropwise with *thorough* shaking after each drop to the solution of the Grignard reagent. Add the benzophenone at such a rate as to maintain the ether at a gentle reflux. Rinse the vial with a few drops of anhydrous ether after all the first solution has been added, and add this rinse to the reaction tube.

Mixing the reaction mixture is very important.

After all the benzophenone has been added, the mixture should be homogeneous. If not, mix it thoroughly, using a stirring rod if necessary. The syringe can be removed, but leave the pressure-relief needle in place. Allow the reaction mixture to stand at room temperature. The reaction apparently is complete when the red color disappears.

At the end of the reaction period, cool the tube in ice, and add to it dropwise with stirring (use a glass rod or a spatula) 2 mL of 3 N hydrochloric acid. A creamy-white precipitate of triphenylmethanol will separate between the layers. Add more ether (it need not be anhydrous) to the reaction tube, and shake the contents to dissolve all the triphenylmethanol. The result should be two perfectly clear layers. Remove a drop of the ether layer for TLC analysis. Any bubbling seen at the interface or in the lower layer is leftover magnesium reacting with the hydrochloric acid. Remove the aqueous layer, and shake the ether layer with an equal volume of saturated aqueous sodium chloride solution in order to remove water and any remaining acid. Carefully remove all the aqueous layer, and then dry the ether layer by adding anhydrous calcium chloride pellets to the reaction tube until the drying agent no longer clumps together. Cork the tube, and shake it from time to time over a 5- or 10-min period to complete the drying.

Using a Pasteur pipette, remove the ether from the drying agent and place it in another tared, dry reaction tube or the centrifuge tube. Use more ether to wash off the drying agent, and combine these ether extracts. Evaporate the ether in a hood by blowing nitrogen or air onto the surface of the solution while warming the tube in a beaker of water or in the hand.

After all the solvent has been removed, determine the weight of the crude product. Note the odor of biphenyl, the product of the side reaction that takes place between bromobenzene and phenylmagnesium bromide during the first reaction.

Trituration (grinding) of the crude product with petroleum ether will remove the biphenyl. Stir the crystals with 0.5 mL of petroleum ether in the ice bath, remove the solvent as thoroughly as possible, add a boiling stick, and recrystallize the residue from boiling 2-propanol (no more than 2 mL). Allow the solution to cool slowly to room temperature, and then cool it thoroughly in ice. Triphenylmethanol crystallizes slowly, so allow the mixture to remain in the ice as long as possible. Stir the ice-cold mixture well, and collect the product by vacuum filtration on the Hirsch funnel. Save the filtrate. Concentration may give a second crop of crystals.

An alternative method for purification of the triphenylmethanol utilizes a mixed solvent. Dissolve the crystals in the smallest possible quantity of warm ether, and add to the solution 1.5 mL of ligroin. Add a boiling stick to the solution, and boil off some of the ether until the solution becomes slightly cloudy, indicating it is saturated. Allow the solution to cool slowly to room temperature. Triphenylmethanol is deposited slowly as large, thick prisms. Cool the solution in ice, and after allowing time for complete crystallization to occur, remove the ether with a Pasteur pipette and wash the crystals once with a few drops of a cold 1 : 4 ether–ligroin mixture. Dry the crystals in the tube under vacuum (Fig. 38.4).

Determine the weight, melting point, and percentage yield of the triphenylmethanol. Analyze the crude and recrystallized product by TLC on silica gel (see Chapter 10), developing the plate with dichloromethane–petroleum ether, 1 : 5. An IR spectrum can be determined in chloroform solution or by preparing a mull or KBr disk (see Chapter 13). Compare the apparatus used in this experiment to the research-type apparatus shown in Fig. 38.5.

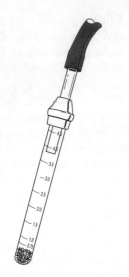

FIG. 38.4 Apparatus for drying crystals in reaction tube under vacuum.

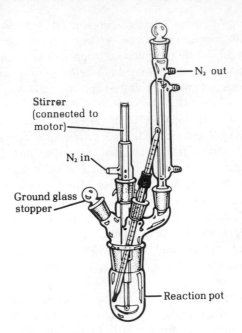

FIG. 38.5 Semimicroscale, research-type apparatus for Grignard reaction, with provision for a motor-driven stirrer and an inlet and outlet for dry nitrogen.

Cleaning Up The acidic aqueous layer and the saturated sodium chloride layer are combined, diluted with water, neutralized with sodium carbonate, and flushed down the drain. Ether is allowed to evaporate from the drying agent in the hood, and then it is discarded in the nonhazardous solid waste container. The petroleum ether and 2-propanol or ether–ligroin mother liquor are placed in the organic solvents container.

MICROSCALE

3. Benzoic Acid

Phenylmagnesium bromide

+ **Carbon dioxide**
MW 44.01
mp − 78.5°C (sublimes)

CO_2

→

$\xrightarrow{\text{HCl}}$

+ MgBrCl

Benzoic acid
MW 122.12
mp 123°C

CAUTION: Handle dry ice with a towel or gloves. Contact with the skin can cause frostbite, because dry ice sublimes at −78.5°C.

Prepare 2 mmol of phenylmagnesium bromide exactly as described in Part 1 of this experiment. Wipe off the surface of a small piece of dry ice (solid carbon dioxide) with a dry towel to remove frost, and place it in a dry 30-mL beaker. Remove the pressure-relief needle from the reaction tube, and then insert a syringe through the septum, turn the tube upside down, and draw into the syringe as much of the reagent solution as possible. Squirt this solution onto the piece of dry ice, and then, using a clean needle, rinse out the reaction tube with a milliliter of anhydrous diethyl ether and squirt this onto the dry ice. Allow excess dry ice to sublime, and then hydrolyze the salt by the addition of 2 mL of 3 M hydrochloric acid.

Transfer the mixture from the beaker to a reaction tube, and shake it thoroughly. Two homogeneous layers should result. Add 1 to 2 mL of acid or of ordinary (not anhydrous) *tert*-butyl methyl ether if necessary. Remove the aqueous layer, and shake the ether layer with 1 mL of water, which is removed and discarded. Then extract the benzoic acid by adding to the ether layer 0.7 mL of 3 M sodium hydroxide solution, shaking the mixture thoroughly, and withdrawing the aqueous layer, which is placed in a very small beaker or vial. The extraction is repeated with another 0.5-mL portion of base and finally 0.5 mL of water. Now that the extraction is complete, the ether, which can be discarded, contains primarily biphenyl, the byproduct formed during the preparation of the phenylmagnesium bromide.

The combined aqueous extracts are heated briefly to about 50°C to drive off dissolved ether from the aqueous solution and then made acidic by the addition of concentrated hydrochloric acid (test with indicator paper). Cool the mixture thoroughly in an ice bath. Collect the benzoic acid on the Hirsch funnel, and wash it with about 1 mL of ice water while on the funnel. A few crystals of this crude material are saved for a melting-point determination, and the remainder of the product is recrystallized from boiling water.

The solubility of benzoic acid in water is 68 g/L at 95°C and 1.7 g/L at 0°C. Dissolve the acid in very hot water. Let the solution cool slowly to room temperature; then cool it in ice for several minutes before collecting the product by vacuum filtration on the Hirsch funnel. Use the ice-cold filtrate in the filter flask to complete the transfer of benzoic acid from the reaction tube. Turn the product out onto a piece of filter paper, squeeze out excess water, and allow it to dry thoroughly. Once dry, weigh it, calculate the percentage yield, and determine the melting point along with the melting point of the crude material. The infrared spectrum may be determined as a solution in chloroform (1 g of benzoic acid dissolves in 4.5 mL of chloroform) or as a mull or KBr disk (see Chapter 13).

Cleaning Up Combine all aqueous layers, dilute with a large quantity of water, and flush the slightly acidic solution down the drain.

MACROSCALE

4. Phenylmagnesium Bromide (Phenyl Grignard Reagent)

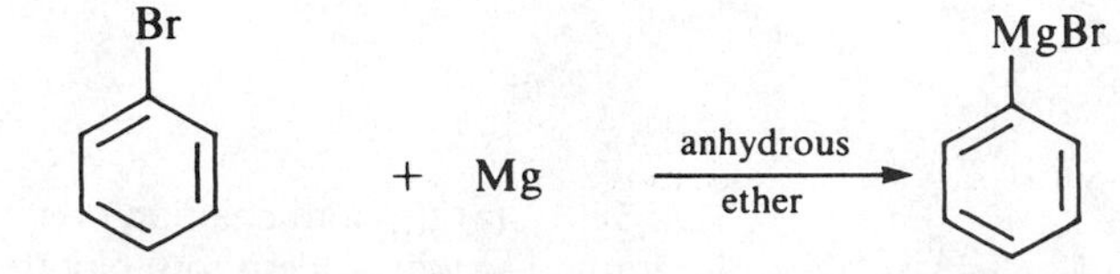

Bromobenzene
MW 157.02
bp 156°C
den. 1.491

Magnesium
At. Wt.
24.31

Phenylmagnesium bromide
not isolated, used *in situ*

All equipment and reagents must be absolutely dry. The Grignard reagent is prepared in a dry 100-mL round-bottomed flask fitted with a long reflux condenser. A calcium chloride drying tube inserted in a cork that will fit either the flask or the top of the condenser is also made ready [Fig. 38.6(a)]. (See also Fig. 38.5.) The flask, condenser, and magnesium (2 g = 0.082 mol of magnesium turnings) should be as dry as possible to begin with, and then should be dried in a 110°C oven for at least 35 min. Alternatively, the magnesium is placed in the flask, the calcium chloride tube is attached directly, and the flask is heated gently but thoroughly with a cool luminous flame. Do not overheat the magnesium. It will become deactivated through oxidation or, if strongly overheated, can burn. The flask on cooling pulls dry air through the calcium chloride. Cool to room

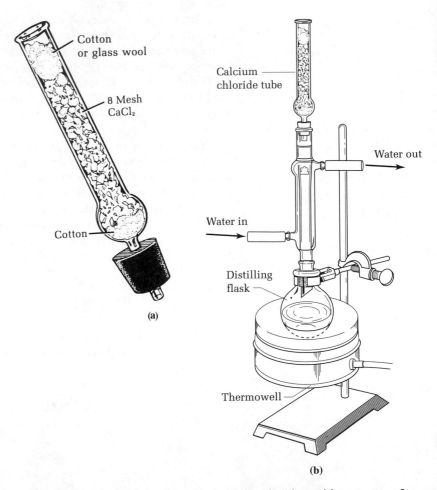

FIG. 38.6 (a) Calcium chloride drying tube fitted with a rubber stopper. Store for future use with cork in top, pipette bulb on bottom. (b) Apparatus for refluxing Grignard reaction.

temperature before proceeding! **Extinguish all flames!** Ether vapor is denser than air and can travel along bench tops and into sinks. Use care.

Make an ice bath ready in case control of the reaction becomes necessary, although this is usually not the case. Remove the drying tube, and fit it to the top of the condenser. Then pour into the flask through the condenser 15 mL of *absolute* ether (absolutely dry, anhydrous) and 9 mL (13.5 g = 0.086 mol) of bromobenzene. Be sure the graduated cylinders used to measure the ether and bromobenzene are absolutely dry. (More ether is to be added as soon as the reaction starts, but at the outset the concentration of bromobenzene is kept high to promote easy starting.) If there is no immediate sign of reaction, insert a *dry* stirring rod with a flattened end and crush a piece of magnesium firmly against the bottom of the flask under the surface of the liquid, giving a twisting motion to the rod. When this is done properly the liquid becomes slightly cloudy, and ebullition commences at the surface of the compressed metal. Be careful not to punch a hole in the bottom of the flask. Attach the condenser at once, swirl the flask to provide fresh surfaces for contact, and, as soon as you are sure that the reaction has started, add 25 mL more absolute ether through the top of the condenser before spontaneous boiling becomes too vigorous (replace the drying tube). Note the volume of ether in the flask. Cool in ice if necessary to slow the reaction, but do not overcool the mixture; the reaction can be stopped by too much cooling. Any difficulty in initiating the reaction can be dealt with by trying the following expedients in succession.

Starting the Grignard reaction

1. Warm the flask with your hands or a beaker of warm water. Then see if boiling continues when the flask (condenser attached) is removed from the warmth.
2. Try further mashing of the metal with a stirring rod.
3. Add a tiny crystal of iodine as a starter (in this case the ethereal solution of the final reaction product should be washed with sodium bisulfite solution to remove the yellow color).
4. Add a few drops of a solution of phenylmagnesium bromide or of methylmagnesium iodide (which can be made in a test tube).
5. Start afresh, taking greater care with the dryness of apparatus, measuring tools, and reagents, and sublime a crystal or two of iodine on the surface of the magnesium to generate Gattermann's "activated magnesium" before beginning the reaction again.

Diethyl ether can be kept anhydrous by storing over Linde 5A molecular sieves. Discard the ether after 90 days because of peroxide formation.

Once the reaction begins, spontaneous boiling in the diluted mixture may be slow or become slow. If so, mount the flask and condenser on the steam bath (one clamp supporting the condenser suffices), and reflux gently until the magnesium has disintegrated and the solution has acquired a cloudy or brownish appearance [Fig. 38.6(b)]. The reaction is complete when only a few small remnants of metal (or metal contaminants) remain. Check to see that the volume of ether has not decreased. If it has, add more anhydrous ether. Since the solution of Grignard reagent deteriorates on standing, the next step should be started at once.

CAUTION: Ether is extremely flammable. Extinguish all flames before using ether.

Specially dried ether is required.

Use minimum steam to avoid condensation on outside of the condenser.

 MACROSCALE

5. Triphenylmethanol from Methyl Benzoate

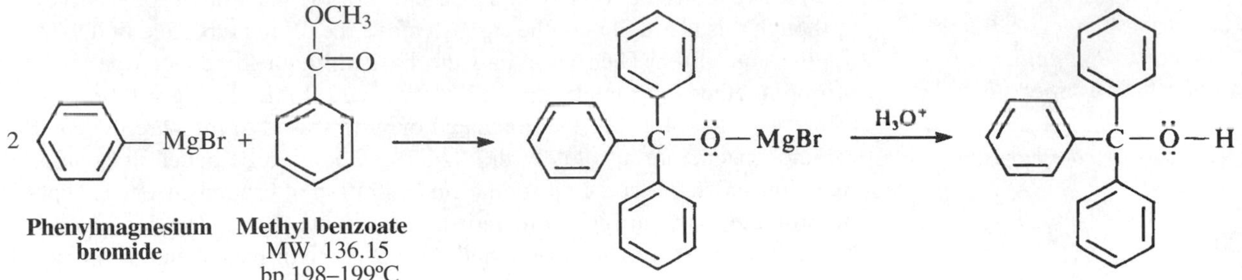

Phenylmagnesium bromide

Methyl benzoate
MW 136.15
bp 198–199°C
21.094

Triphenylmethanol
MW 260.34
mp 164.2°C

Mix 5 g (0.037 mol) of methyl benzoate and 15 mL of absolute ether in a separatory funnel, cool the flask containing phenylmagnesium bromide solution briefly in an ice bath, remove the drying tube, and insert the stem of the separatory funnel into the top of the condenser. Run in the methyl benzoate solution *slowly* with only such cooling as is required to control the mildly exothermic reaction, which affords an intermediate salt that separates as a white solid. Replace the calcium chloride tube; swirl the flask until it is at room temperature and the reaction has subsided.

MACROSCALE

6. Triphenylmethanol from Benzophenone

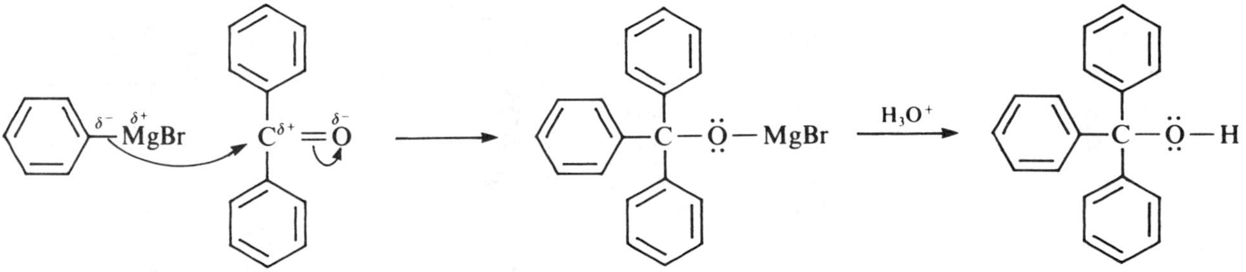

Phenylmagnesium bromide

Benzophenone
MW 182.22
mp 48°C

Triphenylmethanol
MW 260.34
mp 164.2°C

Dissolve 6.75 g (0.037 mol) of benzophenone in 25 mL of absolute ether in a separatory funnel, and cool the flask containing *half* the phenylmagnesium bromide solution (0.041 mol) briefly in an ice bath. (The other half can be used to make benzoic acid.) Remove the drying tube, and insert the stem of the separatory funnel into the top of the condenser. Add the benzophenone solution *slowly* with swirling and only such cooling as is required to control the mildly exothermic reaction, which gives a bright-red solution and then precipitates a white salt. Replace the calcium chloride tube; swirl the flask until it is at room temperature and the reaction has subsided. Go to Part 7.

7. Completion of Grignard Reaction

This is a suitable stopping point.

The reaction is then completed by either refluxing the mixture for one-half hour, or stoppering the flask with the calcium chloride tube and letting the mixture stand overnight (subsequent refluxing is then unnecessary).[2]

Pour the reaction mixture into a 250-mL Erlenmeyer flask containing 50 mL of 10% sulfuric acid and about 25 g of ice, and use both ordinary ether and 10% sulfuric acid to rinse the flask. Swirl well to promote hydrolysis of the addition compound; basic magnesium salts are converted into water-soluble neutral salts, and triphenylmethanol is distributed into the ether layer. An additional amount of ether (ordinary) may be required. Pour the mixture into a separatory funnel (rinse the flask with ether), shake, and draw off the aqueous layer. Shake the ether solution with 10% sulfuric acid to further remove magnesium salts, and wash with saturated sodium chloride solution to remove water that has dissolved in the ether. The amounts of liquid used in these washing operations are not critical. In general, an amount of wash liquid equal to one-third of the ether volume is adequate.

In this part of the experiment, ordinary (not anhydrous) diethyl ether or t-butyl methyl ether may be used.

Saturated aqueous sodium chloride solution removes water from ether.

To effect final drying of the ether solution, pour the ether layer out of the neck of the separatory funnel into an Erlenmeyer flask, add about 5 g of calcium chloride pellets, swirl the flask from time to time, and after 5 min remove the drying agent by gravity filtration through a filter paper held in a funnel into a tared Erlenmeyer flask. Rinse the drying agent with a small amount of ether. Add 25 mL of 66–77°C ligroin, and concentrate the ether-ligroin solutions (steam bath) in an Erlenmeyer flask under an aspirator tube (see Fig. 8.5). Evaporate slowly until crystals of triphenylcarbinol just begin to separate, and then let crystallization proceed, first at room temperature and then at 0°C. The product should be colorless and should melt not lower than 160°C. Concentration of the mother liquor may yield a second crop of crystals. A typical student yield is 5.0 g. Evaporate the mother liquors to dryness, and save the residue for later isolation of the components by chromatography.

Analyze the first crop of triphenylmethanol and the residue from the evaporation of the mother liquors by thin-layer chromatography. Dissolve equal quantities of the two solids (a few crystals) and also biphenyl in equal quantities of dichloromethane (1 or 2 drops). Using a microcapillary, spot equal quantities of material on silica gel TLC plates, and develop the plates in an appropriate solvent system. Try 1:3 dichloromethane–petroleum ether first, and adjust the relative quantities of solvent as needed. The spots can be seen by examining the TLC plate under a fluorescent lamp or by treating the TLC plate with iodine vapor. From this analysis decide how pure each of the solids is and whether it would be worthwhile to attempt to isolate more triphenylmethanol from the mother liquors.

Dispose of recovered and waste solvents in the appropriate containers.

Turn in the product in a vial labeled with your name, the name of the compound, its melting point, and the overall percent yield from benzoic acid.

---

2. A rule of thumb for organic reactions: A 10°C rise in temperature will double the rate of the reaction.

 **MACROSCALE**

8. Benzoic Acid

Wipe the frost from a piece of dry ice, transfer the ice to a cloth towel, and crush it with a hammer. Without delay (so moisture will not condense on the cold solid) transfer about 10 g of dry ice to a 250-mL beaker. Cautiously pour one-half of the solution of phenylmagnesium bromide prepared in Part 1 of this experiment onto the dry ice. A vigorous reaction will ensue. Allow the mixture to warm up; stir it until the dry ice has evaporated. To the beaker add 20 mL of 3 *M* hydrochloric acid, and then heat the mixture over a steam bath in the hood to boil off the ether. Cool the beaker thoroughly in an ice bath, and collect the solid product by vacuum filtration on a Büchner funnel.

Transfer the solid back to the beaker, and dissolve it in the minimum quantity of saturated sodium bicarbonate solution (2.8 *M*). Note that a small quantity of a byproduct remains suspended and floating on the surface of the solution. Note the odor of the mixture. Transfer it to a separatory funnel, and shake it briefly with about 15 mL of *t*-butyl methyl ether. Discard the ether layer, place the clear aqueous layer in the beaker, and heat it briefly to drive off dissolved ether. Carefully add 3 *M* hydrochloric acid to the mixture until the solution tests acid to pH paper. Cool the mixture in ice, and collect the product on a Büchner funnel. Recrystallize it from the minimum quantity of hot water, and isolate it in the usual manner. Determine the melting point and the weight of the benzoic acid, and calculate its yield based on the weight of magnesium used to prepare the Grignard reagent.

Cleaning Up   Combine all aqueous layers, dilute with a large quantity of water, and flush the slightly acidic solution down the drain. The ether–ligroin mother liquor from the crystallization goes in the organic solvents container. The thin-layer chromatography developer, which contains dichloromethane, is placed in the halogenated organic waste container. Calcium chloride from the drying tube should be dissolved in water and flushed down the drain.

Questions

1. Triphenylmethanol also can be prepared from the reaction of ethyl benzoate with phenylmagnesium bromide and by the reaction of diethylcarbonate

   with phenylmagnesium bromide. Write stepwise reaction mechanisms for these two reactions.

2. If the ethyl benzoate used to prepare triphenylmethanol is wet, what byproduct is formed?

3. Exactly what weight of dry ice is needed to react with 2 mmol of phenyl-magnesium bromide?

4. In the synthesis of benzoic acid, benzene is often detected as an impurity. How does this come about?

5. The benzoic acid could have been extracted from the ether layer using

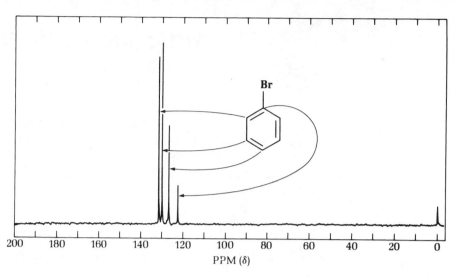

FIG. 38.7 ^{13}C NMR spectrum of bromobenzene (22.6 MHz).

sodium bicarbonate solution. Give equations showing how this might be done and how the benzoic acid would be regenerated. What practical reason makes this extraction less desirable than sodium hydroxide extraction?

6. What is the weight of frost (ice) on the dry ice that will react with all of Grignard reagent used in Experiment 8?

7. How many moles of carbon dioxide are contained in 10 g of dry ice?

8. Just after the dry ice has evaporated from the beaker, what is the white solid remaining?

9. Write an equation for the reaction of the white solid with 3 M hydrochloric acid.

10. Write an equation for the reaction of the product with sodium bicarbonate.

11. Would you expect sodium benzoate to have an odor?

12. What is the odor you detect after the product has dissolved in sodium bicarbonate solution?

13. What is the purpose of the *t*-butyl methyl ether extraction?

14. What is meant by the sentence "Isolate the product in the usual way"?

Surfing the Web

http://odin.chemistry.uakron.edu/organic_lab/grignard/

The synthesis of phenylmagnesium bromide (Experiment 1) is shown in 10 good close-up color photos at this University of Akron site. A further 15 photos show the synthesis of triphenylmethanol from benzophenone. The synthesis of benzoic acid, with three photos, is left, as it is here, as an open-ended experiment.

For updated information visit:
www.mtholyoke.edu/courses/kwilliam/microscale.shtml
or
www.hmco.com/hmco/college/chemistry/Home.html

Prelab Exercise: Account for the fact that the Wittig-Horner reaction of cinnamaldehyde gives almost exclusively the *E,E*-butadiene with very little contaminating *E,Z*-product.

$(C_6H_5)_3P$ + ClH$_2$C—⟨benzene ring⟩ ⟶ $(C_6H_5)_3\overset{+}{P}$CH$_2$—⟨benzene ring⟩ Cl$^-$

$\xrightarrow[\text{(50\% NaOH)}]{\text{Base}}$

Triphenylphosphine
(1)

Benzyl chloride
(2)

Benzyltriphenylphosphonium chloride
(3)

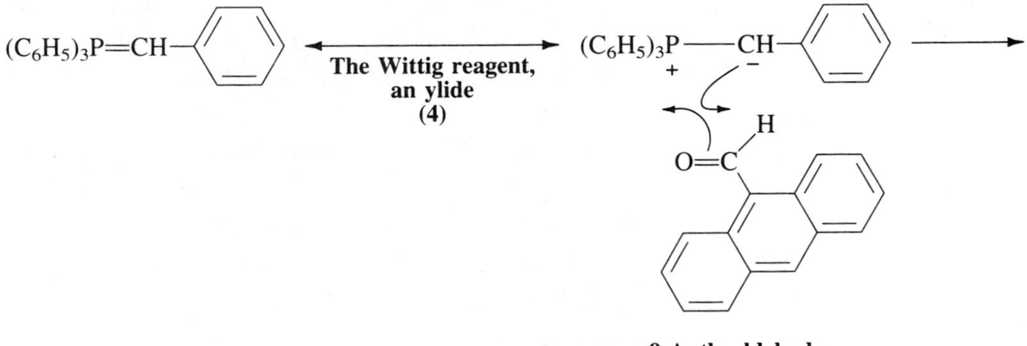

$(C_6H_5)_3P$=CH—⟨benzene ring⟩ $\xleftrightarrow{\hspace{1.2cm}}$ **The Wittig reagent, an ylide (4)** $(C_6H_5)_3\overset{+}{P}$—$\overset{-}{C}$H—⟨benzene ring⟩ ⟶

O=C—H (attached to anthracene ring)

9-Anthraldehyde
(5)

$(C_6H_5)_3P$—CH—⟨benzene ring⟩
│   │
O———C—H (anthracene ring)

⟶ $(C_6H_5)_3P$ +
                 ‖
                 O

Triphenylphosphine oxide
(7)

H—C=C—H (with phenyl and anthracene substituents)

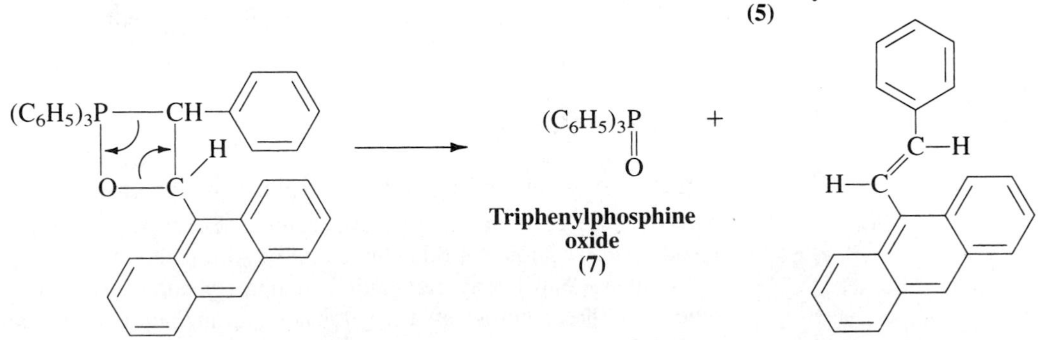

4-Anthranyl-2,2,2,3-tetraphenyl-1,2-oxaphosphetane
(6)

trans-**9-(2-Phenylethenyl)anthracene**
(8)

The Wittig reaction affords an invaluable method for the conversion of a carbonyl compound to an olefin; for example, the conversion of 9-anthraldehyde (**5**) and benzyl chloride (**2**), by means of the Wittig reagent (**4**), through a four-centered intermediate to the product (**8**). Because the active reagent, an ylide, is unstable, it is generated in the presence of the carbonyl compound by dehydrohalogenation of the phosphonium chloride (**3**) with base. Usually a very strong base, like phenyllithium in dry ether, is employed, but in this experiment, 50% sodium hydroxide is employed. The existence of the four-membered ring intermediate was proved by NMR in 1984.[1] The phenylethenylanthracene product produces a green fluorescence with the chemiluminescent substance, Cyalume, in the "light stick" reaction (see Chapter 62).

The light stick reaction

When the halogen compound employed in the first step has an activated halogen atom ($RCH{=}CHCH_2X$, $C_6H_5CH_2X$, XCH_2CO_2H), a simpler procedure known as the *Horner phosphonate modification* of the Wittig reaction is applicable. When benzyl chloride is heated with triethyl phosphite, chloroethane is eliminated from the initially formed phosphonium chloride with the production of diethyl benzylphosphonate. This phosphonate is stable, but in the presence of a strong base, such as the sodium methoxide used here, it condenses with a carbonyl component in the same way that a Wittig ylide condenses. Thus it reacts with benzaldehyde to give *E*-stilbene and with cinnamaldehyde to give *E,E*-1,4-diphenyl-1,3-butadiene.

Simplified Wittig reaction

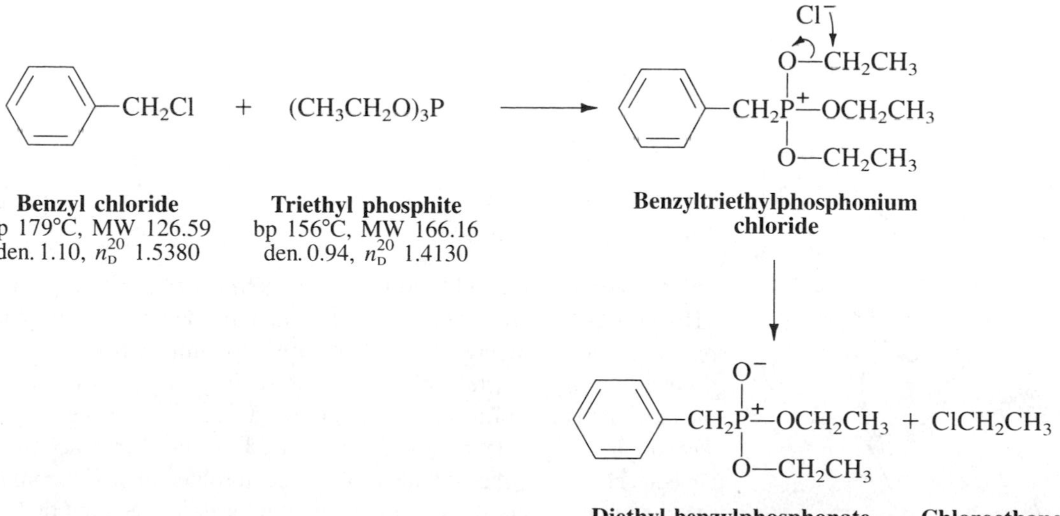

Benzyl chloride
bp 179°C, MW 126.59
den. 1.10, n_D^{20} 1.5380

Triethyl phosphite
bp 156°C, MW 166.16
den. 0.94, n_D^{20} 1.4130

**Benzyltriethylphosphonium
chloride**

Diethyl benzylphosphonate
bp 156°C/9 torr
MW 228.23, den. 1.045
n_D^{20} 1.4970

Chloroethane
bp 12.3°C
MW 64.52, den. 0.891

1. B. E. Maryanoff, A. B. Reitz, and M. S. Mutter, *J. Am. Chem. Soc.* **106,** 1873 (1984).

Diethyl benzylphosphonate + CH$_3$O$^-$Na$^+$ $\longrightarrow$

Sodium methoxide
MW 52.02

Ylide

Ylide + [Cinnamaldehyde structure] $\longrightarrow$ [E,E-1,4-Diphenyl-1,3-butadiene structure] + Na$^+$Ō—P—OCH$_2$CH$_3$

Cinnamaldehyde
MW 132.16, den. 1.11
bp 248°C

E,E-1,4-Diphenyl-1,3-butadiene
mp 153°C, MW 206.27

The energy-minimized conformation of E,E-1,4-diphenyl-1,3-butadiene. Because of steric hindrance, the molecule is not flat.

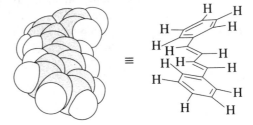

$\equiv$

Experiments

1. Synthesis of *trans*-9-(2-Phenylethenyl)anthracene[2]

Microscale Procedure

To a reaction tube add 200 mg of benzyltriphenylphosphonium chloride, 115 mg of 9-anthraldehyde, 0.6 mL of dichloromethane, and a magnetic stirring bar. With rapid stirring, 0.26 mL of 50% sodium hydroxide solution is added dropwise from a Pasteur pipette. After stirring vigorously for $\frac{1}{2}$ h, 1.5 mL of dichloromethane and 1.5 mL of water are added, and the tube capped and shaken. The organic layer is removed and placed in another reaction tube, and the aqueous layer is extracted with 1 mL of dichloromethane. The combined dichloromethane extracts are dried over calcium chloride pellets, the dichloromethane is removed, the drying agent is washed with more solvent, and the solvent is removed under vacuum in the filter flask (Fig. 39.1). To the solid remaining in the filter flask add 3 mL of 1-propanol, and transfer the hot solution to an Erlenmeyer

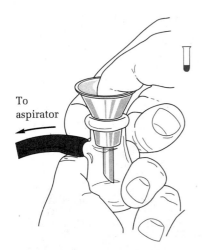

To aspirator

FIG. 39.1 Apparatus for removal of a solvent under a vacuum.

2. For macroscale syntheses see H.-D. Becker and K. Andersson, *J. Org. Chem.* **48,** 4552, 1983; G. Merkl and A. Merz, *Synthesis,* 295, 1973; E. F. Silversmith, *J. Chem. Ed.* **63,** 645, 1986.

flask to crystallize. After cooling spontaneously to room temperature, cool the flask in ice and collect the product on the Hirsch funnel. The triphenylphosphine oxide remains in the propanol solution. The reported melting point of the product is 131–132°C.

Macroscale Procedure

Into a 10-mL round-bottomed flask place 0.97 g of benzyltriphenylphosphonium chloride, 0.57 g of 9-anthraldehyde, 3 mL of dichloromethane, and a stirring bar. Clamp the flask over a magnetic stirrer, and stir the mixture at high speed while adding 1.3 mL of 50% sodium hydroxide solution dropwise from a Pasteur pipette. After the addition is complete, continue the stirring for 30 min, and then transfer the contents of the flask to a 50-mL separatory funnel using 10 mL of water and 10 mL of dichloromethane to complete the transfer. Shake the mixture, remove the organic layer, and then extract the aqueous layer once more with 5 mL of dichloromethane. Dry the combined organic layers with anhydrous calcium chloride pellets, transfer the solution to a 50-mL Erlenmeyer flask, evaporate it to dryness on a steam bath, and recrystallize the yellow partially crystalline residue from 15 mL of 2-propanol. The product crystallizes as thin yellow plates, mp 131–132°C. Save the product for use as a fluorescer in the Cyalume chemiluminescence experiment (Chapter 62). Examine a dilute ethanol solution of the product under an ultraviolet lamp.

MICROSCALE

CAUTION: Take care to keep organophosphorus compounds off the skin.

Use freshly prepared sodium methoxide. Commercial material is often inactive.

CAUTION: Handle benzyl chloride in the hood. Severe lachrymator and irritant of the respiratory tract.

2. Synthesis of 1,4-Diphenyl-1,3-butadiene

The success of this reaction is strongly dependent on the purity of the starting materials.[3] Benzyl chloride and triethyl phosphite are usually pure enough as received from the supplier. Cinnamaldehyde from a new, previously unopened bottle should be satisfactory, but since it is oxidized in air extremely rapidly, it should be distilled, preferably under nitrogen or at reduced pressure, if there is any doubt about its quality. Sodium methoxide, direct from reputable suppliers, has been found on occasion to be completely inactive. No easy method exists for determining its activity; therefore, it is best prepared following the procedure of *Organic Syntheses,* Col. Vol. IV, 651, 1963.[4]

To a 10 mm × 100 mm reaction tube add 316 mg (0.0025 mol) of benzyl chloride (α-chlorotoluene), 415 mg (0.44 mL, 0.0025 mol) of triethyl phosphite, and a boiling chip (Fig. 39.2). Because the triethyl phosphite has an offensive odor, obtain this material from the dispenser that has been calibrated to deliver the correct quantity. Place the reaction tube to a depth of about 1 cm in a sand

---

3. The instructor should try out this experiment before assigning it to a class.

4. For enough sodium methoxide for 50 microscale or 2 macroscale reactions, add 3.5 g of sodium spheres to 50 mL of anhydrous methanol in a 100-mL flask equipped with a condenser. Add the sodium a few pieces at a time through the top of the condenser at such a rate as to keep the reaction under control. It is safest to wait for complete reaction of one portion of sodium before adding the next. After all the sodium has reacted, remove the methanol on a rotary evaporator, first over a steam bath and then over a bath heated to 150°C. The sodium methoxide will be a free-flowing white powder that should keep for several weeks in a desiccator. Yield 8.2 g.

FIG. 39.2 Reaction tube for preparation of phosphonate.

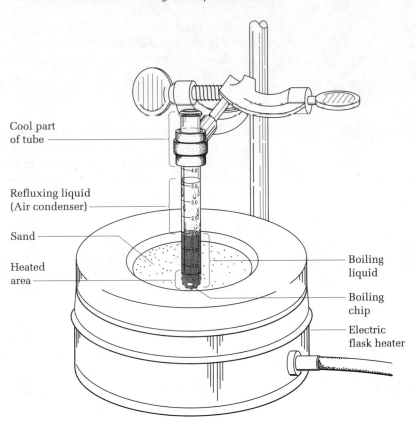

Cool part of tube

Refluxing liquid (Air condenser)

Sand

Heated area

Boiling liquid

Boiling chip

Electric flask heater

O
‖
H—C—N(CH₃)(CH₃)

[structure of DMF: O double bond to C, C bonded to H, C bonded to N, N bonded to two CH₃ groups]

N,N-Dimethylformamide (DMF)
MW 73.10, bp 153°C
n_D^{20} 1.4310

A highly polar solvent capable of dissolving ionic compounds such as sodium methoxide yet miscible with water—avoid skin contact.

bath that has been preheated to 210 to 220°C. Reflux the reaction mixture for 1 h. The vapors condense on the cool upper portion of the reaction tube, so a condenser is not necessary. At an internal temperature of about 140°C (which you need not monitor), ethyl chloride is evolved, and by the end of the reflux period, the temperature of the reaction mixture will be 200 to 220°C. At the end of the reaction period, remove the tube from the sand bath, cool it to room temperature, and then add the contents to 160 mg (0.003 mol) of sodium methoxide in a 10-mL Erlenmeyer flask using 1 mL of dry dimethylformamide to complete the transfer. Cool the mixture in ice, mix the contents of the flask thoroughly, then add 330 mg (0.0025 mol) of freshly distilled cinnamaldehyde in 1 mL of dry dimethylformamide dropwise with thorough mixing in the ice bath. Allow the mixture to come to room temperature. Note the changes that take place over the next few minutes.

After a *minimum* of 10 min, add 2 mL of methanol to the Erlenmeyer flask with thorough stirring. Then almost fill the flask with water, and stir the mixture until it is homogeneous in color. The hydrocarbon precipitates from the reaction mixture upon the addition of the methanol and water. Collect the product by vacuum filtration on the Hirsch funnel (Fig. 39.3), and wash the crystals first with water to remove the red color and then with ice-cold methanol to remove the yellow color. The hydrocarbon is completely insoluble in water and only sparingly soluble in methanol. Weigh the dry product, which should be faint yellow in

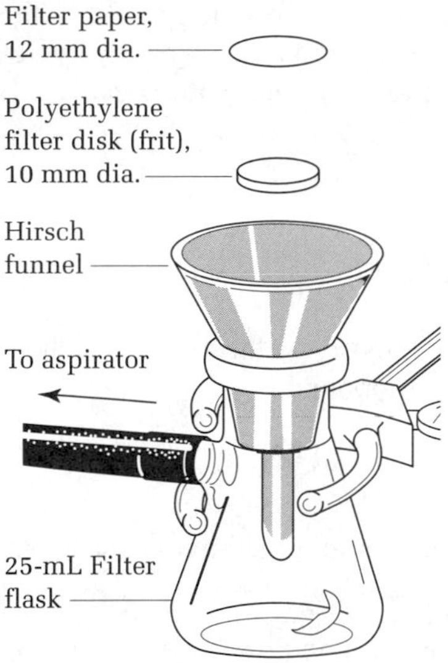

Filter paper,
12 mm dia.

Polyethylene
filter disk (frit),
10 mm dia.

Hirsch
funnel

To aspirator

25-mL Filter
flask

FIG. 39.3   Hirsch funnel for filtration.

color, determine its melting point, and calculate the crude yield. If the melting point is below 150°C, recrystallize the product from methylcyclohexane (10 mL/g), and again determine the melting point and yield. Hand in the product, and give crude and recrystallized weights, melting points, and yields.

Cleaning Up   The filtrate and washings from this reaction are dark, oily, and smell bad. The mixture contains dimethylformamide and could contain traces of all starting materials. Keep the volume as small as possible, and place it in the hazardous waste container for organophosphorus compounds. If methylcyclohexane was used for recrystallization, place the mother liquor in the organic solvents container.

3. Synthesis of 1,4-Diphenyl-1,3-butadiene

The success of this reaction is strongly dependent on the purity of the starting materials.

With the aid of pipettes and a pipetter, measure into a 25 × 150-mm test tube 5 mL of benzyl chloride (α-chlorotoluene) and 7.7 mL of triethyl phosphite. Add a boiling stone, insert a cold finger condenser, and with a flask heater, reflux the liquid gently for 1 h. Alternatively, carry out the reaction in a 25-mL round-bottomed flask equipped with a reflux condenser. (Elimination of ethyl chloride starts at about 130°C, and in the time period specified, the temperature of the liquid rises to 190–200°C.) Let the phosphonate ester cool to room temperature, pour it into a 125-mL Erlenmeyer flask containing 2.4 g of sodium methoxide, and add 40 mL of dimethylformamide, using a part of this solvent to rinse the test tube. Swirl the flask vigorously in a water–ice bath to thoroughly chill the contents, and continue swirling while running in 5 mL of cinnamaldehyde by pipette. The mixture soon turns deep red, and then crystalline hydrocarbon starts to separate. When there is no further change (about 2 min), remove the flask from the cooling bath and let it stand at room temperature for about 10 min. Then add 20 mL of water and 10 mL of methanol, swirl vigorously to dislodge crystals, and finally collect the product on a suction funnel using the red mother liquor to wash the flask. Wash the product with water until the red color of the product is all replaced by yellow. Then wash with methanol to remove the yellow impurity, and continue until the wash liquor is colorless. The yield of the crude, faintly yellow hydrocarbon (mp 150–151°C), should be about 5.7 g. This material is satisfactory for use in Chapter 51 (1.5 g required). A good solvent for crystallization of the remainder of the product is methylcyclohexane (bp 101°C, 10 mL/g; use more if the solution requires filtration). Pure *E,E*-1,4-diphenyl-1,3-butadiene melts at 153°C.

Cleaning Up   The filtrate and washings from this reaction are dark, oily, and smell bad. The mixture contains dimethylformamide and could contain traces of

🧪 MACROSCALE

CAUTION: Keep organophosphorus compounds off the skin. Handle benzyl chloride in the hood. It is a severe lachrymator and respiratory tract irritant.

Use freshly prepared or opened sodium methoxide; keep bottle closed.

Avoid skin contact with dimethylformamide.

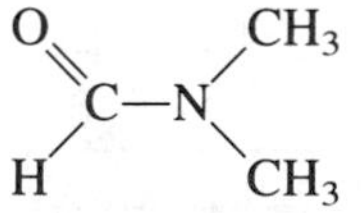

N,N-Dimethylformamide (DMF)
MW 73.10, bp 153°C
n_D^{20} 1.4310

A highly polar solvent capable of dissolving ionic compounds such as sodium methoxide yet miscible with water

all starting materials. Keep the volume as small as possible, and place it in the hazardous waste container for organophosphorus compounds. If methylcyclohexane was used for recrystallization, place the mother liquor in the organic solvents container.

Questions

1. Show how 1,4-diphenyl-1,3-butadiene might be synthesized from benzaldehyde and an appropriate halogenated compound.

2. Explain why the methyl groups of trimethyl phosphite give two peaks in the ^{1}H NMR spectrum (Fig. 39.4).

3. Write the equation for the reaction between sodium methoxide and moist air.

4. The Wittig reaction usually gives a mixture of *cis* and *trans* isomers. Using a molecular mechanics program, calculate the steric energies or heats of formation of both possible products in Experiments 2 and 4. Are these compounds planar? Are the most stable molecules produced in these two experiments?

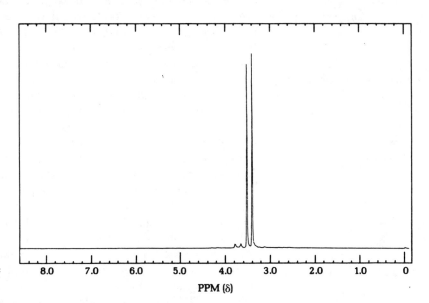

FIG. 39.4 ^{1}H NMR spectrum of trimethyl phosphite (90 MHz).

Esterification and Hydrolysis

Prelab Exercise: Give the detailed mechanism for the acid-catalyzed hydrolysis of methyl benzoate.

The ester group

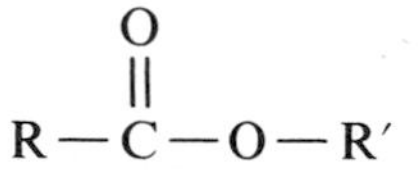

$$R-\overset{\overset{\displaystyle O}{\|}}{C}-O-R'$$

is an important functional group that can be synthesized in a number of different ways. The low molecular weight esters have very pleasant odors and indeed are the major components of the flavor and odor components of a number of fruits. Although the natural flavor may contain nearly a hundred different compounds, single esters approximate the natural odors and are often used in the food industry for artificial flavors and fragrances (Table 40.1).

Flavors and fragrances

Esters can be prepared by the reaction of a carboxylic acid with an alcohol in the presence of a catalyst such as concentrated sulfuric acid, hydrogen chloride, *p*-toluenesulfonic acid, or the acid form of an ion exchange resin:

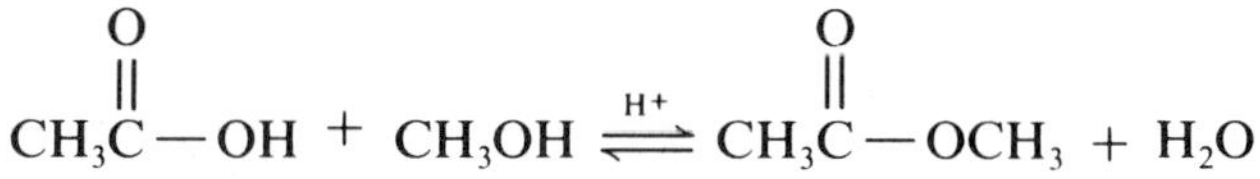

$$\underset{\textbf{Acetic acid}}{CH_3\overset{\overset{\displaystyle O}{\|}}{C}-OH} + \underset{\textbf{Methanol}}{CH_3OH} \underset{}{\overset{H^+}{\rightleftharpoons}} \underset{\textbf{Methyl acetate}}{CH_3\overset{\overset{\displaystyle O}{\|}}{C}-OCH_3} + H_2O$$

Fischer esterification

This Fischer esterification reaction reaches equilibrium after a few hours of refluxing. The position of the equilibrium can be shifted by adding more of the acid or of the alcohol, depending on cost or availability. The mechanism of the reaction involves initial protonation of the carboxyl group, attack by the nucleophilic hydroxyl, a proton transfer, and loss of water followed by loss of the catalyzing proton to give the ester. Each of these steps is completely reversible, so this process is also, in reverse, the mechanism for the hydrolysis of an ester:

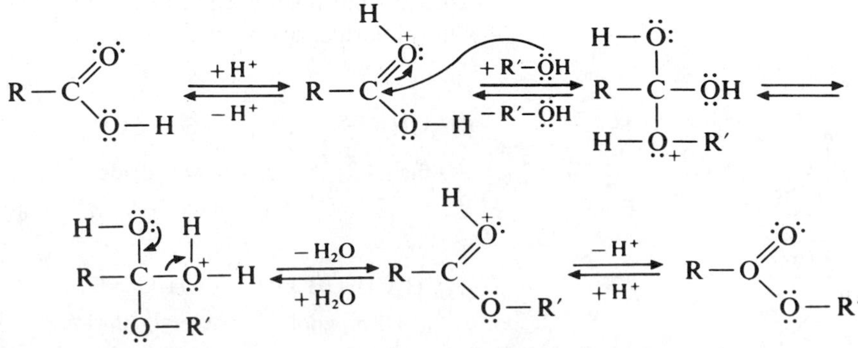

TABLE 40.1 Fragrances and Boiling Points of Esters

Ester	Formula	bp (°C)	Fragrance
2-Methylpropyl formate	$\overset{O}{\overset{\|}{HC}}-OCH_2\overset{CH_3}{\overset{\|}{CH}}CH_3$	98.4	Raspberry
1-Propyl acetate	$CH_3\overset{O}{\overset{\|}{C}}-OCH_2CH_2CH_3$	101.7	Pear
Methyl butyrate	$CH_3CH_2CH_2\overset{O}{\overset{\|}{C}}-OCH_3$	102.3	Apple
Ethyl butyrate	$CH_3CH_2CH_2\overset{O}{\overset{\|}{C}}-OCH_2CH_3$	121	Pineapple
2-Methylpropyl propionate	$CH_3CH_2\overset{O}{\overset{\|}{C}}-OCH_2\overset{CH_3}{\overset{\|}{CH}}CH_3$	136.8	Rum
3-Methylbutyl acetate	$CH_3\overset{O}{\overset{\|}{C}}-OCH_2CH_2\overset{CH_3}{\overset{\|}{CH}}CH_3$	142	Banana
Benzyl acetate	$CH_3\overset{O}{\overset{\|}{C}}-OCH_2-\langle\bigcirc\rangle$	213.5	Peach
Octyl acetate	$CH_3\overset{O}{\overset{\|}{C}}-OCH_2(CH_2)_6CH_3$	210	Orange
Methyl salicylate	$\underset{OH}{\overset{O}{\bigcirc}}\overset{\|}{C}-OCH_3$	222	Wintergreen

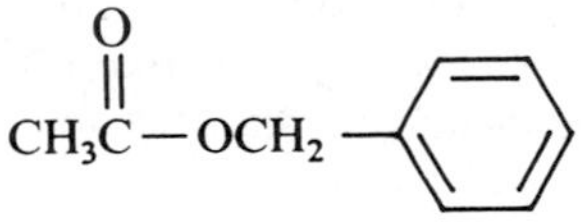

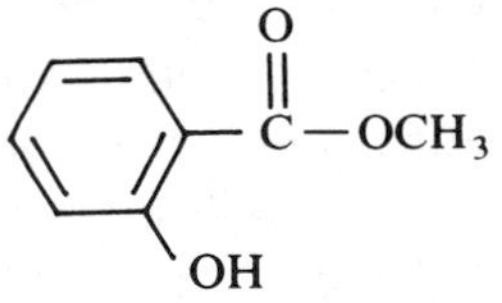

Other methods are available for the synthesis of esters, most of them more expensive but readily carried out on a small scale. For example, alcohols react with anhydrides and with acid chlorides:

$$CH_3CH_2OH + CH_3\overset{O}{\overset{\|}{C}}-O-\overset{O}{\overset{\|}{C}}CH_3 \longrightarrow CH_3\overset{O}{\overset{\|}{C}}-OCH_2CH_3 + CH_3\overset{O}{\overset{\|}{C}}-OH$$

Ethanol   **Acetic anhydride**   **Ethyl acetate**   **Acetic acid**

$$CH_3CH_2CH_2OH + CH_3\overset{O}{\overset{\|}{C}}-Cl \longrightarrow CH_3\overset{O}{\overset{\|}{C}}-OCH_2CH_2CH_3 + HCl$$

1-Propanol   **Acetyl chloride**   ***n*-Propyl acetate**

In the latter reaction, an organic base such as pyridine is usually added to react with the hydrogen chloride.

Other ester syntheses

A number of other methods can be used to synthesize the ester group. Among these are the addition of 2-methylpropene to an acid to form *t*-butyl esters, the addition of ketene to make acetates, and the reaction of a silver salt with an alkyl halide:

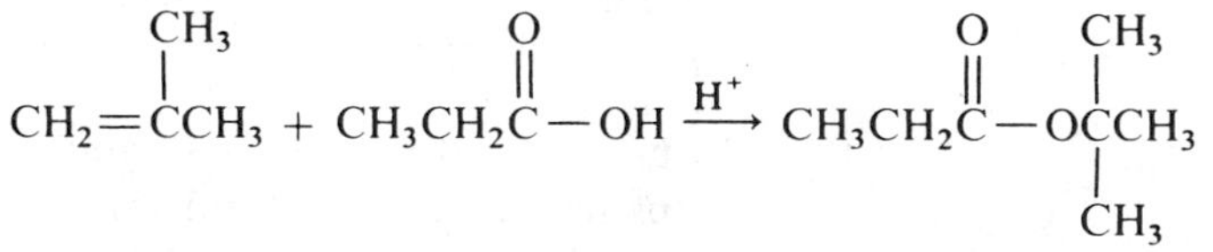

$$CH_2{=}\overset{\overset{\displaystyle CH_3}{|}}{C}CH_3 \;+\; CH_3CH_2\overset{\overset{\displaystyle O}{\|}}{C}{-}OH \;\xrightarrow{\;H^+\;}\; CH_3CH_2\overset{\overset{\displaystyle O}{\|}}{C}{-}\underset{\underset{\displaystyle CH_3}{|}}{\overset{\overset{\displaystyle CH_3}{|}}{O}}CCH_3$$

2-Methylpropene   **Propionic acid**   ***t*-Butyl propionate**
(isobutylene)

$$CH_2{=}C{=}O \;+\; HOCH_2{-}\bigcirc \;\longrightarrow\; CH_3\overset{\overset{\displaystyle O}{\|}}{C}{-}OCH_2{-}\bigcirc$$

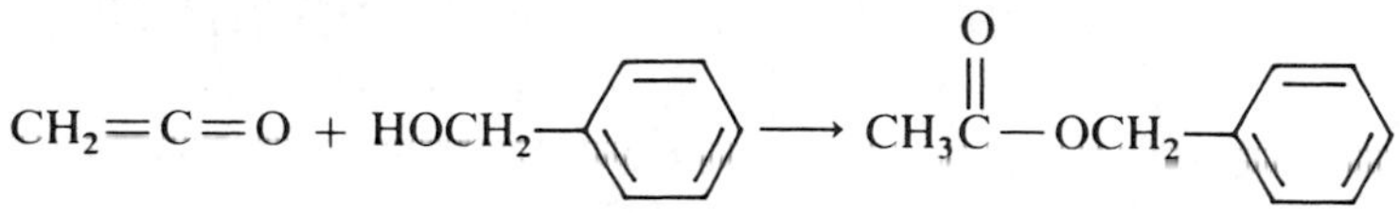

Ketene   **Benzyl alcohol**   **Benzyl acetate**

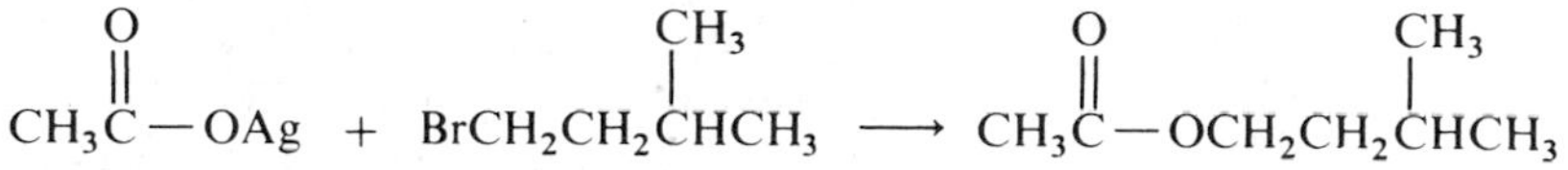

$$CH_3\overset{\overset{\displaystyle O}{\|}}{C}{-}OAg \;+\; BrCH_2CH_2\underset{\underset{\displaystyle CH_3}{|}}{C}HCH_3 \;\longrightarrow\; CH_3\overset{\overset{\displaystyle O}{\|}}{C}{-}OCH_2CH_2\underset{\underset{\displaystyle CH_3}{|}}{C}HCH_3$$

Silver acetate   **1-Bromo-3-methylbutane**   **3-Methylbutyl acetate**

As noted above, Fischer esterification is an equilibrium process. Consider the reaction of acetic acid with 1-butanol to give *n*-butyl acetate:

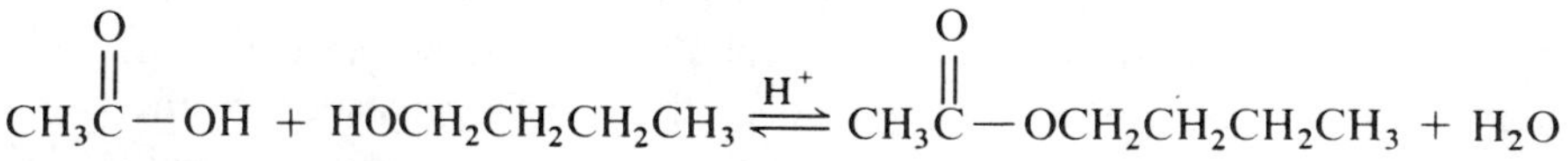

$$CH_3\overset{\overset{\displaystyle O}{\|}}{C}{-}OH \;+\; HOCH_2CH_2CH_2CH_3 \;\underset{}{\overset{H^+}{\rightleftharpoons}}\; CH_3\overset{\overset{\displaystyle O}{\|}}{C}{-}OCH_2CH_2CH_2CH_3 \;+\; H_2O$$

The equilibrium constant is:

$$K_{eq} = \frac{[n\text{-BuOAc}][H_2O]}{[n\text{-BuOH}]\,[HOAc]}$$

For primary alcohols reacting with unhindered carboxylic acids, $K_{eq} \approx 4$. If equal quantities of 1-butanol and acetic acid are allowed to react, at equilibrium the theoretical yield of ester is only 67%. To upset the equilibrium we can, by Le Châtelier's principle, increase the concentration of either the alcohol or acid, as noted above. If either one is doubled, the theoretical yield increases to 85%. When one is tripled it goes to 90%. But note that in the example cited the boiling point of the relatively nonpolar ester is only about 8°C higher than the boiling points of the polar acetic acid and 1-butanol, so a difficult separation

TABLE 40.2   The Ternary Azeotrope of Boiling Point 90.7°C

Compound	Boiling Point of Pure Compound (°C)	Percentage Composition of Azeotrope		
		Vapor Phase	*Upper Layer*	*Lower Layer*
1-Butanol	117.7	8.0	11.0	2.0
n-Butyl acetate	126.7	63.0	86.0	1.0
Water	100.0	29.0	3.0	97.0

problem exists if either starting material is increased in concentration and the product isolated by distillation.

Another way to upset the equilibrium is to remove water. This can be done by adding to the reaction mixture molecular sieves (an artificial zeolite), which preferentially adsorb water. Most other drying agents, such as anhydrous sodium sulfate or calcium chloride pellets, will not remove water at the temperatures used to make esters.

Azeotropic distillation

A third way to upset the equilibrium is to preferentially remove the water as an azeotrope (a constant-boiling mixture of water and an organic liquid). The information in Table 40.2 can be found in a chemistry handbook table of ternary (three-component) azeotropes.

These data tell us that the vapor that distills from a mixture of 1-butanol, *n*-butyl acetate, and water will boil at 90.7°C and the vapor contains 8% alcohol, 63% ester, and 29% water. The vapor is homogeneous, but when it condenses, it separates into two layers. The upper layer is composed of 11% alcohol, 86% ester, and 3% water, but the lower layer consists of 97% water with only traces of alcohol and ester. If some ingenious way to remove the lower layer from the condensate and still return the upper layer to the reaction mixture can be devised, then the equilibrium can be upset and nearly 100% of the ester can be produced in the reaction flask.

Microscale Dean-Stark apparatus

A cork is used instead of a septum so that layer separation can be observed clearly.

The apparatus shown in Fig. 40.1, modeled after that of Dean and Stark, achieves the desired separation of the two layers. The mixture of equimolar quantities of 1-butanol and acetic acid is placed in the flask along with an acid catalyst. Stirring reduces bumping. The vapor, the temperature of which is 90.7°C, condenses and runs down to the sidearm, which is closed with a cork. The layers separate, with the denser water layer remaining in the sidearm while the lighter ester plus alcohol layer runs down into the reaction flask. As soon as the theoretical quantity of water has collected, the reaction is over and the product in the flask should be ester of high purity. The macroscale apparatus is illustrated in Fig. 40.2.

Esterification using a carboxylic acid and an alcohol requires an acid catalyst. In the first experiment, the acid form of an ion-exchange resin is used. This resin, in the form of small beads, is a cross-linked polystyrene that bears sulfonic acid groups on some of the phenyl groups. Essentially it is an immobilized form of *p*-toluenesulfonic acid, an organic-substituted sulfuric acid.

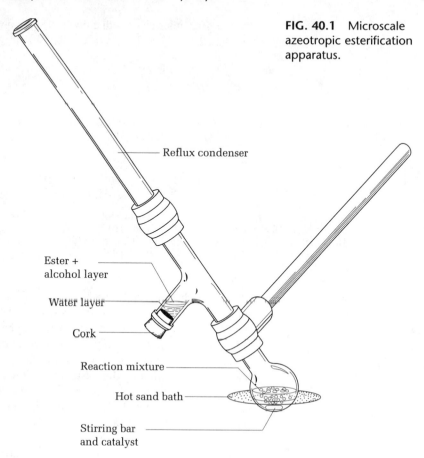

FIG. 40.1   Microscale azeotropic esterification apparatus.

Reflux condenser

Ester + alcohol layer

Water layer

Cork

Reaction mixture

Hot sand bath

Stirring bar and catalyst

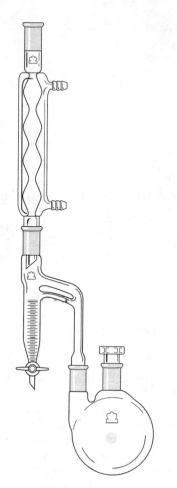

FIG. 40.2   Macroscale azeotropic distillation apparatus, with Dean-Stark trap where water collects.

An ion exchange catalyst

$$\overline{(CH-CH_2-CH-CH_2-CH)_n}$$

$$SO_3{}^-H^+ \quad SO_3{}^-H^+ \quad SO_3{}^-H^+$$

This catalyst has the distinct advantage that at the end of the reaction it can be removed simply by filtration. Immobilized catalysts of this type are becoming more and more common in organic synthesis.

If concentrated sulfuric acid were used as the catalyst, it would be necessary to dilute the reaction mixture with ether; wash the ether layer successively with water, sodium carbonate solution, and saturated sodium chloride solution; and then dry the ether layer with anhydrous calcium chloride pellets before evaporating the ether.

Experiments

MICROSCALE

1. *n*-Butyl Acetate by Azeotropic Distillation of Water

$$CH_3\overset{O}{\underset{\|}{C}}-OH \quad + \quad HOCH_2CH_2CH_2CH_3 \quad \overset{H^+}{\rightleftharpoons} \quad CH_3\overset{O}{\underset{\|}{C}}-OCH_2CH_2CH_2CH_3 + H_2O$$

Acetic acid	**1-Butanol**	***n*-Butyl acetate**	
MW 60.05	MW 74.12	MW 116.16	MW 18
bp 117.9°C, den 1.049	bp 117.7°C, den 0.810	bp 126.5°C, den 0.882	
n_D^{20} 1.3720	n_D^{20} 1.3990	n_D^{20} 1.3940	

Ion-exchange resin catalyst

A cork instead of a septum allows the accumulation of water to be observed.

In a 5-mL short-necked round-bottomed flask, place 0.2 g of Dowex 50X2-100 ion-exchange resin,[1] 0.60 g (0.58 mL) of acetic acid (0.01 mole), 0.74 g (0.91 mL) of 1-butanol (0.01 mole), and a stirring bar. Attach the addition port with the sidearm corked and an empty distilling column as shown in Fig. 40.1, and clamp the apparatus at the angle shown. Heat the flask with stirring on a hot sand bath, and boil the reaction mixture. Stirring the mixture prevents it from bumping. As an option, you might hold a thermometer just above the boiling liquid and note a temperature of about 91°C. Remove the thermometer, and allow the reaction mixture to reflux in such a manner that the vapors condense about one-third of the way up the empty distilling column, which is functioning as an air condenser. Note that the material that condenses is not homogeneous, since droplets of water begin to collect in the upper part of the apparatus. As the sidearm fills with condensate, it is cloudy at first, and then two layers separate. When the volume of the lower aqueous layer does not appear to increase, the reaction is over. This will take about 20 to 30 min. Carefully remove the apparatus from the heat, allow it to cool, and then tip the apparatus very carefully to allow all the upper layer in the sidearm to run back into the reaction flask. Disconnect the apparatus.

Remove the product from the reaction flask with a Pasteur pipette, determine its weight and boiling point, and assess its purity by thin-layer chromatography and IR spectroscopy. The product can be analyzed by gas chromatography on a 10-ft (3-m) Carbowax column. At 152°C, the 1-butanol has a retention time of 2.1 min, the *n*-butyl acetate 2.5 min, and the acetic acid 7.5 min. Look for the presence of hydroxyl and carboxyl absorption bands in the IR spectrum. The product can easily be purified by simple distillation (Fig. 40.3).

Cleaning Up Place the catalyst in the solid hazardous waste container.

---

1. The Dowex resin as received should be washed with water by decantation to remove much of the yellow color. It is then collected by vacuum filtration on a Büchner funnel and returned, in a slightly damp state, to the reagent bottle.

FIG. 40.3 Simple distillation apparatus.

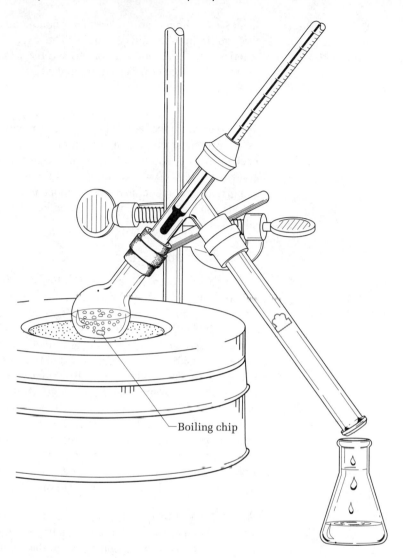

Boiling chip

MICROSCALE

2. Isobutyl Propionate by Fischer Esterification

$$CH_3CH_2\overset{\displaystyle O}{\overset{\|}{C}}OH \ + \ HOCH_2\overset{\displaystyle CH_3}{\overset{|}{C}}HCH_3 \ \underset{\longleftarrow}{\overset{H^+}{\rightleftharpoons}} \ CH_3CH_2\overset{\displaystyle O}{\overset{\|}{C}}OCH_2\overset{\displaystyle CH_3}{\overset{|}{C}}HCH_3 \ + \ H_2O$$

Propanoic acid   **2-Methyl-1-propanol**   **Isobutyl propionate**
  MW 74.08   MW 74.12   MW 130.19
  bp 141°C   bp 108°C   bp 136.8°C

In this experiment, an excess of propanoic acid is employed in order to drive the equilibrium toward the right. The excess propanoic acid is removed by reaction with potassium carbonate, and the water is adsorbed by silica gel when the reaction mixture is chromatographed.

Procedure

To a reaction tube add 112 mg of 2-methyl-1-propanol (isobutyl alcohol), 148 mg of propanoic acid (propionic acid), 50 mg of Dowex 50X2-100 ion-exchange resin, and a boiling chip. Attach the empty distilling column as an air condenser (Fig. 40.4). Reflux the resulting mixture for 1 h or more, cool it to room temperature, remove the product mixture from the resin with a Pasteur pipette, and chromatograph the product (2-methyl-1-propyl propanoate) on a silica-gel column.

Chromatography Procedure

Carry out this part of the experiment in a hood. Dichloromethane is a cancer-suspect agent.

Assemble the column as depicted in Fig. 40.5 being sure it is clamped in a vertical position. Close the valve, and fill the column with dichloromethane to the bottom of the funnel. Prepare a slurry of 1 g of silica gel in 4 mL of dichloromethane in a small beaker. Stir the slurry gently to get rid of air bubbles, and gently swirl, pour, and scrape the slurry into the funnel, which has a capacity of 10 mL. After some of the silica gel has been added to the column, open the stopcock and allow solvent to drain slowly into an Erlenmeyer flask. Use this dichloromethane to rinse the beaker containing the silica gel. As the silica gel is being added, tap the column with a glass rod or pencil so the adsorbent will pack tightly into the column. Continue to tap the column while cycling the dichloromethane through the column once more, and then add 1 g of anhydrous potassium carbonate to the top of the silica gel. The potassium carbonate will remove water from the esterification mixture as well as react with any carboxylic acid present. Run the solvent down to the top of the potassium carbonate layer.

Adding the Sample.   The solvent is drained just to the surface of the potassium carbonate. Using a Pasteur pipette, add the sample to the column and let it run into the adsorbent, stopping when the solution reaches the top of the potassium carbonate. The flask and ion-exchange resin are rinsed twice with 0.5-mL portions of dichloromethane that are run into the column, with the eluent being collected in a tared reaction tube. The elution is completed with 1 mL more dichloromethane.

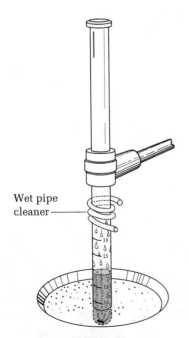

Wet pipe cleaner —

Analysis of the Product.   Evaporate the dichloromethane under a stream of air or nitrogen in the hood, and remove the last traces by connecting the reaction tube to the water aspirator. Since the dichloromethane boils at 40°C and the product at 137°C, separation of the two is easily accomplished. Determine the weight of the product and its boiling point, and calculate the yield. The ester should be a perfectly clear, homogeneous liquid. Obtain an IR spectrum and analyze it for the presence of unreacted alcohol and carboxylic acid. Check the purity of the product by thin-layer and/or gas chromatography.

FIG. 40.4 Esterification apparatus.

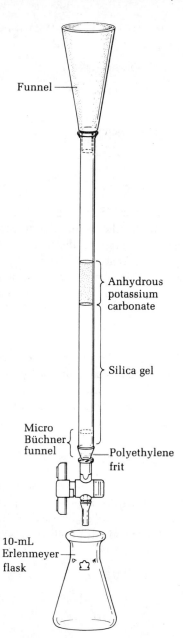

Funnel

Anhydrous
potassium
carbonate

Silica gel

Micro
Büchner
funnel

Polyethylene
frit

10-mL
Erlenmeyer
flask

FIG. 40.5 Chromatographic
column for esters.

Cleaning Up   Place the catalyst in the hazardous waste container. Any dichloromethane should be placed in the halogenated organic waste container. The
contents of the chromatography column, if free of solvent, can be placed in the
nonhazardous solid waste container; otherwise, this material is classified as a
hazardous waste and must go in the container so designated.

MICROSCALE

3. Benzyl Acetate from Acetic Anhydride

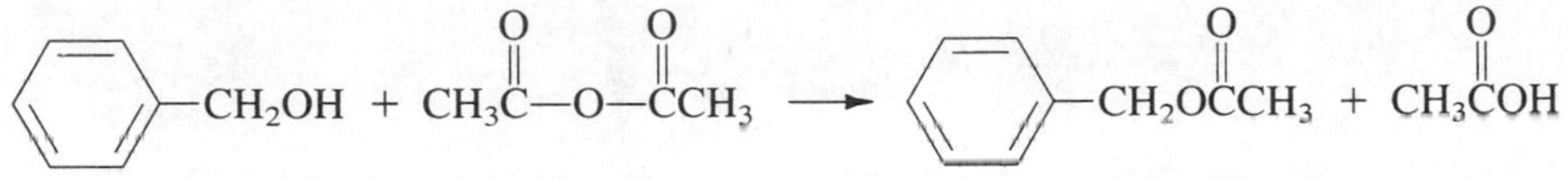

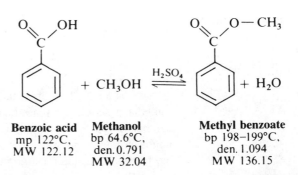

Benzyl alcohol	**Acetic anhydride**	**Benzyl acetate**	**Acetic acid**
MW 108.14	MW 102.09	MW 150.18	MW 60.06
bp 205°C	bp 138–140°C	bp 213.5°C	bp 117–118°C
n_D^{20} 1.5400	n_D^{20} 1.3900	n_D^{20} 1.5020	n_D^{20} 1.3720

To a reaction tube add 108 mg of benzyl alcohol, 102 mg of acetic anhydride, and a boiling chip. Reflux the mixture for at least 1 h, cool the mixture to room temperature, and chromatograph the liquid in exactly the same manner described immediately above. Analyze the product by thin-layer chromatography and by IR spectroscopy as a thin film between salt or silver chloride plates. Note the presence or absence of hydroxyl and carboxyl bands.

   This ester cannot be prepared by Fischer esterification using Dowex. Polymerization seems to occur.

Cleaning Up   Place the catalyst in the hazardous waste container. Any dichloromethane should be placed in the halogenated organic waste container. The contents of the chromatography column, if free of solvent, can be placed in the nonhazardous solid waste container; otherwise, this material is classified as a hazardous waste and must go in the container so designated.

MICROSCALE

4. Other Esterifications

This experiment lends itself to wide-ranging experimentation. All three methods of esterification described above can, in principle, be applied to any unhindered primary or secondary alcohol. These methods work well for most of the esters in Table 40.1 and for hundreds of others as well.

MACROSCALE

5. Methyl Benzoate by Fischer Esterification

Benzoic acid   **Methanol**   **Methyl benzoate**
mp 122°C,   bp 64.6°C,   bp 198–199°C,
MW 122.12   den. 0.791   den. 1.094
                   MW 32.04   MW 136.15

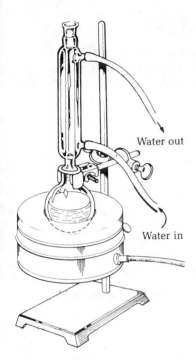

FIG. 40.6 Apparatus for refluxing a reaction mixture.

Place 10.0 g of benzoic acid and 25 mL of methanol in a 125-mL round-bottomed flask, cool the mixture in ice, pour 3 mL of concentrated sulfuric acid *slowly and carefully down the walls of the flask,* and then swirl to mix the components. Attach a reflux condenser, add a boiling chip, and reflux the mixture gently for 1 h. See Fig. 40.6 for a drawing of the apparatus. Cool the solution, decant it into a separatory funnel containing 50 mL of water, and rinse the flask with 35 mL of *t*-butyl methyl ether. Add this ether to the separatory funnel, shake thoroughly, and drain off the water layer, which contains the sulfuric acid and the bulk of the methanol. Wash the ether in the separatory funnel with 25 mL of water followed by 25 mL of 0.5 *M* sodium bicarbonate to remove unreacted benzoic acid. Again shake, with frequent release of pressure by inverting the separatory funnel and opening the stopcock, until no further reaction is apparent; then drain off the bicarbonate layer into a beaker. If this aqueous material is made strongly acidic with hydrochloric acid, unreacted benzoic acid may be observed. Wash the ether layer in the separatory funnel with saturated sodium chloride solution, and dry the solution over anhydrous calcium chloride in an Erlenmeyer flask. Add sufficient calcium chloride pellets so that it no longer clumps together on the bottom of the flask. After 10 min decant the dry ether solution into a flask, wash the drying agent with an additional 5 mL of ether, and decant again.

Remove the ether by simple distillation or by evaporation on the steam bath under an aspirator tube. See Fig. 5.10 or Fig. 3.22, or use a rotary evaporator (Fig. 7.12). When evaporation ceases, add 2–3 g of anydrous calcium chloride pellets to the residual oil and heat for about 5 min longer. Then decant the methyl benzoate into a 50-mL round-bottomed flask, attach a stillhead, dry out the ordinary condenser and use it without water circulating in the jacket, and distill. The boiling point of the ester is so high (199°C) that a water-cooled condenser is liable to crack. Use a tared 25-mL Erlenmeyer as the receiver, and collect material boiling above 190°C. A typical student yield is about 7 g. See Chapter 28 for the nitration of methyl benzoate and Chapter 38 for its use in the Grignard synthesis of triphenylmethanol.

IR and NMR spectra of benzoic acid and methyl benzoate are found at the end of the chapter (see Figs. 40.7 to 40.12).

Tared: previously weighed

Cleaning Up   Pour the sulfuric acid layer into water, combining it with the bicarbonate layer, neutralize it with sodium carbonate, and flush the solution down the drain with much water. The saturated sodium chloride layer can also be flushed down the drain. If the calcium chloride is free of ether and methyl benzoate, it can be placed in the nonhazardous solid waste container; otherwise it must go into the hazardous waste container. *t*-Butyl methyl ether goes into the organic solvents container, along with the pot residues from the final distillation.

6. Hydrolysis (Saponification): The Synthesis of Soap

In general, the reversal of esterification is called *hydrolysis*. In the case of hydrolysis of a fatty acid ester, it is called *saponification*. In this experiment, the saturated fat made from hydrogenated olive oil in Chapter 25 will be saponified to give a soap, which, in this case, will be primarily sodium stearate.

$$
\begin{array}{c}
\text{CH}_2\text{OC(CH}_2)_{16}\text{CH}_3 \\
| \\
\text{CHOC(CH}_2)_{16}\text{CH}_3 \\
| \\
\text{CH}_2\text{OC(CH}_2)_{16}\text{CH}_3
\end{array}
\xrightarrow[\text{H}_2\text{O}]{\text{NaOH}}
\begin{array}{c}
\text{CH}_2\text{OH} \\
| \\
\text{CHOH} \\
| \\
\text{CH}_2\text{OH}
\end{array}
+ \; 3\text{Na}^+\overset{-}{\text{O}}\text{C(CH}_2)_{16}\text{CH}_3
$$

Glycerol tristearate   **Glycerol**   **Sodium stearate (soap)**

Microscale Procedure

Place 0.18 g of the saturated triglyceride made in Chapter 25, Experiment 4, in a 5-mL short-necked, round-bottomed flask. Add 1.5 mL of a 50 : 50 water–ethanol solution that contains 0.18 g of sodium hydroxide. Add an air condenser, and gently reflux the mixture on the sand bath for 30 min, taking care not to boil away the ethanol. At the end of the reaction period, some of the soap will have precipitated. Transfer the mixture to a 10-mL Erlenmeyer flask containing a solution of 0.8 g of sodium chloride in 3 mL of water. Collect the precipitated soap on the Hirsch funnel, and wash it free of excess sodium hydroxide and salt using 4 mL of distilled ice water.

Test the soap by adding a very small piece (about 5–15 mg) to a centrifuge or test tube along with 3 or 4 mL of distilled water. Cap the tube, and shake it vigorously. Note the height and stability of the bubbles. Add a crystal of magnesium chloride or calcium chloride to the tube. Shake the tube again, and note the results. For comparison, do these same tests with a few grains of a detergent instead of the soap.

Macroscale Procedure

In a 100-mL round-bottomed flask, place 5 g of hydrogenated olive oil (Chapter 25) or lard or solid shortening (e.g., Crisco). Add 20 mL of ethanol and a hot solution of 5 g of sodium hydroxide in 20 mL of water. This solution, prepared in a beaker, will become very hot as the sodium hydroxide dissolves. Fit the flask with a water-cooled condenser, and reflux the mixture on the sand bath for 30 min. At the end of the reaction period some of the soap may precipitate from the reaction mixture. Transfer the mixture to a 250-mL Erlenmeyer flask containing an ice-cold solution of 25 g of sodium chloride in 90 mL of distilled water. Collect the precipitated soap on a Büchner funnel, and wash it free of excess sodium hydroxide and salt using no more than 100 mL of distilled ice water.

Test the soap by adding a small piece to a test tube along with about 5 mL of water. Cap the tube, and shake it vigorously. Note the height and stability of the bubbles formed. Add a few crystals of magnesium chloride or calcium chloride to the tube. Shake the tube again, and note the results. For comparison, do these same tests with an amount of detergent equal to that of the soap used.

Questions

1. In the preparation of methyl benzoate, what is the purpose of: (a) Washing the organic layer with sodium bicarbonate solution? (b) Washing the organic layer with saturated sodium chloride solution? (c) Treating the organic layer with anhydrous calcium chloride pellets?

2. Assign the resonances in Fig. 40.9 to specific protons in methyl benzoate.

3. Figures 40.10 and 40.11 each have two resonances that are very small. What do the carbons causing these peaks have in common?

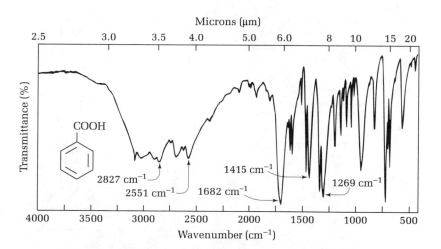

FIG. 40.7 Infrared spectrum of benzoic acid in CS$_2$ (KBr disk).

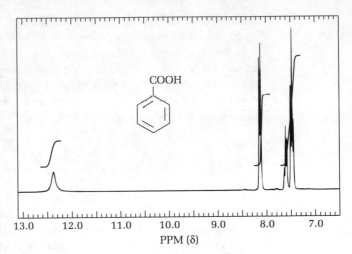

FIG. 40.8 ¹H NMR spectrum of benzoic acid (250 MHz).

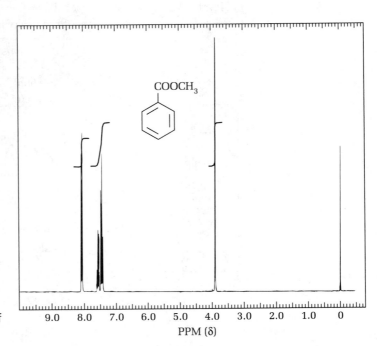

FIG. 40.9 ¹H NMR spectrum of methyl benzoate (250 MHz).

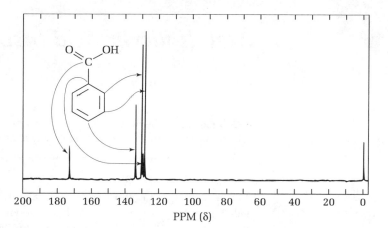

FIG. 40.10 ^{13}C NMR spectrum of benzoic acid (22.6 MHz).

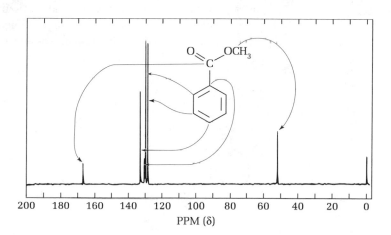

FIG. 40.11 ^{13}C NMR spectrum of methyl benzoate (22.6 MHz).

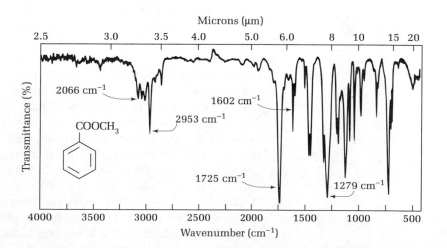

FIG. 40.12   IR spectrum of methyl benzoate.

Acetylsalicylic Acid (Aspirin)

Prelab Exercise: Write detailed mechanisms showing how pyridine and sulfuric acid catalyze the formation of acetylsalicylic acid.

Aspirin is among the most fascinating and versatile drugs known to medicine, and it is among the oldest—the first known use of an aspirinlike preparation can be traced to ancient Greece and Rome. Salicigen, an extract of willow and poplar bark, has been used as a pain reliever (analgesic) for centuries. In the middle of the last century it was found that salicigen is a glycoside formed from a molecule of salicylic acid and a sugar molecule. Salicylic acid is easily synthesized on a large scale by heating sodium phenoxide with carbon dioxide at 150°C under slight pressure (the Kolbe synthesis):

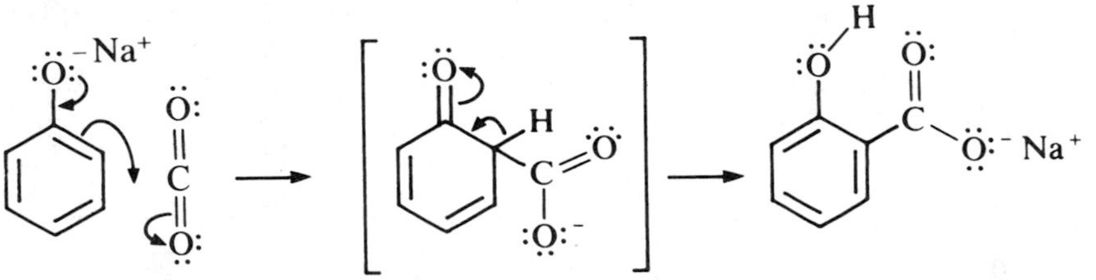

Sodium salicylate

Unfortunately, however, salicylic acid attacks the mucous membranes of the mouth and esophagus and causes gastric pain that may be worse than the discomfort it was meant to cure. Felix Hoffmann, a chemist for Friedrich Bayer, a German dye company, reasoned that the corrosive nature of salicylic acid could be altered by addition of an acetyl group; and in 1893 the Bayer Company obtained a patent on acetylsalicylic acid, despite the fact that it had been synthesized some 40 years previously by Charles Gerhardt. Bayer coined the name Aspirin for its new product to reflect its acetyl nature and its natural occurrence in the Spiraea plant. Over the years they have allowed the term aspirin to fall into the public domain so it is no longer capitalized. The manufacturers of Coke and Sanka work hard to prevent a similar fate befalling their trademarks.

In 1904 the head of Bayer, Carl Duisberg, decided to emulate John D. Rockefeller's Standard Oil Company and formed an "interessen gemeinschaft" (I.G.) of the dye industry (Farbenindustrie). This cartel completely dominated the world dye industry before World War I, and it continued to prosper between the wars, even though some of its assets were seized and sold after World War I. After World War I an American company, Sterling Drug, bought the rights to aspirin for $5.3 million. Sterling was bought by Eastman Kodak in 1988, then sold to SmithKline Beacham.

Because of their involvement at Auschwitz, the top management of I.G. Farbenindustrie was tried and convicted at the Nuremberg trials after World War II, and the cartel broken into three large branches—Bayer, Hoechst, and BASF (Badische Anilin und Sodafabrik)—each of which does more business than DuPont, the largest American chemical company. In 1997 the American rights to the Bayer name and trademark were sold back to Bayer A.G. for $1 billion.

By law, all drugs sold in the United States must meet purity standards set by the U.S. Food and Drug Administration (FDA), and so all aspirin is essentially the same. Each five-grain tablet contains 0.325 g of acetylsalicylic acid held together with a binder. The remarkable difference in price for aspirin is primarily a reflection of the advertising budget of the company that sells it. Bayer has 5% of the painkiller market; lower-priced generic aspirin has 18%.

The gross profit margin on Bayer aspirin is 80%. After advertising costs the profit margin is 60%.

Aspirin is an analgesic (painkiller), an antipyretic (fever reducer), and an anti-inflammatory agent. It is the premier drug for reducing fever, a role for which it is uniquely suited. As an anti-inflammatory, it has become the most widely effective treatment for arthritis. Patients suffering from arthritis must take so much aspirin (several grams per day) that gastric problems may result. For this reason aspirin is often combined with a buffering agent. Bufferin is an example of such a preparation.

The ability of aspirin to diminish inflammation is apparently due to its inhibition of the synthesis of prostaglandins, a group of C-20 molecules that enhance inflammation. Aspirin alters the oxygenase activity of prostaglandin synthetase by moving the acetyl group to a terminal amine group of the enzyme.

If aspirin were a new invention, the FDA would place many hurdles in the path of its approval. It has been implicated, for example, in Reye's syndrome, a brain disorder that strikes children and young people under 18 who take aspirin after flu or chicken pox. It has an effect on platelets, which play a vital role in blood clotting. In newborn babies and their mothers, aspirin can lead to uncontrolled bleeding and problems of circulation for the baby—even brain hemorrhage in extreme cases. This same effect can be turned into an advantage, however. Heart specialists urge potential stroke victims to take aspirin regularly to inhibit clotting in their arteries, and it has been shown that one-half tablet per day will help prevent heart attacks in healthy men.

Aspirin is found in more than 100 common medications, including Alka-Seltzer, Anacin, Coricidin, Excedrin, Midol, and Vanquish. Despite its side effects, aspirin is one of the safest, cheapest, and most effective nonprescription drugs, although acetaminophen (Tylenol, etc.) has 40% and ibuprofen (Advil, etc.) has 26% of the painkiller market in dollar volume ($2.47 billion in 1996). Naproxen (Aleve) has 6% of the market. Aspirin is made commercially employing the same synthesis used here.

The mechanism for the acetylation of salicylic acid is as follows:

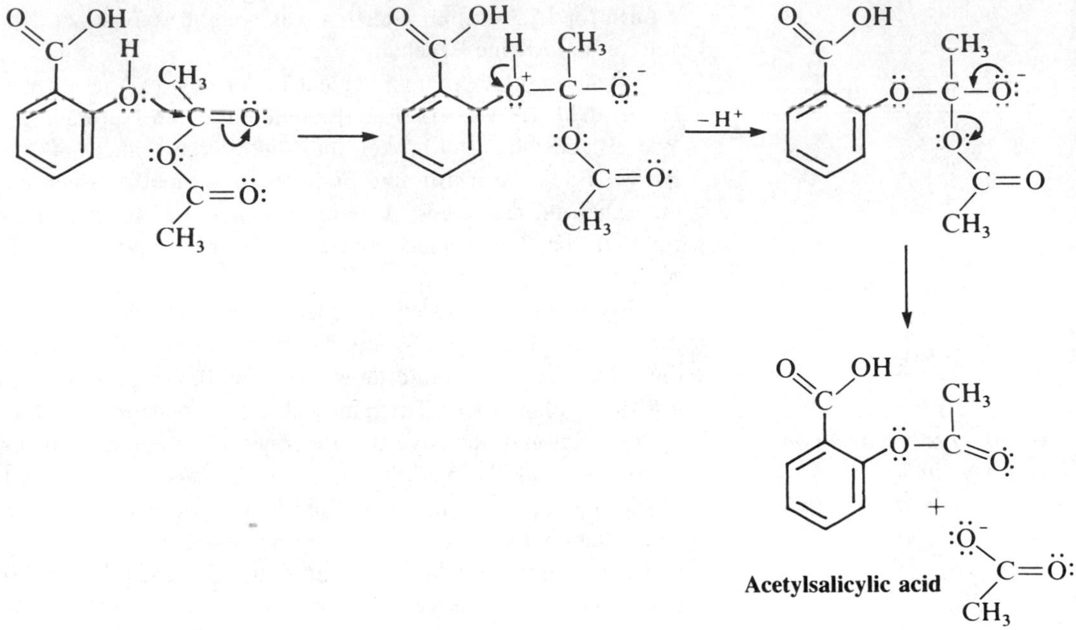

Acetylsalicylic acid

Experiment

Synthesis of Acetylsalicylic Acid (Aspirin)

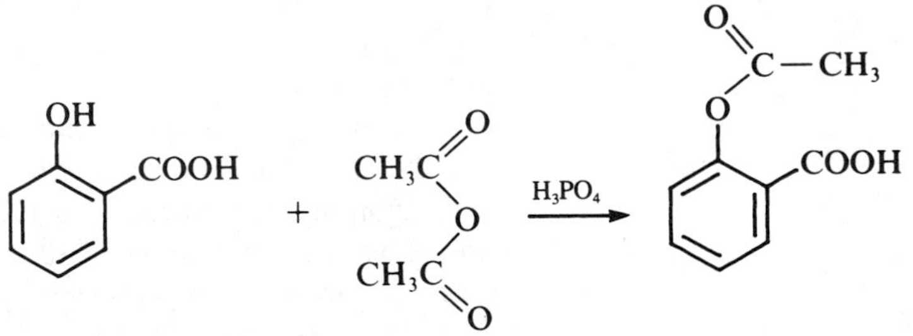

Salicylic acid	**Acetic anhydride**	**Acetylsalicylic acid**
MW 138.12, mp 159°C	MW 102.09, bp 140°C	MW 180.15, mp 128–137°C

Microscale Procedure

To a reaction tube, add 138 mg of salicylic acid, a boiling chip, and one small drop of 85% phosphoric acid followed by 0.3 mL of acetic anhydride, which will serve to wash the reactants to the bottom of the tube. Mix the reactants thoroughly, and then heat the reaction tube on the steam bath or in a beaker of

90°C water for 5 min. Cautiously add 0.2 mL of water to the reaction mixture to decompose excess acetic anhydride. This will be an exothermic reaction. When the reaction is over, add 0.3 mL more water and allow the tube to cool slowly to room temperature. If crystallization of the product does not occur during the cooling process, add a seed crystal or scratch the inside of the tube with a glass stirring rod. Cool the tube in ice until crystallization is complete (at least 10 min), and then remove the solvent with a Pasteur pipette. If the crystals are too fine for this procedure, collect the product by vacuum filtration on the Hirsch funnel. Complete the transfer of the product to the funnel using a very small quantity of ice water. In either case, turn the product out onto a piece of filter paper, and squeeze the crystals between sheets of filter paper to absorb excess water. Allow the product to dry thoroughly in air before determining the weight and calculating the percentage yield. Determine the melting point and the IR spectrum in chloroform solution. Aspirin is hydrolyzed by boiling water, but the reaction is not rapid; if desired, therefore, the product may be recrystallized from a small quantity of very hot water.

Compare a tablet of commercial aspirin with your sample. Test the solubility of the tablet in water and in toluene, and observe if it dissolves completely. Compare its behavior when heated in a melting-point capillary with the behavior of your sample. If an impurity is found, it is probably some substance used as binder for the tablets. Is it organic or inorganic?

Cleaning Up   Dilute the filtrate with water, and flush it down the drain.

Macroscale Procedure

Measure acetic anhydride in the hood, as it is very irritating to breathe.

Place 1 g of salicylic acid in each of four 13 × 100-mm test tubes, and add to each tube 2 mL of acetic anhydride. To the first tube add 0.2 g of anhydrous sodium acetate, note the time, stir the mixture with a glass rod, replace it with a thermometer, and record the time required for a 4°C rise in temperature. Continue to stir occasionally while starting the next acetylation. Obtain a clean thermometer, put it in the second tube, add 5 drops of pyridine, stir, observe as before, and compare with the first results. To the third and fourth tubes add 5 drops of boron trifluoride etherate[1] and 5 drops of concentrated sulfuric acid, respectively. What is the order of activity of the four catalysts as judged by the rates of the reactions?

Acetylation catalysts

1. $CH_3COO^-Na^+$

2.

3. $BF_3 \cdot (C_2H_5)_2O$
4. H_2SO_4

Pyridine has a bad odor, and boron trifluoride etherate and sulfuric acid are corrosive; work in hood.

Put all tubes in hot water (beaker) for 5 min to dissolve solid material and complete the reactions, then pour all the solutions into a 125-mL Erlenmeyer flask containing 50 mL of water, and rinse the tubes with water. Swirl to aid hydrolysis of excess acetic anhydride, then cool thoroughly in ice, scratch the side of the flask with a stirring rod to induce crystallization, and collect the crystalline solid; the yield is about 4 g.

Acetylsalicylic acid melts with decomposition at temperatures reported from 128 to 137°C. It can be crystallized by dissolving it in *t*-butyl methyl ether,

Note for the instructor

1. Commercial reagent, if dark, should be redistilled (bp 126°C, water-white).

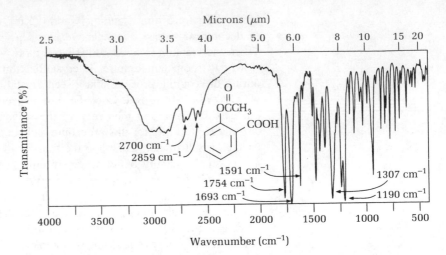

FIG. 41.1 Infrared spectrum of acetylsalicylic acid (aspirin) in CHCl$_3$.

adding an equal volume of petroleum ether, and letting the solution stand undisturbed in an ice bath.

Test the solubility of your sample in toluene and in hot water, and note the peculiar character of the aqueous solution when it is cooled and when it is then rubbed against the tube with a stirring rod. Note also that the substance dissolves in cold sodium bicarbonate solution and is precipitated by addition of an acid. Compare a tablet of commercial aspirin with your sample. Test the solubility of the tablet in water and in toluene, and observe if it dissolves completely. Compare its behavior when heated in a melting point capillary with the behavior of your sample. If an impurity is found, it is probably some substance used as binder for the tablets. Is it organic or inorganic? To interpret your results, consider the mechanism whereby salicylic acid is acetylated.

Note that acetic acid is eliminated during the reaction. What effect would sodium acetate have? How might boron fluoride etherate or sulfuric acid affect the nucleophilic attack of the phenolic oxygen on acetic anhydride? With what might the base, pyridine, associate?

The IR and ^{1}H NMR spectra of acetylsalicylic acid are presented in Figs. 41.1 and 41.2.

Cleaning Up   Combine the aqueous filtrates, dilute with water, and flush the solution down the drain.

t-Butyl methyl ether and petroleum ether are very flammable.

Propose mechanisms to interpret your results.

Questions _____

1. Hydrochloric acid is about as strong a mineral acid as sulfuric acid. Why would it not be a satisfactory catalyst in this reaction?

2. How do you account for the smell of vinegar when an old bottle of aspirin is opened?

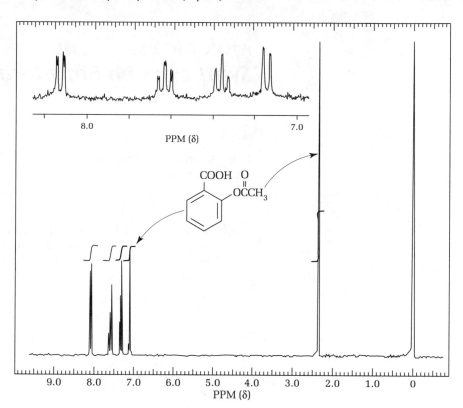

FIG. 41.2 ^{1}H NMR spectrum of acetylsalicylic acid (aspirin) (250 MHz). The COOH proton does not appear on this spectrum.

Malonic Ester Synthesis: Synthesis of a Barbiturate

Prelab Exercise: Which barbiturate discussed in this chapter cannot be synthesized by the acetoacetic ester reaction?

Barbiturates are central nervous system depressants used as hypnotic drugs and anesthetics. They are all derivatives of barbituric acid (R=R′=H), which has no sedative properties. It is called an *acid* because the carbonyl groups render the imide hydrogens acidic:

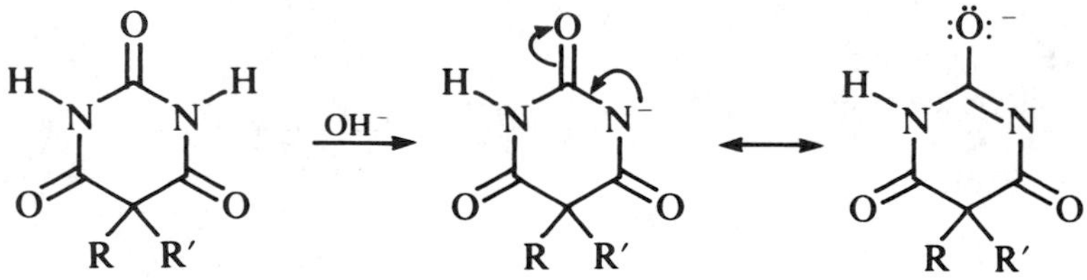

Barbituric acid
R = R′ = H

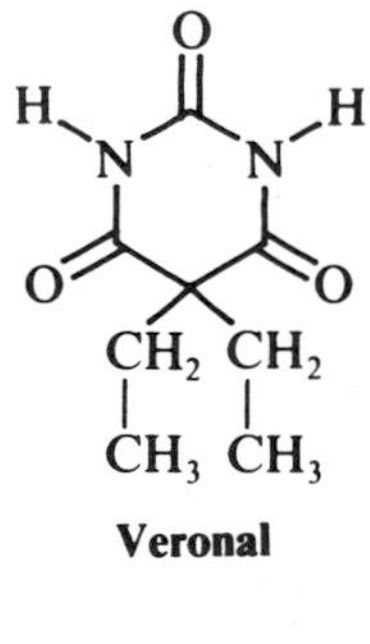

Veronal

Barbituric acid was first synthesized in 1864 by Adolph von Baeyer. It apparently was named at a tavern on St. Barbara's day and is derived from urea. At the turn of the century the great chemist Emil Fischer synthesized the first hypnotic (sleep-inducing) barbiturate, the 5,5-diethyl derivative, at the direction of von Mering. Von Mering, who made the seminal discovery that removal of the pancreas causes diabetes, named the new derivative of barbituric acid Veronal because he regarded Verona as the most restful city on earth.

Barbiturates are the most widely used sleeping pills and are classified according to their duration of action. Because it is quite easy to put almost any conceivable R group onto diethyl malonate, several thousand derivatives of barbituric acid have been synthesized. Studies of these derivatives have shown that as the alkyl chain, R, gets longer or as double bonds are introduced into the chain, the duration of action and the time of onset of action decreases. Maximum sedation occurs when the alkyl chains contain five or six carbons, as found in amobarbital (R=ethyl, R′=3-methylbutyl) and pentobarbital (R=ethyl, R′=1-methylbutyl).

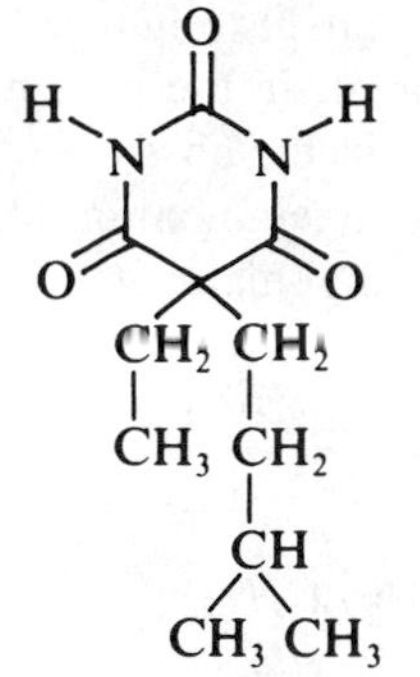

Amobarbital 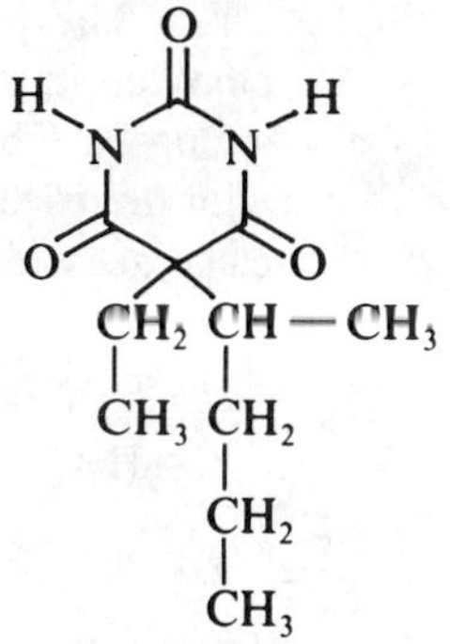   **Pentobarbital** 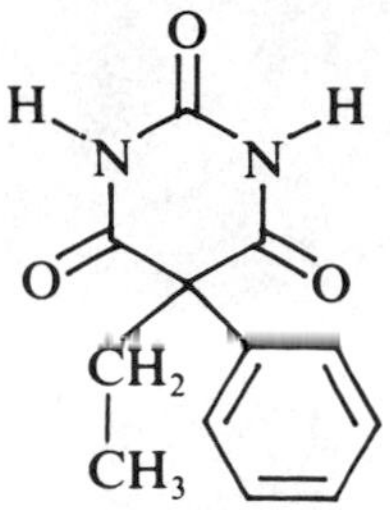   **Phenobarbital**

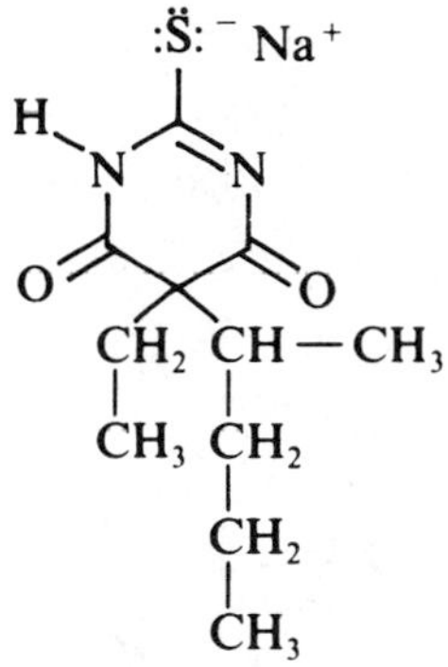

Sodium pentothal

These two molecules illustrate how subtle changes in molecular structure can affect action. Amobarbital requires 30 min to take effect and sedation lasts for 5–6 h, whereas pentobarbital takes effect in 15 min and sedation lasts only 2–3 h. Phenobarbital (R=ethyl, R′=phenyl), on the other hand, requires over an hour to take effect, but sedation lasts for 6–10 h. When the alkyl chains are made much longer, the sedative properties decrease and the substances become anticonvulsants, which are used to treat epileptic seizures. If the alkyl group is too long or is substituted at one of the two nitrogens, convulsants are produced.

The usual dosage is 10–50 mg per pill. Continuous use of barbiturates leads to physiologic dependence (addiction), and withdrawal symptoms are just like those experienced by a heroin addict. Replacing the oxygen atom at carbon-2 with sulfur results in the compound pentothal. Given intravenously as the sodium salt, it is a fast-acting general anesthetic. Like all general anesthetics, the details of its mode of action are unknown. In low, subanesthetic doses sodium pentothal reduces inhibitions and the will to resist and thus functions as the so-called "truth serum," apparently because it takes less mental effort to tell the truth than to prevaricate.

Most barbiturates are made from diethyl malonate. The methylene protons between the two carbonyl groups are acidic and will give a highly stabilized enolate anion.

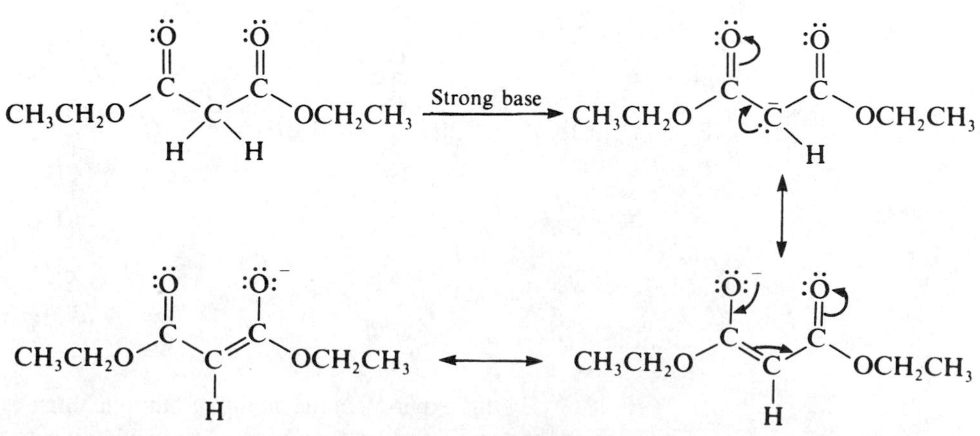

The acidic protons can be removed with a strong base, most often sodium ethoxide in dry ethanol. In the present experiment carbonate functions as the strong base because, in the absence of water, it is not solvated. In association with the tricaprylmethylammonium ion, it is soluble in the organic phase and can react with the diethyl malonate.

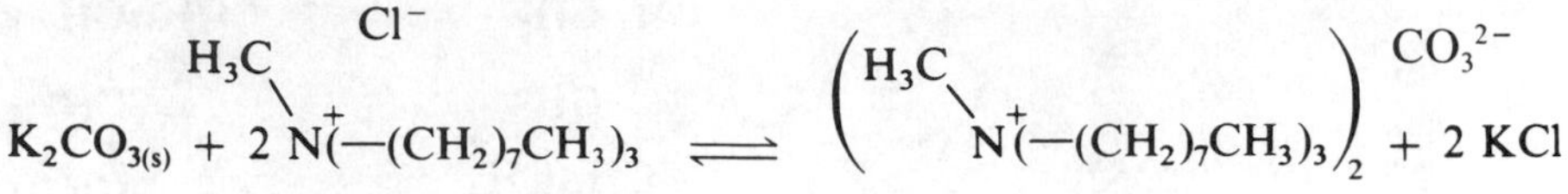

$$K_2CO_{3(s)} + 2\ \overset{\displaystyle H_3C}{\underset{}{N^+}}\overset{Cl^-}{}(-(CH_2)_7CH_3)_3 \rightleftharpoons \left(\overset{\displaystyle H_3C}{\underset{}{N^+}}(-(CH_2)_7CH_3)_3\right)_2 \overset{CO_3^{2-}}{} + 2\ KCl$$

**Tricaprylmethylammonium
chloride**
Phase transfer catalyst

Soluble in
organic phase

The enolate anion of diethyl malonate can be alkylated by an S_N2 displacement of bromide to give diethyl *n*-butylmalonate:

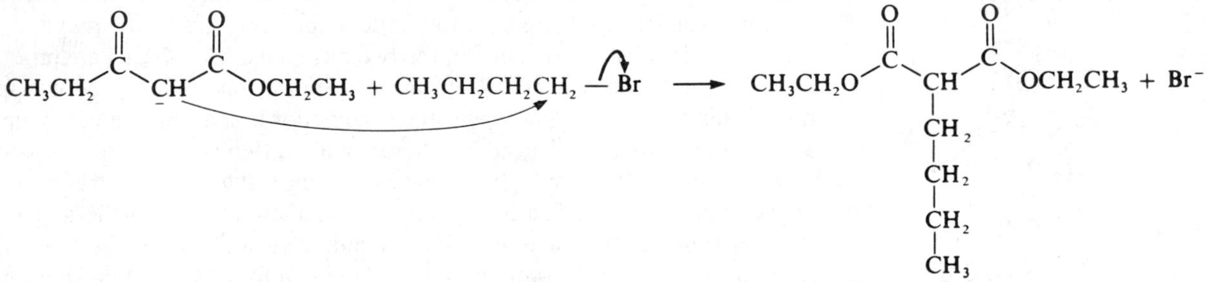

Also, because the product still contains one acidic proton, the process can be repeated using the same or a different alkyl halide. The product above could be hydrolyzed and decarboxylated to give a carboxylic acid:

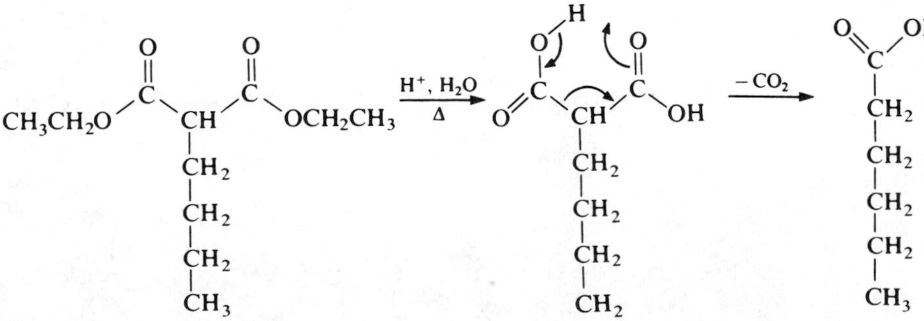

In this experiment the diethyl *n*-butylmalonate is allowed to react with urea in the presence of a strong base, sodium ethoxide, to give the barbiturate:

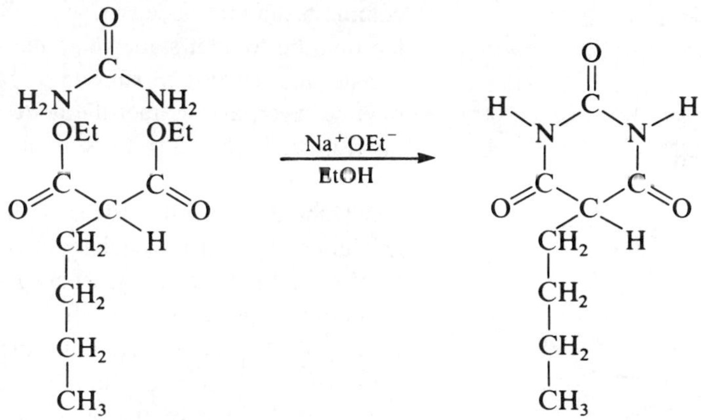

$$\begin{array}{c}
\text{O} \\
\parallel \\
\text{C} \\
\text{H}_2\text{N} \quad \text{NH}_2 \\
\text{OEt} \quad \text{OEt} \\
\text{O}=\text{C} \quad \text{C}=\text{O} \\
\text{C} \\
\text{CH}_2 \quad \text{H} \\
\text{CH}_2 \\
\text{CH}_2 \\
\text{CH}_3
\end{array}
\xrightarrow[\text{EtOH}]{\text{Na}^+\text{OEt}^-}
\begin{array}{c}
\text{O} \\
\parallel \\
\text{C} \\
\text{H}\;\text{N} \quad \text{N}\;\text{H} \\
\text{O}=\text{C} \quad \text{C}=\text{O} \\
\text{C} \\
\text{CH}_2 \quad \text{H} \\
\text{CH}_2 \\
\text{CH}_2 \\
\text{CH}_3
\end{array}$$

n-Butylbarbituric acid

Barbiturates with only one alkyl substituent are rapidly degraded by the body and therefore are physiologically inactive.

Experiments

MICROSCALE

1. Diethyl n-Butylmalonate

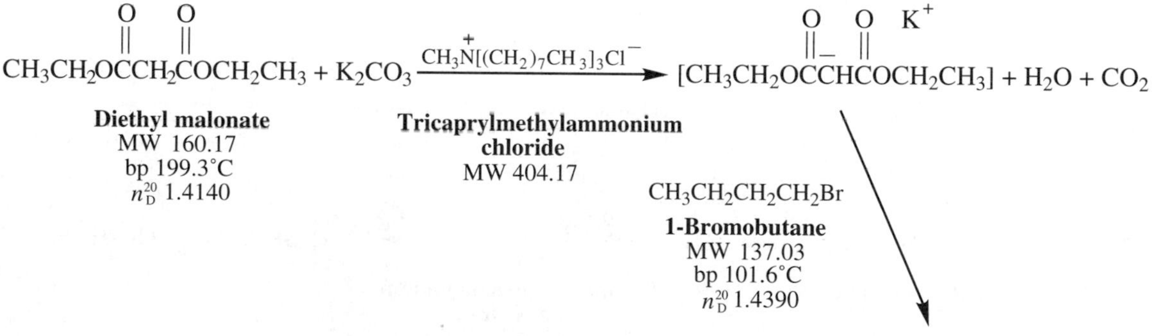

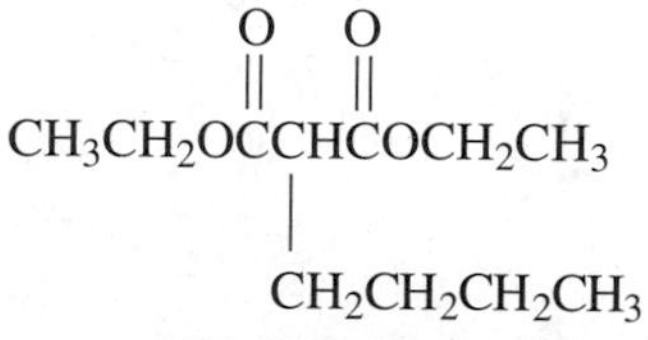

$$\underset{\substack{\text{Diethyl malonate} \\ \text{MW 160.17} \\ \text{bp 199.3}^\circ\text{C} \\ n_\text{D}^{20}\ 1.4140}}{\text{CH}_3\text{CH}_2\text{OCCH}_2\text{COCH}_2\text{CH}_3} + \text{K}_2\text{CO}_3 \xrightarrow{\text{CH}_3\overset{+}{\text{N}}[(\text{CH}_2)_7\text{CH}_3]_3\text{Cl}^-} [\text{CH}_3\text{CH}_2\text{OCCHCOCH}_2\text{CH}_3]\,\overset{\text{K}^+}{} + \text{H}_2\text{O} + \text{CO}_2$$

Tricaprylmethylammonium chloride
MW 404.17

$$\underset{\substack{\text{1-Bromobutane} \\ \text{MW 137.03} \\ \text{bp 101.6}^\circ\text{C} \\ n_\text{D}^{20}\ 1.4390}}{\text{CH}_3\text{CH}_2\text{CH}_2\text{CH}_2\text{Br}}$$

$$\begin{array}{c}
\text{O} \quad \text{O} \\
\parallel \quad \parallel \\
\text{CH}_3\text{CH}_2\text{OCCHCOCH}_2\text{CH}_3 \\
| \\
\text{CH}_2\text{CH}_2\text{CH}_2\text{CH}_3
\end{array}$$

Diethyl n-butylmalonate
MW 216.28
bp 235–240°C
$n_\text{D}^{20}\ 1.4220$

To check reagents, the instructor should run this experiment before assigning it to class.

CAUTION: *Tricaprylmethylammonium chloride (Aliquat 336) is a toxic irritant. Handle with care.*

Into a 5-mL long-necked round-bottomed flask weigh 40 mg of tricaprylmethylammonium chloride, 400 mg of diethyl malonate, 350 mg of 1-bromobutane, and 415 mg of anhydrous potassium carbonate. Attach the empty distillation

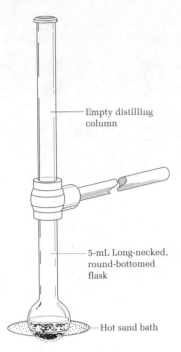

Empty distilling column

5-mL Long-necked, round-bottomed flask

Hot sand bath

FIG. 42.1 Microscale reflux apparatus.

column as an air condenser, and reflux the mixture for 1.5 h (Fig. 42.1). Allow the mixture to cool somewhat, and then cool it in ice. Using 2.5 mL of water, transfer the reaction mixture to a reaction tube, shake the mixture, draw off the organic layer, and extract the aqueous layer with two 1.5-mL portions of *tert*-butyl methyl ether. Dry the combined organic extracts over anhydrous calcium chloride pellets or sodium sulfate, remove the ether solution from the drying agent, and complete the drying of the ether over Drierite (commercial anhydrous calcium sulfate) or molecular sieves. Place the dry ether solutions in a *dry,* tared 5-mL round-bottomed long-necked flask, and remove the ether by distillation or evaporation in the hood. Complete the removal of the ether by connecting the flask to the aspirator. Determine the weight, and calculate the yield of product. Analyze it by thin-layer chromatography on silica gel using dichloromethane as the eluent. Compare the product with a sample of authentic, pure diethyl *n*-butylmalonate and with the starting materials diethyl malonate and bromobutane. The sample must be pure to be used in the next reaction. It must be perfectly clear in appearance and absolutely dry. It can be analyzed by gas chromatography using a Carbowax column at 160°C.

Cleaning Up   The aqueous layer after dilution with water can be flushed down the drain. The drying agents, once free of ether, can be placed in the nonhazardous solid waste container. Ether recovered from distillation is placed in the organic solvents container, and dichloromethane from the thin-layer chromatography goes in the halogenated organic solvents container.

MACROSCALE

2. Diethyl Benzylmalonate

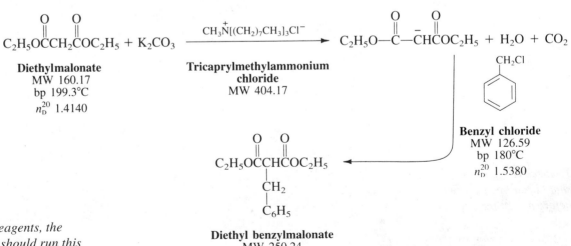

Diethylmalonate
MW 160.17
bp 199.3°C
n_D^{20} 1.4140

Tricaprylmethylammonium chloride
MW 404.17

Benzyl chloride
MW 126.59
bp 180°C
n_D^{20} 1.5380

Diethyl benzylmalonate
MW 250.24
bp 162–163°C/10 mm
n_D^{20} 1.4860

To check reagents, the instructor should run this experiment before assigning it to class.

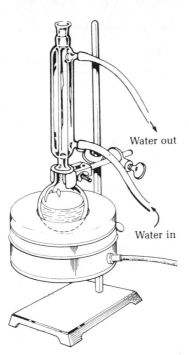

Water out

Water in

FIG. 42.2 Reflux apparatus for macroscale experiments.

Into a 50-mL round-bottomed flask weigh 0.8 g of tricaprylmethylammonium chloride, 8.00 g of diethyl malonate, 13.8 g of benzyl chloride, and 8.3 g of anhydrous potassium carbonate. Attach a water-cooled reflux condenser and reflux the reaction mixture for 1.5 h (Fig. 42.2). Allow the mixture to cool somewhat and then transfer it, using 30 mL of water to rinse out the flask, to a small separatory funnel. Shake the mixture thoroughly, and draw off the aqueous layer after the layers have separated. Place the organic layer in a 50-mL Erlenmeyer flask, and extract the aqueous layer with two 10-mL portions of *t*-butyl methyl ether. Dry the combined extracts over anhydrous calcium chloride pellets (about 10 g), decant the liquid into a 100-mL round-bottomed flask, and evaporate the solvent on the rotary evaporator. Distill the residue from a 25-mL round-bottomed flask, collecting the material that boils from about 150–165°C at 10 mm pressure. Use the apparatus depicted in Fig. 7.1. The product, diethyl benzylmalonate, is reported to boil at 162–163°C/10 mm. See the nomograph (Fig. 7.12) to correct the boiling point to a different pressure.

Cleaning Up The aqueous layer after dilution with water can be flushed down the drain. Calcium chloride, once free of the ether, can be placed in the nonhazardous solid waste container. Ether recovered from distillation is placed in the organic solvents container.

3. n-Butylbarbituric Acid

$$H_2NCNH_2 + CH_3CH_2OCCHCOCH_2CH_3 \xrightarrow{Na^+\bar{O}CH_2CH_3}$$

Urea
MW 60.06
mp 133–135°C

$$\begin{array}{c} CH_2 \\ | \\ CH_2 \\ | \\ CH_2 \\ | \\ CH_3 \end{array}$$

Diethyl n-butylmalonate
MW 216.28
bp 235–240°C

n-Butylbarbituric acid
MW 184.18
mp 209–210°C

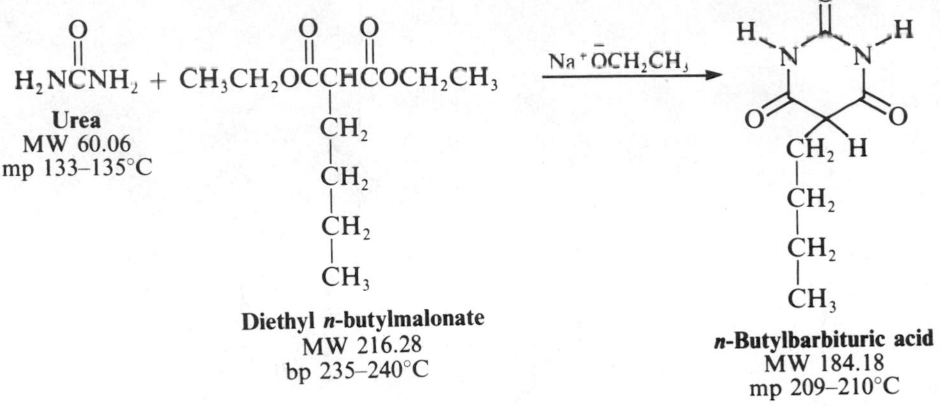

Experimental Considerations

CAUTION: Handle sodium metal with care. It will react violently with water and can cause fire or an explosion. Avoid all contact with moisture in any form.

In this experiment, absolute (anhydrous, 100%) ethanol is reacted with sodium metal to make a solution of sodium ethoxide in the 5-mL flask. (Commercially prepared sodium ethoxide is often not active.) The empty distilling column functions as an air condenser, and moisture is prevented from diffusing into the apparatus by capping the distilling column with a septum and an empty syringe needle.

The sodium metal may come in the form of small spheres or a large block stored under mineral oil. If in block form it must be shaved off with a knife and transferred to a beaker containing light mineral oil or ligroin. Use tweezers to handle the sodium, blot it off, and then transfer it to the beaker for weighing. Blot off each piece as it is added to the reaction flask down the condenser.

Be extremely careful to have no water near the sodium. If a piece of sodium is accidentally dropped, place it in a beaker containing 1-propanol (see "Cleaning Up" below). Do not dispose of it in a trash can because it can easily start a fire due to oxidation.

Microscale Procedure

To a perfectly dry 5-mL round-bottomed long-necked flask add 2.5 mL of absolute ethanol. Add to the flask the empty distilling column, capped with a septum and an empty syringe needle. Add to the ethanol 23 mg of sodium metal. Wait for all the sodium to react, and do not proceed until all traces of the metal have disappeared. To the sodium ethoxide solution add 216 mg of dry n-butyl-malonate [synthesized in the preceding experiment or purchased commercially (Aldrich)]. Immediately add 60 mg of urea that has been dissolved in 1.1 mL of absolute ethanol. Reflux the resulting mixture on a warm sand bath or a steam bath for 2 h. What solid separates during this reaction? It may cause the reaction mixture to bump during the first part of the refluxing period.

At the completion of the reaction, acidify the solution with 1 mL of 3 M hydrochloric acid (check pH with indicator paper), and then reduce the volume in the flask to one-half by distillation of the ethanol or by removal of the solvent on the rotary evaporator. Cool the solution on ice until no more product crystallizes; then collect it by vacuum filtration on the Hirsch funnel. Empty the filter flask, and then wash unreacted diethyl *n*-butylmalonate (detected by its odor) from the crystals with ligroin, press it dry, and then recrystallize the product from boiling water (20 mL of water per gram of product). The acid crystallizes slowly. Cool the solution for at least 30 min before collecting the product by vacuum filtration. Dry the barbituric acid on filter paper until the next laboratory period; then weigh it, calculate the percentage yield, and determine the melting point. Pure *n*-butylbarbituric acid melts at 209 to 210°C.

Wet all paper that has come in contact with sodium with alcohol before disposal. This will decompose traces of sodium.

Cleaning Up   Add scrap sodium metal to 1-propanol in a beaker. After all traces of sodium have disappeared, the alcohol can be diluted with water, neutralized with dilute hydrochloric acid, and flushed down the drain. Combine aqueous filtrates and mother liquors from crystallization, neutralize with sodium carbonate, and flush the solution down the drain. The ligroin wash goes in the organic solvents container.

 MACROSCALE

4. Benzylbarbituric Acid

Experimental Considerations

In this experiment, absolute (anhydrous, 100%) ethanol is reacted with sodium metal to make a solution of sodium ethoxide in the 250-mL flask. (Commercially prepared sodium ethoxide is often not active.) Moisture is prevented from diffusing into the apparatus by capping the condenser with a drying tube. This is prepared by first placing a piece of cotton in the tube, followed by calcium chloride and another firmly placed piece of cotton. At the end of the experiment, dispose of the calcium chloride or cap the drying tube very carefully with corks. The calcium chloride is very hygroscopic.

The sodium metal may come in the form of small spheres or a large block stored under mineral oil. If in block form, it must be shaved off with a knife and transferred to a beaker containing light mineral oil or ligroin. Use tweezers to handle the sodium, blot it off, and then transfer it to the beaker for weighing. Blot off each piece as it is added to the reaction flask down the condenser.

Be extremely careful to have no water near the sodium. If a piece of sodium is accidentally dropped, place it in a beaker containing 1-propanol (see "Cleaning Up" below). Do not dispose of it in a trash can, because it can easily start a fire due to oxidation.

Procedure

To a perfectly dry 250-mL round-bottomed flask add 50 mL of absolute ethanol; then add a water-cooled condenser and a calcium chloride drying tube (Fig. 42.3). Add to the ethanol 0.46 g (0.2 mol) of sodium metal. Warm the flask

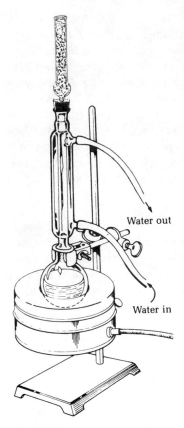

FIG. 42.3 Reflux apparatus with calcium chloride drying tube.

on a steam bath to dissolve all the sodium, and do not proceed until all traces of the metal have disappeared. To the sodium ethoxide solution add 5.00 g (0.2 mol) of dry *n*-benzylmalonate [synthesized in the preceding experiment or purchased commercially (Aldrich)]. Immediately add a hot solution of 1.2 g (0.2 mol) of dry urea that has been dissolved in 22 mL of absolute ethanol. Reflux the resulting mixture on a steam bath or a heating mantle for 2 h. What solid separates during this reaction? It may cause the reaction mixture to bump during the first part of the refluxing period, so clamp the apparatus firmly.

At the completion of the reaction, acidify the solution with 20.0 mL of 3 *M* hydrochloric acid (check with litmus paper) and then reduce the volume to one-half by removing the ethanol either by distillation or evaporation on a rotary evaporator. Cool the solution in an ice bath until no more product crystallizes; then collect it by vacuum filtration on the Büchner funnel. Empty the filter flask, then wash unreacted diethyl benzylmalonate (detected by its odor) from the crystals with ligroin, press it dry, and recrystallize the product from boiling water. The acid crystallizes slowly. Cool the solution for at least 30 min before collecting the product by vacuum filtration. Dry the barbituric acid on filter paper until the next laboratory period; then weigh it, calculate the percentage yield, and determine the melting point. Pure benzylbarbituric acid melts at 206°C.

Cleaning Up   Add scrap sodium metal to 1-propanol in a beaker. After all traces of sodium have disappeared, the alcohol can be diluted with water, neutralized with dilute hydrochloric acid, and flushed down the drain. Combine aqueous filtrates and mother liquors from crystallization, neutralize with sodium carbonate, and flush the solution down the drain. The ligroin wash goes in the organic solvents container.

Questions

1. The anion of diethyl benzylmalonate is often made by reacting the diester with sodium methoxide. What weight of sodium would be required in the present experiment if that method were employed for the experiment?

2. Outline all steps in the synthesis of pentobarbital.

3. What problem might be encountered if $BrCH_2C(CH_3)_3$ were used as the halide in this synthesis?

43

Amines

Prelab Exercise: Explain why an ordinary amide from a primary amine does not dissolve in alkali while the corresponding sulfonamide of the amine does dissolve in alkali.

Amines are weak bases and can be regarded as organic substitution products of ammonia. Just as ammonia reacts with acids to form the ammonium ion, so amines react with acid to form the organoammonium ion:

$$\ddot{N}H_3 + H_3\ddot{O}^+ \rightleftharpoons \overset{+}{N}H_4 + H_2\ddot{O}:$$

Ammonia   **Ammonium ion**

$$CH_3\ddot{N}H_2 + H_3\ddot{O}^+ \rightleftharpoons CH_3\overset{+}{N}H_3 + H_2\ddot{O}:$$

Methylamine   **Methyl-ammonium ion**

When an amine dissolves in water the following equilibrium is established:

$$R\ddot{N}H_2 + H_2O \rightleftharpoons R\overset{+}{N}H_3 + OH^-$$

and from this the basicity constant can be defined as

$$K_b = \frac{[RNH_3^+][OH^-]}{[RNH_2]}$$

Strong bases have the larger values of K_b, meaning the amine has a greater tendency to accept a proton from water to increase the concentration of RNH_3^+ and OH^-.

The basicity constant for ammonia is 1.8×10^{-5}.

$$\ddot{N}H_3 + H_2O \rightleftharpoons NH_4^+ + OH^-$$

$$K_b = 1.8 \times 10^{-5} = \frac{[NH_4^+][OH^-]}{[NH_3]}$$

Alkyl amines such as methylamine, CH_3NH_2, or *n*-propylamine, $CH_3CH_2CH_2NH_2$, are stronger bases than ammonia because the alkyl groups are electron donors and increase the effective electron density on nitrogen. Aromatic amines, on the other hand, are weaker bases than ammonia. Delocalization of the unshared pair of electrons on nitrogen onto the aromatic ring means the electrons are not available to be shared with an acidic proton.

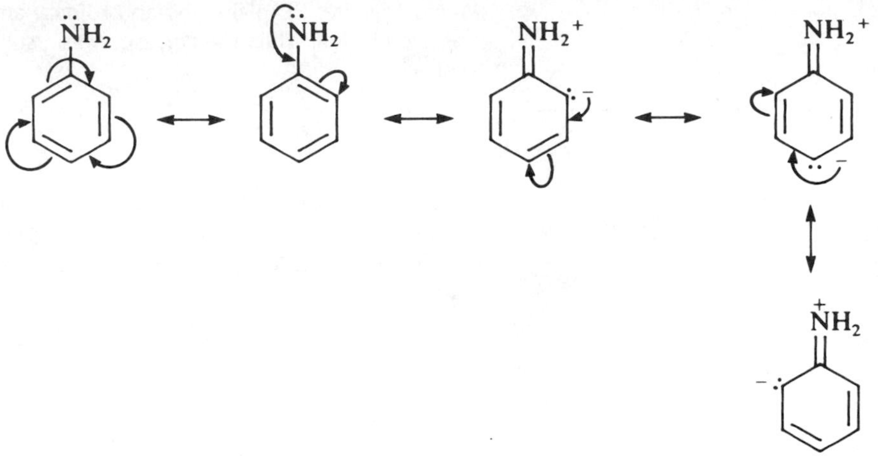

Low molecular weight amines have a powerful fishy odor. Slightly higher molecular weight diamines have names suggestive of their odors: $H_2N(CH_2)_5NH_2$, cadaverine; and $H_2N(CH_2)_4NH_2$, putrescine. The lower molecular weight amines with up to about five carbon atoms are soluble in water. Many amines, especially the liquid aromatic amines, undergo light-catalyzed free-radical air oxidation reactions to give a large variety of highly colored decomposition products. The higher molecular weight amines that are insoluble in water will dissolve in acid to form ionic amine salts. This reaction is useful for both characterization and for separation purposes. The ionic amine salt on treatment with base will regenerate the amine.

Amides are not basic.

Nitriles, $R—C\equiv N$, are not basic and neither are amides or substituted amides. The adjacent electron-withdrawing carbonyl oxygen effectively removes the unshared pair of electrons on nitrogen:

The object of the present experiment is to identify an unknown amine or amine salt. Procedures for solubility tests are given in Experiment 1. Experiment 2 presents the Hinsberg test, a test for distinguishing between primary, secondary, and tertiary amines; Experiment 3 gives procedures for preparation of solid derivatives for melting point characterizations; and Experiment 4 gives spectral characteristics. Apply the procedures to known substances along with the unknown.

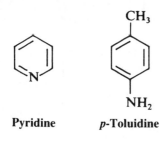

MICROSCALE AND MACROSCALE

Pyridine   ***p*-Toluidine**

Use caution in smelling unknowns, and do it only once.

Experiments

1. Solubility

If necessary, this experiment can be scaled up by a factor of 2 or 3.
Substances to be tested:

Aniline, $C_6H_5NH_2$, bp 184°C

p-Toluidine, $CH_3C_6H_4NH_2$, mp 43°C

Pyridine, C_5H_5N, bp 115°C (a tertiary amine)

Methylamine hydrochloride, dec. (salt of CH_3NH_2, bp -6.7°C)

Aniline hydrochloride, dec. (salt of $C_6H_5NH_2$)

Aniline sulfate, dec.

Amines are often contaminated with dark oxidation products. The amines can easily be purified by a small-scale distillation (Fig. 43.1).

First, see if the substance has a fishy, ammonialike odor; if so, it probably is an amine of low molecular weight. Then test the solubility in water by putting one drop if a liquid or an estimated 10 mg if a solid into a reaction tube or a small glass centrifuge tube, adding 0.1 mL of water (a 5-mm column), and first seeing if the substance dissolves in the cold. If the substance is a solid, rub it well with a stirring rod and break up any lumps before drawing a conclusion.

If the substance is *readily soluble in cold water* and if the odor is suggestive of an amine, test the solution with pH paper and further determine if the odor disappears on addition of a few drops of 10% hydrochloric acid. If the properties are more like those of a salt, add a few drops of 10% sodium hydroxide solution. If the solution remains clear, addition of a little sodium chloride may cause separation of a liquid or solid amine.

If the substance is not soluble in cold water, see if it will dissolve on heating; be careful not to mistake the melting of a substance for dissolving. If it *dissolves in hot water,* add a few drops of 10% alkali, and see if an amine precipitates. (If you are in doubt as to whether a salt has dissolved partially or not at all, pour off the supernatant liquid and make it basic.)

If the substance is *insoluble in hot water,* add 10% hydrochloric acid, heat if necessary, and see if it dissolves. If so, make the solution basic, and see if an amine precipitates.

Cleaning Up All solutions should be combined, made basic to free the amines, extracted with an organic solvent such as ligroin, the aqueous layer flushed down the drain, and the organic layer placed in the organic solvents container.

FIG. 43.1 Apparatus for
instant microscale distillation.
When the refluxing vapor is
drawn into the pipette, it will
condense to a clear liquid,
which is then squirted into the
empty reaction tube. Both tubes
can be held in one hand.

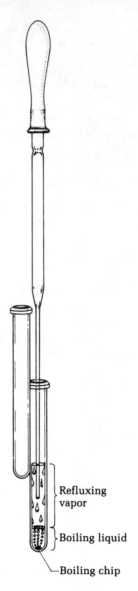

Refluxing
vapor

Boiling liquid

Boiling chip

▯ MICROSCALE

*CAUTION: Handle amines
with care. Many are toxic,
particularly aromatic
amines.*

2. Hinsberg Test

If necessary, this experiment can be doubled in scale.

The procedure for distinguishing amines with benzenesulfonyl chloride is
to be run in parallel on the following substances:

Aniline, $C_6H_5NH_2$ (bp 184°C)
N-Methylaniline, $C_6H_5NHCH_3$ (bp 194°C)
Triethylamine, $(CH_3CH_2)_3N$ (bp 90°C)

Primary and secondary amines react in the presence of alkali with benzenesulfonyl chloride, $C_6H_5SO_2Cl$, to give sulfonamides.

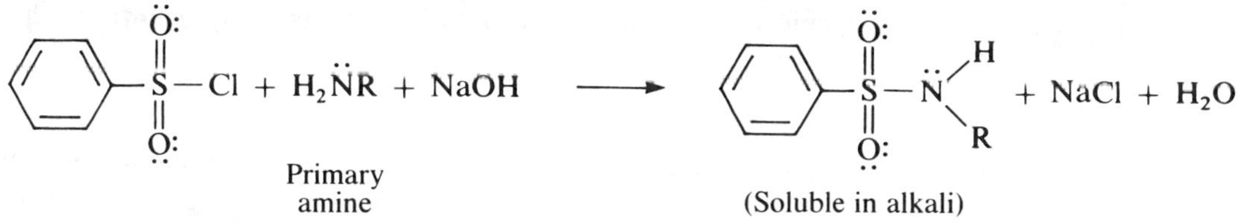

Primary
amine

(Soluble in alkali)   **1**

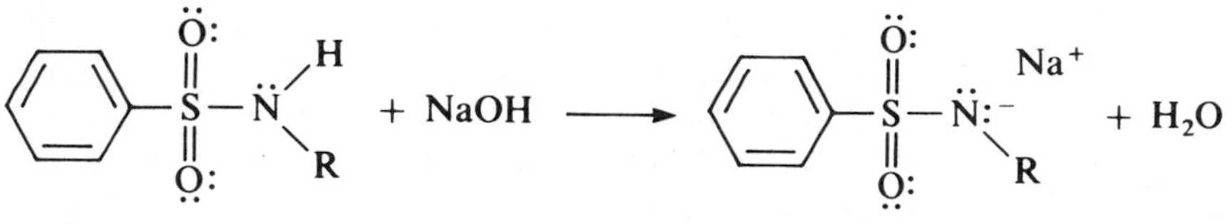

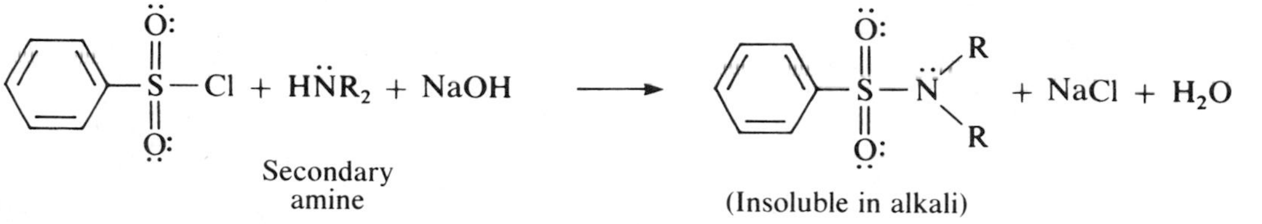

(Insoluble in alkali)   **2**

Secondary
amine

Low molecular weight amines have very bad odors; work in the hood.

The sulfonamides are distinguishable because the derivative from a primary amine has an acidic hydrogen, which renders the product soluble in alkali (reaction **1**); whereas the sulfonamide from a secondary amine is insoluble (reaction **2**). Tertiary amines lack the necessary acidic hydrogen for formation of benzenesulfonyl derivatives.

Procedure

Distinguishing among primary (1°), secondary (2°), and tertiary (3°) amines

CAUTION: *Do not allow benzenesulfonyl chloride to come in contact with the skin.*

To a reaction tube or small centrifuge tube add 50 mg of the amine, 200 mg of benzenesulfonyl chloride, and 1 mL of methanol. Over the hot sand bath or a steam bath heat the mixture to just below the boiling point, cool, and add 2 mL of 6 M sodium hydroxide. Shake the mixture for 5 min, and then allow the tube to stand for 10 more min with occasional shaking. If the odor of benzenesulfonyl chloride is detected, warm the mixture to hydrolyze it (see Question 7). Cool the mixture and acidify it by adding 6 M hydrochloric acid, dropwise, and with stirring. If a precipitate is seen at this point the amine is either primary or secondary. If no precipitate is seen, the amine is tertiary.

If a precipitate is present remove it by filtration on the Hirsch funnel, wash it with 2 mL of water, and transfer it to a reaction tube. Add 2.5 mL of 1.5 M sodium hydroxide solution and warm the mixture to 50°C. Shake the tube vigorously for 2 min. If the precipitate dissolves, the amine is primary. If it does not dissolve it is secondary. The sulfonamide of a primary amine can be recovered by acidifying the alkaline solution. Once dry these sulfonamides can be used to characterize the amine by their melting points.

Cleaning Up The slightly acidic filtrate can be diluted and then flushed down the drain if the unknown is primary or secondary. If the unknown is tertiary, make the solution basic, extract the amine with ligroin, place the ligroin layer in the organic solvents container, and flush the aqueous layer down the drain.

3. Solid Derivatives

Acetyl derivatives of primary and secondary amines are usually solids suitable for melting point characterization and are readily prepared by reaction with acetic anhydride, even in the presence of water. Benzoyl and benzenesulfonyl derivatives are made by reaction of the amine with the appropriate acid chloride in the presence of alkali, as in Experiment 2 (the benzenesulfonamides of aniline and of *N*-methylaniline melt at 110°C and 79°C, respectively).

Solid derivatives suitable for characterization of tertiary amines are the methiodides and picrates:

$$R_3N + CH_3I \longrightarrow R_3\overset{+}{N}CH_3 \; I^-$$

Methiodide

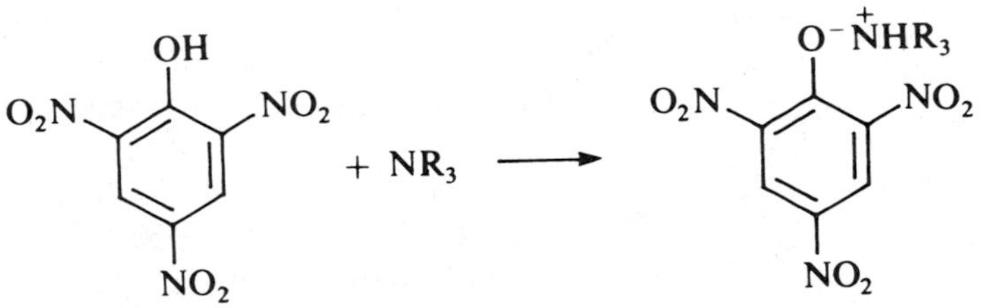

2,4,6-Trinitrophenol
(Picric Acid)

Amine picrate

Typical derivatives are to be prepared, and although determination of melting points is not necessary because the values are given, the products should be saved for possible identification of unknowns.

Acetylation of Aniline with Acetic Anhydride

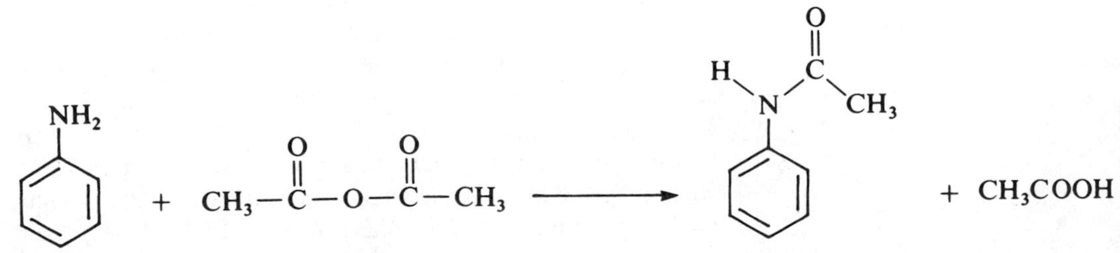

Aniline
MW 93.13
bp 184°C

Acetic anhydride
MW 102.09
bp 138–140°C

Acetanilide
MW 135.17
mp 114°C

In a dry, small test tube place 46 mg (or 5 drops) of freshly distilled aniline, 51 mg (or 5 drops) of pure acetic anhydride, and a small boiling chip. Reflux the mixture for about 8 min, then add water to the mixture dropwise with heating until all of the product is in solution. This will require very little water. Allow the mixture to cool spontaneously to room temperature, and then cool the mixture in ice. Remove the solvent with a Pasteur pipette, and recrystallize the crude material again from boiling water. Isolate the crystals in the same way, cool the tube in ice water, and wash the crystals with a drop or two of ice-cold acetone. Remove the acetone with a Pasteur pipette, and dry the product in the test tube. Once dry, determine the weight, calculate the percent yield, and determine the melting point. With patience this reaction and the recrystallizations can be carried out in a melting point tube on one-tenth the indicated quantities of material using essentially the same techniques, and, if deemed necessary, it can be carried out on 10 to 20 times as much material. On a larger scale, exercise caution when adding water to the reaction mixture, because the exothermic hydrolysis of acetic anhydride is slow at room temperature but can heat up and get out of control.

Acetylation of Aniline Hydrochloride with Acetic Anhydride

If necessary, this experiment can be carried out on 10 to 20 times as much material.

Dissolve 0.26 g of aniline hydrochloride in 2.5 mL of water, and add 0.21 g of acetic anhydride followed immediately by 0.25 g of sodium acetate. Warm the mixture, and cool it to room temperature and then in ice. Recrystallize the product from water, and wash it with acetone as described above. See Chapter 46 for mechanism.

Cleaning Up   The slightly acidic filtrate can be diluted and then flushed down the drain if the unknown is primary or secondary. If the unknown is tertiary, make the solution basic, extract the amine with ligroin, place the ligroin layer in the organic solvents container, and flush the aqueous layer down the drain.

Formation of an Amine Picrate from Picric Acid and Triethylamine

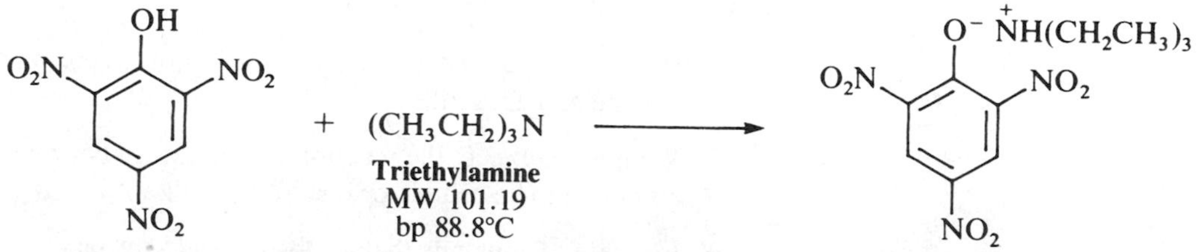

Picric acid
MW 222.11, mp 120–122°C

+ $(CH_3CH_2)_3N$

Triethylamine
MW 101.19
bp 88.8°C

Triethylamine picrate
MW 323.30, mp 171°C

If necessary, this reaction can be doubled in scale.

Dissolve 30 mg of moist (35% water) picric acid in 0.25 mL of methanol, to the warm solution add 10.1 mg of triethylamine, and let the solution stand to

CAUTION: *Picric acid can explode if allowed to become dry. Use only the moist reagent (35% water), and do not allow the reagent to dry out. Your instructor may provide a stock solution containing 3 g of moist picric acid in 25 mL of methanol.*

deposit crystals of the picrate, an amine salt that has a characteristic melting point. This picrate melts at 171°C.

Cleaning Up   The filtrate from this reaction and the product as well can be disposed of by dilution with a large volume of water and flushing down the drain. Larger quantities of moist picric acid (1 g) can be reduced with tin and hydrochloric acid to the corresponding triaminophenol.[1] Dry picric acid is said to be hazardous and should not be handled.

4. The NMR and IR Spectra of Amines

The proton bound to nitrogen can appear between 0.6 and 7.0 ppm on the NMR spectrum, the position depending on solvent, concentration, and structure of the amine. The peak is sometimes extremely broad owing to slow exchange and interaction of the proton with the electric quadrupole of the nitrogen. If addition of a drop of D_2O to the sample causes the peak to disappear, this is evidence for an amine hydrogen; but alcohols, phenols, and enols will also exhibit this exchange behavior. See Fig. 43.2 for the NMR spectrum of aniline, in which the amine hydrogens appear as a sharp peak at 3.3 ppm. Infrared spectroscopy can also be very useful for identification purposes. Primary amines, both aromatic and aliphatic, show a weak doublet between 3300 and 3500 cm^{-1} and a strong absorption between 1560 and 1640 cm^{-1} due to NH bending (Fig. 43.3). Secondary amines show a single peak between 3310 and 3450 cm^{-1}. Tertiary amines have no useful infrared absorptions. In Chapter 14 the characteristic ultraviolet absorption shifts of aromatic amines in the presence and absence of acids were discussed.

5. Unknowns (see Tables 5 and 6 in Chapter 70)

Determine first if the unknown is an amine or an amine salt, and then determine whether the amine is primary, secondary, or tertiary. Complete identification of your unknown may be required.

Questions

1. How could you most easily distinguish between samples of 2-aminonaphthalene and acetanilide?

2. Would you expect the reaction product from benzenesulfonyl chloride and ammonia to be soluble or insoluble in alkali?

3. Is it safe to conclude that a substance is a tertiary amine because it forms a picrate?

4. Why is it usually true that amines that are insoluble in water are odorless?

1. Margaret-Ann Armour, *Hazardous Laboratory Chemicals Disposal Guide,* Lewis Publishers, 1996.

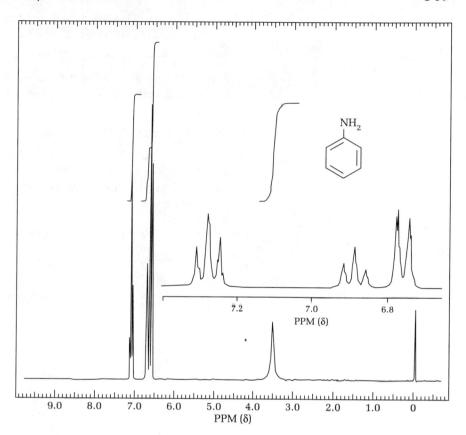

FIG. 43.2   The ¹H NMR spectrum of aniline (250 MHz).

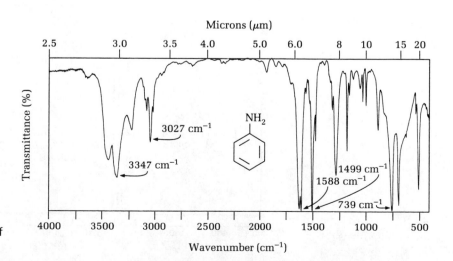

FIG. 43.3   Infrared spectrum of aniline.

5. Technical dimethylaniline contains traces of aniline and of methylaniline. Suggest a method for elimination of these impurities.

6. How would you prepare aniline from aniline hydrochloride?

7. Write a balanced equation for the reaction of benzenesulfonyl chloride with sodium hydroxide solution in the absence of an amine. What solubility would you expect the product to have in acid and base?

8. Assign the peak at 3347 cm^{-1} in the IR spectrum of aniline (Fig. 43.3).

9. In the proton noise-decoupled ^{13}C NMR spectra of benzenesulfonyl chloride (Fig. 43.4) and benzoyl chloride (Fig. 43.5), why is the peak of the carbon bearing the substituent so small?

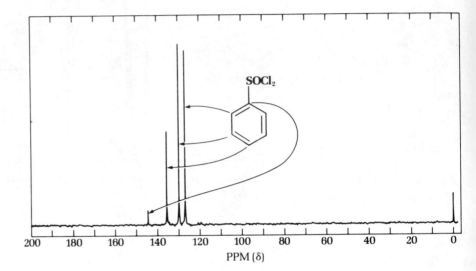

FIG. 43.4 ^{13}C NMR spectrum of benzenesulfonyl chloride (22.6 MHz).

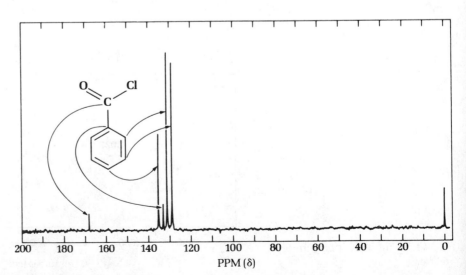

FIG. 43.5 ^{13}C NMR spectrum of benzoyl chloride (22.6 MHz).

Separation and Purification of the Components of an Analgesic Tablet: Aspirin, Caffeine, and Acetaminophen

Prelab Exercise: Prepare a detailed flow sheet for the separation of aspirin, caffeine, and acetaminophen from an analgesic tablet.

This experiment puts into practice the techniques learned in several previous chapters for separating and purifying, on a very small scale, the components in a common analgesic tablet. It is presumed that thin-layer or high-performance liquid chromatography has already shown that the tablet does indeed contain aspirin, caffeine, and acetaminophen.

Many pharmaceutical tablets are held together with a binder to prevent the components from crumbling on storage or while being swallowed. A close reading of the contents on the package will disclose the nature of the binder. Starch is commonly used, as is microcrystalline cellulose and silica gel. All of these have one property in common: They are insoluble in water and common organic solvents.

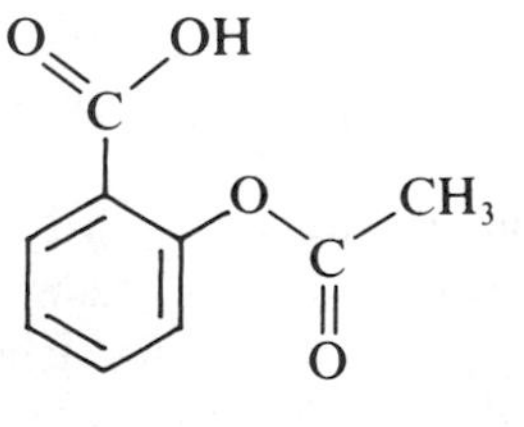

This experiment has been placed here because the chemistry of amines is usually discussed late in the term. It could very well follow Chapter 8, "Extraction."

| Aspirin (Acetylsalicylic acid) mp 135°C | Acetaminophen (*p*-Hydroxyacetanilide) mp 169–170.5°C | Caffeine mp 238°C |

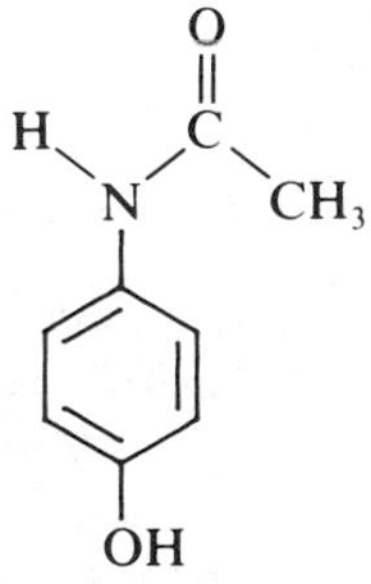

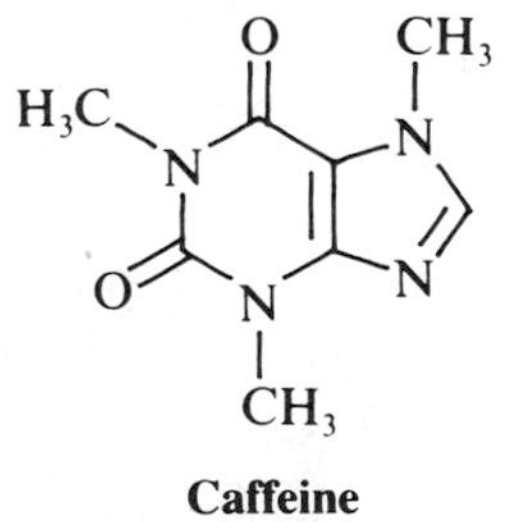

Inspection of the structures of caffeine, acetylsalicylic acid, and acetaminophen reveals that one is a base, one a strong organic acid, and one a weak organic acid. One might be tempted to separate this mixture using exactly the same procedure as employed in Chapter 8 to separate benzoic acid, 2-naphthol, and 1,4-dimethoxybenzene, i.e., dissolve the mixture in dichloromethane; remove the benzoic acid (the strong acid) by reaction with bicarbonate ion, a

TABLE 44.1   Solubilities

	Water	Ethanol	Chloroform	Diethyl ether	Ligroin
Aspirin	0.33 g/100 mL 25°C 1 g/100 mL 37°C	1 g/5 mL	1 g/17 mL	1 g/13 mL	
Acetaminophen	v. sl. sol. cold sol. hot	sol.	ins.	sl. sol.	ins.
Caffeine	1 g/46 mL 25°C 1 g/5.5 mL 80°C 1 g/1.5 mL 100°C	1 g/66 mL 25°C 1 g/22 mL 60°C	1 g/5.5 mL	1 g/530 mL	sl. sol.

weak base; then remove the naphthol, the weak acid, by reaction with hydroxide, a strong base. This process would leave the neutral compound, 1,4-dimethoxybenzene, in the dichloromethane solution.

In the present experiment the solubility data (see Table 44.1) reveal that the weak acid, acetaminophen, is not soluble in ether, chloroform, or dichloromethane, so it cannot be extracted by strong base. We can take advantage of this lack of solubility by dissolving the other two components, caffeine and aspirin, in dichloromethane, and removing the acetaminophen by filtration. The binder is also insoluble in dichloromethane, but treatment of the solid mixture with ethanol will dissolve the acetaminophen and not the binder. They can then be separated by filtration and the acetaminophen isolated by evaporation of the ethanol.

This experiment is a test of technique. It is not easy to separate and crystallize a few milligrams of a compound that occurs in a mixture.

Experiment

📥 MICROSCALE

Extraction and Purification Procedure

Handle dichloromethane in the hood. It is a carcinogen.

In a mortar, grind an Extra Strength Excedrin tablet to a very fine powder. The label states that this analgesic contains 250 mg of aspirin, 250 mg of acetaminophen, and 65 mg of caffeine per tablet. Place 300 mg of this powder in a reaction tube, and add to it 2 mL of dichloromethane. Warm the mixture briefly, and note that a large part of the material does not dissolve. Filter the mixture on the microscale Büchner funnel (Fig. 44.1) into another reaction tube. This is done by transferring the slurry to the funnel with a Pasteur pipette and completing the transfer with a small portion of dichloromethane. This filtrate is solution 1. The Hirsch funnel or the Wilfilter also can be used for this procedure. See Fig. 44.2 for the apparatus. Pressure filtration is another alternative.

The binder can be starch, microcrystalline cellulose, or silica gel.

Transfer the powder on the filter to a reaction tube, add 1 mL of ethanol, and heat the mixture to boiling on the sand bath (boiling stick). Not all the material will go into solution. That which does not is the binder. Filter the mixture on the same microscale Büchner funnel into a tared reaction tube, and complete the transfer and washing using a few drops of hot ethanol.

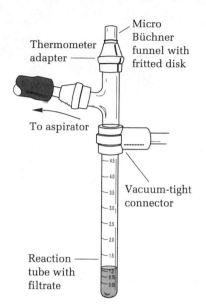

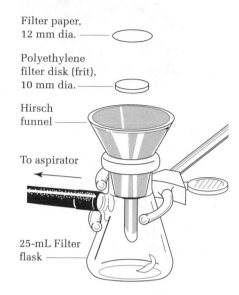

FIG. 44.1 Microscale Büchner funnel assembly.

FIG. 44.2 Hirsch funnel with integral adapter, polyethylene frit, and 25-mL filter flask.

Acetaminophen

Check product purity by TLC using 25 : 1 ethyl acetate–acetic acid to elute the silica gel plates.

Evaporate about two-thirds of the filtrate by boiling off the ethanol or, better, by warming the solution and blowing a stream of air into the reaction tube. Heat the residue to boiling (add a boiling stick to prevent bumping), and add more ethanol, if necessary, to bring the solid into solution. Allow the saturated solution to cool slowly to room temperature to deposit crystals of acetaminophen, which is reported to melt at 169 to 170.5°C. After the mixture has cooled to room temperature, cool it in ice for several minutes, remove the solvent with a Pasteur pipette, wash the crystals once with 2 drops of ice-cold ethanol, remove the ethanol, and dry the crystals under aspirator vacuum while heating the tube on a steam bath.

Alternatively, the original ethanol solution can be evaporated to dryness and the residue recrystallized from boiling water. The crystals can be collected on a Hirsch funnel (Fig. 44.2) or by use of the Wilfilter (Fig. 44.3). Once the crystals are dry, determine their weight and melting point. Thin-layer chromatographic analysis and the melting points of these crystals and the two other components of this mixture will indicate their purity.

The dichloromethane filtered from the binder and acetaminophen mixture (solution 1) should contain caffeine and aspirin. These can be separated by extraction either with acid, which will remove the caffeine as a water-soluble salt, or with base, which will remove the aspirin as a water-soluble salt. We shall use the latter procedure.

To the dichloromethane solution in a reaction tube, add 1 mL of 3 M sodium hydroxide solution and shake the mixture thoroughly. Remove the aqueous layer,

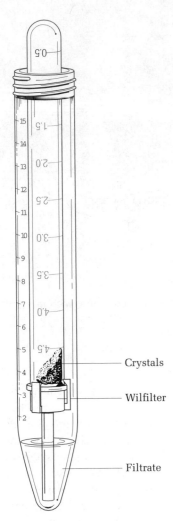

FIG. 44.3 Wilfilter filtration apparatus. See page 52 in Chapter 3.

Caffeine

See Fig. 3.19 for drying crystals under vacuum.

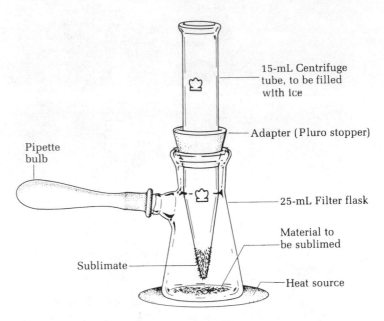

FIG. 44.4 Sublimation apparatus.

add 0.2 mL more water, shake the mixture thoroughly, and again remove the aqueous layer, which is combined with the first aqueous extract.

To the dichloromethane, add calcium chloride pellets until the drying agent no longer clumps together. Shake the mixture over a 5- to 10-min period to complete the drying process; then remove the solvent, wash the drying agent with more solvent, and evaporate the combined extracts to dryness under a stream of air to leave crude caffeine.

The caffeine can be purified by sublimation (Fig. 44.4), as has been done in the experiment in which it was extracted from tea (Chapter 8), or it can be purified by crystallization. Recrystallize the caffeine by dissolving it in the minimum quantity of 30% ethanol in tetrahydrofuran. It also can be crystallized by dissolving the product in a minimum quantity of hot toluene or acetone and adding to this solution ligroin until the solution is cloudy while at the boiling point. In any case, allow the solution to cool slowly to room temperature, then cool the mixture in ice, and remove the solvent from the crystals with a Pasteur pipette. Remove the remainder of the solvent under aspirator vacuum, and determine the weight of the caffeine and its melting point.

The aqueous hydroxide extract contains aspirin as the sodium salt of the carboxylic acid. To the aqueous solution, add 3 M hydrochloric acid dropwise until the solution tests strongly acid to indicator paper, and then add 2 more drops of acid. This will give a suspension of white acetylsalicylic acid in the aqueous solution. It could be filtered off and recrystallized from boiling water, but this would entail losses in transfer. An easier procedure is to heat the aqueous solution that contains the precipitated aspirin.

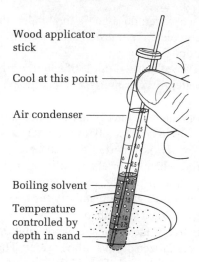

Wood applicator
stick

Cool at this point

Air condenser

Boiling solvent

Temperature
controlled by
depth in sand

FIG. 44.5   Crystallization in the reaction tube.

Add a boiling stick, and heat the mixture to boiling (Fig. 44.5), at which time the aspirin should dissolve completely. If it does not, add more water. Long boiling will hydrolyze the aspirin to salicylic acid (mp 157–159°C). Once completely dissolved, the aspirin should be allowed to crystallize slowly as the solution cools to room temperature in an insulated container. Once the tube has reached room temperature, it should be cooled in ice for several minutes and then the solvent removed with a Pasteur pipette. Wash the crystals with a few drops of ice-cold water, and then isolate them with the Wilfilter or scrape them out onto a piece of filter paper. Squeezing the crystals between sheets of filter paper will hasten drying. Once they are completely dry, determine the weight of the acetylsalicylic acid and its melting point.

Aspirin

Alternative procedure: Use the Hirsch funnel (Fig. 3.18) or the Wilfilter to isolate the aspirin.

Cleaning Up   Place any dichloromethane-containing solutions in the halogenated organic waste container and the other organic liquids in the organic solvents container. The aqueous layers should be diluted and neutralized with sodium carbonate before being flushed down the drain. After it is free of solvent, the calcium chloride can be placed in the nonhazardous solid waste container.

🖤 MACROSCALE

Extraction and Purification Procedure

Handle dichloromethane in the hood. It is a cancer suspect agent.

The binder can be starch, microcrystalline cellulose, or silica gel.

In a mortar, grind two Extra Strength Excedrin tablets to a very fine powder. The label states that this analgesic contains 250 mg of aspirin, 250 mg of acetaminophen, and 65 mg of caffeine per tablet. Place this powder in a test tube, and add to it 7.5 mL of dichloromethane. Warm the mixture briefly, and note that a large part of the material does not dissolve. Filter the mixture into another test tube. This can be done by transferring the slurry to a funnel equipped with a piece of filter paper. Use a Pasteur pipette, and complete the transfer with a small portion of dichloromethane. This filtrate is solution 1.

Transfer the powder on the filter to a test tube, add 4 mL of ethanol, and heat the mixture to boiling (boiling stick). Not all of the material will go into solution. That which does not is the binder. Filter the mixture into a tared test tube, and complete the transfer and washing using a few drops of hot ethanol. This is solution 2.

Evaporate about two-thirds of solution 2 by boiling off the ethanol (boiling stick) or, better, by warming the solution and blowing a stream of air into the test tube. Heat the residue to boiling (add a boiling stick to prevent bumping), and add more ethanol if necessary to bring the solid into solution. Allow the saturated solution to cool slowly to room temperature to deposit crystals of acetaminophen, which is reported to melt at 169–170.5°C. After the mixture has cooled to room temperature, cool it in ice for several minutes, remove the solvent with a Pasteur pipette, wash the crystals once with two drops of ice-cold ethanol, remove the ethanol, and dry the crystals under aspirator vacuum while heating the tube on a steam bath.

Alternatively, the original ethanol solution is evaporated to dryness and the residue recrystallized from boiling water. The crystals are best collected and dried on a Hirsch funnel (Fig. 44.2). Once the crystals are dry determine their weight and melting point. Thin-layer chromatographic analysis and the melting points of these crystals and the two other components of this mixture will indicate their purity.

The dichloromethane filtered from the binder and acetaminophen mixture (solution 1) should contain caffeine and aspirin. These can be separated by extraction either with acid, which will remove the caffeine as a water-soluble salt, or with base, which will remove the aspirin as a water-soluble salt. We shall use the latter procedure.

To the dichloromethane solution in a test tube add 4 mL of 3 M sodium hydroxide solution, and shake the mixture thoroughly. Remove the aqueous layer, add 1 mL more water, shake the mixture thoroughly, and again remove the aqueous layer, which is combined with the first aqueous extract.

To the dichloromethane add anhydrous calcium chloride pellets until the drying agent no longer clumps together. Shake the mixture over a 5- or 10-min period to complete the drying process, then remove the solvent, wash the drying agent with more solvent, and evaporate the combined extracts to dryness under a stream of air to leave crude caffeine.

The caffeine can be purified by sublimation as has been done in the experiment in which it was extracted from tea (Fig. 44.4), or it can be purified by crystallization. Recrystallize the caffeine by dissolving it in the minimum quantity of 30% ethanol in tetrahydrofuran. It can also be crystallized by dissolving the product in a minimum quantity of hot toluene or acetone and adding to this solution ligroin (hexanes) until the solution is cloudy while at the boiling point. In any case, allow the solution to cool slowly to room temperature, then cool the mixture in ice, and remove the solvent from the crystals with a Pasteur pipette. Remove the remainder of the solvent under aspirator vacuum, and determine the weight of the caffeine and its melting point.

The aqueous hydroxide extract contains aspirin as the sodium salt of the carboxylic acid. To the aqueous solution add 3 M hydrochloric acid dropwise until the solution tests strongly acid to indicator paper, and then add two more drops of acid. This will give a suspension of white acetylsalicylic acid in the

Acetaminophen

Check product purity by TLC using 25 : 1 ethyl acetate–acetic acid to elute the silica gel plates.

Caffeine

Sodium acetylsalicylate (soluble in water)

aqueous solution. It could be filtered off and recrystallized from boiling water, but this would entail losses in transfer. An easier procedure is to simply heat the aqueous solution that contains the precipitated aspirin and allow it to crystallize on slow cooling.

Aspirin

Add a boiling stick and heat the mixture to boiling, at which time the aspirin should dissolve completely. If it does not, add more water. Long boiling will hydrolyze the aspirin to salicylic acid, mp 157–159°C. Once completely dissolved, the aspirin should be allowed to crystallize slowly as the solution cools to room temperature in an insulated container. Once the tube has reached room temperature it should be cooled in ice for several minutes and then the solvent removed with a Pasteur pipette. The crystals are to be washed with a few drops of ice-cold water and then scraped out onto a piece of filter paper. Squeezing the crystals between sheets of the filter paper will hasten drying. Once they are completely dry, determine the weight of the acetylsalicylic acid and its melting point.

Alternative procedure: Use the Hirsch funnel (Fig. 3.8) to isolate the aspirin.

Cleaning Up   Place any dichloromethane-containing solutions in the halogenated organic waste container and the other organic liquids in the organic solvents container. The aqueous layers should be diluted and neutralized with sodium carbonate before being flushed down the drain. After it is free of solvent, the sodium sulfate can be placed in the nonhazardous solid waste container.

Questions

1. Write equations showing how caffeine could be extracted from an organic solvent and subsequently isolated.

2. Write equations showing how acetaminophen might be extracted from an organic solvent such as an ether, if it were soluble.

3. Write detailed equations showing the mechanism by which aspirin is hydrolyzed in boiling, slightly acidic water.

The Sandmeyer Reaction:
1-Bromo-4-chlorobenzene,
2-Iodobenzoic Acid, and 4-Chlorotoluene

Prelab Exercise: Outline the steps necessary to prepare 4-bromotoluene, 4-iodotoluene, and 4-fluorotoluene from benzene.

The Sandmeyer reaction is a versatile means of replacing the amine group of a primary aromatic amine with a number of different substituents:

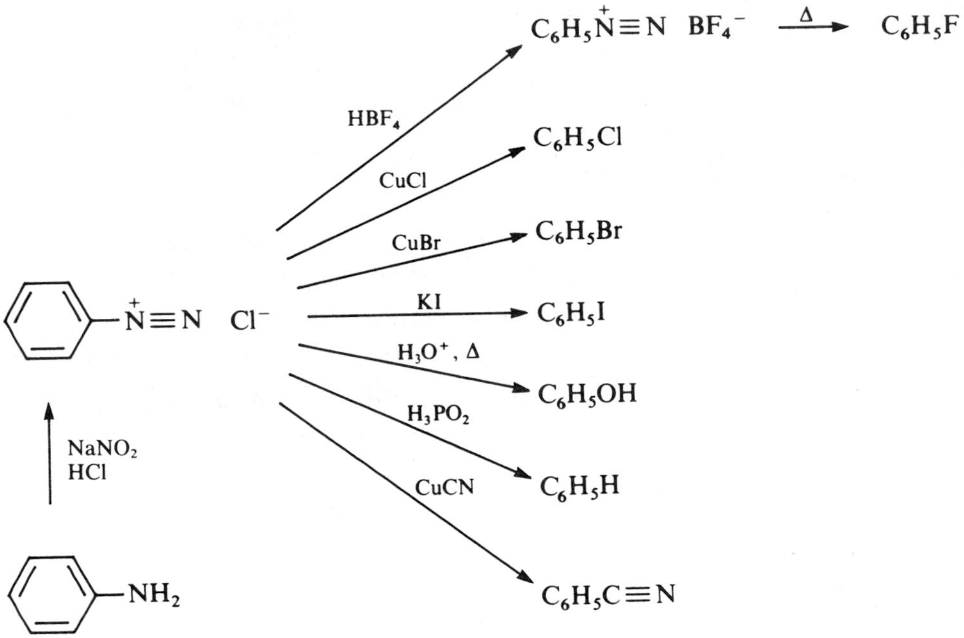

$$C_6H_5\overset{+}{N}\equiv N \;\; BF_4^- \;\; \xrightarrow{\;\Delta\;} \;\; C_6H_5F$$

HBF$_4$

CuCl → C_6H_5Cl

CuBr → C_6H_5Br

KI → C_6H_5I

H_3O^+, Δ → C_6H_5OH

H_3PO_2 → C_6H_5H

CuCN → $C_6H_5C\equiv N$

NaNO$_2$
HCl

The diazonium salt is formed by the reaction of nitrous acid with the amine in acid solution. Nitrous acid is not stable and must be prepared *in situ;* in strong acid it dissociates to form nitroso ions, $^+$NO, which attack the nitrogen of the amine. The intermediate so formed loses a proton, rearranges, and finally loses water to form the resonance-stabilized diazonium ion.

$$NaNO_2 + HCl \rightleftharpoons HO\ddot{N}O + Na^+Cl^-$$

Sodium nitrite   **Nitrous acid**

$$H_3O^+ + HO\ddot{N}O \rightleftharpoons H_2O + H_2\overset{+}{\underset{..}{O}}NO \rightleftharpoons 2\,H_2O + \overset{+}{:}\!N{=}O$$

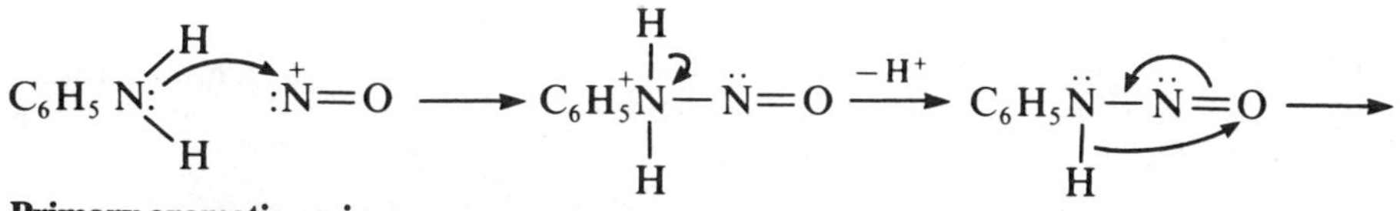

Primary aromatic amine

$$C_6H_5{-}\ddot{N}{=}\ddot{N}{-}OH \overset{H^+}{\rightleftharpoons} C_6H_5{-}\ddot{N}{=}\overset{+}{N}{-}\underset{..}{\overset{..}{O}}H_2 \overset{-H_2O}{\rightleftharpoons} [C_6H_5{-}\overset{+}{N}{\equiv}N \leftrightarrow C_6H_5{-}\ddot{N}{=}\overset{+}{\ddot{N}}]$$

Diazonium ion

The diazonium ion is reasonably stable in aqueous solution at 0°C; on warming it will form the phenol, as seen on the opposite page. A versatile functional group, it will undergo all the reactions depicted there as well as couple to aromatic rings activated with substituents such as amino and hydroxyl groups to form the huge class of azo dyes (see Chapter 66).

| **Benzenediazonium chloride** | **N,N-Dimethylaniline** | **A diazo compound** |

Diazonium salts are not ordinarily isolated, because the dry solid is explosive.

Three Sandmeyer Reactions

Experiments 1, 2, and 3

$$NaNO_2 + HCl \longrightarrow Na^+Cl^- + HONO$$

Sodium nitrite   **Nitrous acid**
MW 69.01

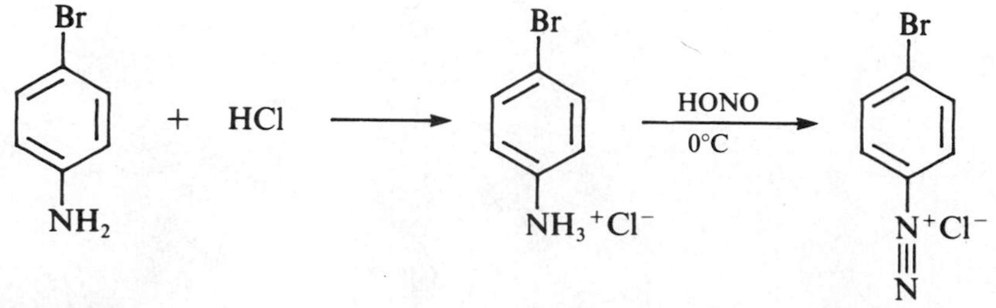

4-Bromoaniline   **4-Bromoaniline hydrochloride**   **4-Bromobenzenediazonium**
MW 172.03, mp 62–64°C   **chloride**

Experiment 3

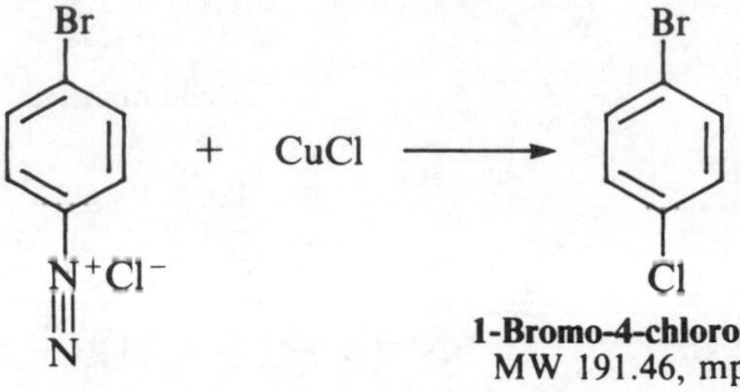

1-Bromo-4-chlorobenzene
MW 191.46, mp 68°C

In the first microscale procedure (Experiments 1, 2, and 3), after the copper(I) chloride is prepared, 4-bromoaniline is dissolved in the required amount of hydrochloric acid, two more equivalents of acid are added, and the mixture cooled in ice to produce a paste of the crystalline amine hydrochloride. When this salt is treated at 0 to 5°C with one equivalent of sodium nitrite, nitrous acid is generated and reacts to produce the diazonium salt. The excess hydrochloric acid (beyond the two equivalents required to form the amine hydrochloride and react with sodium nitrite) maintains acidity sufficient to prevent formation of the diazo–amino compound and rearrangement of the diazonium salt. The diazonium salt reacts with copper(I) chloride to give the easily sublimed 1-bromo-4-chlorobenzene.

Experiments 6, 7, and 8

$$NaNO_2 + HCl \xrightarrow{0°C} HONO + NaCl$$

Sodium nitrite   **Nitrous**
MW 69.01   **acid**

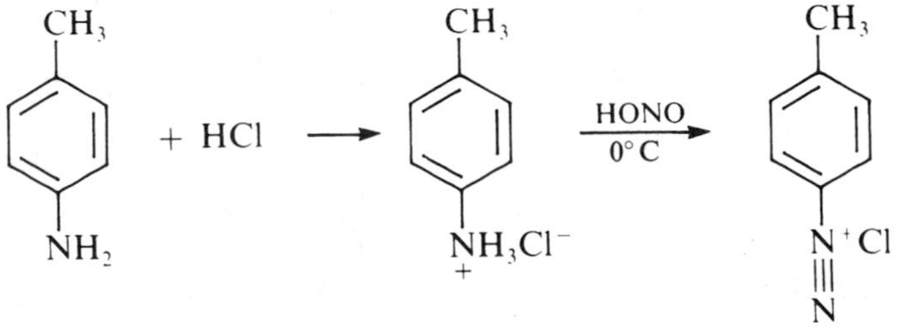

p-**Toluidine**   *p*-**Toluidine**   *p*-**Methylbenzenediazonium**
MW 107.15, mp 45 C   **hydrochloride**   **chloride**

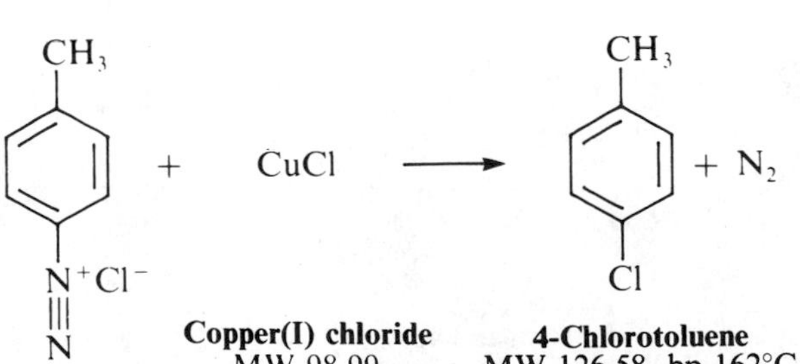

Copper(I) chloride   **4-Chlorotoluene**
MW 98.99   MW 126.58, bp 162°C

The second procedure (Experiment 4 microscale, Part 5 macroscale) employs potassium iodide, which reacts with a diazonium salt to give an iodobenzoic acid.

The third procedure is a macroscale synthesis of 4-chlorotoluene (Experiments 6, 7, and 8). *p*-Toluidine is dissolved in the required amount of hydrochloric acid, two more equivalents of acid are added, and the mixture cooled in ice to produce a paste of the crystalline amine hydrochloride. When this salt is treated at 0 to 5°C with one equivalent of sodium nitrite, nitrous acid is liberated and reacts to produce the diazonium salt. The excess hydrochloric acid, beyond the two equivalents required to form the amine hydrochloride and react with sodium nitrite, maintains acidity sufficient to prevent formation of the diazo–amino compound and rearrangement of the diazonium salt. The diazonium salt reacts with copper(I) chloride to give liquid 4-chlorotoluene.

Copper(I) chloride is made by reduction of copper(II) sulfate with sodium sulfite (which is produced as required from the cheaper sodium bisulfite). The white solid is left covered with the reducing solution for protection against air oxidation until it is to be used and then dissolved in hydrochloric acid. On addition of the diazonium salt solution, a complex forms and rapidly decomposes to give *p*-chlorotoluene and nitrogen. The mixture is very discolored, but steam distillation leaves most of the impurities and all salts behind and gives material substantially pure except for the presence of a trace of yellow pigment, which can be eliminated by distillation of the dried oil.

Experiments

MICROSCALE

1. Copper(I) Chloride Solution

$$2 \ CuSO_4 \cdot 5 \ H_2O \ + \ 4 \ NaCl \ + \ NaHSO_3 \ + \ NaOH \ \longrightarrow \ 2 \ CuCl \ + \ 3 \ Na_2SO_4 \ + \ 2 \ HCl \ + \ 10 \ H_2O$$

MW 249.71   MW 58.45   MW 104.97   MW 40.01   MW 99.02

NaHSO₃, sodium bisulfite,
sodium hydrogen sulfite, not
Na₂S₂O₄, sodium hydrosulfite

Do not measure hydrochloric
acid in a syringe that has a
metal needle.

In a reaction tube dissolve 0.30 g of copper(II) sulfate crystals ($CuSO_4 \cdot 5 \ H_2O$) in 1.0 mL of water by boiling and then add 140 mg of sodium chloride, which may give a small precipitate of basic copper(II) chloride. Prepare a solution of sodium sulfite from 70 mg of sodium bisulfite and 0.4 mL of 3 M sodium hydroxide solution, or use 57 mg of sodium sulfite in 0.4 mL of water. Add this solution dropwise to the hot copper(II) sulfate solution. Rinse the tube that contained the sulfite with a couple drops of water, and add it to the copper(I) chloride mixture. Shake the reaction mixture well, and then allow it to cool in ice while diazotizing the amine. When you are ready to use the copper(I) chloride, remove the water above the solid, wash the white solid once with water, remove the water, and dissolve the solid in 0.45 mL of concentrated hydrochloric acid. The solution is susceptible to air oxidation and should not stand for an appreciable time before use.

Cleaning Up   Combine the filtrate and washings, neutralize with sodium carbonate, and flush down the drain.

2. Diazotization of 4-Bromoaniline

In a 10-mL Erlenmeyer flask place 172 mg of 4-bromoaniline, and 0.50 mL of 3 M hydrochloric acid. Warm the mixture on the sand bath to dissolve the amine and ensure transformation into hydrochloride. Cool the flask in ice (the hydrochloride will crystallize), and add an ice-cold solution of 70 mg of fresh sodium nitrite in 0.2 mL of water. Use a drop of water to complete the transfer of the nitrite solution. Mix the solid and the sodium sulfite thoroughly. The result should be a homogeneous yellow solid/liquid mixture. Do not allow this mixture to warm above 0°C.

The extraction can be carried out in the microscale separatory funnel (Fig. 8.5).

3. Sandmeyer Reaction: 1-Bromo-4-chlorobenzene

Cool the copper(I) chloride solution in ice, and add it to the diazonium chloride solution dropwise with thorough mixing using a Pasteur pipette. Rinse the tube that contained the copper(I) chloride with a drop of water. Allow the reaction mixture to warm to room temperature, and observe the reaction mixture closely. It bubbles as nitrogen gas is evolved. Heat the mixture on the steam bath to complete the reaction. Cool the mixture thoroughly in ice, and collect the dark solid on the Hirsch funnel. Squeeze the product between sheets of filter paper to dry it, then sublime the bromochlorobenzene on the steam bath. After initial sublimation, scrape the residue in the filter flask to sublime more product.

Another way to isolate the product is to extract the product with three 0.5-mL portions of *tert*-butyl methyl ether and wash the ether extract with 0.2 mL of 3 M sodium hydroxide solution, which will remove any 4-bromophenol, and then with 0.5 mL of saturated sodium chloride solution. Dry the ether solution over anhydrous calcium chloride pellets. Add sufficient drying agent so that it no longer clumps together. After 5 or 10 min, remove the ether with a Pasteur pipette, wash the drying agent with ether, and evaporate the ether solutions in a small, tared filter flask. Use care in this evaporation because the product is volatile. Determine the weight of the crude product, and then purify it by sublimation at atmospheric pressure.

The 15-mL centrifuge tube (Fig. 45.1) is filled with ice water while the filter flask is heated on a sand or steam bath. Close the sidearm with a rubber pipette bulb. Roll the flask in the sand and/or use a heat gun to drive product from the sides of the flask to the centrifuge tube. Replace the ice water in the centrifuge tube with room-temperature water before removing it so that moisture will not condense on the tube. Scrape the product onto a piece of glazed paper, determine the weight and the melting point, and analyze it for purity by thin-layer chromatography and IR spectroscopy in a chloroform solution.

Cleaning Up   Combine all aqueous filtrates and washings, neutralize with sodium carbonate, and flush down the drain. Allow the ether to evaporate from the drying agent before it is placed in the nonhazardous solid waste container.

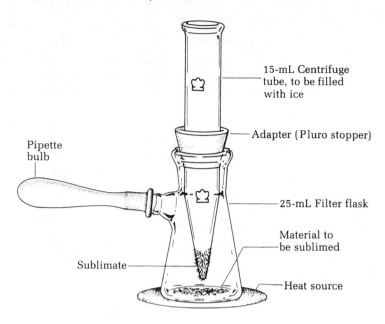

FIG. 45.1 Sublimation apparatus.

15-mL Centrifuge tube, to be filled with ice

Adapter (Pluro stopper)

Pipette bulb

25-mL Filter flask

Material to be sublimed

Sublimate

Heat source

📋 **MICROSCALE**

4. 2-Iodobenzoic Acid

COOH
I

+ N$_2$ + K$^+$Cl$^-$

KI

2-Iodobenzoic acid
MW 248.02
mp 126–163°C

COOH
NH$_2$

2-Aminobenzoic acid
MW 137.14, mp 144–148°C

+ NaNO$_2$ + 2 HCl ⟶

COOH Cl$^-$
N$^+$≡N

+ Na$^+$Cl$^-$ + 2 H$_2$O

Do not measure hydrochloric acid in a syringe that has a metal needle.

To a reaction tube containing 137 mg of 2-aminobenzoic acid (anthranilic acid) add 1 mL of water and 0.25 mL of concentrated hydrochloric acid. Heat to dissolve the amino acid and form the hydrochloride, and then cool the tube in ice. Cap the tube with a septum, and then, with the tube in an ice bath, add a solution of 75 mg of sodium nitrite dissolved in 0.3 mL of water using a syringe. This addition should be made dropwise with thorough agitation or magnetic stirring of the reaction. Rinse the syringe with 0.1 mL of water, which is also injected into the reaction mixture. After 5 min, a solution of 0.17 g of potassium iodide in 0.25 mL of water is added when a brown complex partially separates.

The empty distilling column is added to the top of the reaction tube to catch any foam. The mixture is let stand without cooling for 5 min and then cautiously warmed to 40°C. At this point a vigorous reaction begins, with nitrogen gas evolution, foaming, and separation of a tan solid. After reacting for 10 min, the mixture is heated in a beaker of boiling water for 10 min and then cooled in ice. A few milligrams of sodium sulfite are added to destroy any iodine present, and the granular tan product is collected and washed with water.

The still-moist product is dissolved in 0.7 mL of ethanol, 0.35 mL of water is added, and the hot, brown solution is treated with enough granular Norit to remove most of the color. The solution is transferred to another reaction tube to remove the decolorizing charcoal, diluted at the boiling point with 0.35 to 0.40 mL of water, and let stand. 2-Iodobenzoic acid separates in large, slightly yellow needles. The yield should be about 150 mg with a melting point near 164°C.

Cleaning Up   The reaction mixture filtrate and mother liquor from the crystallization are combined, neutralized with sodium carbonate, and flushed down the drain. Norit is placed in the nonhazardous solid waste container.

MACROSCALE

5. 2-Iodobenzoic Acid

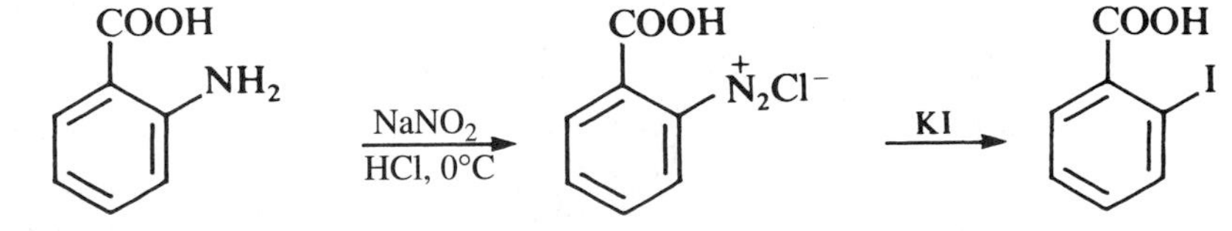

2-Aminobenzoic acid
(Anthranilic acid)
MW 137.14, mp 144–148°C

2-Iodobenzoic acid
MW 248.02, mp 162–164°C

A 100-mL round-bottomed flask containing 3.4 g of anthranilic acid, 25 mL of water, and 6 mL of concentrated hydrochloric acid is heated until the solid is dissolved. The mixture is then cooled in ice while bubbling in nitrogen to displace the air. When the temperature reaches 0–5°C, a solution of 1.8 g of sodium nitrite in 6 mL of water is added slowly. After 5 min, a solution of 4.25 g of potassium iodide in 6 mL of water is added when a brown complex partially separates. The mixture is let stand without cooling for 5 min (under nitrogen) and then warmed to 40°C, at which point a vigorous reaction ensues (gas evolution, separation of a tan solid). After reacting for 10 min, the mixture is heated on the steam bath for 10 min and then cooled in ice. A pinch of sodium bisulfite is added to destroy any iodine present, and the granular tan product is collected and washed with water. The still-moist product is dissolved in 18 mL of 95% ethanol; 9 mL of hot water is added, and the brown solution is treated with decolorizing charcoal, filtered, diluted at the boiling point with 9–10 mL of water, and let stand. 2-Iodobenzoic acid separates in large, slightly yellow needles (mp 164°C); yield is approximately 4.3 g (72%).

Cleaning Up   The reaction mixture, filtrate, and mother liquor from the crystallization are combined, neutralized with sodium carbonate, and flushed down the drain with a large excess of water. Decolorizing charcoal is placed in the nonhazardous solid waste container.

 **MACROSCALE**

6. Copper(I) Chloride Solution

$$2 \, CuSO_4 \cdot 5 \, H_2O + 4 \, NaCl + NaHSO_3 + NaOH \;\rightarrow\; 2 \, CuCl + 3 \, Na_2SO_4 + 2 \, HCl + 10 \, H_2O$$

MW 249.71   MW 58.45 MW 104.97 MW 40.01   MW 99.02

NaHSO₃, sodium bisulfite, sodium hydrogen sulfite, not Na₂S₂O₄, sodium hydrosulfite

One may stop here.

In a 250-mL round-bottomed flask (to be used later for steam distillation) dissolve 15 g of copper(II) sulfate crystals ($CuSO_4 \cdot 5H_2O$) in 50 mL of water by boiling, and then add 7 g of sodium chloride, which may give a small precipitate of basic copper(II) chloride. Prepare a solution of sodium sulfite from 3.5 g of sodium bisulfite, 2.25 g of sodium hydroxide, and 25 mL of water; and add this, not too rapidly, to the hot copper(II) sulfate solution (rinse flask and neck). Shake well, put the flask in a pan of cold water in a slanting position favorable for decantation, and let the mixture stand to cool and settle during the diazotization. When you are ready to use the copper(I) chloride, decant the supernatant liquid, wash the white solid once with water by decantation, and dissolve the solid in 23 mL of concentrated hydrochloric acid. The solution is susceptible to air oxidation and should not stand for an appreciable time before use.

 **MACROSCALE**

7. Diazotization of *p*-Toluidine

CAUTION: p-Toluidine is toxic. Handle with care.

Put 5.5 g of *p*-toluidine and 7.5 mL of water in a 50-mL Erlenmeyer flask. Measure 12.5 mL of concentrated hydrochloric acid, and add 5 mL of it to the flask. Heat over a hot plate, and swirl to dissolve the amine and hence ensure that it is all converted into the hydrochloride. Add the remaining acid, cool thoroughly in an ice bath, and let the flask stand in the bath while preparing a solution of 3.5 g of sodium nitrite in 10 mL of water. To maintain a temperature of 0 to 5°C during diazotization, add a few pieces of ice to the amine hydrochloride suspension and add more later as the first ones melt. Pour in the nitrite solution in portions during 5 min with swirling in the ice bath. The solid should dissolve to a clear solution of the diazonium salt. After 3 to 4 min, test for excess nitrous acid: Dip a stirring rod in the solution, touch off the drop on the wall of the flask, put the rod in a small test tube, and add a few drops of water. Then insert a strip of starch-iodide paper; an instantaneous deep blue color due to a starch-iodine complex indicates the desirable presence of a slight excess of nitrous acid. (The sample tested is diluted with water because strong hydrochloric acid alone produces the same color on starch-iodide paper, after a slight induction period.) Leave the solution in the ice bath.

🧴 MACROSCALE

8. Sandmeyer Reaction: 4-Chlorotoluene

Complete the preparation of copper(I) chloride solution, cool it in the ice bath, pour in the solution of diazonium chloride through a long-stemmed funnel, and rinse the flask. Swirl occasionally at room temperature for 10 min, and observe initial separation of a complex of the two components and its decomposition with liberation of nitrogen and separation of an oil. Arrange for steam distillation (Fig. 6.4), or generate steam *in situ* simply by boiling the flask contents with a Thermowell using the apparatus for simple distillation (Fig. 5.10) (add more water during the distillation). Do not start the distillation until bubbling in the mixture has practically ceased and an oily layer has separated. Then steam distill and note that *p*-chlorotoluene, although lighter than the solution of inorganic salts in which it was produced, is heavier (den 1.07) than water. Extract the distillate with a little *tert*-butyl methyl ether, wash the extract with 3 M sodium hydroxide solution to remove any *p*-cresol present, then wash with saturated sodium chloride solution; dry the ether solution over about 2.5 g of anhydrous calcium chloride pellets and filter or decant it into a tared flask, evaporate the ether, and determine the yield and percentage yield of product (your crude yield should be about 4.5 g). Pure *p*-chlorotoluene, the IR and NMR spectra of which are shown in Figs. 45.2 through 45.4, is obtained after simple distillation of this crude product. Analyze your crude and distilled product by TLC (hexane/CH_2Cl_2), 3:1 IR spectroscopy, and GC using a Carbowax column. Is it pure?

Dispose of copper salts and solutions in the container provided.

CAUTION: *Use aspirator tube.*

Cleaning Up Combine the pot residue from the steam distillation with the aqueous washings, neutralize with sodium carbonate, dilute with water, and flush down the drain. Allow the ether to evaporate from the drying agent in the hood, and then place it in the nonhazardous solid waste container.

Questions _____

1. Nitric acid is generated by the action of sulfuric acid on sodium nitrate. Nitrous acid is prepared by the action of hydrochloric acid on sodium nitrite. Why is nitrous acid prepared *in situ,* rather than obtained from the reagent shelf?

2. What byproduct would be obtained in high yield if the diazotization of *p*-toluidine were carried out at 30°C instead of 0–5°C?

3. How would 4-bromoaniline be prepared from benzene?

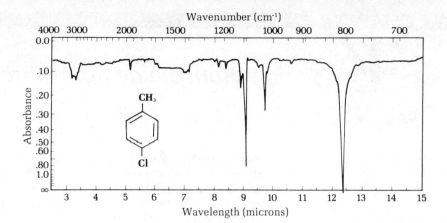

FIG. 45.2 IR spectrum of *p*-chlorotoluene in CS$_2$.

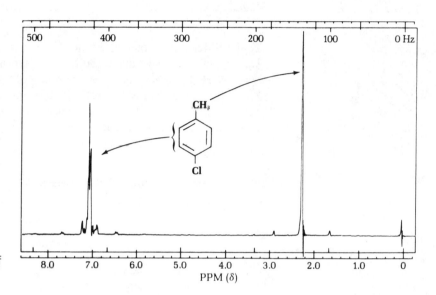

FIG. 45.3 ^{1}H NMR spectrum of *p*-chlorotoluene (90 MHz).

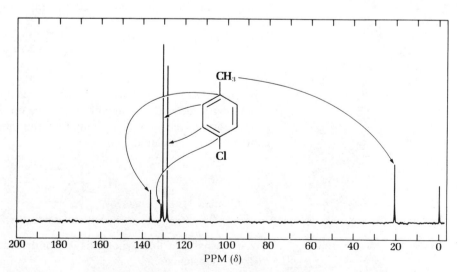

FIG. 45.4 ^{13}C NMR spectrum of *p*-chlorotoluene (22.6 MHz).

Sulfanilamide from Nitrobenzene

Prelab Exercise: Prepare detailed flow sheets for the three reactions. Look up the solubilities of all reagents, and indicate in separation steps in which layer you would expect to find the desired product. Calculate the theoretical amount of ammonium hydroxide needed to react with 5 g of *p*-acetaminobenzenesulfonyl chloride to form sulfanilamide.

Paul Ehrlich, the father of immunology and chemotherapy, discovered Salvarsan, an arsenical "magic bullet" (a favorite phrase of his) used to treat syphilis. He hypothesized at the beginning of this century that it might be possible to find a dye that would selectively stain, or dye, a bacterial cell and thus destroy it. In 1932 I.G. Farbenindustrie patented a new azo dye, Prontosil, which they put through routine testing for chemotherapeutic activity when it was noted it had particular affinity for protein fibers like silk.

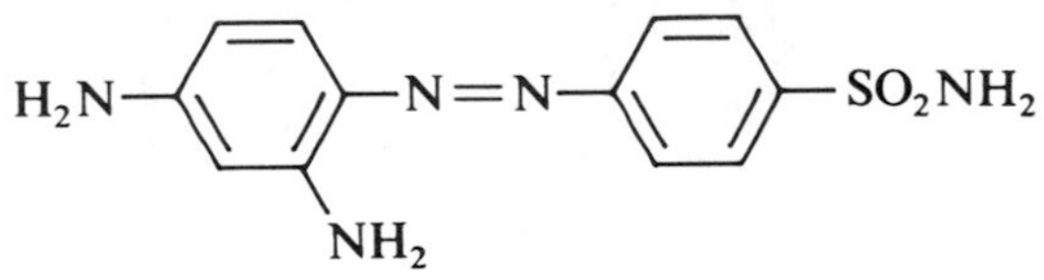

Prontosil

The dye was found to be effective against streptococcal infections in mice, but somewhat surprisingly, ineffective *in vitro* (outside the living animal). A number of other dyes were tested, but only those having the group

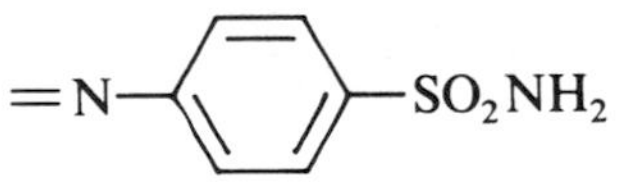

were effective. French workers hypothesized that the antibacterial activity had nothing to do with the identity of the compounds as dyes, but rather with the reduction of the dyes in the body to *p*-aminobenzenesulfonamide, known commonly as sulfanilamide.

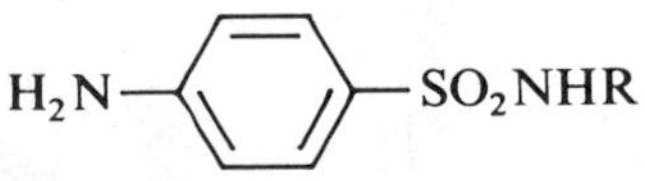

Sulfanilamide, R=H

On the basis of this hypothesis, sulfanilamide was tested and found to be the active substance.

Because sulfanilamide had been synthesized in 1908, its manufacture could not be protected by patents, so the new drug and thousands of its derivatives were rapidly synthesized and tested. When the R group in sulfanilamide is replaced with a heterocyclic ring system—pyridine, thiazole, diazine, merazine, etc.—the sulfa drug so produced is often faster acting or less toxic than sulfanilamide. Although they have been supplanted for the most part by antibiotics of microbial origin, these drugs still find wide application in chemotherapy.

Unlike that of most drugs, the mode of action of the sulfa drugs is now completely understood. Bacteria must synthesize folic acid in order to grow. Higher animals, like humans, do not synthesize this vitamin and hence must acquire it in their food. Sulfanilamide inhibits the formation of folic acid, stopping the growth of bacteria; and because the synthesis of folic acid does not occur in humans, only the bacteria are affected.

A closer look at these events reveals that bacteria synthesize folic acid using several enzymes, including one called dihydropteroate synthetase, which catalyzes the attachment of *p*-aminobenzoic acid to a pteridine ring system. When sulfanilamide is present it competes with the *p*-aminobenzoic acid (note the structural similarity) for the active site on the enzyme. This activity makes it a competitive inhibitor. Once this site is occupied on the enzyme, folic acid synthesis stops and bacterial growth stops. Folic acid can also be synthesized in the laboratory.[1]

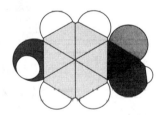

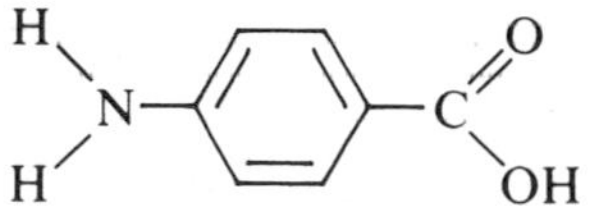

Sulfanilamide

***p*-Aminobenzoic acid**

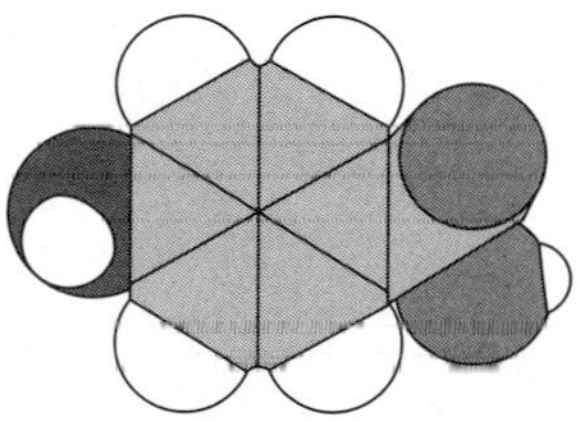

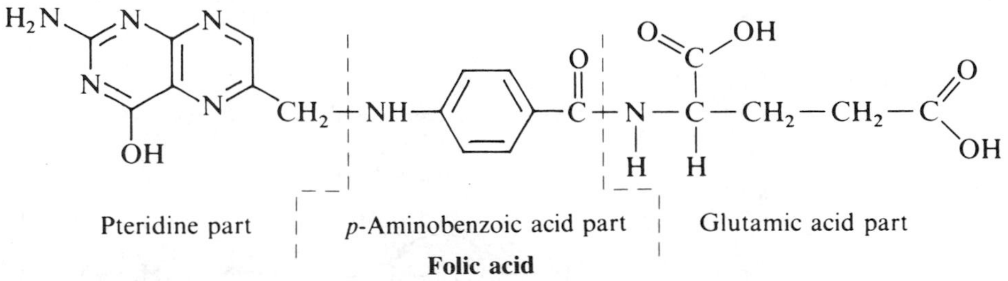

Folic acid

Pteridine part | *p*-Aminobenzoic acid part | Glutamic acid part

This experiment is a multistep synthesis of sulfanilamide starting with nitrobenzene. Nitrobenzene is reduced with tin and hydrochloric acid to give the anilinium ion, which is converted to aniline with base:

1. L. T. Plante, K. L. Williamson, and E. J. Pastore, *Methods in Enzymology*, Vol. 66, p. 533, D. B. McCormick and L. D. Wright, eds., Academic Press, New York, 1980.

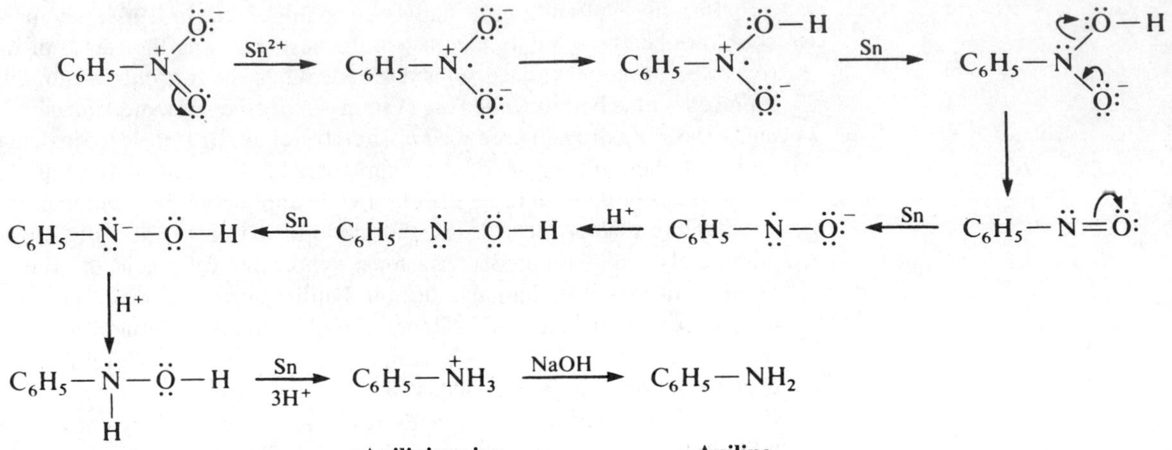

$$C_6H_5-\overset{+}{N}\overset{\ddot{O}:^-}{\underset{\ddot{O}:}{}} \xrightarrow{Sn^{2+}} C_6H_5-\overset{+}{N}\overset{\ddot{O}:^-}{\underset{\ddot{O}:^-}{\cdot}} \longrightarrow C_6H_5-\overset{+}{N}\overset{:\ddot{O}-H}{\underset{\ddot{O}:^-}{\cdot}} \xrightarrow{Sn} C_6H_5-\overset{:\ddot{O}-H}{\underset{\ddot{O}:^-}{N}}$$

$$C_6H_5-\overset{..}{\underset{}{N}}{}^-\!-\ddot{O}-H \xleftarrow{Sn} C_6H_5-\overset{..}{\underset{}{N}}-\ddot{O}-H \xleftarrow{H^+} C_6H_5-\overset{..}{\underset{}{N}}-\ddot{O}:^- \xleftarrow{Sn} C_6H_5-\overset{..}{N}=\ddot{O}:$$

$$\downarrow H^+$$

$$C_6H_5-\underset{\underset{H}{|}}{\overset{..}{N}}-\ddot{O}-H \xrightarrow[3H^+]{Sn} C_6H_5-\overset{+}{N}H_3 \xrightarrow{NaOH} C_6H_5-NH_2$$

$$\textbf{Anilinium ion} \qquad\qquad \textbf{Aniline}$$

The acetylation of aniline in aqueous solution to give acetanilide serves to protect the amine group from reaction with chlorosulfonic acid. The acetylation takes place readily in aqueous solution. Aniline reacts with acid to give the water-soluble anilinium ion. The anilinium ion reacts with acetate ion to set up an equilibrium that liberates a small quantity of aniline:

$$C_6H_5-\overset{..}{N}H_2 + H^+ \longrightarrow C_6H_5-\overset{+}{N}H_3$$

$$\textbf{Aniline} \qquad\qquad\qquad \textbf{Anilinium ion}$$

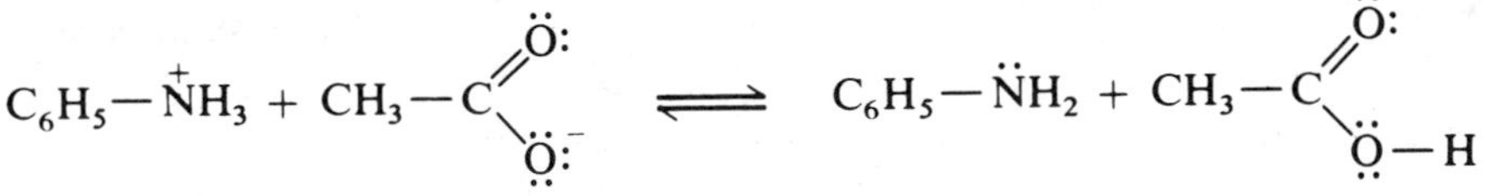

$$C_6H_5-\overset{+}{N}H_3 + CH_3-C\overset{\ddot{O}:}{\underset{\ddot{O}:^-}{}} \rightleftharpoons C_6H_5-\overset{..}{N}H_2 + CH_3-C\overset{\ddot{O}:}{\underset{\ddot{O}-H}{}}$$

This aniline reacts with acetic anhydride to give water-insoluble acetanilide:

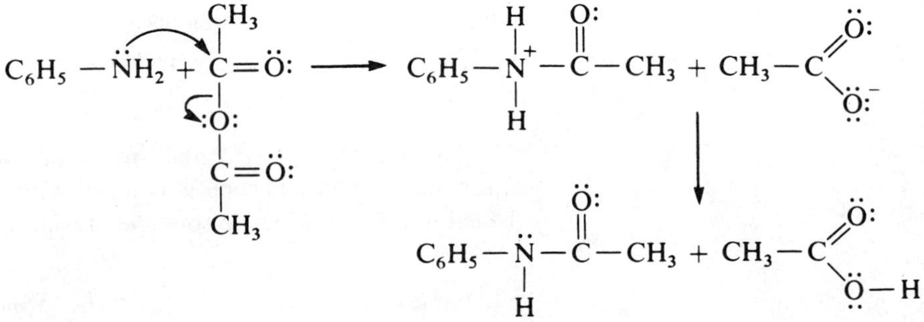

$$C_6H_5-\overset{..}{N}H_2 + \underset{\underset{CH_3}{|}}{\overset{\overset{CH_3}{|}}{C}}\!=\!\ddot{O}: \longrightarrow C_6H_5-\underset{\underset{H}{|}}{\overset{\overset{H}{|}}{N}}{}^+\!-\overset{\overset{\ddot{O}:}{||}}{C}-CH_3 + CH_3-C\overset{\ddot{O}:}{\underset{\ddot{O}:^-}{}}$$

$$\downarrow$$

$$C_6H_5-\underset{\underset{H}{|}}{\overset{..}{N}}-\overset{\overset{\ddot{O}:}{||}}{C}-CH_3 + CH_3-C\overset{\ddot{O}:}{\underset{\ddot{O}-H}{}}$$

$$\textbf{Acetanilide}$$

This upsets the equilibrium, releasing more aniline, which can then react with acetic anhydride to give more acetanilide.

Acetanilide reacts with chlorosulfonic acid in an electrophilic aromatic substitution:

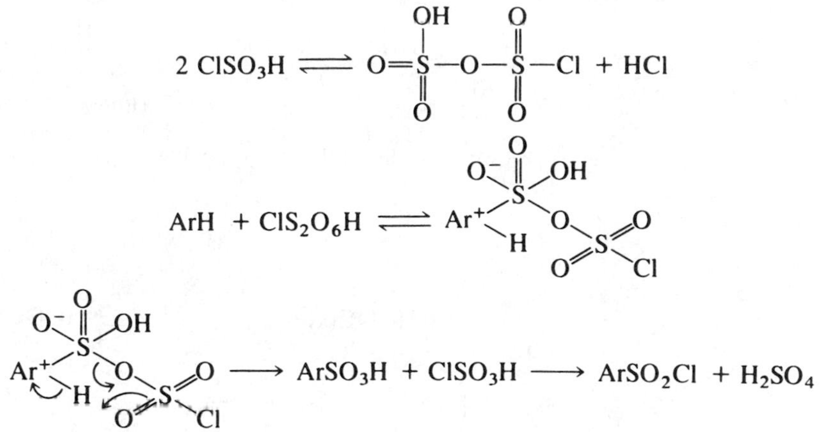

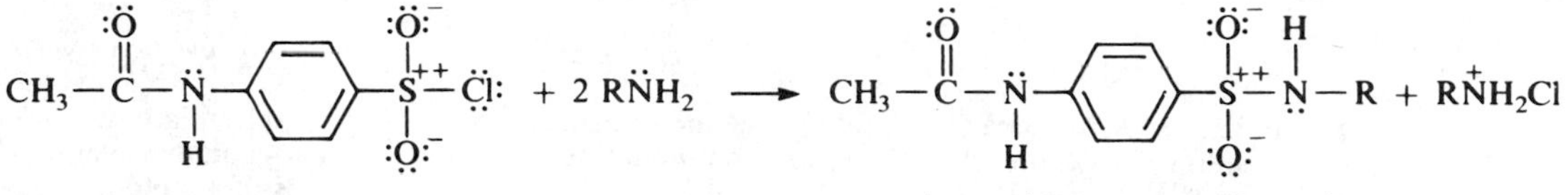

$$2\ ClSO_3H \rightleftharpoons O{=}\overset{OH}{\underset{O}{S}}{-}O{-}\overset{O}{\underset{O}{S}}{-}Cl\ +\ HCl$$

$$ArH\ +\ ClS_2O_6H \rightleftharpoons Ar^+{-}H\,\overset{O^-}{\underset{}{S}}{-}OH \cdots O{=}\overset{O}{\underset{Cl}{S}}$$

$$Ar^+{-}H\,\overset{O^-}{S}{-}OH \cdots \overset{O}{\underset{Cl}{S}}{=}O \longrightarrow ArSO_3H\ +\ ClSO_3H \longrightarrow ArSO_2Cl\ +\ H_2SO_4$$

$$CH_3{-}\overset{:\ddot{O}:}{\underset{}{C}}{-}\overset{}{\underset{H}{\ddot{N}}}{-}\!\!\bigcirc\!\!{-}\overset{:\ddot{O}:^-}{\underset{:\ddot{O}:^-}{\overset{++}{S}}}{-}\ddot{C}l:\ +\ 2\ R\ddot{N}H_2 \longrightarrow CH_3{-}\overset{:\ddot{O}:}{\underset{}{C}}{-}\overset{}{\underset{H}{\ddot{N}}}{-}\!\!\bigcirc\!\!{-}\overset{:\ddot{O}:^-}{\underset{:\ddot{O}:^-}{\overset{++}{S}}}{-}\overset{H}{\underset{}{N}}{-}R\ +\ R\overset{+}{N}H_2Cl$$

p-Acetamidobenzenesulfonamide

The protecting amide group is removed from the *p*-acetamidobenzenesulfonamide by acid hydrolysis. The amide group is more easily hydrolyzed than the sulfonamide group:

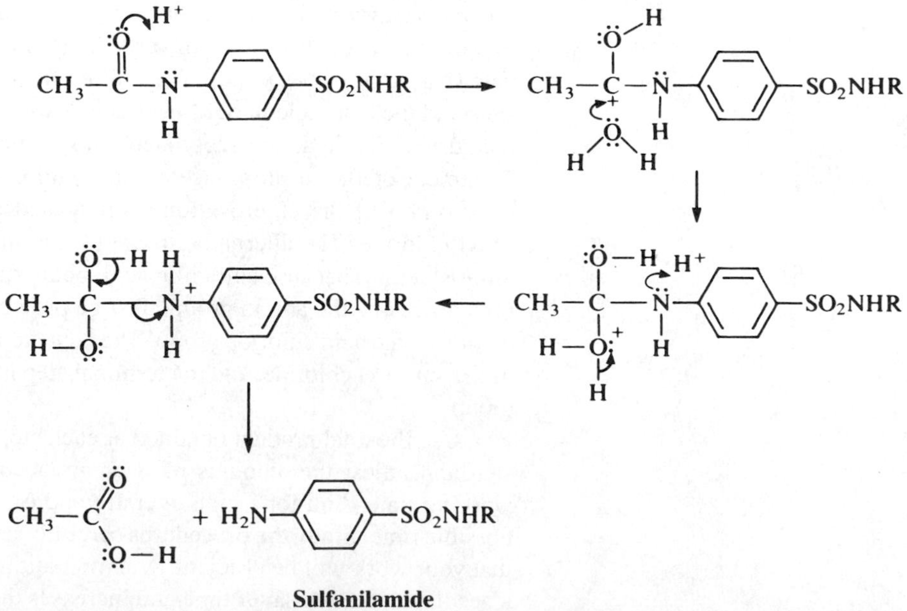

$$CH_3{-}\overset{:\ddot{O}\curvearrowleft H^+}{\underset{\underset{H}{\ddot{N}}}{C}}{-}\!\!\bigcirc\!\!{-}SO_2NHR \longrightarrow CH_3{-}\overset{:\ddot{O}{-}H}{\underset{\underset{\overset{\ddot{O}}{\underset{H\ \ H}{}}}{\overset{+}{\underset{H}{\ddot{N}}}}}{\overset{+}{C}}}{-}\!\!\bigcirc\!\!{-}SO_2NHR$$

$$CH_3{-}\overset{:\ddot{O}{-}H\ H}{\underset{H{-}\overset{+}{\ddot{O}}}{C}}{-}\overset{}{\underset{H}{\overset{+}{N}}}{-}\!\!\bigcirc\!\!{-}SO_2NHR \longleftarrow CH_3{-}\overset{:\ddot{O}{-}H\ H^+}{\underset{H{-}\overset{+}{\ddot{O}}}{C}}{-}\overset{}{\underset{H}{N}}{-}\!\!\bigcirc\!\!{-}SO_2NHR$$

$$CH_3{-}\overset{:\ddot{O}}{\underset{:\ddot{O}{-}H}{C}}\ +\ H_2N{-}\!\!\bigcirc\!\!{-}SO_2NHR$$

Sulfanilamide

The Synthetic Sequence

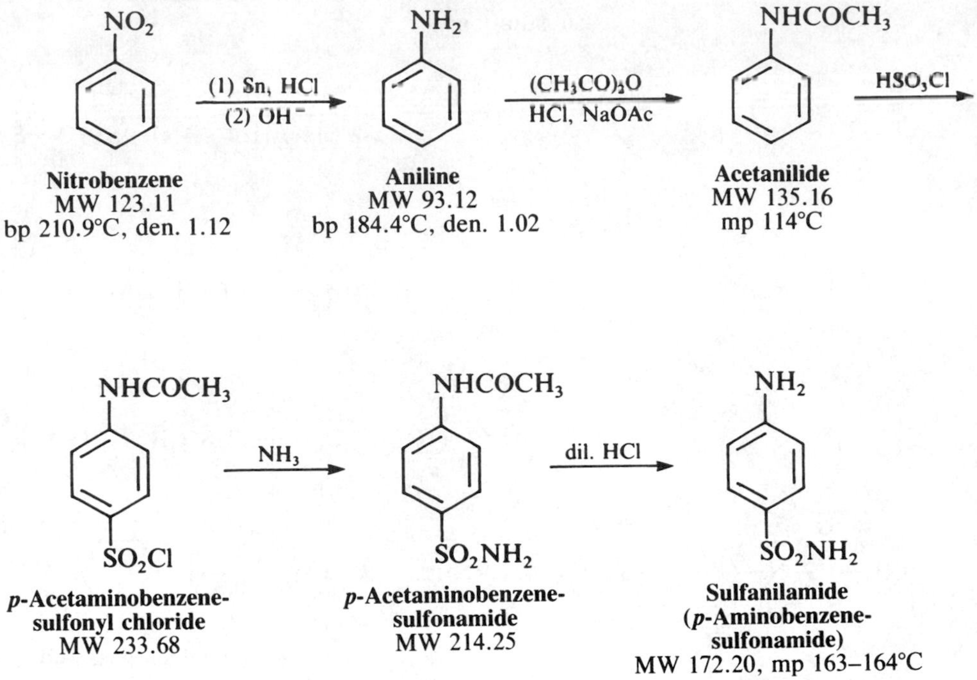

NO$_2$ $\xrightarrow[\text{(2) OH}^-]{\text{(1) Sn, HCl}}$ NH$_2$ $\xrightarrow[\text{HCl, NaOAc}]{\text{(CH}_3\text{CO)}_2\text{O}}$ NHCOCH$_3$ $\xrightarrow{\text{HSO}_3\text{Cl}}$

Nitrobenzene
MW 123.11
bp 210.9°C, den. 1.12

Aniline
MW 93.12
bp 184.4°C, den. 1.02

Acetanilide
MW 135.16
mp 114°C

NHCOCH$_3$ / SO$_2$Cl $\xrightarrow{\text{NH}_3}$ NHCOCH$_3$ / SO$_2$NH$_2$ $\xrightarrow{\text{dil. HCl}}$ NH$_2$ / SO$_2$NH$_2$

**p-Acetaminobenzene-
sulfonyl chloride**
MW 233.68

**p-Acetaminobenzene-
sulfonamide**
MW 214.25

**Sulfanilamide
(p-Aminobenzene-
sulfonamide)**
MW 172.20, mp 163–164°C

In the present experiment nitrobenzene is reduced to aniline by tin and hydrochloric acid. A double salt with tin having the formula $(C_6H_5NH_3)_2SnCl_4$ separates partially during the reaction, and at the end it is decomposed by addition of excess alkali, which converts the tin into water-soluble stannite or stannate $(Na_2SnO_2$ or $Na_2SnO_3)$. The aniline liberated is separated from inorganic salts and the insoluble impurities derived from the tin by steam distillation. It is then dried, distilled, and acetylated with acetic anhydride in aqueous solution. Treatment of the resulting acetanilide with excess chlorosulfonic acid effects substitution of the chlorosulfonyl group and affords p-acetaminobenzenesulfonyl chloride. The alternative route to this intermediate via sulfanilic acid is unsatisfactory, because sulfanilic acid, being dipolar, is difficult to acetylate. In both processes the amino group must be protected by acetylation to permit formation of the acid chloride group. The next step in the synthesis is ammonolysis of the sulfonyl chloride, and the terminal step is removal of the protective acetyl group.

Use the total product obtained at each step as starting material for the next step and adjust the amounts of reagents accordingly. Keep a record of your working time. Aim for a high overall yield of pure final product in the shortest possible time. Study the procedures carefully before coming to the laboratory so that your work will be efficient. A combination of consecutive steps that avoids a needless isolation saves time and increases the yield.

Experiments _____

1. Microscale Preparation of Aniline

Nitrobenzene   **Aniline**

Measure nitrobenzene (toxic) in the hood. Do not breathe the vapors.

Handle aniline with care. It may be a carcinogen.

Reaction time: 0.5 h

Handle the hot solution of sodium hydroxide with care; it is very corrosive.

The aniline can be isolated (see next page) or converted directly to acetanilide (Experiment 3).

In a 5-mL round-bottomed long-necked flask place 1.25 g of granulated tin, 600 mg of nitrobenzene, 2.75 mL of concentrated hydrochloric acid, and a magnetic stirring bar. Have an ice bath ready to control the reaction, insert a thermometer, and let the mixture react until the temperature reaches 60°C; then cool it briefly, if necessary, so that the temperature does not rise above 62°C. Swirl the reaction mixture, and maintain the temperature in the range of 55 to 60°C for 15 min. Remove the thermometer, rinse it with water, and fit the flask with an empty fractionating column that will function as an air condenser. Additional cooling can, if necessary, be furnished by a damp pipe cleaner (Fig. 46.1). Heat the reaction mixture on the steam bath or boiling water bath with frequent swirling or magnetic stirring until droplets of nitrobenzene are absent from the condenser and the color due to an intermediate reduction product is gone (about 15 min). During this period, dissolve 2 g of sodium hydroxide in 5 mL of water, and cool the solution to room temperature.

At the end of the reduction period, transfer the acid solution to the 15-mL centrifuge tube, cool the mixture in ice, and add the sodium hydroxide solution dropwise with stirring and ice cooling. The alkali neutralizes the aniline hydrochloride, releasing aniline. Gently extract the mixture with three 1.5-mL portions of *tert*-butyl methyl ether, and place the ether extracts in the 5-mL short-necked round-bottomed flask, which need not be clean or dry. If intractable emulsions form, centrifuge the tube for 1 min to cause the layers to separate. Transfer any emulsion layer to the flask along with the ether layer.

Construct the apparatus shown in Fig. 46.2, add a boiling chip, and remove the ether (bp 55°C) by distillation. To the dirty residue of aniline and water add 2 mL of water through the addition port using a syringe. Heat the contents of the flask to boiling. A mixture of water and aniline will steam distill (see Chapter 6), and the distillate will be cloudy as the two layers, aniline and water, separate. As the distillation proceeds, add water through the addition port, using a syringe, to keep the volume in the distilling flask constant. Since aniline is fairly soluble in water (3.6 g/100 g of water at 18°C), distillation should be continued somewhat beyond the point at which the distillate has lost its original turbidity (2.5–3 mL more). Make an accurate estimate of the volume of the distillate, for example, by pouring it into a 10-mL graduated cylinder. The ^{13}C NMR spectrum of nitrobenzene (Fig. 46.6) and the IR (Fig. 46.7) and ^{1}H NMR spectra of aniline (Fig. 46.8) are found at the end of the chapter.

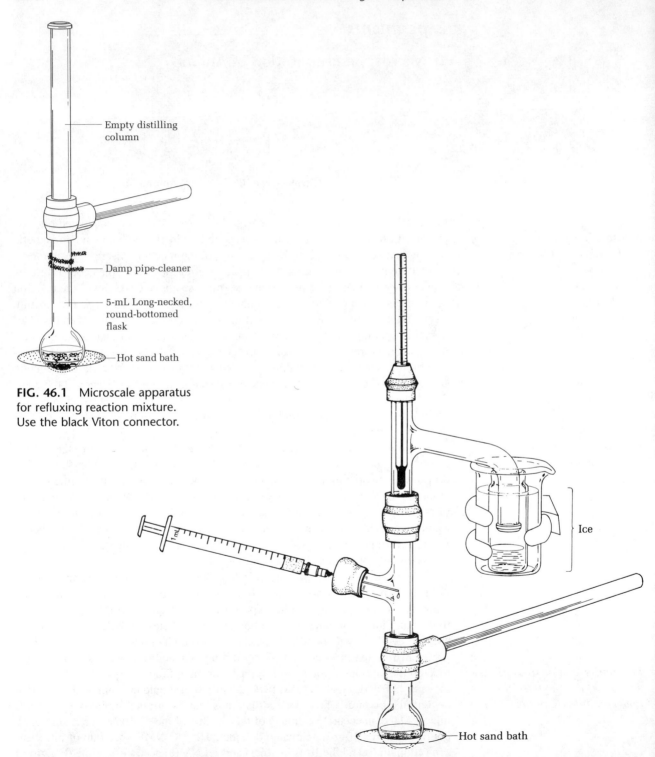

Empty distilling column

Damp pipe-cleaner

5-mL Long-necked, round-bottomed flask

Hot sand bath

FIG. 46.1 Microscale apparatus for refluxing reaction mixture. Use the black Viton connector.

Ice

Hot sand bath

FIG. 46.2 Small-scale steam distillation apparatus. Add water via syringe.

Cleaning Up   Filter the pot residue from the steam distillation through a layer of Celite filter aid on a Büchner funnel. Discard the solid in the nonhazardous solid waste container. The filtrate, after neutralization with hydrochloric acid, can be flushed down the drain.

2. Macroscale Preparation of Aniline

Measure nitrobenzene (toxic) in the hood. Do not breathe the vapors.

The reduction of the nitrobenzene is carried out in a 500-mL round-bottomed flask suitable for steam distillation of the reaction product. Put 25 g of granulated tin and 12.0 g of nitrobenzene in the flask, make an ice-water bath ready, add 55 mL of concentrated hydrochloric acid, insert a thermometer, and swirl well to promote reaction in the three-phase system. Let the mixture react until the temperature reaches 60°C, and then cool briefly in ice just enough to prevent a rise above 60°C, so the reaction will not get out of hand. Continue to swirl, cool as required, and maintain the temperature in the range 55–60°C for 15 min. Remove the thermometer, rinse it with water, fit the flask with a reflux condenser, and heat on the steam bath with frequent swirling until droplets of nitrobezene are absent from the condenser and the color due to an intermediate reduction product is gone (about 15 min). During this period, dissolve 40 g of sodium hydroxide in 100 mL of water and cool to room temperature.

Handle aniline with care. It may be a carcinogen.

Handle the hot solution of sodium hydroxide with care; it is very corrosive.

Suitable point of interruption

At the end of the reduction reaction, cool the acid solution in ice (to prevent volatilization of aniline) during gradual addition of the solution of alkali. This alkali neutralizes the aniline hydrochloride, releasing aniline, which will now be volatile in steam. Attach a stillhead with steam-inlet tube, condenser, adapter, and receiving Erlenmeyer flask (Fig. 6.4); heat the flask with a microburner to prevent the flask from filling with water from condensed steam, and proceed to steam distill. Because aniline is fairly soluble in water (3.6 g/100 g$^{18°}$) distillation should be continued somewhat beyond the point where the distillate has lost its original turbidity (50–60 mL more). Make an accurate estimate of the volume of distillate by filling a second flask with water to the level of liquid in the receiver and measuring the volume of water.

The ^{13}C NMR spectrum of nitrobenzene (Fig. 46.6) and the IR (Fig. 46.7) and ^{1}H NMR (Fig. 46.8) spectra of aniline are found at the end of the chapter.

MICROSCALE AND MACROSCALE

Isolation of Aniline—Alternative Choices

Salting out aniline

At this point aniline can be isolated. You could reduce the solubility of aniline by dissolving in the steam distillate 0.2 g of sodium chloride per milliliter, extracting the aniline with 2–3 portions of dichloromethane, drying the extract, distilling the dichloromethane (bp 41°C), and then distilling the aniline (bp 184°C). Or the aniline can be converted directly to acetanilide. The procedure calls for pure aniline, but note that the first step is to dissolve the aniline in water and hydrochloric acid. Your steam distillate is a mixture of aniline and water, both of which have been distilled. Are they not both water-white and presumably pure? Hence, an attractive procedure would be to assume that the steam distillate contains the theoretical amount of aniline and to add to it, in turn, appropriate amounts of hydrochloric acid, acetic anhydride, and sodium acetate, calculated from the quantities given in Experiment 2.

Cleaning Up   Filter the pot residue from the steam distillation through a layer of Celite filter aid on an Büchner funnel. Discard the tin salts in the non-hazardous solid waste container. The filtrate, after neutralization with hydrochloric acid, can be flushed down the drain.

MICROSCALE

3. Acetanilide: Acetylation in Aqueous Solution

Handle aniline and acetic anhydride under hood; avoid contact with skin.

Aniline   Acetic anhydride   Acetanilide

From the volumes of reagents given for this experiment, calculate the numbers of grams, the numbers of moles, and the mole ratios. What is the limiting reagent? Why is sodium acetate in this mixture?

Add to the steam distillate, which consists of a mixture of aniline and water, 0.43 mL of concentrated hydrochloric acid, and make up the volume to 13 mL. Add 0.620 mL of acetic anhydride, and also prepare a solution of 0.53 g of anhydrous sodium acetate in 3 mL of water. Add the acetic anhydride to the solution of aniline hydrochloride with stirring, and at once add the sodium acetate solution. Stir, cool in ice, and collect the product by suction filtration on the Hirsch funnel. It should be colorless, and, when dry, the melting point should be close to 114°C. Since the acetanilide *must be completely dry* for use in the next step, it is advisable to put the material in a tared 25-mL Erlenmeyer flask and to heat it on the steam bath under evacuation until the weight is constant (Fig. 46.3). It could then be used to make sulfanilamide.

FIG. 46.3 Drying of acetanilide under reduced pressure. Heat flask on steam bath.

^{1}H NMR (Fig. 46.9) and IR (Fig. 46.10) spectra of acetanilide are found at the end of the chapter.

Cleaning Up   The aqueous filtrate can be flushed down the drain.

MACROSCALE

4. Acetanilide

Acetylation in Aqueous Solution

Handle aniline and acetic anhydride under hood; avoid contact with skin.

Dissolve 5.0 g (0.054 mol) of aniline in 135 mL of water and 4.5 mL (0.054 mol) of concentrated hydrochloric acid, and, if the solution is colored, filter it by suction through a pad of decolorizing charcoal. Measure out 6.2 mL (0.065 mol) of acetic anhydride, and also prepare a solution of 5.3 g (0.065 mol) of anhydrous sodium acetate in 30 mL of water. Add the acetic anhydride to the solution of aniline hydrochloride with stirring, and at once add the sodium acetate solution. Stir, cool in ice, and collect the product. It should be colorless and the mp close to 114°C. Since the acetanilide *must be completely dry* for use in the next step, it is advisable to put the material in a tared 125-mL Erlenmeyer flask and to heat this on the steam bath under evacuation until the weight is constant. (See Fig. 46.3.)

^{1}H NMR (Fig. 46.9) and IR (Fig. 46.10) spectra of acetanilide are found at the end of the chapter.

Cleaning Up   The aqueous filtrate can be flushed down the drain with a large excess of water.

5. Sulfanilamide

MICROSCALE

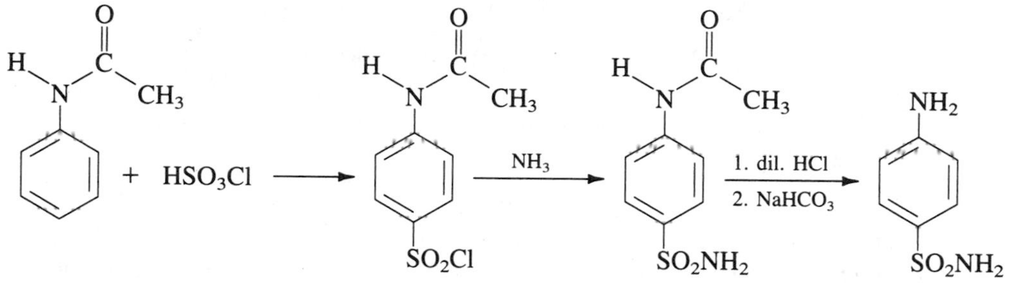

| Acetanilide | Chloro-sulfonic acid | *p*-Acetaminobenzene-sulfonyl chloride | *p*-Acetaminobenzene-sulfonamide | Sulfanilamide |

The chlorosulfonation of acetanilide in the preparation of sulfanilamide is conducted without solvent in the 25-mL Erlenmeyer flask used for drying the precipitated acetanilide from Experiment 3. Alternatively, commercial acetanilide can be employed.

Fit the Erlenmeyer flask with a septum connected to a short length of polyethylene tubing leading into a reaction tube that contains a small piece of damp cotton to trap hydrogen chloride vapors (Fig. 46.4). Add 0.625 mL of chlorosulfonic acid a few drops at a time to 0.25 g of acetanilide from a capped vial in the hood using a Pasteur pipette (not a syringe with metal needle). Connect the flask to the gas trap between additions. In 5 to 10 min the reaction subsides, and only a few small pieces of acetanilide remain undissolved.

When this point has been reached, heat the mixture on the steam bath for 10 min to complete the reaction, cool the flask in ice, and deliver the oily product by drops with a Pasteur pipette while stirring it into 3.5 mL of ice water contained in a 10-mL Erlenmeyer flask in the hood. Use extreme caution when adding the oil to ice water and when rinsing out any containers that have held chlorosulfonic acid. Rinse the flask with cold water, and stir the precipitated *p*-acetaminobenzenesulfonyl chloride for a few minutes until an even suspension of granular white solid is obtained. Collect and wash the solid on a Hirsch funnel.

After pressing and draining the filter cake, transfer the solid to the rinsed

HSO_3Cl

Chlorosulfonic acid

Caution: *Chlorosulfonic acid is a corrosive chemical and reacts violently with water. Withdraw with pipette and pipetter. Neutralize any spills and drips immediately. The wearing of gloves and a face shield is advised.*

$$HSO_3Cl + H_2O \rightarrow$$
$$H_2SO_4 + HCl$$

FIG. 46.4 Hydrogen chloride trap.

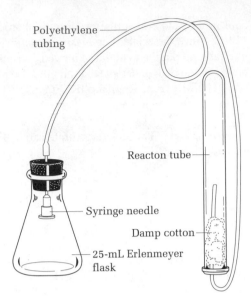

Polyethylene tubing

Reacton tube

Syringe needle

Damp cotton

25-mL Erlenmeyer flask

Do not let the mixture stand before addition of ammonia.

reaction flask, add 0.75 mL of concentrated aqueous ammonia solution (ammonium hydroxide) and 0.75 mL of water, and heat the mixture over a hot sand bath to just below the boiling point with occasional swirling in the hood. Heat the mixture in this manner for 5 min. During this treatment, a change can be noted as the sulfonyl chloride undergoes transformation to a more pasty suspension of the amide. Cool the suspension well in an ice bath, collect the *p*-acetaminobenzenesulfonamide by suction filtration, press the cake on the Hirsch funnel, and allow it to drain thoroughly. Any excess water will unduly dilute the acid used in the next step.

Transfer the still-moist amide to the well-drained reaction flask, add 0.25 mL of concentrated hydrochloric acid and 0.5 mL of water, boil the mixture gently until the solid has all dissolved (5–10 min), and then continue the heating at the boiling point for 10 min longer (do not evaporate to dryness). The solution, when cooled to room temperature, should deposit no solid amide, but if it is deposited, heating should be continued for a further period. The cooled solution of sulfanilamide hydrochloride is shaken with granulated decolorizing charcoal and filtered by removal of the solution with a Pasteur pipette.

Place the solution in a 30-mL beaker, and cautiously add an aqueous solution of 0.25 g of sodium bicarbonate while stirring to neutralize the hydrochloride. After the foam has subsided, test the suspension with indicator paper, and if it is still acidic, add more bicarbonate until the neutral point is reached. Cool thoroughly in ice and collect the granular, white precipitate of sulfanilamide. The crude product (mp 161–163°C) on crystallization from alcohol or water affords pure sulfanilamide (mp 163–164°C) with about 90% recovery.

The IR spectrum (Fig. 46.11) of sulfanilamide is found at the end of this chapter.

Cleaning Up   Rinse the cotton in the trap with water, add this rinse to the combined aqueous filtrates from all reactions, and neutralize the solution by adding either 3 M hydrochloric acid or sodium carbonate. Flush the neutral solution down the drain. Any spilled drops of chlorosulfonic acid should be covered with sodium carbonate and the powder collected in a beaker, dissolved in water, and flushed down the drain.

MACROSCALE

HSO_3Cl

Chlorosulfonic acid

CAUTION: Chlorosulfonic acid is a corrosive chemical and reacts violently with water. Withdraw with pipette and pipetter. Neutralize any spills and drips immediately. The wearing of gloves and a face shield is advised.

$$HSO_3Cl + H_2O \rightarrow$$
$$H_2SO_4 + HCl$$

6. Sulfanilamide

The chlorosulfonation of acetanilide in the preparation of sulfanilamide is conducted without solvent in the 125-mL Erlenmeyer flask used for drying the precipitated acetanilide from Experiment 2. Because the reaction is most easily controlled when the acetanilide is in the form of a hard cake, the dried solid is melted by heating the flask over a hot plate; as the melt cools, the flask is swirled to distribute the material as it solidifies over the lower walls of the flask. Let the flask cool while making provision for trapping the hydrogen chloride evolved in the chlorosulfonation. Fit the Erlenmeyer flask with a stopper connected by a section of rubber tubing to a glass tube fitted with a stopper into the neck of a 250-mL filter flask half-filled with water. The tube should be about 1 cm above the surface of the water and *must not dip into the water.* See Fig. 46.5.

Cool the flask containing the acetanilide thoroughly in an ice-water bath, and for 5.0 g of acetanilide measure 12.5 mL of chlorosulfonic acid in a graduate (supplied with the reagent and kept away from water). Add the reagent in 1–2 mL portions with a capillary dropping tube, and connect the flask to the gas trap. The flask is now removed from the ice bath and swirled until a part of the solid has dissolved and the evolution of hydrogen chloride is proceeding rapidly. Occasional cooling in ice may be required to prevent too brisk a reaction.

In 5–10 min, the reaction subsides and only a few lumps of acetanilide remain undissolved. When this point has been reached, heat the mixture on the steam bath for 10 min to complete the reaction, cool the flask under the tap, and

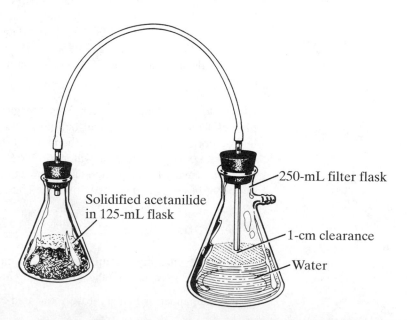

Solidified acetanilide in 125-mL flask

250-mL filter flask

1-cm clearance

Water

FIG. 46.5 Chlorosulfonation apparatus fitted with HCl gas trap.

deliver the oil by drops with a capillary dropper while stirring it into 75 mL of ice water contained in a beaker cooled in an ice bath (hood). Use extreme caution when adding the oil to ice water and when rinsing out any containers that have held chlorosulfonic acid. Rinse the flask with cold water, and stir the precipitated *p*-acetaminobenzenesulfonyl chloride for a few minutes until an even suspension of granular white solid is obtained. Collect and wash the solid on a Büchner funnel.

Do not let the mixture stand before addition of ammonia.

After pressing and draining the filter cake, transfer the solid to the rinsed reaction flask, add (for 5 g of aniline) 15 mL of concentrated aqueous ammonia solution and 15 mL of water, and heat the mixture over a flame with occasional swirling (hood). Maintain the temperature of the mixture just below the boiling point for 5 min. During this treatment a change can be noted as the sulfonyl chloride undergoes transformation to a more pasty suspension of the amide. Cool the suspension well in an ice bath, collect the *p*-acetaminobenzenesulfonamide by suction filtration, press the cake on the funnel, and allow it to drain thoroughly. Any excess water will unduly dilute the acid used in the next step.

Transfer the still-moist amide to the well-drained reaction flask, add 5 mL of concentrated hydrochloric acid and 10 mL of water (for 5 g of aniline), boil the mixture gently until the solid has all dissolved (5–10 min), and then continue the heating at the boiling point for 10 min longer (do not evaporate to dryness). The solution when cooled to room temperature should deposit no solid amide, but, if it is deposited, heating should be continued for a further period. The cooled solution of sulfanilamide hydrochloride is shaken with decolorizing charcoal and filtered by suction.

Place the solution in a beaker, and cautiously add an aqueous solution of 5 g of sodium bicarbonate with stirring to neutralize the hydrochloride. After the foam has subsided, test the suspension with litmus, and, if it is still acidic, add more bicarbonate until the neutral point is reached. Cool thoroughly in ice and collect the granular, white precipitate of sulfanilamide. The crude product (mp 161–163°C) on crystallization from alcohol or water affords pure sulfanilamide, mp 163–164°C, with about 90% recovery. Determine the mp and calculate the overall yield from the starting material.

The IR spectrum (Fig. 46.11) of sulfanilamide is found at the end of the chapter.

Cleaning Up   Add the water from the gas trap to the combined aqueous filtrates from all reactions, and neutralize the solution by adding either 10% hydrochloric acid or sodium carbonate. Flush the neutral solution down the drain with a large excess of water. Any spilled drops of chlorosulfonic acid should be covered with sodium carbonate and the powder collected in a beaker, dissolved in water, and flushed down the drain.

Questions

1. Why is an acetyl group added to aniline (making acetanilide) and then removed to regenerate the amine group in sulfanilamide?

2. What happens when chlorosulfonic acid comes in contact with water?

3. Acetic anhydride, like any anhydride, reacts with water to form a carboxylic acid. How then is it possible to carry out an acetylation in aqueous
   solution? What is the purpose of the hydrochloric acid and the sodium
   acetate in this reaction?

4. What happens when *p*-acetaminobenzenesulfonyl chloride is allowed to
   stand for some time in contact with water?

5. Assign the groups of peaks at 7.0, 7.2, and 7.4 ppm to specific protons in
   the NMR spectrum of acetanilide (Fig. 46.10).

6. Assign the peaks at 3470 cm^{-1} and 1619 cm^{-1} in the IR spectrum of sulfanilamide (Fig. 46.11).

7. At 98°C the vapor pressure of water is 720 mm Hg and that of aniline is 40
   mm Hg. How much aniline steam distills with each gram of water collected?

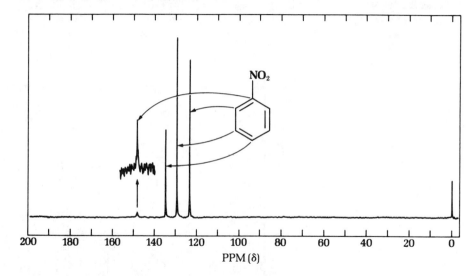

FIG. 46.6 ^{13}C NMR spectrum of nitrobenzene (22.6 MHz).

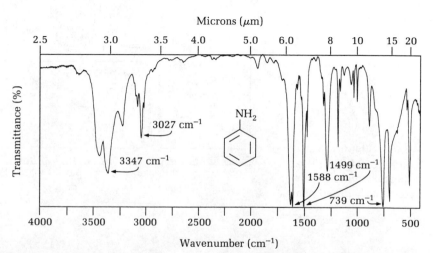

FIG. 46.7   Infrared spectrum of aniline.

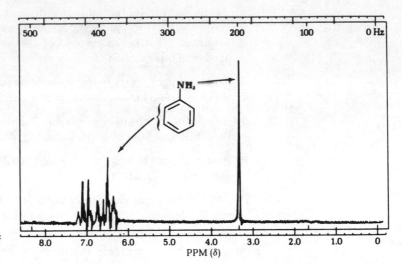

FIG. 46.8 ¹H NMR spectrum of aniline (60 MHz).

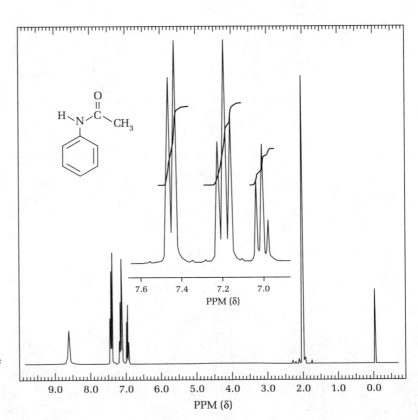

FIG. 46.9 ¹H NMR spectrum of acetanilide (250 MHz). The amide proton shows a characteristically broad peak.

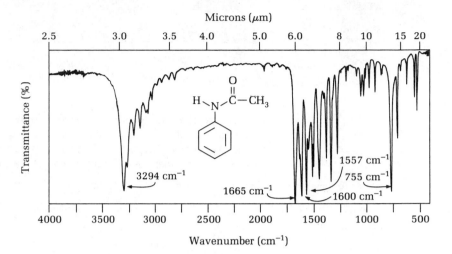

FIG. 46.10 IR spectrum of acetanilide (KBr disk).

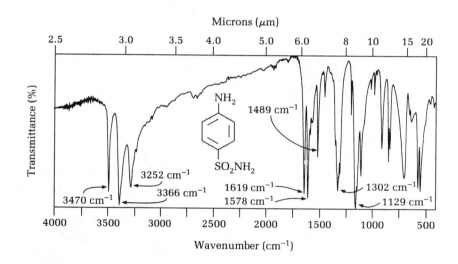

FIG. 46.11 IR spectrum of sulfanilamide (KBr disk).

Dyes and Dyeing[1]

Prelab Exercise: Operating on the simple hypothesis that the intensity of a dye on a fiber will depend on the number of strongly polar or ionic groups in the fiber molecule, predict the relative intensities produced by Methyl Orange when it is used to dye a variety of different fibers, such as are found in the Multifiber Fabric.

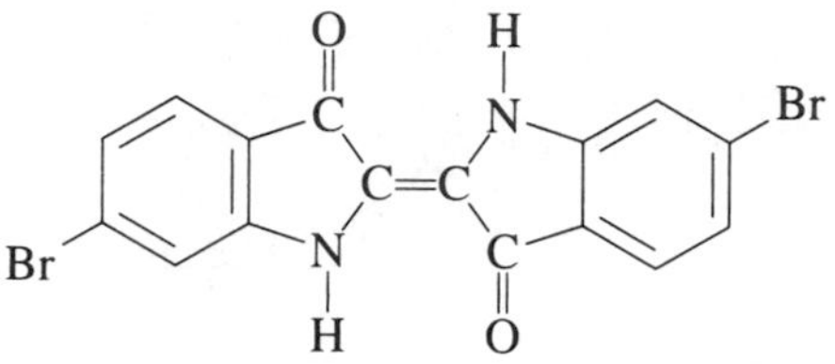

Tyrian Purple
(6,6'-Dibromoindigo)

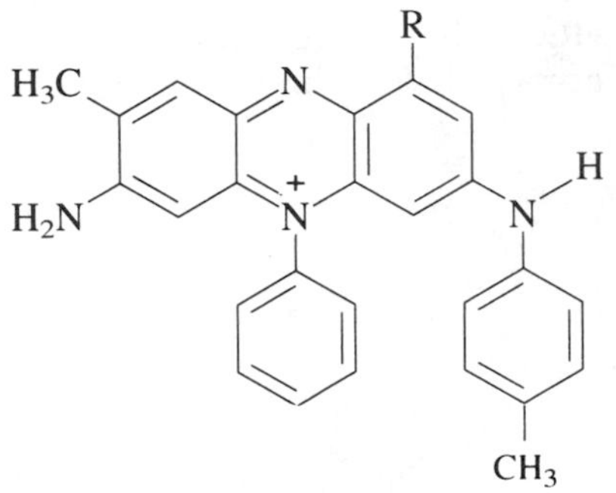

Perkin's Mauve[2]
R = H or CH$_3$

Since prehistoric times humans have been dyeing cloth. The "wearing of the purple" has long been synonymous with royalty, attesting to the cost and rarity of Tyrian purple, a dye derived from the sea snail *Murex brandaris*. The organic chemical industry originated with William Henry Perkin's discovery of the first synthetic dye, Perkin's Mauve, in 1856. The correct structure of this dye has only recently been elucidated.[2]

Dyes are absorbed from solution, usually aqueous, by the fiber. A natural fiber such as cotton has a surface area of about 5 acres per pound (4.4 hectares per kilogram). The dye penetrates the pores in the fiber and is bound to the fiber by electrostatic forces, by van der Waals attraction, by hydrogen bonding, and, in the case of fiber-reactive dyes, by covalent bonds. A good dye must be fast to light, heat, and washing; that is, it must not fade, sublime away, or come off during washing.

In this experiment several dyes are synthesized, and these and other dyes are used to dye a representative group of natural and synthetic fibers. You will

---

1. For a detailed discussion of the chemistry of dyes and dyeing, see *Topics in Organic Chemistry*, by Louis F. Fieser and Mary Fieser, Reinhold Publishing Corp., New York, 1963, pp. 357–417.

2. Otto Meth-Cohn and Mandy Smith, *J. Chem. Soc. Perkin 1*, 5 (1994).

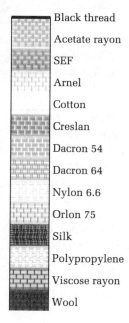

Black thread

Acetate rayon

SEF

Arnel

Cotton

Creslan

Dacron 54

Dacron 64

Nylon 6.6

Orlon 75

Silk

Polypropylene

Viscose rayon

Wool

FIG. 47.1   Multifiber Fabric.

receive several pieces of Multifiber Fabric[3] (Fig. 47.1), which has 13 strips of different fibers woven into it. Below the black thread at the top, the fibers are acetate rayon (cellulose di- or triacetate), SEF (Monsanto's modacrylic Self-Extinguishing Fiber), Arnel (cellulose triacetate), cotton, Creslan (polyacrylonitrile), Dacron 54 and 64 (polyester without and with a brightener), nylon 6.6 (polyamide), Orlon 75 (polyacrylonitrile), silk (polyamide), polypropylene, viscose rayon (regenerated cellulose), and wool (polyamide).

Acetate rayon is cellulose (from any source) in which about two of the hydroxyl groups in each unit have been acetylated. This renders the polymer soluble in acetone from which it can be spun into fiber. The smaller number of hydroxyl groups in acetate rayon compared to cotton makes direct dyeing of rayon more difficult than cotton.

Cotton is pure cellulose. Nylon is a polyamide and made by polymerizing adipic acid and hexamethylenediamine. The nylon polymer chain can be prepared with one acid and one amine group at the termini or with both acids or both amines. Except for these terminal groups, there are no polar centers in nylon, and consequently it is difficult to dye. Similarly Dacron, a polyester made by polymerizing ethylene glycol and terephthalic acid, has few polar centers within the polymer and consequently is difficult to dye. Even more difficult to dye is Orlon, a polymer of acrylonitrile. Wool and silk are polypeptides crosslinked with disulfide bridges. The acidic and basic amino acids (e.g., glutamic acid and lysine) provide many polar groups in wool and silk to which a dye can bind, making these fabrics easy to dye. In this experiment note the marked differences in shade produced by the same dye on different fibers.

Cellulose (Cotton, R=H)
Acetylated cellulose (Acetate rayon, R=OAc)

Wool (R=amino acid residue)

**Polyethylene glycol terephthalate
(Dacron)**

Nylon

**Polyacrylonitrile
(Orlon)**

Polypropylene

---

3. See *Instructor's Guide.*

Part 1. Dyes

The most common dyes are the azo dyes, formed by coupling diazotized amines to phenols. The dye can be made in bulk, or, as we shall see, the dye molecule can be developed on and in the fiber by combining the reactants in the presence of the fiber.

One dye, Orange II, is made by coupling diazotized sulfanilic acid with 2-naphthol in alkaline solution; another, Methyl Orange, is prepared by coupling the same diazonium salt with *N,N*-dimethylaniline in a weakly acidic solution. Methyl Orange is used as an indicator, because it changes color at pH 3.2–4.4. The change in color is due to transition from one chromophore (azo group) to another (quinonoid system).

You are to prepare one of these two dyes and then exchange samples with a neighbor and do the tests with both dyes. Both substances dye wool, silk, and skin, and you must work carefully to avoid getting them on your hands or clothes. The dye will eventually wear off your hands, or you can clean your hands by soaking them in warm, slightly acidic (H_2SO_4) permanganate solution until heavily stained with manganese dioxide and then removing the stain in a bath of warm, dilute bisulfite solution.

Experiments

1. Diazotization of Sulfanilic Acid

Microscale Procedure

Sulfanilic acid
MW 173.19

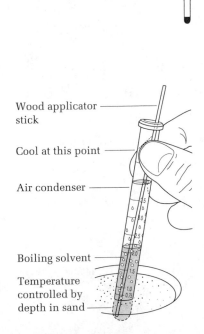

Wood applicator stick

Cool at this point

Air condenser

Boiling solvent

Temperature controlled by depth in sand

FIG. 47.2 Reaction tube with boiling stick to promote even boiling.

In a 10×100 mm reaction tube (Fig. 47.2) dissolve, by boiling, 120 mg of sulfanilic acid monohydrate in 1.25 mL of 2.5% sodium carbonate solution (or use 35 mg of anhydrous sodium carbonate and 1.25 mL of water). Cool the solution to room temperature, add 50 mg of sodium nitrite, and stir until it is dissolved. Cool the tube in ice, and add to it with thorough stirring a mixture of 0.75 g of ice and 0.125 mL of concentrated hydrochloric acid. In a minute or two a powdery white precipitate of the diazonium salt should separate, and the material is then ready for use. The product is not collected but is used in the preparation of the dyes Orange II and/or Methyl Orange while in suspension. It is more stable than most diazonium salts and will keep for a few hours.

Macroscale Procedure

Avoid skin contact with diazonium salts. Some diazonium salts are explosive when dry. Always use in solution.

In a 50-mL Erlenmeyer flask dissolve, if necessary by boiling, 2.4 g of sulfanilic acid monohydrate in 25 mL of 2.5% sodium carbonate solution (or use 0.66 g of anhydrous sodium carbonate and 25 mL of water). Cool the solution under the tap, add 0.95 g of sodium nitrite, and stir until it is dissolved. Pour the solution into a flask containing about 15 g of ice and 2.5 mL of concentrated hydrochloric acid. In a minute or two a powdery white precipitate of the diazonium salt should separate; the material is then ready for use. The product is not collected but is used in the preparation of the dye Orange II and/or Methyl Orange while in suspension. It is more stable than most diazonium salts and will keep for a few hours.

MICROSCALE

2. Orange II (1-*p*-Sulfobenzeneazo-2-naphthol Sodium Salt)

$$N=N-\!\!\!\!\!\!\bigcirc\!\!\!-SO_3^-\overset{+}{Na}$$

OH

Orange II

Handle 2-naphthol with care, in the hood. Do not breathe the dust or allow skin contact. Carcinogen.

In a 10-mL Erlenmeyer flask dissolve 90 mg of 2-naphthol in 0.5 mL of 3 M sodium hydroxide solution, and transfer to this solution, with *thorough* stirring, the suspension of diazotized sulfanilic acid prepared in the preceding experiment. Rinse all the diazonium salt into the naphthol solution with a few drops of cold water. Coupling occurs very rapidly, and the dye, being a sodium salt, separates easily from the solution because a considerable excess of sodium ion from the carbonate, the nitrite, and the alkali is present. Stir the crystalline paste thoroughly to effect good mixing, and after 5 to 10 min, heat the mixture in a beaker of boiling water until the solid dissolves. Add 0.25 g of sodium chloride to further decrease the solubility of the product, and bring this all into solution by heating and stirring. Allow the flask to cool to near room temperature undisturbed, and then cool it in ice. Collect the product on the Hirsch funnel bearing a filter paper (Fig. 47.3). Use saturated sodium chloride solution rather than water to rinse out the flask and to wash the filter cake free of the dark-colored mother liquor. The filtration is somewhat slow, and transfer to the Hirsch funnel is difficult.

The product dries slowly, and it contains about 20% sodium chloride. The

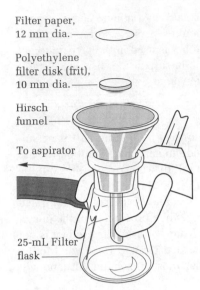

Filter paper, 12 mm dia.

Polyethylene filter disk (frit), 10 mm dia.

Hirsch funnel

To aspirator

25-mL Filter flask

FIG. 47.3 Filter flask with Hirsch funnel, polyethylene frit, and filter paper.

MACROSCALE

Choice of 2 or 3

CAUTION: Handle 2-naphthol with care, in the hood. Do not breathe the dust or allow skin contact. Carcinogen.

crude yield is thus not significant, and the material need not be dried before being purified. This particular azo dye is too soluble to be crystallized from water; it can be obtained in a fairly satisfactory form by adding saturated sodium chloride solution to a hot, filtered solution in water and cooling, but the best crystals are obtained from aqueous ethanol. Transfer the filter cake to a reaction tube, wash the material from the filter paper and funnel with 1 mL of water, and bring all the solid into solution at the boiling point. It may be necessary to add another 0.25 mL (no more) of water in the course of filtering this hot solution through the micro Hirsch funnel equipped with a polyethylene frit (see Fig. 3.17). Collect the filtrate in a reaction tube, add 2.5 to 3 mL of ethanol, and allow crystallization to proceed as the tube cools slowly to room temperature. Cool well in ice before collecting the product. Rinse the tube with some of the filtrate, and, finally, wash the product with a small quantity of ethanol. The yield of pure, crystalline material should be about 150 mg. Orange II separates from aqueous alcohol as the dihydrate, containing two molecules of water of crystallization, and allowance for this should be made in calculating the yield. If the water of hydration is eliminated by drying at 120°C, the material becomes fiery red.

Cleaning Up The filtrate from the reaction, although highly colored, contains little dye but is very soluble in water. It can be diluted with a large quantity of water and flushed down the drain; or, with the volume kept as small as possible, it can be placed in the aromatic amines hazardous waste container; or it can be reduced with tin(II) chloride (see Experiment 7). The crystallization filtrate should go into the organic solvents container. Wash the Hirsch funnel *thoroughly*.

3. Orange II (1-*p*-Sulfobenzeneazo-2-naphthol Sodium Salt)

$$N=N--SO_3^-Na^+$$
OH

Orange II

In a 250-mL beaker, dissolve 1.8 g of 2-naphthol in 10 mL of cold 3 *M* sodium hydroxide solution, and pour into this solution, with stirring, the suspension of diazotized sulfanilic acid from Experiment 1. Rinse the Erlenmeyer flask with a small amount of water, and add it to the beaker. Coupling occurs very rapidly, and the dye, being a sodium salt, separates easily from the solution because a considerable excess of sodium ion from the carbonate, the nitrite, and the alkali is present. Stir the crystalline paste thoroughly to effect good mixing, and, after 5–10 min, heat the mixture until the solid dissolves. Add 5 g of sodium chloride to further decrease the solubility of the product, bring this all into solution by heating and stirring, set the beaker in a pan of ice and water, and let the solution cool undisturbed. When near room temperature, cool further by stirring and collect the product on a Büchner funnel. Use saturated sodium chloride solution rather than water for rinsing the material out of the beaker and for washing the filter cake free of the dark-colored mother liquor. The filtration is somewhat slow.[4]

The product dries slowly; it contains about 20% of sodium chloride. The crude yield is thus not significant, and the material need not be dried before being purified. This particular azo dye is too soluble to be crystallized from water; it can be obtained in a fairly satisfactory form by adding saturated sodium chloride solution to a hot, filtered solution in water and cooling, but the best crystals are obtained from aqueous ethanol. Transfer the filter cake to a beaker, wash the material from the filter paper and funnel with water, and bring the cake into solution at the boiling point. Avoid a large excess of water, but use enough to prevent separation of solid during filtration (use about 25 mL). Filter by suction through a Büchner funnel that has been preheated on the steam bath. Pour the filtrate into an Erlenmeyer flask, rinse the filter flask with a small quantity of water, add it to the flask, estimate the volume, and if greater than 30 mL evaporate by boiling. Cool to 80°C, add 50–60 mL of ethanol, and allow crystallization to proceed. Cool the solution well before collecting the product. Rinse the beaker with mother liquor, and wash finally with a little ethanol. The yield of pure, crystalline material is about 3.4 g. Orange II separates from aqueous alcohol with two molecules of water of crystallization, and allowance for this should be made in calculating the yield. If the water of hydration is eliminated by drying at 120°C, the material becomes fiery red.

Extinguish flames!

Cleaning Up   The filtrate from the reaction, although highly colored, contains little dye but is very soluble in water. It can be diluted with a large quantity of water and flushed down the drain; or, with the volume kept as small as possible, it can be placed in the aromatic amines hazardous waste container or it can be reduced with tin(II) chloride (see Experiment 5). The crystallization filtrate should go into the organic solvents container.

4. Methyl Orange (*p*-Sulfobenzeneazo-4-dimethylaniline Sodium Salt)

Diazotized sulfanilic acid
4-Diazobenzenesulfonic acid

N,N-**Dimethylaniline**
MW 121.18
bp 194°C

Methyl Orange
4-[4-(Dimethylamino)phenylazo] benzenesulfonic acid,
sodium salt
MW 327.34
λ_{max} 507 nm

---

4. If the filtration must be interrupted, fill the funnel, close the rubber suction tubing (while the aspirator is still running) with a screw pinchclamp placed close to the filter flask, and then disconnect the tubing from the trap. Set the unit aside; thus, suction will be maintained, and filtration will continue.

Microscale Procedure

In a 10-mL Erlenmeyer flask, mix 75 mg of dimethylaniline and 65 mg of acetic acid. Add the suspension of diazotized sulfanilic acid, with stirring, to the solution of dimethylaniline acetate, rinsing out the last portions with a few drops of water. Stir the mixture thoroughly, and within a few minutes, the red, acid-stable form of the dye should separate. A stiff paste should result in 5 to 10 min, at which time 1 mL of 3 M sodium hydroxide is added to produce the orange sodium salt. Heat the mixture to the boiling point with constant stirring (to avoid bumping), when a large part of the dye should dissolve. This might better be done by placing the flask in a small beaker of boiling water. Allow the flask to cool slowly to room temperature, and then cool it thoroughly in ice before collecting the product by vacuum filtration on the Hirsch funnel bearing a piece of filter paper. Use saturated sodium chloride solution rather than water to rinse the flask and to wash the dark mother liquor from the filter cake.

The crude product need not be dried but can be crystallized from water after making preliminary solubility tests to determine the proper conditions. The yield should be about 125 to 150 mg. Recrystallized material need not necessarily be used for the dyeing process.

Cleaning Up   The highly colored filtrates from the reaction and crystallization are very water soluble. After dilution with a large quantity of water, they can be flushed down the drain, because the amount of solid is small. Alternatively, the combined filtrates should be placed in the hazardous waste container, or the mixture can be reduced with tin(II) chloride (see Experiment 5). Wash the Hirsch funnel *thoroughly*.

Macroscale Procedure

In a test tube, thoroughly mix 1.6 mL of dimethylaniline and 1.25 mL of glacial acetic acid. To the suspension of diazotized sulfanilic acid from Experiment 1 contained in a 250-mL beaker add, with stirring, the solution of dimethylaniline acetate. Rinse the test tube with a small quantity of water, and add it to the beaker. Stir and mix thoroughly, and within a few minutes the red, acid-stable form of the dye should separate. A stiff paste should result in 5–10 min; 18 mL of 3 *M* sodium hydroxide solution is then added to produce the orange sodium salt. Stir well, and heat the mixture to the boiling point, when a large part of the dye should dissolve. Place the beaker in a pan of ice and water, and allow the solution to cool undisturbed. When cooled thoroughly, collect the product on a Büchner funnel, using saturated sodium chloride solution rather than water to rinse the flask and to wash the dark mother liquor from the filter cake.

The crude product need not be dried but can be crystallized from water after making preliminary solubility tests to determine the proper conditions. The yield is about 2.5–3 g.

Methyl Orange is an acid/base indicator.

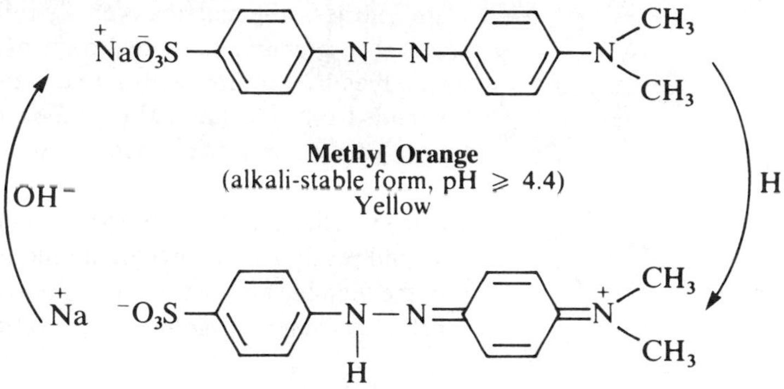

Methyl Orange
(alkali-stable form, pH ≥ 4.4)
Yellow

OH⁻   H⁺

Methyl Orange
(acid-stable form, pH ≤ 3.2)
Red

Cleaning Up   The highly colored filtrates from the reaction and crystallization are very water soluble. After dilution with a large quantity of water, they can be flushed down the drain because the amount of solid is small. Alternatively, the combined filtrates should be placed in the hazardous waste container, or the mixture can be reduced with tin(II) chloride (see Experiment 5).

MICROSCALE AND MACROSCALE   *For Further Investigation*

Following the procedure for the synthesis of Orange II, couple diazotized sulfanilic acid with one or more of the following naphthols: 4-amino-5-hydroxy-2,7-naphthalenedisulfonic acid, monosodium salt; 6-amino-4-hydroxy-2-naphthalenesulfonic acid; 7-amino-4-hydroxy-2-naphthalenesulfonic acid; 4-hydroxy-1-naphthalenesulfonic acid, sodium salt; or 6-hydroxy-2-naphthalenesulfonic acid, sodium salt. Despite the apparent complexity of these molecules, they are not expensive; they are used for the commercial synthesis of dyes. The many sulfonic acid groups have been added to make the dyes water soluble; addition of solid sodium chloride "salts out" the dye, rendering it insoluble.

MICROSCALE AND MACROSCALE   **5. Tests**

Solubility and Color

Compare the solubility in water of Orange II and Methyl Orange, and account for the difference in terms of structure. Treat the first solution with alkali, and note the change in shade due to salt formation; to the other solution alternately add acid and alkali.

Reduction

Characteristic of an azo compound is the ease with which the molecule is cleaved at the double bond by reducing agents to give two amines. Since amines are colorless, the reaction is easily followed by the color change. The reaction is of use in preparation of hydroxyamino and similar compounds, in analysis of azo dyes by titration with a reducing agent, and in identification of azo compounds from an examination of the cleavage products.

This reaction can, if necessary, be run on five times the indicated scale. Dissolve about 0.1 g of tin(II) chloride in 0.2 mL of concentrated hydrochloric acid, add a small quantity of the azo compound (20 mg), and heat. A colorless solution should result, and no precipitate should form on adding water. The aminophenol or the diamine products are present as the soluble hydrochlorides; the other product of cleavage, sulfanilic acid, is sufficiently soluble to remain in solution.

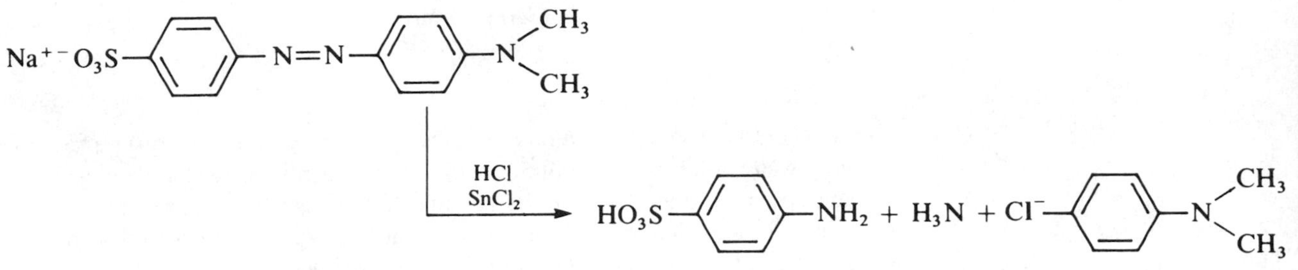

$$Na^{+\,-}O_3S-\!\!\left\langle\bigcirc\right\rangle\!\!-N\!=\!N-\!\!\left\langle\bigcirc\right\rangle\!\!-N\!\!\begin{array}{c}CH_3\\ \diagdown\\ CH_3\end{array}$$

$$\xrightarrow[\text{SnCl}_2]{\text{HCl}} \quad HO_3S-\!\!\left\langle\bigcirc\right\rangle\!\!-NH_2 \;+\; H_3N \;+\; Cl^- \!\!-\!\!\left\langle\bigcirc\right\rangle\!\!-N\!\!\begin{array}{c}CH_3\\ \diagdown\\ CH_3\end{array}$$

Cleaning Up   Dilute the reaction mixture with water, neutralize with sodium carbonate, and remove the solids by vacuum filtration. The solids go in the aromatic amines hazardous waste container, and the filtrate can be flushed down the drain.

Part 2. Dyeing

Experiments

The following experiments can, if necessary, be run on ten times the indicated amounts. This is one experiment where disposable gloves (polyethylene or vinyl) might be advisable.

MICROSCALE AND MACROSCALE

With good laboratory technique your hands will not be dyed; use care.

1. Direct Dyes

The sulfonate groups on the Methyl Orange and Orange II molecules are polar and thus enable these dyes to combine with polar sites in the fibers. Wool and silk have many polar sites on their polypeptide chains and hence bind strongly to a dye of this type. Martius Yellow, picric acid, and eosin are also highly polar dyes and thus dye directly to wool and silk.

Orange II or Methyl Orange

The dye bath is prepared from 50 mg of Orange II or Methyl Orange, 0.5 mL of 3 M sodium sulfate solution, 15 mL of water, and 5 drops of 3 M sulfuric acid in a 30-mL beaker. Place a piece of test fabric, a strip 3/4-in. wide, in the bath for 5 min at a temperature near the boiling point. Remove the fabric from the dye bath, allow it to cool, and then wash it thoroughly with soap under running water before drying it.

Dye untreated test fabric and one or more of the pieces of test fabric that have been treated with a mordant following this same procedure. See Experiment 3 in this part for application of mordants.

Picric acid

Picric Acid or Martius Yellow

Martius Yellow

Dissolve 50 mg of one of these acidic dyes in 15 mL of hot water to which a few drops of dilute sulfuric acid have been added. Heat a piece of test fabric in this bath for 1 min, then remove it with a stirring rod, rinse well, scrub with soap and water, and dry. Describe the results.

Eosin

Dissolve 10 mg of sodium eosin in 20 mL of water, and dye a piece of test fabric by heating it with the solution for about 10 min. Eosin is the dye used in red ink. Also dye pieces of mordanted cloth in eosin (see Experiment 3). Wash and rinse the dyed cloth in the usual way.

Eosin A
(λ_{max} 516, 483 nm)

Cleaning Up   In each case dilute the dye bath with a large quantity of water and flush it down the drain.

**MICROSCALE
AND MACROSCALE**

2. Substantive Dyes: Congo Red

Cotton and the rayons do not have the anionic and cationic carboxyl and amine groups of wool and silk and hence do not dye well with direct dyes; but they can be dyed with substances of rather high molecular weight showing colloidal properties. Such dyes probably become fixed to the fiber by hydrogen bonding. Such a dye is Congo Red, a substantive dye.

Dissolve 10 mg of Congo Red in 40 mL of water, add about 0.1 mL each of 3 M solutions of sodium carbonate and sodium sulfate, heat to a temperature just below the boiling point, and introduce a piece of test fabric. At the end of 10 min, remove the fabric, and wash in warm water with soap as long as the dye continues to be removed. Place pieces of the dyed material in very dilute hydrochloric acid solution, and observe the result. Rinse and wash the material with soap.

Congo Red
(λ_{max} 497 nm)

Cleaning Up   The dye bath should be diluted with water and flushed down the drain.

3. Mordant Dyes

One of the oldest known methods of producing wash-fast colors involves the use of metallic hydroxides, which form a link, or mordant (L. *mordere,* to bite), between the fabric and the dye. Other substances, such as tannic acid, also function as mordants. The color of the final product depends on both the dye used and the mordant. For instance, the dye Turkey Red (alizarin) is red with an aluminum mordant, violet with an iron mordant, and brownish-red with a chromium mordant. Some important mordant dyes possess a structure based on triphenylmethane, as do Crystal Violet and Malachite Green.

*Chromium functioning as
a mordant*

Alizarin
1,2-Dihydroxyanthraquinone

Applying Mordants—Tannic Acid, Fe, Sn, Cr, Cu, Al

Caution: $K_2Cr_2O_2$ *is toxic and the dust a cancer-suspect agent.*

Mordant pieces of test fabric or wool yarn by allowing them to stand in a hot (nearly boiling) solution of 0.1 g of tannic acid in 50 mL of water for 30 min. The tannic acid mordant must now be fixed to the cloth; otherwise it would wash out. For this purpose, transfer the cloth or yarn to a hot bath made from 20 mg of potassium antimonyl tartrate (tartar emetic) in 20 mL of water. After 5 min, wring the cloth and dry it as much as possible over a warm hot plate.

Mordant 1/2-in. strips of test cloth or yarn in the following mordants, which are 0.1 M solutions of the indicated salts. Immerse pieces of cloth in the solutions, which are kept near the boiling point, for about 15 to 20 min or longer. The mordants are ferrous sulfate, stannous chloride, potassium dichromate, copper sulfate, and potassium aluminum sulfate (alum). The alum and dichromate solutions should also contain 0.05 M oxalic acid. These mordanted pieces of cloth or yarn can then be dyed with alizarin (1,2-dihydroxy-anthraquinone) and either Methyl Orange or Orange II in the usual way.

Cleaning Up   Mix the mordant baths. The Fe^{2+} and Sn^{2+} will reduce the Cr^{6+} to Cr^{3+}. The mixture can then be diluted with water and flushed down the drain because the quantity of metal ions is extremely small. Alternatively, precipitate them as the hydroxides, collect by vacuum filtration, and place the solid in the hazardous waste container.

N(CH₃)₂

Malachite Green
(λ_{max} 617 nm)

Dyeing with a Triphenylmethane Dye—Crystal Violet or Malachite Green

A dye bath is prepared by dissolving 10 mg of either Crystal Violet or Malachite Green in 20 mL of boiling water. Dye the mordanted cloth or yarn in this bath for 5–10 min at a temperature just below the boiling point. Dye another piece of cloth that has not been mordanted, and compare the two. In each case allow as much of the dye to drain back into the beaker as possible, and then, using glass rods, wash the dyed cloth or yarn under running water with soap, blot, and dry.

Cleaning Up   The stains on glass produced by triphenylmethane dyes can be removed with a few drops of concentrated hydrochloric acid and washing with water, as HCl forms a di- or trihydrochloride more soluble in water than the original monosalt.

The dye bath and acid washings are diluted with water and flushed down the drain because the quantity of dye is extremely small.

Crystal Violet
(λ_{max} 591,540 nm)

MICROSCALE AND MACROSCALE

4. Developed Dyes

A superior method of applying azo dyes to cotton, patented in England in 1880, is that in which cotton is soaked in an alkaline solution of a phenol and then in an ice-cold solution of a diazonium salt; the azo dye is developed directly on the fiber. The reverse process (ingrain dyeing) of impregnating cotton with an amine, which is then diazotized and developed by immersion in a solution of the phenol, was introduced in 1887. The first ingrain dye was Primuline Red, obtained by coupling the sulfur dye Primuline, after application to the cloth and diazotization, with 2-naphthol. Primuline (substantive to cotton) is a complex thiazole, prepared by heating *p*-toluidine with sulfur and then introducing a solubilizing sulfonic acid group.

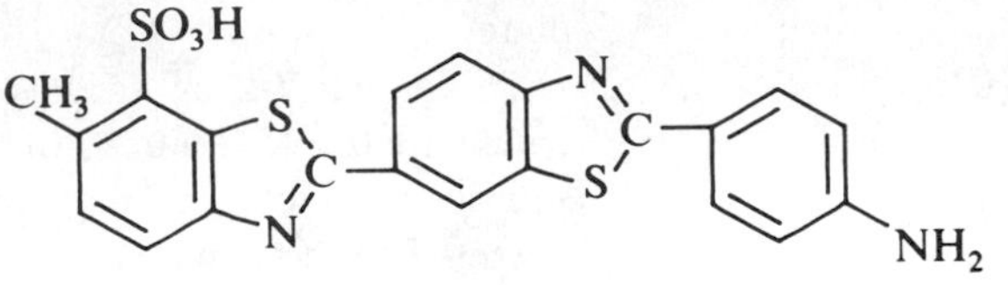

Primuline

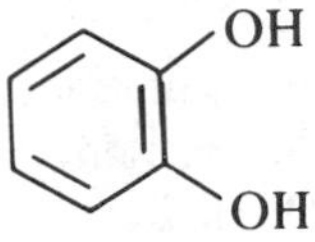

Primuline Red

Resorcinol

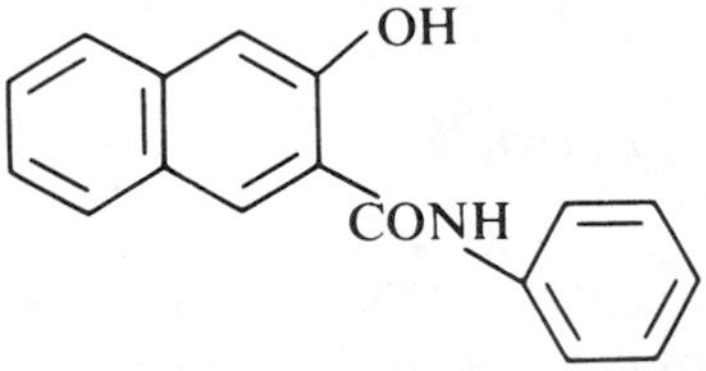

Naphthol AS

Dye three pieces of cotton cloth in a solution of 20 mg of Primuline and 0.5 mL of sodium carbonate solution in 50 mL of water at a temperature just below the boiling point for 15 min. Wash the cloth twice in about 50 mL of water. Prepare a diazotizing bath by dissolving 20 mg of sodium nitrite in 50 mL of water containing a little ice, and, just before using the bath, add 0.5 mL of concentrated hydrochloric acid. Allow the cloth dyed with Primuline to stay in this diazotizing bath for about 5 min. Now prepare three baths for the coupling reaction. Dissolve 10 mg of 2-naphthol in 0.2 mL of 5% sodium hydroxide solution and dilute with 10 mL of water; prepare similar baths from phenol, resorcinol, Naphthol AS, or other phenolic substances.

Transfer the cloth from the diazotizing bath to a beaker containing about 50 mL of water and stir. Put one piece of cloth in each of the developing baths, and allow them to stay for 5 min. Primuline coupled to 2-naphthol gives the dye called Primuline Red. Draw the structure of the dye.

Para Red, an Ingrain Color

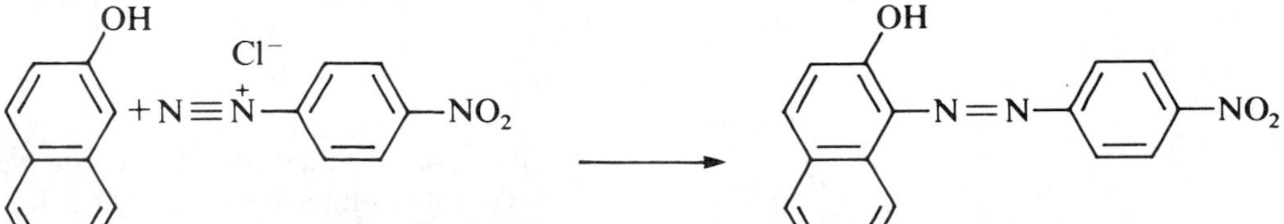

2-Naphthol   **4-Nitrobenzene diazonium chloride**   **Para Red**

A solution is prepared by suspending 300 mg of 2-naphthol in 10 mL of water, stirring well, and adding 3 *M* sodium hydroxide solution, a drop at a time, until the naphthol just dissolves. Do not add excess alkali. The material to be dyed is first soaked in or painted with this solution and then dried, preferably in an oven.

Prepare a solution of 4-nitrobenzenediazonium chloride as follows: Dissolve 140 mg of 4-nitroaniline in a mixture of 3 mL of water and 0.6 mL of 3 *M* hydrochloric acid by heating. Cool the solution in ice (the hydrochloride of the amine may crystallize), add all at once a solution of 80 mg of sodium nitrite in about 0.5 mL of water, and stir. In about 10 min a clear solution of the diazonium salt will be obtained. Just before developing the dye on the cloth add a solution of 80 mg of sodium acetate in 0.5 mL of cold water. Stir in the acetate well, add 30 mL of water, and immediately add the cloth. The diazonium chloride solution may also be painted onto the cloth.

This is the red dye used for the American flag.

Good results can be obtained by substituting Naphthol-AS for 2-naphthol; in this case it is necessary to warm the Naphthol-AS with alkali and to break the lumps with a flattened stirring rod in order to bring the naphthol into solution.

Cleaning Up   The dye baths should be diluted with water and flushed down the drain because the quantity of dye is extremely small.

MICROSCALE AND MACROSCALE

5. Dyeing with Indigo—The Denim Dye

$$\text{Na}_2\text{S}_2\text{O}_4 + \text{NaOH} \rightleftharpoons \text{O}_2 \text{ (air)}$$

Indigo
Insoluble in water
Deep blue

Indigo white
The leuco form, soluble in water
Colorless or light yellow

Look about you. No matter where in the world you are carrying out these experiments, you will find people wearing cotton clothes dyed with indigo. And this, "the king of dyes," has been used for more than 3500 years, first in India and then in ancient Egypt. It is a unique dye in that it can be abraded from the surface of a fabric, where other dyes penetrate the fiber. This explains why the knees and other parts of blue jeans (dyed exclusively with indigo) will gradually turn white. Advantage of this fact has been taken by manufacturers who sell "stone-washed" denim blue jeans that have been tumbled with pumice to wear off some of the dye (which can then be reused).

Denim, de Nimes, from Nimes, in southern France

The dye originally came from the leaves of the indigo plant, but now it is produced by chemical synthesis. It is one of the *vat dyes,* a term applied to dyes that are reduced to a colorless (leuco) form that is then oxidized on the surface of the fiber. Formerly, the reduction was carried out by fermentation of the plant leaves; now, chemical reducing agents, most commonly sodium hydrosulfite, are used.

Experimental Considerations

Indigo is a dark-blue powder that is completely insoluble in water and is not easily wet by water. To disperse the dye in water, a bit of soap is added to the dye bath. The liquid will appear blue, but the dye is not dissolved, merely suspended. Sodium hydrosulfite ($Na_2S_2O_4$) is then added to the hot dye bath. Because this reducing agent decomposes on storage, it is not easy to state exactly how much should be used. Add the required amount, stir, and then look through the side of the beaker. If the solution is not transparent and light yellow in color, add more of the reducing agent. From the top the liquid will still appear to be dark blue because of a film of the oxidized dye on the surface of the solution.

Procedure

Use 100 mg of indigo and a drop of detergent or a pinch of soap powder. Boil the dye with 50 mL of water, 2.5 mL of 3 *M* sodium hydroxide solution, and about 0.5 g of sodium hydrosulfite until the dye is reduced. At this point a clear solution will be seen through the side of the beaker. Add more sodium hydrosulfite if necessary. Introduce a piece of cloth, and boil the solution gently for 10 min. Rinse the cloth well in water, and allow it to dry. To increase the intensity of the dye, repeat the process several times with no drying. Describe what happens during the drying process.

Other dyes that can be used in this procedure are Indanthrene Brilliant Violet and Indanthrene Yellow.

Cleaning Up   Add household bleach (5.25% sodium hypochlorite solution) to the dye bath to oxidize it to the starting material. The mixture can be diluted with water and flushed down the drain or the small amount of solid removed by filtration and placed in the aromatic amines hazardous waste container.

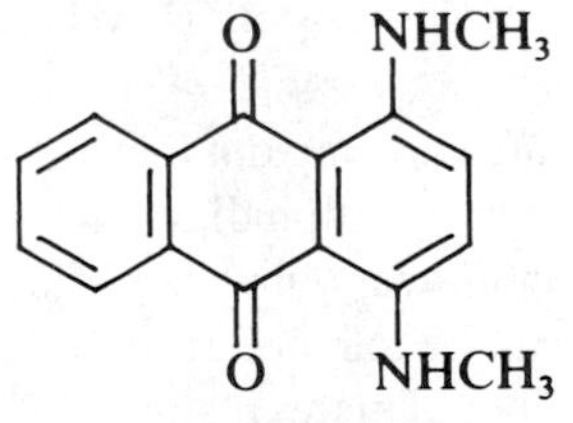

Celliton Fast Blue B

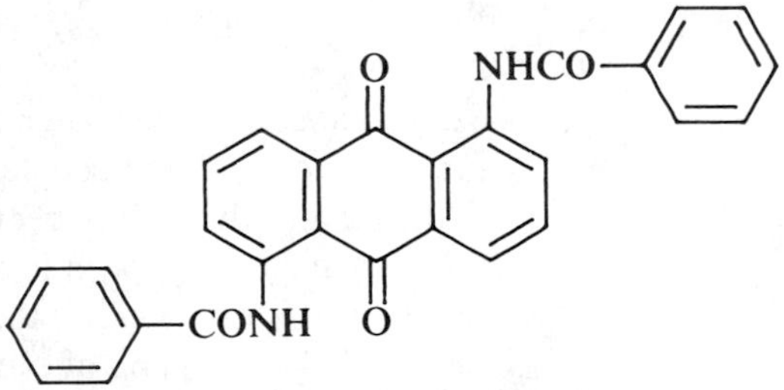

Indanthrene Yellow

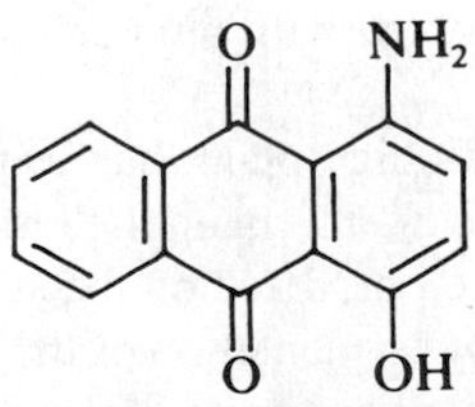

Celliton Fast Pink B

6. Disperse Dyes: Disperse Red

Fibers such as Dacron, acetate rayon, nylon, and polypropylene are difficult to dye with conventional dyes because they contain so few polar groups. These fibers are dyed with substances that are insoluble in water but that, at elevated temperatures (pressure vessels), are soluble in the fiber as true solutions. They are applied to the fiber in the form of a dispersion of finely divided dye (hence the name). The Cellitons are typical disperse dyes.

In this experiment Disperse Red, a brilliant red dye used commercially, is synthesized.

MICROSCALE

Diazotization of 2-Amino-6-methoxybenzothiazole

2-Amino-6-methoxybenzothiazole
MW 180.23
mp 165–167°C

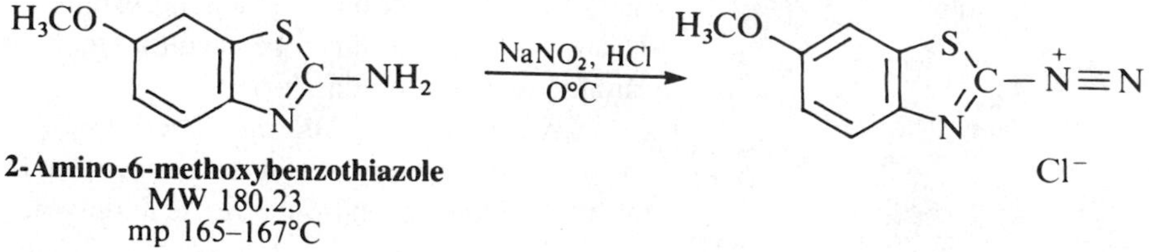

To 135 mg (0.75 mmol) of 2-amino-6-methoxybenzothiazole in 1.5 mL of water in a test tube, add 0.175 mL of concentrated hydrochloric acid, then cool the solution to 0–5°C. To this mixture add, dropwise, an *ice-cold* solution of 55 mg of sodium nitrite that has been dissolved in 0.75 mL of water. The reaction mixture changes color and some of the diazonium salt crystallizes out, but it should not foam. Foaming, caused by the evolution of nitrogen, is an indication the mixture is too warm. Keep the mixture ice-cold until used in the coupling reaction.

MICROSCALE

Disperse Red

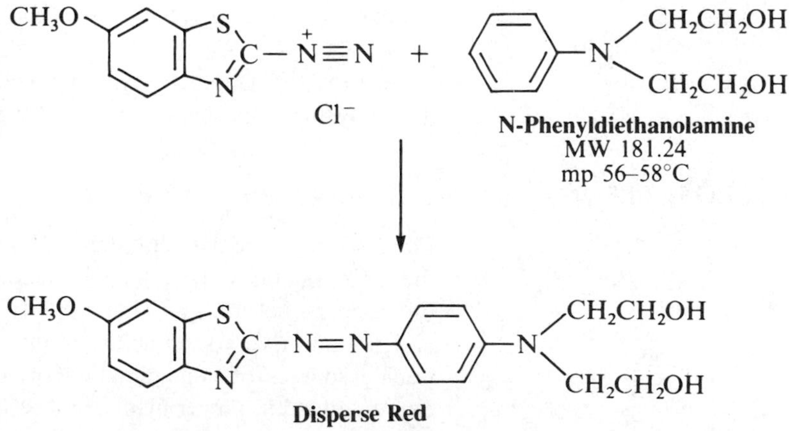

N-Phenyldiethanolamine
MW 181.24
mp 56–58°C

Disperse Red

To 135 mg (0.75 mmol) of *N*-phenyldiethanolamine in 0.75 mL of hot water, add just enough 3 *M* hydrochloric acid to bring the amine into solution. This amount is less than 0.5 mL. Cool the resulting solution to 0°C in ice, and add to it, dropwise and with *very thorough mixing,* the diazonium chloride solution. Mix the solution well by drawing into the Pasteur pipette and expelling it into the cold reaction tube. Allow the mixture to come to room temperature over a period of 10 min, then add 225 mg of sodium chloride, and heat the mixture to boiling. The sodium chloride decreases the solubility of the product in water. Allowing the hot solution to cool slowly to room temperature should afford easily filterable crystals. Collect the dye on the Hirsch funnel, wash it with a few drops of saturated sodium chloride solution, and press it dry on the funnel.

Often the reaction gives a noncrystalline product that looks like purple tar. This is the dye, and it can be used to dye the multifiber test cloth, so don't discard the reaction mixture.

Save the filtrate. Add the material on the filter paper to 50 mL of boiling water, and dye a piece of test cloth for 5 min. Do the same with the filtrate. Wash the cloth with soap and water, rinse and dry it, and compare the results.

Cleaning Up   The dye baths should be diluted with water and flushed down the drain because the quantity of dye is extremely small.

MICROSCALE AND MACROSCALE

7. Test Identification Stain

In the dye industry, a proprietary mixture of three dyes called Test Identification Stain (obtained from Kontes) is used to dye cloth. Prepare a dye bath by dissolving 25 mg of the dye in 25 mL of water, adding a drop of acetic acid, and bring the mixture to a boil. Dye a piece of test cloth in the mixture for 5 min, remove it, immediately wash it under running water, and then scrub it *thoroughly* with soap and water before rinsing and drying. Describe the rather extraordinary result, and suggest how this mixture could be used industrially. Several pieces of test cloth can be dyed in this one dye bath. Analyze the dye mixture by thin-layer chromatography, using 40 : 60 ethanol/hexane as the eluent.

Cleaning Up   Dilute the dye bath extensively with water, and flush it down the drain. Dyeing cloth removes much of the dye added initially.

MICROSCALE

8. Fluorescein and Eosin

Fluorescein, as the name implies, is fluorescent; in fact, it is so fluorescent that the sodium salt in water can be detected under ultraviolet light at concentrations of 0.02 parts per million. This property makes it useful for tracing the paths of underground rivers such as are found in Mammoth Cave in Kentucky or for tracing sources of contamination of drinking water, leaks in the huge condensers found in power plants, and so on. It is also used to visualize scratches on the cornea of the eye. The tetrabromo derivative is eosin, a dye used in lipstick, nail polish, red ink, and as a biological stain.

The synthesis of fluorescein is very straightforward; however, in this experiment we have deliberately omitted the equations for the synthesis, the mechanism of the reaction, the structures of the molecules in acid and base, and the structures of the tetrabromo derivatives. These are left for you to solve or look up in the library.[5]

Synthesis of Fluorescein

In a reaction tube heat 100 mg of zinc chloride in a hot sand bath until no more moisture comes from the tube, then add 200 mg of resorcinol and 140 mg of phthalic anhydride. Put a thermometer into the mixture, and heat it to 180°C. Stir and continue heating for 10 min, at which time the mixture should cease bubbling and become stiff. Cool the mixture to below 100°C, add 2 mL of water and 0.2 mL of concentrated hydrochloric acid, then stir and grind the solid (the process is called *trituration*) while heating it on a steam bath or boiling water bath (not on the sand bath—it will bump). Remove the liquid with a Pasteur pipette, and, after again triturating with another similar portion of water and acid, collect the solid on the Hirsch funnel, wash it well with water, and allow it to dry. A typical yield is 240 mg.

Fluorescence

Add 50 mg of fluorescein to 0.5 mL of 3 *M* sodium hydroxide and 0.5 mL of water; then dilute the mixture to 100 mL. Examine the solution by transmitted light and then by reflected sunlight or ultraviolet light. Make the solution acid, and again examine it under sunlight or UV light.

Tetrabromofluorescein, Eosin: The Red Ink Dye

In a reaction tube place 100 mg of fluorescein and 0.4 mL of ethanol. To this mixture add 215 mg of bromine in 0.5 mL of carbon tetrachloride, dropwise, mixing thoroughly after each drop. After half the bromine is added, a solution of soluble dibromofluorescein results. When all the bromine has been added, the tetrabromo compound will precipitate. After about 15 min, cool the reaction mixture in ice, and remove the solvent with a Pasteur pipette. Wash the product in the reaction tube with a few drops of ice-cold ethanol, and dry it under vacuum. The yield should be about 160 mg.

Prepare the ammonium salt of eosin by exposing the salt to ammonia vapors. Moisten about 80 mg of fluorescein with a few drops of ethanol on a small filter paper. Fold up the paper, and push it into a reaction tube above 0.5 mL of ammonium hydroxide. Cap the tube with a septum. After about an hour, a test portion of the derivative should be completely soluble in water. Dye a piece of test cloth with this dye. Eosin is used commercially to dye silk.

Before you perform this experiment, answer the following questions:

---

5. If you would like to read about the original synthesis, see *J. Prakt. Chem.* **104**, 123, 1922, which will require a reading knowledge of German, and the structure proof in *Comp. Rendu.* **205**, 864, 1937, which will require a reading knowledge of French. Further searching in the library may disclose other papers in English.

1. Write a balanced equation for the reaction of resorcinol with phthalic anhydride.
2. Calculate the number of moles of starting material, determine the limiting reagent, and calculate the theoretical yield.
3. In view of the balanced equation, why is it necessary to dehydrate zinc chloride, a very hygroscopic substance?
4. To what general category of substances does zinc chloride belong that is relevant to its use in this reaction? What two functions does zinc chloride serve?
5. Write a possible mechanism for the synthesis of fluorescein.
6. Write an equation showing the reaction of fluorescein with sodium hydroxide. *Hint:* A quinoid structure is involved.
7. How do you account for the changes in fluorescence in going from an acidic to a basic medium?
8. Compare the color and fluorescence changes of fluorescein with those of the related compound, phenolphthalein.
9. Write a balanced equation for the bromination of fluorescein to give first the dibromo and then the tetrabromo compound. Explain your choices for the positions of bromination.
10. Write a balanced equation for the reaction of tetrabromofluorescein with ammonia.

9. Fiber-Reactive Dyes

Among the newest of the dyes are the fiber-reactive compounds, which form a covalent link to the hydroxyl groups of cellulose. The reaction involves an amazing and little-understood nucleophilic displacement of a chloride ion from the triazine part of the molecule by the hydroxyl groups of cellulose, yet the reaction occurs in aqueous solution.

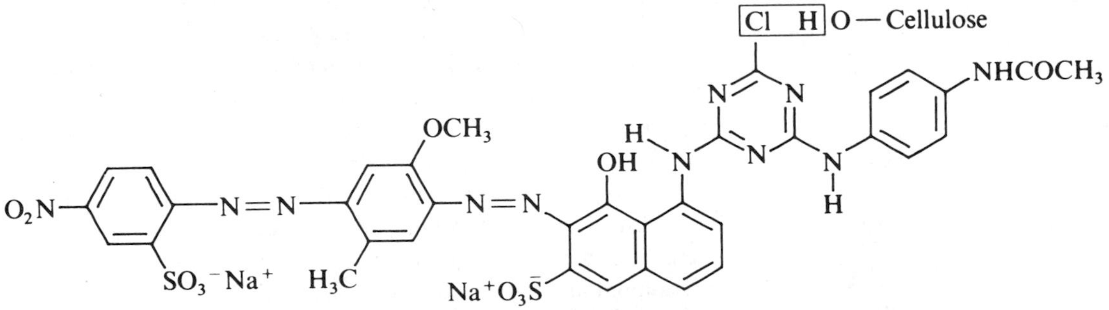

Chlorantin Light Blue 8G

MICROSCALE AND MACROSCALE

10. Optical Brighteners—Fluorescent White Dyes

Most modern detergents contain a blue-white fluorescent dye that is adsorbed on the cloth during the washing process. These dyes fluoresce, that is, absorb ultraviolet light and reemit light in the visible blue region of the spectrum. This blue color counteracts the pale-yellow color of white goods, which develops because of a buildup of insoluble lipids. The modern-day use of optical brighteners has replaced a past custom of using bluing (ferriferrocyanide).

Dyeing with Detergents

Immerse a piece of test fabric in a hot solution (0.5 g of detergent, 200 mL of water) of a commercial laundry detergent that you suspect may contain an

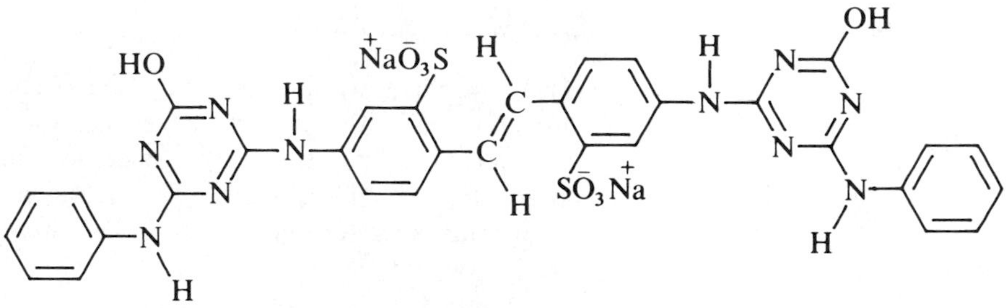

Blankophor B
An optical brightener

optical brightener (e.g., Tide and Cheer) for 15 min. Rinse the fabric thoroughly, dry, and compare with an untreated fabric sample under an ultraviolet lamp.

Cleaning Up   The solution should be diluted with water and flushed down the drain.

Questions _____

1. Write reactions showing how nylon can be synthesized such that it will react with (a) basic dyes and (b) acidic dyes.

2. Draw the resonance form of dimethylaniline that is most prone to react with diazotized sulfanilic acid.

3. Draw a resonance form of indigo that would be present in base.

4. Draw a resonance form of indigo that has been reduced and is therefore colorless.

48

Martius Yellow

Prelab Exercise: Prepare a time line for this experiment, indicating clearly which experiments can be carried out simultaneously.

This experiment was introduced by Louis Fieser of Harvard University over half a century ago. Although the experiment starts with macroscale quantities of materials, the later steps often become microscale experiments and benefit greatly from the use of microscale equipment. It has long been the basis for an interesting laboratory competition. The present author was a winner over 40 years ago.

Starting with 5 g of 1-naphthol, a skilled chemist familiar with the procedures can prepare pure samples of the seven compounds in 3–4 h. In a first trial of the experiment, a particularly competent student, who plans his or her work in advance, can complete the program in two laboratory periods (6 h).

The first compound of the series, Martius Yellow, a mothproofing dye for wool (1 g of Martius Yellow dyes 200 g of wool) discovered in 1868 by Karl Alexander von Martius, is the ammonium salt of 2,4-dinitro-1-naphthol (**1**), shown below. Compound **1** in the series of reactions is obtained by sulfonation of 1-naphthol with sulfuric acid and treatment of the resulting disulfonic acid with nitric acid in aqueous medium. The exchange of groups occurs with remarkable ease, and it is not necessary to isolate the disulfonic acid. The advantage of introducing the nitro groups in this indirect way is that 1-naphthol is very sensitive to oxidation and would be partially destroyed on direct nitration. Martius Yellow is prepared by reaction of the acidic phenolic group of **1** with ammonia to form the ammonium salt. A small portion of this salt (Martius Yellow) is converted by acidification and crystallization into pure 2,4-dinitro-1-naphthol (**1**), a sample of which is saved. The rest is suspended in water and reduced to diaminonaphthol with sodium hydrosulfite according to the equation:

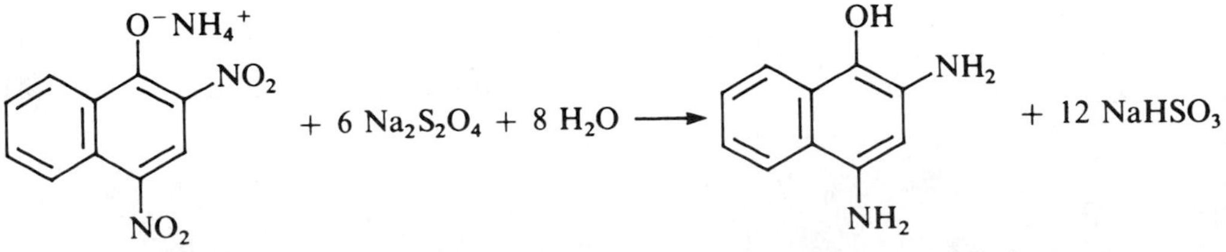

Martius Yellow $+ 6\ Na_2S_2O_4 + 8\ H_2O \longrightarrow$ **2,4-Diamino-1-naphthol** $+ 12\ NaHSO_3$

The diaminonaphthol separates in the free condition, rather than as an ammonium salt, because the diamine, unlike the dinitro compound, is a very weakly acidic substance.

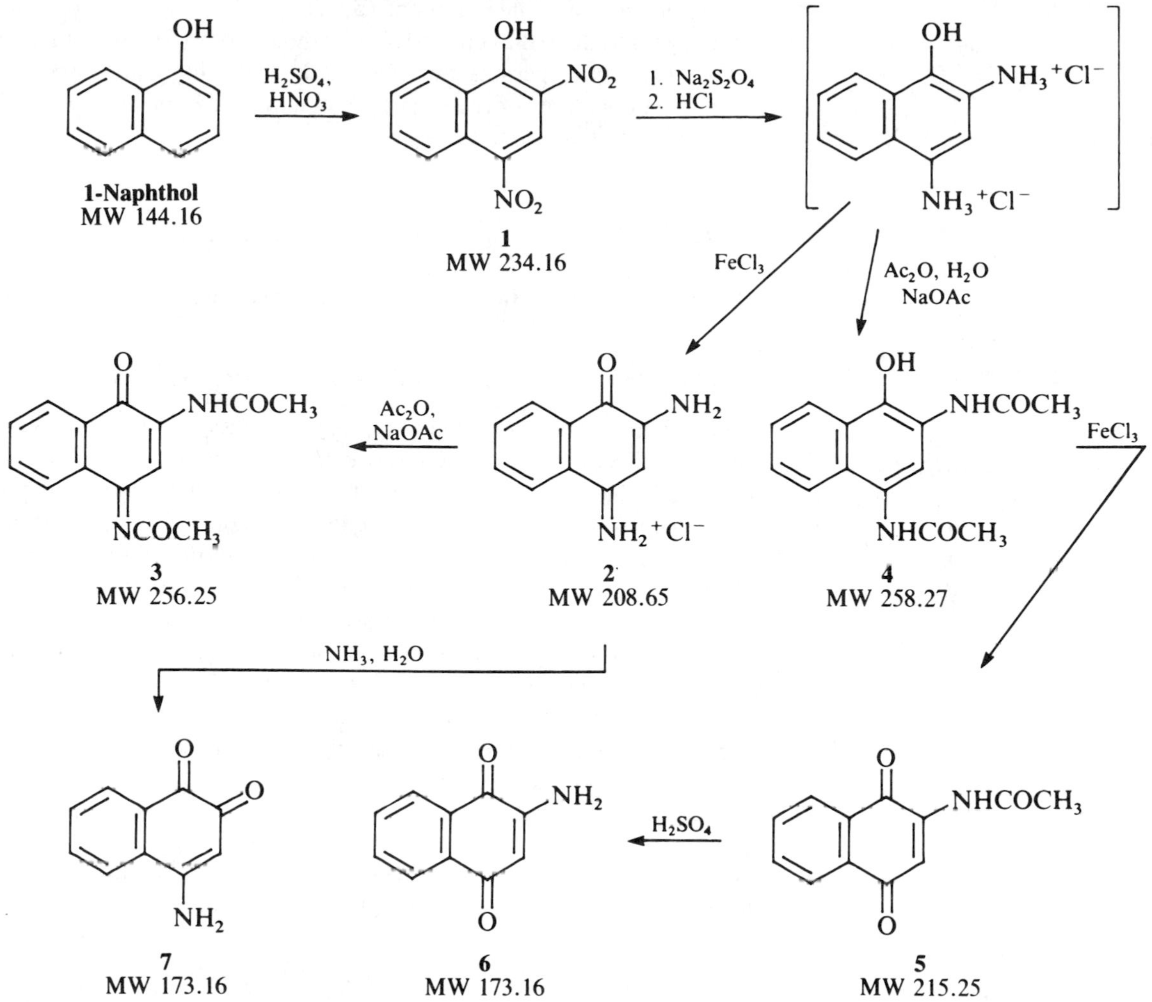

1-Naphthol
MW 144.16

1
MW 234.16

3
MW 256.25

2
MW 208.65

4
MW 258.27

7
MW 173.16

6
MW 173.16

5
MW 215.25

Since 2,4-diamino-1-naphthol is exceedingly sensitive to air oxidation as the free base, it is at once dissolved in dilute hydrochloric acid. The solution of diaminonaphthol dihydrochloride is clarified with decolorizing charcoal and divided into equal parts. One part on oxidation with iron(III) chloride affords the fiery-red 2-amino-1,4-naphthoquinonimine hydrochloride (**2**). Since this substance, like many other salts, has no melting point, it is converted for identification to the yellow diacetate, **3**. Compound **2** is remarkable in that it is stable enough to be isolated. On hydrolysis it affords the orange 4-amino-1,2-naphthoquinone (**7**).

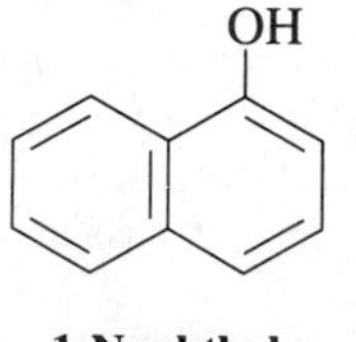

The other part of the diaminonaphthol dihydrochloride solution is treated with acetic anhydride and then sodium acetate; the reaction in aqueous solution effects selective acetylation of the amino groups and affords 2,4-diacetylamino-1-naphthol (**4**). Oxidation of **4** by Fe^{3+} and oxygen from the air is attended with cleavage of the acetylamino group at the 4-position, and the product is

2-acetylamino-1,4-naphthoquinone (**5**). This yellow substance is hydrolyzed by sulfuric acid to the red 2-amino-1,4-naphthoquinone (**6**), the last member of the series. The reaction periods are brief and the yields high; however, remember to scale down quantities of reagents and solvents if the quantity of starting material is less than that called for.[1]

Caution: *Toxic, irritant*

Experiments

If necessary, the quantities in the following seven experiments can be doubled.

MICROSCALE AND MACROSCALE

Use care when working with hot concentrated sulfuric and nitric acids.

1

Avoid contact of the yellow product and its orange NH_4^- *salt with the skin.*

1. Preparation of 2,4-Dinitro-1-naphthol(1)

Place 2.5 g of pure 1-naphthol[2] in a 50-mL Erlenmeyer flask, add 5 mL of concentrated sulfuric acid, and heat the mixture with swirling on the steam bath for 5 min, when the solid should have dissolved and an initial red color should be discharged. Cool in an ice bath, add 13 mL of water, and cool the solution rapidly to 15°C. Measure 3 mL of concentrated nitric acid into a test tube, and transfer it with a Pasteur pipette in small portions (0.25 mL) to the chilled aqueous solution while keeping the temperature in the range 15 to 20°C by swirling the flask vigorously in the ice bath. When the addition is complete and the exothermic reaction has subsided (1–2 min), warm the mixture gently to 50°C (1 min), when the nitration product should separate as a stiff yellow paste. Apply the full heat of the steam bath for 1 min more, fill the flask with water, break up the lumps and stir to an even paste, collect the product (**1**) on a small Büchner funnel, wash it well with water, and then wash it into a 250-mL beaker with water (50 mL). Add 75 mL of hot water and 2.5 mL of concentrated ammonium hydroxide solution (den 0.90), heat to the boiling point, and stir to dissolve the solid. Filter the hot solution by suction if it is dirty, add 5 g of

---

1. This series of reactions lends itself to a laboratory competition, the rules for which might be as follows: (1) No practice or advance preparation is allowable except collection of reagents not available at the contestant's bench (ammonium chloride, sodium hydrosulfite, iron(III) chloride solution, acetic anhydride). (2) The time scored is the actual working time, including that required for cleaning up and cleaning the apparatus and bench; labels for ziplock bags can be prepared out of the working period. (3) Time is not charged during an interim period (overnight) when solutions are let stand to crystallize or solids are let dry, on condition that during this period no adjustments are made and no cleaning or other work is done. (4) Melting-point and color-test characterizations are omitted. (5) Successful completion of the contest requires preparation of authentic and macroscopically crystalline samples of all seven compounds. (6) Judgment of the winners among the successful contestants is based on quality and quantity of samples, technique and neatness, and working time. (Superior performance 3 to 4 h.)

2. If the 1-naphthol is dark, it can be purified by distillation at atmospheric pressure in the hood. The colorless distillate is most easily pulverized (also in the hood) before it has completely cooled and hardened.

Martius Yellow

ammonium chloride to the filtrate to salt out the ammonium salt (Martius Yellow), cool in an ice bath, collect the orange salt, and wash it with water containing 1% to 2% of ammonium chloride. The salt does not have to be dried (dry weight 3.8 g, 88%).

Set aside an estimated 150 mg of the moist ammonium salt. This sample is to be dissolved in hot water, the solution acidified (HCl), and the free 2,4-dinitro-1-naphthol (**1**) crystallized from methanol or ethanol (use decolorizing charcoal if necessary); it forms yellow needles (mp 138°C).

The NMR and IR spectra of 1-naphthol are shown in Figs. 48.1 and 48.2 at the end of the chapter.

Cleaning Up   Combine aqueous filtrates, dilute with water, neutralize with sodium carbonate, and flush the solution down the drain. Recrystallization solvents go in the organic solvents container.

Preparation of Unstable 2,4-Diamino-1-naphthol

2,4-Diamino-1-naphthol dihydrochloride

$Na_2S_2O_4$

Sodium hydrosulfite

Wash the rest of the ammonium salt into a beaker with a total of about 100 mL of water, add 2.0 g of sodium hydrosulfite, stir until the original orange color has disappeared and a crystalline tan precipitate has formed (5–10 min), and cool in ice. Make ready a solution of 1 g of sodium hydrosulfite in 50 mL of water for use in washing and a 250-mL beaker containing 3 mL of concentrated hydrochloric acid and 12 mL of water. In collecting the precipitate by suction filtration on a small Büchner funnel, use the hydrosulfite solution for rinsing and washing, avoid even briefly sucking air through the cake after the reducing agent has been drained away, and wash the solid at once into the beaker containing the dilute hydrochloric acid and stir to convert all the diamine to the dihydrochloride.

The acid solution, often containing suspended sulfur and filter paper, is clarified by filtration by suction through a moist charcoal bed made by shaking 1 g of powdered decoloring carbon with 13 mL of water in a stoppered flask to produce a slurry and pouring this on the paper of a 50-mm Büchner funnel. Pour the water out of the filter flask, and then filter the solution of dihydrochloride. Divide the pink or colorless filtrate into approximately two equal parts, and at once add the reagents for conversion of one part to **2** and the other part to **4**.

Cleaning Up   Neutralize the filtrate with sodium carbonate, dilute with water, and then, in the hood, cautiously add household bleach (5.25% sodium hypochlorite solution) to the mixture until a test with 5% silver nitrate shows that no more hydrosulfite is present (absence of a black precipitate). Neutralize the solution, and filter through Celite to remove suspended solids. The filtrate should be diluted with water and flushed down the drain. The solid residue goes in the nonhazardous solid waste container.

MICROSCALE
AND MACROSCALE

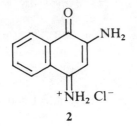

2

2. Preparation of 2-Amino-1,4-naphthoquinonimine Hydrochloride (2)

To one-half of the diamine dihydrochloride solution add 12.5 mL of 1.3 M iron(III) chloride solution,[3] cool in ice, and, if necessary, initiate crystallization by scratching. Rub the liquid film with a glass stirring rod at a single spot slightly above the surface of the liquid. If efforts to induce crystallization are unsuccessful, add more hydrochloric acid. Collect the red product, and wash with dilute HCl. Dry weight is 1.2 to 1.35 g.

Divide the moist product into three equal parts, and spread out one part to dry for conversion to **3**. The other two parts can be used while still moist for conversion to **7** and for recrystallization. Dissolve one part by warming in a little water containing 2 to 3 drops of hydrochloric acid, shake for a minute or two with decolorizing charcoal, filter by suction, and add concentrated hydrochloric acid to decrease the solubility. Collect the product by suction filtration.

Cleaning Up   Neutralize the filtrate with sodium carbonate, and collect the iron hydroxide by vacuum filtration through Celite on a Büchner funnel. The solid goes into the nonhazardous solid waste container, while the filtrate is diluted with water and flushed down the drain.

**MICROSCALE
AND MACROSCALE**

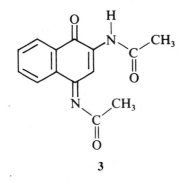

3

3. Preparation of 2-Amino-1,4-naphthoquinonimine Diacetate (3)

A mixture of 0.25 g of the dry quinonimine hydrochloride (**2**), 0.25 g of sodium acetate (anhydrous), and 1.5 mL of acetic anhydride is stirred in a reaction tube and warmed gently on a sand or steam bath. With thorough stirring, the red salt should soon change into yellow crystals of the diacetate. The solution may appear red, but as soon as particles of red solid have disappeared, the mixture can be poured into about 5 mL of water. Stir until the excess acetic anhydride has either dissolved or become hydrolyzed, collect and wash the product (dry weight 250 mg), and (drying is unnecessary) crystallize it from ethanol or methanol; yellow needles result (mp 189°C).

Cleaning Up   The filtrate should be diluted with water and flushed down the drain. Crystallization solvent goes in the organic solvents container.

**MICROSCALE
AND MACROSCALE**

4. Preparation of 2,4-Diacetylamino-1-naphthol (4)

To one-half of the diaminonaphthol dihydrochloride solution saved from Experiment 1 add 1.5 mL of acetic anhydride, stir vigorously, and add a solution of 1.5 g of sodium acetate (anhydrous) and about 50 mg of sodium hydrosulfite in 10 to 15 mL of water. The diacetate may precipitate as a white powder, or it

Note for the instructor

---

3. Dissolve 45 g of $FeCl_3 \cdot 6\ H_2O$ (MW 270.32) in 50 mL of water and 50 mL of concentrated hydrochloric acid by warming, cool, and filter (124 mL of solution).

may separate as an oil that solidifies when chilled in ice and rubbed with a rod. Collect the product, and to hydrolyze any triacetate present, dissolve it in 2.5 mL of 3 M sodium hydroxide and 25 mL of water by stirring at room temperature. If the solution is colored, a few milligrams of sodium hydrosulfite may bleach it. Filter by suction, and acidify by gradual addition of well-diluted hydrochloric acid (1 mL of concentrated acid). The diacetate tends to remain in supersaturated solution; and hence, either to initiate crystallization or to ensure maximum separation, it is advisable to stir well, rub the walls with a rod, and cool in ice. Collect the product, wash it with water, and divide it into thirds (dry weight 1–1.3 g).

Two-thirds of the material can be converted without drying into **5** and the other third used for preparation of a crystalline sample. Dissolve the third reserved for crystallization (moist or dry) in enough hot acetic acid to bring about solution, add a solution of a small crystal of tin(II) chloride in a few drops of dilute hydrochloric acid to inhibit oxidation, and dilute gradually with 5 to 6 volumes of water at the boiling point. Crystallization may be slow, and cooling and scratching may be necessary. The pure diacetate forms colorless prisms (mp 224°C, dec.).

Cleaning Up   Combine all filtrates, including the acetic acid used to crystallize the product, dilute with water, neutralize with sodium carbonate, and flush the solution down the drain.

MICROSCALE AND MACROSCALE

5. Preparation of 2-Acetylamino-1,4-napthoquinone (5)

Dissolve 0.75 g of the moist diacetylaminonaphthol (**4**) in 5 mL of acetic acid (hot), dilute with 10 mL of hot water, and add 5 mL of 0.13 M iron(III) chloride solution. The product separates promptly in flat, yellow needles, which are collected (after cooling) and washed with a little alcohol; the yield is usually 0.6 g. Dry one-half of the product for conversion to 6, and crystallize the rest from 95% ethanol (mp 204°C).

Cleaning Up   Dilute the filtrate with water, neutralize it with sodium carbonate, and flush it down the drain. A negligible quantity of iron is disposed of in this way.

MICROSCALE AND MACROSCALE

6. Preparation of 2-Amino-1,4-napthoquinone (6)

To 0.25 g of dry 2-acetylamino-1,4-naphthoquinone (**5**) contained in a reaction tube add 1 mL of concentrated sulfuric acid, and heat the mixture on the steam bath with swirling to promote rapid solution (1–2 min). After 5 min, cool the deep-red solution, dilute extensively with water, and collect the precipitated product; wash it with water, and crystallize the moist sample (dry weight about 200 mg) from alcohol or alcohol–water; red needles (mp 206°C).

Cleaning Up   The filtrate is neutralized with sodium carbonate, diluted with water, and flushed down the drain.

**MICROSCALE
AND MACROSCALE**

Moist 2 is satisfactory.

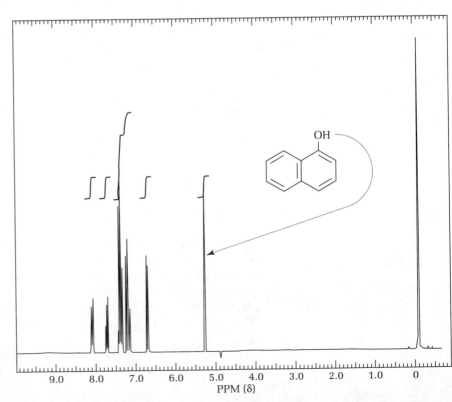

7

7. Preparation of 4-Amino-1,2-naphthoquinone (7)

Dissolve 0.5 g of the aminonaphthoquinonimine hydrochloride (**2**) reserved from Experiment 2 in 12 mL of water, add 1 mL of concentrated ammonium hydroxide solution (den 0.90), and boil the mixture for 5 min. The free quinonimine initially precipitated is hydrolyzed to a mixture of the aminoquinone (**7**) and the isomer **6**. Cool, collect the precipitate, suspend it in about 25 mL of water, and add 12.5 mL of 3 M sodium hydroxide solution. Stir well, remove the small amount of residual 2-amino-1,4-naphthoquinone (**6**) by filtration, and acidify the filtrate with acetic acid. The orange precipitate of **7** is collected, washed, and crystallized while still wet from 250 to 300 mL of hot water (the separation is slow). The yield of orange needles (dec about 270°C) is about 200 mg.

Cleaning Up　　The filtrate is neutralized with sodium carbonate, diluted with water, and flushed down the drain.

Questions _____

1. Write a balanced equation for the preparation of 2-amino-1,4-naphthoquinonimine hydrochloride (**2**).

2. Assign the peak at 3300 cm^{-1} in the IR spectrum of 1-naphthol (Fig. 48.2).

FIG. 48.1　^{1}H NMR spectrum of α-naphthol (250 MHz).

FIG. 48.2 IR spectrum of
1-naphthol (KBr disk).

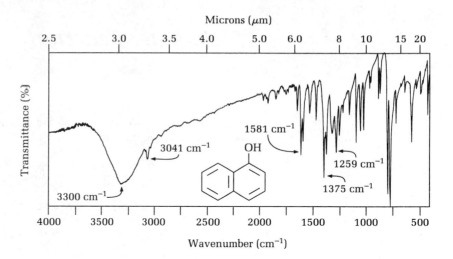

Diels-Alder Reaction

Prelab Exercise: Describe in detail the laboratory operations, reagents, and solvents you would employ to prepare

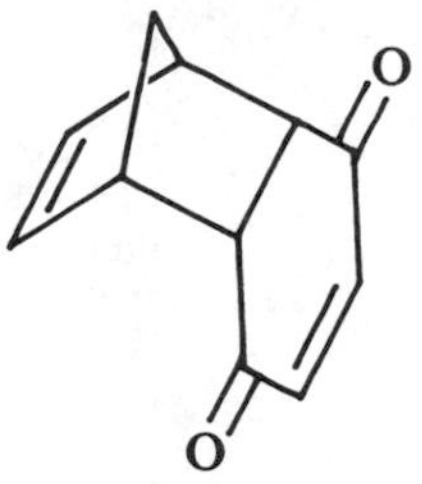

Otto Diels and his pupil Kurt Alder received the Nobel Prize in 1950 for their discovery and work on the reaction that bears their names. Its great usefulness lies in its high yield and high stereospecificity. A cycloaddition reaction, it involves the 1,4-addition of a conjugated diene in the s-*cis* conformation to an alkene in which two new σ (sigma) bonds are formed from two π (pi) bonds.

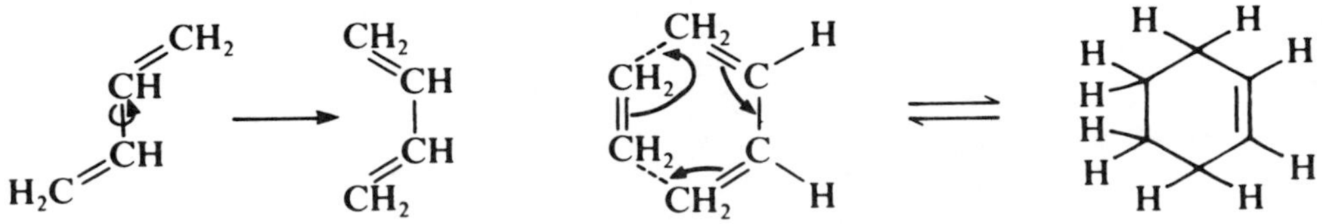

s-*trans*   s-*cis*

The adduct is a six-membered ring alkene. The diene can have the two conjugated bonds contained within a ring system, as with cyclopentadiene or cyclohexadiene, or the molecule can be an acyclic diene that must be in the *cis* conformation about the single bond before reaction can occur.

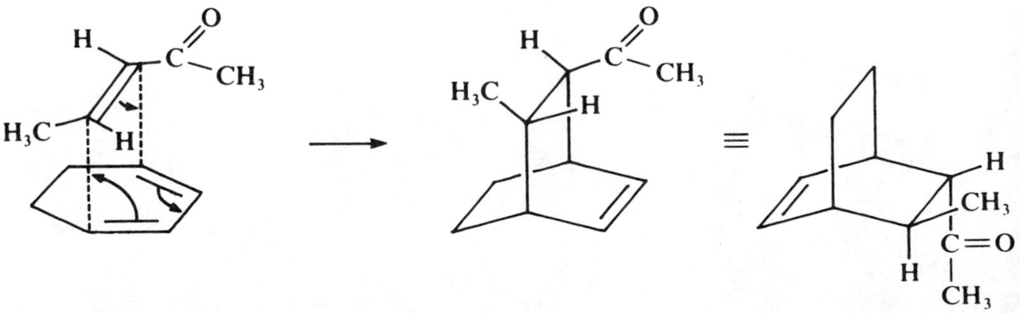

The reaction works best when there is a marked difference between the electron densities in the diene and the alkene with which it reacts, the dienophile. Usually the dienophile has electron-attracting groups attached to it, while the diene is electron rich, for example, as in the reaction of methyl vinyl ketone with 1,3-butadiene.

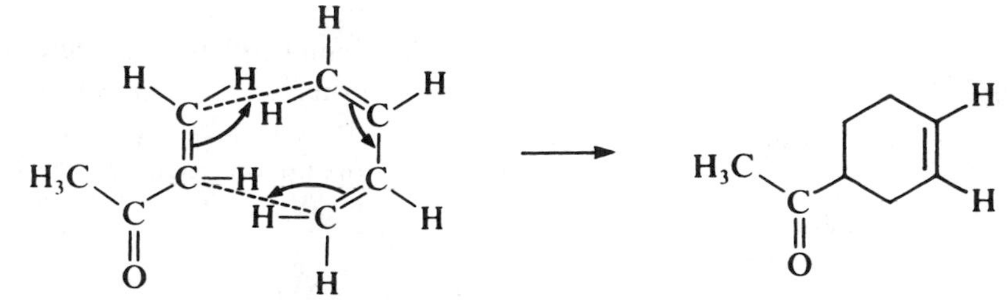

Methyl vinyl ketone   1,3-Butadiene

Retention of the configurations of the reactants in the products implies that both new σ bonds are formed almost simultaneously. If not, then the intermediate with a single new bond could rotate about that bond before the second σ bond is formed, thus destroying the stereospecificity of the reaction.

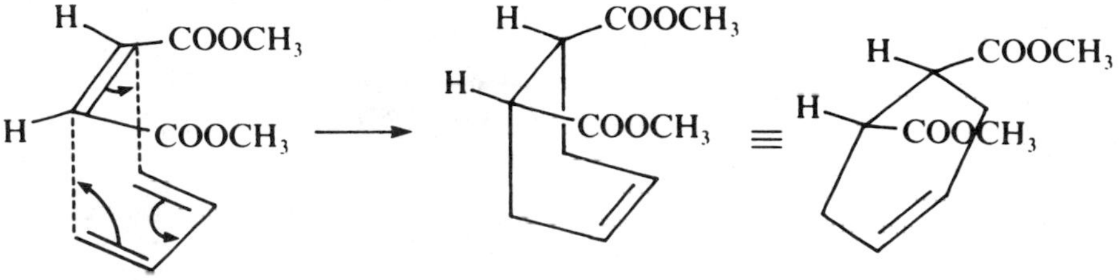

Dimethyl maleate
+
1,3-butadiene

cis **Isomer**

This does not happen:

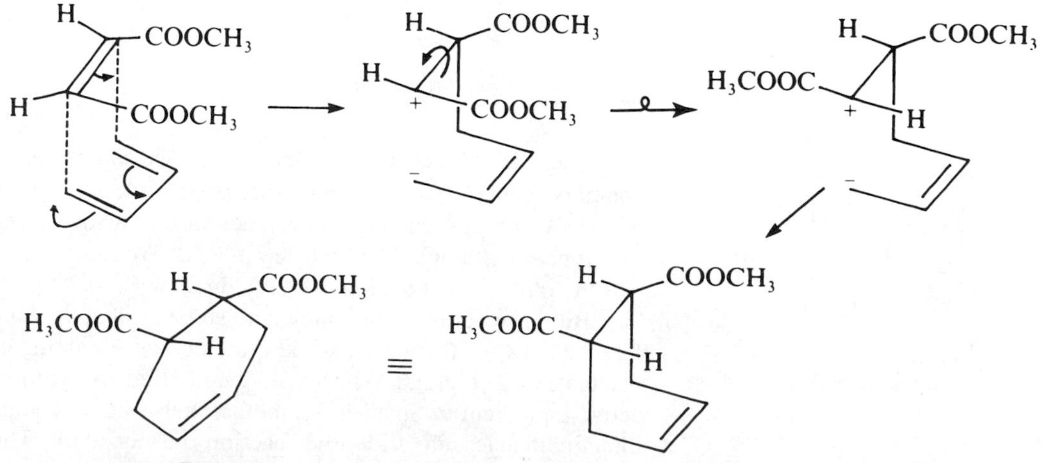

trans **Isomer**

A highly stereospecific reaction

This reaction is not polar in that no charged intermediates are formed. Neither is it a radical reaction, because no unpaired electrons are involved. It is instead known as a *concerted reaction,* or one in which several bonds in the transition state are simultaneously made and broken. When a cyclic diene and a cyclic dienophile react with each other as in the present reaction, more than one stereoisomer may be formed. The isomer that predominates is the one that involves maximum overlap of π electrons in the transition state. The transition state for the formation of the *endo* isomer in the present reaction involves a sandwich with the diene directly above the dienophile. To form the *exo* isomer, the diene and dienophile would need to be arranged in a stair-step fashion.

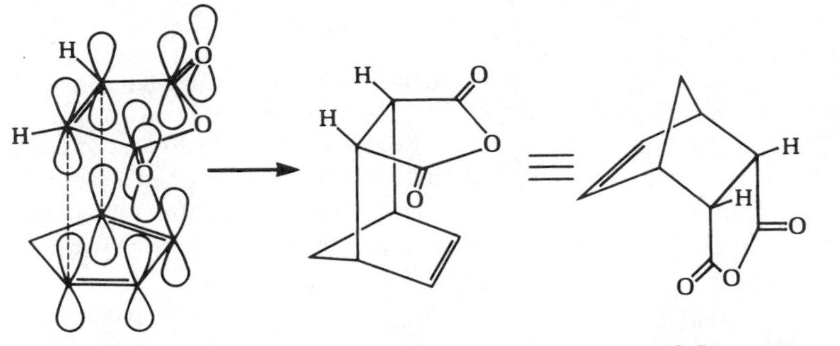

**Maximum overlap
of π electrons**

endo **Isomer
predominant product**

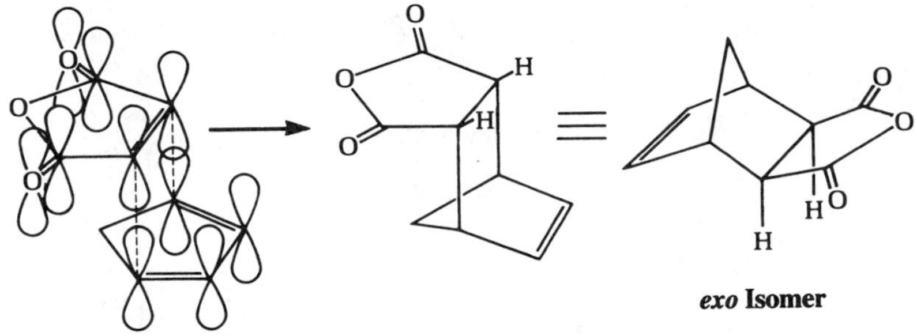

π **Electron overlap not so large**

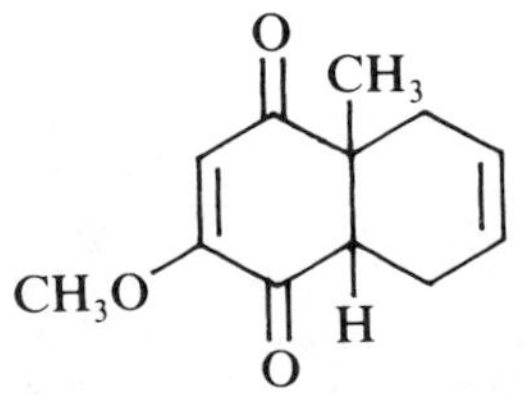

exo **Isomer**

Woodward's Diels-Alder adduct

The Diels-Alder reaction has been used extensively in the synthesis of complex natural products because it is possible to exploit the formation of a number of chiral centers in one reaction and also the regioselectivity of the reaction. For example, the first step in R. B. Woodward's synthesis of cortisone was the formation of a Diels-Alder adduct.

But the reaction is also subject to steric hindrance, especially when the difference between the electron-withdrawing and -donating characters of the two reactants is not great. When Woodward tried to synthesize cantharidin, the active ingredient in Spanish fly, by the Diels-Alder condensation of furan with dimethylmaleic anhydride, the reaction did not work. The reaction possesses

Cantharidin

$-\Delta V^*$ (it proceeds with a net decrease in volume). High pressure should overcome this problem, but this reaction will not proceed even at 600,000 lb/in.^{2} (4.1 × 10^{10} dynes/cm^2). A closely related reaction will proceed at 300,000 lb/in.^{2} and has been used to synthesize this molecule.[1] Cantharidin is a powerful vesicant (blister-former).

R. B. Woodward and Roald Hoffmann formulated the theoretical rules involving the correlation of orbital symmetry, which govern the Diels-Alder and other electrocyclic reactions.

Cyclopentadiene is obtained from the light oil from coal tar distillation but exists as the stable dimer, dicyclopentadiene, which is the Diels-Alder adduct from two molecules of the diene. Thus generation of cyclopentadiene by pyrolysis of the dimer represents a reverse Diels-Alder reaction. See Figs. 49.5 and 49.6 at the end of the chapter for NMR and IR spectra, respectively, of dicyclopentadiene.

In the Diels-Alder addition of cyclopentadiene and maleic anhydride, the two molecules approach each other in the orientation shown in the top drawing on the previous page, because this orientation provides maximal overlap of π bonds of the two reactants and favors formation of an initial π complex and then the final *endo* product. Dicyclopentadiene also has the *endo* configuration.

The infrared spectrum of dicyclopentadiene appears in Fig. 49.6 and the ^{13}C NMR spectrum in Fig. 49.7 at the end of the chapter.

Experiments

1. Microscale Cracking of Dicyclopentadiene

$$ \xrightleftharpoons[\text{room temperature}]{\sim 160°C} $$

Dicyclopentadiene
den 0.98
MW 132.20
bp 170°C
n_D^{20} 1.5100

Cyclopentadiene
bp 41°C, den 0.80
MW 66.10

Half fill with mineral oil the short-necked 5-mL round-bottomed flask equipped with an addition port bearing a septum on the sidearm and topped with a distillation head and thermometer (Fig. 49.1). Heat the flask on a sand bath. Start the heating with the thermometer down in the oil. Heat it to about 250°C, and then raise the thermometer to the position shown in Fig. 49.1. Lubricate the thermometer with a drop of oil to make it slide easily.

1. W. G. Dauben, C. R. Kessel, and K. H. Takemura, *J. Am. Chem. Soc.,* **102,** 6893 (1980).

FIG. 49.1 Apparatus for microscale cracking of dicyclopentadiene. Add dicyclopentadiene dropwise via syringe so that distillate temperature does not exceed 45°C.

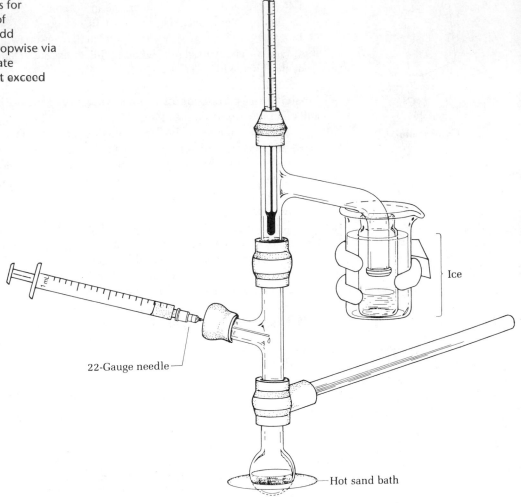

Ice

22-Gauge needle

Hot sand bath

Set controller on the flask heater at half maximum. Temperature can be controlled by piling up or scraping away sand from the flask. Turn down heat after finishing reaction.

Place a small tared collection vial in an ice-filled 30-mL beaker at the end of the distilling head, taking care to keep water out of the vial. Using a syringe, draw 0.6 mL of dicyclopentadiene (Fig. 49.2) from the septum-capped storage container after first injecting 0.6 mL of air into the container to overcome the vacuum. Stick the needle of the filled syringe into a rubber stopper or cork to avoid loss of the contents until they are used.

When the mineral oil is hot (250°C), inject the dicyclopentadiene through the septum on the addition port. Add it dropwise at such a rate that the temperature of the thermometer never exceeds 45°C. The boiling point of cyclopentadiene is 41°C. Add the dimer over a 10-min period. If the dimer is added too slowly, the yield will be lower. Once the dicyclopentadiene has all been added, remove

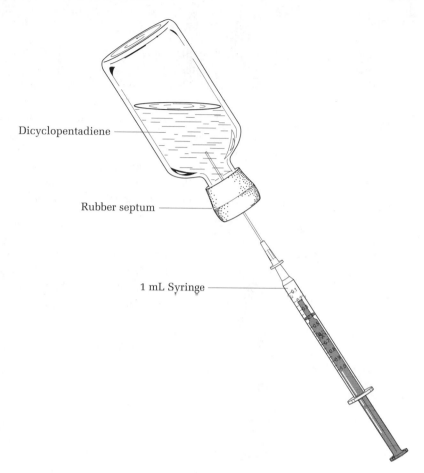

Dicyclopentadiene

Rubber septum

1 mL Syringe

FIG. 49.2   Dicyclopentadiene has a very bad odor and so is dispensed from a closed container. Remove 0.6 mL from the septum-capped bottle after injecting 0.6 mL of air.

Don't disassemble the apparatus until you are sure you have 0.3 mL of product.

the syringe, and weigh the product after closing the vial. Rinse the syringe with acetone in the hood. Calculate the percentage yield of cyclopentadiene. If the product is cloudy, add a small quantity of anhydrous calcium chloride pellets to dry it if the maleic anhydride experiment is being done next. It need not be dry to make ferrocene. Keep this cyclopentadiene on ice, and use it the same day it is prepared.

Cleaning Up   Place the mineral oil from the reaction flask and any unused dicyclopentadiene in the organic solvents container. If calcium chloride was used, it should be freed of cyclopentadiene by evaporation in the hood and then placed in the nonhazardous solid waste container.

2. Macroscale Cracking of Dicyclopentadiene

The infrared spectrum of dicyclopentadiene appears in Fig. 49.6 at the end of the chapter.

Measure 20 mL of dicyclopentadiene into a 100-mL flask, and arrange for fractional distillation into an ice-cooled receiver (Fig. 19.2). Heat the dimer with an electric flask heater until it refluxes briskly and at such a rate that the monomeric diene begins to distill in about 5 min and soon reaches a steady boiling point in the range 40–42°C. Apply heat continuously to promote rapid distillation without exceeding the boiling point of 42°C. Distillation for 45 min should provide the 12 mL of cyclopentadiene required for two preparations of the adduct; continued distillation for another half hour gives a total of about 20 mL of monomer.

Reverse Diels-Alder

Check old dicyclopentadiene for peroxides.

Cleaning Up Pour the pot residue of dicyclopentadiene and unused cyclopentadiene into the recovered dicyclopentadiene container. This recovered material can, despite its appearance, be cracked in the future to give cyclopentadiene. If the pot residue is not to be recycled, place it in the organic solvents container.

2. *cis*-Norbornene-5,6-*endo*-dicarboxylic Anhydride

Maleic anhydride
mp 53°C, MW 98.06

cis-**Norbornene-5,6-*endo*-dicarboxylic anhydride**
mp 165°C, MW 164.16

Microscale Procedure

Ligroin = hexane(s)

Mixing of the reactants is very important. Pull the reaction mixture into a pipette, and then expel it into the reaction tube or blow air from a pipette through the reaction mixture.

Dissolve 0.20 g of powdered maleic anhydride in 1 mL of ethyl acetate in a tared 10 × 100 mm reaction tube, and then add 1 mL bp 60–80°C ligroin. This combination of solvents is used because the product is too soluble in pure ethyl acetate and not soluble enough in pure ligroin. To the solution of maleic anhydride add 0.20 mL (0.160 g) of dry cyclopentadiene, mix the reactants, and observe the reaction. Allow the tube to cool to room temperature, during which time crystallization of the product should occur. If crystallization does not occur, scratch the inside of the test tube with a stirring rod at the liquid–air interface. The scratch marks on the inside of the tube often form the nuclei on which crystallization starts. Should crystallization occur very rapidly at room temperature, the crystals will be very small. If so, save a seed crystal, heat the mixture until the product dissolves, seed it, and allow it to cool slowly to room temperature. You will be rewarded with large platelike crystals. Remove the solvent from the crystals with a Pasteur pipette that is forced to the bottom of the tube, wash the crystals with one portion of cold ligroin, and remove the solvent (see Fig. 3.13). Scrape the product onto a piece of filter paper, allow the crystals to dry in air, determine their weight, and calculate the yield of the product. Determine the melting point of the product, and turn in any material not used in the next experiment. Thin-layer chromatography of the product is hardly necessary; it is quite pure. The IR spectrum of the anhydrite appears in Fig. 49.8, the ^{1}H NMR spectrum in Fig. 49.9, and the ^{13}C NMR spectrum in Fig. 49.10.

Cleaning Up   Place the crystallization solvent mixture in the organic solvents container. It contains a very small quantity of the product.

Macroscale Procedure

Place 6 g of maleic anhydride in a 125-mL Erlenmeyer flask, and dissolve the anhydride in 16 mL of ethyl acetate by heating on a hot plate or steam bath. Add 16 mL of ligroin (bp 60–90°C) or hexane, cool the solution thoroughly in an ice-water bath, and leave it in the bath (some anhydride may crystallize).

Cyclopentadiene is flammable.

     The distilled cyclopentadiene may be slightly cloudy because of the condensation of moisture in the cooled receiver and water in the starting material. Add about 1 g of calcium chloride pellets to remove the moisture. It will redimerize in a few hours; use it immediately. Measure 6 mL of dry cyclopentadiene, and add it to the ice-cold solution of maleic anhydride. Swirl the solution in the ice bath for a few minutes until the exothermic reaction is over

Rapid addition at 0°C

and the adduct separates as a white solid. Then heat the mixture on a hot plate or steam bath until the solid is all dissolved.[2] If you let the solution stand undisturbed, you will be rewarded with a beautiful display of crystal formation. The anhydride crystallizes in long spars (mp 164–165°C); a typical yield is 8.2 g.[3] The IR spectrum of the anhydride appears in Fig. 49.8, the ^{1}H NMR spectrum in Fig. 49.9, and the ^{13}C NMR spectrum in Fig. 49.10 at the end of the chapter.

---

2. In case moisture has gotten into the system, a little of the corresponding diacid may remain undissolved at this point and should be removed by filtration of the hot solution.

3. The student need not work up the mother liquor but may be interested in learning the result. Concentration of the solution to a small volume is not satisfactory because of the presence of dicyclopentadiene, formed by dimerization of excess monomer; the dimer has high solvent power. Hence the bulk of the solvent is evaporated on the steam bath, and the flask is connected to the water pump with a rubber stopper and glass tube and heated under vacuum on the steam bath until dicyclopentadiene is removed and the residue solidifies. Crystallization from 1:1 ethyl acetate–ligroin affords 1.3 g adduct (mp 156–158°C); total yield is 95%.

3. *cis*-Norbornene-5-6,-*endo*-dicarboxylic Acid

$$H_2O \ + \qquad\qquad\qquad \longrightarrow$$

endo,cis-**Diacid**

Microscale Procedure

To 0.2 g (200 mg) of the anhydride from the preceding experiment, add 2.5 mL of water and a boiling stick in a 10 × 100 mm reaction tube. Heat the mixture to boiling by immersing the tube in a hot sand bath (Fig. 49.3). The anhydride may appear to melt and form globules on the bottom of the tube. As the reaction proceeds, the anhydride will react with the water, and the diacid, which is soluble in boiling water, will be formed. Continue to heat for about 2 min after the last globule disappears. Remove the boiling stick from the hot solution, and allow the mixture to cool to room temperature.

If crystallization of the diacid does not occur, follow exactly the same procedure used for the anhydride. On slow cooling with simultaneous crystal growth the solution will deposit long needlelike crystals. Again, cool the mixture in ice, allow sufficient time for crystal growth to occur, and then collect the product by filtration on the Hirsch funnel. Use the filtrate (ice cold) to complete the transfer. Wash the crystals once with a small quantity of ice water, and place the product on a piece of filter paper to dry. Do not discard the filtrate until you have weighed the product. More material can be recovered by concentration of the filtrate and allowing it to cool to give a second crop of crystals. This is a general strategy. Weigh the diacid, and determine the melting point and percentage yield. The melting point depends on the rate of heating as the anhydride reforms and water splits out. IR and ^{1}H NMR spectra are in Figs. 49.11 and 49.12 at the end of the chapter.[4]

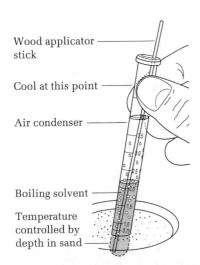

Wood applicator stick

Cool at this point

Air condenser

Boiling solvent

Temperature controlled by depth in sand

FIG. 49.3 Hydrolysis of the anhydride to the *endo,cis*-diacid.

Macroscale Procedure

$$H_2O \; + \qquad \longrightarrow$$

endo,cis-Diacid

For preparation of the *endo,cis*-diacid, place 4.0 g of the anhydride from Experiment 2 and 50 mL of distilled water in a 125-mL Erlenmeyer flask, grasp this with a clamp, swirl the flask over a hot plate, and bring the contents to the boiling point, at which point the solid partly dissolves and partly melts. Continue to heat until the oil is all dissolved, and then let the solution stand undisturbed. Because the diacid has a strong tendency to remain in supersaturated solution, allow half an hour or more for the solution to cool to room temperature, and then drop in a boiling stone or touch the surface of the liquid once or twice with a stirring rod. Observe the stone and its surroundings carefully, waiting several minutes before applying the more effective method of making one scratch with a stirring rod on the inner wall of the flask at the air–liquid interface. Let crystallization proceed spontaneously to give large needles; then cool the solution in ice and collect the product. The melting point depends on the rate of heating as the anhydride reforms and water splits out. IR and ^{1}H NMR spectra of the diacid are found in Figs. 49.11 and 49.12 at the end of the chapter.[4]

The temperature of decomposition is variable.

Cleaning Up   The aqueous filtrate from the crystallization contains a very small quantity of the diacid. It can be flushed down the drain.

4. Synthesis of Compound X[5]

To a tared 10 × 100 mm reaction tube add 0.15 g of the *endo,cis*-diacid from the preceding experiment, followed by 0.25 mL of concentrated sulfuric acid. Warm the mixture on the steam bath or in a beaker of boiling water for about

---

4. The *endo,cis*-diacid is stable to alkali but can be isomerized to the *trans*-diacid (mp 192°C) by conversion to the dimethyl ester (3 g acid, 10 mL methanol, 0.5 mL concentrated H_2SO_4; reflux 1 h). This ester is equilibrated with sodium ethoxide in refluxing ethanol for 3 days and saponified. For an account of a related epimerization and discussion of the mechanism, see J. Meinwald and P. G. Gassman, *J. Am. Chem. Soc.*, **82**, 5445, 1960. See also K. L. Williamson, Y.-F. Li, R. Lacko, and C. H. Youn, *J. Am. Chem. Soc.*, **91**, 6129, 1969; and K. L. Williamson and Y.-F. Li, *J. Am. Chem. Soc.*, **92**, 7654, 1970.

5. Introduced by James A. Deyrup.

If you do not have the necessary quantity of diacid, scale down the amounts of reactants and solvents to match the quantity of starting material. This is a general rule.

Do not put a wood boiling stick into this solution. The sulfuric acid will attack the wood.

2 min to allow the anhydride to dissolve/react, cool, and then *cautiously* add 0.70 mL of water to the test tube. This should be done dropwise with vigorous mixing of the contents after the addition of each drop. The product will crystallize as a fine powder, often gray in color.

Save a seed crystal, and heat the tube on a hot sand bath until the crystals redissolve. Seed the solution, and allow it to cool slowly to room temperature. Compound X will crystallize in platelike crystals. The crystallization process for this compound is fairly slow; allow at least 10 min for the solution to come to room temperature and a further 10 min in the ice bath before collecting the crystals on the Hirsch funnel or with the Wilfilter (Fig. 49.4). The longer you wait to collect the product the higher your yield will be. Wash the crystals with one small portion of ice water, and scrape the product onto a piece of filter paper. Squeeze the crystals between sheets of filter paper to complete the drying process, and then determine the weight, yield, and melting point of the product.

Cleaning Up   Dilute the aqueous filtrate with water, neutralize it with sodium carbonate, and flush the resulting solution down the drain. It contains an extremely small quantity of compound X.

 ### Macroscale Procedure

For preparation of compound X, place 1 g of the *endo,cis*-diacid and 5 mL of concentrated sulfuric acid in a 50-mL Erlenmeyer flask, and heat gently on the hot plate for a minute or two until the crystals are all dissolved. Then cool in an ice bath, add a small piece of ice, swirl to dissolve, and add further ice until the volume is about 20 mL. Heat to the boiling point, and let the solution simmer on the hot plate for 5 min. Cool well in ice, scratch the flask (see Chapter 3) to induce crystallization, and allow for some delay in complete separation. The crystals will often be gray in color. Collect, wash with water, and crystallize from water. Compound X (about 0.7 g) forms large prisms (mp 203°C). To run an NMR spectrum of compound X, the compound must be dissolved in deutero-dimethyl sulfoxide (DMSO d_6).

Cleaning Up   Dilute the aqueous filtrate with water, neutralize it with sodium carbonate, and flush the resulting solution down the drain. It contains a small quantity of compound X.

 **MICROSCALE AND MACROSCALE**

7. Structure Determination of Compound X

To determine the formula for compound X, which is an isomer of the diacid, try to answer the following questions: What intermediate is formed when the diacid dissolves in concentrated sulfuric acid? Why is the ^{1}H NMR spectrum of X (Fig. 49.13) so much more complex than the anhydride and diacid spectra (Figs. 49.9 and 49.12)? What functional group is missing from X that is seen in Figs. 49.9 and 49.12? Write formulas for possible structures of X, and devise tests to distinguish among them. What functional group is present in X that is not found in the diacid as determined by an analysis of the infrared spectrum of X (Fig. 49.14)?

Computational Chemistry

Would you predict that compound X is more or less stable than the isomeric diacid from which it is formed? Test your prediction by carrying out a heat of formation calculation on each of the isomers. To do this, first find the conformation of each isomer with the minimum steric energy using a molecular mechanics program. Then submit each of these to a heat of formation calculation at the AM1 or higher level of calculation.

Questions _____

1. In the cracking of dicyclopentadiene, why is it necessary to distill the product very slowly?

2. Draw the products of the following reactions:

(a) H₃C ... + ... CN ... $\xrightarrow{\Delta}$

(b) ... + ... $\xrightarrow{\Delta}$

(c) ... O + ... CN ... $\xrightarrow{\Delta}$

3. What starting material would be necessary to prepare the following compound by the Diels-Alder reaction?

4. If the Diels-Alder reaction between dimethylmaleic anhydride and furan had worked, would cantharidin have been formed?

5. Determine the heats of formation or the steric energies of the *exo-* and

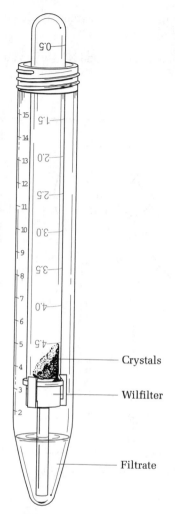

FIG. 49.4   The Wilfilter filtration apparatus.

Crystals

Wilfilter

Filtrate

endo-anhydride adducts using a molecular mechanics program. What do these energies tell you about the mechanism of the reaction?

6. Which molecule, norbornene dicarboxylic acid or compound X, would you predict to be the more stable? Why?

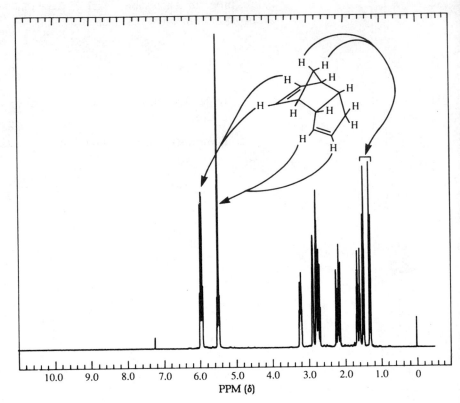

FIG. 49.5 ¹H NMR spectrum of dicyclopentadiene (250 MHz).

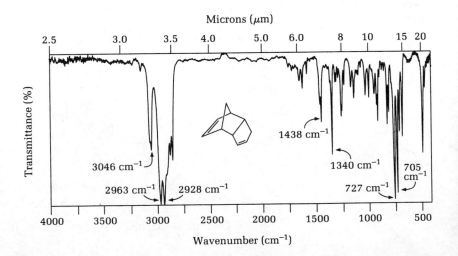

FIG. 49.6 IR spectrum of dicyclopentadiene (thin film).

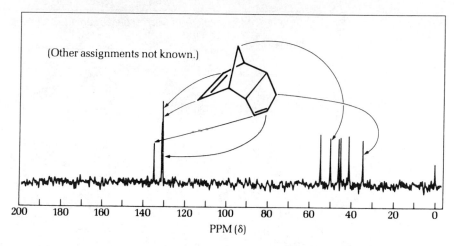

FIG. 49.7 ¹³C NMR spectrum of dicyclopentadiene.

(Other assignments not known.)

200   180   160   140   120   100   80   60   40   20   0
PPM (δ)

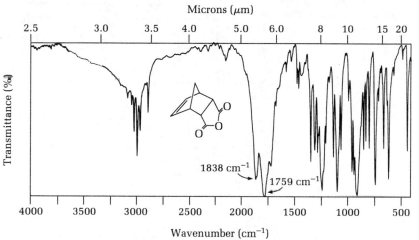

FIG. 49.8 Infrared spectrum of *cis*-norbornene-5,6-*endo*-dicarboxylic anhydride.

Microns (μm)
2.5   3.0   3.5   4.0   5.0   6.0   8   10   15   20

Transmittance (%)

1838 cm⁻¹
1759 cm⁻¹

4000   3500   3000   2500   2000   1500   1000   500
Wavenumber (cm⁻¹)

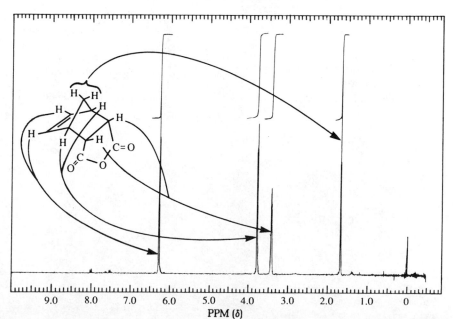

FIG. 49.9 ¹H NMR spectrum of *cis*-norbornene-5,6-*endo*-dicarboxylic anhydride. (250 MHz).

9.0   8.0   7.0   6.0   5.0   4.0   3.0   2.0   1.0   0
PPM (δ)

585

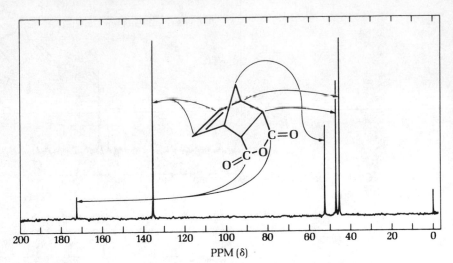

FIG. 49.10 ^{13}C NMR spectrum of *cis*-norbornene-5,6-*endo*-dicarboxylic anhydride (22.6 MHz).

PPM (δ)

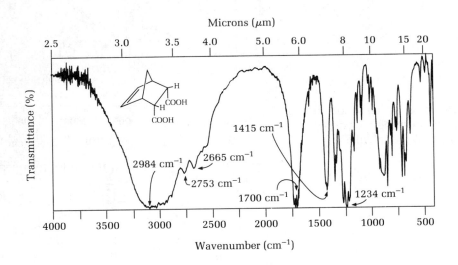

FIG. 49.11 Infrared spectrum of *cis*-5,6-*endo*-norbornene dicarboxylic acid (KBr disk).

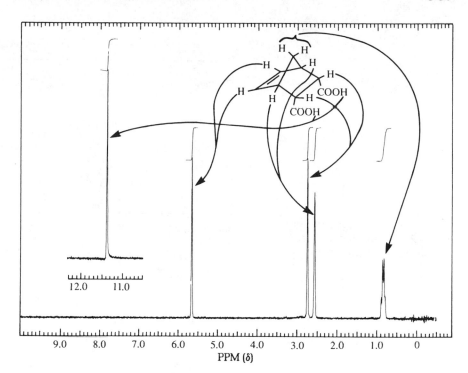

FIG. 49.12 ^{1}H NMR spectrum of *cis*-norbornene-5,6-*endo*-dicarboxylic acid (250 MHz).

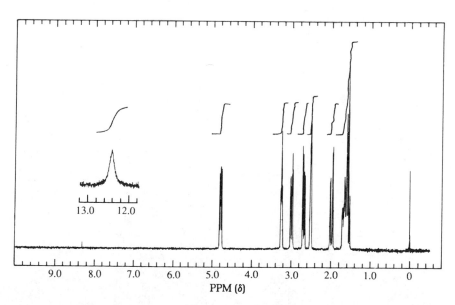

FIG. 49.13 ^{1}H NMR spectrum of compound X (250 MHz), dissolved in DMSO d$_6$.

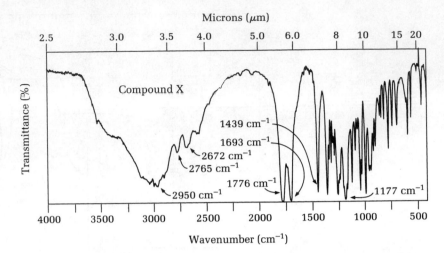

FIG. 49.14 Infrared spectrum of compound X (KBr disk).

Ferrocene [Bis(cyclopentadienyl)iron]

Prelab Exercise: Propose a detailed outline of the procedure for the synthesis of ferrocene, paying particular attention to the time required for each step.

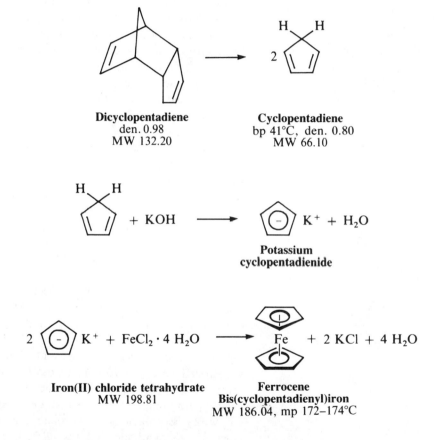

Dicyclopentadiene
den. 0.98
MW 132.20

Cyclopentadiene
bp 41°C, den. 0.80
MW 66.10

**Potassium
cyclopentadienide**

Iron(II) chloride tetrahydrate
MW 198.81

**Ferrocene
Bis(cyclopentadienyl)iron**
MW 186.04, mp 172–174°C

The Grignard reagent is a classic organometallic compound. The magnesium ion in Group IIA of the periodic table needs to lose two and only two electrons to achieve the inert gas configuration. This metal has a strong tendency to form ionic bonds by electron transfer:

$$RBr + Mg \longrightarrow \overset{\delta-}{R}\!-\!\overset{\delta+}{Mg}Br$$

In the transition metals, the situation is not so simple. Consider the bonding between iron and carbon monoxide in $Fe(CO)_5$:

$$\overset{\displaystyle \diagdown \mid}{\underset{\diagup \mid}{Fe}} \longleftarrow :C \equiv \overset{+}{\underset{}{O}:} \longrightarrow \overset{\displaystyle \diagdown \mid}{\underset{\diagup \mid}{Fe}} = C = \ddot{O}:$$

The pair of electrons on the carbon atom is shared with iron to form a σ bond between the carbon and iron. The π bond between iron and carbon is formed from a pair of electrons in the d-orbital of iron. The π bond is thus formed by the overlap of a d orbital of iron with the p-π bond of the carbonyl group. This mutual sharing of electrons results in a relatively nonpolar bond.

Iron has 6 electrons in the $3d$ orbital, 2 in the $4s$, and none in the $4p$ orbital. The inert gas configuration requires 18 electrons—ten $3d$, two $4s$, and six $4p$ electrons. Iron pentacarbonyl enters this configuration by accepting two electrons from each of the five carbonyl groups, a total of 18 electrons. Back-bonding of the d-π type distributes the excess electrons among the five carbon monoxide molecules.

Early attempts to form σ-bonded derivatives linking alkyl carbon atoms to iron were unsuccessful, but P. L. Pauson in 1951 succeeded in preparing a very stable substance, ferrocene, $C_{10}H_{10}Fe$, by reacting 2 mol of cyclopentadienyl-magnesium bromide with anhydrous ferrous chloride. Another group of chemists—Wilkinson, Rosenblum, Whiting, and Woodward—recognized that the properties of ferrocene (its remarkable stability to water, acids, and air and its ease of sublimation) could only be explained if it had the structure depicted and that the bonding of the ferrous iron with its 6 electrons must involve all 12 of the π-electrons on the two cyclopentadiene rings, with a stable 18-electron inert gas structure as the result.

In the present experiment ferrocene is prepared by reaction of the anion of cyclopentadiene with iron(II) chloride. Abstraction of one of the acidic allylic protons of cyclopentadiene with base gives the aromatic cyclopentadienyl anion. It is considered aromatic because it conforms to the Hückel rule in having $4n + 2\pi$-electrons (where n is 1). Two molecules of this anion will react with iron(II) to give ferrocene, the most common member of the class of metal-organic compounds referred to as metallocenes. In this centrosymmetric sandwich-type π-complex, all carbon atoms are equidistant from the iron atom, and the two cyclopentadienyl rings rotate more or less freely with respect to each other. The extraordinary stability of ferrocene (stable to 500°C) can be attributed to the sharing of the 12 π-electrons of the two cyclopentadienyl rings with the 6 outer shell electrons of iron(II) to give the iron a stable 18-electron inert gas configuration. Ferrocene is soluble in organic solvents, can be dissolved in concentrated sulfuric acid and recovered unchanged, and is resistant to other acids and bases as well (in the absence of oxygen). This behavior is consistent with that of an aromatic compound; ferrocene is found to undergo electrophilic aromatic substitution reactions with ease.

Ferrocene: soluble in organic solvents, stable to 500°C

Cyclopentadiene readily dimerizes at room temperature by a Diels-Alder reaction to give dicyclopentadiene. This dimer can be "cracked" by heating (an example of the reversibility of the Diels-Alder reaction) to give low-boiling cyclopentadiene. In most syntheses of ferrocene the anion of cyclopentadiene is prepared by reaction of the diene with metallic sodium. Subsequently, this anion is allowed to react with anhydrous iron(II) chloride. In the present experiment the anion is generated using powdered potassium hydroxide, which functions as both a base and dehydrating agent.

The anion of cyclopentadiene rapidly decomposes in air, and iron(II) chloride, although reasonably stable in the solid state, is readily oxidized to the iron(III) (ferric) state in solution. Consequently this reaction must be carried out in the absence of oxygen, accomplished by bubbling nitrogen gas through the solutions to displace dissolved oxygen and to flush air from the apparatus. In research laboratories rather elaborate apparatus is used to carry out an experiment in the absence of oxygen. In the present experiment, because no gases are evolved, no heating is necessary, and the reaction is only mildly exothermic, very simple apparatus is used.

Experiments

1. Microscale Synthesis of Ferrocene

$CH_3OCH_2CH_2OCH_3$

1,2-Dimethoxyethane
(Ethylene glycol dimethyl ether, monoglyme), bp 85°C
Completely miscible with water

Potassium hydroxide is extremely corrosive and hygroscopic. Immediately wash any spilled powder or solutions from the skin, and wipe up all spills. Keep containers tightly closed. Work in the hood.

To a 5-mL short-necked, round-bottomed flask add a magnetic stirring bar and then quickly add 0.75 g of finely powdered potassium hydroxide,[1] followed by 1.25 mL of dimethoxyethane. The funnel that is a part of the chromatography column makes a convenient addition funnel. Cap the flask with a good septum, and pass nitrogen into the flask or, better, through the solution for about 1 min. This is done by connecting a tank of nitrogen via a rubber tube to a 22-gauge needle and adjusting the nitrogen flow to a few milliliters per minute by bubbling it under a liquid such as acetone. With the nitrogen flow adjusted, insert an empty syringe needle through the septum of the flask as an outlet, and then insert the nitrogen inlet needle (Fig. 50.1). Remove the needles, and shake the flask to dislodge the solid from the bottom and to help dissolve some of it. If possible, then stir the mixture magnetically.

To a 10 × 100 mm reaction tube add 0.35 g of finely powdered green iron(II) chloride tetrahydrate and 1.5 mL of dimethyl sulfoxide. Cap the tube with a good rubber septum, insert an empty syringe needle through the septum, and pass nitrogen into the tube for about 1 min to displace the oxygen present. Remove the needles, and then shake the vial vigorously to dissolve all the iron chloride. Some warming may be needed.

---

1. Potassium hydroxide is easily ground to a fine powder in 25-g batches in 1 min, employing an ordinary food blender (e.g., Waring, Osterizer). The finely powdered base is transferred in a hood to a bottle with a tightly fitting cap. Alternatively, grind about 1 g potassium hydroxide in a mortar, and transfer it rapidly to the reaction flask.

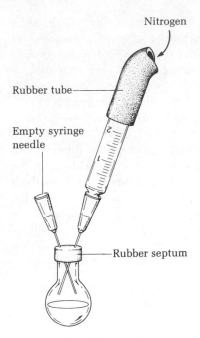

Nitrogen

Rubber tube

Empty syringe
needle

Rubber septum

FIG. 50.1 Apparatus for
flushing air from reaction flask.
Check the nitrogen flow rate by
allowing it to bubble through an
organic solvent before inserting
the needle through the septum.

*Dimethyl sulfoxide is rapidly
absorbed through the skin. Wash
off spills with water. Wear
disposable gloves when shaking
the apparatus.*

Dimethyl sulfoxide, DMSO
(Methyl sulfoxide), bp 189°C
Completely miscible with water

Isolation of ferrocene

Purification by sublimation

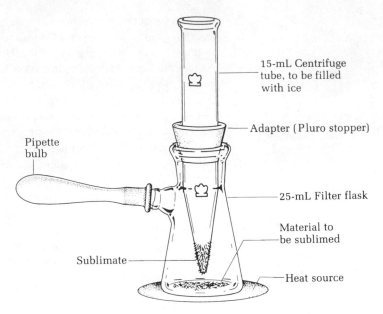

Pipette
bulb

15-mL Centrifuge
tube, to be filled
with ice

Adapter (Pluro stopper)

25-mL Filter flask

Material to
be sublimed

Sublimate

Heat source

FIG. 50.2 Apparatus for the sublimation of ferrocene.

Using an accurate syringe, inject 0.300 mL of freshly prepared cyclopentadiene (see Chapter 49 for the preparation of cyclopentadiene) into the flask containing the potassium hydroxide. Do not grasp the body of the syringe, because
the heat of your hand will cause the cyclopentadiene to volatilize. Stir the mixture vigorously, and note the color change as the potassium cyclopentadienide is
formed. After waiting about 5 min for the anion to form, pierce the septum with
an empty needle for pressure relief and inject the iron(II) chloride solution contained in the reaction tube in six 0.25-mL portions over a 10-min period. Stir the
mixture well with the magnetic stirring bar. If the bar is immobilized, then
between injections remove both needles from the septum and shake the flask vigorously. After all the iron(II) chloride solution has been added, rinse the reaction
tube with 0.25 mL more dimethyl sulfoxide, and add this to the flask. Continue
to stir or shake the solution for about 15 min to complete the reaction.

To isolate the ferrocene, pour the dark slurry onto a mixture of 4.5 mL of 6 M
hydrochloric acid and 5 g of ice in a 30-mL beaker. Stir the contents of the beaker
thoroughly to dissolve and neutralize all the potassium hydroxide. Collect the crystalline orange ferrocene on a Hirsch funnel, wash the crystals well with water, press
out excess water, squeeze the product between sheets of filter paper to complete the
drying, and then purify the ferrocene by sublimation. The filtrate is blue because of
dissolved ferrocinium ion. It can be reduced with a mild reducing agent, such as
ascorbic acid, to regenerate ferrocene. The amount produced is usually negligible.

To sublime the ferrocene, add the crude dry product to a 25-mL filter flask
equipped with a neoprene filter adapter (Pluro stopper) and a 15-mL centrifuge
tube that is pushed to within 5 mm of the bottom of the flask (Fig. 50.2). Put
a rubber bulb on the side arm of the flask, and then add ice to the centrifuge
tube and heat the flask on the sand bath to sublime the product. Tilting and

FIG. 50.3 Evacuation of a melting point capillary prior to sealing.

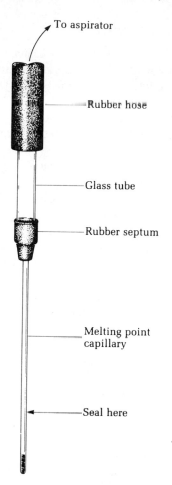

To aspirator

Rubber hose

Glass tube

Rubber septum

Melting point capillary

Seal here

rolling the filter flask in the hot sand will help drive ferrocene onto the centrifuge tube. Remove the flask from the sand bath, and use a heat gun to drive the last of the ferrocene from the sides of the flask to the centrifuge tube. Ferrocene sublimes nicely at atmospheric pressure. Vacuum sublimation is not needed.

When sublimation is complete, cool the flask, remove the ice water from the centrifuge tube, and replace it with room temperature water (to prevent moisture from collecting on the tube). Transfer the product to a tared stoppered vial, determine the weight, and calculate the percentage yield. Determine the melting point in an evacuated capillary, since the product sublimes at the melting point (Fig. 50.3). See Chapter 4 for this technique.

Cleaning Up   The filtrate should be slightly acidic. Neutralize it with sodium carbonate, dilute it with water, and flush it down the drain. Place any unused cyclopentadiene in the recovered dicyclopentadiene or the organic solvents container. Add 0.4 mL concentrated nitric acid to the sublimation flask to clean it. After at least 24 hours, dilute the acid, neutralize it with sodium carbonate, and flush the solution down the drain.

2. Macroscale Synthesis of Ferrocene

Following the procedure described in Chapter 49, prepare 3 mL of cyclopentadiene. It need not be dry. While this distillation is taking place, rapidly weigh 12.5 g of finely powdered potassium hydroxide[2] into a 50-mL Erlenmeyer flask, add 30 mL of dimethoxyethane ($CH_3OCH_2CH_2OCH_3$), and immediately cool the mixture in an ice bath. Swirl the mixture in the ice bath for a minute or two, then bubble nitrogen through the solution for about 2 min. Quickly stopper the flask, and shake the mixture to dislodge the cake of potassium hydroxide from the bottom of the flask and to dissolve as much of the base as possible (much will remain undissolved).

Grind 3.5 g of iron(II) chloride tetrahydrate to a fine powder, and then add 3.5 g of the green salt to 12.5 mL of dimethyl sulfoxide (DMSO) in a 25-mL Erlenmeyer flask. Pass nitrogen through the DMSO mixture for about 2 min, stopper the flask, and shake it vigorously to dissolve all the iron(II) chloride. Gentle warming of the flask on a steam bath may be necessary to dissolve the last traces of iron(II) chloride. Transfer the solution rapidly to a 60-mL separatory funnel equipped with a cork or stopper to fit the 50-mL Erlenmeyer flask, flush air from the funnel with a stream of nitrogen, and stopper it.

Transfer 3 mL of the freshly distilled cyclopentadiene to the slurry of potassium hydroxide in dimethoxyethane. Shake the flask vigorously, and note the color change as the potassium cyclopentadienide is formed. After waiting about 5 min for the anion to form, replace the cork or stopper on the Erlenmeyer flask with the separatory funnel quickly (to avoid admission of air to the flask). (See Fig. 50.4.) Fig. 50.5 depicts research-quality apparatus that would be used for this experiment.

Add the iron(II) chloride solution to the base dropwise over a period of 20 min with vigorous swirling and shaking. Dislodge the potassium hydroxide should it cake on the bottom of the flask. The shaking will allow nitrogen to pass from the Erlenmeyer flask into the separatory funnel as the solution leaves the funnel.[3] Continue to shake and swirl the solution for 10 min after all the iron(II) chloride is added, then pour the dark slurry onto a mixture of 45 mL of 6 *M* hydrochloric acid and 50 g of ice in a 250-mL beaker. Stir the contents of the beaker thoroughly to dissolve and neutralize all the potassium hydroxide. Collect the crystalline orange ferrocene on a Büchner funnel, wash the crystals with water, press out excess water, and allow the product to dry on a watch glass overnight.

---

2. Potassium hydroxide is easily ground to a fine powder in 75-g batches in 1 min employing an ordinary food blender (e.g., Waring, Osterizer). The finely powdered base is transferred in a hood to a bottle with a tightly fitting cap. If a blender is unobtainable, crush and then grind 27 g of potassium hydroxide pellets in a large mortar and quickly weigh 25 g of the resulting powder into the 125-mL Erlenmeyer flask.

3. If the particular separatory funnel being used does not allow nitrogen to pass from the flask to the funnel, connect the two with a rubber tube leading to a glass tube and stopper at the top of the separatory funnel and to a syringe needle that pierces the rubber stopper in the flask (suggestion of D. L. Fishel).

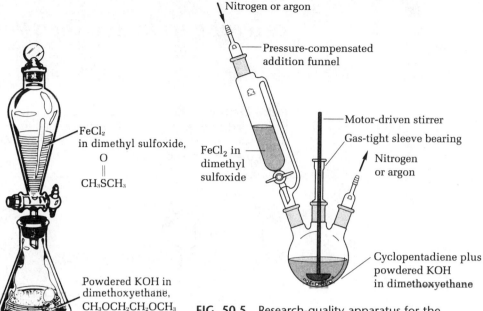

FIG. 50.4 Apparatus for ferrocene synthesis.

FIG. 50.5 Research-quality apparatus for the preparation of ferrocene.

Recrystallize the ferrocene from methanol or, better, from ligroin. It is also very easily sublimed. In a hood place about 0.5 g of crude ferrocene on a watch glass on a hot plate set to about 150°C. Invert a glass funnel over the watch glass. Ferrocene will sublime in about 1 h, leaving nonvolatile impurities behind. Pure ferrocene melts at 172–174°C. Determine the melting point in an evacuated capillary (Fig. 50.3) since the product sublimes at the melting point. Compare the melting points of your sublimed and recrystallized materials.

Cleaning Up   The filtrate from the reaction mixture should be slightly acidic. Neutralize it with sodium carbonate, dilute it with water, and flush it down the drain. Place any unused cyclopentadiene in the recovered dicyclopentadiene or the organic solvents container. If the ferrocene has been crystallized from methanol or ligroin, place the mother liquor in the organic solvents container.

Questions

1. If ferrocene is stable to air and all of the reagents are stable to air before the reaction begins, why must air be so carefully excluded from this reaction?

2. What special properties do the solvents dimethoxyethane and dimethyl sulfoxide have compared to diethyl ether, for example, that make them particularly suited for this reaction?

3. What is it about ferrocene that allows it to sublime easily where many other compounds do not?

p-Terphenyl by the Diels-Alder Reaction

Prelab Exercise: Why does the decarboxylate, **5,** undergo double decarboxylation to give terphenyl, while the diacid formed from the reaction of cyclopentadiene with maleic anhydride (Chapter 49) does not undergo decarboxylation under the same reaction conditions?

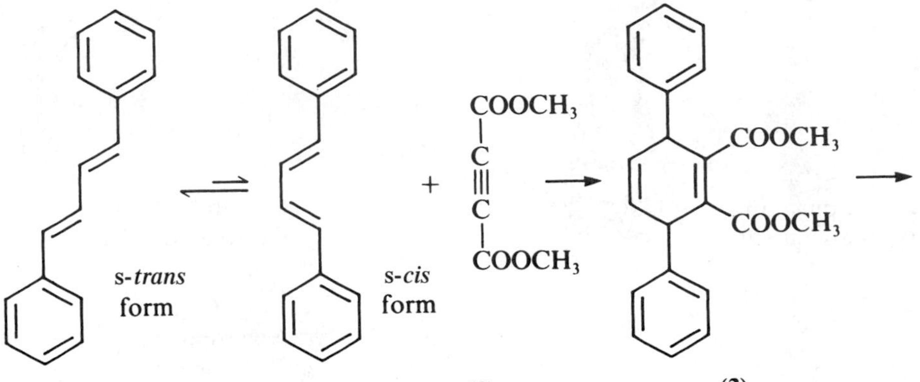

(1)
E,E-**1,4-Diphenyl-1,3-butadiene**
MW 206.27

s-*trans*
form

s-*cis*
form

(2)
MW 142.11

(3)
MW 348.38, mp 98°C

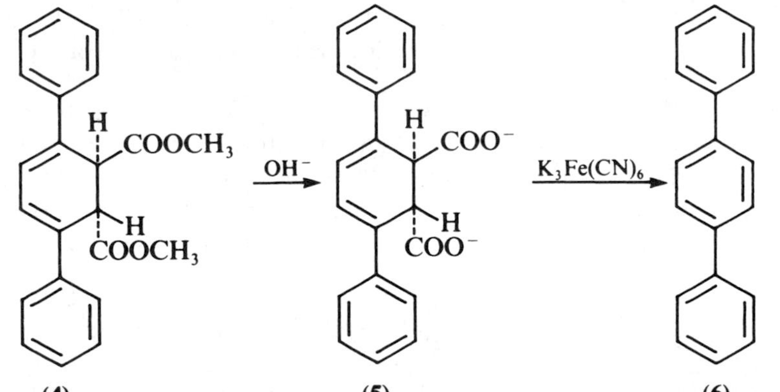

(4)
mp 170°C

(5)

(6)
p-**Terphenyl**
MW 230.29, mp 211°C

E-E-1,4-Diphenyl-1,3-butadiene (**1**) is most stable in the s-*trans* form, but at a suitably elevated temperature the s-*cis* form present in the equilibrium adds to dimethyl acetylenedicarboxylate (**2**) to give dimethyl 1,4-diphenyl-1,4-dihy-

Diels-Alder reaction

drophthalate (**3**). This low-melting ester is obtained as an oil and, when warmed briefly with methanolic potassium hydroxide, is isomerized to the high-melting *E*-ester (**4**). The free *E*-acid can be obtained in 86% yield by refluxing the suspension of (**3**) in methanol for 4 h; but in the recommended procedure the isomerized ester is collected, washed to remove dark mother liquor, and hydrolyzed by brief heating with potassium hydroxide in a high-boiling solvent. The final step, an oxidative decarboxylation, is rapid and nearly quantitative. It probably involves reaction of the oxidant with the dianion (**5**) with removal of two electrons and formation of a diradical, which loses carbon dioxide with formation of *p*-terphenyl (**6**).

Experiments

1. Microscale Synthesis of *p*-Terphenyl

CAUTION: Handle dimethyl acetylenedicarboxylate with care.

In a 10 × 100 mm reaction tube place 100 mg of *E,E*-1,4-diphenyl-1,3-butadiene prepared by the Wittig reaction in Chapter 39. Add 74 mg of dimethyl acetylenedicarboxylate (*caution:* skin irritant[1]), followed by 0.35 mL of triethylene glycol dimethyl ether (triglyme, bp 222°C) and a boiling chip. Reflux the mixture on a hot sand bath for 30 min.

Reaction time: 30 min

Cool the yellowish solution in cold water, add 2.5 mL of *tert*-butyl methyl ether, and extract the solution three times with 1.5-mL portions of water to remove the high-boiling solvent, shake the ethereal solution once with saturated sodium chloride solution, and dry the solution over calcium chloride pellets. Add the drying agent until it no longer clumps together. Remove the ether solution with a Pasteur pipette, and place it in a tared reaction tube. Wash off the calcium chloride pellets with more ether. Evaporate the solvent under a stream of nitrogen or air, removing the last traces under aspirator vacuum.

$$CH_3OCH_2CH_2OCH_2$$
$$|$$
$$CH_3OCH_2CH_2OCH_2$$

**Triethylene glycol
dimethyl ether (triglyme)**
Miscible with water; bp 222°C

CAUTION: Forms peroxides. Discard 90 days after container is opened.

While the evaporation is in progress, dissolve 35 mg of potassium hydroxide in 0.35 mL of methanol by heating and stirring. Crystallization of the yellow oil containing **3** can be initiated by cooling and scratching; this provides assurance that the reaction has proceeded properly. Transfer the methanolic potassium hydroxide, and heat with stirring on the hot sand bath for about 1 min until a stiff paste of crystals of the isomerized ester (**4**) appears. Cool, thin the mixture with methanol, collect the product on the Hirsch funnel, wash it free of dark mother liquor, and spread it thinly on a paper for rapid drying. The yield of pure, white ester (**4**) is about 120 mg. Solutions in methanol are strongly fluorescent.

Rapid hydrolysis of the hindered ester (4)

Potassium hydroxide is easily powdered in a blender.

Place the ester **4** in a reaction tube, add 50 mg of potassium hydroxide, and add 0.35 mL of triethylene glycol. Stir the mixture with a thermometer, and heat, raising the temperature to 140°C in the course of about 5 min. By intermittent heating, keep the temperature close to 140°C for 5 min longer, and then cool the mixture under the tap. Pour into a 4-in. test tube, and rinse the reaction

---

1. This ester is a powerful lachrymator (tear producer) and vesicant (blistering agent) and should be dispensed from a bottle provided with a pipette and a pipetter. Even a trace of ester on the skin should be washed off promptly with methanol, followed by soap and water.

tube with about 3.3 mL of water. Heat to boiling, and in case there is a small precipitate or the solution is cloudy, add a little pelletized Norit decolorizing charcoal, swirl, and filter the alkaline solution. Then add 225 mg of potassium ferricyanide, and heat on the hot sand bath for about 5 min to dissolve the oxidant and to coagulate the white precipitate, which soon separates and is collected by filtration on the Hirsch funnel. The product can be air-dried overnight or dried to constant weight by heating in an evacuated reaction tube on the steam bath. The yield of colorless *p*-terphenyl (mp 209–210°C) is 50 to 60 mg.

Cleaning Up   The aqueous layer from the reaction after dilution can be flushed down the drain. Allow the ether to evaporate from the calcium chloride, and then discard it in the nonhazardous solid waste container. The aqueous reaction mixture after dilution with water is neutralized with dilute hydrochloric acid and then flushed down the drain.

2. Macroscale Synthesis of *p*-Terphenyl

CAUTION: Handle dimethyl acetylenedicarboxylate with care.

$$CH_3OCH_2CH_2OCH_2$$
$$|$$
$$CH_3OCH_2CH_2OCH_2$$

Triethylene glycol dimethyl ether (triglyme)
Miscible with water. bp 222°C

Reaction time: 30 min; reflux over a flame

CAUTION: Extinguish all flames before working with ether.

Potassium hydroxide is easily powdered in a Waring blender.

Place 1.5 g of *E,E*-1,4-diphenyl-1,3-butadiene (from the Wittig reaction, Chapter 39) and 1.0 mL (1.1 g) of dimethyl acetylenedicarboxylate (*caution:* skin irritant[2]) in a 25 × 150-mm test tube, and rinse down the walls with 5 mL of triethylene glycol dimethyl ether (triglyme) (bp 222°C). Clamp the test tube in a vertical position, introduce a cold finger condenser, and reflux the mixture gently for 30 min. Alternatively, carry out the experiment in a 25-mL round-bottomed flask equipped with a reflux condenser. Cool the yellowish solution under the tap, pour into a separatory funnel, and rinse out the reaction vessel with a total of about 50 mL of *t*-butyl methyl ether. Extract twice with water (50–75 mL portions) to remove the high-boiling solvent, shake the ethereal solution with saturated sodium chloride solution, and dry the ether layer over anhydrous calcium chloride pellets. Filter or decant the ether solution into a tared 125-mL Erlenmeyer flask, and evaporate the *t*-butyl methyl ether on the steam bath, eventually with evacuation at the aspirator, until the weight of yellow oil is constant; yield is about 2.5–2.8 g.

While evaporation is in progress, dissolve 0.5 g of potassium hydroxide (about 5 pellets) in 10 mL of methanol by heating and swirling; the process is greatly hastened by crushing the lumps with a stirring rod with a flattened head. Crystallization of the yellow oil containing (**3**) can be initiated by cooling and scratching; this provides assurance that the reaction has proceeded properly. Pour in the methanolic potassium hydroxide, and heat with swirling on the hot plate for about 1 min until a stiff paste of crystals of the isomerized ester (**4**) appears. Cool, thin the mixture with methanol, collect the product, wash it free of dark mother liquor, and spread it thinly on a paper for rapid drying. The yield of pure, white ester (**4**) is 1.7–1.8 g. Solutions in methanol are strongly fluorescent.

---

2. This ester is a powerful lachrymator (tear producer) and vesicant (blistering agent) and should be dispensed from a bottle provided with a pipette and a pipetter. Even a trace of ester on the skin should be washed off promptly with methanol, followed by soap and water.

*Rapid hydrolysis of the
hindered ester (4)*

Place the ester (**4**) in a 25 × 150-mm test tube, add 0.7 g of potassium hydroxide (7–8 pellets), and pour in 5 mL of triethylene glycol. Stir the mixture with a thermometer, and heat, raising the temperature to 140°C in the course of about 5 min. By intermittent heating, keep the temperature close to 140°C for 5 min longer, and then cool the mixture under the tap. Pour into a 125-mL Erlenmeyer flask, and rinse the tube with about 50 mL of water. Heat to boiling, and, in case there is a small precipitate or the solution is cloudy, add a little pelletized Norit, swirl, and filter the alkaline solution by gravity. Then add 3.4 g of potassium ferricyanide, and heat on the hot plate with swirling for about 5 min to dissolve the oxidant and to coagulate the white precipitate that soon separates. The product can be air-dried overnight or dried to constant weight by heating in an evacuated Erlenmeyer flask on the steam bath. The yield of colorless *p*-terphenyl, mp 209–210°C, is 0.7–0.8 g.

Cleaning Up   The aqueous layer from the reaction after dilution can be flushed down the drain. Allow the ether to evaporate from the calcium chloride, and then discard it in the nonhazardous solid waste container. The aqueous reaction mixture after dilution with water is neutralized with dilute hydrochloric acid and then flushed down the drain.

Computational Chemistry

Using a molecular mechanics program, calculate the steric energies of all possible *cis-* and *trans-* isomers as well as all s-*cis-* and s-*trans-* isomers. There are six possible combinations of these. Which is the most stable?

Questions _____

1. What is the driving force for the isomerization of **3** to **4**?

2. Why is the *trans* and not the *cis* diester **4** formed in the isomerization of **3** to **4**?

3. Why does hydrolysis of **4** in methanol require 4 h, whereas hydrolysis in triethylene glycol requires only 10 min?

Prelab Exercise: Write a detailed mechanism for the formation of tetraphenylcyclopentadienone from benzil and 1,3-diphenylacetone. To which general class of reactions does this condensation belong?

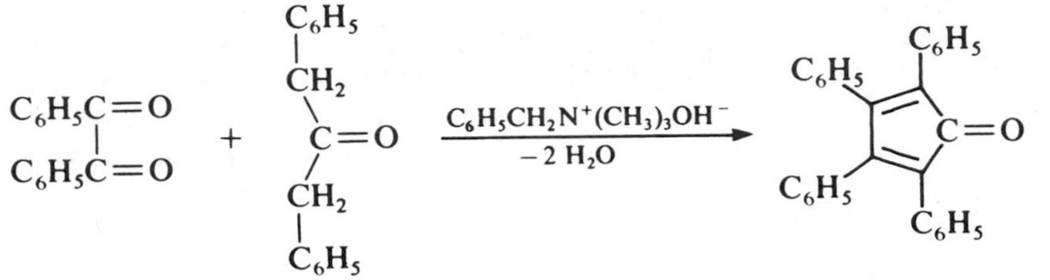

$$C_6H_5C=O \atop C_6H_5C=O \quad + \quad {C_6H_5 \atop \underset{C_6H_5}{\overset{CH_2}{\underset{CH_2}{C=O}}}} \quad \xrightarrow[-2\ H_2O]{C_6H_5CH_2N^+(CH_3)_3OH^-} \quad \text{Tetraphenylcyclopentadienone}$$

Benzil
MW 210.22
mp 96°C

1,3-Diphenylacetone
MW 210.26
mp 35°C

Tetraphenylcyclopentadienone
MW 384.45 mp 219°C

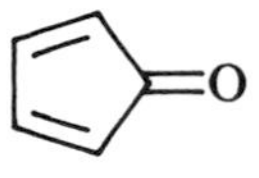

Cyclopentadienone

Louis Fieser introduced the idea of using very high-boiling solvents to speed up this and many other reactions.

Cyclopentadienone is an elusive compound that has been sought for many years but with little success. Molecular orbital calculations predict that it should be highly reactive, and so it is; it exists only as the dimer. The tetraphenyl derivative of this compound is synthesized in this experiment. This derivative is stable and reacts readily with dienophiles. It is used not only for the synthesis of highly aromatic, highly arylated compounds, but also for examination of the mechanism of the Diels-Alder reaction itself. Tetraphenylcyclopentadienone has been carefully studied by means of molecular orbital methods in attempts to understand its unusual reactivity, color, and dipole moment.

The literature procedure for condensation of benzil with 1,3-diphenylacetone in ethanol with potassium hydroxide as basic catalyst suffers from the low boiling point of the alcohol and the limited solubility of both potassium hydroxide and the reaction product in this solvent. Triethylene glycol is a better solvent and permits operation at a higher temperature. In the procedure that follows, the glycol is used with benzyltrimethylammonium hydroxide, a strong base readily soluble in organic solvents, which serves as catalyst.

In the next chapter tetraphenylcyclopentadienone will be used to synthesize hexphenylbenzene and dimethyl tetraphenylphthalate. The mechanism for the formation of tetraphenycyclopentadienone is as follows:

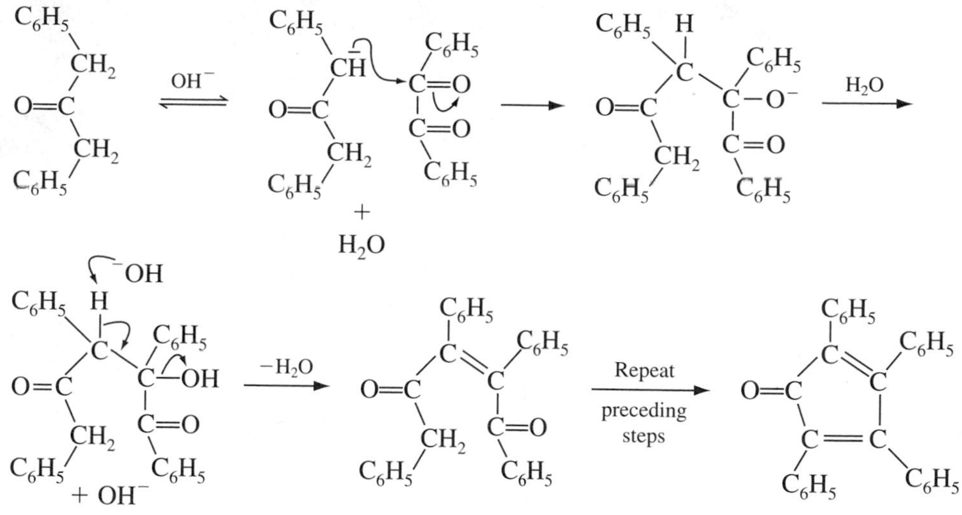

Experiments

1. Tetraphenylcyclopentadienone

For preparation of benzil, see Chapter 55. Into a 10×100 mm reaction tube place 42 mg of pure benzil (free of benzoin), 42 mg of 1,3-diphenylacetone, and 0.4 mL of triethylene glycol, using the solvent to wash the walls of the tube. Clamp the tube over a hot sand bath, insert a thermometer, and heat the solution until the benzil is dissolved. Remove the tube from the heat, and then, using a 1-mL syringe, add to the solution 0.20 mL of a 40% solution of benzyltrimethylammonium hydroxide in methanol (Triton B) when the temperature of the solution reaches exactly 100°C. Stir once to mix. Crystallization usually starts in 10 to 20 s. Let the mixture cool to near room temperature, and then cool it in cold water. Add 0.5 mL of methanol, cool the tube in ice, and collect the product.

Short reaction period

CAUTION: *Triton B is toxic and corrosive.*

If the crystals are large enough, collection can be done by inserting a Pasteur pipette into the tube and removing the solvent between the tip of the pipette and the bottom of the tube (Fig. 52.1). If the crystals are small, then collect them on the Wilfilter, the Hirsch funnel, or the microscale Büchner funnel (Fig. 52.2). In any case, wash the crystals with cold methanol until the washings are purple-pink, not brown. The yield of deep purple crystals is about 60 mg. If either the crystals are not well formed or the melting point is low, place the material in a reaction tube, add 0.6 mL of triethylene glycol, stir with a thermometer, and raise the temperature to 220°C to bring the solid into solution. Let it stand for crystallization (if initially pure material is recrystallized, the recovery is about 90%).

Cleaning Up   Since the filtrate and washings from the reaction contain Triton B, they should be placed in the hazardous waste container. Crystallization solvent should be diluted with water and flushed down the drain.

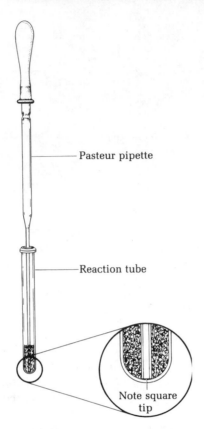

FIG. 52.1 Pasteur pipette filtration technique. Solvent is removed between pipette tip and bottom of tube, leaving crystals in reaction tube.

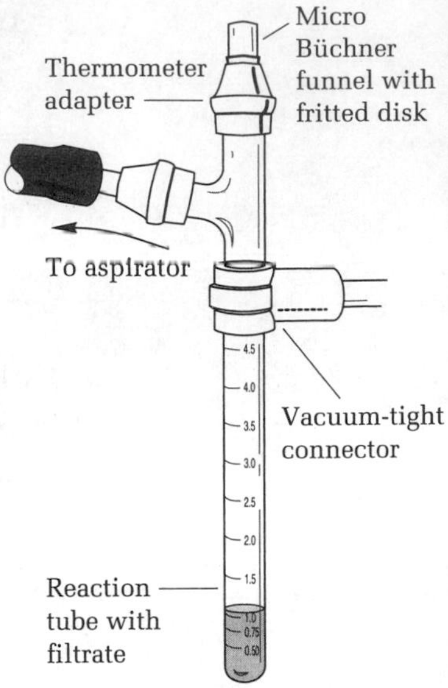

FIG. 52.2 Apparatus for filtration on a microscale Büchner funnel.

🍶 **MACROSCALE**

Short reaction period

CAUTION: Triton B is toxic and corrosive.

2. Tetraphenylcyclopentadienone

Measure into a 25 × 150 mm test tube 2.1 g of benzil, 2.1 g of 1,3-diphenylacetone, and 10 mL of triethylene glycol, using the solvent to wash the walls of the test tube. Support the test tube in a hot sand bath, stir the mixture with a thermometer, heat until the benzil is dissolved, and then remove it from the sand. Measure 1 mL of a commercially available 40% solution of benzyltrimethylammonium hydroxide (Triton B) in methanol into a 10 × 75-mm test tube, adjust the temperature of the solution to exactly 100°C, remove from heat, add the catalyst, and stir once to mix. Crystallization usually starts in 10–20 sec. Let the temperature drop to about 80°C, and then cool under the tap, add 10 mL of methanol, stir to a thin crystal slurry, collect the product, and wash it with methanol until the filtrate is purple-pink, not brown. The yield of deep purple crystals is 3.3–3.7 g. If either the crystals are not well formed or the melting point is low, place 1 g of material and 10 mL of triethylene glycol in a vertically supported test tube, stir with a thermometer, raise the temperature to 220°C to bring the solid into solution, and let stand for crystallization. (If initially pure material is recrystallized, the recovery is about 90%.)

Cleaning Up   Since the filtrate and washings from the reaction contain Triton B, they should be placed in the hazardous waste container. Crystallization solvent should be diluted with water and flushed down the drain.

Question

Draw the structure of the dimer of cyclopentadienone. Why doesn't tetraphenylcyclopentadienone undergo dimerization? This will become much clearer if an energy-minimized structure is generated using a molecular mechanics program.

Hexaphenylbenzene and Dimethyl Tetraphenylphthalate

Prelab Exercise: Explain the driving force behind the loss of carbon monoxide from the intermediates formed in these two reactions.

This experiment illustrates two examples of the Diels-Alder reaction, a means of synthesizing molecules that would be extremely difficult to synthesize in any other way. Both reactions employ as the diene the tetraphenylcyclopentadienone prepared in the previous chapter. Although the Diels-Alder reaction is reversible, the intermediate in each of these reactions spontaneously loses carbon monoxide (why?) to form the products.

Hexaphenylbenzene melts at 465°C without decomposition. Few completely covalent organic molecules have higher melting points. Lead melts at 327.5°C. As is often the case, high melting point also means limited solubility. The solvent used to recrystallize hexaphenylbenzene, diphenyl ether, is very high boiling (bp 259°C) and has superior solvent power.

In the first experiment the dienone is condensed with dimethyl acetylenedicarboxylate using as the solvent 1,2-dichlorobenzene. This solvent is chosen for its solvent properties as well as its high boiling point, which guarantees the reaction is complete in a minute or two.

Experiments

1. Dimethyl Tetraphenylphthalate

Microscale Procedure

CAUTION: Work in a hood. Dimethyl acetylenedicarboxylate is corrosive and a lachrymator. 1,2-Dichlorobenzene is a toxic irritant.

Measure into a reaction tube 50 mg of tetraphenylcyclopentadienone (prepared in Chapter 52), 0.4 mL of 1,2-dichlorobenzene, and 27.5 mg of dimethyl acetylenedicarboxylate. Clamp the tube over a hot sand bath, insert a thermometer, and raise the temperature to the boiling point (180–185°C). Boil gently until there is no further color change, and let the rim of condensate rise just high enough to wash the walls of the tube. The pure adduct is colorless, and if the starting ketone is adequately pure, the color changes from purple to pale tan. A 5-min boiling period should be sufficient. Cool the tube to 100°C, slowly stir in 0.6 mL of 95% ethanol, and let crystallization proceed. After the mixture has cooled to near room temperature, cool it in ice. Then collect the product, either by removal of the solvent with a Pasteur pipette or, if the crystals are too small for this technique, on

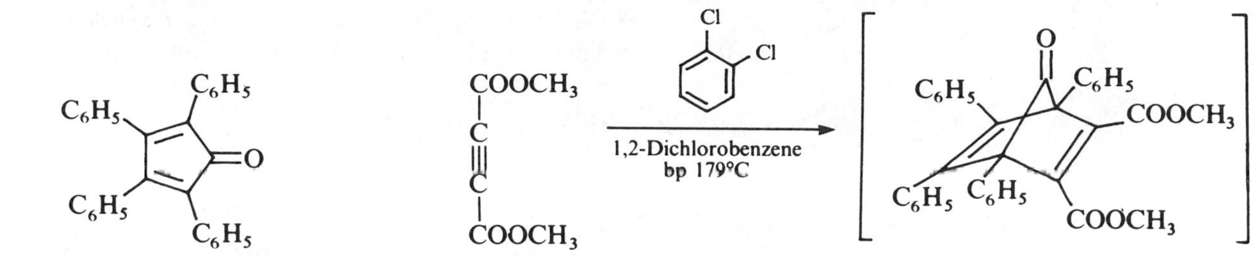

Tetraphenylcyclopentadienone
MW 384.45, mp 219°C

Dimethyl acetylenedicarboxylate
MW 142.11, bp ~300°C

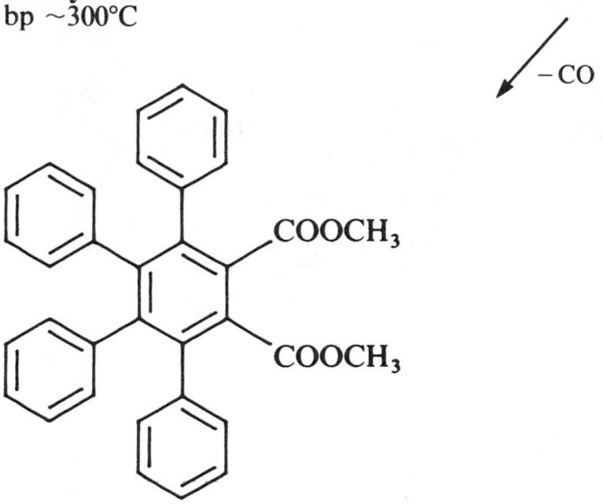

1,2-Dichlorobenzene
bp 179°C

− CO

Dimethyl tetraphenylphthalate
MW 474.53, mp 258°C

the Hirsch funnel or Wilfilter. Wash the crystals with cold methanol. The yield of dimethyl tetraphenylphthalate will be about 50 mg.

Cleaning Up   Since the filtrate from this reaction contains 1,2-dichlorobenzene, it should be placed in the halogenated organic solvents container.

 Macroscale Procedure

CAUTION: Work in a hood. Dimethyl acetylenedicarboxylate is corrosive and a lachrymator. 1,2-Dichlorobenzene is a toxic irritant.

Measure into a 25 × 150-mm test tube 2 g of tetraphenylcyclopentadienone, 10 mL of 1,2-dichlorobenzene, and 1 mL (1.1 g) of dimethyl acetylenedicarboxylate. Clamp the test tube in a hot sand bath, insert a thermometer, and raise the temperature to the boiling point (180–185°C). Boil gently until there is no further color change, and let the rim of condensate rise just high enough to wash the walls of the tube. The pure adduct is colorless, and if the starting ketone is adequately pure the color changes from purple to pale tan. A 5-min boiling period should be sufficient. Cool to 100°C, slowly stir in 15 mL of 95% ethanol, and let crystallization proceed. Cool under the tap, collect the product, and rinse the tube with methanol. The yield of colorless crystals should be in the range of 2.1–2.2 g.

Reaction time: about 5 min

Cleaning Up Since the filtrate from this reaction contains 1,2-dichlorobenzene, it should be placed in the halogenated organic solvents container.

2. Hexaphenylbenzene

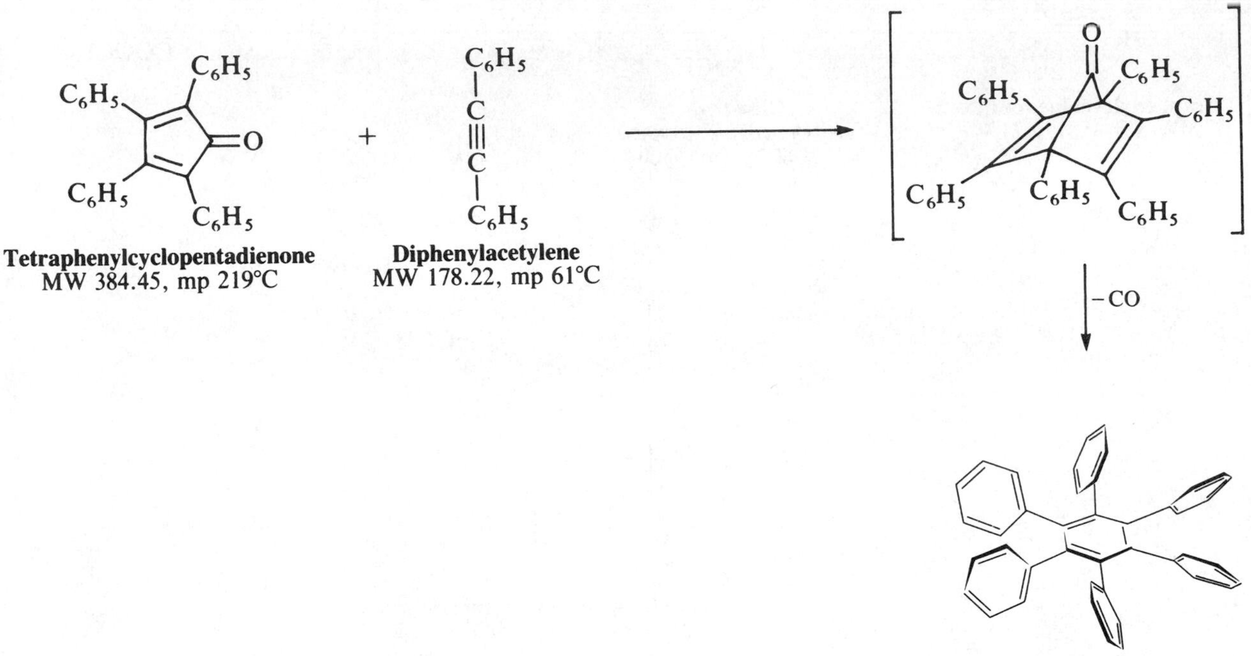

Tetraphenylcyclopentadienone
MW 384.45, mp 219°C

Diphenylacetylene
MW 178.22, mp 61°C

Hexaphenylbenzene
MW 534.66
mp 465°C
Energy-minimized conformation

Diphenylacetylene is a less reactive dienophile than dimethylacetylenedicarboxylate; but when heated with tetraphenylcyclopentadienone without solvent a temperature (ca. 380–400°C) suitable for reaction can be attained. In the following procedure the dienophile is taken in large excess to serve as solvent. Since refluxing diphenylacetylene (bp about 300°C) keeps the temperature below the melting point of the product, removal of the diphenylacetylene lets the reaction mixture melt, which ensures completion of the reaction.

Microscale Procedure

Place 50 mg each of tetraphenylcyclopentadienone (prepared in Chapter 52) and diphenylacetylene (prepared in Chapter 59) in a reaction tube, clamp it upright, and heat the mixture with the flame of a microburner. This is the only experiment in this text that requires a flame to conduct a microscale reaction. Use care to ensure that no flammable solvents are nearby. Soon after the reactants have melted with strong bubbling, the white product becomes visible. Let the diphenylacetylene reflux briefly on the walls of the tube, and then remove some of this excess diphenylacetylene by inserting a Pasteur pipette into the vapors above the solid and drawing the hot vapors into the pipette. Repeat this operation until it is possible, through strong heating with the flame, to melt the mixture completely. Then let the melt cool and solidify. Add 0.25 mL of diphenyl ethel (bp 259°C), heat the mixture *carefully* to dissolve the solid, and then let the product crystallize slowly. Extinguish the flame, and when the tube is cold, add 0.5 mL of toluene to dilute the mixture. Collect the product using the Pasteur pipette method or on the Hirsch funnel by vacuum filtration, and wash it with toluene. The yield of colorless plates is about 60 mg. Using a Mel-Temp apparatus *equipped with a 500°C thermometer,* determine the melting point. The product should melt at 465°C.

Cleaning Up   Place all filtrates in the organic solvent container.

 ### Macroscale Procedure

A hydrocarbon of very high mp. Although the small amount of carbon monoxide produced probably presents no hazard, work in a hood.

Place 0.5 g each of tetraphenylcyclopentadienone and diphenylacetylene in a 25 × 150 mm test tube supported by a clamp, and heat the mixture strongly with the free flame of a microburner held in the hand (do not insert a thermometer into the test tube; the temperature will be too high). Soon after the reactants have melted with strong bubbling, white masses of the product become visible. Let the diphenylacetylene reflux briefly on the walls of the tube, and then remove some of the diphenylacetylene by letting it condense for a minute or two on a cold finger condenser filled with water, but without fresh water running through it (Fig. 53.1). Remove the flame, withdraw the cold finger, and wipe it with a towel. Repeat the operation until you are able, by strong heating, to melt the mixture completely. Then let the melt cool and solidify. Add 10 mL of diphenyl ether, using it to rinse the walls. Heat *carefully* over a free flame to dissolve the solid, and then let the product crystallize. When cold, add 10 mL of toluene to thin the mixture, collect the product, and wash with toluene. The yield of colorless plates, mp 465°, is 0.6–0.7 g.

Notes:

1. The melting point of the product can be determined with a Mel-Temp apparatus and a 500°C thermometer. To avoid oxidation, seal the sample in an evacuated capillary tube.
2. In case the hexaphenylbenzene is contaminated with insoluble material, crystallization from a filtered solution can be accomplished as follows: Place 10 mL of diphenyl ether in a 25 × 150-mm test tube, pack the sample of hexaphenylbenzene into a 10-mm extraction thimble, and suspend this in the test tube with two nichrome wires, as shown in Fig. 53.1. Insert a cold finger condenser supported by a filter adapter, and adjust the

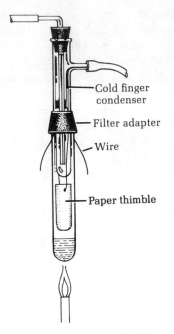

- Cold finger condenser
- Filter adapter
- Wire
- Paper thimble

FIG. 53.1 Soxhlet-type extractor.

FIG. 53.2 Research-quality Soxhlet extractor. Solvent vapors from the flask rise through A and up into the condenser. As they condense the liquid just formed returns to B, where sample to be extracted is placed. (Bottom of B is sealed at C.) Liquid rises in B to level D, at which time the automatic siphon, E, starts. Extracted material accumulates in the flask as more pure liquid vaporizes to repeat the process.

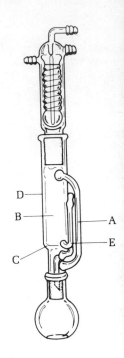

condenser and the wires so that condensing liquid will drop into the thimble. Let the diphenyl ether reflux until the hexaphenylbenzene in the thimble is dissolved, and then let the product crystallize, add toluene, collect the product, and wash with toluene as described previously. A larger version of this apparatus is shown in Fig. 53.2.

Cleaning Up   Place all filtrates in the organic solvents container.

Computational Chemistry

Bicyclo[2,2,1]hepta-2,5-dien-7-one   **Benzene**   **Carbon monoxide**

Calculate, using the AM1 semiempirical method, the heats of formation of bicyclo[2,2,1]hepta-2,5-dien-7-one and of benzene and carbon monoxide in order to confirm theoretically that a reaction might occur.

Questions

1. What two factors probably contribute to the very high melting points of these two hexasubstituted benzenes?

2. What volume of carbon monoxide, measured at STP, is produced by the decomposition of 38 mg of tetraphenylcyclopentadienone?

3. Is the energy-minimized conformation of hexaphenylbenzene a chiral molecule?

Derivatives of 1,2-Diphenylethane: A Multistep Synthesis[1]

Procedures are given in the next eight chapters for rapid preparation of small samples of twelve related compounds starting with benzaldehyde and phenylacetic acid. The quantities of reagents specified in the procedures are often such as to provide somewhat more of each intermediate than is required for completion of subsequent steps in the sequence of reactions. If the experiments are dovetailed, the entire series of preparations can be completed in very short working time. For example, one can start the preparation of benzoin (record the time of starting; do not rely on memory), and during the reaction period start the preparation of α-phenylcinnamic acid; this requires refluxing for 35 min, and while it is proceeding the benzoin preparation can be stopped when the time is up and the product allowed to crystallize. The α-phenylcinnamic acid mixture can be let stand (and cooled) until one is ready to isolate it. Also, while a crystallization is proceeding, one may want to observe the crystals occasionally but should utilize most of the time for other operations.

Points of interest concerning stereochemistry and reaction mechanisms are discussed in the introductions to the individual chapters. Because several of the compounds have characteristic ultraviolet or infrared absorption spectra, pertinent spectroscopic constants are recorded, and brief interpretations of the data are presented. Molecular mechanics calculations give insight into the conformations of several of these compounds.

Question

Starting with 150 mg of benzaldehyde and assuming an 80% yield on each step, what yield of diphenylacetylene, in grams, might you expect? From the information given and assuming an 80% yield on the last two reactions, what yield of stilbene dibromide would you expect employing 150 mg of benzaldehyde?

Note for the instructor

1. If the work is well organized and proceeds without setbacks, the experiments can be completed in about four laboratory periods. The instructor may elect to name a certain number of periods in which the student is to make as many of the compounds as possible; the instructor may also decide to require submission only of the end products in each series.

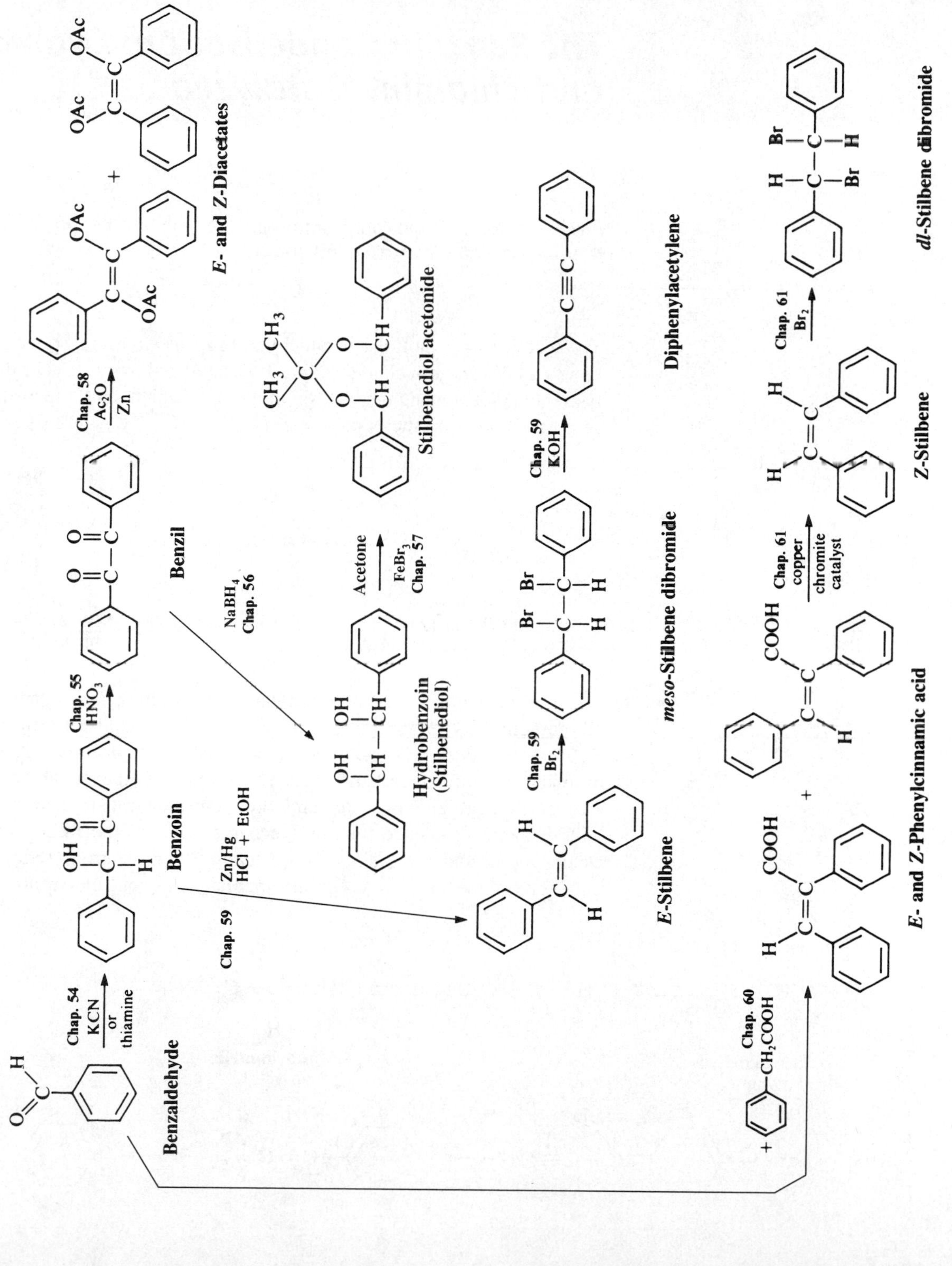
Benzaldehyde
Chap. 54
KCN
or
thiamine
Benzoin
Chap. 55
HNO3
Benzil
Chap. 58
Ac2O
Zn
E- and Z-Diacetates
NaBH4
Chap. 56
Hydrobenzoin
(Stilbenediol)
Acetone
FeBr3
Chap. 57
Stilbenediol acetonide
Chap. 59
Zn/Hg
HCl + EtOH
E-Stilbene
Chap. 59
Br2
meso-Stilbene dibromide
Chap. 59
KOH
Diphenylacetylene
Chap. 60
E- and Z-Phenylcinnamic acid
Chap. 61
copper
chromite
catalyst
Z-Stilbene
Chap. 61
Br2
dl-Stilbene dibromide

The Benzoin Condensation: Cyanide Ion and Thiamine Catalyzed

Prelab Exercise: What purposes does the sodium hydroxide serve in the thiamine-catalyzed benzoin condensation?

The reaction of 2 mol of benzaldehyde to form a new carbon–carbon bond is known as the *benzoin condensation.* It is catalyzed by two rather different catalysts—cyanide ion and the vitamin thiamine—which, on close examination, are seen to function in exactly the same way.

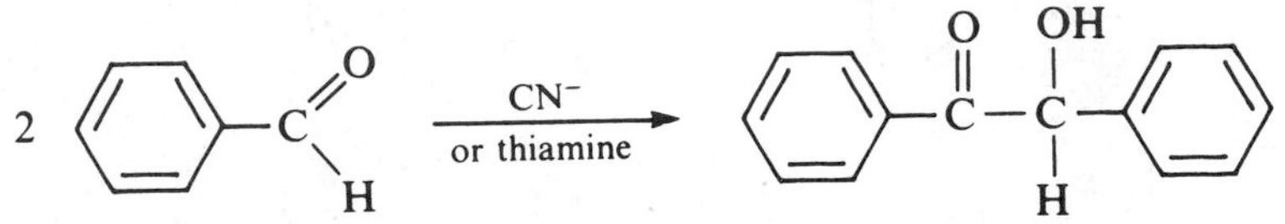

Benzaldehyde
MW 106.12
bp 178°C, den. 1.044

Benzoin
MW 212.24
mp 135°C

Consider first the cyanide ion-catalyzed reaction. The cyanide ion attacks the carbonyl oxygen to form a stable cyanohydrin, mandelonitrile, a liquid of bp 170°C that under the basic conditions of the reaction loses a proton to give a resonance-stabilized carbanion, **A.** The carbanion attacks another molecule of benzaldehyde to give **B,** which undergoes a proton transfer and loses cyanide to give benzoin. Evidence for this mechanism lies in the failure of 4-nitrobenzaldehyde to undergo the reaction, because the nitro group reduces the nucleophilicity of the anion in **A.** On the other hand, a strong electron-donating group

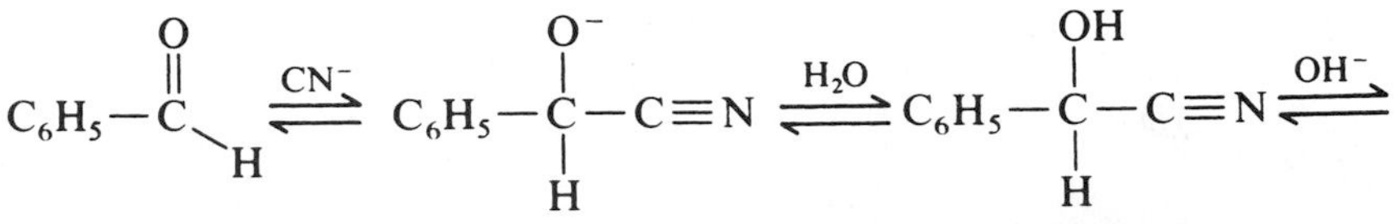

Benzaldehyde

Mandelonitrile
bp 170°C

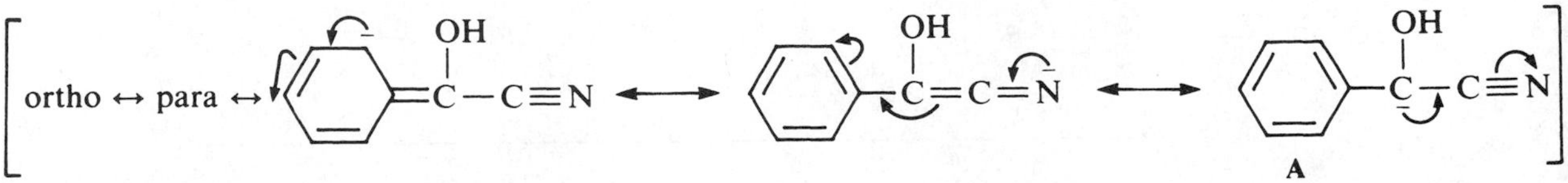

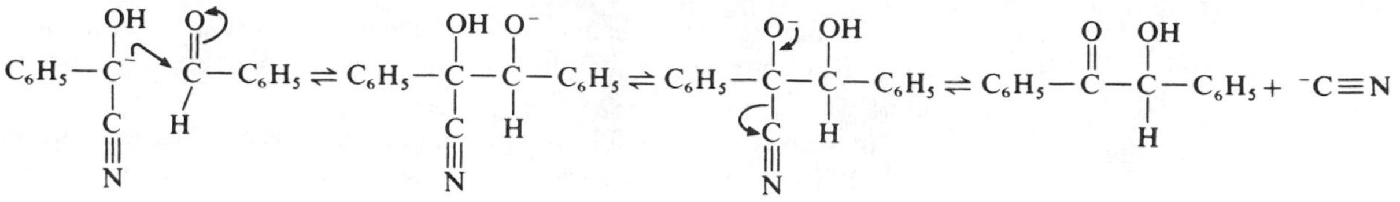

B   **Benzoin**

in the 4-position of the phenyl ring makes the loss of the proton from the cyanohydrin very difficult, and thus 4-dimethylaminobenzaldehyde also does not undergo the benzoin condensation with itself.

A number of biochemical reactions bear a close resemblance to the benzoin condensation but are not, obviously, catalyzed by the highly toxic cyanide ion. Some 30 years ago Breslow proposed that vitamin B_1, thiamine hydrochloride, in the form of the coenzyme thiamine pyrophosphate, can function in a manner completely analogous to cyanide ion in promoting reactions like the benzoin condensation. The resonance-stabilized conjugate base of the thiazolium ion, thiamine, and the resonance-stabilized carbanion, **C**, which it forms, are again the keys to the reaction. Like the cyanide ion, the thiazolium ion has just the right balance of nucleophilicity, ability to stabilize the intermediate anion, and good leaving group qualities.

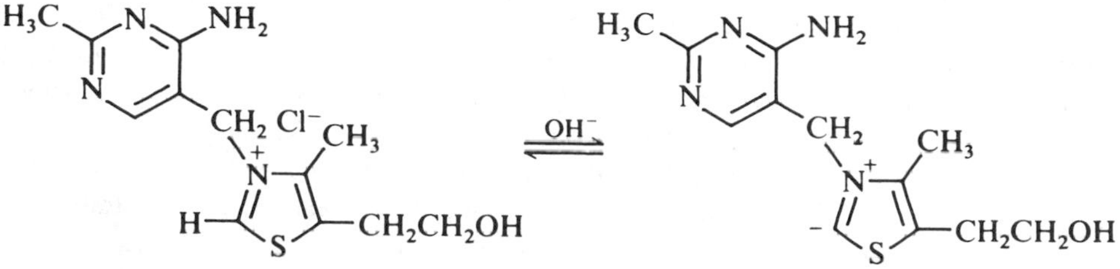

Cyanide ion binds irreversibly to hemoglobin, rendering it useless as a carrier of oxygen.

Thiamine hydrochloride   **Thiamine**

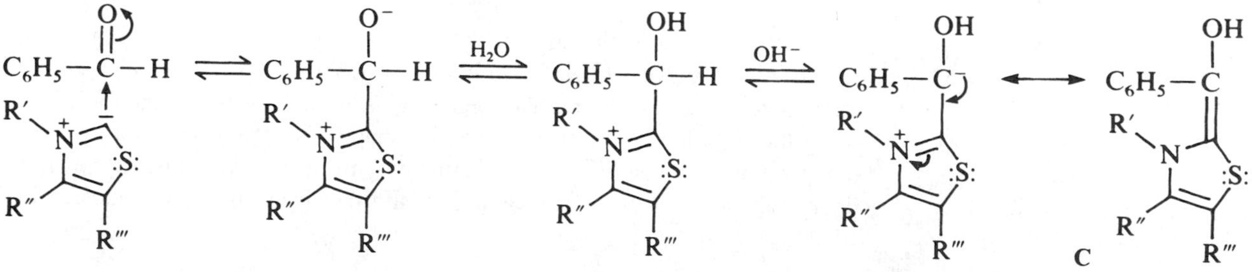

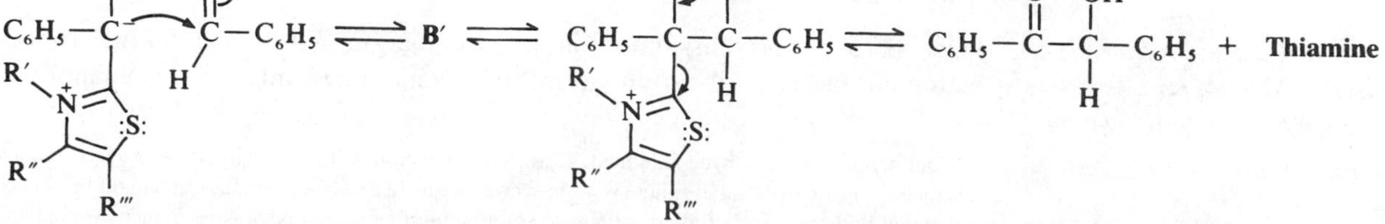

The importance of thiamine is evident in that it is a vitamin, an essential substance that must be provided in the diet to prevent beriberi, a nervous system disease.

In the reactions that follow, cyanide ion functions as a fast and efficient catalyst, although in large quantities it is highly toxic. The amount of potassium cyanide used in the present experiment (15 mg) is about eight times less than the average fatal dose, a difference that underlines the advantage of carrying out organic experiments on a microscale.

The thiamine-catalyzed reaction is much slower, but the catalyst is edible.

Experiments

1. Cyanide Ion–Catalyzed Benzoin Condensation

Microscale Procedure

In a reaction tube place 15 mg of potassium cyanide (**poison!**), dissolve it in 0.15 mL of water, add 0.30 mL of 95% ethanol and from a small, accurate syringe 0.15 mL (or 157 mg) of pure benzaldehyde,[1] introduce a boiling chip, and reflux the solution gently on a warm sand bath or a steam bath for 30 min (Fig. 54.1). Remove the tube, cool it in an ice bath, and if no crystals appear within a few minutes, withdraw a drop on a stirring rod and rub it against the inside of the tube to induce crystallization. When crystallization is complete, remove the solvent using a Pasteur pipette, and while keeping it on ice, wash the crystals thoroughly with 1 mL of a 1 : 1 mixture of 95% ethanol and water. Mix the crystals with the wash solvent, and then isolate them by filtration on the Hirsch funnel or on the Wilfilter. This material is usually colorless and has a melting point of 134 to 135°C, indicating that it is pure. The usual yield is 100 to 120 mg.

Cleaning Up   Add the aqueous filtrate to 10 mL of a 1% sodium hydroxide solution. Add 10 mL of household bleach (5.25% sodium hypochlorite) to oxidize the cyanide ion. The resulting solution can be tested for cyanide using the Prussian blue test described in Chapter 70, Part 3a. When cyanide is not present, the solution can be diluted with water and flushed down the drain. Ethanol used in crystallization should be placed in the organic solvents container.

Macroscale Procedure

Place 0.75 g of potassium cyanide (see margin note) in a 50-mL round-bottomed flask, dissolve it in 7.5 mL of water, add 15 mL of 95% ethanol and

Potassium cyanide
POISON: *Do not handle if you have open cuts on your hands. Never acidify a cyanide solution (HCN gas is evolved). Wash hands after handling cyanide.*

CAUTION: *Potassium cyanide is poisonous. Do not handle if you have open cuts on your hands.*

Note for the instructor

---

1. Commercial benzaldehyde inhibited against autoxidation with 0.1% hydroquinone is usually satisfactory. If the material available is yellow or contains benzoic acid crystals, it should be shaken with equal volumes of 5% sodium carbonate solution until carbon dioxide is no longer evolved and the upper layer dried over calcium chloride. Distillation is not necessary.

FIG. 54.1   Microscale reflux apparatus.

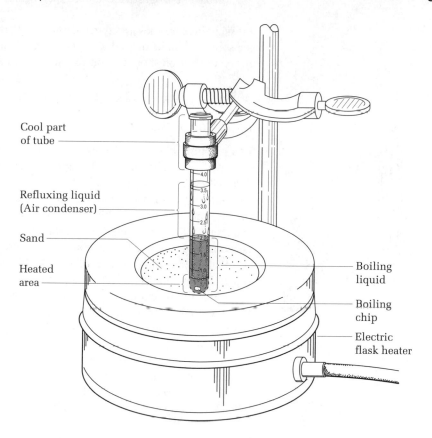

Cool part of tube

Refluxing liquid (Air condenser)

Sand

Heated area

Boiling liquid

Boiling chip

Electric flask heater

Never acidify a cyanide solution (HCN gas is evolved). Wash hands after handling cyanide.

7.5 mL of pure benzaldehyde,[2] introduce a boiling stone, attach a short condenser, and reflux the solution gently on the flask heater for 30 min (Fig. 54.2). Remove the flask, cool it in an ice bath, and, if no crystals appear within a few minutes, withdraw a drop on the stirring rod and rub it against the neck of the flask to induce crystallization. When crystallization is complete, collect the product and wash it free of yellow mother liquor with a 1:1 mixture of 95% ethanol and water. Usually this first-crop material is colorless and of satisfactory melting point (134–135°C); usual yield 5–6 g.[3]

Cleaning Up   Add the aqueous filtrate to 5 mL of a 1% sodium hydroxide solution. Add 25 mL of household bleach (5.25% sodium hypochlorite) to oxidize the cyanide ion. The resulting solution can be tested for cyanide using the

---

2. Commercial benzaldehyde inhibited against autoxidation with 0.1% hydroquinone is usually satisfactory. If the material available is yellow or contains benzoic acid crystals, it should be shaken with equal volumes of 5% sodium carbonate solution until carbon dioxide is no longer evolved and the upper layer dried over calcium chloride and distilled (bp 178–180°C), with avoidance of exposure of the hot liquid to air. The distillation step can be omitted if the benzaldehyde is colorless.

3. Concentration of the mother liquor to a volume of 10 mL gives a second crop (1.4 g, mp 133–134.5°C); best total yield 6.8 g (87%). Recrystallization can be accomplished with either methanol (11 mL/g) or 95% ethanol (7 mL/g) with 90% recovery in the first crop.

Prussian blue test described in Chapter 70, Part 3a. When cyanide is not present, the solution can be diluted with water and flushed down the drain. Ethanol used in crystallization should be placed in the organic solvents container.

2. Thiamine-Catalyzed Benzoin Condensation

Microscale Procedure

In a reaction tube place 26 mg of thiamine hydrochloride, dissolve it in a drop of water, add 0.30 mL of 95% ethanol, and cool the solution in an ice bath. Add 0.05 mL of 3 M sodium hydroxide, followed by 0.15 mL (157 mg) of pure benzaldehyde,[4] stir well, add the distilling column as an air condenser, and heat the mixture in a water bath at 60°C for 1 to 1.5 h. The progress of the reaction should be followed by thin-layer chromatography. Use benzaldehyde in one lane, benzoin in another, and the reaction mixture in the center. Elute with 50 : 50 petroleum ether–ethyl acetate. Alternatively, the reaction mixture can be stored at room temperature for at least 24 h, although a week will do no harm.

Cool the reaction mixture in an ice bath. If crystallization does not occur, withdraw a drop of solution on a stirring rod and rub it against the inside surface of the tube to induce crystallization. Remove the solvent using a Pasteur pipette, and while keeping the mixture on ice, wash the crystals with a 1 : 1 ice-cold mixture of 95% ethanol and water. The product should be colorless and of sufficient purity (mp 134–135°C) to use in subsequent reactions; the usual yield is 100 to 120 mg. If desired, the moist product can be recrystallized from 95% ethanol (7 mL/g) or methanol (11 mL/g) with 90% recovery.

Cleaning Up   The aqueous filtrate, after neutralization with dilute hydrochloric acid, is diluted with water and flushed down the drain. Ethanol used in crystallization should be placed in the organic solvents container.

Macroscale Procedure

Place 1.3 g of thiamine hydrochloride in a 50-mL Erlenmeyer flask, dissolve it in 4 mL of water, add 15 mL of 95% ethanol, and cool the solution in an ice bath. Add 2.5 mL of 3 M sodium hydroxide dropwise with swirling such that the temperature of the solution does not rise above 20°C. To the yellow solution add 7.5 mL of pure benzaldehyde[5] and heat the mixture at 60°C for 1–1.5 h. The progress of the reaction can be followed by thin-layer chromatography. Alternatively, the reaction mixture can be stored at room temperature for at least 24 h. (The rate of most organic reactions doubles for each 10°C rise in temperature.)

Cool the reaction mixture in an ice bath. If crystallization does not occur, withdraw a drop of solution on a stirring rod and rub it against the inside surface of the flask to induce crystallization. Collect the product by suction filtration, and wash it free of yellow mother liquor with a 1 : 1 mixture of 95% ethanol and

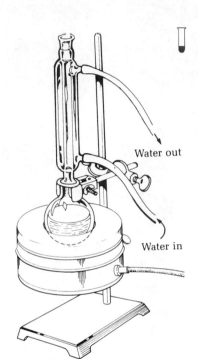

FIG. 54.2 Macroscale reflux apparatus.

Reaction time: 30 min

---

4. See footnote 1.

5. See footnote 1.

water. The product should be colorless and of sufficient purity (mp 134–135°C) to use in subsequent reactions; usual yield is 5–6 g. If desired, the moist product can be recrystallized from 95% ethanol (8 mL/g).

Cleaning Up   The aqueous filtrate, after neutralization with dilute hydrochloric acid, is diluted with water and flushed down the drain. Ethanol used in crystallization should be placed in the organic solvents container.

Questions

1. Speculate on the structure of the compound formed when 4-dimethylaminobenzaldehyde is condensed with 4-chlorobenzaldehyde.

2. Why might the presence of benzoic acid be deleterious to the benzoin condensation?

3. How many π-electrons are in the thiazoline ring of thiamine hydrochloride? of thiamine?

4. Locate the CH and OH protons in the NMR spectrum of benzoin (Fig. 54.3).

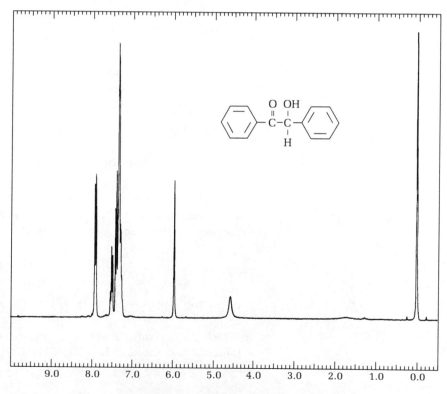

FIG. 54.3 ¹H NMR spectrum of benzoin.

PPM (δ)

Nitric Acid Oxidation. Preparation of Benzil from Benzoin. Synthesis of a Heterocycle: Diphenylquinoxaline

Prelab Exercise: Write a detailed mechanism for the formation of 2,3-dimethylquinoxaline.

Benzoin can be oxidized to the α-diketone, benzil, very efficiently by nitric acid or by copper(II) sulfate in pyridine. On oxidation with sodium dichromate in acetic acid, the yield is lower because some material is converted into benzaldehyde by cleavage of the bond between two oxidized carbon atoms that is activated by both phenyl groups (**a**). Similarly, hydrobenzoin on oxidation with dichromate or permanganate yields chiefly benzaldehyde and only a trace of benzil (**b**).

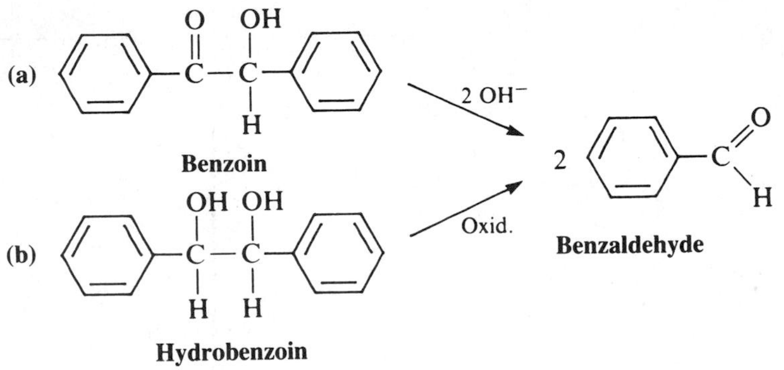

Ultraviolet spectroscopy is used to help characterize aromatic molecules such as benzoin. In Fig. 55.1, the absorption band at 247 nm is attributable to the presence of the phenyl ketone group,

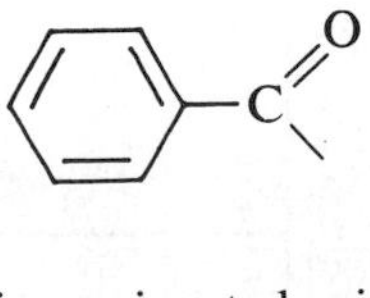

in which the carbonyl group is conjugated with the benzene ring. Aliphatic α,β-unsaturated ketones, R—CH=CH—C=O, show selective absorption of ultraviolet light of comparable wavelength. See the IR spectrum of benzoin in Fig. 55.2.

FIG. 55.1   The ultraviolet spectrum of benzoin. λ_{max}^{EtOH} 247 nm (ε = 13,200). Concentration: 12.56 mg/L = 5.92 × 10^{-5} mol/L. See Chapter 14 (Ultraviolet Spectroscopy) for the relationship between the extinction coefficient, ε, absorbance, A, and concentration, C. The absorption band at 247 nm is attributable to the presence of the phenyl ketone group,

in which the carbonyl group is conjugated with the benzene ring. Aliphatic α,β-unsaturated ketones, R—CH=CH—C=O, show selective absorption of ultraviolet light of comparable wavelength.

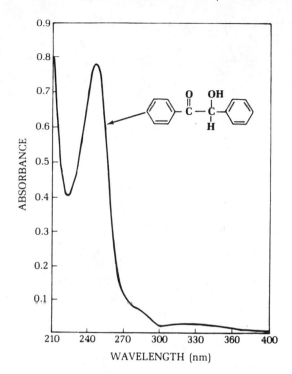

FIG. 55.2   IR spectrum of benzoin (KBr disk).

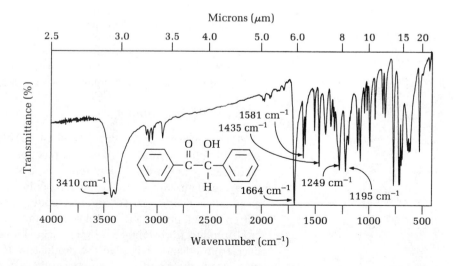

Experiments

1. Nitric Acid Oxidation of Benzoin

Benzoin
MW 212.24, mp 135°C   **Benzil**
                                       MW 210.23, mp 94–95°C

Microscale Procedure

Heat a mixture of 100 mg of benzoin and 0.35 mL of concentrated nitric acid on the steam bath or in a small beaker of boiling water for about 11 min. Carry out the reaction in the hood, or use an aspirator tube near the top of the tube to remove nitrogen oxides. Be sure all the benzoin gets washed down inside the tube and is oxidized. Add 2 mL of water to the reaction mixture, cool to room temperature, and stir the mixture for a minute or two to coagulate the precipitated product; remove the solvent with a Pasteur pipette, and wash the solid with 2 mL more water. Dissolve the solid in 0.5 mL of hot ethanol, and add water dropwise to the hot solution until the solution appears to be cloudy, indicating it is saturated. Heat to bring the product completely into solution, and allow it to cool slowly to room temperature. Cool the tube in ice, and isolate the product using the Wilfilter or the Hirsch funnel. Scrape the benzil onto a piece of filter paper, squeeze out excess solvent, and allow the solid to dry. Record the percentage yield, crystalline form, color, and melting point of the product.

Cleaning Up   The aqueous filtrate should be neutralized with sodium carbonate, diluted with water, and flushed down the drain. Ethanol used in crystallization should be placed in the organic solvents container.

Macroscale Procedure

Reaction time: 10 min

Heat a mixture of 4 g of benzoin and 14 mL of concentrated nitric acid on the steam bath for 11 min. Carry out the reaction under a hood, or use an aspirator tube near the top of the flask to remove nitrogen oxides. Add 75 mL of water to the reaction mixture, cool to room temperature, and swirl for a minute or two to coagulate the precipitated product; collect and wash the yellow solid on a Hirsch funnel, pressing the solid well on the filter to squeeze out the water. This crude product (dry weight 3.7–3.9 g) need not be dried but can be crystallized at once from ethanol. Dissolve the product in 10 mL of hot ethanol, add water dropwise to the cloud point, and set aside to crystallize. Record the yield, crystalline form, color, and mp of the purified benzil. The IR and ^{1}H NMR spectra are found at the end of the chapter (Figs. 55.5 and 55.6).

Test for the Presence of Unoxidized Benzoin.   Dissolve about 0.5 mg of crude or purified benzil in 0.5 mL of 95% ethanol or methanol, and add one drop of 3 *M* sodium hydroxide. If benzoin is present, the solution soon acquires a purplish color owing to a complex of benzil with a product of autoxidation of benzoin. If no color develops in 2–3 min, an indication that the sample is free from benzoin, add a small amount of benzoin, observe the color that develops, and note that if the test tube is stoppered and shaken vigorously the color momentarily disappears; when the solution is then let stand, the color reappears.

Cleaning Up   The aqueous filtrate should be neutralized with sodium carbonate, diluted with water, and flushed down the drain. Ethanol used in crystallization should be placed in the organic solvents container.

2. Preparation of Benzil Quinoxaline

A reaction that characterizes benzil as an α-diketone is a condensation reaction with 1,2-phenylenediamine to the quinoxaline derivative. The aromatic heterocyclic ring formed in the condensation is fused to a benzene ring to give a bicyclic system analogous to naphthalene.

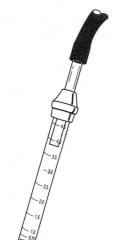

FIG. 55.3   Apparatus for vacuum sublimation of phenylene diamine.

Benzil
MW 210.22, mp 96°C

1,2-Phenylenediamine
MW 108.14
mp 103°C

2,3-Diphenylquinoxaline
MW 282.33, mp 126°C

Microscale Procedure

Handle 1,2-phenylenediamine with care. It is classified as a cancer-suspect agent. Similar compounds (hair dyes) are mild carcinogens. Carry out reaction in hood.

Commercial 1,2-phenylenediamine is usually badly discolored (air oxidation) and gives a poor result unless purified as follows. Place 100 mg of material in a reaction tube, evacuate the tube at full aspirator suction, clamp it, and heat the bottom of the tube with a hot sand bath to distill or sublime colorless 1,2-phenylenediamine from the dark residue into the upper half of the tube (Fig. 55.3). Let the tube cool until the melt has solidified, and scrape out the white solid.

Reaction time: 10 min

Weigh 105 mg of benzil and 54 mg of your purified 1,2-phenylenediamine into a reaction tube, and heat in a steam bath for 10 min, which changes the initially molten mixture to a light-tan solid. Dissolve the solid in hot methanol (about 2.5 mL), and let the solution stand undisturbed. If crystallization does not occur within 10 min, reheat the solution and dilute it with a little water to the point of saturation. The crystals should be filtered on the Hirsch funnel as soon as formed, for brown oxidation products accumulate on standing. The quinoxaline forms colorless needles (mp 125–126°C); yield is about 90 mg.

Cleaning Up The residues from the distillation (sublimation) of 1,2-phenylenediamine (the tube can be rinsed with acetone) and the solvent from the reaction should be placed in the aromatic amines hazardous waste container, because the diamine may be a carcinogen.

 ## Macroscale Procedure

Handle o-phenylenediamine with care. Similar compounds (hair dyes) are mild carcinogens. Carry out reaction in hood.

Commercial *o*-phenylenediamine is usually badly discolored (air oxidation) and gives a poor result unless purified as follows. Place 200 mg of material in a 20 × 150 mm test tube, evacuate the tube at full aspirator suction, clamp it in a horizontal position, and heat the bottom of the tube with a hot sand bath to sublime colorless *o*-phenylenediamine from the dark residue into the upper half of the tube. Let the tube cool in position until the melt has solidified, and scrape out the white solid.

Reaction time: 10 min

Weigh 0.20 g of benzil (theory = 210 mg) and 0.10 g of your purified *o*-phenylenediamine (theory = 108 mg) into a 20 × 150-mm test tube, and heat in a steam bath for 10 min, which changes the initially molten mixture to a light tan solid. Dissolve the solid in hot methanol (about 5 mL), and let the solution stand undisturbed. If crystallization does not occur within 10 min, reheat the solution and dilute it with a little water to the point of saturation. The crystals should be filtered as soon as formed, for brown oxidation products accumulate on standing. The quinoxaline forms colorless needles, mp 125–126°C; yield is 185 mg. The UV spectrum is shown in Fig. 55.4.

Cleaning Up The residues from the distillation of 1,2-phenylenediamine and the solvent from the reaction should all be placed in the aromatic amines hazardous waste container because the diamine may be a carcinogen.

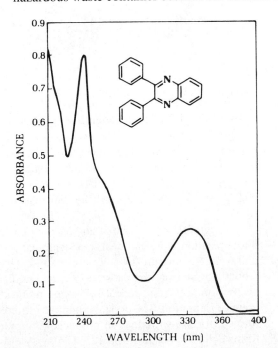

FIG. 55.4 Ultraviolet spectrum of quinoxaline derivative. λ_{max}^{EtOH} 244 nm (ε = 37,400), 345 nm (ε = 12,700).

Questions

1. Assign the peak at 1646 cm^{-1} in the IR spectrum of benzil (Fig. 55.5).

2. Assign the peaks at 3410 cm^{-1} and 1664 cm^{-1} in the infrared spectrum of benzoin (see Fig. 55.2).

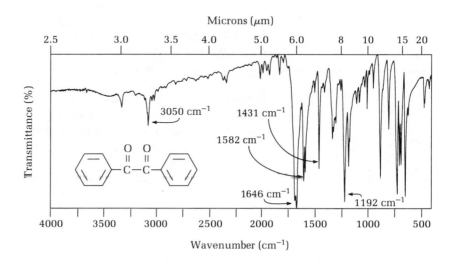

FIG. 55.5   IR spectrum of benzil (KBr disk).

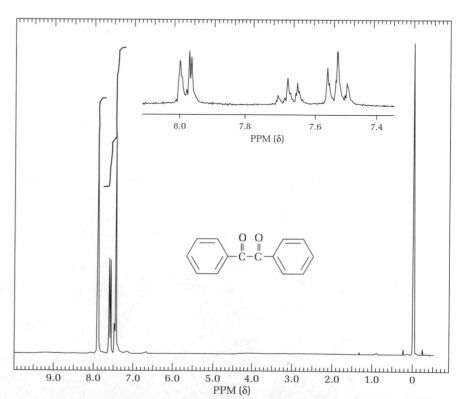

FIG. 55.6 ^{1}H NMR spectrum of benzil (250 MHz).

Borohydride Reduction of a Ketone: Hydrobenzoin from Benzil

Prelab Exercise: Compare the reductive abilities of lithium aluminum hydride with those of sodium borohydride.

Sodium borohydride was discovered in 1943 by H. I. Schlesinger and H. C. Brown. Brown devoted his entire scientific career to this reagent, making it and other hydrides the most useful and versatile of reducing reagents. He received a Nobel prize for his work.

Considering the extreme reactivity of most hydrides (such as sodium hydride and lithium aluminum hydride) toward water, sodium borohydride is somewhat surprisingly sold as a stabilized aqueous solution 14 molar in sodium hydroxide containing 12% sodium borohydride. Unlike lithium aluminum hydride, sodium borohydride is insoluble in ether and soluble in methanol and ethanol.

Sodium borohydride is a mild and selective reducing reagent. In ethanol solution it reduces aldehydes and ketones rapidly at 25°C, esters very slowly, and is inert toward functional groups that are readily reduced by lithium aluminum hydride: carboxylic acids, epoxides, lactones, nitro groups, nitriles, azides, amides, and acid chlorides.

The present experiment is a typical sodium borohydride reduction. These same conditions and isolation procedures could be applied to hundreds of other ketones and aldehydes.

Experiment

MICROSCALE AND MACROSCALE

Sodium Borohydride Reduction of Benzil

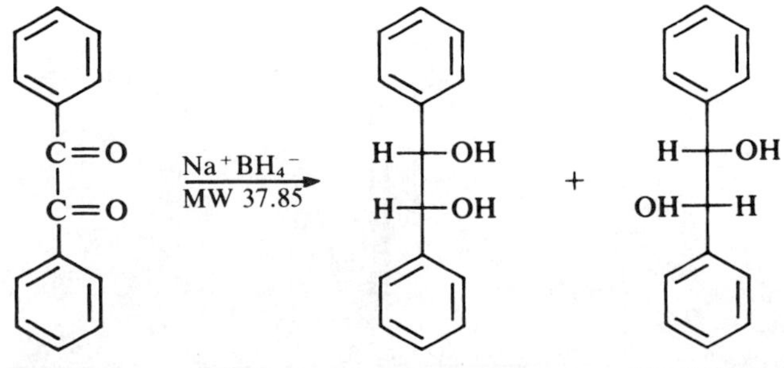

$$\xrightarrow[\text{MW 37.85}]{\text{Na}^+\text{BH}_4^-}$$

Benzil
MW 210.22

(1*R*,2*S*)-(*meso*)-Hydrobenzoin
mp 137°C, MW 214.25

(1*R*,2*R*) and (1*S*,2*S*)-Hydrobenzoin
mp 120°C

Addition of two atoms of hydrogen to benzoin or of four atoms of hydrogen to benzil gives a mixture of stereoisomeric diols, of which the predominant isomer is the nonresolvable (1*R*,2*S*)-hydrobenzoin, the *meso* isomer, accompanied by the enantiomeric (1*R*,2*R*) and (1*S*,2*S*) compounds. The reaction proceeds rapidly at room temperature; the intermediate borate ester is hydrolyzed with water to give the product alcohol.

$$4 R_2C{=}O + Na^+BH_4^- \longrightarrow (R_2CHO)_4B^-Na^+$$

$$(R_2CHO)_4B^-Na^+ + 2 H_2O \longrightarrow 4 R_2CHOH + Na^+BO_2^-$$

The procedure that follows specifies use of benzil rather than benzoin, because you can then follow the progress of the reduction by the discharge of the yellow color of the benzil.

See Fig. 55.5 for the IR spectrum and 55.6 for the ^{1}H NMR spectrum of benzil. The UV spectrum is shown in Fig. 56.2 at the end of this chapter.

Microscale Procedure

In a 10 × 100 mm reaction tube, dissolve 50 mg of benzil in 0.5 mL of 95% ethanol, and cool the solution in ice to produce a fine suspension. Add to this suspension 10 mg of sodium borohydride (a large excess). The benzil dissolves, the mixture warms up, and the yellow color disappears in 2 to 3 min. After a total of 10 min, add 0.5 mL of water, heat the solution to the boiling point, filter the solution in case it is not clear (usually not necessary), and dilute the hot solution with hot water to the point of saturation (cloudiness), a process requiring about 1 mL of water. (1*R*,2*S*)-Hydrobenzoin separates in lustrous thin plates (mp 136–137°C) and is best isolated by withdrawing some of the solvent using a Pasteur pipette and then using the Wilfilter (Fig. 56.1). The yield is about 35 mg. It also can be collected on the microscale Büchner funnel.

Cleaning Up   The aqueous filtrate should be diluted with water and neutralized with acetic acid (to destroy borohydride) before flushing the mixture down the drain.

Macroscale Procedure

In a 50-mL Erlenmeyer flask, dissolve 0.5 g of benzil in 5 mL of 95% ethanol, and cool the solution under the tap to produce a fine suspension. Then add 0.1 g of sodium borohydride (large excess). The benzil dissolves, the mixture warms up, and the yellow color disappears in 2–3 min. After a total of 10 min, add 5 mL of water, heat to the boiling point, filter in case the solution is not clear, dilute to the point of saturation with more water (10 mL), and set the solution aside to crystallize. *meso*-Hydrobenzoin separates in lustrous thin plates, mp 136–137°C; yield is about 0.35 g. A second crop of material can be obtained

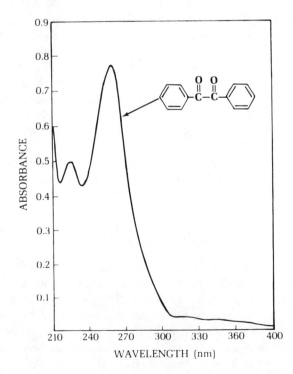

by adding solid sodium chloride to the filtrate until no more of the salt will dissolve.

Cleaning Up   The aqueous filtrate should be diluted with water and neutralized with acetic acid (to destroy borohydride) before flushing the mixture down the drain.

Questions

1. Draw and name, using the *R,S* system of nomenclature, all of the isomers of hydrobenzoin.

2. Calculate the theoretical weight of sodium borohydride needed to reduce 50 mg of benzil.

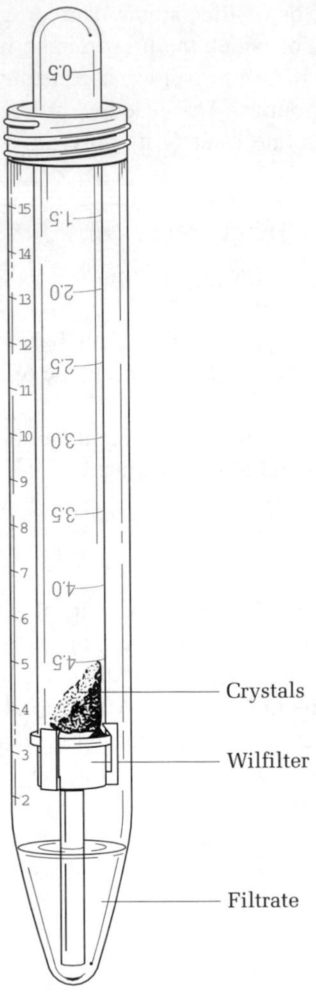

FIG. 56.1   Wilfilter filtration apparatus. See page 52 in Chapter 3.

Crystals

Wilfilter

Filtrate

FIG. 56.2   Ultraviolet spectrum of benzil. λ_{max}^{EtOH} 260 nm ($\varepsilon = 19{,}800$). One-centimeter cells and 95% ethanol have been employed for the UV spectrum.

57

Synthesis of 2,2-Dimethyl-1,5-dioxolane. The Acetonide Derivative of a Vicinal Diol

Prelab Exercise: Draw the most stable structures of *meso-* and *racemic*-1,2-stilbenediol using Newman projections. Draw the same two isomers with the hydroxyl groups eclipsed.

In the previous experiment, yellow benzil was reduced with sodium borohydride to the colorless diol. The melting point of the product, 137°C, indicates, according to the experiment, that (1*R*,2*S*)-(*meso*)-hydrobenzoin has been produced. But how do we *know* that is the isomer produced? The object of the present experiment is to demonstrate, using logic based on nuclear magnetic resonance (NMR) spectral data, that the product is indeed the *meso* and not the *racemic* isomer.

The diol is reacted with acetone to form a crystalline acetonide (a ketal) that will have either structure **1** or **2**. Examination of the NMR spectrum of this derivative will reveal whether the two methyl groups of the acetonide are equivalent or not and thus whether the diol was *meso* or *racemic*. Rather than simply explaining each step in this process and the reasoning behind it, we have elected to let you discover them for yourself by answering the questions at the end of this experiment. By considering each question in turn you will see how it is possible to determine the stereochemistry of the starting diol.

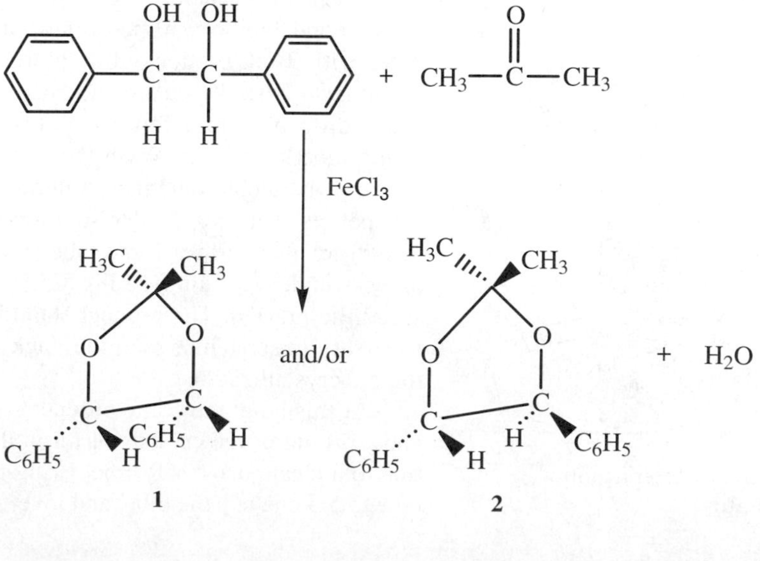

1   **2**

Experiment

Microscale Procedure

Dissolve 100 mg of the diol in 3 mL of acetone, and add 30 mg of anhydrous iron(III) chloride. The iron(III) chloride is very hygroscopic and therefore should be weighed from the sealed storage container into a vial that can be closed. Reflux the mixture for 20 min using the apparatus depicted in Fig. 57.1.

Transfer the reaction mixture to a 15 mL centrifuge tube containing 4 mL of water and 1 mL of 3 M potassium carbonate solution. Extract the product three times with 1 mL portions of dichloromethane that are transferred to a reaction tube. Wash the combined extracts with 2.5 mL of water, and then dry the solution over a few pieces of granular anhydrous calcium chloride. Transfer the dry solution to a clean reaction tube, and carefully evaporate it to dryness.

Dissolve the crude product in 0.3 to 0.5 mL of hexane, and cool the solution in ice, whereupon the product will crystallize. Collect the product using the Pasteur pipette method or the Wilfilter (Fig. 57.2). Dry the product under vacuum, if necessary, and then record the yield, melting point, and IR spectrum of the product. Record the ^{1}H NMR spectrum of a solution in CDCl$_3$, and assign the stereochemistry of the product and thus of the diol. Store the acetonide in the refrigerator before analysis; it slowly decomposes at room temperature.

Macroscale Procedure

Dissolve 0.30 g of *meso*-hydrobenzoin (*meso*-1,2-stilbenediol) from the previous experiment in 9 mL of acetone, and add 0.1 g of anhydrous iron(III) chloride. The iron(III) chloride is very hygroscopic and therefore should be weighed from the sealed storage container into a vial that can be closed. Reflux the mixture for 20 min using the apparatus depicted in Fig. 25.3 (or 40.6). Be sure to add a boiling chip to the flask.

Transfer the reaction mixture to a small separatory funnel that contains 30 mL of water and 3 mL of 3 *M* potassium carbonate solution. Extract the product three times with 3-mL portions of dichloromethane that are transferred to a 10-mL Erlenmeyer flask. If you are uncertain which layer is the dichloromethane layer, add a drop of it to a few drops of water. If two phases are seen, it is the dichloromethane layer. Wash the combined extracts with 8 mL of water, and then dry the dichloromethane solution over granular anhydrous calcium chloride pellets. Transfer the dry solution to a clean 25-mL filter flask using more dichloromethane to wash the tube and drying agent. Evaporate the solution to dryness under vacuum (see Fig. 25.9). Warm the filter flask with your hand to speed the process. The product should crystallize in the filter flask. It may be necessary to scratch or seed the thick liquid that sometimes results in order to initiate crystallization.

At this point the crude product could be used for NMR analysis. Dissolve about 50 mg of the crystals or oil in deuterochloroform, and transfer the solution to a clean, dry NMR tube. Bring the level of the deuterochloroform in the tube to 5.5 cm, cap the tube, and invert it several times to mix the solution.

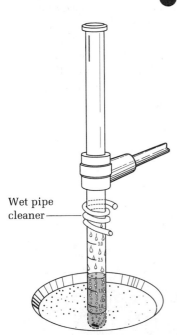

Wet pipe cleaner

FIG. 57.1 Esterification apparatus.

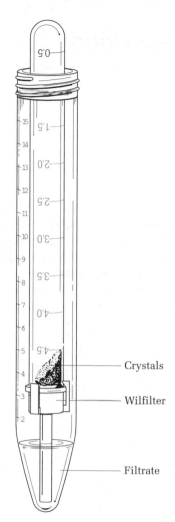

FIG. 57.2 Wilfilter filtration apparatus. See page 52 in Chapter 3.

To purify the crude product, dissolve it in the minimum quantity (0.9 to 1.5 mL) of dichloromethane, transfer the solution to a reaction tube or small test tube, evaporate the dichloromethane, seed the resulting oil and then add 0.9 to 1.5 mL of hexane, stir the mixture, and allow crystallization to proceed. Cool the solution in ice. Collect the acetonide on a small Hirsch funnel. Dry it under vacuum if necessary (Fig. 3.22) and then record the yield, melting point, and IR spectrum of the product.

Record the ^{1}H NMR spectrum of a solution in CDCl$_3$, and assign the stereochemistry of the product and thus of the diol. Store the acetonide in the refrigerator before analysis; it slowly decomposes at room temperature.

Cleaning Up Place any excess or recovered dichloromethane in the halogenated waste container. Other organic solutions go in the organic solvents container. The aqueous phases from the extractions can, after dilution with water, be washed down the drain. The drying agent, calcium chloride pellets, after removal of solvent, can be placed in the nonhazardous solid waste container.

Questions

1. Write the mechanism and discuss the stereochemistry of the sodium borohydride reduction of diphenylethanedione (benzil).

2. Give the mechanism of acetonide formation. What is the role of the iron(III) chloride?

3. Carefully draw the structures, using Newman projections, of (1R,2S)-(*meso*)-hydrobenzoin and (1R,2R) and (1S,2S)-(*racemic*)-hydrobenzoin with the hydroxyl groups eclipsed. Then draw the corresponding acetonides of these two isomers.

4. Is acetonide **1,** above, from the *meso* or the *racemic* diol?

5. Consider the symmetry of the two different acetonides formed. How many methyl peaks would you expect to observe in the NMR spectrum for the acetonide of the *meso*-isomer? For the *racemic*-isomer?

6. Compare and contrast the infrared spectra of the dione, the diol, and the acetonide. Assign major peaks to functional groups.

1,4-Addition: Reductive Acetylation of Benzil

Prelab Exercise: Write the complete mechanism for the reaction of acetic anhydride with an alcohol.

Ac = acetyl group

$$-OAc = -OCCH_3$$

$$CH_3C-O-CCH_3$$

Ac_2O *is acetic anhydride.*

In one of the first demonstrations of the phenomenon of 1,4-addition, Johannes Thiele (1899) established that reduction of benzil with zinc dust in a mixture of acetic anhydride–sulfuric acid involves 1,4-addition of hydrogen to the α-diketone grouping and acetylation of the resulting enediol before it can undergo ketonization to benzoin. The process of reductive acetylation results in a mixture of the *E*- and *Z*-isomers **1** and **2**. Thiele and subsequent investigators isolated the more soluble, lower melting *Z*-stilbenediol diacetate (**2**) only in impure form, mp 110°C. Separation of the two isomers by chromatography is not feasible because they are equally adsorbable on alumina. However, separation is possible by fractional crystallization (described in the following procedure), and both isomers can be isolated in pure condition. In the method prescribed here for the preparation of the isomer mixture, hydrochloric acid is substituted for sulfuric acid because the latter acid gives rise to colored impurities and is reduced to sulfur and to hydrogen sulfide.[1]

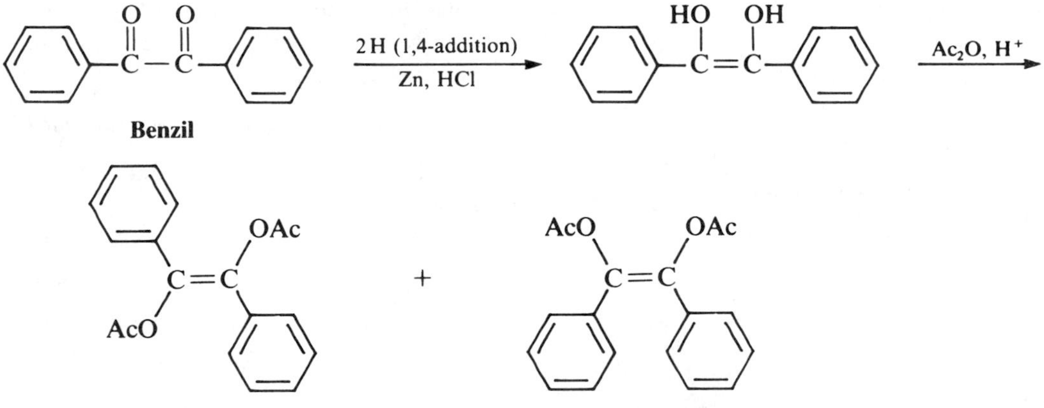

Benzil

$$\xrightarrow[\text{Zn, HCl}]{2\,H\ (1,4\text{-addition})}$$

$$\xrightarrow{Ac_2O,\ H^+}$$

E-Stilbenediol diacetate (1)
mp 155°C, λ_{max}^{EtOH} 271 nm (ϵ = 23,400)
MW 296.31
1

Z-Stilbenediol diacetate (2)
mp 119°C, λ_{max}^{EtOH} 265 nm (ϵ = 12,800)
MW 296.31
2

---

1. If acetyl chloride (2 mL) is substituted for the hydrochloric acid–acetic anhydride mixture in the procedure, the *Z*-isomer is the sole product.

FIG. 58.1 Ultraviolet spectra of *Z*- and *E*-stilbene diacetate. *Z*: λ_{max}^{EtOH} 223 nm ($\varepsilon = 20{,}500$), 269 nm ($\varepsilon = 10{,}800$); *E*: λ_{max}^{EtOH} 272 nm ($\varepsilon = 20{,}800$). In the *Z*-diacetate the two phenyl rings cannot be coplanar, which prevents overlap of the *p*-orbitals of the phenyl groups with those of the central double bond. This steric inhibition of resonance accounts for the diminished intensity of the *Z*-isomer relative to the *E*. Both spectra were run at the same concentration.

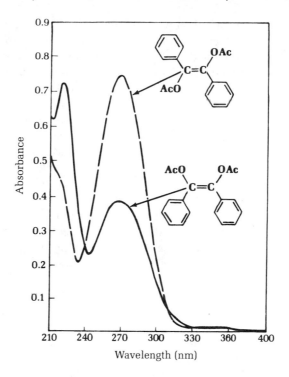

Assignment of configuration

The configurations of this pair of geometrical isomers remained unestablished for over 50 years, but the tentative inference that the higher melting isomer has the more symmetrical *E* configuration eventually was found to be correct. Evidence of infrared spectroscopy is of no avail; the spectra are nearly identical in the interpretable region ($4000–1200$ cm^{-1}) characterizing the acetoxyl groups but differ in the fingerprint region ($1200–400$ cm^{-1}). However, the isomers differ markedly in ultraviolet absorption (Fig. 58.1); and, in analogy to *E*- and *Z*-stilbene, the conclusion is justified that the higher melting isomer, because it has an absorption band at longer wavelength and higher intensity than its isomer, does indeed have the configuration **1**.

Experiments

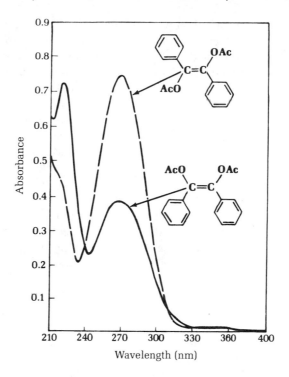 **MICROSCALE**

1. Reductive Acetylation of Benzil

Place one reaction tube containing 0.7 mL of acetic anhydride and another containing 0.1 mL of concentrated hydrochloric acid in an ice bath, and, when both are thoroughly chilled, transfer the acid to the anhydride dropwise by means of a Pasteur pipette. Transfer the acidic anhydride solution into a reaction tube containing 100 mg of pure benzil (prepared in Chapter 55) and 100 mg of zinc dust, and stir the mixture for 2 to 3 min in an ice bath. Remove the tube, and allow it to warm up slowly, cooling in ice if necessary.

Use fresh zinc dust.

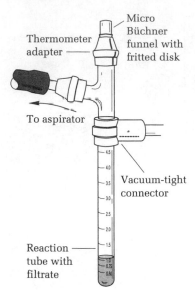

Thermometer adapter

Micro Büchner funnel with fritted disk

To aspirator

Vacuum-tight connector

Reaction tube with filtrate

FIG. 58.2 Microscale filtration assembly.

When there is no further exothermic effect, let the mixture stand for 5 min, and then add 2.5 mL of water. Swirl, break up any lumps of product, and allow a few minutes for hydrolysis of excess acetic anhydride. Then collect the mixture of product and zinc dust on the Hirsch funnel, wash with water, and press and apply suction to the filter cake until there is no further drip. Digest the solid in a 25-mL Erlenmeyer flask (drying is not necessary) with 7.0 mL of diethyl ether to dissolve the organic material, add about 1 g of anhydrous calcium chloride pellets, and swirl briefly; using a Pasteur pipette, transfer the ether solution into another 10-mL Erlenmeyer flask, wash the drying agent, and concentrate the ether solution (steam bath, boiling stone, water aspirator) to a volume of approximately 1.5 mL, transfer to a reaction tube, and let the tube stand, corked and undisturbed.

The E-diacetate (**1**) soon begins to separate in prismatic needles, and, after 20 to 25 min, crystallization appears to stop. Remove the mother liquor from the reaction tube with a Pasteur pipette or use the Wilfilter (Fig. 3.15) or the microscale filtration apparatus shown in Fig. 58.2, and evaporate it to dryness in another reaction tube.

Dissolve the white solid left in the second reaction tube in 1.0 mL of methanol, let the solution stand undisturbed for about 10 min, and drop in one tiny crystal of the E-diacetate. This should give rise, in 20 to 30 min, to a second crop of the E-diacetate. Remove the mother liquor from these crystals and concentrate it to 0.7 mL, let cool to room temperature as before, and again seed with a crystal of E-diacetate; this usually affords a third crop of the E-diacetate. The three crops might be 30 mg (mp 154–156°C), 5 mg (mp 153–154°C), and 5 mg (mp 153–155°C).

At this point the mother liquor should be rich enough in the more soluble Z-diacetate (**2**) for its isolation. Concentrate the methanol mother liquor and washings from the third crop of **1** to a volume of 0.4 mL, stopper the tube, and let the solution stand undisturbed overnight. The Z-diacetate (**2**) sometimes separates spontaneously in large rectangular prisms of great beauty. If the solution remains supersaturated, addition of a seed crystal of **2** causes prompt separation of the Z-diacetate in a paste of small crystals [e.g., 20 mg (mp 118–119°C), then a second crop, 7 mg (mp 116–117°C)]. This experiment points up the great convenience of column chromatography in isomer separations when it can be employed. Separating compounds by fractional crystallization is rarely necessary.

Cleaning Up   Spread the zinc out on a watch glass for about 20 min before wetting it and placing it in the nonhazardous solid waste container. Sometimes zinc dust from a reaction like this is pyrophoric (spontaneously flammable in air) because it is so finely divided and has such a large, clean surface area able to react with air. The aqueous filtrate should be neutralized with sodium carbonate, zinc salts removed by vacuum filtration, and the filtrate diluted with water and flushed down the drain. The zinc salts are placed in the nonhazardous solid waste container. The filtrate from the fractional crystallization should be placed in the organic solvents container. Allow the solvent to evaporate from the drying agent in the hood, and then place it in the nonhazardous solid waste container.

**MACROSCALE**

Reaction time: 10 min

Carry out reaction in hood.

Use fresh zinc dust.

Handle acetic anhydride and hydrochloric acid with care.

2. Reductive Acetylation of Benzil

Place one test tube (20 × 150 mm) containing 7 mL of acetic anhydride and another (13 × 100 mm) containing 1 mL of concentrated hydrochloric acid in an ice bath. When both are thoroughly chilled, transfer the acid to the anhydride dropwise in not less than 1 min by means of a capillary dropping tube. Wipe the test tube dry, pour the chilled solution into a 50-mL Erlenmeyer flask containing 1 g of pure benzil and 1 g of zinc dust, and swirl for 2–3 min in an ice bath. Remove the flask, and hold it in the palm of your hand; if it begins to warm up, cool further in ice. When there is no further exothermic effect, let the mixture stand for 5 min, and then add 25 mL of water. Swirl, break up any lumps of product, and allow a few minutes for hydrolysis of excess acetic anhydride. Then collect the mixture of product and zinc dust by vacuum filtration, wash with water, and press and apply suction to the cake until there is no further drip. Digest the solid (drying is not necessary) with 70 mL of *t*-butyl methyl ether to dissolve the organic material, add about 4 g of calcium chloride pellets and swirl briefly; filter the solution into a 125-mL Erlenmeyer flask, concentrate the filtrate (steam bath, boiling stone, water aspirator) to a volume of approximately 15 mL,[2] and let the flask stand, corked and undisturbed.

The *E*-diacetate (**1**) soon begins to separate in prismatic needles, and after 20 to 25 min, crystallization appears to stop. Remove the crystals by filtration on the Hirsch funnel. Wash them once with ether, and then evaporate the ether to dryness. Dissolve the residue in 10 mL of methanol, and transfer the hot solution to a 25-mL Erlenmeyer flask.

Let the solution stand undisturbed for about 10 min, and drop in one tiny crystal of the *E*-diacetate (**1**). This should give rise, in 20 to 30 min, to a second crop of the *E*-diacetate. Then concentrate the mother liquor and washings to a volume of 7 to 8 mL, let cool to room temperature as before, and again seed with a crystal of *E*-diacetate; this usually affords a third crop of the *E*-diacetate. The three crops might be 300 mg (mp 154–156°C), 50 mg (mp 153–156°C), and 50 mg (mp 153–155°C). The products also can be collected using the Wilfilter or the microscale Büchner funnel.

At this point the mother liquor should be rich enough in the more soluble *Z*-diacetate (**2**) for its isolation. Concentrate the methanol mother liquor and washings from the third crop of **1** to a volume of 4 to 5 mL, stopper the flask, and let the solution stand undisturbed overnight. The *Z*-diacetate (**2**) sometimes separates spontaneously in large rectangular prisms of great beauty. If the solution remains supersaturated, addition of a seed crystal of (**2**) causes prompt separation of the *Z*-diacetate in a paste of small crystals [e.g., 215 mg (mp 118–119°C); then 70 mg (mp 116–117°C)].

Cleaning Up   Spread the zinc out on a watch glass for about 20 min before wetting it and placing it in the nonhazardous solid waste container. Sometimes

---

2. Measure 15 mL of a solvent into a second flask of the same size, and compare the levels in the two flasks.

zinc dust from a reaction like this is pyrophoric (spontaneously flammable in air) because it is so finely divided and has such a large, clean surface area able to react with air. The aqueous filtrate should be neutralized with sodium carbonate, zinc salts removed by vacuum filtration, and the filtrate diluted with water and flushed down the drain. The zinc salts are placed in the nonhazardous solid waste container. The filtrate from the fractional crystallization should all be placed in the organic solvents container. Allow the solvent to evaporate from the calcium chloride pellets in the hood and then place it in the nonhazardous solid waste container.

Computational Chemistry

Construct *Z*- and *E*-stilbene diacetate, and carry out an energy minimization on the resulting molecules using a molecular mechanics program. Repeat the calculation using the semiempirical molecular orbital program AM1. From the resulting conformations can you explain why the ultraviolet extinction coefficient ε for the peak at 269 nm is smaller in the *Z*-isomer than in the *E*-isomer?

Synthesis of an Alkyne from an Alkene. Bromination and Dehydrobromination: Stilbene and Diphenylacetylene

Prelab Exercise: Calculate the theoretical quantities of thionyl chloride and of sodium borohydride needed to convert benzoin to *E*-stilbene.

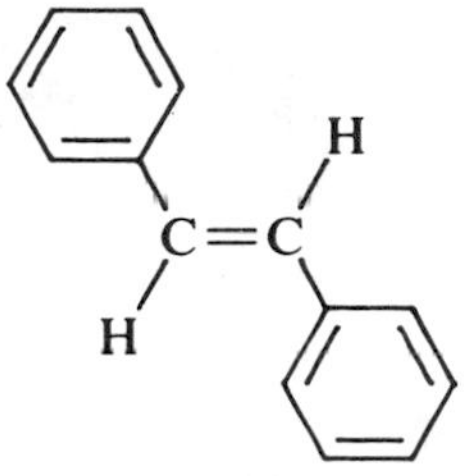

E-Stilbene
mp 125°C, MW 180.24
λ_{max}^{EtOH} 301 nm (ϵ = 28,500)
226 nm (ϵ = 17,700)

Heat of hydrogenation, -20.1 kcal/mol

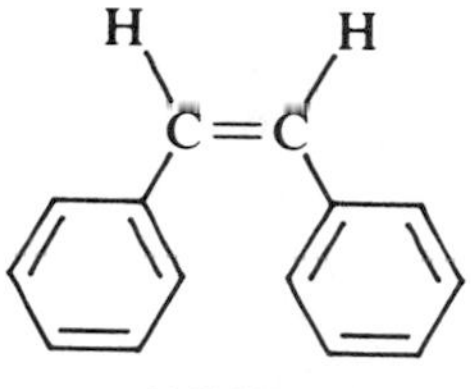

Z-Stilbene
mp 6°C, MW 180.24
λ_{max}^{EtOH} 280 nm (ϵ = 13,500)
223 nm (ϵ = 23,500)

Heat of hydrogenation, -25.8 kcal/mol

In this experiment, benzoin, prepared in Chapter 54, is converted to the alkene *trans*-stilbene (*E*-stilbene), which is in turn brominated and dehydrobrominated to form the alkyne diphenylacetylene.

One method of preparing *E*-stilbene is reduction of benzoin with zinc amalgam in ethanol-hydrochloric acid, presumably through an intermediate:

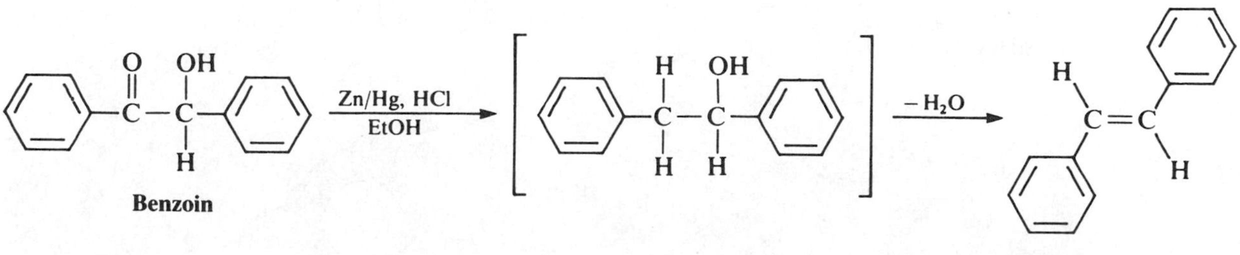

Benzoin

E-Stilbene

The procedure that follows is quick and affords very pure hydrocarbon. It involves three steps: (1) replacement of the hydroxyl group of benzoin by chlorine to form desyl chloride, (2) reduction of the keto group with sodium borohydride to give what appears to be a mixture of the two diastereoisomeric chlorohydrins, and (3) elimination of the elements of hypochlorous acid with zinc and acetic acid. The last step is analogous to the debromination of an olefin dibromide.

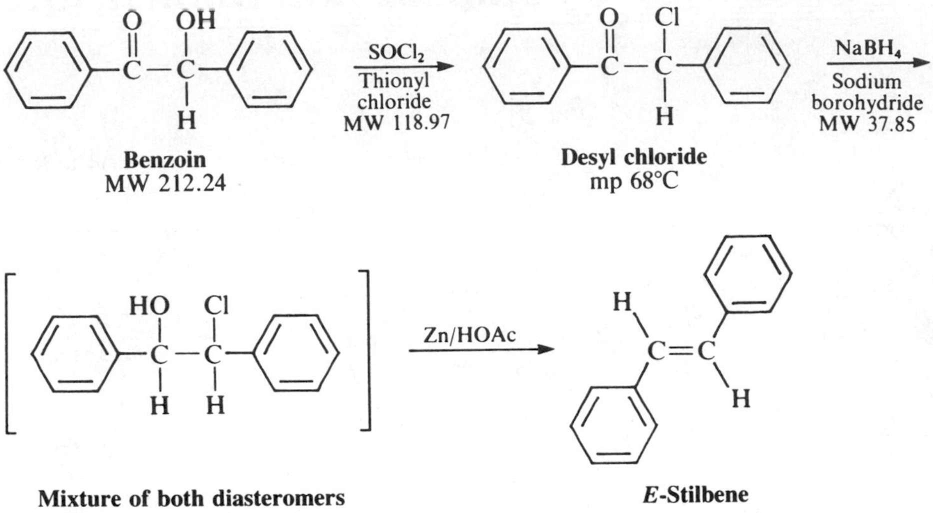

Benzoin
MW 212.24

SOCl₂
Thionyl chloride
MW 118.97

Desyl chloride
mp 68°C

NaBH₄
Sodium borohydride
MW 37.85

Mixture of both diasteromers

Zn/HOAc

E-Stilbene

The minimum energy conformations of *E*- and *Z*-stilbene are shown in Figs. 59.1 and 59.2, respectively. They have been calculated using Spartan's molecular mechanics routine. Note that the *E*-isomer is planar, while the *Z*-isomer is markedly distorted from planarity.

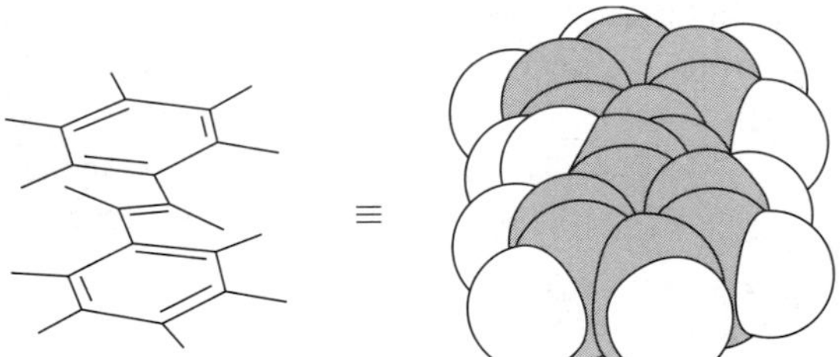

FIG. 59.1 Minimum energy conformation of *E*-stilbene.

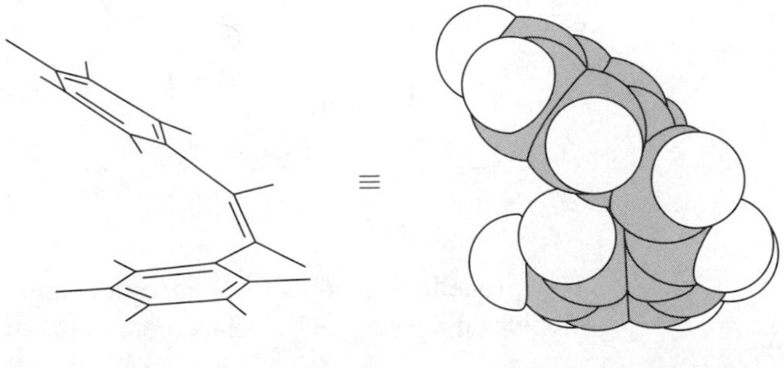

FIG. 59.2 Minimum energy conformation of *Z*-stilbene.

Experiments

1. Stilbene

Place 100 mg of benzoin (crushed to a powder) in a reaction tube, cover it with 0.15 mL of thionyl chloride, and add a boiling chip and a rubber septum bearing a polyethylene tube leading to another reaction tube bearing a plug of damp cotton (Fig. 59.3). This will serve to trap the hydrogen chloride and sulfur dioxide evolved in this reaction. Warm the reaction mixture gently on the steam bath or on the warm sand bath until the solid has dissolved and then more strongly for 5 min.

Caution: If the mixture of benzoin and thionyl chloride is let stand at room temperature for an appreciable time before being heated, an undesired reaction intervenes,[1] and the synthesis of *E*-stilbene is spoiled.

To remove excess thionyl chloride (bp 77°C), connect the reaction tube to the aspirator for a few minutes, add 0.5 mL of petroleum ether (bp 30–60°C), boil it off, and evacuate again. Desyl chloride is thus obtained as a viscous, pale-yellow oil (it will solidify if let stand). Dissolve it in 1 mL of 95% ethanol, cool under the tap, and add 9 mg of sodium borohydride (an excess is harmful). Stir, break up any lumps of the borohydride, and after 10 min add to the solution

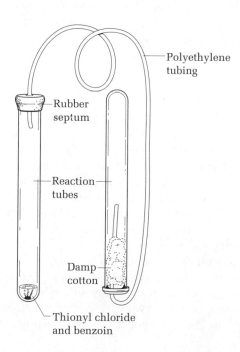

FIG. 59.3 Trap for hydrogen chloride and sulfur dioxide. Warm the tubing in a steam bath in order to bend it permanently.

1. $C_6H_5C{=}O$
   $|$
   C_6H_5CHOH
   $\xrightarrow{\text{SOCl}_2}$
   $C_6H_5C{-}O$
   $\|SO$
   $C_6H_5C{-}O$
   $\xrightarrow{\text{NaBH}_4}$
   C_6H_5CO
   $|$
   $C_6H_5CH_2$
   Desoxybenzoin

FIG. 59.4 Pasteur pipette filtration technique.

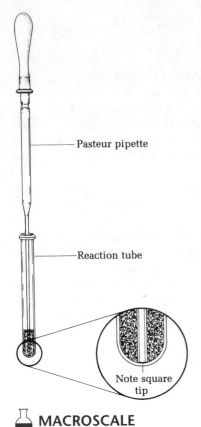

— Pasteur pipette

— Reaction tube

Note square tip

⚗ **MACROSCALE**

Carry out procedure in hood.

of chlorohydrins 60 mg of zinc dust and 0.15 mL of acetic acid and reflux for 1 h. Then cool under the tap. When white crystals separate, add 2 mL of *tert*-butyl methyl ether, and then wash the ether solution once with an equal volume of water containing a drop of concentrated hydrochloric acid (to dissolve basic zinc salts) and then with water. Dry the ether over calcium chloride pellets, which is added until it no longer clumps together. Remove the ether from the drying agent with a Pasteur pipette (Fig. 59.4), wash the drying agent with ether, evaporate the ether to dryness, dissolve the residue in the minimum amount of hot 95% ethanol (about 1 mL), and let the product crystallize. *E*-Stilbene separates in diamond-shaped iridescent plates (mp 124–125°C); the yield is about 50 mg. It can easily be isolated using the Pasteur pipette technique (Fig. 59.4) or the Wilfilter (Fig. 3.15).

Cleaning Up   Combine washings from the cotton in the trap and all aqueous layers, neutralize with sodium carbonate, remove zinc salts by vacuum filtration, and flush the filtrate down the drain with excess water. The zinc salts are placed in the nonhazardous solid waste container. Allow the ether to evaporate from the calcium chloride pellets in the hood, and then place them in the nonhazardous solid waste container. Ethanol mother liquor goes in the organic solvents container. Any zinc isolated should be spread out on a watch glass to dry and air oxidize. It should then be wetted with water and placed in the nonhazardous solid waste container.

2. Stilbene

Place 2 g of benzoin (crushed to a powder) in a 50-mL round-bottomed flask, cover it with 4 mL of thionyl chloride,[2] warm gently on the steam bath (hood) until the solid has all dissolved, and then more strongly for 5 min.

Caution: If the mixture of benzoin and thionyl chloride is left standing at room temperature for an appreciable time before being heated, an undesired reaction intervenes[3] and the synthesis of *E*-stilbene is spoiled.

To remove excess thionyl chloride (bp 77°C), evacuate at the aspirator for a few minutes, add 5 mL of petroleum ether (bp 30–60°C), boil it off (hood), and evacuate again. Desyl chloride is thus obtained as a viscous, pale yellow oil (it will solidify if let stand). Dissolve this in 20 mL of 95% ethanol, cool under the tap, and add 180 mg of sodium borohydride (an excess is harmful). Stir, break up any lumps of the borohydride, and after 10 min add to the solution of chlorohydrins 1 g of zinc dust and 2 mL of acetic acid and reflux for 1 h. Then cool under the tap. When white crystals separate, add 25 mL of

---

2. The reagent can be dispensed from a burette or measured by pipette; in the latter case the liquid should be drawn into the pipette with a pipetter, *not by mouth*.

3. $C_6H_5C{=}O$
   $\quad\quad|$
   $\quad C_6H_5CHOH$
   $\xrightarrow{\text{SOCl}_2}$
   $C_6H_5C{-}O$
   $\quad\quad\|\quad\quad\quad SO$
   $C_6H_5C{-}O$
   $\xrightarrow{\text{NaBH}_4}$
   C_6H_5CO
   $\quad\quad$
   $C_6H_5CH_2$
   Desoxybenzoin

t-butyl methyl ether, and decant the solution from the bulk of the zinc into a separatory funnel. Wash the solution twice with an equal volume of water containing 0.5–1 mL of concentrated hydrochloric acid (to dissolve basic zinc salts) and then, in turn, with sodium carbonate solution and saturated sodium chloride solution. Dry the ether over anhydrous calcium chloride pellets (2 g), filter to remove the drying agent, evaporate the filtrate to dryness, dissolve the residue in the minimum amount of hot 95% ethanol (15–20 mL), and let the product crystallize. *E*-Stilbene separates in diamond-shaped iridescent plates, mp 124–125°C; yield is about 1 g. The UV spectrum is shown at the end of the chapter (Fig. 59.5).

Cleaning Up   Combine washings from the cotton in the trap and all aqueous layers, neutralize with sodium carbonate, remove zinc salts by vacuum filtration, and flush the filtrate down the drain with excess water. The zinc salts are placed in the nonhazardous solid waste container. Allow the ether to evaporate from the sodium sulfate in the hood, and then place it in the nonhazardous solid waste container. Ethanol mother liquor goes in the organic solvents container. Any zinc isolated should be spread out on a watch glass to dry and air oxidize. It should then be wetted with water and placed in the nonhazardous solid waste container.

MICROSCALE

3. *meso*-Stilbene Dibomide

E-Stilbene reacts with bromine predominantly by the usual process of *trans*-addition and affords the optically inactive, nonresolvable *meso*-dibromide; the much lower melting enantiomeric mixture of dibromides is a very minor product of the reaction.

In this experiment, the brominating agent will be pyridinium hydrobromide perbromide,[4] a crystalline, nonvolatile, odorless complex of high molecular weight (319.86), which, in the presence of a bromine acceptor such as an alkene, dissociates to liberate 1 mol of bromine. For microscale experiments it is much more convenient and agreeable to measure and use than free bromine.

Procedure

Pyridine + HBr + Br$_2$

↓

Pyridinium hydrobromide perbromide

CAUTION: Pyridinium hydrobromide perbromide is corrosive and a lachrymator.

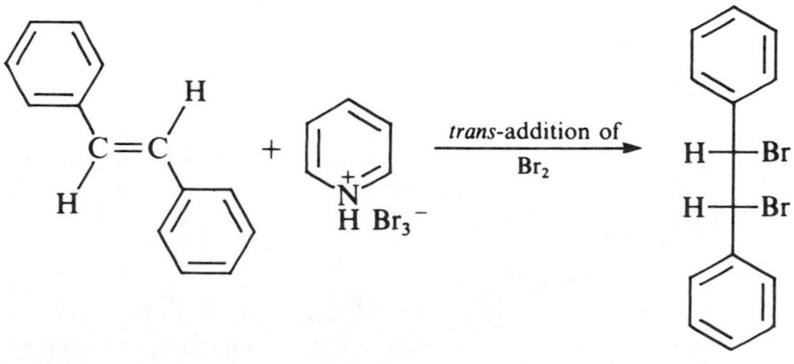

E-Stilbene
MW 180.24

Pyridinium hydrobromide perbromide
MW 319.86

(1 R,2S)-Stilbene dibromide (*meso* isomer)
mw 238°C, MW 340.07

4. Can be purchased from Aldrich, or see Chapter 18 for synthesis.

Total time required: 10 min

Carry out procedure in hood.

In a reaction tube, dissolve 50 mg of *E*-stilbene in 1 mL of acetic acid, by heating on the steam bath or a hot water bath, and then add 100 mg of pyridinium hydrobromide perbromide. Mix by swirling; if necessary, rinse crystals of reagent down the walls of the flask with a little acetic acid, and continue the heating for 1 to 2 min longer. The dibromide separates in small plates. Cool the mixture under the tap, collect the product on the Wilfilter (Fig. 3.15) or on the Hirsch funnel, and wash it with methanol to remove any color; yield of colorless crystals (mp 236–237°C) is 80 mg. Use this material for the preparation of diphenyl-acetylene after determining the percentage yield and melting point.

Cleaning Up   To the filtrate add sodium bisulfite (until a negative test with starch-iodide paper is observed) to destroy any remaining perbromide, neutralize with sodium carbonate, and extract the pyridine released with ether, which goes in the organic solvents container. The aqueous layer can then be diluted with water and flushed down the drain.

MACROSCALE

4. *meso*-Stilbene Dibromide

E-Stilbene reacts with bromine predominantly by the usual process of *trans*-addition and affords the optically inactive, nonresolvable *meso*-dibromide; the much lower melting *dl*-dibromide (Chapter 61) is a very minor product of the reaction.

Procedure

Pyridine + HBr + Br$_2$
↓
Pyridinium hydrobromide perbromide

CAUTION: *Corrosive, lachrymator*

Total time required: 10 min
Carry out procedure in hood.

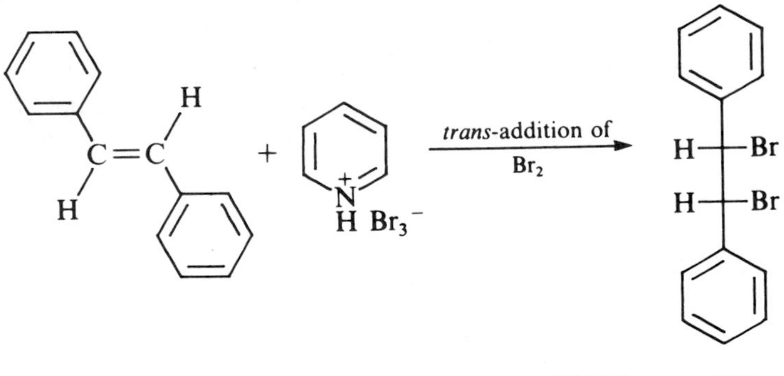

E-Stilbene	**Pyridinium hydro-bromide perbromide**	**(1R,2S)-*meso*-Stilbene dibromide**
MW 180.24	MW 319.86	mp 238°C, MW 340.07

In a 50-mL Erlenmeyer flask dissolve 1 g of *E*-stilbene in 20 mL of acetic acid, by heating on the steam bath, and then add 2 g of pyridinium hydrobromide per-bromide.[5] Mix by swirling; if necessary, rinse crystals of reagent down the walls

5. Can be purchased from Aldrich, or see Chapter 18 for synthesis.

of the flask with a little acetic acid; and continue the heating for 1–2 min longer. The dibromide separates almost at once in small plates. Cool the mixture under the tap, collect the product, and wash it with methanol; yield of colorless crystals, mp 236–237°C, about 1.6 g. Use 0.5 g of this material for the preparation of diphenylacetylene, and turn in the remainder.

Cleaning Up To the filtrate add sodium bisulfite (until a negative test with starch-iodide paper is observed) to destroy any remaining perbromide, neutralize with sodium carbonate, extract the pyridine released during the reaction with ether, which goes in the organic solvents container. The aqueous layer can then be diluted with water and flushed down the drain.

Diphenylacetylene

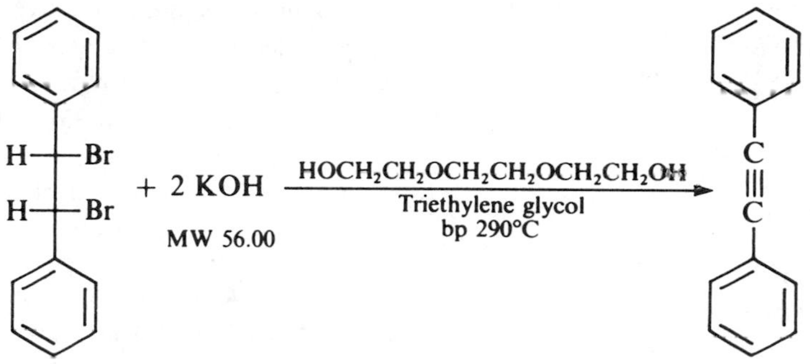
H
Br
H
Br
+ 2 KOH
MW 56.00
$HOCH_2CH_2OCH_2CH_2OCH_2CH_2OH$
Triethylene glycol
bp 290°C
C
C

(1*R*,2*S*)-(*meso*)Stilbene dibromide
MW 340.07

Diphenylacetylene
mp 61°C, MW 178.22

One method for the preparation of diphenylacetylene involves oxidation of benzil dihydrazone with mercuric oxide; the intermediate diazo compound loses nitrogen as formed to give the hydrocarbon:

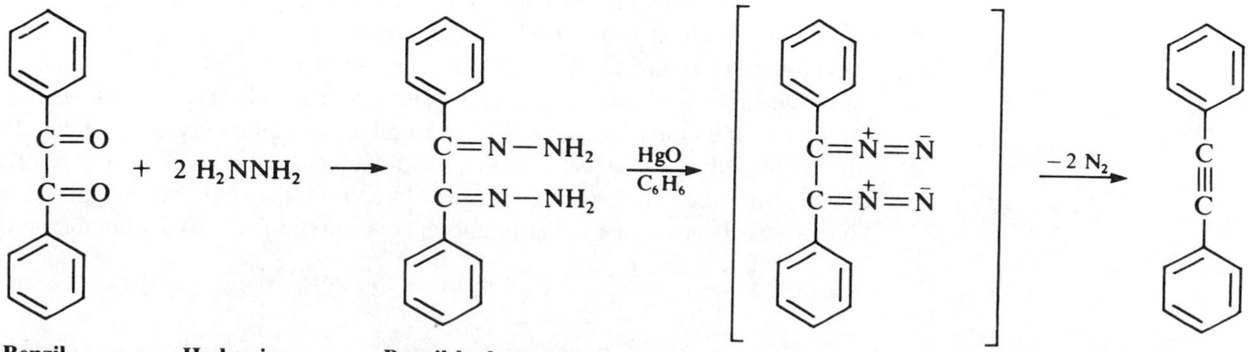
C=O
C=O
+ 2 H_2NNH_2
C=N−NH_2
C=N−NH_2
HgO
C_6H_6
$C=\overset{+}{N}=\overset{-}{N}$
$C=\overset{+}{N}=\overset{-}{N}$
− 2 N_2
C
C

Benzil **Hydrazine** **Benzil hydrazone** **Diphenylacetylene**

The method used in the following procedure involves dehydrohalogenation of *meso*-stilbene dibromide. An earlier procedure called for refluxing the dibromide with 43% ethanolic potassium hydroxide in an oil bath at 140°C for 24 h. In the following procedure the reaction time is reduced to a few minutes by use of the high-boiling triethylene glycol as solvent to permit operation at a higher reaction temperature, a technique introduced by Louis Fieser.

5. Synthesis of Diphenylacetylene

Microscale Procedure

Reaction time: 5 min

CAUTION: *Potassium hydroxide is corrosive to skin. Handle ether away from flames.*

In a reaction tube place 80 mg of *meso*-stilbene dibromide, 40 mg of potassium hydroxide,[6] and 0.5 mL of triethylene glycol. Heat the mixture on a hot sand bath to a temperature of 160°C, when potassium bromide begins to separate. By intermittent heating, keep the mixture at 160 to 170°C for 5 min more, and then cool to room temperature, remove the thermometer, and add 2 mL of water. The diphenylacetylene that separates as a nearly colorless, granular solid is collected using the Pasteur pipette. The crude product need not be dried but can be crystallized directly from 95% ethanol. Let the solution stand undisturbed in order to observe the formation of beautiful, very large spars of colorless crystals. After a first crop has been collected, the mother liquor on concentration affords a second crop of pure product; total yield is 35 mg (mp 60–61°C).

Cleaning Up   Combine the crystallization mother liquor with the filtrate from the reaction, dilute with water, neutralize with 3 M hydrochloric acid, and flush down the drain.

Macroscale Procedure

Reaction time: 5 min

CAUTION: *Potassium hydroxide is corrosive to skin.*

In a 20 × 150-mm test tube place 0.5 g of *meso*-stilbene dibromide, 3 pellets of potassium hydroxide[6] (250 mg), and 2 mL of triethylene glycol. Insert a thermometer into a 10 × 75-mm test tube containing enough triethylene glycol to cover the bulb, and slip this assembly into the larger tube. Clamp the tube in a vertical position in a hot sand bath, and heat the mixture to a temperature of 160°C, when potassium bromide begins to separate. By intermittent heating, keep the mixture at 160–170°C for 5 min more, then cool to room temperature, remove the thermometer and small tube, and add 10 mL of water. The diphenylacetylene that separates as a nearly colorless, granular solid is collected by suction filtration. The crude product need not be dried but can be crystallized directly from 95% ethanol. Let the solution stand undisturbed in order to observe the formation of beautiful, very large spars of colorless crystals. After a first crop has been collected, the mother liquor on concentration affords a second crop of

---

6. Potassium hydroxide pellets are 85% KOH and 15% water.

pure product; total yield, about 0.23 g; mp 60–61°C. The UV spectrum is shown at the end of the chapter (Fig. 59.6).

Cleaning Up   Combine the crystallization mother liquor with the filtrate from the reaction, dilute with water, neutralize with 10% hydrochloric acid, and flush down the drain.

Questions

1. Using a molecular mechanics program, calculate the energy difference or the difference in heats of formation between *Z*- and *E*-stilbene. How do your computed values compare with the difference in heats of hydrogenation of the two isomers (see the first page of this experiment)?

2. Why is *Z*-stilbene not planar?

3. How do you account for the big difference in the extinction coefficients, ϵ, for the long-wavelength peaks in *Z*- and *E*-stilbene (Fig. 59.6)?

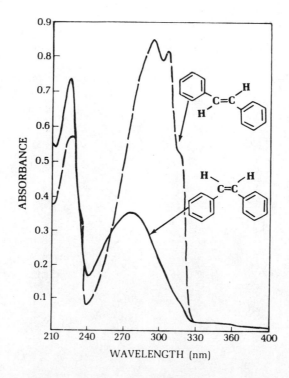

FIG. 59.5   Ultraviolet spectra of *Z*- and *E*-stilbene. *Z:* λ_{max}^{EtOH} 224 nm ($\epsilon = 23,300$), 279 nm ($\epsilon = 11,100$); *E:* λ_{max}^{EtOH} 226 nm ($\epsilon = 18,300$), 295 nm ($\epsilon = 27,500$). Like the diacetates, steric hindrance and lack of coplanarity in these hydrocarbons cause the long-wavelength absorption of the *Z*-isomer to be of diminished intensity relative to the *E*-isomer.

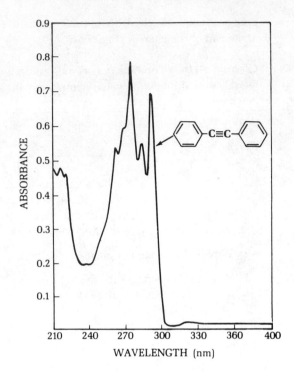

FIG. 59.6 Ultraviolet spectrum of diphenylacetylene. λ_{max}^{EtOH} 279 nm ($\varepsilon = 31{,}400$). This spectrum is characterized by considerable fine structure (multiplicity of bands) and a high extinction coefficient.

The Perkin Reaction: Synthesis of α-Phenylcinnamic Acid

Prelab Exercise: At the end of this reaction the products are present as mixed anhydrides. What are these, how are they formed, and how are they converted to product?

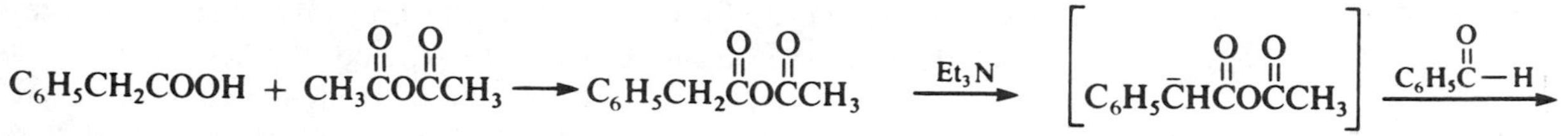

$$C_6H_5CH_2COOH + CH_3\overset{O}{\overset{\|}{C}}O\overset{O}{\overset{\|}{C}}CH_3 \longrightarrow C_6H_5CH_2\overset{O}{\overset{\|}{C}}O\overset{O}{\overset{\|}{C}}CH_3 \xrightarrow{Et_3N} \left[C_6H_5\overset{-}{C}H\overset{O}{\overset{\|}{C}}O\overset{O}{\overset{\|}{C}}CH_3 \right] \xrightarrow{C_6H_5\overset{O}{\overset{\|}{C}}-H}$$

Phenylacetic acid
mp 77°C, bp 265°C
MW 136.14

Acetic anhydride
bp 138–140°C
MW 102.09
n_D^{20} 1.3900

Triethylamine
bp 89.5°C, den. 0.729
MW 101.19
n_D^{20} 1.4000

Benzaldehyde
bp 179°C, den. 1.046
MW 106.12
n_D^{20} 1.5450

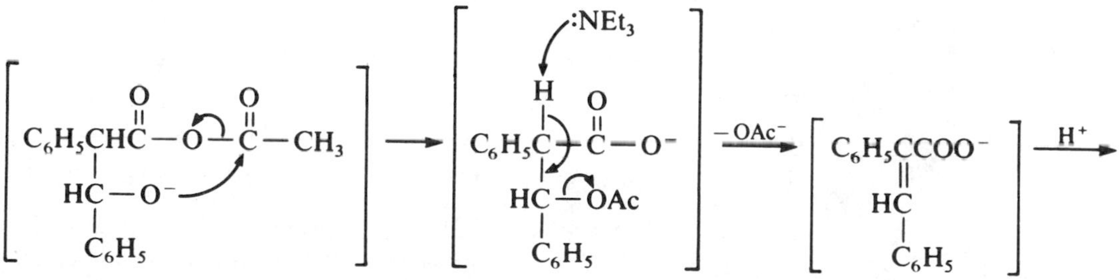

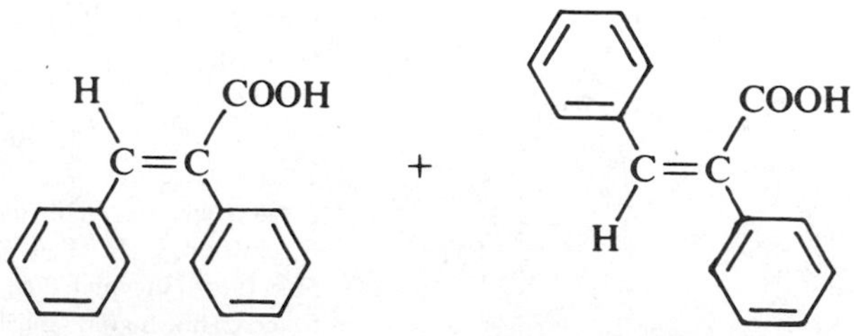

$$\left[\begin{array}{c} C_6H_5\overset{O}{\overset{\|}{CHC}}-O-\overset{O}{\overset{\|}{C}}-CH_3 \\ HC-O^- \\ C_6H_5 \end{array} \right] \longrightarrow \left[\begin{array}{c} :NEt_3 \\ H \quad O \\ C_6H_5\overset{|}{C}-\overset{\|}{C}-O^- \\ HC-OAc \\ C_6H_5 \end{array} \right] \xrightarrow{-OAc^-} \left[\begin{array}{c} C_6H_5CCOO^- \\ \| \\ HC \\ | \\ C_6H_5 \end{array} \right] \xrightarrow{H^+}$$

E-α-Phenylcinnamic acid[1]
mp 174°C, pK_a 6.1
MW 224.25

Z-α-Phenylcinnamic acid
mp 138°C, pK_a 4.8
MW 224.25

---

1. pK_a measured in 60% ethanol.

The reaction of benzaldehyde with phenylacetic acid to produce a mixture of the α-carboxylic acid derivatives of *Z*- and *E*-stilbene,[2] a form of aldol condensation known as the *Perkin reaction,* is effected by heating a mixture of the components with acetic anhydride and triethylamine. In the course of the reaction the phenylacetic acid is probably present both as anion and as the mixed anhydride resulting from equilibration with acetic anhydride. A reflux period of 5 h specified in an early procedure has been shortened by a factor of 10 by restriction of the amount of the volatile acetic anhydride, use of an excess of the less expensive, high-boiling aldehyde component, and use of a condenser that permits some evaporation and consequent elevation of the reflux temperature.

E-Stilbene is a byproduct of the condensation, but experiment has shown that neither the *E*- nor *Z*-acid undergoes decarboxylation under the conditions of the experiment.

At the end of the reaction the α-phenylcinnamic acids are present in part as the neutral mixed anhydrides, but these can be hydrolyzed by addition of excess hydrochloric acid. The organic material is taken up in ether and the acids extracted with alkali. Neutralization with acetic acid (pK_a 4.76) then causes precipitation of only the less acidic *E*-acid (see pK_a values under the formulas); the *Z*-acid separates on addition of hydrochloric acid.

Whereas *Z*-stilbene is less stable and lower melting than *E*-stilbene, the reverse is true of the α-carboxylic acids, and in this preparation the more stable, higher melting *E*-acid is the predominant product. Evidently the steric interference between the carboxyl and phenyl groups in the *Z*-acid is greater than that between the two phenyl groups in the *E*-acid. Steric hindrance is also evident from the fact that the *Z*-acid is not subject to Fischer esterification (ethanol and an acid catalyst), whereas the *E*-acid is.

1. Microscale Synthesis of α-Phenylcinnamic Acid

Reflux time: 35 min

Measure into a reaction tube 250 mg of phenylacetic acid, 0.30 mL of benzaldehyde, 0.2 mL of triethylamine, and 0.2 mL of acetic anhydride. Insert a boiling stone, and reflux the mixture for 35 min (Fig. 60.1). Cool the yellow melt, add 0.4 mL of concentrated hydrochloric acid, and mix, whereupon the mixture sets to a stiff paste. Add 2.5 mL of *t*-butyl methyl ether, warm to dissolve the bulk of the solid, wash the ethereal solution twice with water, and then extract

2. Stereochemistry of alkenes can be designated by the *E,Z*-system of nomenclature [see J. L. Blackwood, C. L. Gladys, K. L. Loening, A. E. Petrarca, and J. E. Rush, *J. Am. Chem. Soc.,* **90,** 509 (1968)] in which the groups attached to the double bond are given an order of priority according to the Cahn, Ingold, and Prelog system [see R. S. Cahn, C. K. Ingold, and V. Prelog, *Experentia,* **12,** 81 (1956)]. The atom of highest atomic number attached directly to the alkene is given highest priority. If two atoms attached to the alkene are the same, one goes to the second or third atom, etc., away from the alkene carbons. When the two groups of highest priority are on adjacent sides of the double bond, the stereochemistry is designated as *Z* (German *zusammen,* together). When the two groups are on opposite sides of the double bond, the stereochemistry is designated *E* (German *entgegen,* opposed).

FIG. 60.1 Apparatus for refluxing reaction mixture.

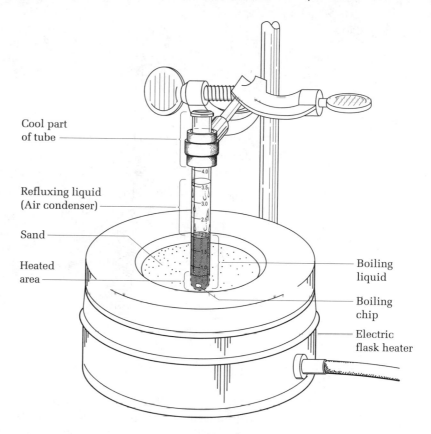

Cool part of tube

Refluxing liquid (Air condenser)

Sand

Heated area

Boiling liquid

Boiling chip

Electric flask heater

it five times with 2-mL portions of 0.5 M potassium hydroxide solution.[3] Discard the dark-colored ethereal solution.[4] Acidify the combined, colorless alkaline extract to pH 6 by adding 0.4 mL of acetic acid, collect the E-acid that precipitates, and save the filtrate and washings. The yield of E-acid (mp 163–166°C) is usually about 290 mg. Crystallize 30 mg of material by dissolving it in 0.6 mL of *tert*-butyl methyl ether, adding 0.8 mL of petroleum ether (bp 30–60°C), heating briefly to the boiling point, and letting the solution stand. Silken needles form (mp 173–174°C). These can be easily collected on the Wilfilter (Fig. 3.15).

Addition of 0.5 mL of concentrated hydrochloric acid to the aqueous filtrate from precipitation of the E-acid produces a cloudy emulsion, which on standing for about one-half hour coagulates to crystals of Z-acid; yield is 30 mg (mp 136–137°C).[5]

---

3. If stronger alkali is used, the potassium salt may separate.

4. For isolation of stilbene, wash this ethereal solution with saturated sodium bisulfite solution for removal of benzaldehyde, dry, evaporate, and crystallize the residue from a little methanol. Large, slightly yellow spars (mp 122–124°C) separate (9 mg).

5. The E-acid can be recrystallized by dissolving 30 mg in 0.5 mL of *tert*-butyl methyl ether, filtering if necessary from a trace of sodium chloride, adding 1 mL of petroleum ether (bp 30–60°C), and evaporating to a volume of 0.5 mL; the acid separates as a hard crust of prisms (mp 138–139°C).

FIG. 60.2 Reflux apparatus.

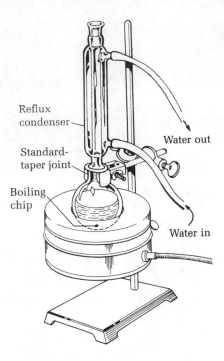

Reflux
condenser

Standard-
taper joint

Boiling
chip

Water out

Water in

Cleaning Up   The dark-colored ether solution and the mother liquor from crystallization are placed in the organic solvents container. After the Z-acid has been removed from the reaction mixture, the acidic solution should be diluted with water and flushed down the drain.

2. Macroscale Synthesis of α-Phenylcinnamic Acid

Reflux time: 35 min

Measure into a 25-mL round-bottomed flask 2.5 g of phenylacetic acid, 3 mL of benzaldehyde, 2 mL of triethylamine, and 2 mL of acetic anhydride. Add a water-cooled condenser and a boiling stone, and reflux the mixture for 35 min (see Fig. 60.2). Cool the yellow melt, add 4 mL of concentrated hydrochloric acid, and swirl, whereupon the mixture sets to a stiff paste. Add *t*-butyl methyl ether, warm to dissolve the bulk of the solid, and transfer to a separatory funnel with use of more ether. Wash the ethereal solution twice with water, and then extract it with a mixture of 25 mL of water and 5 mL of 3 *M* sodium hydroxide solution.[6] Repeat the extraction twice more and discard the dark-colored ethereal solution.[7] Acidify the combined, colorless, alkaline extract to pH 6 by adding 5 mL of acetic acid, collect the *E*-acid that precipitates, and save the filtrate and

---

6. If stronger alkali is used, the sodium salt may separate.

7. For isolation of stilbene, wash this ethereal solution with saturated sodium bisulfite solution for removal of benzaldehyde, dry, evaporate, and crystallize the residue from a little methanol. Large, slightly yellow spars, mp 122–124°C, separate (9 mg).

FIG. 60.3 Ultraviolet spectra of Z- and E-α-phenylcinnamic acid, run at identical concentrations. E: λ_{max}^{EtOH} 222 nm (ε = 18,000), 282 nm (ε = 14,000); Z: λ_{max}^{EtOH} 222 nm (ε = 15,500), 292 nm (ε = 22,300).

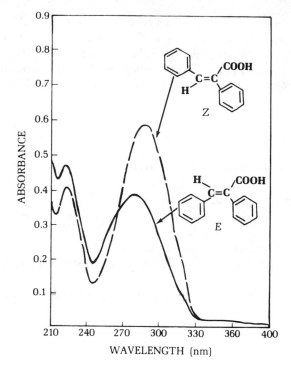

washings. The yield of E-acid, mp 163–166°C, is usually about 2.9 g. Crystallize 0.3 g of material by dissolving it in 8 mL of t-butyl methyl ether, adding 8 mL of petroleum ether (bp 30–60°C) or pentane, heating briefly to the boiling point, and letting the solution stand. Silken needles form, mp 173–174°C.

Addition of 5 mL of concentrated hydrochloric acid to the aqueous filtrate from precipitation of the E-acid produces a cloudy emulsion, which, on standing for about one-half hour, coagulates to crystals of Z-acid; yield, about 0.3 g, mp 136–137°C.[8]

Cleaning Up   The dark-colored ether solution and the mother liquor from crystallization are placed in the organic solvents container. After the Z-acid has been removed from the reaction mixture, the acidic solution should be diluted with water and flushed down the drain.

Computational Chemistry

Construct Z- and E-α-phenylcinnamic acids and carry out energy minimizations. Be sure to arrange the carboxyl group to give the lowest steric energy,

8. The E-acid can be recrystallized by dissolving 0.3 g in 5 mL of ether, filtering if necessary from a trace of sodium chloride, adding 10 mL of petroleum ether (bp 30–60°C), and evaporating to a volume of 5 mL; the acid separates as a hard crust of prisms, mp 138–139°C.

which you should note. Then carry out AM1 semiempirical molecular orbital heats of formation calculations for the two isomers. Do the calculations confirm that the Z-isomer is the more stable one? Would simple inspection of the space-filling model have led to the same conclusion? Can you rationalize the absorbances in the UV spectra of the two isomers? (See Fig. 60.3.)

Surfing the Web

http://www.ch.ic.ac.uk/motm/perkin.html
This Imperial College site has a brief biography of William Henry Perkin, for whom this reaction is named. It also includes the story of his discovery of mauvine, the dye that became the world's first industrial chemical.

For updated information visit:
www.mtholyoke.edu/courses/kwilliam/microscale.shtml
or
www.hmco.com/hmco/college/chemistry/Home.html

61

Decarboxylation: Synthesis of cis-Stilbene

Prelab Exercise: Calculate the volume (measured at STP) of carbon dioxide evolved in this reaction.

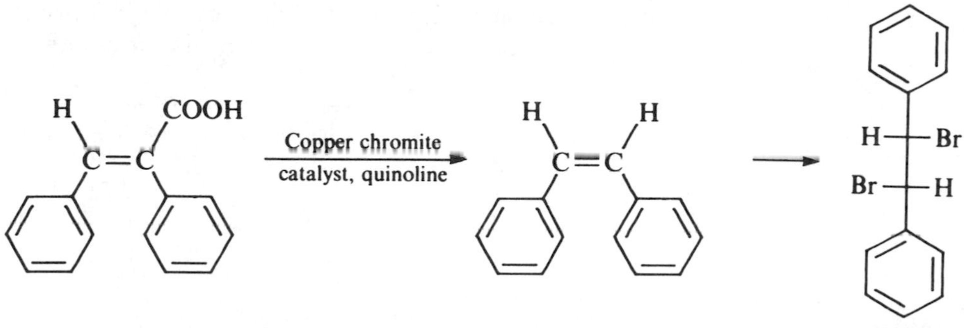

E-α-Phenylcinnamic acid
MW 224.25

Z-Stilbene
mp 4°C, bp 82–84°C/0.4 mm
MW 180.24

1R,2R and 1S,2S(±)-
Stilbene dibromide
mp 114°C
MW 340.07

The catalyst for this reaction is copper chromite, $2\ CuO \cdot Cr_2O_3$, a relatively inexpensive commercially available catalyst used for both hydrogenation and dehydrogenation as well as decarboxylation.

Decarboxylation of *E*-α-phenylcinnamic acid is effected by refluxing the acid in quinoline in the presence of a trace of copper chromite catalyst; both the basic properties and boiling point (237°C) of quinoline make it a particularly favorable solvent. *Z*-Stilbene, a liquid at room temperature, can be characterized by *trans* addition of bromine to give the crystalline ±-dibromide. A little *meso*-dibromide derived from *E*-stilbene in the crude hydrocarbon starting material is easily separated by virtue of its sparing solubility.

Although free rotation is possible around the single bond connecting the chiral carbon atoms of the stilbene dibromides and hydrobenzoins, evidence from dipole-moment measurements indicates that the molecules tend to exist predominantly in the specific shape or conformation in which the two phenyl groups repel each other and occupy positions as far apart as possible. The optimal conformations of the + or − dibromide and the *meso*-dibromide are represented in Fig. 61.1 by Newman projection formulas, in which the molecules are viewed along the axis of the bond connecting the two chiral carbon atoms. In the *meso*-dibromide the two repelling phenyl groups are on opposite sides of the

FIG. 61.1 Favored conformations of stilbene dibromide.

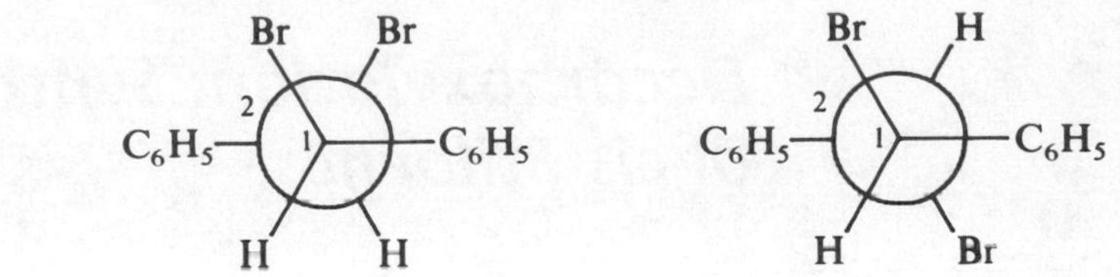

(**1R,2R** or **1S,2S**) + (or −)**Dibromide**   (**1R,2S**)*meso*-**Dibromide**

molecule, and so are the two large bromine atoms. Hence, the structure is much more symmetrical than that of the + (or −) dibromide. X-ray diffraction measurements of the dibromides in the solid state confirm the conformations indicated in Fig. 61.1. The Br–Br distances found are *meso*-dibromide, 4.50 nm; ±-dibromide, 3.85 nm. The difference in symmetry of the two optically inactive isomers accounts for the marked contrast in properties:

	mp (°C)	Solubility in ether (18°C)
±-Dibromide	114	1 part in 3.7 parts
meso-Dibromide	237	1 part in 1025 parts

▯ MICROSCALE

2CuO · Cr₂O₃
Copper chromite

Reaction time in first step: 10 min

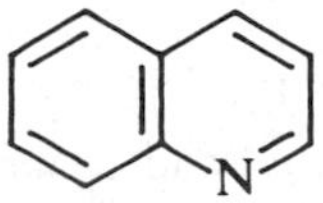

Quinoline
bp 237°C

1. Stilbene Dibromide

Because a trace of moisture causes troublesome spattering, the reactants and catalyst are dried prior to decarboxylation. Stuff 250 mg of crude, "dry" *E-α*-phenylcinnamic acid and 40 mg of copper chromite catalyst into a reaction tube, add 0.3 mL of quinoline[1] (bp 237°C), and let it wash down the solids. Make connection to the aspirator, and turn it on full force. Make sure that you have a good vacuum (pressure gauge), and heat the tube strongly on the steam bath with most of the rings removed. Heat and evacuate for 5 to 10 min to remove all traces of moisture. Then wipe the outside walls of the tube dry, insert a thermometer, clamp the tube over a hot sand bath, raise the temperature to 230°C, and note the time. Then maintain a temperature close to 230°C for 10 min. Cool the yellow solution containing suspended catalyst to 25°C, add 3 mL of *tert*-butyl methyl ether, and filter the solution by gravity (use more ether for rinsing). Transfer the solution to a reaction tube, and remove the quinoline by extraction twice with about 1.5 mL of water containing 0.35 mL of concentrated hydrochloric acid. Then shake the ethereal solution well with water containing a little sodium hydroxide solution, draw off the alkaline liquor, and acidify it. A substantial precipitate will show that decarboxylation was incomplete, in which case the starting material can be recovered and the reaction repeated. If there is only a trace of precipitate, shake the ethereal solution with saturated sodium chloride solution for preliminary drying, dry the ethereal solution over calcium

---

1. Material that has darkened in storage should be redistilled over a little zinc dust.

chloride pellets, remove the ether from the drying agent with a Pasteur pipette, and evaporate the ether. The residual brownish oil (about 150 mg) is crude *Z*-stilbene containing a little *E*-isomer formed by rearrangement during heating.

Dissolve the crude *Z*-stilbene (for example, 150 mg) in 1 mL of acetic acid, and in subdued light, add double the weight of pyridinium hydrobromide perbromide (for example, 300 mg). Warm on the steam bath until the reagent is dissolved, and then cool under the tap and scratch to effect separation of a small crop of plates of the *meso*-dibromide. Filter the solution by suction if necessary, dilute extensively with water, and extract with ether. Wash the solution twice with water and then with saturated sodium bicarbonate solution (1.3 M) until neutral; shake with saturated sodium chloride solution, dry over calcium chloride pellets, and evaporate to a volume of about 1 mL. If a little more of the sparingly soluble *meso*-dibromide separates, remove it and then evaporate the remainder of the solvent. The residual ±-dibromide is obtained as a dark oil that readily solidifies when rubbed with a rod. Dissolve it in a small amount of methanol, and let the solution stand to crystallize. Isolate the product on the Wilfilter (Fig. 61.2). The ±-dibromide separates as colorless prismatic plates (mp 113–114°C); yield is about 60 mg.

Cleaning Up   The catalyst removed by filtration should be placed in the nonhazardous solid waste container. The combined aqueous layers from all parts of the experiment containing quinoline and pyridine should be treated with a small quantity of bisulfite to destroy any bromine, neutralized with sodium carbonate, the quinoline and pyridine released extracted into ligroin, and the ligroin solution placed in the organic solvents container. The aqueous layer should be diluted with water and flushed down the drain. After ether is allowed to evaporate from the calcium chloride pellets in the hood, they can be placed in the nonhazardous solid waste container. Methanol from the crystallization goes in the organic solvents container.

 **MACROSCALE**

Stilbene Dibromide

Because a trace of moisture causes troublesome spattering, the reactants and catalyst are dried prior to decarboxylation. Stuff 2.5 g of crude, "dry" *E*-α-phenylcinnamic acid and 0.2 g of copper chromite catalyst[2] into a 20 × 150-mm test tube, add 3 mL of quinoline[3] (bp 237°C), and let it wash down the solids. Make connection with a rubber stopper to the aspirator, and turn it on full force. Make sure that you have a good vacuum (pressure gauge), and heat the tube strongly on the steam bath with most of the rings removed. Heat and evacuate for 5–10 min to remove all traces of moisture. Then wipe the outside walls of the test tube dry, insert a thermometer, clamp the tube over a microburner, raise the temperature to 230°C, and note the time. Then maintain a temperature close

Second step requires about one-half hour.

Carry out procedure in hood.

Copper chromite is toxic; weigh it in hood.

$$2CuO \cdot Cr_2O_3$$
Copper chromite

Reaction time in first step: 10 min

---

2. The preparations described in *J. Am. Chem. Soc.*, **54**, 1138 (1932) and *ibid.*, **72**, 2626 (1950) are both satisfactory.

3. Material that darkened in storage should be redistilled over a little zinc dust.

FIG. 61.2 Wilfilter filtration apparatus.

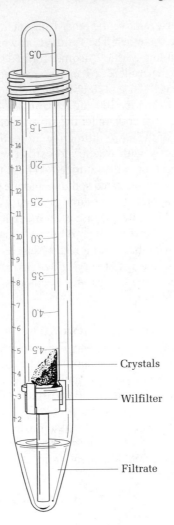

— Crystals

— Wilfilter

— Filtrate

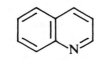

Quinoline, bp 237°C

to 230°C for 10 min. Cool the yellow solution containing suspended catalyst to 25°C, add 30 mL of *t*-butyl methyl ether, and filter the solution by gravity (use more ether for rinsing). Transfer the solution to a separatory funnel, and remove the quinoline by extraction twice with about 15 mL of water containing 3–4 mL of concentrated hydrochloric acid. Then shake the ethereal solution well with water containing a little sodium hydroxide solution, draw off the alkaline liquor, and acidify it. A substantial precipitate will show that decarboxylation was incomplete, in which case the starting material can be recovered and the reaction repeated. If there is only a trace of precipitate, shake the ethereal solution with saturated sodium chloride solution for preliminary drying, dry the ethereal solution over anhydrous calcium chloride pellets, remove the drying agent by filtration, and evaporate the ether. The residual brownish oil (1.3–1.8 g) is crude *Z*-stilbene containing a little *E*-isomer formed by rearrangement during heating.

Second step requires about one-half hour.

Carry out procedure in hood.

Dissolve the crude Z-stilbene (e.g., 1.5 g) in 10 mL of acetic acid and, in subdued light, add double the weight of pyridinium hydrobromide perbromide (e.g., 3.0 g). Warm on the steam bath until the reagent is dissolved, and then cool under the tap and scratch to effect separation of a small crop of plates of the *meso*-dibromide (10–20 mg). Filter the solution by suction, dilute extensively with water, and extract with *t*-butyl methyl ether. Wash the solution twice with water and then with 5% sodium bicarbonate solution until neutral; shake with saturated sodium chloride solution, dry over sodium sulfate, and evaporate to a volume of about 10 mL. If a little more of the sparingly soluble *meso*-dibromide separates, remove it by gravity filtration and then evaporate the remainder of the solvent. The residual *dl*-dibromide is obtained as a dark oil that readily solidifies when rubbed with a rod. Dissolve it in a small amount of methanol, and let the solution stand to crystallize. The *dl*-dibromide separates as colorless prismatic plates, mp 113–114°C; yield, about 0.6 g.

Cleaning Up   The catalyst removed by filtration should be placed in the nonhazardous solid waste container. The combined aqueous layers from all parts of the experiment containing quinoline and pyridine should be treated with a small quantity of bisulfite to destroy any bromine, neutralized with sodium carbonate, the quinoline and pyridine released extracted into ligroin, and the ligroin solution placed in the organic solvents container. The aqueous layer should be diluted with water and flushed down the drain. After the ether is allowed to evaporate from the calcium chloride pellets in the hood, it can be placed in the nonhazardous solid waste container. Methanol from the crystallization goes in the organic solvents container.

Computational Chemistry

The Br–Br distances in 1*R*,2*R*- or 1*S*,2*S*-stilbene dibromide and 1*R*,2*S*-stilbene dibromide are 4.50 and 3.85 nm, respectively, as determined by X-ray diffraction on the crystals. Do simple molecular mechanics energy-minimized structures give these same intramolecular distances once the global minimum is found?

Questions

1. Draw the mechanism that shows how the bromination of *E*-stilbene produces *meso*-stilbene dibromide.

2. Can ^{1}H NMR spectroscopy be used to distinguish between *meso*- and ±-stilbene dibromide?

3. Why is Z-stilbene not planar?

4. Draw a Newman projection of 1*R*,2*S*-stilbene dibromide (the *meso*-isomer).

Chemiluminescence: Syntheses of Cyalume and Luminol

Prelab Exercise: Write a balanced equation for the reduction of nitrophthalhydrazide to aminophthalhydrazide using sodium hydrosulfite, which is oxidized to bisulfite.

Chemiluminescence is the process whereby light is produced by a chemical reaction with the evolution of little or no heat. The periodic flashes of the male firefly in quest of a mate and the glow of light seen in the wake of a boat, under a rotten log, or among the many organisms found at great depth in the ocean are examples of natural chemiluminescence. The interaction of luciferin from the firefly, the enzyme luciferase, adenosine triphosphate (ATP), and molecular oxygen is the most carefully studied of these reactions. In this experiment, both Cyalume, an invention of the American Cyanamid Company, and luminol are synthesized.

"Light sticks" sold at entertainment events and as toys are solutions of Cyalume. Bending the light stick activates it (by mixing hydrogen peroxide with the Cyalume) to give off a bright glow in a variety of colors. These sticks are also used as emergency lights on life vests, as roadside markers, on fishing lures, and even in golf balls that light up at night! A variety of different colors can be produced by allowing your synthetic Cyalume to interact with a variety of fluorescers, including one you can prepare by the Wittig reaction. Luminol is not a commercial product.[1]

Cyalume

Cyalumes are the most efficient nonenzyme fluorescing substances known, with quantum efficiencies of up to 26%.

In this experiment, 2,4,6-trichlorophenol (**1**) reacts with the diacid chloride, oxalyl chloride (**2**), in the presence of triethylamine to give the desired oxalic acid ester (**3**) and the amine hydrochloride. A number of different oxalic acid esters will work in the luminescent reaction.

In the light-producing reaction, the ester undergoes nucleophilic attack by hydrogen peroxide to produce a peroxyoxalic acid, which, by an intramolecular

---

1. For the chemiluminescence of luminol, see E. H. Huntress, L. N. Stanley, and A. S. Parker, *J. Chem. Ed.,* **11,** 142 (1934). The mechanism of the reaction has been investigated by Emil H. White and coworkers, *J. Am. Chem. Soc.,* **86,** 940 and 942 (1964). Peroxyoxalate chemistry is discussed by A. G. Mohan and N. J. Turro, *J. Chem. Ed.,* **51,** 528 (1974).

displacement reaction, generates a small intermediate molecule, perhaps the 1,2-dioxetanedione (**4**). This molecule transmits its energy to a fluorescer, such as 9,10-diphenylanthracene (**5**), through a charge-transfer complex that produces the excited singlet state of the fluorescer when it decomposes to two molecules of carbon dioxide. *trans*-9-(2-Phenylethenyl)anthracene, synthesized by the Wittig reaction in Chapter 39, is also an excellent fluorescer. The observed color depends on the structure of the fluorescer.

The structure of the intermediate molecule, written here as **4**, has not been established with certainty. The intermediate is known to be volatile; carbon dioxide is a product of the reaction, but carbon monoxide is not. Efforts to detect this intermediate by Fourier transform infrared, carbon NMR, and mass spectroscopy have all been unsuccessful.

The light-producing reaction of the oxalate esters differs from luminol in that the ester decomposes to a small molecule that transfers its energy to a different fluorescent molecule, while in luminol, the molecule itself is fluorescent and gives off light from the excited state. In addition, the luminol reaction is over in a few minutes, but the peroxyoxalate reaction can last for a day or more.

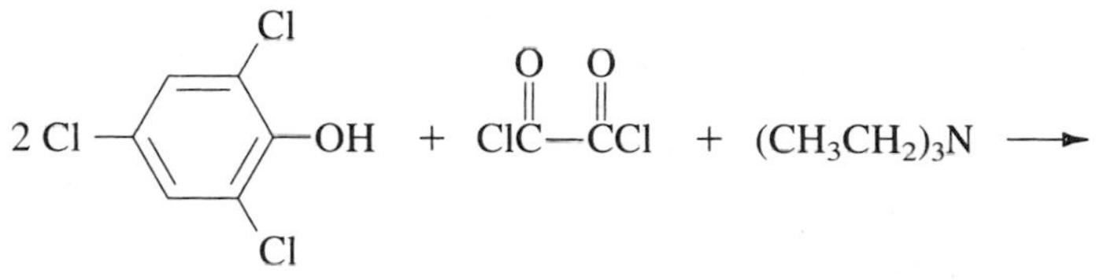

2,4,6-Trichlorophenol
MW 197.45
mp 64–66°C

1

Oxalyl chloride
MW 126.93
bp 63–64°C
den. 1.455

2

Triethylamine
MW 101.19
bp 89°C
den. 0.726

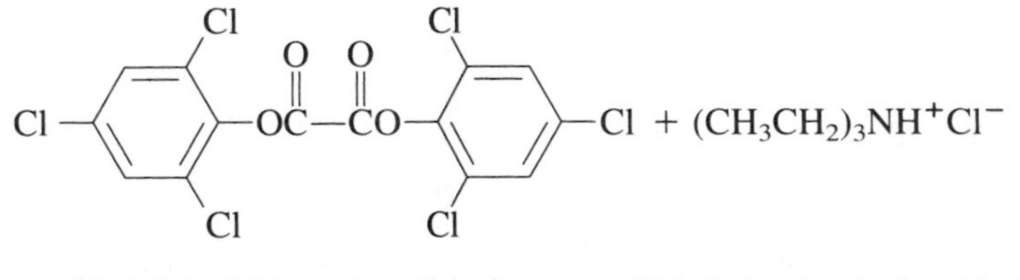

Bis(2,4,6-trichlorophenyl)oxalate
(A Cyalume)
MW 448.9
mp 162–163°C

3

Triethylamine hydrochloride

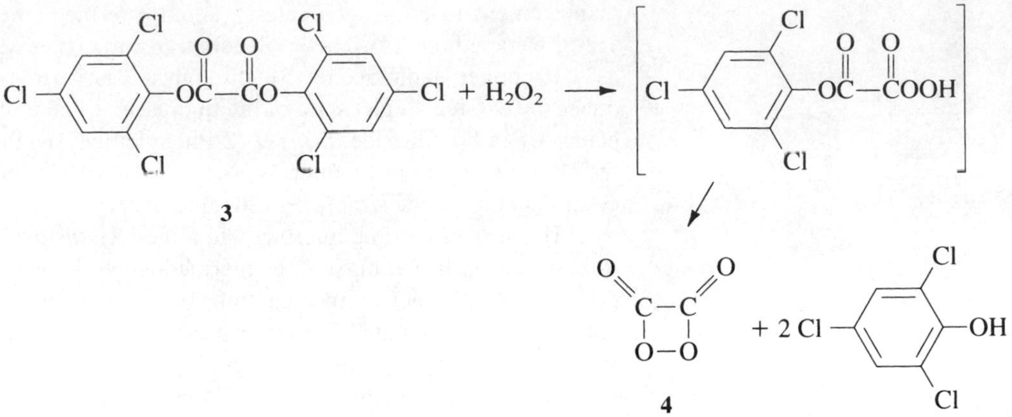

3

4

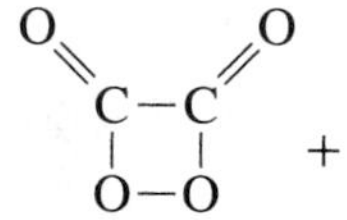

1,2-Dioxetanedione (?)
4

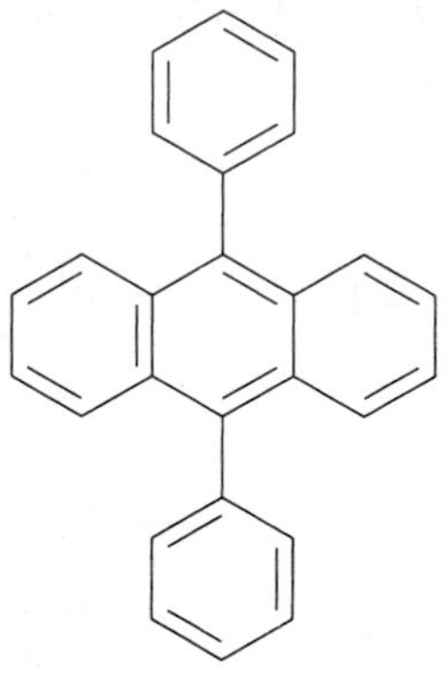

9,10-Diphenylanthracene
MW 330.43
mp 245–248°C
5

or

Charge transfer complex

$2 CO_2$ + **9,10-Diphenylanthracene**
Singlet excited state

$h\nu$ + **9,10-Diphenylanthracene**
Ground state

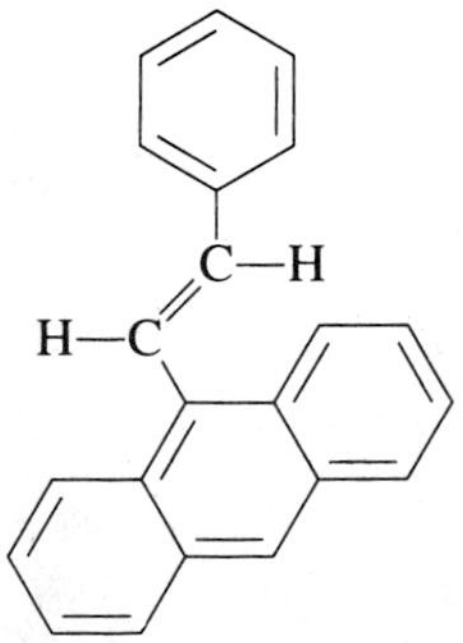

trans-**9-(2-Phenylethenyl)anthracene**
(from Chapter 39)

CHEMILUMINESCENCE OF LUMINOL

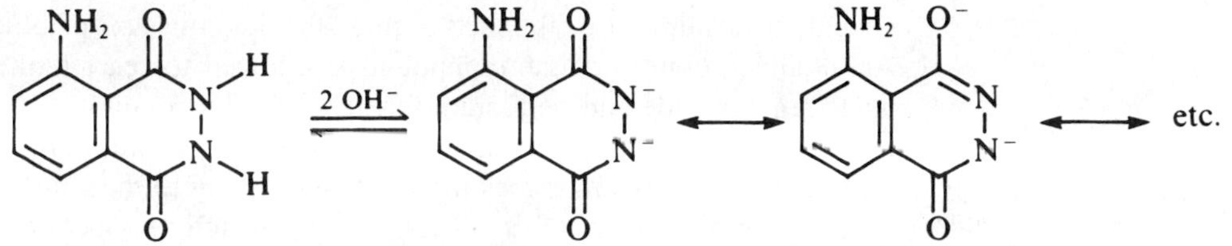
NH₂ O
H
N
N
H
O
2 OH⁻
NH₂ O
N⁻
N⁻
O
NH₂ O⁻
N
N⁻
O
etc.

6 7

Luminol
(3-Aminophthalhydrazide)

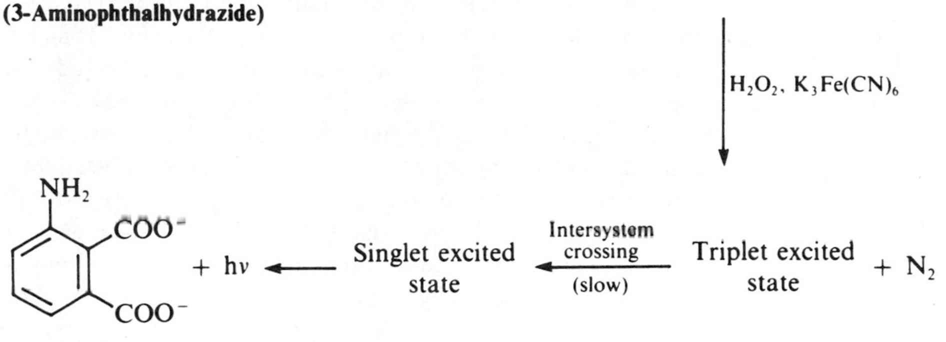
H₂O₂, K₃Fe(CN)₆
NH₂
COO⁻
COO⁻
+ hν
Singlet excited state
Intersystem crossing
(slow)
Triplet excited state
+ N₂

Ground state

8 9 10

SYNTHESIS OF LUMINOL

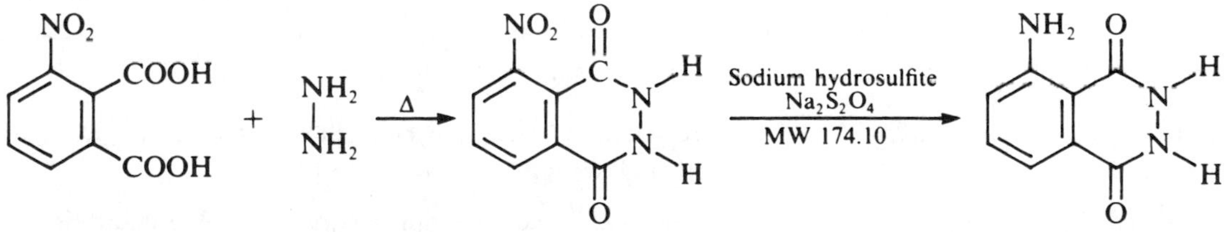
NO₂
COOH
COOH
+
NH₂
NH₂
Δ
NO₂ O
C
N
H
N
H
O
Sodium hydrosulfite
Na₂S₂O₄
MW 174.10
NH₂ O
N
H
N
H
O

11 12 13 6

3-Nitrophthalic acid
MW 211.13
mp 222°C

Hydrazine

Luminol
MW 177.16
mp 332°C

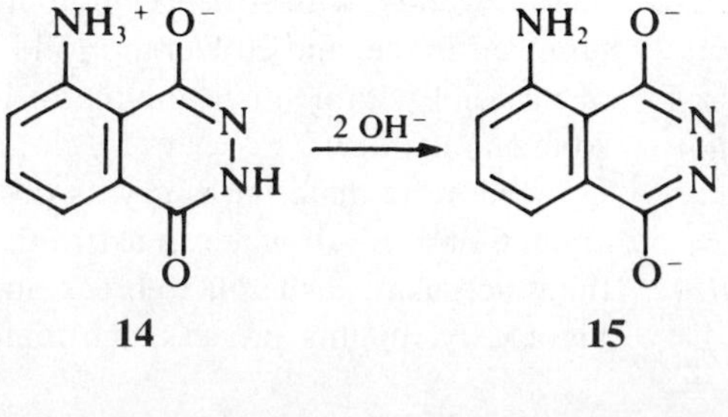
NH₃⁺ O⁻
N
NH
O
2 OH⁻
NH₂ O⁻
N
N
O⁻

14 15

Luminol

Oxidation of luminol is attended with a striking emission of blue-green light. An alkaline solution of the compound is allowed to react with a mixture of hydrogen peroxide and potassium ferricyanide. The dianion (**7**) is oxidized to the triplet excited state (two unpaired electrons of like spin) of the amino phthalate ion. This slowly undergoes intersystem crossing to the singlet excited state (two unpaired electrons of opposite spin) (**9**), which decays to the ground state ion (**8**) with the emission of one quantum of light (a photon) per molecule. (See Chapter 63 for a more detailed discussion of photochemistry.)

Luminol (**6**) is made by reduction of the nitro derivative (**13**) formed on thermal dehydration of a mixture of 3-nitrophthalic acid (**11**) and hydrazine (**12**). An earlier procedure for effecting the first step called for addition of hydrazine sulfate to an alkaline solution of the acid, evaporation to dryness, and baking the resulting mixture of the hydrazine salt and sodium sulfate at 165°C; it required a total of 4.5 h for completion. Louis Fieser reduced this working time drastically by adding high-boiling triethylene glycol (bp 290°C) to an aqueous solution of the hydrazine salt, distilling the excess water, and raising the temperature to a point where dehydration to (**13**) is complete within a few minutes. Nitrophthalhydrazide (**3**) is insoluble in dilute acid but soluble in alkali, by virtue of enolization; it is conveniently reduced to luminol (**6**) by sodium hydrosulfite (sodium dithionite) in alkaline solution. In dilute, weakly acidic or neutral solution luminol exists largely as the dipolar ion (**14**), which exhibits beautiful blue fluorescence.

An alkaline solution contains the doubly enolized anion (**15**) and displays particularly marked chemiluminescence when oxidized with a combination of hydrogen peroxide and potassium ferricyanide.

MICROSCALE AND MACROSCALE

1. Synthesis of Bis(2,4,6-trichlorophenyl) Oxalate, a Cyalume

For a macroscale reaction, triple or quadruple the quantities. Add the thionyl chloride in portions, with cooling if necessary.

Dissolve 0.395 g of 2,4,6-trichlorophenol in 3 mL of toluene in a 5-mL round-bottomed flask equipped with a magnetic stirring bar. Add to this solution 0.28 mL (0.2 g) of triethylamine, cool the container in ice, and add 0.10 mL (0.14 g) of oxalyl chloride. There will be an immediate voluminous precipitate of triethylamine hydrochloride. Add an air condenser to the flask, and reflux the mixture gently, with stirring, for 30 min to complete the reaction. Cool the mixture well in ice, and collect the solid on a Hirsch funnel. Complete the transfer of material with a small volume of hexane(s), which is also used to wash the solid on the filter.

Press the material as dry as possible on the filter, and then suspend it in about 5 or 6 mL of water in a 10-mL Erlenmeyer flask. Stir the solid well with the water, using a spatula to break up any lumps, then cap the flask and shake it vigorously. In this process, the amine hydrochloride dissolves in water and

CAUTION: 2,4,6-Trichlorophenol is a carcinogen. Oxalyl chloride is corrosive and a lachrymator. Triethylamine has a very bad odor. Carry out the measurement and transfer of reagents in the hood.

Magnet stirring of this reaction mixture is desirable but not mandatory. The precipitate is viscous and does not stir well.

leaves the product as a suspension. Empty the filtrate in the filter flask into the flammable waste container, and then collect the solid product from the aqueous solution on the Hirsch funnel. Wash the solid well with water, and press it as dry as possible on the filter. Spread the product out to dry, determine its weight, and then recrystallize it from the minimum volume of boiling toluene. Allow the product to crystallize slowly, and then cool the tube in ice. Collect the product on the Hirsch funnel (use hexane to complete the transfer of material), and wash the crystals with a small volume of hexane. Determine the weight and melting point of the dry material, and calculate the percentage yield.

Cleaning Up   Discard the aqueous filtrate down the drain and the organic filtrates in the flammable waste container.

MICROSCALE AND MACROSCALE

2. The Chemiluminescent Reaction

Dissolve 50 mg (or less, if that is all you have) of bis(2,4,6-trichloro)oxalate and about 3 mg of 9-(2-phenylethenyl)anthracene (Chapter 39) or 9,10-diphenylanthracene in 5 mL of diethyl phthalate by warming on the steam bath. In another container, shake 0.2 mL of 30% hydrogen peroxide (or, less satisfactory, 2 mL of 3% peroxide) with 5 mL of diethyl phthalate. Add the peroxide suspension dropwise to the solution of the oxalate ester and the fluorescer in a darkened room. Try cooling and warming the mixture after initial mixing, and note the effect of temperature on the rate of a reaction. Also try adding one or two more milligrams of the fluorescer (9,10-diphenylanthracene is expensive) and more hydrogen peroxide.

Cleaning Up   Dispose of the mixture in the halogenated waste container.

Computational Chemistry

Using a molecular mechanics program calculate the energy, after minimization, of the dimer of carbon dioxide, 1,2-dioxetanedione (**4**), the molecule thought to be the intermediate in Cyalume chemiluminescence. Note the bond lengths and angles. This molecule, with only 6 atoms, is small enough to carry out some of the more advanced calculational methods in a reasonable length of time. Using the energy-minimized structure as input, calculate the optimal geometry of **4** using the AM1 semiempirical molecular orbital method. Again note the bond lengths and angles.

Using the program Spartan on an Apple PowerMac 120 computer this calculation took 8 sec. Using this AM1 geometry as input again calculate the optimal geometry and heat of formation with an *ab initio* program. At the 6-31G(*) level this calculation took 17 min. The energy is given by the program in hartrees (1 hartree = 365.5 kcal/mol). Note the changes in energy and in geometry as the calculational method supposedly gets better (and more lengthy). *Ab initio* calculations are, due to this time constraint, limited to molecules with fewer than 100 atoms. The time needed to carry out the calculation increases at more than the fourth power of the number of atoms in the molecule.

Experiments

MICROSCALE

Hydrazine is a carcinogen. Handle with care. Wear gloves and carry out this experiment in the hood.

HO(CH$_2$CH$_2$O)$_3$H

Triethylene glycol
bp 290°C

The two-step synthesis of a chemiluminescent substance can be completed in 25 min.

Na$_2$S$_2$O$_4$ · 2 H$_2$O

Sodium hydrosulfite dihydrate
(Sodium dithionite)
MW 210.15

CAUTION: Contact of solid sodium hydrosulfite with moist combustible material may cause fire.

1. Synthesis of Luminol

First heat a tube containing 3 mL of water on the steam bath. Then heat a mixture of 200 mg of 3-nitrophthalic acid and 0.4 mL of an 8% aqueous solution of hydrazine[2] (*Caution!*) in a 10 × 100 mm reaction tube over a hot sand bath until the solid is dissolved. Add 0.6 mL of triethylene glycol, and clamp the tube in a vertical position above the hot sand bath. Insert a boiling chip and a thermometer, and boil the solution vigorously to distill the excess water. Intermittently remove the thermometer, and replace it with an aspirator tube to facilitate this. There will be a period during which the solution will boil at 110°C, and then over a 3- or 4-min period it will rise to 215°C. Lift the tube from the hot sand, and by intermittent gentle heating maintain a temperature of 215 to 220°C for 2 min. Remove the tube, cool to about 100°C (crystals of the product often appear), add the 3 mL of hot water, cool the tube in cold water, and collect the light-yellow granular nitro compound (**8**) by vacuum filtration on the Hirsch funnel.[3] The dry weight should be about 140 mg.

The nitro compound need not be dried and can be transferred at once, for reduction, to the uncleaned tube in which it was prepared. Add 1.0 mL of 3 M sodium hydroxide solution, stir with a rod, and to the resulting deep brown-red solution add 0.6 g of fresh sodium hydrosulfite dihydrate (*not* sodium hydrogen sulfite or sodium bisulfite). Wash the solid down the walls with a little water. Heat to the boiling point, stir, and keep the mixture hot for 5 min, during which time some of the reduction product may separate. Then add 0.4 mL of acetic acid, cool the tube in a beaker of cold water, and stir; collect the resulting precipitate of light-yellow luminol (**1**) by vacuum filtration on the Hirsch funnel. The filtrate on standing overnight usually deposits a further crop of luminol (20–40 mg).

Cleaning Up   Combine the filtrate from the first and second reactions, dilute with a few milliliters of water, neutralize with sodium carbonate, add 3 mL of household bleach (5.25% sodium hypochlorite solution), and heat the mixture to 50°C for 1 h. This will oxidize any unreacted hydrazine and hydrosulfite. Dilute the mixture, and flush it down the drain.

MACROSCALE

Hydrazine is a carcinogen. Handle with care. Wear gloves, and carry out this experiment in the hood.

3. Synthesis of Luminol

First heat a flask containing 15 mL of water on the steam bath. Then heat a mixture of 1 g of 3-nitrophthalic acid and 2 mL of an 8% aqueous solution of hydrazine[4] (*Caution!*) in a 20 × 150-mm test tube over a Thermowell until the solid is dissolved, add 3 mL of triethylene glycol, and clamp the tube in a verti-

---

2. Dilute 3.12 g of the commercial 64% hydrazine solution to a volume of 25 mL.

3. The reason for adding hot water and then cooling rather than adding cold water is that the solid is then obtained in more easily filterable form.

4. Dilute 3.12 g of the commercial 64% hydrazine solution to a volume of 25 mL.

cal position in a hot sand bath. Insert a thermometer, a boiling chip, and an aspirator tube connected to an aspirator, and boil the solution vigorously to distill the excess water (110–130°C). Let the temperature rise rapidly until (3–4 min) it reaches 215°C. Remove the burner, note the time, and by intermittent gentle heating maintain a temperature of 215–220°C for 2 min. Remove the tube, cool to about 100°C (crystals of the product often appear), add the 15 mL of hot water, cool under the tap, and collect the light-yellow granular nitro compound **(13)**. Dry weight 0.7 g.[5]

The two-step synthesis of a chemiluminescent substance can be completed in 25 min.

$Na_2S_2O_4 \cdot 2\,H_2O$

**Sodium hydrosulfite dihydrate
(Sodium dithionite)
MW 210.15**

The nitro compound need not be dried and can be transferred at once, for reduction, to the uncleaned test tube in which it was prepared. Add 5 mL of 3 *M* sodium hydroxide solution, stir with a rod, and to the resulting deep brown-red solution add 3 g of fresh sodium hydrosulfite dihydrate. Wash the solid down the walls with a little water. Heat to the boiling point, stir, and keep the mixture hot for 5 min, during which time some of the reduction product may separate. Then add 2 mL of acetic acid, cool under the tap, and stir; collect the resulting precipitate of light-yellow luminol **(6)**. The filtrate on standing overnight usually deposits a further crop of luminol (0.1–0.2 g).

CAUTION: Contact of solid sodium hydrosulfite with moist combustible material may cause fire.

Cleaning Up   Combine the filtrate from the first and second reactions, dilute with a few milliliters of water, neutralize with sodium carbonate, add 40 mL of household bleach (5.25% sodium hypochlorite solution), and heat the mixture to 50°C for 1 h. This will oxidize any unreacted hydrazine and hydrosulfite. Dilute the mixture, and flush it down the drain.

 **MICROSCALE
AND MACROSCALE**

4. The Light-Producing Reaction

This reaction can be run on a scale five times larger. Dissolve the first crop of moist luminol (dry weight about 40–60 mg) in 2 mL of 3 *M* sodium hydroxide solution and 18 mL of water; this is stock solution A. Prepare a second stock solution, B, by mixing 4 mL of 3% aqueous potassium ferricyanide, 4 mL of 3% hydrogen peroxide, and 32 mL of water. Now dilute 5 mL of solution A with 35 mL of water, and in a dark place, pour this solution and solution B simultaneously into an Erlenmeyer flask. Swirl the flask and, to increase the brilliance, gradually add further small quantities of alkali and ferricyanide crystals.

Ultrasonic sound also can be used to promote this reaction. Prepare stock solutions A and B again, but omit the hydrogen peroxide. Place the combined solutions in an ultrasonic cleaning bath, or immerse an ultrasonic probe into the reaction mixture. Spots of light are seen where the ultrasonic vibrations produce hydroxyl radicals.

And the sanguinary-minded can mix solutions A and B, omitting the ferricyanide from solution B. Light can be generated by adding blood dropwise to the reaction mixture.

Cleaning Up   Add 2 mL of 3 *M* hydrochloric acid, dilute the solution with water, and flush the mixture down the drain.

---

5. The reason for adding hot water and then cooling rather than adding cold water is that the solid is then obtained in more easily filterable form.

Photochemistry: The Synthesis of Benzopinacol

Prelab Exercise: Draw a mechanism for the base-catalyzed cleavage of benzopinacol.

Photochemistry, as the name implies, is the chemistry of reactions initiated by light. A molecule can absorb light energy and then undergo isomerization, fragmentation, rearrangement, dimerization, or hydrogen atom abstraction. The last reaction is the subject of this experiment. On irradiation with sunlight, benzophenone abstracts a proton from the solvent 2-propanol and becomes reduced to benzopinacol.

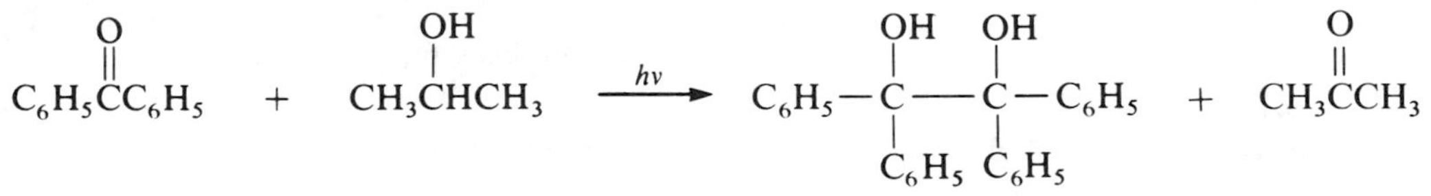

$$
\underset{\substack{\textbf{Benzophenone}\\ \text{MW } 182.21\\ \text{mp } 48°C}}{\overset{\overset{\displaystyle O}{\parallel}}{C_6H_5CC_6H_5}}
\;+\;
\underset{\substack{\textbf{2-Propanol}\\ \text{MW } 60.09\\ \text{bp } 82°C\\ n_D^{20}\ 1.3770}}{\overset{\overset{\displaystyle OH}{\mid}}{CH_3CHCH_3}}
\;\xrightarrow{h\nu}\;
\underset{\substack{\textbf{Benzopinacol}\\ \text{MW } 366.44\\ \text{mp } 189°C}}{C_6H_5-\overset{\overset{\displaystyle OH}{\mid}}{\underset{\underset{\displaystyle C_6H_5}{\mid}}{C}}-\overset{\overset{\displaystyle OH}{\mid}}{\underset{\underset{\displaystyle C_6H_5}{\mid}}{C}}-C_6H_5}
\;+\;
\underset{\substack{\textbf{Acetone}\\ \text{MW } 58.08\\ \text{bp } 56°C\\ n_D^{20}\ 1.3590}}{\overset{\overset{\displaystyle O}{\parallel}}{CH_3CCH_3}}
$$

The energy of light varies with its frequency according to this equation:

$$\Delta E = h\nu = h\frac{c}{\lambda}$$

where h is Planck's constant, ν is the frequency of the light, λ is its wavelength, and c is the speed of light. The fact that benzophenone is colorless means it does not absorb visible light, yet irradiation of benzophenone in alcohol results in a chemical change. Pyrex glass is not transparent to ultraviolet light with a wavelength shorter than 290 nm, so some wavelength between 290 and 400 nm (the edge of the visible region) must be responsible. The UV spectrum of benzophenone indicates an absorption band centered at about 355 nm. Light of that wavelength is absorbed by benzophenone.

What, in a general sense, occurs when a molecule absorbs light? A photon is absorbed only if its energy (wavelength) corresponds exactly to the difference between two electronic energy levels in the molecule. In benzophenone the electrons most loosely held and thus most easily excited are the two pairs of nonbonded electrons on the carbonyl oxygen, called the *n*-electrons.

$$\text{\textbackslash}C=\ddot{O}: \qquad \text{\textit{n}-electrons}$$

These electrons have paired spins $\searrow C=O \uparrow\downarrow$ and only one is excited by the photon of light. The electron goes into the lowest unoccupied excited energy level, the π^*. Each electronic energy level of benzophenone has within it many vibrational and rotational energy levels. An electron can reside in many of these in the lower electronic energy level, the ground state S_0, and it can be promoted to many vibrational and rotational energy levels in the first excited state (S_1). Hence the ultraviolet spectrum does not appear as a single sharp peak but as a band made up of many peaks that arise from transitions between these many energy levels. See Fig. 63.1.

The spin of the electron cannot change in going from the ground state S_0 to the excited singlet state S_1 (conservation of angular momentum). Once in a higher vibrational or rotational state it can drop to S_1 by vibrational relaxation, losing some of its energy as heat. It then undergoes either fluorescence or inter-system crossing. As seen in Fig. 63.1, the rate for fluorescence, a light-emitting process in which the electron returns to the ground state, is 10^4 times slower than intersystem crossing; so benzophenone does not fluoresce. The electron flips its orientation during intersystem crossing so that it has the same orientation as the electron with which it was paired in the ground state. In this new state, called the *triplet*, it can lose energy as light; but this process, phosphorescence, is a slow one. The lifetime of the triplet state is long enough for chemical

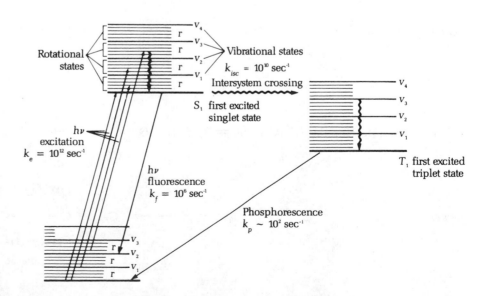

FIG. 63.1 Electronic energy levels and possible transitions.

reactions to take place. The triplet can also lose energy as heat in a radiationless transition, but the probability of this happening is relatively low. This situation can be represented diagrammatically:

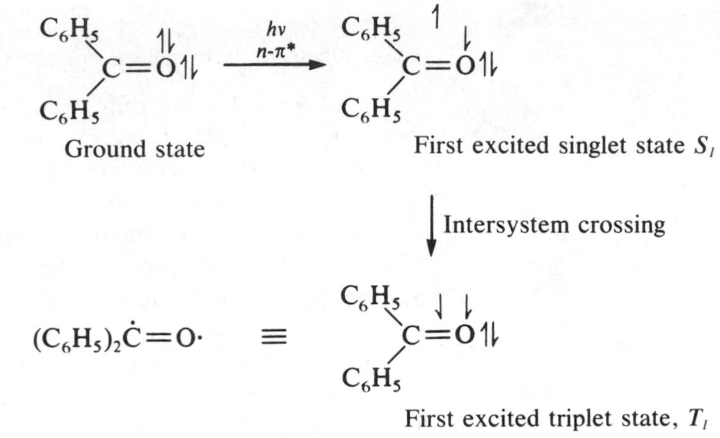

$$(C_6H_5)_2\dot{C}=O\cdot \quad \equiv \quad \text{First excited triplet state, } T_1$$

This T_1 state is a diradical and can abstract a methine proton from the solvent, then a hydroxyl proton to give acetone and diphenyl hydroxy radical, which then dimerizes to give the product.

Removal of methine proton

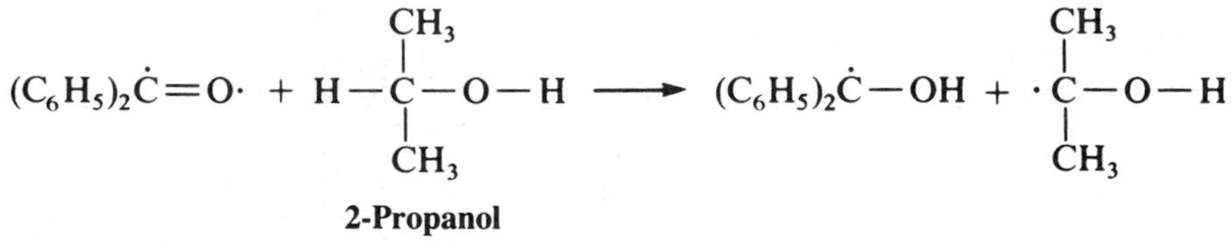

$$(C_6H_5)_2\dot{C}=O\cdot \; + \; H-\underset{\underset{CH_3}{|}}{\overset{\overset{CH_3}{|}}{C}}-O-H \longrightarrow (C_6H_5)_2\dot{C}-OH \; + \; \cdot\underset{\underset{CH_3}{|}}{\overset{\overset{CH_3}{|}}{C}}-O-H$$

2-Propanol

Removal of hydroxyl proton

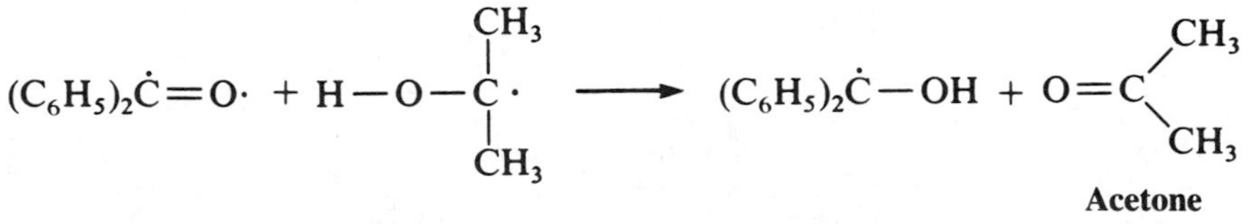

$$(C_6H_5)_2\dot{C}=O\cdot \; + \; H-O-\underset{\underset{CH_3}{|}}{\overset{\overset{CH_3}{|}}{C}}\cdot \longrightarrow (C_6H_5)_2\dot{C}-OH \; + \; O=C\overset{CH_3}{\underset{CH_3}{}}$$

Acetone

Dimerization

$$2\,(C_6H_5)_2\dot{C}-OH \longrightarrow C_6H_5-\underset{\underset{C_6H_5}{|}}{\overset{\overset{OH}{|}}{C}}-\underset{\underset{C_6H_5}{|}}{\overset{\overset{OH}{|}}{C}}-C_6H_5$$

Benzopinacol

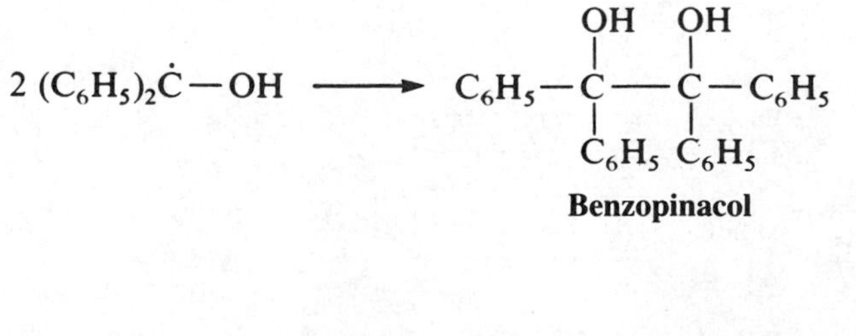

Experiments

1. Benzopinacol

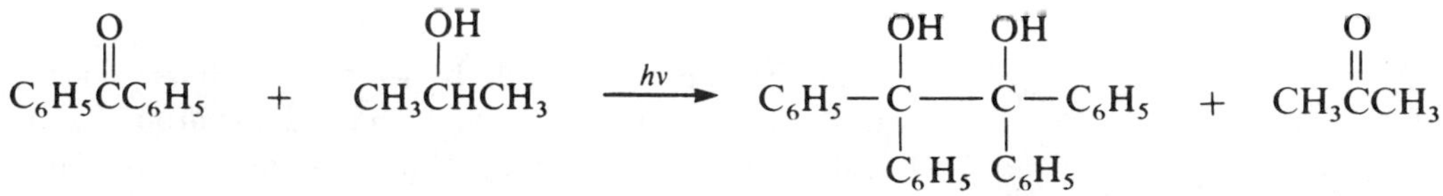

$$C_6H_5\overset{\displaystyle O}{\overset{\|}{C}}C_6H_5 \;+\; \underset{\displaystyle CH_3CHCH_3}{\overset{\displaystyle OH}{\overset{|}{}}} \xrightarrow{h\nu} \; C_6H_5-\underset{\displaystyle C_6H_5}{\overset{\displaystyle OH}{\overset{|}{C}}}-\underset{\displaystyle C_6H_5}{\overset{\displaystyle OH}{\overset{|}{C}}}-C_6H_5 \;+\; CH_3\overset{\displaystyle O}{\overset{\|}{C}}CH_3$$

Benzophenone	**2-Propanol**	**Benzopinacol**	**Acetone**
MW 182.21	MW 60.09	MW 366.44	MW 58.08
mp 48°C	bp 82°C	mp 189°C	bp 56°C

This experiment should be done when there is good prospect for several hours of bright sun. The benzopinacol is cleaved by alkali to benzhydrol and benzophenone (Experiment 3), and it is rearranged in acid to benzopinacolone (Experiment 4).

Microscale Procedure

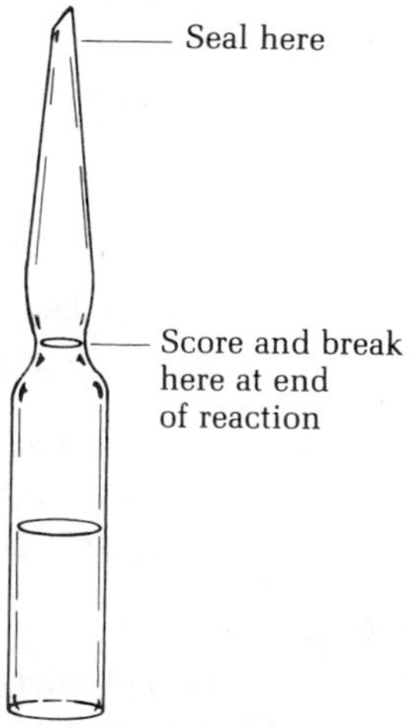

In a 2-mL or 1-dram ampule dissolve 100 mg of benzophenone in 1 mL of isopropyl alcohol by warming on a sand bath. Add a microdrop of acetic acid, cool the ampule in dry ice, and seal the neck of the ampule in a flame (Fig. 63.2). Set the ampule outside in bright sunlight. On a larger scale, this experiment can take as long as a week because of absorption of ultraviolet radiation by the thick glass walls of a flask and by self-absorption of the thick layers of solution. On a microscale, the glass of an ampule is very thin and the thickness of the entire solution is small, so the reaction proceeds very rapidly. A rough parabolic reflector fashioned from aluminum foil can speed the reaction even more.

Since benzopinacol is but sparingly soluble in the isopropyl alcohol, its formation is followed by the separation of small, colorless crystals (benzophenone forms large, thick prisms) from around the walls of the ampule. If the reaction mixture is exposed to direct sunlight throughout the daylight hours, the first crystals separate in less than an hour, and the reaction is complete in a day. In winter the reaction will take longer, and any benzophenone that crystallizes must be brought into solution by warming on the steam bath. When the reaction appears to be over, chill the tube if necessary and collect the product. The material should be pure (mp 188–189°C). If the yield is low, more material can be obtained by further exposure of the mother liquor to sunlight.

FIG. 63.2 Microscale photolysis in 1-dram (1.77-mL) ampule.

Seal here

Score and break here at end of reaction

Reaction time: 1 day to 1 week

Cleaning Up   Dilute the isopropyl alcohol filtrate with water, and flush the solution down the drain. Should any unreacted benzophenone precipitate, collect it by vacuum filtration, discard the filtrate down the drain, and place the recovered solid in the nonhazardous solid waste container.

Macroscale Procedure

The experiment should be done when there is good prospect for long hours of bright sunshine for several days. The benzopinacol then can be cleaved by alkali to benzhydrol and benzophenone in Experiment 3, and it can be rearranged in acid to benzopinacolone.

In a 50-mL round-bottomed flask dissolve 5 g of benzophenone in 30–35 mL of isopropyl alcohol by warming on the steam bath, fill the flask to the neck with more of this alcohol, and add one drop of glacial acetic acid. (If the acid is omitted enough alkali may be derived from the glass of the flask to destroy the reaction product by the alkaline cleavage described in Experiment 3.) Stopper the flask with a well-rolled, tight-fitting cork or a lightly greased glass stopper, which is then wired in place. Invert the flask in a 100-mL beaker placed where the mixture will be most exposed to direct sunlight for some time. Since benzopinacol is but sparingly soluble in alcohol, its formation can be followed by the separation from around the walls of the flask of small, colorless crystals (benzophenone forms large, thick prisms). If the reaction mixture is exposed to direct sunlight throughout the daylight hours, the first crystals separate in about 5 h, and the reaction is practically complete (95% yield) in 4 days. In winter the reaction may take as long as 2 weeks, and any benzophenone that crystallizes must be brought into solution by warming on the steam bath. When the reaction appears to be over, chill the flask if necessary and collect the product. The material should be pure, mp 188–189°C. If the yield is low, more material can be obtained by further exposure of the mother liquor to sunlight.

Reaction time: 4 days to 2 weeks

Cleaning Up   Dilute the isopropyl alcohol filtrate with water, and flush the solution down the drain. Should any unreacted benzophenone precipitate, collect it by vacuum filtration, discard the filtrate down the drain, and place the recovered solid in the nonhazardous solid waste container.

MICROSCALE AND MACROSCALE

2. Alkaline Cleavage

Suspend a small test sample of benzopinacol in alcohol, and heat to boiling on a steam bath, making sure that the amount of solvent is not sufficient to dissolve the solid. Add a drop of sodium hydroxide solution, heat for a minute or two, and observe the result. The solution contains equal parts of benzhydrol and benzophenone, formed by the following reaction:

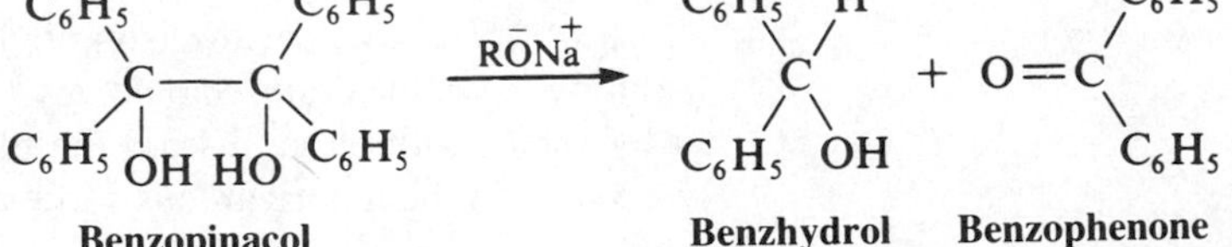

$$
\underset{\substack{\text{Benzopinacol}\\ \text{mp 189°C}}}{\underset{C_6H_5}{\overset{C_6H_5}{\big|}}\underset{\text{OH HO}}{\overset{}{C}}-\underset{C_6H_5}{\overset{C_6H_5}{\big|}}C} \xrightarrow{\text{RO}^-\text{Na}^+} \underset{\substack{\text{Benzhydrol}\\ \text{mp 68°C}}}{\underset{C_6H_5}{\overset{C_6H_5}{\big|}}\underset{\text{OH}}{\overset{H}{C}}} + \underset{\substack{\text{Benzophenone}\\ \text{mp 48°C}}}{O=\underset{C_6H_5}{\overset{C_6H_5}{C}}}
$$

The low-melting products resulting from the cleavage are much more soluble than the starting material. Analyze the mixture by TLC.

Benzophenone can be converted into benzhydrol in nearly quantitative yield by following the procedure outlined above for the preparation of benzopinacol, modified by addition of a *very* small piece of sodium (5 mg) instead of the acetic acid. The reaction is complete when, after exposure to sunlight, the greenish-blue color disappears. To obtain the benzhydrol the solution is diluted with water, acidified, and evaporated. Benzopinacol is produced as before by photochemical reduction, but it is at once cleaved by the sodium alkoxide. The benzophenone formed by cleavage is converted into more benzopinacol, cleaved, and eventually consumed. The IR and ^{1}H NMR spectra of benzophenone are shown at the end of the chapter (Figs. 63.3 and 63.4).

3. Pinacolone Rearrangement

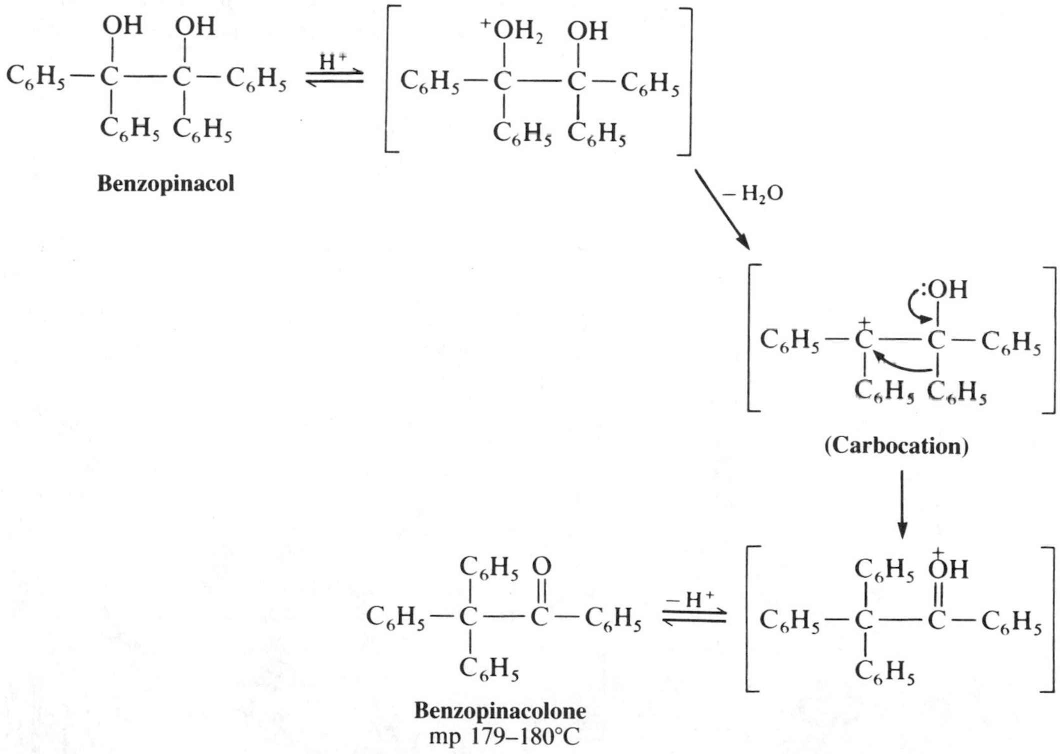

This acid-catalyzed carbonium ion rearrangement is characterized by rapidity and by the high yield.

Microscale Procedure

In a reaction tube place 50 mg of benzopinacol, 0.25 mL of acetic acid, and one *very* small crystal of iodine (0.0005 g). Heat to boiling for a minute or two on the sand bath until the crystals are dissolved, and then reflux the red solution for 5 min. On cooling, the pinacolone separates as a stiff paste. Thin the paste with alcohol, collect the product by vacuum filtration on the Hirsch funnel, and wash it free of iodine with cold 95% ethanol. The material should be pure; the expected yield is 95%.

Cleaning Up   Dilute the filtrate with water, neutralize with sodium carbonate, and flush down the drain.

Macroscale Procedure

In a 100-mL round-bottomed flask place 5 g of benzopinacol, 25 mL of acetic acid, and two or three very small crystals of iodine (0.05 g). Heat to the boiling point for a minute or two under a reflux condenser until the crystals are dissolved, and then reflux the red solution for 5 min. On cooling, the pinacolone separates as a stiff paste. Thin the paste with alcohol, collect the product, and wash it free from iodine with alcohol. The material should be pure; yield is 95%.

Cleaning Up   Dilute the filtrate with water, neutralize with sodium carbonate, and flush down the drain.

Questions _____

1. Would the desired reaction occur if ethanol or *t*-butyl alcohol were used instead of isopropyl alcohol in the attempted photochemical dimerization of benzophenone?

2. Assign the peak at 1639 cm^{-1} in the IR spectrum of benzophenone (Fig. 63.3).

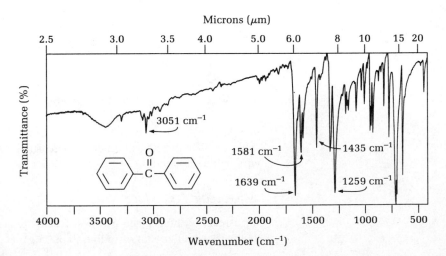

FIG. 63.3   Infrared spectrum of benzophenone (KBr disk).

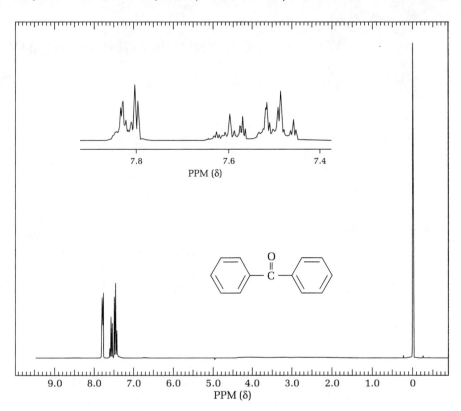

FIG. 63.4   ^{1}H NMR spectrum of benzophenone (250 MHz).

CHAPTER 64

Carbohydrates and Sweeteners

Prelab Exercise: Draw a flow sheet showing the sequence of tests you would conduct to identify an unknown carbohydrate.

Carbohydrates, the direct product of the photosynthetic combination of carbon dioxide and water, are, by weight, the most common organic compounds on earth. Since most have the empirical formula $C_nH_{2n}O_n = C_n(H_2O)_n$, it is easy to see why early workers considered these molecules to be hydrates of carbon. They are polyhydroxyaldehydes and ketones and exist as hemi- and full acetals and ketals. Carbohydrates are classified as mono-, di-, and polysaccharides.

Sweeteners

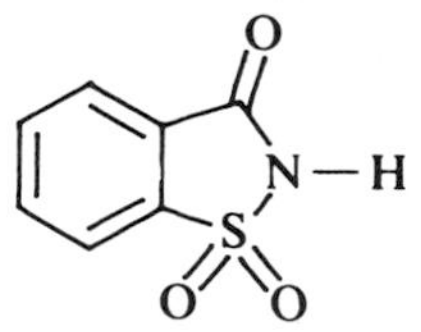

Saccharin

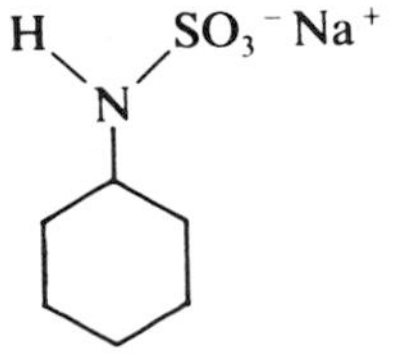

Sodium cyclamate

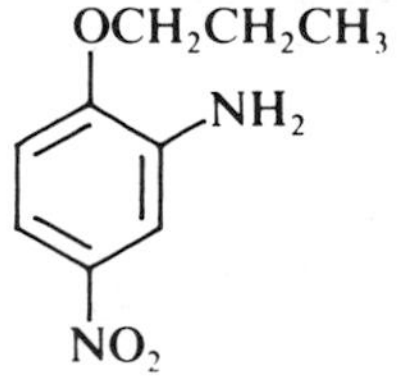

2-Amino-4-nitro-*n*-propoxybenzene

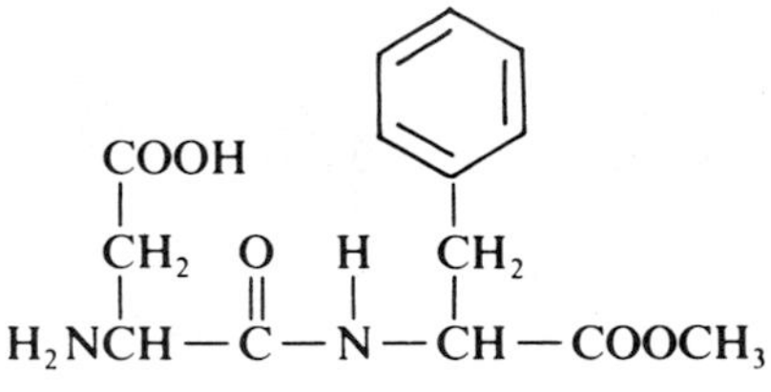

Aspartame

L-Aspartyl-L-phenylalanine methyl ester

SYNTHETIC SUGAR SUBSTITUTES

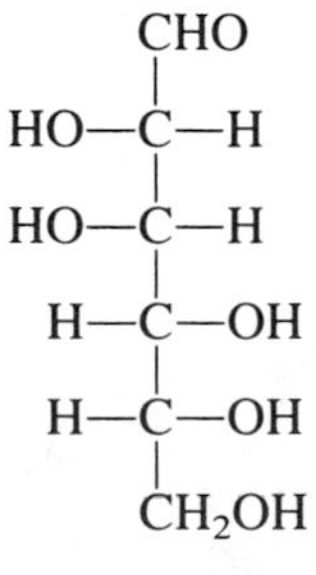

D-Mannose

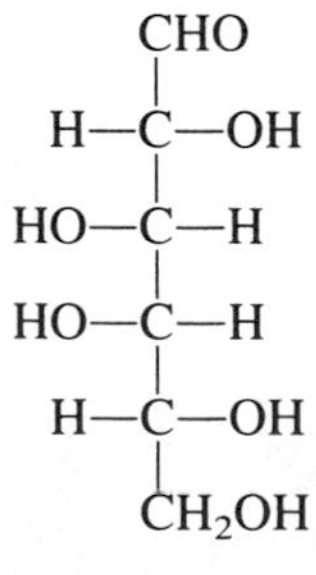

D-Galactose

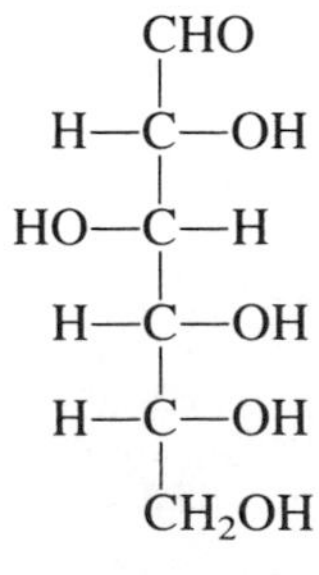

D-Glucose

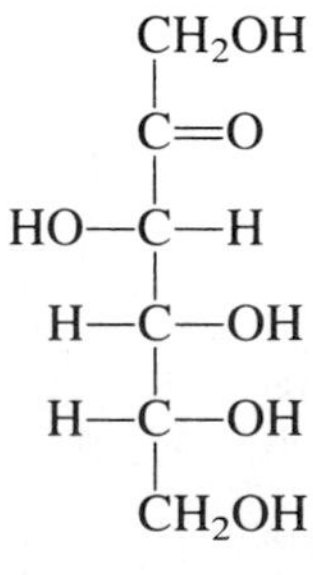

D-Fructose

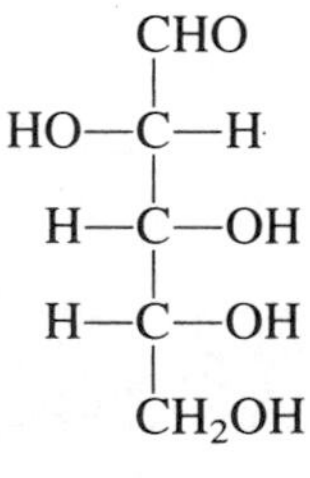

D-Arabinose

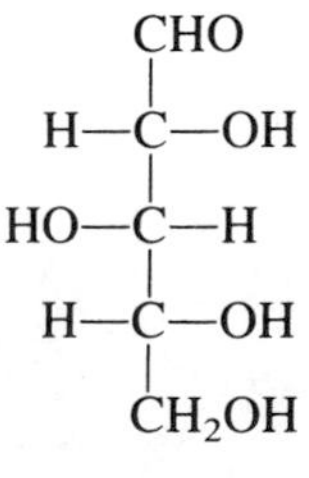

D-Xylose

ALDOHEXOSES | KETOHEXOSE | ALDOPENTOSES

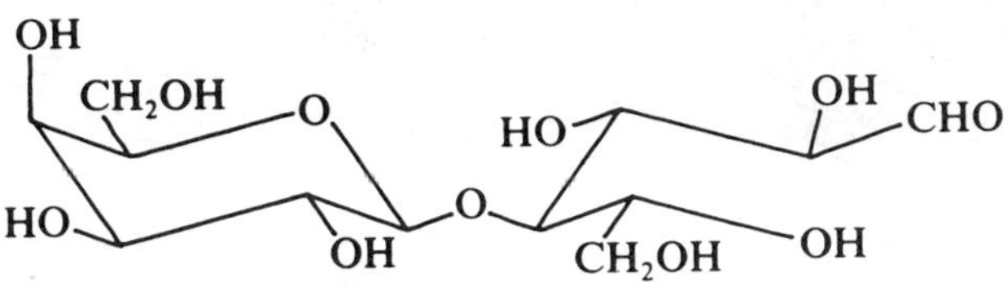

Lactose (aldehyde form)

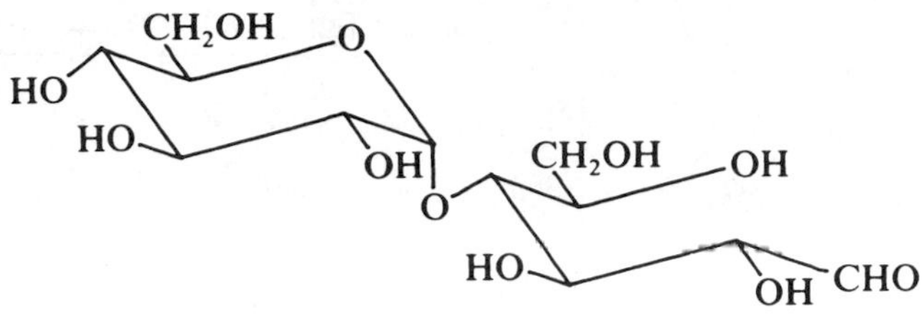

Maltose (aldehyde form)

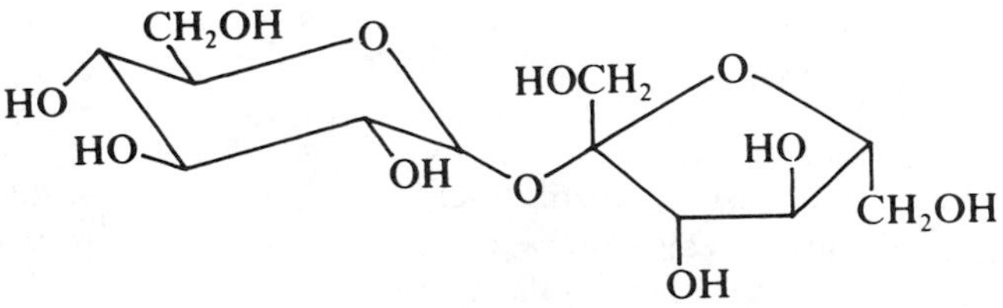

Sucrose

DISACCHARIDES

The term *sugar* applies to mono-, di-, and oligosaccharides, which are all soluble in water and thereby distinguished from polysaccharides. Many natural sugars are sweet, but data in Table 64.1 show that sweetness varies greatly with stereochemical configuration and is exhibited by compounds of widely differing structural type. Note that D-fructose is 50% sweeter than sucrose, which accounts for the presence of high-fructose corn syrup in many commercial foods and drinks.

The synthesis and sale of the artificial, low-calorie sweetener aspartame (NutraSweet) is a multibillion-dollar industry. It is interesting to note that the discovery of all artificial sweeteners has been accidental, which says something about how careless chemists have been. The exact physiological basis for sweetness, and therefore the relationship between organic structure and sweetness, is not understood.

One gram of sucrose (common table sugar) dissolves in 0.5 mL of water at 25°C and in 0.2 mL at the boiling point, but the substance has marked, atypical crystallizing properties. Despite the high solubility, it can be obtained in beautiful, large crystals (rock candy). More typical sugars are obtainable in crystalline form only with difficulty, particularly in the presence of a trace of impurity, and even then, they give small, ill-formed crystals. Alcohol is often added to a water solution to decrease solubility and thus to induce crystallization. The amounts of 95% ethanol required to dissolve 1-g samples at 25°C are sucrose, 170 mL; glucose, 60 mL; and fructose, 15 mL. Some sugars have never been obtained in crystalline condition and are known only as viscous syrups.

Table 64.1 Relative Sweetness of Sugars and Sugar Substitutes

	Sweetness	
Compound	*To humans*	*To bees*
Monosaccharides		
D-Fructose	1.50	+
D-Glucose	0.55	+
D-Mannose	Sweet, then bitter	−
D-Galactose	0.55	−
D-Arabinose	0.70	−
D-Xylose	Very sweet	+
Disaccharides		
Sucrose (glucose, fructose)	1.0	+
Maltose (2 glucose)	0.3	+
Lactose (glucose, galactose)	0.2	−
Cellobiose (2 glucose)	Indifferent	−
Gentiobiose (2 glucose)	Bitter	−
Polysaccharides		
Soluble starch suspension	Not sweet	−
Soluble glycogen suspension	Not sweet	−
Synthetic sugar substitutes		
Aspartame (NutraSweet)	80	
Saccharin	550	
2-Amino-4-nitro-1-*n*-propoxybenzene	4000	

Analysis of Carbohydrates

In this experiment, unknown carbohydrates will be identified using a series of classification tests. The Molisch test (Experiment 1) is a general test for carbohydrates. If the test is positive and the unknown is water soluble (which distinguishes it from polysaccharides such as cellulose), then it can be tested with iodine, which will give a positive test with the partially water-soluble polysaccharide, starch.

If the unknown is a sugar, it can be a mono-, di-, or, less commonly, an oligosaccharide. If it is a monosaccharide, it usually has either five carbons or six carbons, and its carbonyl group can be either an aldehyde or a ketone.

To distinguish aldehydo- from ketosugars, take advantage of the reducing ability of the aldehyde group. Either the traditional Fehling's or Benedict's tests will give a positive reaction with any reducing sugar; in this experiment, we will employ the much more sensitive red tetrazolium (RT) test. If the RT test (Experiment 2) is positive, then Barfoed's test (Experiment 3) allows us to make a distinction between reducing mono- and disaccharides.

To distinguish between pentoses and hexoses, run the Bial test (Experiment 4). The Seliwanov test (Experiment 5) allows us to make a distinction between aldohexoses and ketohexoses.

Osazones

With phenylhydrazine, many sugars form beautiful crystalline derivatives called *osazones* (Experiment 6). Osazones are much less soluble in water than the parent sugars because the molecular weight is increased by 178 units and the number of hydroxyl groups is reduced by one. It is easier to isolate an osazone than to isolate the sugar, and sugars that are syrups often give crystalline osazones. Osazones of the more highly hydroxylic disaccharides are notably more soluble than those of monosaccharides.

Some disaccharides do not form osazones, but a test for formation or nonformation of the osazone is ambiguous because the glycosidic linkage may suffer hydrolysis in a boiling solution of phenylhydrazine and acetic acid with formation of an osazone derived from a component sugar and not from the disaccharide. If a sugar has reducing properties (a positive RT test), it is also capable of osazone formation; hence an unknown sugar should be tested for reducing properties before preparation of an osazone is attempted.

Experiments

1. Molisch Test for Carbohydrates

MICROSCALE AND MACROSCALE

CAUTION: Handle concentrated sulfuric acid with care. It is very corrosive. Do not mix the layers in the tube because the heat of mixing can form steam, thus causing the solution to erupt from the tube. Use great care in pouring the reaction mixture into a beaker for disposal.

Label a number of reaction tubes or 10×75-mm culture tubes. To 1 mL of a 1% solution of the carbohydrates in Table 64.1 (or a representative selection), add a drop of Molish reagent[1] and mix thoroughly. Place just 1 mL of water in another tube as a control. Then, while tilting each tube, carefully introduce 1 mL of concentrated sulfuric acid down the inside. The dense sulfuric acid will form a lower layer. Note the appearance of each tube. Do not mix the layers. A purple color at the interface between the two layers is a positive test for carbohydrates.

Cleaning Up   Pour the contents of each tube into a large beaker (*Caution!*), and neutralize with sodium bicarbonate. Dilute the neutral solution with water, and flush down the drain.

MICROSCALE AND MACROSCALE

2. Red Tetrazolium[2]

The reagent (red tetrazolium) is a nearly colorless, water-soluble substance that oxidizes aldoses and ketoses, as well as other α-ketols, and is thereby reduced.

---

1. In 25 mL of ethanol, dissolve 1.25 g of 1-naphthol.

2. 2,3,5-Triphenyl-2H-tetrazolium chloride. Available from Aldrich. Light-sensitive. Freshly prepared aqueous solutions should be used in tests. Any unused solution should be acidified and discarded.

The reduced form is a water-insoluble, intensely colored pigment, a diformazan.

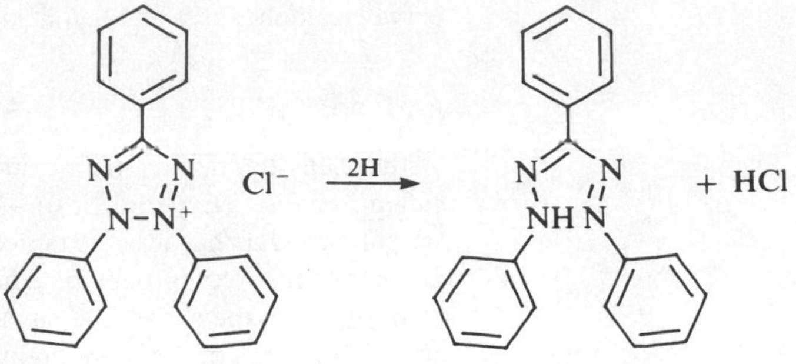

Red tetrazolium (RT)   **RT-Diformazan**

A sensitive test for reducing sugars

Red tetrazolium affords a highly sensitive test for reducing sugars, and it distinguishes between α-ketols and simple aldehydes more sharply than Fehling's and Tollens' tests, two older tests that were used for this purpose.

Procedure

Put one drop of each of the 1% test solutions in the cleaned, marked test tubes, and, to each tube, add 1 mL of a 0.5% aqueous solution of red tetrazolium[3] and one drop of 3 *M* sodium hydroxide solution. Put the tubes in the beaker of hot water, note the time, and note the time at which each tube develops color.

To estimate the sensitivity of the test, use the substance that you regard as the most reactive of the five studied. Dilute 1 mL of the 1% solution with water to a volume of 100 mL, and run a test with RT on 0.2 mL of the diluted solution.

Cleaning Up   The test solutions should be diluted with water and flushed down the drain.

MICROSCALE AND MACROSCALE

3. Barfoed's Test

Barfoed's test is used to distinguish between reducing mono- and disaccharides. In labeled reaction tubes or 10 × 75-mm culture tubes, place 1 mL of 1% solutions of a representative selection of reducing mono- and disaccharide solutions chosen from Table 64.1. To each, add 0.5 mL of Barfoed's reagent[4] and mix. Heat the tubes in a hot-water bath for 15 min. Note the time needed for a red precipitate to appear in each tube.

Cleaning Up   The test solutions should be diluted with water and flushed down the drain.

---

3. Dissolve 0.5 g of red tetrazolium in 100 mL of water.

4. Dissolve 6.7 g of copper(II) acetate and 0.9 mL of acetic acid in 100 mL of water.

MICROSCALE AND MACROSCALE

4. Bial's Test

Bial's test is used to distinguish between pentoses and hexoses. In labeled reaction tubes or 10×75-mm culture tubes, place 1 mL of 1% solutions of a representative selection of pentoses and hexoses taken from Table 64.1. In one tube, place 1 mL of distilled water to serve as a reference or control solution.

To each tube, add 1 mL of Bial's reagent[5] and a boiling chip. Heat each tube to boiling on the sand bath, and note the color that develops. In tubes that have weak or no color, the product of the reaction can be concentrated by diluting the mixture with 2 mL of water, adding 0.5 mL of cyclohexanol, and then extracting the colored product into the alcohol layer by shaking the tube vigorously.

Cleaning Up   The test solutions should be diluted with water and flushed down the drain.

MICROSCALE AND MACROSCALE

5. Seliwanov's Test

Seliwanov's test is used to distinguish between aldohexoses and ketohexoses. To a number of reaction tubes or 10×75-mm culture tubes, add 1 mL of the Seliwanov reagent.[6] To each labeled tube, add 5 drops of a 1% solution of the sugars to be tested. After mixing, heat the tubes in a boiling water bath for 2 min. Examine the tubes. The intensity of the color is proportional to the α-ketol concentration.

Cleaning Up   The test solutions should be diluted with water and flushed down the drain.

MICROSCALE AND MACROSCALE

6. Osazones

This experiment can be carried out on a scale two or three times larger than is specified here.

Put 0.33-mL portions of phenylhydrazine reagent[7] into each of four cleaned, numbered reaction tubes. Add 1-mL portions of 1% solutions of glucose, fructose, lactose, and maltose, and heat the tubes in the beaker of hot water for 20 min. Shake or flick the tubes occasionally to relieve supersaturation. Note the times at which the osazones separate. If, after 20 min, no product has separated, cool and scratch the test tube to induce crystallization.

CAUTION: Conduct this procedure in hood. Avoid skin contact with phenylhydrazine.

---

5. In 100 mL of concentrated hydrochloric acid, dissolve 300 mg of 3,5-dihydroxytoluene, and then add 0.3 mL of a 10% iron(III) chloride solution.

6. The reagent is prepared by dissolving 50 mg of 1,3-dihydroxybenzene (resorcinol) in a mixture of 33 mL of concentrated hydrochloric acid and 67 mL of distilled water.

7. Neutralize phenylhydrazine (3 mL) with 9 mL of acetic acid in a 50-mL Erlenmeyer flask, add 15 mL of water, transfer the mixture to a graduated cylinder, and make up the volume to 30 mL.

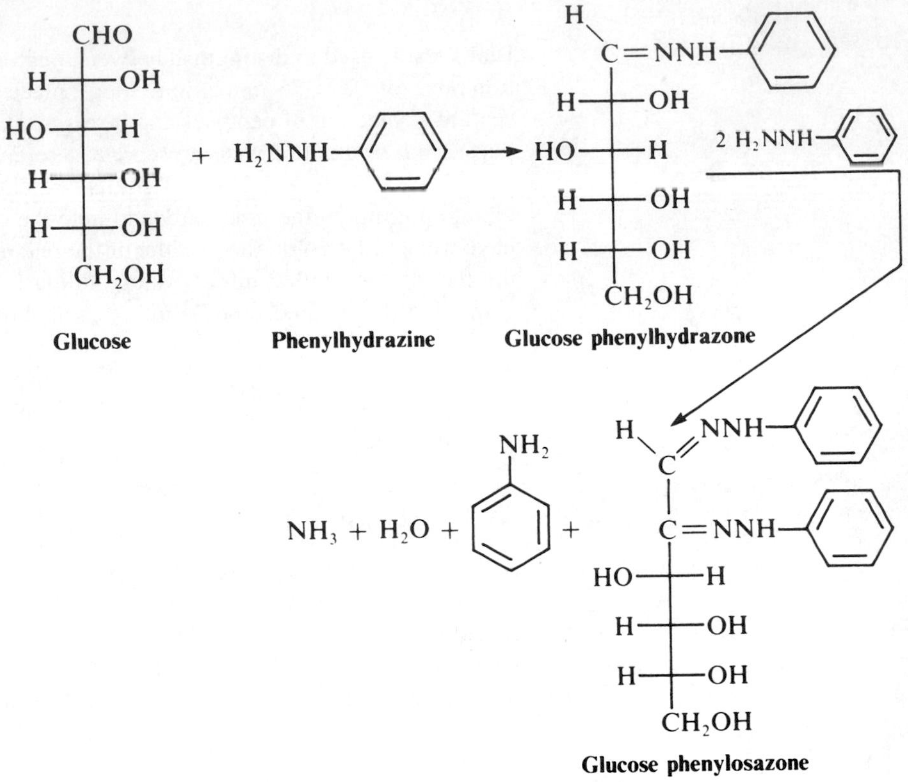

Collect and save the products for possible use in later identification of unknowns. Since osazones melt with decomposition, heat the bath at a standard rate (0.5°C/sec) when determining the melting point.

Cleaning Up   The filtrate can be diluted with water and flushed down the drain. If you must destroy the phenylhydrazine, neutralize the solution and add 2 mL of laundry bleach (5.25% sodium hypochlorite) for each 1 mL of the reagent. Heat the mixture to 45–50°C for 2 h to oxidize the amine, then cool the mixture, and flush it down the drain.

Questions

1. What do you conclude is the order of relative reactivity in the RT test of the compounds studied?

2. Which test do you regard as the most reliable for distinguishing reducing from nonreducing sugars? Which is most reliable for differentiating an α-ketol from a simple aldehyde?

3. Write a mechanism for the acid-catalyzed hydrolysis of the disaccharide sucrose. Take care to draw the stereochemistry of the products.

4. Will the anomeric forms of glucose give different phenylosazones?

65

Biosynthesis of Ethanol

Prelab Exercise: List the essential chemical substances, the solvent, and conditions for converting glucose to ethanol.

Fermentation

Human beings have been preparing fermented beverages for more than 5000 years. Materials excavated from Egyptian tombs dating to the third millennium B.C. demonstrate the operations used in making beer and leavened bread. The history of fermentation, whereby sugar is converted to ethanol by the action of yeast, is also a history of chemistry. The word "gas" was coined by van Helmont in 1610 to describe the bubbles produced in fermentation. Leeuwenhoek observed and described the cells of yeast with his newly invented microscope in 1680. Joseph Black in 1754 discovered carbon dioxide and showed it to be a product of fermentation, the burning of charcoal, and respiration. Lavoisier in 1789 showed that sugar gives ethanol and carbon dioxide and made quantitative measurements of the amounts consumed and produced.

Glucose $\rightarrow$ 2 C_2H_5OH + 2 CO_2

Once the mole concept was established, Gay-Lussac in 1815 could show that 1 mol of glucose gives exactly 2 mol of ethanol and 2 mol of carbon dioxide. But the process of fermentation stumped some great chemists. The little-known Kutzing wrote in 1837, on the basis of microscopic observation, "It is obvious that chemists must now strike yeast off the roll of chemical compounds, since it is not a compound but an organized body, an organism." On the other side were chemists such as Berzelius, who believed that yeast had a catalytic action, and Liebig, who put forth a "theory of motion of the elements within a compound which caused a disturbance of equilibrium which was communicated to the elements of the substance with which it came in contact thus forming new compounds."

Pasteur's contribution

It remained for Pasteur to show that fermentation was a physiologic action associated with the life processes of yeast. Through his microscope he could see the yeast cells that grew naturally on the surface of grapes. He showed that grape juice carefully extracted from the center of a grape and exposed to clean air would not ferment. In his classic paper of 1857, he described fermentation as the action of a living organism; but because the conversion of glucose to ethanol and carbon dioxide is a balanced equation, other chemists disputed his findings. They searched for the substance in yeast that might cause the reaction. The search lasted for 40 years, eventually ended by a clever experiment by Eduard Büchner. He made a cell-free extract of yeast that would still cause the conversion of sugar to alcohol. This cell-free extract contained the catalysts, which we now call enzymes, that were necessary for fermentation—a discovery that earned him the 1907 Nobel prize. In 1905 Harden discovered that inorganic phosphate added to the enzymes increased the rate of fermentation and was

Enzymes: fermentation catalysts

itself consumed. This result led him to eventually isolate fructose 1,6-diphosphate. Clearly the history of biochemistry is intimately associated with the study of alcoholic fermentation.

Ancient peoples discovered many of the essential reactions of alcoholic fermentation completely by accident. That crushed grapes would soon begin to froth and bubble and produce a pleasant beverage is a discovery lost in time. But what of those who lived in colder climates where the grape did not grow? How did they discover that the starch of wheat or barley could be converted to sugar by the enzymes in malt? When grain germinates, enzymes are produced that turn the starch into sugar. The process of malting involves letting the grain start to germinate and then heating and drying the sprouts to stop the process before the enzymes are used up. The color of the malt depends on the temperature of the drying. The darkest is used for stout and porter, the lighter for brown, amber, and pale ale. Because of a discovery some time ago that the resulting beverage did not spoil as rapidly if hops were added, we now also have beer.

Sources of enzymes:
  Grape skins
  Malt
  Saliva
  Yeast

Other sources exist for the amylases that catalyze the conversion of starch to glucose. The Peruvian campasinos (peasants) make a drink called "chicha" from masticated wheat, which is dried in small cakes. When water, yeast, and more ground wheat are added, the resulting mixture ferments to a beerlike beverage. The enzyme salivary amylase is the catalyst for this starch-to-glucose conversion.

Bakers make use of fermentation by taking advantage of the gas released to leaven the bread. In the present experiment baker's yeast is used to convert sucrose, ordinary table sugar, into ethanol and carbon dioxide with the aid of some 14 enzymes as catalysts, in addition to adenosine triphosphate (ATP), phos-

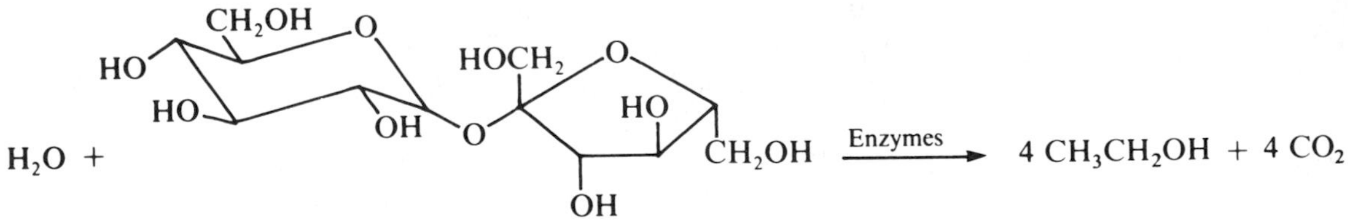

$$H_2O + \text{Sucrose} \xrightarrow{\text{Enzymes}} 4\ CH_3CH_2OH + 4\ CO_2$$

Sucrose

phate ion, thiamine pyrophosphate, magnesium ion, and reduced nicotinamide adenine dinucleotide (NADH), all present in yeast. The fermentation process—known as the Emden-Meyerhof-Parnas scheme—involves the hydrolysis of sucrose to glucose and fructose, which, as their phosphates, are cleaved to two three-carbon fragments. These fragments, as their phosphates, eventually are converted to pyruvic acid, which is decarboxylated to give acetaldehyde. Acetaldehyde, in turn, is reduced to ethanol in the final step. Each step requires a specific enzyme as a catalyst and often inorganic ions, such as magnesium and, of course, phosphate. Thirty-one kilocalories of heat are released per mole of glucose consumed in this sequence of anaerobic reactions.

This same sequence of reactions, up to the formation of pyruvic acid, occurs in the human body in times of stress when energy is needed, but not enough oxygen is available for normal aerobic oxidation. The pyruvic acid under these conditions is converted to lactic acid. It is the buildup of lactic acid in the muscles that is partly responsible for the feeling of fatigue.

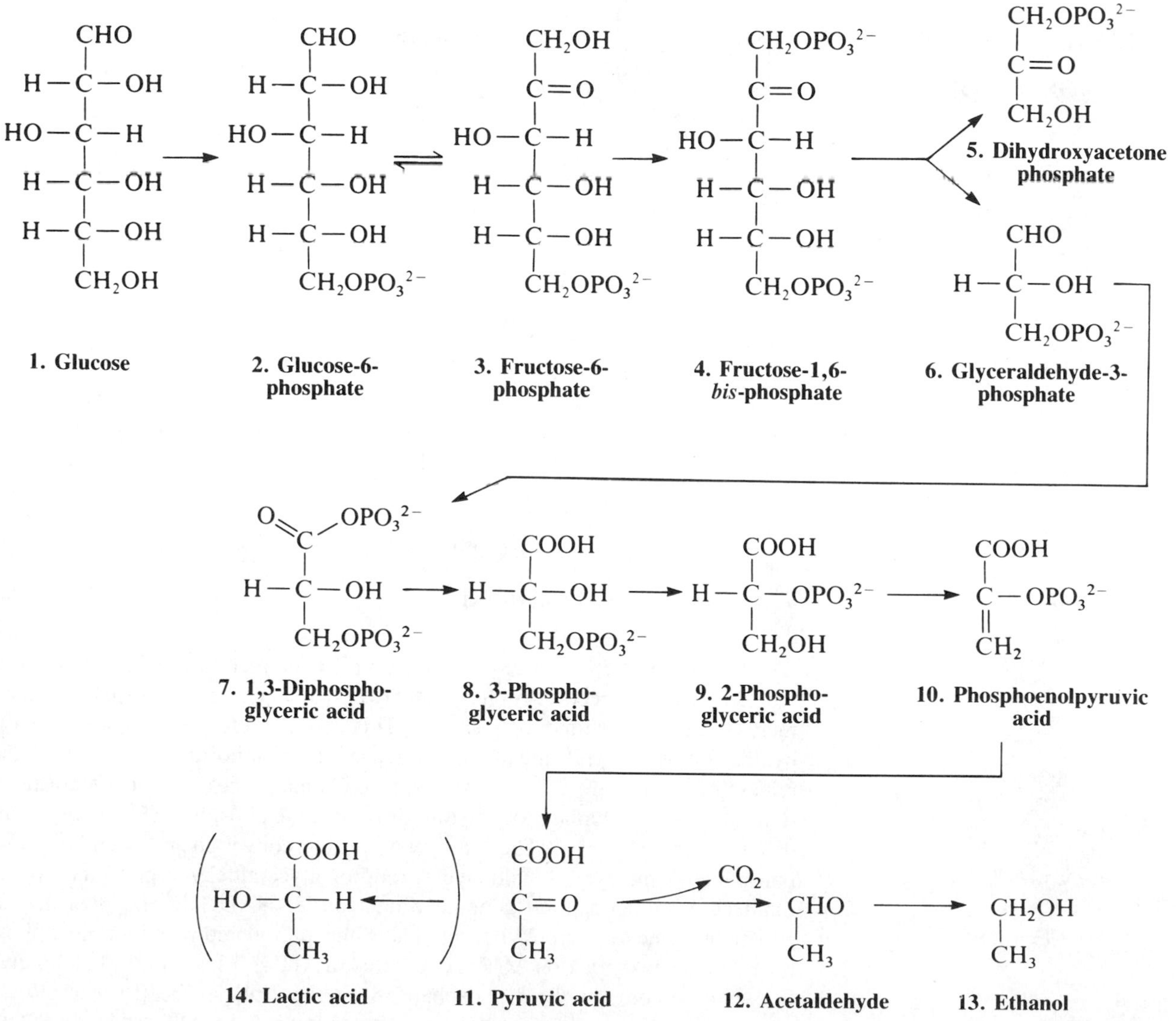

1. Glucose 2. Glucose-6-phosphate 3. Fructose-6-phosphate 4. Fructose-1,6-*bis*-phosphate 5. Dihydroxyacetone phosphate 6. Glyceraldehyde-3-phosphate 7. 1,3-Diphospho-glyceric acid 8. 3-Phospho-glyceric acid 9. 2-Phospho-glyceric acid 10. Phosphoenolpyruvic acid 14. Lactic acid 11. Pyruvic acid 12. Acetaldehyde 13. Ethanol

The first step in the sequence is the formation of glucose-6-phosphate (**2**) from glucose (**1**). The reaction requires adenosine triphosphate (ATP), which is converted to the diphosphate (ADP) by catalysis with the enzyme glucokinase, which requires magnesium ion to function. This conversion is one of the reactions in which energy is released. In the living organism this energy can be used to do work; in fermentation it simply creates heat.

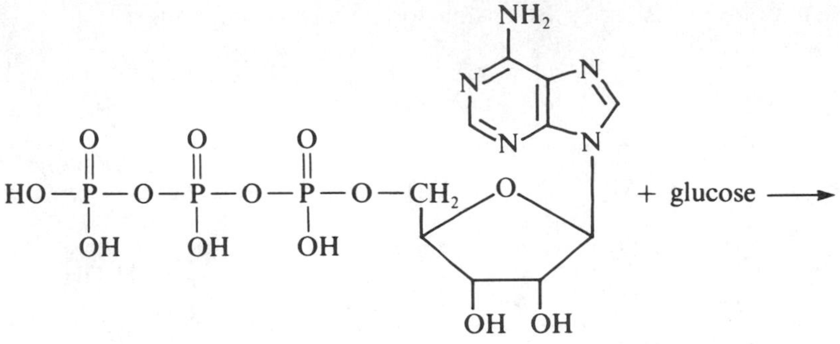

Adenosine triphosphate (ATP)

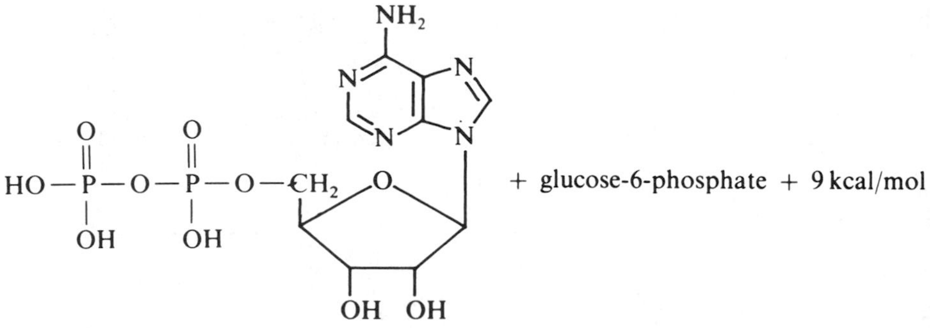

Adenosine diphosphate (ADP)

In the next step glucose-6-phosphate (**2**) is converted through the enol of the aldehyde to fructose-6-phosphate (**3**) by the enzyme phosphoglucoisomerase. The fructose monophosphate (**3**) is converted to the *bis*-phosphate (**4**) by the action of ATP under the influence of phosphofructokinase with the release of more energy. This *bis*-phosphate (**4**) undergoes a reverse aldol reaction catalyzed by aldolase to give dihydroxyacetone phosphate (**5**) and glyceraldehyde-3-phosphate (**6**). The latter two are interconverted by means of triosphosphate isomerase. The aldehyde group of glyceraldehyde phosphate (**6**) is oxidized by nicotinamide adenine dinucleotide (NAD⁺) in the presence of another enzyme to a carboxyl group that is phosphorylated with inorganic phosphate. In the next reaction ADP is converted to ATP as **7** loses phosphate to give **8**. A mutase converts the 3-phosphate (**8**) to the 2-phosphate (**9**). An enolase converts **9** to **10**, and a kinase converts **10** to pyruvic acid (**11**). A decarboxylase converts pyruvic acid (**11**) to acetaldehyde (**12**) in the fermentation process. Yeast alcohol dehydrogenase (YAD), a well-studied enzyme, catalyzes the reduction of acetaldehyde to ethanol. The reducing agent is reduced nicotinamide adenine dinucleotide, NADH.

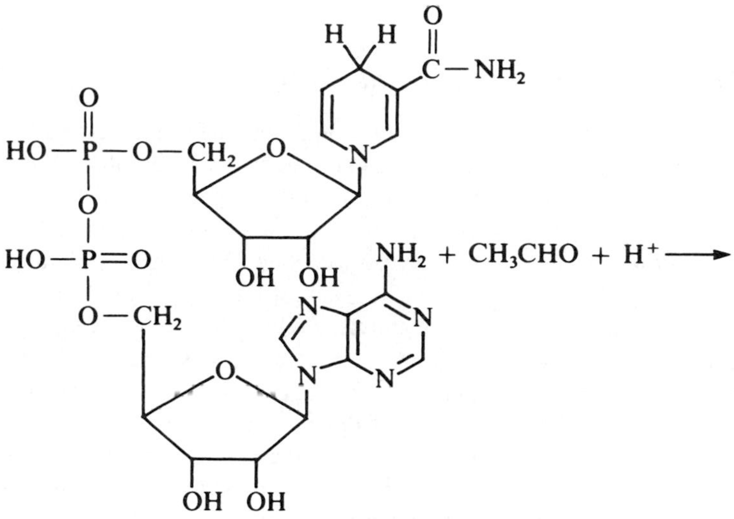

Reduced nicotinamide adenine dinucleotide, NADH

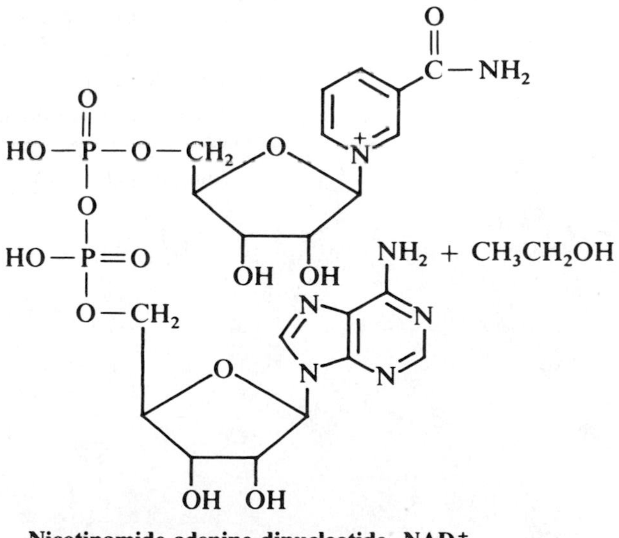

Nicotinamide adenine dinucleotide, NAD⁺

Enzymes are labile

Enzymes are remarkably efficient catalysts, but they are also labile (sensitive) to such factors as heat and cold, changes in pH, and various specific inhibitors. In the first experiment of this chapter you will have an opportunity to observe the biosynthesis of ethanol and to test the effects of various agents on the enzyme system.

This experiment involves the fermentation of ordinary cane sugar using baker's yeast. The resulting dilute solution of ethanol, after removal of the yeast by filtration, can be distilled according to the procedures of Chapter 5.

Experiments

▯ MICROSCALE

Protect the fermenting reaction from exposure to oxygen and contaminating materials. Under aerobic conditions acetobacter bacteria can convert ethanol to acetic acid.

Reaction time: 1 week

1. Fermentation of Sucrose

To a 5-mL round-bottomed long-necked flask add 90 mg of dry yeast and 1.25 mL of warm (up to 50°C) water. Shake the mixture thoroughly until it is more or less homogeneous in appearance, and then add to it 9 mg of disodium hydrogen phosphate, 1.30 g of sucrose, and an additional 3.75 mL of water, which should be warmed to about 45°C. Shake this mixture to ensure complete mixing, and then fit the neck of the flask with a septum that is connected to an 8-in. length of polyethylene tubing (Fig. 65.1). Lead this tubing beneath the surface of about 2 mL of a saturated aqueous solution of calcium hydroxide (limewater) in a reaction tube. The tube in the limewater will act as a seal to prevent air and unwanted enzymes from entering the flask but will allow gas to escape. Place the assembly in a warm spot (the optimal temperature for the reaction is 35°C) for a week, at which time the evolution of carbon dioxide will have ceased. What is the precipitate in the limewater?

On a small scale, it will be necessary to provide external heat to maintain the fermentation or to group several reactions closely together in an insulated container. On a larger scale, since the fermentation is an exothermic reaction, there is enough heat evolved by the reaction to keep the mixture warm and promote the biosynthesis of ethanol.

On completion of fermentation, add about 0.25 g of Celite filter aid (diatomaceous earth, face powder) to the flask and shake it vigorously. Celite is

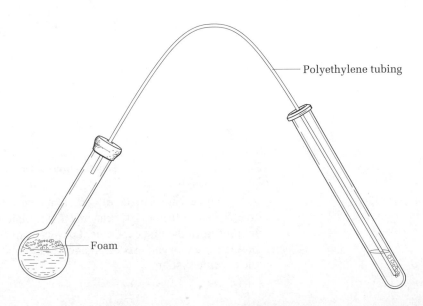

FIG. 65.1 Fermentation apparatus. See Fig. 18.2 for the technique of threading a polyethylene tube through a septum.

Polyethylene tubing

Foam

added to make it possible to filter the solution. Otherwise the yeast cells will clog the filter paper. Filter the mixture on the Hirsch funnel into a 25-mL filter flask that is attached to the water aspirator through a trap by vacuum tubing (see Fig. 3.14). Because the small filter flask will easily tip over, clamp it to a ring stand. Moisten the filter paper with water, apply gentle suction (water supply to aspirator turned on full force, valve to trap partly open), and slowly pour the reaction mixture onto the filter. Wash out the flask with a milliliter of water, and rinse the filter cake with this water. Full vacuum in this filtration will evaporate some of the desired ethanol.

The filtrate, which is a dilute solution of ethanol contaminated with a few bits of cellular material and other organic compounds (acetic acid if you are not careful), is saved in a stoppered container until it is distilled following the procedure outlined in Chapter 5.

MACROSCALE

2. Fermentation of Sucrose

Macerate (grind) one-half cake of yeast or half an envelope of dry yeast in 50 mL of water in a beaker, add 0.35 g of disodium hydrogen phosphate, and transfer this slurry to a 500-mL round-bottomed flask. Add a solution of 51.5 g of sucrose in 150 mL of water, and shake to ensure complete mixing. Fit the flask with a one-hole rubber stopper containing a bent glass tube that dips below the surface of a saturated aqueous solution of calcium hydroxide (limewater) in a 6-in. test tube (Fig. 65.2). The tube in limewater will act as a seal to prevent air and unwanted enzymes from entering the flask, but will allow gas to escape.[1] Place the assembly in a warm spot in your desk (the optimum temperature for the reaction is 35°C) for one week, at which time the evolution of carbon dioxide will have ceased. What is the precipitate in the limewater?

Upon completion of fermentation add 10 g of Celite filter aid (diatomaceous earth, face powder) to the flask, shake vigorously, and filter. Use a 5.5-cm

Protect the fermenting reaction from exposure to oxygen and contaminating materials. Under aerobic conditions acetobacter bacteria can convert ethanol to acetic acid.

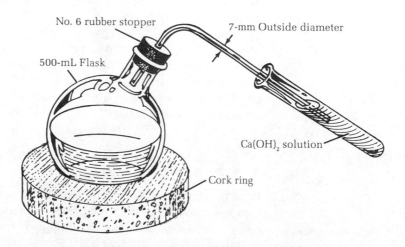

No. 6 rubber stopper

7-mm Outside diameter

500-mL Flask

Ca(OH)$_2$ solution

Cork ring

FIG. 65.2 Apparatus used in the fermentation of sucrose experiment.

---

1. The tube should be about 0.5 cm below the limewater so that limewater will not be sucked back into the flask should the pressure change.

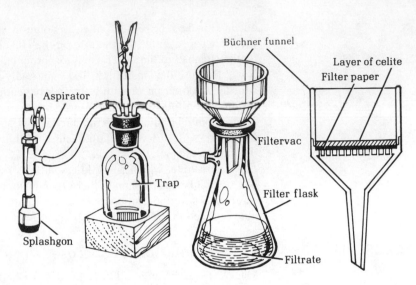

FIG. 65.3 Vacuum filtration apparatus.

Reaction time: 1 week

Filter aid

Büchner funnel placed on a neoprene adapter or Filtervac atop a 500-mL filter flask that is attached to the water aspirator through a trap by vacuum tubing (Fig. 65.3). Since the apparatus is top-heavy, clamp the flask to a ring stand. Moisten the filter paper with water, apply gentle suction (water supply to aspirator turned on full force, clothespin on trap partially closed), and slowly pour the reaction mixture onto the filter. Wash out the flask with a few milliliters of water from your wash bottle, and rinse the filter cake with this water. The filter aid is used to prevent the pores of the filter paper from becoming clogged with cellular debris from the yeast.

The filtrate, which is a dilute solution of ethanol contaminated with bits of cellular material and other organic compounds (acetic acid if you are not careful), is saved in a stoppered flask until it is distilled following the procedure outlined in Chapter 5.

Cleaning Up Since sucrose, yeast, and ethanol are all natural products, all solutions produced in this experiment contain biodegradable material and can be flushed down the drain after dilution with water. The limewater can be disposed of in the same way. The Celite filter aid can be placed in the nonhazardous solid waste container.

▌ MICROSCALE

3. Effect of Various Reagents and Conditions on Enzymatic Reactions

CAUTION: Handle the sodium fluoride solution with great care—it is very poisonous if ingested.

Prepare a fermentation mixture exactly as described in the first part of this experiment. After about 15 min, when the fermentation should be progressing nicely, split the mixture into five equal parts in reaction tubes. To one tube add 1.0 mL of water, to the next add 1.0 mL of 95% ethanol, and to the next add 1.0 mL of 0.5 M sodium fluoride solution. Heat the next tube for 5 min in a steam bath or boiling water bath, and cool the next tube for 5 min in ice. Add to each tube 2 drops of mineral oil. The oil will float on the aqueous solution and prevent exposure to oxygen, which is necessary because fermentation is an anaerobic process. Place the tubes in a beaker of water at a temperature of 30°C for 15 min; then connect them one at a time while in the beaker to an inverted 1.0-mL graduated pipette that has been plugged at one end as shown in Fig. 65.4. Record the volume of gas in the pipette every minute for 5 min, or count the bubbles evolved per minute. Plot a graph of volume of gas against time for each of the five reactions. What conclusions can you draw from the results of these five reactions?

MACROSCALE

CAUTION: Handle the sodium fluoride solution with great care—it is very poisonous if ingested.

4. Effect of Various Reagents and Conditions on Enzymatic Reactions

About 15 min after mixing the yeast, sucrose, and phosphate, remove 20 mL of the mixture and place 4 mL in each of five test tubes. To one tube add 1.0 mL of water, to the next add 1.0 mL of 95% ethanol, to the next add 1.0 mL of 0.5 *M* sodium fluoride. Heat the next tube for 5 min in a steam bath, and cool the next tube for 5 min in ice. Add 10 to 15 drops of mineral oil on top of the reaction mixture in each tube (to exclude air, since the process is anaerobic). Place tubes in a beaker of water at a temperature of 30°C for 15 min, then take them one at a time and connect each to the manometer as shown in Fig. 65.5. Allow about 30 sec for temperature equilibration, then clamp the vent tube and read the manometer. Record the height of the manometer fluid in the open arm of the U-tube every minute for 5 min or until the fluid reaches the top of the manometer. Plot a graph of the height of the manometer fluid against time for each of the four reactions. What conclusions can you draw from the results of these five reactions?

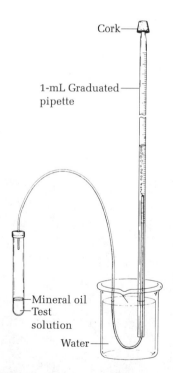

Cork

1-mL Graduated pipette

Mineral oil
Test solution

Water

FIG. 65.4 Apparatus for measuring carbon dioxide evolved. See Fig. 18.2 for the technique of threading a polyethylene tube through a septum.

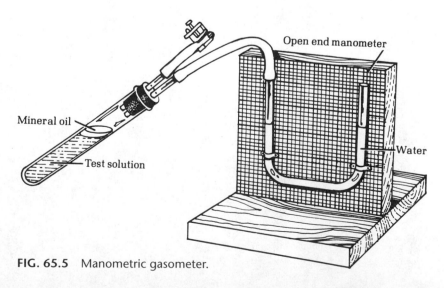

Open end manometer

Mineral oil

Test solution

Water

FIG. 65.5 Manometric gasometer.

Cleaning Up   Combine all reaction mixtures, remove the mineral oil from the top, and place it in the organic solvents container. The aqueous solutions, after diluting with about 50 volumes of water, can be flushed down the drain. Solutions containing fluoride ion are very toxic and should not be placed in the sewer system but neutralized and treated with excess calcium chloride to precipitate calcium fluoride. The latter is separated by filtration and put in the non-hazardous solid waste container.

Questions

1. Using yeast, can glucose be converted to ethanol? Can fructose be converted to ethanol?

2. Write the equation for the formation of the precipitate formed in the test tube containing calcium hydroxide.

3. In this experiment could 90% ethanol be made by adding more sugar to the fermentation flask?

Enzymatic Reactions: A Chiral Alcohol from a Ketone and Enzymatic Resolution of DL-Alanine

Prelab Exercise: Study the biochemistry of the conversion of glucose to ethanol (Chapter 65). Which enzyme might be responsible for the reduction of ethyl acetoacetate? Discuss at least three other methods for resolving DL-alanine into its enantiomers. Judging from its name, would you expect acylase to be equally effective on the acetyl derivatives of other amino acids? Explain why the enzyme acylase will hydrolyze only one of a pair of enantiomers and how a mixture of L-alanine and *N*-acetyl-D-alanine can be separated at the end of the reaction.

Part 1. Enzymatic Reduction of a Ketone to a Chiral Alcohol

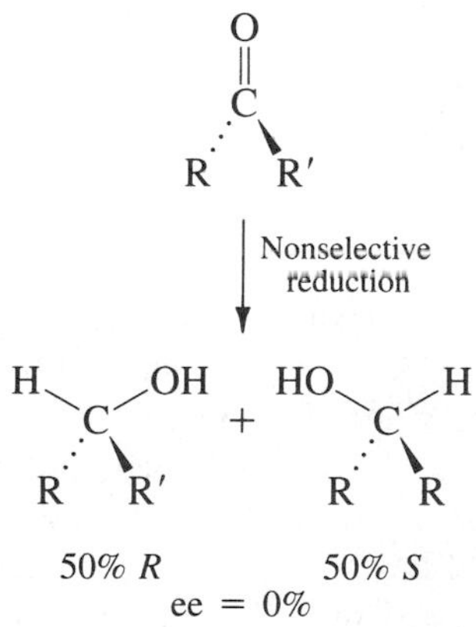

Reduction of an achiral ketone with the usual laboratory reducing agents, such as sodium borohydride or lithium aluminum hydride, will not give a chiral alcohol, because the chances for attack on two sides of the planar carbonyl group are equal. However, if the reducing agent is chiral, there is the possibility of obtaining a chiral alcohol. Organic chemists in recent years have devised a number of such chiral reducing agents, but few of them are as efficient as those found in nature.

In this experiment, we will use the enzymes found in baker's yeast[1] to reduce ethyl acetoacetate to S(+)-ethyl 3-hydroxybutanoate. This compound is a very useful synthetic building block. Many chiral natural product syntheses are based on this hydroxyester.[2]

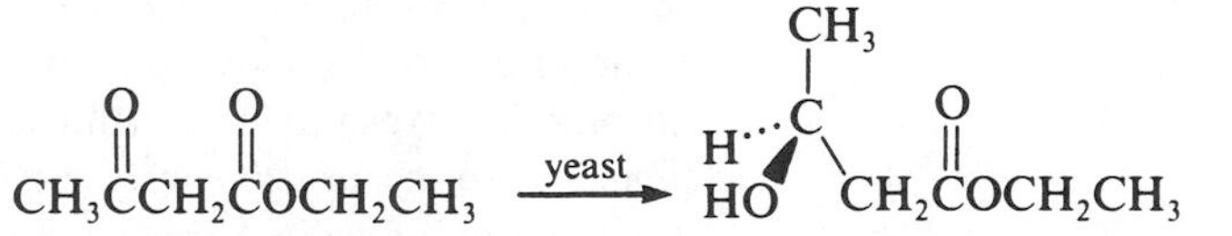

50% *R*   50% *S*

ee = 0%

ee = enantiomeric excess

Ethyl acetoacetate
MW 130.14
bp 181°C, den. 1.021
n_D^{20} 1.4190

S(+)-Ethyl 3-hydroxybutanoate
MW 132.16
bp 180–182°C (71–73°C/12 mm)
n_D^{20} 1.4210

1. S. Servi, *Synthesis* 1–25 (1990).

2. R. Amstutz, E. Hungerbühler, and D. Seebach, *Helv. Chim. Acta* **64,** 1796 (1981); and D. Seebach, *Tetrahedron Lett.* 159 (1982).

A large number of different enzymes are present in yeast. The primary ones responsible for the conversion of glucose to ethanol are discussed in Chapter 65. While this fermentation reaction is taking place, certain ketones can be reduced to chiral alcohols.

Whenever a chiral product is produced from an achiral starting material, the chemical yield as well as the optical yield is important—in other words, how stereoselective the reaction has been. The usual method for recording this is to calculate the enantiomeric excess (ee). A sodium borohydride reduction will produce 50% *R* and 50% *S* alcohol with no enantiomer in excess. If 93% *S*(+) and 7% *R*(−) isomer are produced, then the enantiomeric excess is 86%. In the present yeast reduction of ethyl acetoacetate, various authors have reported enantiomeric excesses ranging from 70 to 97%. This optical yield is distinct from the chemical yield, which depends on how much material is isolated from the reaction mixture.

The use of enzymes to carry out stereospecific chemical reactions is not new, but it is not always possible to predict if an enzymatic reaction (unlike a purely chemical reaction) will occur or how stereospecific it will be if it does. Because this experiment is easily carried out, it might be an interesting research project to explore the range of possible ketones that yeast will reduce to chiral alcohols. For example, butyrophenone can be reduced to the corresponding chiral alcohol. For a review see Sih and Rosazza.[3] The present experiment is based on the work of Seebach,[4] Mori,[5] and Ridley.[6] See also the work of Bucciarelli *et al.*[7]

Experiments

1. Enzymatic Resolution

Microscale Procedure

In a 25-mL flask (see Chapter 59) dissolve 2.3 g of sucrose and 15 mg of disodium hydrogen phosphate in 8.5 mL of warm (35°C) tap water. Add 0.5 g of dry yeast, and swirl to suspend the yeast throughout the solution. After 15 min, while fermentation is progressing vigorously, add 150 mg of ethyl acetoacetate. Store the flask in a warm place, ideally at 30 to 35°C, for at least 48 h (a longer time will do no harm). At the end of this time, add 0.5 g of Celite filtration aid, and remove the yeast cells by filtration on a Hirsch funnel (see Chapter 59). Wash the cells with 1.5 mL of water; then saturate the filtrate with sodium chloride to decrease the solubility of the product. See a handbook for the solubility of sodium chloride in water to decide approximately how much to

Keep the mixture warm; the reaction is slow at low temperatures.

3. C. J. Sih and J. P. Rosazza, *Application of Biochemical Systems in Organic Chemistry,* Part 1 (J. B. Jones, C. J. Sih, and D. Perlman, eds.), pp. 71–78, Wiley, New York, 1976.

4. D. Seebach, *Tetrahedron Lett.* 159 (1982).

5. K. Mori, *Tetrahedron* **37,** 1341 (1981).

6. D. D. Ridley and M. Stralow, *Chem. Commun.* 400 (1975).

7. M. Bucciarelli, A. Forni, I. Moretti, and G. Torre, *Synthesis* 897 (1983).

use. Extract the resulting solution five times with 1.5-mL portions of *tert*-butyl methyl ether in a test tube, taking care to shake hard enough to mix the layers but not so hard as to form an emulsion between the ether and water. Addition of a small amount of methanol may help to break up emulsions. Dry the ether layer by adding anhydrous sodium sulfate or calcium chloride pellets until the drying agent no longer clumps together. After approximately 15 min, remove the ether solution, and evaporate it in portions in a tared reaction tube. The residue should weigh about 100 mg. It should, unlike the starting material, give a negative iron(III) chloride test (see p. 752). It can be analyzed by TLC (use dichloromethane as the solvent) to determine whether unreacted ethyl acetoacetate is present. Infrared spectroscopy should show the presence of the hydroxyl group and may show unreduced methyl ketone. The NMR spectrum of the product is easily distinguished from the starting material.

A 15-mL centrifuge tube can be used for the extraction, and the emulsions can be broken in the centrifuge.

Cleaning Up   The aqueous layer can be diluted with water and flushed down the drain. After the ether evaporates from the drying agent in the hood, it can be placed in the nonhazardous solid waste container. Dichloromethane (the TLC solvent) goes in the halogenated organic solvents container. Ether distillate goes in the organic solvents container.

Macroscale Procedure

In a 500-mL flask (see Chapter 65) dissolve 80 g of sucrose and 0.5 g of disodium hydrogen phosphate in 300 mL of warm (35°C) tap water. Add two packets (16 g) of dry yeast, and swirl to suspend the yeast throughout the solution. After 15 min, while fermentation is progressing vigorously, add 5 g of ethyl acetoacetate. Store the flask in a warm place, ideally at 30–35°C, for at least 48 h (a longer time will do no harm). At the end of this time, add 20 g of Celite filtration aid, and remove the yeast cells by filtration on a 10-cm Büchner funnel (see Chapter 65). Wash the cells with 50 mL of water; then saturate the filtrate with sodium chloride in order to decrease the solubility of the product. See a handbook for the solubility of sodium chloride in water to decide approximately how much to use. Extract the resulting solution five times with 50-mL portions of *t*-butyl methyl ether, taking care to shake the separatory funnel hard enough to mix the layers but not so hard as to form an emulsion between the ether and water. Addition of a small amount of methanol may help to break up emulsions. Dry the ether layer by adding anhydrous sodium sulfate or calcium chloride pellets until the drying agent no longer clumps together. After approximately 15 min, decant the ether solution into a tared distilling flask, and remove the ether by distillation or by evaporation. The residue should weigh about 3.5 g. It should, unlike the starting material, give a negative iron(III) chloride test (see Chapter 70). It can be analyzed by TLC (use dichloromethane as the solvent) to determine whether unreacted ethyl acetoacetate is present. Infrared spectroscopy should show the presence of the hydroxyl group and may show unreduced methyl ketone. The NMR spectrum of the product is easily distinguished from that of starting material. NMR and IR spectra of starting material and *racemic* products are shown in Figs. 66.1, 66.2, and 66.3.

Keep the mixture warm; the reaction is slow at low temperatures.

Extinguish all flames when working with ether.

FIG. 66.1 ¹H NMR spectrum of ethyl acetoacetate (250 MHz).

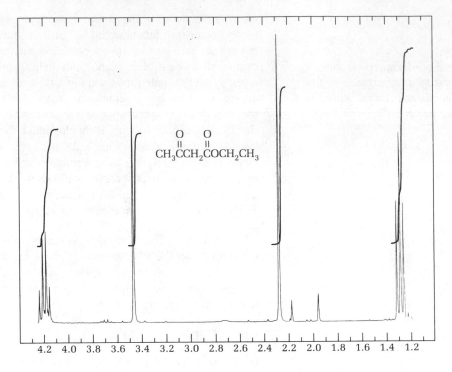

Cleaning Up The aqueous layer can be diluted with water and flushed down the drain. After the ether evaporates from the drying agent in the hood, it can be placed in the nonhazardous solid waste container. Dichloromethane (the TLC solvent) goes in the halogenated organic solvents container. Ether distillate goes in the organic solvents container.

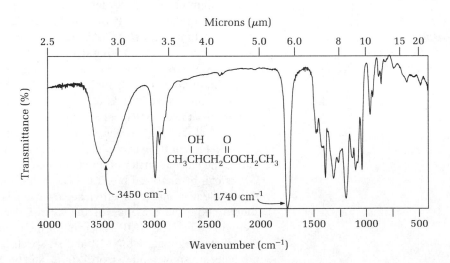

FIG. 66.2 IR spectrum of ethyl 3-hydroxybutanoate (thin film).

FIG. 66.3 ^{1}H NMR spectra of racemic (±)-ethyl 3-hydroxy-butanoate in 0.3 mL of carbon tetrachloride and 0.2 mL deuterochloroform (250 MHz). (a) Pure. (b) With 30 mg of shift reagent. (c) With 50 mg of shift reagent.

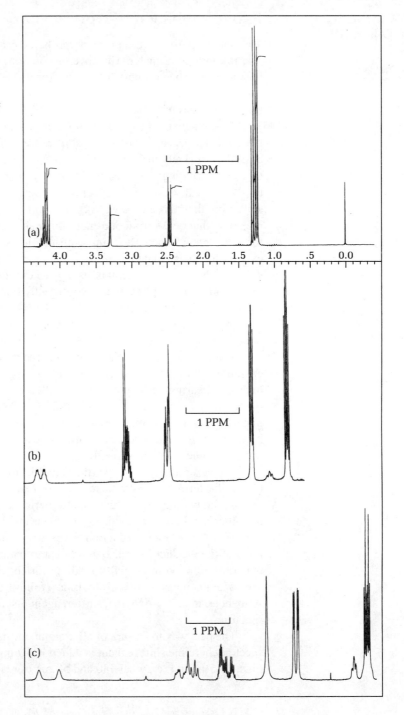

2. Determination of Optical Purity

The optical purity of the product can be determined by measuring the optical rotation in a polarimeter. The specific rotation, $[\alpha]_D^{25}$, of $S(+)$-ethyl 3-hydroxybutanoate has been reported to vary from $+31.3°$ to $41.7°$ in chloroform. The specific rotation, $[\alpha]_D^{25}$, of $37.2°$ (chloroform, C 1.3) corresponds to an enantiomeric excess of 85%.

A better method for determining optical purity of such a small quantity of material is to use an NMR chiral shift reagent.[8] The use of shift reagents is discussed extensively in Chapter 13. A shift reagent will complex with a basic center (the hydroxyl group in ethyl 3-hydroxybutanoate) and cause the NMR peaks to shift, usually downfield. Protons nearest the shift reagent/hydroxyl group shift more than those far away. A chiral shift reagent forms diastereomeric complexes so that peaks from one enantiomer shift downfield more than peaks from the other enantiomer. By comparing the areas of the peaks, the enantiomeric excess can be calculated.

The NMR experiment is easily accomplished by adding 10- to 20-mg increments of tris[3-(heptafluoropropylhydroxymethylene)-(+)-camphorato]-europium(III) shift reagent (Aldrich 16,474–7) to a solution of 30 mg of pure dry ethyl 3-hydroxybutanoate in a mixture of 0.3 mL of carbon tetrachloride and 0.2 mL of deuterochloroform.[9]

CAUTION: Handle carbon tetrachloride in the hood.

The analysis of the NMR spectra is not completely straightforward. For example, in Fig. 66.3(a), can you locate the peaks from the proton on the hydroxyl-bearing carbon atom? (Pay attention to the integrals.) And note that the methylene hydrogens on carbon-2 [$\delta = 2.45$ ppm in Fig. 66.3(a)] are not magnetically (or chemically) equivalent. They are said to be diasteriotopic because diasteriomers could be formed by replacing either one or the other proton with some other substituent.

As the optically active shift reagent is added, sets of peaks begin to double as diasteriomeric complexes are formed [Figs. 66.3(b) and (c)]. By integrating the pairs of peaks, the enantiomeric excess of one isomer in relation to the other can be determined. In this experiment, only one set of peaks can be seen if the enzymatic reduction gives only one of the two possible enantiomers of the product, so it is a good idea to compare your product with racemic material, either commercially made (Aldrich) or made by reduction of ethylacetoacetate by sodium borohydride (follow the procedure in Chapter 26). Also bear in mind that the shift reagent itself will contribute peaks to the spectrum.

The best way to be sure of all assignments is to run and integrate a series of spectra where the shift reagent is added in 10-mg increments, as is illustrated in Figs. 13.5(b) and (c), where 30 and 50 mg of reagent have been added.

---

8. K. B. Lipkowitz and J. L. Mooney, *J. Chem. Ed.* **64,** 985 (1987).

9. *Ibid.*

Cleaning Up Place the contents of the NMR tube in the halogenated solvents waste container.

MICROSCALE 4. Preparation of 3,5-Dinitrobenzoate

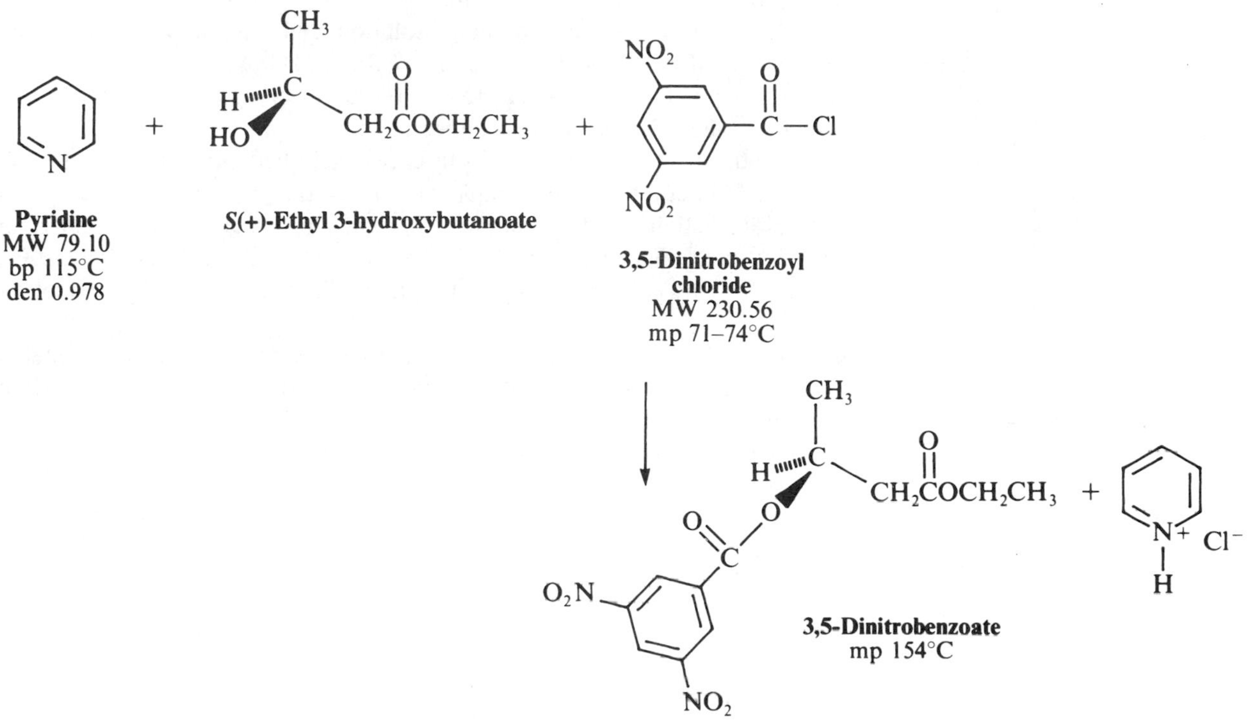

Pyridine
MW 79.10
bp 115°C
den 0.978

S(+)-**Ethyl 3-hydroxybutanoate**

**3,5-Dinitrobenzoyl
chloride**
MW 230.56
mp 71–74°C

3,5-Dinitrobenzoate
mp 154°C

3,5-Dinitrobenzoates are common alcohol derivatives (see Chapter 70); they are easily prepared, the dinitrophenyl group adds considerably to the molecular weight, and they are easily crystallized.

The 3,5-dinitrobenzoates of a racemic mixture of *R*- and *S*-ethyl 3-hydroxybutanoate, like almost all racemates, crystallize together to give crystals that in this case melt at 146°C. No amount of recrystallization causes one enantiomer to crystallize out while the other remains in solution, since they are, after all, mirror images of each other. However, if one enantiomer is in large excess, it is possible to effect a separation by crystallization. In the present case, the *S* enantiomer predominates and it crystallizes out, leaving most of the *R* + *S* racemate in solution. Repeated crystallization increases the melting point and the purity to the point at which the optical purity of the crystalline product is almost 100%. At this point the 3,5-dinitrobenzoate has a melting point of 154°C. Treatment of the derivative with excess acidified ethanol regenerates 100% ee *S*(+)-ethyl 3-hydroxybutanoate.

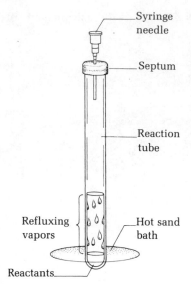

Syringe needle

Septum

Reaction tube

Refluxing vapors

Hot sand bath

Reactants

FIG. 66.4 Apparatus for refluxing reaction mixture.

Procedure

To 100 mg of the *S*(+)-ethyl 3-hydroxybutanoate in a reaction tube (Fig. 66.4) add 175 mg of pure[10] 3,5-dinitrobenzoyl chloride and 1 mL of pyridine. Reflux the mixture for 15 min, and then transfer it with a Pasteur pipette to 3.5 mL of water in another reaction tube. Remove the solvent from the crystals, and shake the crystals with 2 mL of 0.5 *M* sodium bicarbonate solution to remove dinitrobenzoic acid. Remove the bicarbonate solution, and recrystallize the derivative from ethanol. Collect the product on the Wilfilter. Dry a portion, and determine the melting point. If it is not near 154°C, repeat the crystallization. Dry the pure derivative, and calculate the percentage yield.

Treatment of the derivative with excess acidified ethanol will regenerate 100% ee *S*(+)-ethyl 3-hydroxybutanoate. If several crops of crude product are pooled, distillation at reduced pressure (see Chapter 7) can give chemically pure ethyl 3-hydroxybutanoate (bp 71–73°C/12 mm). The *R* and *S* enantiomers have the same boiling point, so the optical purity will not be changed by distillation.

Cleaning Up   The pyridine solvent goes in the organic solvents container. The aqueous layer should be diluted with water and flushed down the drain. Ethanol from the crystallization goes in the organic solvents container, along with the pot residue if the final product is distilled.

---

10. Check the melting point of the 3,5-dinitrobenzoyl chloride. If it is below 70°C, it should be recrystallized from dichloromethane.

Part 2. Enzymatic Resolution of DL-Alanine

DL-Alanine (1) mp 295°C MW 89.10

***N*-Acetyl-DL-alanine (2)** mp 137°C MW 131.13

L-Alanine (3) mp 297°C α_D + 14.4°

***N*-Acetyl-D-alanine (4)** mp 125°C α_D + 66.5°

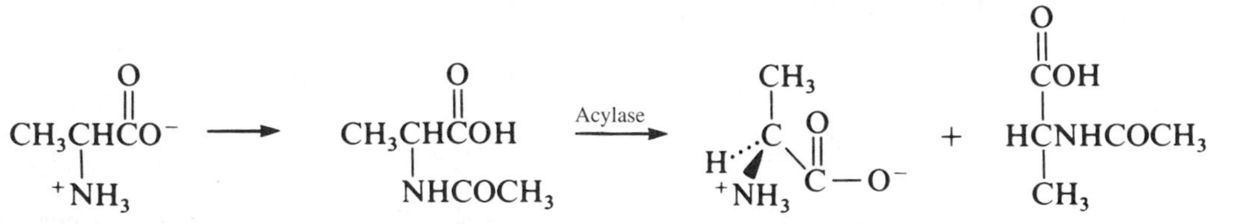

Resolution of DL-alanine **(1)** is accomplished by heating the *N*-acetyl derivative **(2)** in weakly alkaline solution with acylase, a proteinoid preparation from porcine kidney containing an enzyme that promotes rapid hydrolysis of *N*-acyl derivatives of natural L-amino acids but acts only immeasurably slowly on the unnatural D-isomers. *N*-Acetyl-DL-alanine **(2)** can thus be converted into a mixture of L-(+)-alanine **(3)** and *N*-acetyl-D-alanine **(4).** The mixture is easily separable into the components, because the free amino acid **(3)** is insoluble in ethanol and the *N*-acetyl derivative **(4)** is readily soluble in this solvent. Note that, in contrast to the weakly levorotatory D-(−)-alanine (α_D − 14.4°), its acetyl derivative is strongly dextrorotatory (α_D + 66.5°).

(4) **(5)** **(6)**

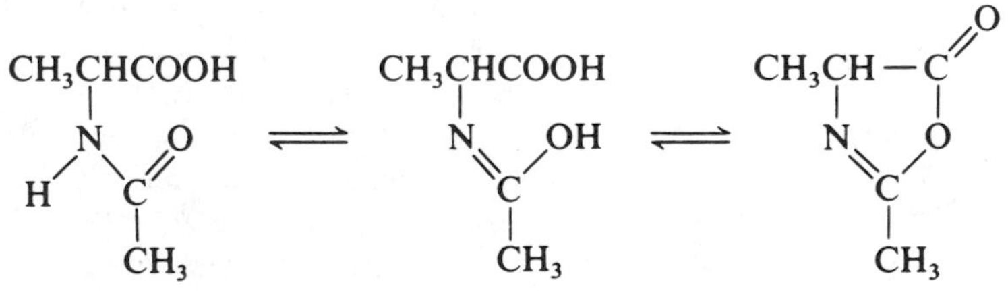

The acetylation of an α-amino acid presents the difficulty that, if the conditions are too drastic, the *N*-acetyl derivative **(4)** is converted in part through the enol **(5)** to the azlactone **(6).**[11] However, under critically controlled conditions of concentration, temperature, and reaction time, *N*-acetyl-DL-alanine can be prepared easily in high yield.

11. The azlactone of DL-alanine is known only as a partially purified liquid.

Experiments

1. Acetylation of DL-Alanine

$$CH_3CHCO^- \quad + \quad CH_3COCCH_3 \quad \xrightarrow{CH_3COOH} \quad CH_3CHCOH \quad + \quad CH_3COOH$$

DL-Alanine (1)	Acetic anhydride	N-Acetyl-DL-alanine (2)
MW 89.10	MW 102.09	MW 131.13
mp 295°C	bp 140°C	mp 137°C

Check the pressure gauge.

FIG. 66.5 Drying *N*-acetyl-DL-alanine under reduced pressure.

Place 200 mg of DL-alanine and 0.5 mL of acetic acid in a reaction tube, insert a thermometer, and clamp the tube in a vertical position. Measure 0.3 mL of acetic anhydride, which is to be added when the alanine–acetic acid mixture is at exactly 100°C. Heat the tube on a hot sand bath with stirring until the temperature of the suspension has risen a little above 100°C. Stir the suspension, let the temperature gradually fall, and, when it reaches 100°C, add the 0.3-mL portion of acetic anhydride and note the time. In the course of 1 min the temperature falls (91–95°C, cooled by added reagent), rises (100–103°C, the acetylation is exothermic), and begins to fall with the solid largely dissolving. Stir to facilitate reaction of a few remaining particles of solid, let the temperature drop to 80°C, pour the solution into a tared 10-mL Erlenmeyer flask, and rinse the thermometer and test tube with a little acetone. Add 1 mL of water to react with excess anhydride, connect the flask to the aspirator operating at full force (Fig. 66.5), put the flask *inside* the rings of the steam bath, and wrap the flask with a towel. Evacuation and heating for about 5 to 10 min should remove most of the acetic acid and water and leave an oil or thick syrup. Add 2 mL of cyclohexane, and evacuate and heat as before for 5 to 10 min. Traces of water in the syrup are removed as a cyclohexane–water azeotrope. If the product has not yet separated as a white solid or semisolid, determine the weight of the product, add 2 mL more cyclohexane, and repeat the process. When the weight becomes constant, the yield of acetyl DL-alanine should be close to the theoretical amount. The product has a pronounced tendency to remain in supersaturated solution and hence does not crystallize readily.

2. Enzymatic Resolution of *N*-Acetyl-DL-Alanine

This reaction can be run on ten times the indicated scale. Add 1.0 mL of distilled water[12] to the Erlenmeyer flask, grasp this with a clamp, swirl the mixture over a hot sand bath to dissolve all the product, and cool under the tap. Remove a drop of the solution on a stirring rod, add it to 0.5 mL of a 0.3% solution of ninhydrin in water, and heat to boiling. If any unacetylated DL-alanine is present,

---

12. Tap water may contain sufficient heavy metal ion to deactivate the enzyme.

a purple color will develop and should be noted. Pour the solution into a reaction tube, and rinse the flask with a little water. Add 0.15 mL of concentrated ammonia solution, stir to mix, check the pH with Hydrion paper, and if necessary adjust to pH 8 by addition of more ammonia with a capillary dropping tube. Add 5 mg of commercial acylase,[13] mix with a stirring rod, rinse the rod with distilled water, and make up the volume until the tube is about half full. Then stopper the tube, mark it for identification, and let the mixture stand at room temperature overnight or at 37°C[14] for 4 h.

At the end of the incubation period, add 0.3 mL of acetic acid to denature the enzyme, and if the solution is not as acidic as pH 5, add more acid. Rinse the cloudy solution into a 10-mL Erlenmeyer flask, add 10 mg of granulated decolorizing carbon, heat and swirl over a hot sand bath for a few moments to coagulate the protein, and filter the solution. Transfer the solution to a 25-mL round-bottomed flask and evaporate on a rotary evaporator under vacuum (Fig. 66.6), or add 2 mL of cyclohexane (to prevent frothing) and a boiling stone and evaporate on the steam bath under vacuum to remove water and acetic acid as completely as possible. Remove the last traces of water and acid by adding 2 mL of cyclohexane and evaporating again to remove water and acetic acid as azeotropes.

Workup time 30–40 min

The mixture of L-alanine and acetyl-D-alanine separates as a white scum on the walls. Add 1.5 mL of 95% ethanol, digest on the steam bath, and dislodge some of the solid with a spatula. Cool well in ice for a few minutes, and then scrape as much of the crude L-alanine as possible onto a suction funnel and wash it with ethanol. Save the ethanol mother liquor.[15] To recover the L-alanine retained by the flask, add 0.2 mL of water and warm on the steam bath until the solid is all dissolved; then transfer the solution to a reaction tube by means of a Pasteur pipette, rinse the flask with 0.2 mL more water, and transfer in the same way. Add the filtered L-alanine, dissolve by warming, and filter the solution into a reaction tube using the pressure apparatus. Rinse the flask and funnel with 0.1 mL of water and then with 0.5 mL of warm 95% ethanol. Then heat the filtrate on the steam bath, and add more 95% ethanol (1–1.5 mL) in portions until crystals of L-alanine begin to separate from the hot solution. Let crystallization proceed. Collect the crystals,

Pool several products to make the optical rotation measurement.

and wash with ethanol. The yield of colorless needles of L-alanine, α_D + 13.7 to + 14.4°[16] (in 1 N hydrochloric acid) varies from 40 to 50 mg, depending on the activity of the enzyme.

Note for the instructor

---

13. Commercial porcine kidney acylase is available from Schwarz/Mann, Division of Becton, Dickinson and Co., Orangeburg, NY 10962, or from Sigma Chemical Co., P.O. Box 14503, St. Louis, MO 63172.

14. A reasonably constant heating device that will hold 30 or more tubes is made by partially filling a 1-L beaker with water, adjusting to 37°C, and maintaining this temperature by the heat of a 250-W infrared drying lamp shining horizontally on the beaker from a distance of about 40 cm.

15. In case the yield of L-alanine is low, evaporation of this mother liquor may reveal the reason. If the residue solidifies readily and crystallizes from acetone to give acetyl-DL-alanine, mp 130°C or higher, the acylase preparation is recognized as inadequate in activity or amount. Acetyl-D-alanine is much more soluble and slow to crystallize.

16. Determination of optical activity can be made in the student laboratory with a Zeiss pocket polarimeter, which requires no monochromatic light source and no light shield. For construction of a very inexpensive polarimeter, see W. H. R. Shaw, *J. Chem. Ed.* **32**, 10, 1955.

FIG. 66.6   Rotary evaporation.

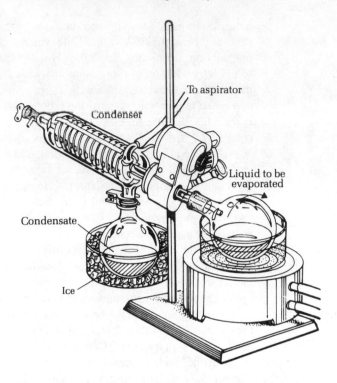

Cleaning Up   The decolorizing carbon goes in the nonhazardous solid waste container. Ethanol used in the crystallization can be diluted with water and flushed down the drain.

MACROSCALE

3. Acetylation of DL-Alanine

$$\underset{\substack{\text{DL-Alanine (1)}\\ \text{MW 89.10}\\ \text{mp 295°C}}}{\overset{\displaystyle O}{\underset{\displaystyle \overset{|}{\underset{+NH_3}{CH_3CHCO^-}}}{\parallel}}} + \underset{\substack{\text{Acetic anhydride}\\ \text{MW 102.09}\\ \text{bp 140°C}}}{\overset{\displaystyle O\quad O}{CH_3COCCH_3}} \xrightarrow{CH_3COOH} \underset{\substack{N\text{-Acetyl-DL-alanine (2)}\\ \text{MW 131.13}\\ \text{mp 137°C}}}{\overset{\displaystyle O}{\underset{\displaystyle \underset{NHCOCH_3}{CH_3CHCOH}}{\parallel}}} + CH_3COOH$$

Check the pressure gauge.

Place 2 g of DL-alanine and 5 mL of acetic acid in a 25 × 150-mm test tube, insert a thermometer, and clamp the tube in a vertical position. Measure 3 mL of acetic anhydride, which is to be added when the alanine/acetic acid mixture is at exactly 100°C. Heat the test tube cautiously, with stirring, until the temperature of the suspension has risen a little above 100°C. Stir the suspension, let the temperature gradually fall, and when it reaches 100°C add the 3-mL portion of acetic anhydride and note the time. In the course of 1 min the temperature falls (91–95°C, cooled by added reagent), rises (100–103°C, the acetylation is exothermic), and begins to fall with the solid largely dissolving. Stir to facilitate reaction of a few remaining particles of solid, let the temperature drop to 80°C,

FIG. 66.7 Drying *N*-acetyl-DL-alanine under reduced pressure.

pour the solution into a tared 100-mL round-bottomed flask, and rinse the thermometer and test tube with a little acetone. Add 10-mL of water to react with excess anhydride, connect the flask to the aspirator operating at full force (Fig. 66.7), put the flask *inside* the rings of the steam bath, and wrap the flask with a towel. Evacuation and heating for about 5–10 min should remove most of the acetic acid and water and leave an oil or thick syrup. Add 10 mL of cyclohexane and evacuate and heat as before for 5–10 min. Traces of water in the syrup are removed as a cyclohexane/water azeotrope. If the product has not yet separated as a white solid or semisolid, determine the weight of the product, add 10 mL more cyclohexane, and repeat the process. When the weight becomes constant, the yield of acetyl DL-alanine should be close to the theoretical amount. The product has a pronounced tendency to remain in supersaturated solution and hence does not crystallize readily.

4. Enzymatic Resolution of *N*-Acetyl-DL-Alanine

Add 10 mL of distilled water[17] to the reaction flask, grasp this with a clamp, swirl the mixture over a free flame to dissolve all the product, and cool under the tap. Remove a drop of the solution on a stirring rod, add it to 0.5 mL of a 0.3% solution of ninhydrin in water, and heat to boiling. If any unacetylated DL-alanine is present, a purple color will develop and should be noted. Pour the solution into a 20 × 150-mm test tube, and rinse the flask with a little water. Add 1.5 mL of concentrated ammonia solution, stir to mix, check the pH with Hydrion paper, and, if necessary, adjust to pH 8 by addition of more ammonia with a Pasteur pipette. Add 10 mg of commercial acylase powder, or 2 mL of fresh acylase solution,[18] mix with a stirring rod, rinse the rod with distilled water, and make up the volume until the tube is about half full. Then stopper the tube, mark it for identification (Fig. 66.8), and let the mixture stand at room temperature overnight, or at 37°C[19] for 4 h.

At the end of the incubation period add 3 mL of acetic acid to denature the enzyme, and, if the solution is not as acidic as pH 5, add more acid. Rinse the cloudy solution into a 125-mL Erlenmeyer flask, add 100 mg of decolorizing carbon (0.5-cm column in a 13 × 100-mm test tube), heat and swirl over a free flame for a few moments to coagulate the protein, and filter the solution by suction. Transfer the solution to a 100-mL round-bottomed flask and evaporate on a rotary evaporator under vacuum (Fig. 66.9), or add 20 mL of cyclohexane (to prevent frothing) and a boiling stone and evaporate on the steam bath under vacuum to remove water and acetic acid as completely as possible. Remove the last traces of water and acid by adding 15 mL of cyclohexane and evaporating again

FIG. 66.8 Filter paper identification marker.

Note for the instructor

---

17. Tap water may contain sufficient heavy metal ion to deactivate the enzyme.

18. Commercial porcine kidney acylase is available from Schwarz/Mann, Division of Becton, Dickinson and Co., Orangeburg, NY 10962, or from Sigma Chemical Co., P.O. Box 14503, St. Louis, MO 63172.

19. A reasonably constant heating device that will hold 15 tubes is made by filling a 1-L beaker with water, adjusting to 37°C, and maintaining this temperature by the heat of a 250-watt infrared drying lamp shining horizontally on the beaker from a distance of about 40 cm. The capacity can be tripled by placing other beakers on each side of the first one and a few centimeters closer to the lamp.

FIG. 66.9 Rotary evaporation.

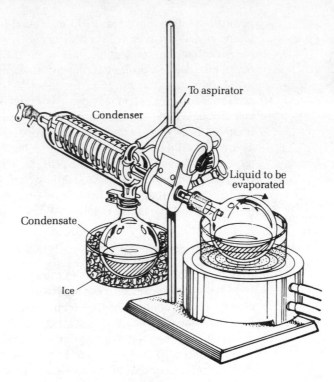

Workup time: $\frac{1}{2}-\frac{3}{4}$ h

to remove water and acetic acid as azeotropes. The mixture of L-alanine and acetyl D-alanine separates as a white scum on the walls. Add 15 mL of 95% ethanol, digest on the steam bath, and dislodge some of the solid with a spatula. Cool well in ice for a few minutes, and then scrape as much of the crude L-alanine as possible onto a suction funnel, and wash it with ethanol. Save the ethanol mother liquor.[20] To recover the L-alanine retained by the flask, add 2 mL of water and warm on the steam bath until the solid is all dissolved, then transfer the solution to a 25-mL Erlenmeyer flask by means of a Pasteur pipette, rinse the flask with 2 mL more water, and transfer in the same way. Add the filtered L-alanine, dissolve by warming, and filter the solution by gravity into a 50-mL Erlenmeyer flask (use the dropping tube to effect the transfer of solution to filter). Rinse the flask and funnel with 1 mL of water and then with 5 mL of warm 95% ethanol. Then heat the filtrate on the steam bath, and add more 95% ethanol (10–15 mL) in portions until crystals of L-alanine begin to separate from the hot solution. Let crystallization proceed. Collect the crystals and wash with ethanol. The yield of colorless needles of L-alanine, α_D + 13.7 to + 14.4°[21]

---

20. In case the yield of L-alanine is low, evaporation of this mother liquor may reveal the reason. If the residue solidifies readily and crystallizes from acetone to give acetyl-DL-alanine, mp 130°C or higher, the acylase preparation is recognized as inadequate in activity or amount. Acetyl-D-alanine is much more soluble and slow to crystallize.

21. Determination of optical activity can be made in the student laboratory with a Zeiss Pocket Polarimeter, which requires no monochromatic light source and no light shield. For construction of a very inexpensive polarimeter, see W. H. R. Shaw, *J. Chem. Ed.,* **32,** 10 (1955).

(in 1 *N* hydrochloric acid) varies from 0.40 to 0.56 g, depending on the activity of the enzyme.

Cleaning Up   The decolorizing carbon goes in the nonhazardous solid waste container. Ethanol used in the crystallization can be diluted with water and flushed down the drain.

Questions

1. In Fig. 66.1 the integrals of the NMR peaks are not exactly 2:2:3:3, as one might expect from the structure written for ethyl acetoacetate. Assign the four groups of integrated peaks, and explain the discrepancy.

2. Look up the prices of racemic ethyl 3-hydroxybutanoate and each of the enantiomers in a catalog; for example, Aldrich. How do you explain the rather different prices?

3. Assign the peaks at 3450 and 1740 cm^{-1} in the IR spectrum of ethyl 3-hydroxybutanoate (Fig. 66.2).

4. The melting point of *N*-acetyl-DL-alanine is 137°C, and that of *N*-acetyl-D-alanine is 125°C. What would you expect the melting point of *N*-acetyl-L-alanine to be, or is this impossible to predict?

5. Would the ^{13}C NMR spectrum of D-alanine differ from that of L-alanine (Fig. 66.10)?

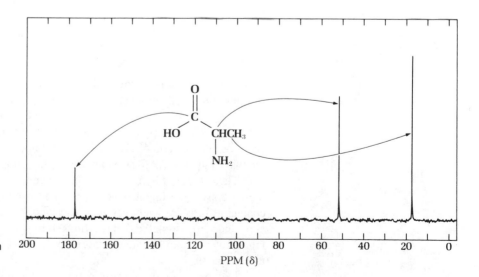

FIG. 66.10 ^{13}C NMR spectrum of L-alanine.

Isolation of Lycopene and β-Carotene

Prelab Exercise: The primary reason for low yields in this experiment is oxidation of the product during isolation. Once crystalline, it is reasonably stable. Speculate on the primary products formed by photochemical air oxidation of lycopene.

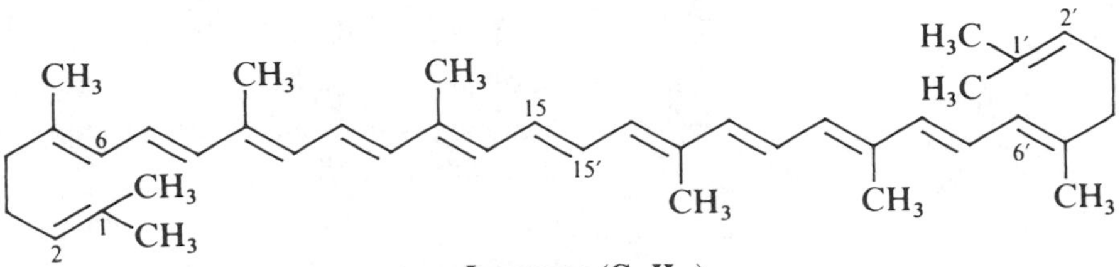

Lycopene ($C_{40}H_{56}$)
MW 536.85
mp 173°C, λ_{max}^{hexane} 475 nm

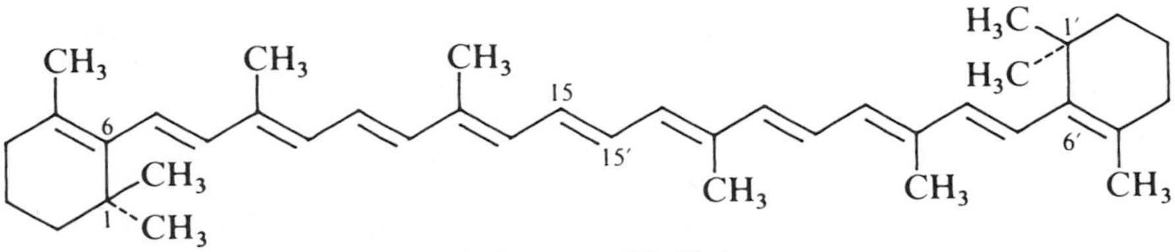

β-Carotene ($C_{40}H_{56}$)
mp 183°C, λ_{max}^{hexane} 451 nm

Lycopene, the red pigment of the tomato, is a C_{40}-carotenoid made up of eight isoprene units. β-Carotene, the yellow pigment of the carrot, is an isomer of lycopene in which the double bonds at C_1—C_2 and C_1'—C_2' are replaced by bonds extending from C_1 to C_6 and from C_1' to C_6' to form rings. The chromophore in each case is a system of eleven all-*trans* conjugated double bonds; the closing of the two rings renders β-carotene less highly pigmented than lycopene.

These colored hydrocarbons have been encountered in the thin-layer chromatography experiment (Chapter 9). The isolation procedure described in this chapter affords sufficient carotene and lycopene to carry out analytical spectroscopy and some isomerization reactions. It might be of interest if some students isolate carotene from strained carrot baby food while others isolate lycopene from tomato paste.

Lycopene is not only responsible for the red color of tomatoes, but also of red grapefruit and flamingos. If flamingos do not receive lycopene in their diet, they will be white.

Until recently lycopene was thought to have no utility until a study showed a lower incidence of prostate cancer among men who consumed such foods as spaghetti and pizza (but not tomato juice).[1] It is theorized that the lycopene, which is insoluble in water, is dissolved in the fat of pasta sauce and pizza and thus absorbed through the intestine. Carotene does not have the same anticancer effect.

Lycopene is the predominant carotenoid in blood plasma and prostate tissue. It is not converted into vitamin A as carotene is, but is a powerful antioxidant, being an efficient scavenger of singlet oxygen.

Carotenoids are very sensitive to photochemical air oxidation. Protect solutions and solids from undue exposure to light and heat, and work as rapidly as possible. Do not heat solutions when evaporating solvents, and, if possible, flush apparatus with nitrogen to exclude oxygen. Research workers would isolate these compounds in dimly lit rooms and/or wrap all containers and chromatographic columns in aluminum foil and carry out extractions and crystallizations using solvents that have been deoxygenated.

Experiment

MICROSCALE AND MACROSCALE

Dehydration and Extraction of Tomato or Carrot Paste

Add 5 g of tomato or carrot paste to a 15-mL centrifuge tube or 25 × 150-mm test tube, then add about 7 mL of acetone and stir the paste for several minutes until it is no longer gummy. This acetone treatment removes most of the water from the cellular mixture. Filter the mixture on a small Büchner funnel. Scrape out the tube with a spatula, let it drain thoroughly, and squeeze as much liquid as possible out of the solid residue in the funnel with a spatula. Discard the yellow filtrate. Then return the solid residue to the centrifuge tube, and add 5 mL of dichloromethane, to effect extraction. Cap the tube, and shake the mixture vigorously. Filter the mixture on the Büchner funnel once more, repeat the extraction and filtration with two or three further 5-mL portions of dichloromethane, clean the tube thoroughly, and place the filtrates in it. Dry the solution over anhydrous calcium chloride pellets, filter the solution into a small flask, and evaporate the solution to dryness under vacuum. Apply heat only with the palm of the hand.

Determine the weight of the crude material. It will be very small. If the residue is dry, as it should be, add to it just enough dichloromethane to dissolve the residue. Save a drop of this solution in order to carry out a thin-layer chromatographic analysis (using dichloromethane as the eluent on silica gel plates; see Chapter 9). Then add 200 mg of alumina to the remaining dichloromethane solution, and evaporate the mixture to dryness, again without heat.

1. E. Giovannucci, A. Ascherio, E. B. Rimm, M. J. Stampfer, G. A. Colditz, and W. C. Willett, *J. Natl. Cancer Inst.*, **87,** 1767 (1995).

MICROSCALE AND MACROSCALE

Column Chromatography

The crude carotenoid is to be chromatographed on an 8-cm column of basic or neutral alumina, prepared with hexane as solvent (see Chapter 10). Run out excess solvent, or remove it from the top of the chromatography column with a Pasteur pipette and add the 200 mg of alumina that has the crude carotenoids absorbed on it. Add a few drops of hexane to wash down the inside of the chromatography column and to consolidate the carotenoid mixture at the top of the column. Elute the column with hexane, discard the initial colorless eluate, and collect all yellow or orange eluates together. Place a drop of solution on a microscope slide, and evaporate the rest to dryness. Examination of the material spotted on the slide may reveal crystallinity. If you are using tomato paste, a small amount of yellow β-carotene will come off first followed by lycopene. Collect the red lycopene separately by eluting with a mixture of 10% acetone in hexane, and evaporate that solution to dryness also.

Finally, dissolve the samples obtained by evaporating petroleum ether in the least possible amount of dichloromethane, and carry out thin-layer chromatography of the two products in order to ascertain their purity. Combine your purified products with those of several other students, evaporate the solution to dryness, dissolve the residue in $CDCl_3$, and determine the ^1H NMR spectrum. Also obtain an infrared spectrum and a visible spectrum (in hexane).

Look up the current prices of commercial lycopene[2] and β-carotene.[3] Note that β-carotene is in demand as a source of vitamin A and is manufactured by an efficient synthesis. Until very recently no use for lycopene had been found.

Cleaning Up   Recovered and unused dichloromethane should be placed in the halogenated organic waste container, the solvents used for TLC should be placed in the organic solvents container, and the drying agent, once free of solvent, can be placed in the nonhazardous solid waste container along with the used plant material and TLC plates.

Research Projects

The carotenoids of any leaf can be isolated in the manner described in this experiment. Grind the leaf material (about 10 g) in a mortar with some sand, and then follow the above procedure. Waxy leaves do not work well. The carotenoids are present in the leaf during its whole life, so a green leaf from a maple or euonymus known to turn bright red in the fall will show lycopene even when the leaf is green. In the fall the chlorophyll decomposes before the carotenoids, so the leaves appear in a variety of orange and red hues.

It is of interest to investigate the carotenoids of the tomato, of which there are some 80 varieties. The orange-colored tangerine tomato contains an isomer of lycopene. If a hexane solution of the prolycopene from this tomato is treated

2. Pfaltz and Bauer, Flushing, NY 11368.

3. Aldrich Chemical Co., Milwaukee, WI 53233.

with a drop of a very dilute solution of iodine in hexane and then exposed to bright light, the solution will turn deep-orange in color, indicating that a *cis*-double bond has isomerized to the *trans* form. The product is, however, still not identical to natural lycopene.

Isomerization

Prepare a hexane solution of either carotene or lycopene, save a drop for TLC, treat the solution with a very dilute solution of iodine in hexane, expose the resulting mixture to strong light for a few minutes, and then carry out thin-layer chromatography on the resulting solution. Also compare the visible spectrum before and after isomerization.

   Iodine serves as a catalyst for the light-catalyzed isomerization of some of the *trans*-double bonds to an equilibrium mixture containing *cis*-isomers.

Surfing the Web _____

http://wellweb.com/PROSTATE/tomatoes.htm

   Prostate cancer and tomatoes.

http://www.vrp.com/Tsauce.htm

   Lycopene and Prostate Cancer: Tomato Sauce to the Rescue?

For updated information visit:
www.mtholyoke.edu/courses/kwilliam/microscale.shtml
or
www.hmco.com/hmco/college/chemistry/Home.html

CHAPTER

68

The Synthesis of Natural Products: Pseudopellitierene and Camphor

Prelab Exercise: Look up and write out Willstätter's synthesis of tropinone.

Part 1. Synthesis of Pseudopellitierene

A principal focus of organic chemistry and certainly of its more specialized branch, biochemistry, is an understanding of the life process. There has long been a quest to understand and then to reproduce in the laboratory the reactions that take place in living organisms without the intervention of catalysts (enzymes) that have been isolated from living organisms.

In the present experiment, we attempt to reproduce one of the classic syntheses of a complex natural product by a biogeneticlike synthesis, a synthesis of pseudopellitierene. This alkaloid is a naturally occurring substance isolated from the bark of the pomegranate tree. Several isomeric compounds were isolated in 1878; the structures of none were known at the time, hence the names pellitierene, isopellitierene, and pseudopellitierene.

Tropinone is the simplest representative of a very similar class of alkaloids that includes atropine (used by ophthalmologists to dilate the pupil of the eye), scopolamine (used to combat seasickness), and cocaine (a component of Coca-Cola at the turn of the century). Tropinone was synthesized by Willstätter in 1903 in a classic 15-step synthesis. It also was the object of the first biogenetic-like synthesis of a naturally occurring molecule. Robert Robinson (later Sir Robert), while at the University of Liverpool in 1917, synthesized tropinone by the Mannich-type condensation of succindialdehyde, methyl amine, and acetone dicarboxylic acid. For his many contributions to synthetic organic chemistry, Robinson was awarded the Nobel Prize in 1974. Willstätter received the Nobel Prize in 1920.

Experiment

The Biomimetic Synthesis of Pseudopellitierene

In this experiment, we will synthesize pseudopellitierene using essentially the same procedure employed by Robinson: simply stirring a mixture of the three reactants at room temperature. Pseudopellitierene, with a higher melting point, is somewhat easier to isolate than tropinone is.

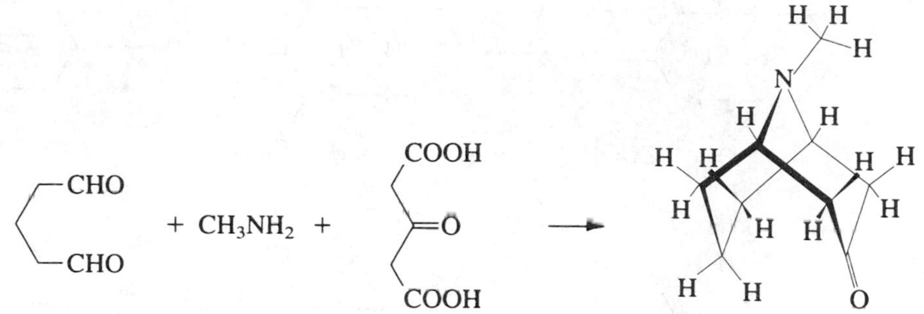

| Glutaraldehyde | Methyl amine | Acetone-dicarboxylic acid | Pseudopellitierene |

The mechanism for the glutaraldehyde precursor hydrolysis and for the biogeneticlike synthesis of pseudopellitierene are given below:

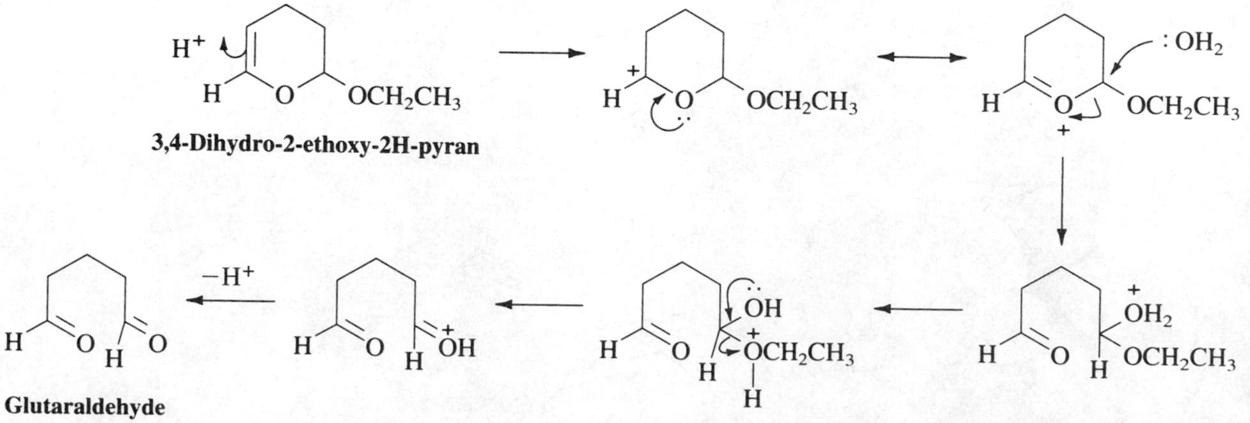

3,4-Dihydro-2-ethoxy-2H-pyran

Glutaraldehyde

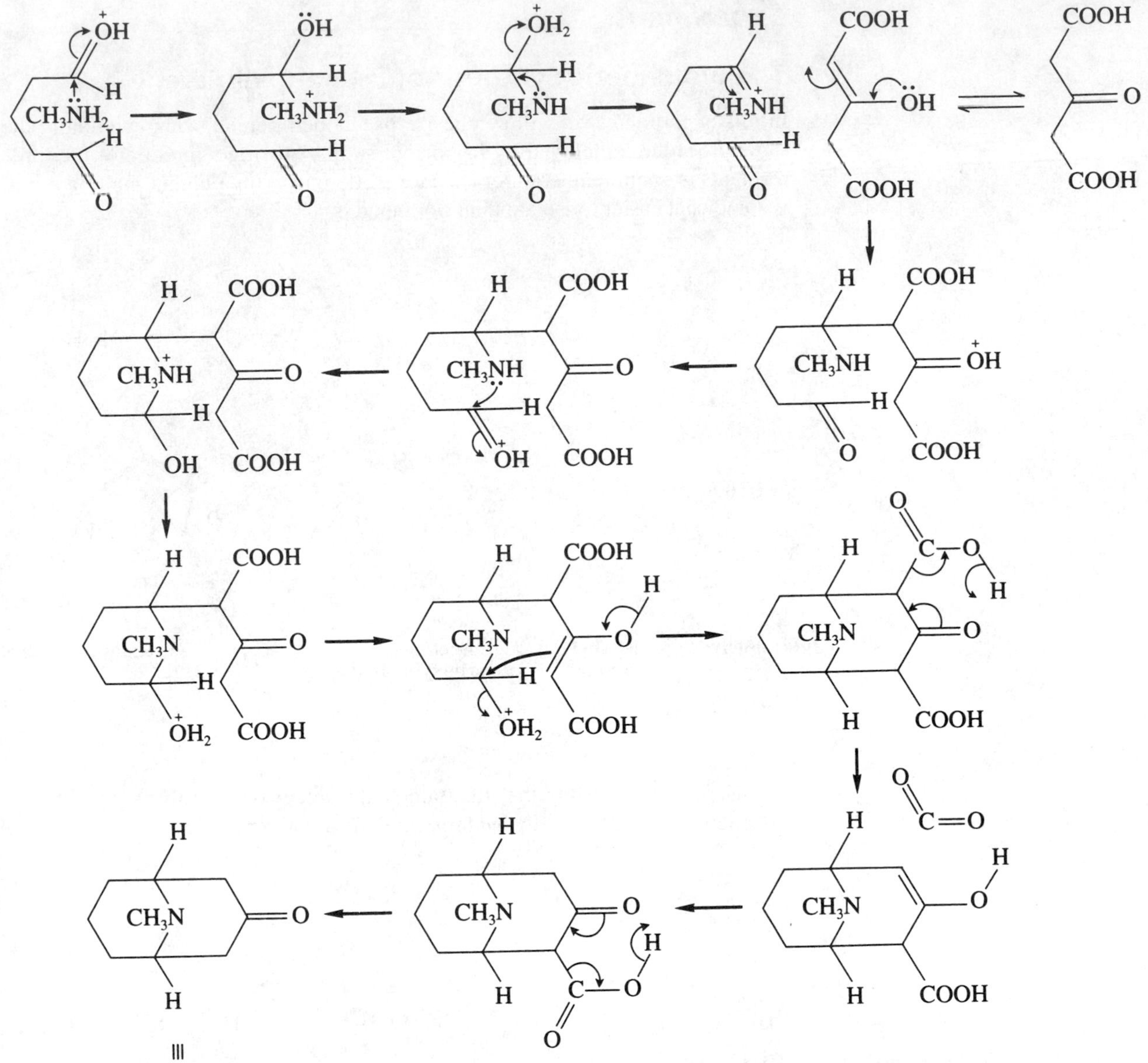

III

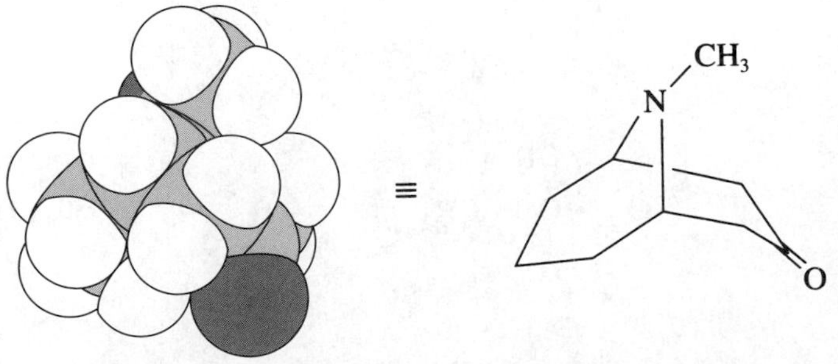

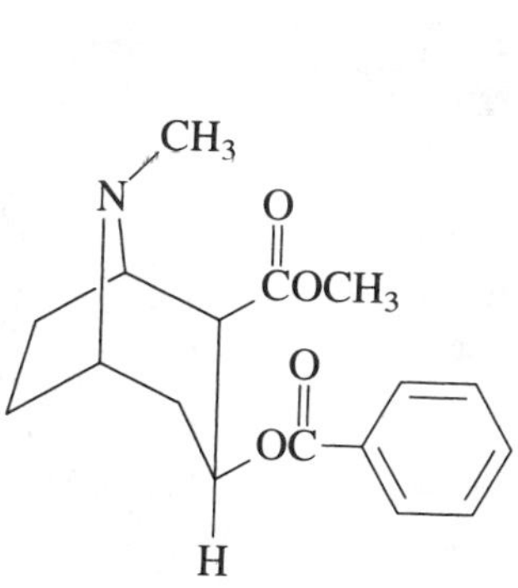

Cocaine

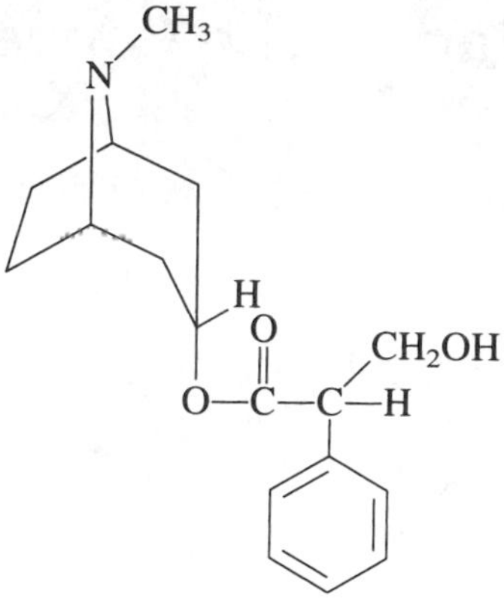

Atropine

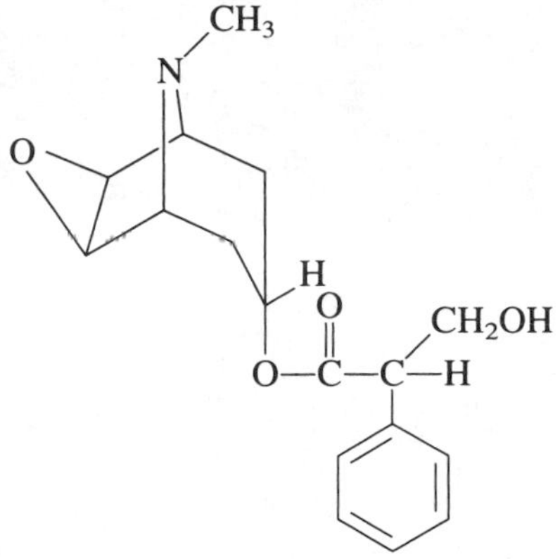

Scopolamine

Procedure

To a 250-mL round-bottomed flask equipped with a magnetic stirring bar, add 6.6 mL of deoxygenated water, 0.88 mL of concentrated hydrochloric acid, and 2.56 g of 3,4-dihydro-2-ethoxy-2H-pyran. The mixture is stirred rapidly for 20 min and then allowed to stand for 1 h. To the yellow solution, add 14.0 mL of water, 2.0 g of methylamine hydrochloride dissolved in 20 mL of water, 3.32 g of acetonedicarboxylic acid (freshly recrystallized from ethyl acetate) dissolved in 32 mL of water and 1.42 g of disodium hydrogen phosphate, and 0.44 g of potassium hydroxide dissolved in 8 mL of water by heating. Measure the pH. The solution is stirred slowly under a nitrogen atmosphere for at least 24 h (again measure the pH), then 1.32 mL of concentrated hydrochloric acid is added, and the mixture is heated on the steam bath for 1 h. After cooling to room temperature, 4.4 g of potassium hydroxide dissolved in 4 mL of water is added. The mixture is extracted eight times with 12-mL portions of dichloromethane in the hood, and the extract is dried with anhydrous sodium sulfate (not calcium chloride). The combined dichloromethane extract is carefully evaporated, and the residue is sublimed under aspirator (or oil pump) vacuum to give pseudopellitierene. Resublimation may be necessary to give purer material. The melting point reported for pseudopellitierene is 54°C.

Part 2. The Conversion of Camphene to Camphor

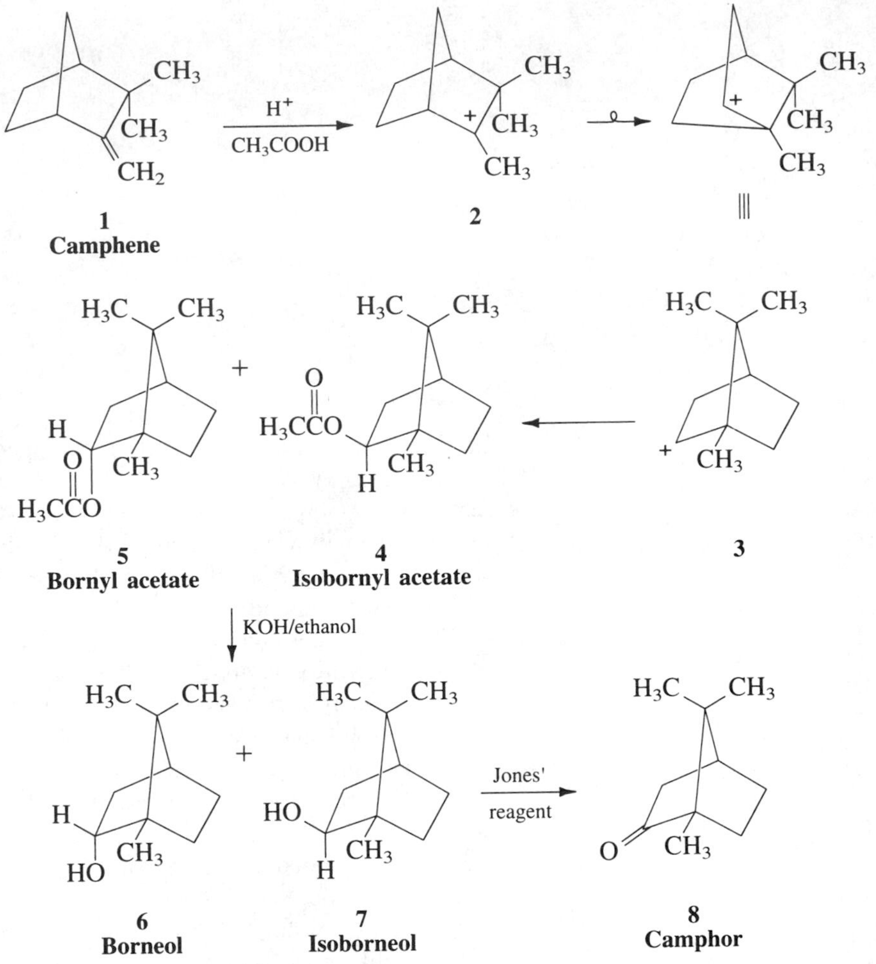

In these experiments, 2,2-dimethyl-3-methylenebicyclo[2.2.1]heptane, otherwise known as *camphene* (**1**), will be converted to 1,7,7-trimethylbicyclo-[2.2.1]heptan-2-one, which is camphor (**8**) through the intermediate isomeric alcohols borneol (**6**) and isoborneol (**7**) *endo-* and *exo-*1,7,7-trimethylbicyclo-[2.2.1]heptan-2-ol.

Camphene (**1**), as its odor will indicate, can be isolated from turpentine and a number of other naturally occurring oils. Upon reaction with acetic acid and a catalytic amount of sulfuric acid, it forms a carbocation (**2**) that undergoes rearrangement (Wagner-Meerwein rearrangement) to the secondary carbocation (**3**), with subsequent solvolysis (reaction with the solvent) to form predominantly the *exo*-acetate (**4**) with some of the *endo*-acetate (**5**). Even though **2** is a tertiary carbocation and more stable than **3**, the acetate (**4**) is formed because acetate ion can more easily attack the unhindered cation (**3**) than the hindered cation (**2**). In Experiment 3 of this section, computational chemistry is used to predict and/or confirm that the *exo*-acetate (**4**) will be formed in preference to the *endo*-isomer (**5**).

Isobornyl acetate (**4**), with the acetate group in the *exo* position, is accompanied by a small amount of the isomeric *endo* compound, bornyl acetate (**5**). These isomers are liquids and not easily purified. They can be hydrolyzed to the corresponding alcohols, borneol (**6**) and isoborneol (**7**), which are solids.

This mixture of isomers can be oxidized to camphor (**8**). All these compounds have a roughly spherical shape. Intermolecular interactions are relatively small, and therefore they sublime easily. When you look up the physical properties of isoborneol, note the small difference between its melting point and its boiling point.

In carrying out this series of reactions, bear in mind that one of the goals of the experiment is to prepare enough of the final product to purify and to characterize by thin-layer chromatography, melting point, and infrared and NMR spectroscopy. For characterization you will need about 20 mg. With skill, you should end up with a lot more than that.

Experiments _____

MICROSCALE

1. Laboratory Conversion of Camphene to Camphor

Camphene (1) to Isobornyl Acetate (4)

In a 5-mL round-bottomed flask place a stirring bar, 3 mL of acetic acid, and 1.36 g of camphene (0.01 mol), and to the resulting solution add a solution of 0.25 g of concentrated sulfuric acid in 0.3 mL of water. Since the acetic acid is the solvent and sulfuric acid is a catalyst, the amounts of these reagents need not be measured with great accuracy. Heat the mixture in a boiling water bath for 15 min with stirring. Alternatively, the mixture can simply be heated over the steam bath with frequent swirling of the contents. Do not reflux the mixture directly on the sand bath because that would be too hot and cause undesired side reactions. The reaction mixture will turn dark in color.

After the heating period, place the cooled reaction mixture in a 15-mL centrifuge tube, complete the transfer with about 2 mL of *tert*-butyl methyl ether, and extract the mixture with two 4-mL portions of water. This extraction will remove most of the acetic acid. The separations are achieved by drawing off the aqueous layer using a 9-in. Pasteur pipette. Wash the ether layer with about 2 mL of 10% sodium carbonate or bicarbonate solution, and dry the solution over calcium chloride pellets for about 5 min. Transfer the solution to the tared 25-mL filter flask, and remove the ether under reduced pressure (Fig. 68.1).

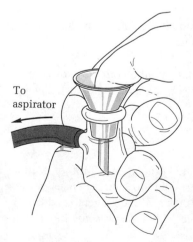

To aspirator

FIG. 68.1 Evaporation of solvent under reduced pressure.

This will take just a minute or so because of the low boiling point of the ether. Save about 50 mg of the product for analysis. It will be a brown liquid mixture of isobornyl acetate with some bornyl acetate. In this product isolation the exact amounts of ether, water, and carbonate solutions are not important, so there is no need to measure them precisely.

Cleaning Up   Neutralize the aqueous washes with bicarbonate, and then flush the resulting solution down the sink with lots of water, or combine these washes with the filtrate from the next experiment, neutralize, and flush down the sink. The calcium chloride can be placed in the nonhazardous waste container or dissolved in water and flushed down the drain.

MICROSCALE

Isobornyl Acetate (4) [and Bornyl Acetate (5)] to Isoborneol (7) [and Borneol (6)]

In a 5-mL round-bottomed flask place a boiling chip and a solution of 0.45 g of potassium hydroxide pellets (the pieces are 85% potassium hydroxide and 15% water) dissolved in 0.7 mL water to which is added 2.3 mL of ethanol. To this basic solution add 1.0 g of isobornyl/bornyl acetate, add an air condenser, and reflux the mixture on the sand bath for 1 h. With careful adjustment of the amount of heat, it is easy to reflux this mixture using an air condenser. Just make sure that the top of the condenser is not hot, which would indicate that ethanol may be escaping. Wrap a wet pipe cleaner around the air condenser if more cooling is needed. At the end of the hour, pour the solution onto about 5 g of ice in a 10-mL Erlenmeyer flask. Swirl the mixture until the product is completely solidified; then collect it on the Hirsch funnel. Wash it well with water, and then carefully press the product onto the funnel using a spatula. It will have the consistency of light-brown sugar. Try to squeeze out as much of the water as possible. Save about 60 mg of this sample for analysis. It sublimes easily and will give pure white crystals. Determine the melting point of the sublimed product in a sealed melting-point capillary (see p. 70 for this technique). Explain the melting point in terms of the expected composition of the product.

Cleaning Up   Neutralize the filtrate with dilute hydrochloric acid, or combine it with the washings from the previous experiment, neutralize, and flush down the drain with excess water.

MICROSCALE

Borneols (5 and 6) to Camphor (7)

In a 15-mL centrifuge tube dissolve 0.6 g of the borneols in 1.2 mL of acetone, add a magnetic stirring bar, and then add to the solution 1 mL of Jones' reagent dropwise with thorough stirring and cooling.[1] After all the reagent is added, the mixture should be homogeneous in appearance. If not, stir it some more. After

---

1. Jones' reagent is prepared by dissolving 13.5 g of chromium trioxide in 11.5 mL of concentrated sulfuric acid and then adding, with great care, enough water to bring the volume up to 50 mL.

FIG. 68.2 Sublimation apparatus.

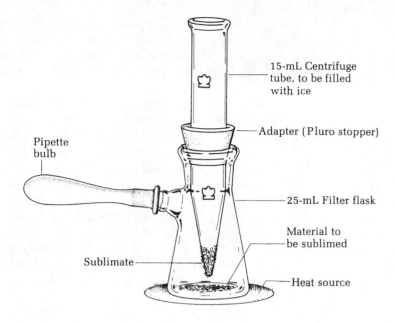

15-mL Centrifuge tube, to be filled with ice

Adapter (Pluro stopper)

Pipette bulb

25-mL Filter flask

Material to be sublimed

Sublimate

Heat source

CAUTION: *Jones' reagent is very corrosive because it is a powerful oxidizing agent prepared in concentrated sulfuric acid. Wash up spills immediately. If spilled on the skin, flush with water until the skin is neutral. Cr^{6+} as a dust is a carcinogen if inhaled. It is unlikely that you would inhale this reagent; however, exercise care when preparing the reagent from solid chromium trioxide.*

MACROSCALE

about 30 min, add water dropwise to the reaction mixture, diluting it to about 12 mL. Collect the solid by filtration on the Hirsch funnel, squeeze it as dry as possible, and then sublime the camphor at atmospheric pressure (Fig. 68.2). Determine the melting point, analyze the product by thin-layer chromatography and infrared and NMR spectroscopy, and then submit the product in a labeled vial (not a plastic bag).

Cleaning Up   Complete the reduction of any remaining chromic ion in the filtrate by adding solid sodium thiosulfate until the solution becomes cloudy and blue. Neutralize with sodium carbonate, and filter the flocculent precipitate of $Cr(OH)_3$ on a Hirsch funnel. The filtrate can be diluted with water and flushed down the drain, while the precipitate on the filter paper should be placed in the heavy metals hazardous waste container.

2. Laboratory Conversion of Camphene to Camphor

Camphene (1) to Isobornyl Acetate (4)

In a 25-mL round-bottomed flask place a stirring bar, 12 mL of acetic acid, and 5.44 g of camphene (0.01 mol), and to the resulting solution add a solution of 1.0 g of concentrated sulfuric acid in 1.2 mL of water. Since the acetic acid is the solvent and sulfuric acid is a catalyst, the amounts of these reagents need not be measured with great precision. Heat the mixture in a boiling water bath for 15 min with stirring. Alternatively, the mixture can simply be heated over the steam bath with frequent swirling of the contents. Do not reflux the mixture directly on the sand bath because that would be too hot and cause undesired side reactions. The reaction mixture will turn dark in color.

After the heating period, place the cooled reaction mixture in a small separatory funnel, complete the transfer with about 8 mL of *t*-butyl methyl ether, and extract the mixture with two 15-mL portions of water. This extraction will remove most of the acetic acid. Wash the ether layer with about 8 mL of 10% sodium carbonate or bicarbonate solution, and dry the solution over calcium chloride pellets for about 5 min. Transfer the solution to a tared 100-mL round-bottomed flask, and remove the ether with a rotary evaporator. This will take just a minute or so because of the low boiling point of the ether. Save about 200 mg of the product for analysis. It will be a brown liquid mixture of isobornyl acetate with some bornyl acetate. In this product isolation, the exact amounts of ether, water, and carbonate solutions are not important, so there is no need to measure them precisely.

Cleaning Up   Neutralize the aqueous washes with bicarbonate, and then flush the resulting solution down the sink with lots of water, or combine these washes with the filtrate from the next experiment, neutralize, and flush down the sink. The calcium chloride can be placed in the nonhazardous waste container or dissolved in water and flushed down the drain.

MACROSCALE

Isobornyl Acetate (4) [and Bornyl Acetate (5)] to Isoborneol (7) [and Borneol (6)]

In a 25-mL round-bottomed flask place a boiling chip and a solution of 1.8 g of potassium hydroxide pellets (the pieces are 85% potassium hydroxide and 15% water) dissolved in 2.8 mL water to which is added 9.2 mL of ethanol. To this basic solution add 4.0 g of isobornyl/bornyl acetate, add a condenser, and reflux the mixture on the sand bath for 1 h. At the end of the hour, pour the solution onto about 20 g of ice in a 50-mL Erlenmeyer flask. Swirl the mixture until the product is completely solidified; then collect it on the Hirsch funnel. Wash it well with water, and then carefully press the product onto the funnel using a spatula. It will have the consistency of light-brown sugar. Try to squeeze out as much of the water as possible. Save about 250 mg of this sample for analysis. It sublimes easily and will give pure white crystals. Determine the melting point of the sublimed product in a sealed melting-point capillary. Explain the melting point in terms of the expected composition of the product.

Cleaning Up   Neutralize the filtrate with dilute hydrochloric acid, or combine it with the washings from the previous experiment, neutralize, and flush down the drain with excess water.

MACROSCALE

Borneols (5 and 6) to Camphor (8)

In a 25-mL Erlenmeyer flask, dissolve 2.4 g of the borneols in 4.8 mL of acetone, add a magnetic stirring bar, and then add to the solution 4.0 mL of Jones' reagent dropwise with thorough stirring and cooling.[2] After all the reagent is

---

2. Jones' reagent is prepared by dissolving 13.5 g of chromium trioxide in 11.5 mL of concentrated sulfuric acid and then adding, with great care, enough water to bring the volume up to 50 mL.

CAUTION: *Jones' reagent is very corrosive because it is a powerful oxidizing agent prepared in concentrated sulfuric acid. Wash up spills immediately. If spilled on the skin, flush with water until the skin is neutral. Cr^{6+} as a dust is a carcinogen if inhaled. It is unlikely that you would inhale this reagent; however, exercise care when preparing the reagent from solid chromium trioxide.*

added, the mixture should be homogeneous in appearance. If not, stir it some more. After about 30 min, add water dropwise to the reaction mixture, diluting it to about 50 mL. Collect the solid by filtration on the Hirsch funnel, squeeze it as dry as possible, and then sublime the camphor at atmospheric pressure (Fig. 68.2). Determine the melting point, analyze the product by thin-layer chromatography and infrared and NMR spectroscopy (Fig. 68.3), and then submit the product in a labeled vial (not a plastic bag).

Cleaning Up   Complete the reduction of any remaining chromic ion in the filtrate by adding solid sodium thiosulfate until the solution becomes cloudy and blue. Neutralize with sodium carbonate, and filter the flocculent precipitate of $Cr(OH)_3$ on a Hirsch funnel. The filtrate can be diluted with water and flushed down the drain, while the precipitate on the filter paper should be placed in the heavy metals hazardous waste container.

3. Computational Chemistry: The Theoretical Conversion of Camphene to Camphor

Molecular mechanics calculations have been dealt with previously (Chapter 15), but these calculations represent just a part of a larger topic, computational chemistry, an important new part of organic chemistry. Just a few years ago this area of chemistry was relegated to the specialist, the "theoretical chemist," who laboriously encoded the Cartesian coordinates of all the atoms in a molecule and then submitted the calculation to a large mainframe computer for solution. The results came back in enormous piles of digital printout, interpretable only by the expert. In the last few years, however, workstations and now even desktop computers that are graphically oriented can handle these computations, while the input and output can be manipulated and displayed on a computer monitor as a drawing of the molecule. For example, the distribution of charge density can be color encoded and superimposed on the space-filling model of the

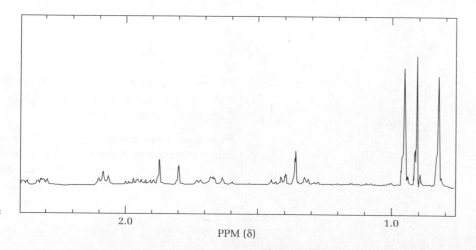

FIG. 68.3 ¹H NMR spectrum of camphor (250 MHz).

2.0

1.0

PPM (δ)

molecule. Thus computational chemistry has passed from the hands of the specialist to the practicing organic chemist.

In the example presented here, we try to predict by computation which isomer, isobornyl acetate (**4**) or bornyl acetate (**5**), is favored when camphene is treated with acetic acid and a catalytic amount of sulfuric acid.

As seen in the drawing, camphene (**1**), on protonation, will give a carbocation (**3**), which is attacked by the nucleophile, acetic acid, to give either **4** or **5**, the isomeric acetates. The carbocation, an sp^2-hybridized carbon, is planar and has an empty p orbital. We know that the acetic acid must attack this from "above" or "below," along the axis of the p orbital. The reactants must come within bonding distance for the product to form, but there is steric hindrance from other parts of the molecule preventing this. It is not obvious from scrutiny of the drawing whether there is more steric hindrance from the "top" or the "bottom" of the ion in **3.** Even when molecular models are held in the hand, it looks as if the bridge methyl group is going to get in the way of the acetic acid, so we would predict that attack by acetic acid would take place from the bottom side of the cation.

To approach this problem from a calculational standpoint, we first assemble the framework of the cation on the computer and then minimize its energy. We then proceed to carry out a molecular orbital calculation. In this case we use a semiempirical method because it is fast and will give us the type of information we need. Among the well-known methods are modified neglect of diatomic overlap (MNDO), Austin Method 1 (AM1),[3] and MNDO parameterization method 3 (MNDO/PM 3 or simply PM3).

Using a computer program such as Spartan, which runs on a workstation such as a Silicon Graphics Indigo, the lowest unoccupied molecular orbital (LUMO) of the cation (**3**) can be calculated and superimposed on the space-filling model of the cation. Through color coding, you can see on the computer monitor that there is more of the empty molecular orbital exposed on the top (*exo*) side of the cation than on the bottom (*endo*) side. If you have the facilities, try to reproduce this calculation. This same procedure is used to predict the direction of attack of the borohydride anion on 2-methylcyclohexanone (Chapter 22).

Questions

1. The molal freezing point depression constant for camphor is 39.7°C. What does this mean, and what effect might it have on your observed melting point for camphor?

2. Determine, using a molecular mechanics program, whether the most stable conformation of pseudopellitierene is shown on the third page of this chapter. Pay particular attention to the locations of the carbonyl group and the methyl group.

---

3. Its originator, Michael J. S. Dewar, was a professor at the University of Texas at Austin.

CHAPTER 69

Polymers: Synthesis and Recycling

Prelab Exercise: In the preparation of nylon by interfacial polymerization sebacoyl chloride is synthesized from decanedioic acid and thionyl chloride. What volume of hydrogen chloride is produced in this reaction? What volume of sulfur dioxide is produced?

Part 1. Synthesis of Polymers

We use more synthetic polymers than steel, aluminum, and copper combined.

Polymers are ubiquitous. Natural polymers such as proteins (polyamino acids), DNA (polynucleotides), and cellulose (polyglucose) are the basic building blocks of plant and animal life. Synthetic organic polymers, or plastics, are now among our most common structural materials. In the United States we make and use more synthetic polymers than we do steel, aluminum, and copper combined—at present 60 billion pounds, worth $25 billion each year.

Polymers, from the Greek meaning "many parts," are high molecular weight molecules made up of repeating units of smaller molecules. Most polymers consist of long chains held together by hydrogen bonds, van der Waals forces, and the tangling of the long chains. When heated, the covalent bonds of some polymers, which are thermoplastic, do not break, but the chains slide over one another to adopt new shapes. These shapes can be films, sheets, extrusions, or molded parts in a myriad of forms.

Nitrocellulose

The first synthetic plastic was nitrocellulose, made in 1862 by nitrating the natural polymer, cellulose. Nitrocellulose, when mixed with a plasticizer such as camphor to make it more workable, was originally used as a replacement for ivory in billiard balls and piano keys and to make Celluloid collars. This material, from which the first movie film was made, is notoriously flammable.

Cellulose acetate

Cellulose acetate, made by treating cellulose with acetic acid and acetic anhydride, was originally used as a waterproof varnish to coat the fabric of airplanes during World War I. It later became important as a photographic film base and as acetate rayon.

Bakelite

The first completely synthetic organic polymer was Bakelite, named for its discoverer Leo Baekeland. He was a Belgian chemistry professor who invented the first successful photographic paper, Velox. He came to America at the age of 35 and sold his invention to George Eastman for $1 million in 1899. He then turned his attention to finding a replacement for shellac, which comes from the Asian lac beetle. At the time, shellac was coming into great demand in the fledgling electrical industry as an insulator. The polymer Baekeland produced is still used for electrical plugs, switches, and the black handles and knobs on pots and pans. It has superior electrical insulating properties and very high heat resistance. It is made by the base-catalyzed reaction of excess formaldehyde with phenol. In a low molecular weight form it is used to glue together the plies of plywood or mixed with a filler such as sawdust. When it is heated to a high temperature, crosslinking occurs as the polymer "cures."

Thermoplastic polymers

Most polymers are amorphous, linear macromolecules that are thermoplastic and soften at high temperature. In Bakelite the polymer crosslinks to form a three-dimensional network, and the polymer becomes a dark, insoluble, infusible substance. Such polymers are said to be *thermosetting*. Natural rubber is thermoplastic. It becomes a thermosetting polymer when heated with sulfur, as Charles Goodyear discovered. With 2% sulfur the rubber becomes crosslinked but is still elastic; at 30% sulfur it can be made into bowling balls. Some other important thermosetting polymers are urea-formaldehyde resins and melamine-formaldehyde resins. The latter are among the hardest of polymers and take on a high-gloss finish. Melamine is used extensively to manufacture plastic dinnerware.

**Melamine,
a thermosetting polymer**

Vinyl chloride

Even though vinyl chloride was discovered in 1835, polyvinyl chloride was not produced until 1912. It is now one of our most common polymers; production in 1997 was over 20 million metric tons. The monomer is made by the pyrolysis of 1,2-dichloroethane, formed by chlorination of ethylene. Free radical polymerization follows Markovnikov's rule to give the head-to-tail polymer with high specificity:

$$n\text{CH}_2\!=\!\text{CHCl} \longrightarrow -(\text{CH}_2\text{CHCl})_n-$$

Vinyl chloride   Polyvinyl chloride

Plasticizers

Pure polyvinyl chloride (PVC) is an extremely hard polymer. It, along with some other polymers, can be modified by the addition of plasticizers; the greater the ratio of plasticizer to polymer, the greater the flexibility of the polymer. PVC pipe is rigid and contains little plasticizer, while shower curtains contain a large percentage. The most common plasticizer is di(2-ethylhexyl) phthalate, which can be added in concentrations of up to 50%.

Di(2-ethylhexyl) phthalate (Dioctyl phthalate)

PVC is used for raincoats, house siding, and artificial leather for handbags, briefcases, and inexpensive shoes. It is found in garden hose, floor covering, swimming pool liners, and automobile upholstery. When vinyl upholstery is exposed to high temperatures, as in the interior of an automobile, the plasticizer distills out. The result is an opaque, difficult-to-remove film on the insides of the windows, and upholstery that is hard and brittle.

Polymerization methods

Monomers can be polymerized in the gas phase, in bulk, as suspensions, and as emulsions. The most common method of making PVC is by emulsifying the monomer, vinyl chloride, in water with surfactants (soaps), water-soluble catalysts, and heat. The monomer is polymerized to solid particles, which are suspended in the aqueous phase. This product can be centrifuged and dried or used as such. Chemists can control the average molecular weight, which can become very high in emulsion polymerization. A high molecular weight means a more rigid and stronger polymer, but also one that is more difficult to work with. An emulsion of polyvinyl acetate is sold as latex paint. When the vehicle, water, evaporates, the polymer is left as a hard film. A thicker emulsion of polyvinyl acetate is an excellent adhesive, the familiar white glue. When vinyl acetate and vinyl chloride are polymerized together, a copolymer results that has properties all of its own. This copolymer is particularly good for detailed moldings and is used to make phonograph records.

Polyvinylidene chloride is primarily extruded as a film that has low permeability to water vapor and air and is therefore used as the familiar clinging plastic food wrap, Saran Wrap.

Teflon

Polytetrafluoroethylene, Teflon, another of the halogenated polymers, has a number of unique properties. It has a very high melting point, 327°C, it does not dissolve in any solvent, and nothing sticks to it. It is also an extremely good electrical insulator. A product of the DuPont company, Teflon is one of the densest of the polymers and also one of the most expensive. The surface of the polymer must be etched with metallic sodium to form free radicals, to which glue can adhere. The polymer has an extremely low coefficient of friction, which makes it useful for bearings. Its chemical inertness makes it an ideal liner for chemical reagent bottle caps, and its no-stick property is ideal for the coating on the inside of frying pans. At 380°C Teflon is still so viscous that it cannot be injection-molded. Instead it is molded by pressing the powdered polymer at high temperature and pressure, a process called *sintering.*

$$—(CH_2CHCl)_n— \quad —(CH_2CCl_2)_n— \quad\quad —(CF_2CF_2)_n— \quad\quad —(CH_2CH_2)_n—$$

Polyvinyl chloride   Polyvinylidene chloride   Polytetrafluoroethylene   Polyethylene

The polymer produced in highest volume is polyethylene. Invented by the British, who call it *polythene,* and put into production in 1939, it could for a long time only be produced by the oxygen-catalyzed polymerization of ethylene at pressures near 40,000 lb/in.^{2}. Such pressures are expensive and dangerous to maintain on an industrial scale. The polyethylene produced has a low density and is used primarily to make film for bags of all types—from sandwich bags to trash can liners. The opaque appearance of polyethylene is due to crystallites, regions of order in the polymer that resemble crystals. In the 1950s Karl Ziegler and Giulio Natta developed catalysts composed of $TiCl_4$, alkyl aluminum, and transition metal halides with which ethylene can be polymerized at pressures of

High-density polyethylene

just 450 lb/in.^{2}. The resulting product has a higher density and a 20°C higher softening temperature than the low-density material. The catalysts which Ziegler and Natta developed, and for which they received the 1963 Nobel Prize, cause stereoregular polymerization and thus a crystalline product. High-density polyethylene is as rigid as polystyrene and yet has high impact resistance. It is used to mold very large articles such as luggage, the cases for domestic appliances, trash cans, and soft drink crates.

Polystyrene

Polystyrene is a brilliantly clear, high-refractive-index polymer familiar in the form of disposable drinking glasses. It is brittle and produces sharp, jagged edges when fractured. It softens in boiling water and it burns readily with a very smoky flame. But it foams readily and makes a very good insulator; witness the disposable white hot-drink cup. It is used extensively for insulation when properly protected from ignition. The addition of a small quantity of butadiene to the styrene makes a polymer that is no longer transparent but that has high impact resistance. Blends of acrylonitrile, butadiene, and styrene (ABS) have excellent molding properties and are used to make car bodies. One formulation can be chrome-plated for automobile grills and bumpers.

$$CH_2{=}CHCN \quad C_6H_5CH{=}CH_2 \quad CH_2{=}CH{-}CH{=}CH_2$$

Acrylonitrile   **Styrene**   **1,3-Butadiene**

Rubber

Joseph Priestley, the discoverer of oxygen, named *rubber* for its ability to remove lead pencil marks. Rubber is an *elastomer,* defined as a substance that can be stretched to at least twice its length and return to its original size. The Germans, cut off from a supply of natural rubber, began manufacturing synthetic rubber during World War I. Called "buna" for *bu*tadiene and sodium, *Na,* the polymerization catalyst, it was not the ideal substitute. Cars with buna tires had to be jacked up when not in use because their tires would develop flat spots. The addition of about 25% styrene greatly improved the qualities of the product;

Synthetic rubber; a styrene-butadiene copolymer

styrene-butadiene synthetic rubber now dominates the market, a principal outlet being automobile tires. Addition of 30% acrylonitrile to butadiene produces nitrile rubber, which is used to make conveyor belts, tank liners, rubber hose, and gaskets.

The chemist classifies polymers in several ways. There are thermosetting plastics such as Bakelite and melamine and the much larger category of thermoplastic materials, which can be molded, blown, and formed after polymerization. There are the arbitrary distinctions made among plastics, elastomers, and fibers. And there are the two broad categories formed by the polymerization reaction

Addition polymers
Condensation polymers

itself: (1) addition polymers (e.g., vinyl polymerizations), in which a double bond of a monomer is transformed into a single bond between monomers, and (2) condensation polymers (e.g., Bakelite), in which a small molecule, such as water or alcohol, is split out as the polymerization reaction occurs.

Nylon

One of the most important condensation polymers is nylon, a name so ingrained into our language that it has lost trademark status. It was developed by Wallace Carothers, director of organic chemicals research at DuPont, and was the outgrowth of his fundamental research into polymer chemistry. Introduced in 1938, it was the first totally synthetic fiber. The most common form of nylon is the polyamide formed by the condensation of hexamethylene diamine and adipic acid:

$$n\text{H}_2\text{N}(\text{CH}_2)_6\text{NH}_2 \; + \; n\text{HOOC}(\text{CH}_2)_4\text{COOH}$$

Hexamethylene diamine | **Adipic acid**

$$n\overset{+}{\text{H}_3}\text{N}(\text{CH}_2)_6\overset{+}{\text{N}}\text{H}_3 \; + \; n\overset{-}{\text{O}}\overset{\overset{\text{O}}{\|}}{\text{C}}(\text{CH}_2)_4\overset{\overset{\text{O}}{\|}}{\text{C}}\text{O}^{-}$$

Pressure, 280°C

$$\text{H} \!\!+\!\! \text{NH}(\text{CH}_2)_6\text{NH}\overset{\overset{\text{O}}{\|}}{\text{C}}(\text{CH}_2)_4\overset{\overset{\text{O}}{\|}}{\text{C}} \!\!+\!\!_n \text{OH} \; + \; (2n - 1)\text{H}_2\text{O}$$

Nylon 6.6

The reactants are mixed together to form a salt that melts at 180°C. This is converted into the polyamide by heating to 280°C under pressure, which eliminates water. Nylon 6.6 is used to make textiles, while nylon 6.10, from the 10-carbon diacid, is used for bristles and high-impact sports equipment. Nylon can also be made by interfacial and by ring-opening polymerization, both of which are used in the following experiments.

The condensation polymer made by reacting ethylene glycol with 1,4-benzene dicarboxylic acid (terephthalic acid) produces a polymer that is almost exclusively converted into the fiber Dacron. The polymerization is run as an ester interchange reaction using the methyl ester of terephthalic acid:

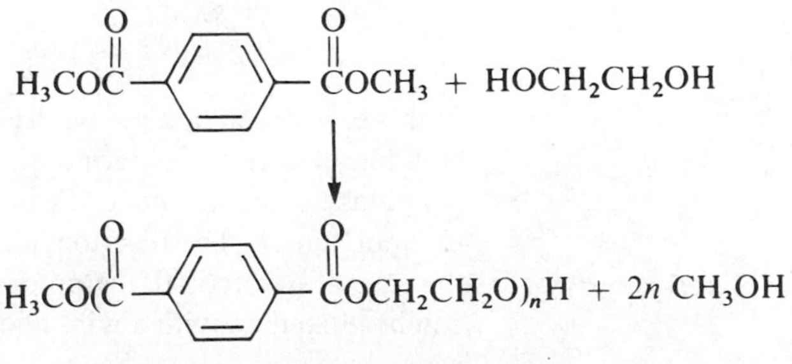

$$\text{H}_3\text{CO}\overset{\overset{\text{O}}{\|}}{\text{C}} \!\!-\!\! \langle\text{benzene}\rangle \!\!-\!\! \overset{\overset{\text{O}}{\|}}{\text{C}}\text{OCH}_3 \; + \; \text{HOCH}_2\text{CH}_2\text{OH}$$

$$\text{H}_3\text{CO}(\overset{\overset{\text{O}}{\|}}{\text{C}} \!\!-\!\! \langle\text{benzene}\rangle \!\!-\!\! \overset{\overset{\text{O}}{\|}}{\text{C}}\text{OCH}_2\text{CH}_2\text{O})_n\text{H} \; + \; 2n\,\text{CH}_3\text{OH}$$

Polyethylene glycol terephthalate (Dacron)

The structure of a polymer is not simple. For example, polymerization of styrene produces a chiral carbon at each benzyl position. We can ask whether the phenyl rings are all on the same side of the long carbon chain, whether they alternate positions, or whether they adopt some random configuration. In the case of copolymers we can ask whether the two components alternate: ABABABABAB . . . ; whether they adopt a random configuration: AABBBABAAB . . . ; or whether they polymerize as short chains of one and then the other: AAAABBBBBBAAABBBB. . . . These questions are important because the physical properties of the resulting polymer depend on them. The polymer chemist is concerned with finding the answers and discovering catalysts and reaction conditions that can control these parameters.

In the experiments that follow, you will find that the preparation of nylon by interfacial polymerization is a spectacular and reliable experiment easily carried out in one afternoon. The synthesis of Bakelite works well; it requires overnight heating in an oven to complete the polymerization. Nylon by ring-opening polymerization requires skill and care because of the high temperatures involved. The polymerization of styrene also requires care, but is somewhat easier to carry out.

In the recycling experiments you will depolymerize a PET bottle to short-chain polyols and then use that material to make fiberglass and polyurethane foam.

Experiments

**MICROSCALE
AND MACROSCALE**

1. Nylon by Interfacial Polymerization[1]

$$SOCl_2 \ + \ HOOC(CH_2)_8COOH \ \longrightarrow \ \overset{\overset{O}{\|}}{Cl}C(CH_2)_8\overset{\overset{O}{\|}}{C}Cl \ + \ HCl \ + \ SO_2$$

Thionyl chloride	**Sebacic acid**	**Sebacoyl chloride**
MW 118.97	MW 202.25	MW 239.14
bp 79°C	mp 137°C	bp 168°C/12 mm

$$\overset{\overset{O}{\|}}{Cl}C(CH_2)_8\overset{\overset{O}{\|}}{C}Cl \ + \ H_2N(CH_2)_6NH_2 \ \longrightarrow \ \underset{}{+}\overset{\overset{O}{\|}}{C}(CH_2)_8\overset{\overset{O}{\|}}{C}NH(CH_2)_6NH\underset{}{+}_n \ + \ HCl$$

Hexane-1,6-diamine   **Nylon 6.10**

MW 116.21
mp 45–46°C

In this experiment a diamine dissolved in water is carefully floated on top of a solution of a diacid chloride dissolved in an organic solvent. Where the two solutions come in contact (the interface), an S_N2 reaction occurs to form a film of a polyamide. The reaction stops there unless the polyamide is removed. In the case of nylon 6.10, the product of this reaction, the film is so strong that it can be picked up with a wire hook and continuously removed in the form of a rope.

This reaction works because the diamine is soluble in both water and dichloromethane, the organic solvent used. As the diamine diffuses into the organic layer, reaction occurs immediately to give the insoluble polymer. The HCl produced

---

1. P. W. Morgan and S. L. Kwolek, *J. Chem. Ed.*, **36**, 182 (1959).

reacts with the sodium hydroxide in the aqueous layer. The acid chloride does not hydrolyze before reacting with the amine because it is not very soluble in water. The acid chloride is conveniently prepared using thionyl chloride.

Microscale Procedure

In a reaction tube fitted with a gas trap (Fig. 69.1) place 0.25 g of sebacic acid (1,8-octane dicarboxylic acid, decanedioic acid), 0.25 mL of thionyl chloride and 12 mg (a small drop) of *N,N*-dimethylformamide. Heat the tube to 60 to 70°C in a water bath in the hood. This is best accomplished by putting very hot water in a beaker and placing the beaker on a steam bath to keep it at 60 to 70°C. As the reaction proceeds, the product forms a liquid layer on the bottom of the tube. Flick the tube to use this liquid to wash down unreacted acid (if necessary) as the reaction proceeds. When the acid has all reacted and gas evolution has ceased (about 10–15 min), the product should be a clear liquid. Add some dichloromethane to the reaction tube, and transfer the solution to a 50-mL beaker using a total of 12 mL of dichloromethane. To prevent the polymer from sticking to the beaker, coat the inside with a thin layer of silicone vacuum grease or silicone oil. Carefully pour onto the top of this dichloromethane solution 0.25 g of hexane-1,6-diamine (hexamethylenediamine) that has been dissolved in 6 mL of water containing 1 mL of 3 M sodium hydroxide. Immediately pick up the polymer film at the center with a copper wire hook and lead it over the outer surface of a bottle, beaker, or graduated cylinder as it is removed. Wind

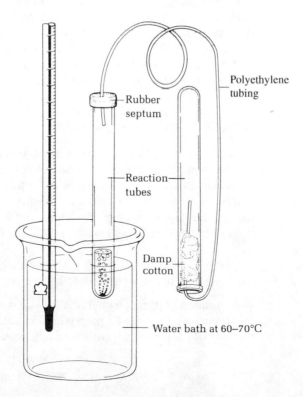

Polyethylene tubing

Rubber septum

Reaction tubes

Damp cotton

Water bath at 60–70°C

FIG. 69.1 Apparatus for acid chloride synthesis. HCl and SO_2 are trapped in the damp cotton.

up as much of the polymer as possible, wash it thoroughly in water, and press it as dry as possible. After the polymer has dried, determine its weight and calculate the yield. Try dissolving the polymer in two or three solvents. Attach a piece of the polymer to your laboratory report.

Cleaning Up   Add the cotton from the trap to the used reaction mixture, and stir the mixture vigorously to cause nylon to precipitate. Decant the water and dichloromethane, and then squeeze the solid as dry as possible; place it in the nonhazardous solid waste container. The dichloromethane should be placed in the halogenated organic solvents container, and the aqueous layer, after neutralization, should be diluted with water and flushed down the drain.

Macroscale Procedure[2]

In a 15-mL tapered 14/20 flask, fitted with a condenser and gas trap like the one shown in Fig. 18.3, place 2 g of sebacic acid (1,8-octane dicarboxylic acid, decanedioic acid), 2 mL of thionyl chloride, and 0.1 mL of *N,N*-dimethylformamide. Heat the tube to 60–70°C in a water bath in the hood. This is best accomplished by putting very hot water in a beaker and placing the beaker on a steam bath to keep it at 60–70°C. As the reaction proceeds the product forms a liquid layer on the bottom of the tube. Use this liquid to wash down unreacted acid as the reaction proceeds. When the acid has all reacted and gas evolution has ceased (about 10–15 min), transfer the product, which should be a clear liquid at this point, to a 250-mL beaker using 50 mL of dichloromethane. Carefully pour onto the top of this dichloromethane solution 2 g of hexane-1,6-diamine (hexamethylenediamine) that has been dissolved in 50 mL of water containing 1 g of sodium hydroxide. Pick up the polymer film at the center with a copper wire and lead it over the outer surface of a bottle, beaker, or graduated cylinder as it is removed. Remove as much of the polymer as possible, wash it thoroughly in water, and press it as dry as possible. After the polymer has dried determine its weight and calculate the yield. Try dissolving the polymer in two or three solvents. Attach a piece of the polymer to your laboratory report.

Cleaning Up   Add the cotton from the trap to the used reaction mixture, and stir the mixture vigorously to cause nylon to precipitate. Decant the water and dichloromethane, and then squeeze the solid as dry as possible; place it in the nonhazardous solid waste container. The dichloromethane should be placed in the halogenated organic solvents container, and the aqueous layer, after neutralization, should be diluted with water and flushed down the drain.

MICROSCALE AND MACROSCALE

2. The Condensation Polymerization of Phenol and Formaldehyde: Bakelite

Condensation of phenol with formaldehyde is a base-catalyzed process in which one resonance form of the phenoxide ion attacks formaldehyde. The resulting trimethylol phenol is then crosslinked by heat, presumably by dehydration with

2. G. C. East and S. Hassell, *J. Chem. Ed.*, **60,** 69 (1983).

the intermediate formation of benzylcarbocations. The resulting polymer is Bakelite. Since the cost of phenol is relatively high and the polymer is somewhat brittle, it is common practice to add an extender such as sawdust to the material before crosslinking. The mixture is placed in molds and heated to form the polymer. The resulting polymer, like other thermosetting polymers, is not soluble in any solvent and does not soften when heated. It is the plastic used to make the black handles on kitchen pots and pans that withstand the heat of an oven.

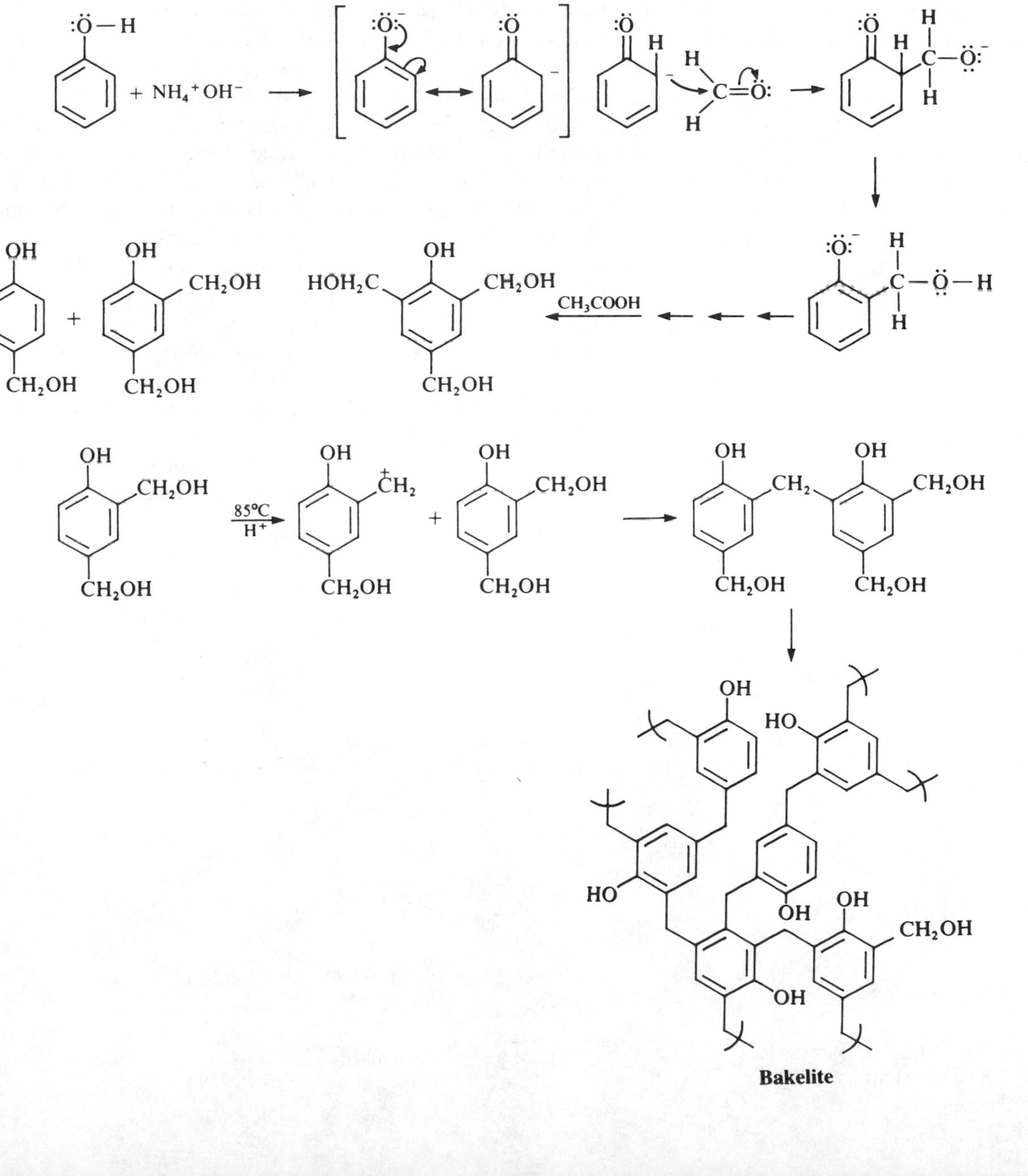

Bakelite

Microscale Procedure

In a 5-mL short-necked, round-bottomed flask place 0.75 g of phenol and 2.5 mL of 37% by weight aqueous formaldehyde solution. The formaldehyde solution contains 10% to 15% methanol, which has been added as a stabilizer to prevent the formaldehyde from polymerizing. Add 0.37 mL of concentrated ammonium hydroxide to the solution, attach an empty distilling column as an air condenser (Fig. 69.2), and reflux it for 5 min beyond the point at which the solution turns cloudy, a total reflux time of about 10 min. In the hood pour the warm solution into a small, disposable tube, for example, a 10 × 75 mm soft glass test tube or vial. *Immediately* clean the flask with a small amount of acetone. Heat the tube and contents inside a steam bath for 10 min to complete the reaction. Allow the mixture to stand at room temperature for 10–20 min. As the mixture cools, droplets of product will sink to the bottom, giving a clear upper layer that is then poured off. If this is made prematurely the yield of product will be much lower. Warm the yellow-orange lower layer on the steam bath, and add acetic acid dropwise (0.5 mL maximum) with thorough mixing until the layer is clear, even when the polymer is cooled to room temperature. Heat the tube on a water bath at 60 to 65°C for 30 min, and then, after placing a wood stick in the polymer to use as a handle, leave the tube, with your name attached, in an 85°C oven overnight or until the next laboratory period. To free the polymer, the tube may need to be broken (Fig. 69.3). A filler such as sawdust or a small piece of cloth can be used to make a stronger polymer. Attach a piece of the polymer to your lab report.

Cleaning Up   The acetone wash should be placed in the organic solvents container. Should it be necessary to destroy formaldehyde, add 25 mL of house-

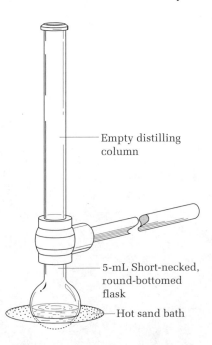

FIG. 69.2 Apparatus for Bakelite synthesis.

Empty distilling column

5-mL Short-necked, round-bottomed flask

Hot sand bath

hold bleach for each milliliter of the solution. After 20 min, flush the solution down the drain. Discard the used glass tube in the nonhazardous solid waste container.

 Macroscale Procedure

CAUTION: Formaldehyde solution is a cancer-suspect agent. Handle the solution with care. Make measurements and carry out the experiment in the hood. The oven should be well ventilated or in a hood.

In a 25-mL round-bottomed flask place 3.0 g of phenol and 10 mL of 37% by weight aqueous formaldehyde solution. The formaldehyde solution contains 10–15% methanol, which has been added as a stabilizer to prevent the formaldehyde from polymerizing. Add 1.5 mL of concentrated ammonium hydroxide to the solution, and reflux it for 5 min beyond the point at which the solution turns cloudy, a total reflux time of about 10 min. In the hood pour the warm solution into a disposable test tube or culture tube, and draw off the upper layer. Immediately clean the flask with a small amount of acetone. Warm the viscous milky lower layer on the steam bath, and add acetic acid dropwise with thorough mixing until the layer is clear, even when the polymer is cooled to room temperature. Heat the tube on a water bath at 60–65°C for 30 min. Then, after placing a wood stick in the polymer to use as a handle, leave the tube, with your name attached, in an 85°C oven overnight or until the next laboratory period. To free the polymer the tube may need to be broken. Attach a piece of the polymer to your lab report.

MICROSCALE AND MACROSCALE

3. Nylon by Ring-Opening Polymerization[3]

$$\underset{\mathbf{1}}{\overset{}{\text{N}^-}} + \underset{\mathbf{2}}{\overset{}{\text{N}-\text{COCH}_3}} \longrightarrow \text{N}-\text{CO}-(\text{CH}_2)_5-\bar{\text{N}}-\text{COCH}_3$$

$$\text{N}-\text{CO}-(\text{CH}_2)_5-\bar{\text{N}}-\text{COCH}_3 + \overset{}{\text{NH}} \longrightarrow$$

$$\text{N}-\text{CO}-(\text{CH}_2)_5-\text{NH}-\text{COCH}_3 + \underset{\mathbf{4}}{\overset{}{\text{N}^-}}$$
$$\mathbf{3}$$

While interfacial polymerization in the manner described above is not a commercial process, the ring-opening of caprolactam is. The nylon produced, nylon 6, is used extensively in automobile tire cord and for gears and bearings in small mechanical devices.

3. L. J. Mathias, R. A. Vaidya, and J. B. Canterbury, *J. Chem. Ed.,* **61,** 805 (1984).

FIG. 69.3 To free the polymer, turn the thumbscrew to break the tube. Wear glasses. You may wish to wrap the tube in a piece of cloth.

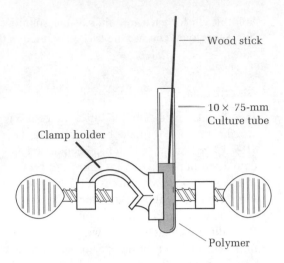

Wood stick

10 × 75-mm Culture tube

Clamp holder

Polymer

The catalyst used in this reaction is sodium hydride; therefore, this is referred to as an *anionic polymerization.* The sodium hydride removes the acidic lactam proton to form an anion **(1)** that attacks the coinitiator, acetylcaprolactam **(2),** which has an electron-attracting acetyl group attached to the nitrogen. The ring of the acetylcaprolactam is attacked by the anion, and the acetylcaprolactam ring opens, forming a substituted caprolactam **(3)** that still has an electron-attracting group attached to nitrogen. A proton transfer reaction occurs, generating a new caprolactam anion **(4),** and **4** then attacks **3,** and so on.

Ordinarily anionic polymerizations must be run in the absence of oxygen, but the addition of polyethylene glycol serves to complex with the sodium ion just as 18-crown-6 does and enhances the catalytic activity of the sodium hydride.

Microscale Procedure

Into a disposable 10 × 75 mm test tube place 1 g of caprolactam, 60 mg of polyethylene glycol, and 1 small drop of *N*-acetylcaprolactam. Heat the mixture, and as soon as it has melted, remove it from the heat and add 20 mg of gray (not white) sodium hydride (50% dispersion in mineral oil). Mix the catalyst with the reactants by stirring with a Pasteur pipette, and heat the mixture *rapidly* to boiling (200–230°C). This should take place over a 2-min period. Polymerization takes place rapidly, as indicated by an increase in viscosity. If polymerization has not occurred within 3 min, remove the tube from the heat, cool it somewhat, and add another 15 mg of sodium hydride. When the solution is so viscous that it will barely flow, insert a wood stick and with help from a neighbor draw fibers from the melt. After it cools, the nylon 6 can usually be removed from the tube as one cylindrical piece. Try dissolving a piece of the nylon or the fibers in various solvents. Test the physical properties of the fibers by stretching them to the breaking point. Describe your observations, and attach a piece of fiber to your report. Do not forget to turn off the heater.

Cleaning Up   Place the used tube, if coated with polymer, in the nonhazardous solid waste container. Should it be necessary to destroy sodium hydride, add it to excess 1-butanol (38 mL/g of hydride). After the reaction has ceased, cautiously add water; then dilute with more water, and flush the solution down the drain.

Macroscale Procedure

Set the heater to its maximum setting on your sand bath. The sand must be quite hot before beginning this experiment. Heating can also be done over a small Bunsen burner flame.

Into a disposable 4-in. test tube place 4 g of caprolactam, 0.25 g of polyethylene glycol, and 2 drops of *N*-acetylcaprolactam. Heat the mixture, and as soon as it has melted remove it from the heat and add 50 mg of gray (not white) sodium hydride (50% dispersion in mineral oil). Mix the catalyst with the reactants by stirring with a Pasteur pipette, and heat the mixture *rapidly* to boiling (200–230°C). This should take place over a 2-min period. Polymerization takes place rapidly, as indicated by an increase in viscosity. If polymerization has not occurred within 3 min remove the tube from the heat, cool it somewhat, and add another 50 mg of sodium hydride. When the solution is so viscous that it will barely flow, insert a wood stick and with help from a neighbor draw fibers (monofilament line) several feet long from the melt. After it cools, the nylon-6 can usually be removed from the tube as one cylindrical piece. Try dissolving a piece of the nylon or the fibers in various solvents. Test the physical properties of the fibers by stretching them to the breaking point. Describe your observations, and attach a piece of fiber to your report. Don't forget to turn off the heater.

Cleaning Up   Place the used tube, if coated with polymer, in the nonhazardous solid waste container. Should it be necessary to destroy sodium hydride, add it to excess 1-butanol (38 mL/g of hydride). After the reaction has ceased, cautiously add water, then dilute with more water and flush the solution down the drain.

 **MICROSCALE**

4. Polystyrene by Free-Radical Polymerization

Polystyrene, the familiar crystal-clear brittle plastic used to make disposable drinking glasses and, when foamed, the lightweight white cups for hot drinks, is usually made by free-radical polymerization. Commercially an initiator is not used because polymerization begins spontaneously at elevated temperatures. At lower temperatures a variety of initiators could be used [e.g., 2,2′-azobis-(2-methylpropionitrile), which was used in the free-radical chlorination of 1-chlorobutane]. In this experiment we use benzoyl peroxide as the initiator. On mild heating it splits into two benzoyloxy radicals

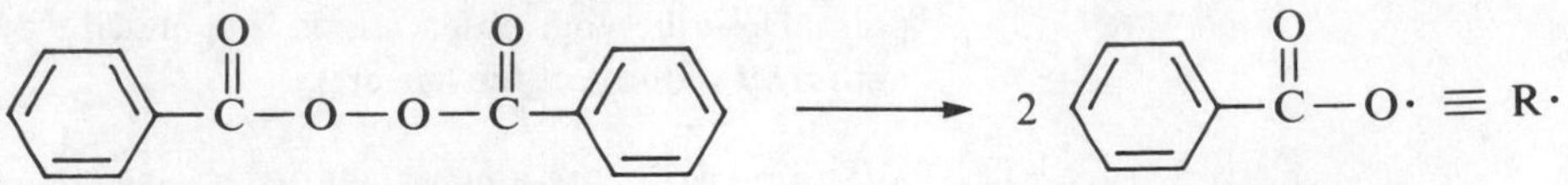

Benzoyl peroxide
MW 242.23
dec 106°C

Benzoyloxy radical

which react with styrene through initiation, propagation, and termination steps to form polystyrene:

Initiation:

$$R\cdot + CH_2{=}CH(C_6H_5) \longrightarrow RCH_2CH(C_6H_5)\cdot$$

Propagation:

$$RCH_2CH(C_6H_5)\cdot + CH_2{=}CH(C_6H_5) \longrightarrow RCH_2CH(C_6H_5){-}CH_2CH(C_6H_5)\cdot + CH_2{=}CH(C_6H_5) \longrightarrow RCH_2CH(C_6H_5){-}CH_2CH(C_6H_5){-}CH_2CH(C_6H_5)\cdot\text{, etc.}$$

Termination:

$$2\,R\cdot \longrightarrow R{-}R,\quad RCH_2CH(C_6H_5)\cdot + R\cdot \longrightarrow RCH_2CH(C_6H_5){-}R,\quad 2\,RCH_2CH(C_6H_5)\cdot \longrightarrow RCH_2CH(C_6H_5){-}CH(C_6H_5)CH_2R$$

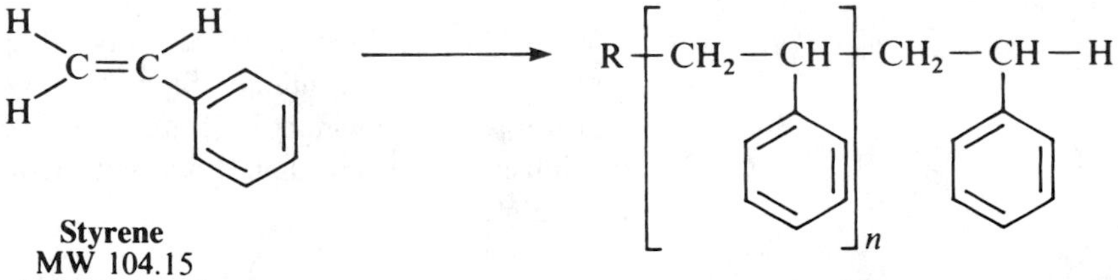

Styrene
MW 104.15
bp 145–146°C

Polystyrene
MW 300,000–25,000,000

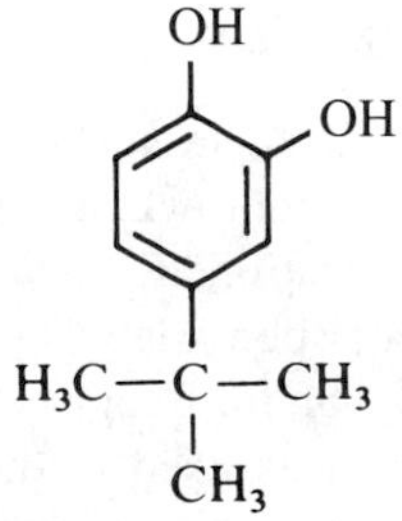

4-t-Butylcatechol

The final polymer has about 3000 monomer units in a single chain, but it can be made with up to 240,000 monomers per chain.

To prevent styrene from polymerizing in the bottle in which it is sold, the manufacturer adds 10 to 15 parts per million of 4-*tert*-butylcatechol, a radical inhibitor (a particularly good chain terminator). This must be removed by passing the styrene through a column of alumina before the styrene can be polymerized.

Synthesis of Polystyrene

Procedure

Styrene is flammable, an irritant, and has a bad odor. Work with it in the hood.

CAUTION: Benzoyl peroxide is flammable and may explode on heating or on impact. There is no need for more than a gram or two in the laboratory at any one time. It should be stored in and dispensed from waxed paper containers and not in metal or glass containers with screw-cap lids.

In a Pasteur pipette loosely place a very small piece of cotton, then fill the pipette half full with alumina. Add to the top of the pipette 1.5 mL of styrene, and collect 1 mL in a disposable 10 × 75-mm test tube. Add to the tube 50 mg of benzoyl peroxide and a wood boiling stick, and heat the tube over a hot sand bath. When the temperature reaches about 135°C, polymerization begins, and, because it is an exothermic process, the temperature rises. Keep the reaction under control by cautious heating. The temperature rises, perhaps to 180°C, well above the boiling point of styrene (145°C); the viscosity also increases. Pull the wood stick from the melt from time to time to form fibers; when a cool fiber is found to be brittle, remove the tube from the heat. Alternatively, heat the tube on a steam bath for 90 min with the wood stick in place. Allow the tube to cool; then cool it in ice. Often the polymer will shrink enough so that it can be pulled from the tube. Alternatively, break the tube. Should the polymer be sticky the polymerization can be completed in an oven overnight at a temperature of about 85°C.

Cleaning Up   Shake the alumina out of the pipette, and place it in the styrene/alumina hazardous waste container. Clean up spills of benzoyl peroxide immediately. It can be destroyed by reacting each gram with 1.4 g of sodium iodide in 28 mL of acetic acid. After 30 min the brown solution should be neutralized with sodium carbonate, diluted with water, and flushed down the drain.

Part 2. Recycling Plastics

2.2 lbs (pounds) =
1 kg (kilogram)

Garbage. Trash. Hardly the concern of the organic chemistry lab. Or is it? In this country, we generate 320 billion pounds of trash a year. Into dumps goes 85% of this, and these landfills are filling up, fast. Plastics make up 7% of this waste. In any given year in the United States, we produce about 55 billion pounds of plastic, of which 22 billion pounds get discarded. More than half of this plastic has been used in packaging. By volume, this waste plastic makes up 30% of municipal solid waste, but only 1% of this is recycled. Some plastic can be recycled simply by melting it and remolding it; other polymers must be treated chemically.

Plastics must be carefully sorted before recycling because some are incompatible with others. The Society of the Plastics Industry has adopted a coding system in which manufacturers place an alphanumeric identification code at or near the bottom of each container to identify the type of plastic resin being used. PET is the leading recycled plastic.

In this experiment, a plastic soft-drink bottle is converted to fiberglass-reinforced polyester laminates and polyurethane foam.

Polyethylene terephthalate (PET) is the plastic from which clear plastic bottles for soft drinks are made by blow-molding. It is made of repeating units of ethylene glycol and terephthalic acid. Each repeating unit has a molecular weight of 192, with the number of repeating units more than 100, giving the polymer a molecular weight of about 20,000. When this polymer is formed into sheets, it is called *Mylar,* and when it is spun into fibers, it is called *Dacron.* In the United Kingdom it is called *Terylene.* Many millions of pounds of it are produced each year, and most of this ends up in waste dumps wherever it is not recycled. As we become increasingly aware of the environmental toll such waste entails, we are turning to recycling. This experiment demonstrates the commercial processes by which PET bottles are chemically recycled.

The process is fairly straightforward. The polymer is broken down (depolymerized) into small fragments that can then be used as the starting material for making a new polymer. In the case of PET, the new polymer could be fiberglass or polyurethane foam. It cannot be blow-molded again to make new beverage bottles because of possible contamination.

The chemical structure of polyethylene terephthalate (PET) is

Polyethylene
terephthalate
PETE

High-density
polyethylene
HDPE

Polyvinyl
chloride
V

Low-density
polyethylene
LDPE

Polypropylene
PP

Polystyrene
PS

Other
OTHER

$$\left[\!\!-CH_2CH_2O\overset{\displaystyle O}{\overset{\displaystyle \|}{C}}\!-\!\!\bigcirc\!\!-\overset{\displaystyle O}{\overset{\displaystyle \|}{C}}O-\!\!\right]_n$$

It is an ester. Low-molecular-weight esters are prepared and hydrolyzed according to the following equilibrium reaction, which can be either acid- or base-catalyzed (see Chapter 40):

$$\overset{O}{\underset{\|}{RCOH}} + R'OH \underset{}{\overset{H^+}{\rightleftharpoons}} \overset{O}{\underset{\|}{RCOR'}} + H_2O$$

If you put small pieces of PET in a flask with either aqueous acid or base and reflux the mixture, there will be no apparent reaction. It is remarkably resistant to hydrolysis. Try it. Esters undergo a related reaction called *transesterification* that can be used to depolymerize PET:

$$\overset{O}{\underset{\|}{RCOR'}} + R''OH \underset{}{\overset{cat}{\rightleftharpoons}} \overset{O}{\underset{\|}{RCOR''}} + R'OH$$

When applied to PET, the reaction produces a mixture containing three esters when the alcohol is diethylene glycol:

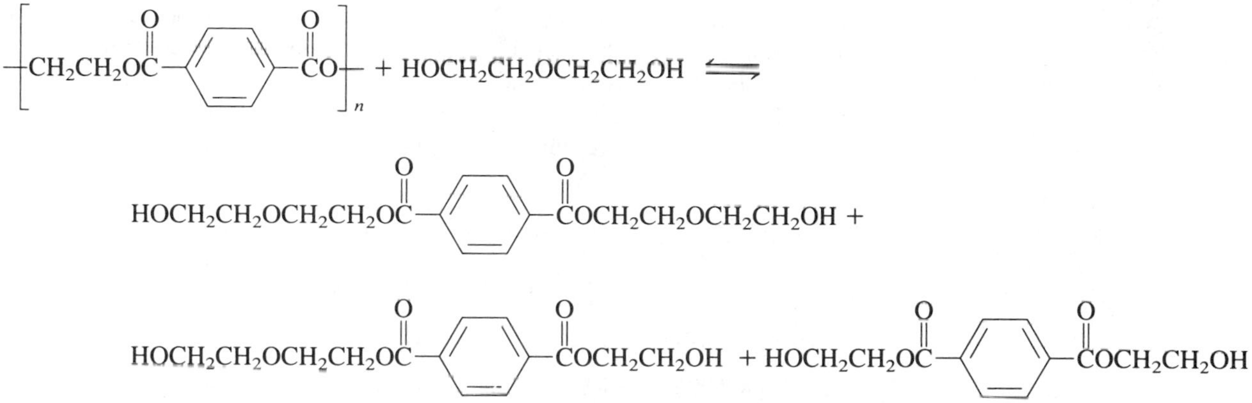

$$\left[CH_2CH_2O\overset{O}{\underset{\|}{C}} - \langle \bigcirc \rangle - \overset{O}{\underset{\|}{C}}O \right]_n + HOCH_2CH_2OCH_2CH_2OH \rightleftharpoons$$

$$HOCH_2CH_2OCH_2CH_2O\overset{O}{\underset{\|}{C}} - \langle \bigcirc \rangle - \overset{O}{\underset{\|}{C}}OCH_2CH_2OCH_2CH_2OH +$$

$$HOCH_2CH_2OCH_2CH_2O\overset{O}{\underset{\|}{C}} - \langle \bigcirc \rangle - \overset{O}{\underset{\|}{C}}OCH_2CH_2OH + HOCH_2CH_2O\overset{O}{\underset{\|}{C}} - \langle \bigcirc \rangle - \overset{O}{\underset{\|}{C}}OCH_2CH_2OH$$

At the end of the reaction, ethylene glycol and unreacted diethylene glycol are also present. This reaction is catalyzed by a mixture of manganese acetate and manganese carbonate.

Polyurethane Foam from Plastic Beverage Bottles

The familiar foam-rubber cushions that you sit on are more than likely made of polyurethane. It is also widely used in the building trades as a rigid insulating material in the form of lightweight boards.

A urethane (also called a carbamate) is a functional group that looks like an amide on one side and an ester on the other. It is made by reacting an isocyanate with an alcohol.

$$R-N=C=O + R'-OH \rightleftharpoons R-NH-\overset{O}{\underset{\|}{C}}-O-R'$$

An isocyanate   **A urethane**

Polyurethanes, like many other polymers, are made from difunctional starting materials:

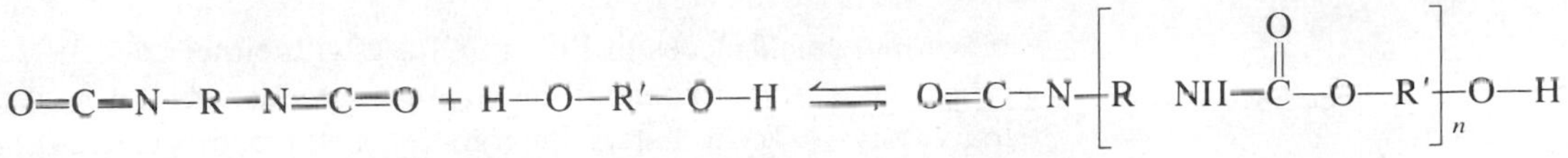

$$O=C=N-R-N=C=O \ + \ H-O-R'-O-H \ \rightleftharpoons \ O=C=N \left[R \quad NII-\overset{\displaystyle O}{\overset{\|}{C}}-O-R' \right]_n O-H$$

A polyurethane

Isocyanates react with water to form an amine and carbon dioxide:

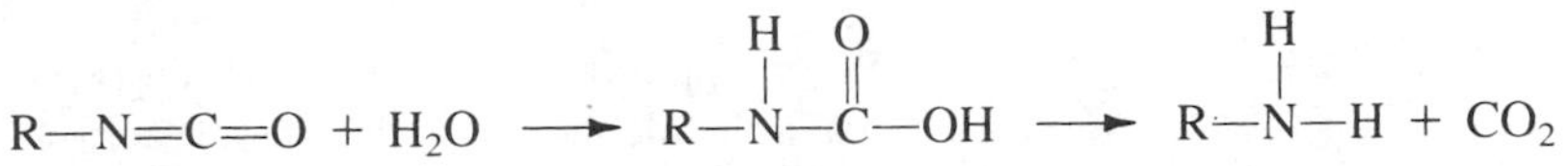

$$R-N=C=O \ + \ H_2O \ \longrightarrow \ R-\overset{\displaystyle H}{\underset{\displaystyle}{N}}-\overset{\displaystyle O}{\overset{\|}{C}}-OH \ \longrightarrow \ R-\overset{\displaystyle H}{\underset{\displaystyle}{N}}-H \ + \ CO_2$$

For this reason, a small amount of water can be added during the polymerization step. The carbon dioxide is one means of forming the bubbles in the foam.

As noted earlier, commercial polyurethane can be rigid or flexible, depending on the nature of the R and R′ group. In this experiment, the foamed polymer will be rigid. The most common diisocyanate is tolylene diisocyanate.

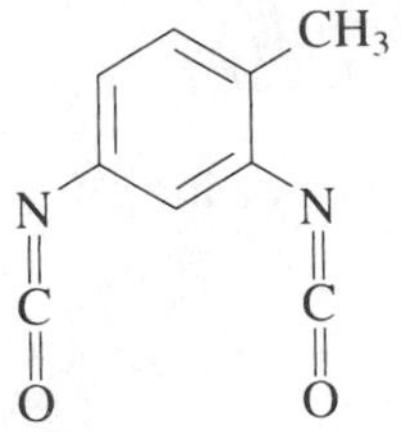

Tolylene 2,4-diisocyanate

Experiments

1. Depolymerization of PET with Diethylene Glycol

Microscale Procedure

Into a 13×100 mm test tube, weight 1.74 g of PET from a clear soft-drink bottle that has been cut into $\frac{1}{4}$-in. squares or comes from a commercial shredder.[4] Add 3 mg of $Mn(OAc)_2$ and 3 mg of $MnCO_3$ followed by 1.26 g of diethylene glycol. Add a boiling chip, and reflux the mixture. Note the time at which all the chips of PET disappear. If the mixture is cooled at this point, it will have a cloudy white appearance, because the material still has a high molecular weight and comes out of solution. Continue to reflux for a total of 45 min, at the end of which time the mixture should remain clear after cooling. At this time, the glycolysis (transesterification) is finished, and the tube contains the mixture of three esters shown above. This material will be used for making fiberglass and polyurethane foam in the next two experiments.

Look up information on PET. How much is produced each year? Are Mylar and Dacron simply different forms of this polymer? See if you can discover anything about the use of manganese acetate and carbonate to catalyze a reaction of this type.

---

4. In places where a deposit on bottles is required, there is often a machine that will shred the bottles and return the deposit. Obtain this shredded material from the grocer, and treat it first with toluene to remove the label glue, and then, after drying, separate it by flotation in water. The paper and black polyethylene base of the bottle will float, the aluminum cap will sink, and the PET (density about 1) will be suspended in the water.

Macroscale Procedure

Into a 20 × 150-mm test tube, weight 7.5 g of PET from a clear soft-drink bottle that has been cut into $\frac{1}{4}$-in. squares or comes from a commercial shredder.[5] Add 18 mg of $Mn(OAc)_2$ or $Sb(OAc)_3$ and 18 mg of $MnCO_3$ followed by 5.0 g of diethylene glycol. Add a boiling chip, and reflux the mixture. Poke the shredded plastic down into the liquid as the reaction proceeds. Note the time at which all the chips of PET disappear. If the mixture is cooled at this point, it will have a cloudy white appearance, because the material still has a high molecular weight and comes out of solution. Continue to reflux for a total of 45 min, at the end of which time the mixture should remain clear after cooling. At this time, the glycolysis (transesterification) is finished, and the tube contains the mixture of three esters shown earlier. This material will be used for making fiberglass and polyurethane foam in the next two experiments.

Look up information on PET. How much is produced each year? Are Mylar and Dacron simply different forms of this polymer? See if you can discover anything about the use of manganese acetate and carbonate or antimony acetate to catalyze a reaction of this type.

 **MACROSCALE**

2. Prepolymer for Polyurethane Foam

$$CH_2 - O - \overset{\overset{\text{O}}{\|}}{C}(CH_2)_7CH = CHCH_2\overset{\overset{\text{OH}}{|}}{CH}(CH_2)_5CH_3$$
$$CH - O - \overset{\overset{\text{O}}{\|}}{C}(CH_2)_7CH = CHCH_2\overset{\overset{\text{OH}}{|}}{CH}(CH_2)_5CH_3$$
$$CH_2 - O - \overset{\overset{\text{O}}{\|}}{C}(CH_2)_7CH = CHCH_2\overset{\overset{\text{OH}}{|}}{CH}(CH_2)_5CH_3$$

Castor oil
Mostly the triglyceride
of ricinoleic acid

CAUTION: Do not carry out this experiment unless good hood facilities are available!

Into a 100-mL round-bottomed flask, place 21.5 g of castor oil and 7.5 g of transesterified PET from the preceding experiment. Add a magnetic stirring bar and thermometer, and with vigorous stirring in the hood, add 27.5 g of tolylene 2,4-diisocyanate (*Caution!*) all at once from an Erlenmeyer flask. Over about the next 15 min, the temperature of the mixture rises to 80°C. Some foaming takes place. Once the temperature starts to drop, place the flask in a Thermowell

---

5. In places where a deposit on bottles is required, there is often a machine that will shred the bottles and return the deposit. Obtain this shredded material from the grocer, and treat it first with toluene to remove the label glue, and then, after drying, separate it by flotation in water. The paper and black polyethylene base of the bottle will float, the aluminum cap will sink, and the PET (density about 1) will be suspended in the water.

and heat the mixture at 120°C for 1 h. Remove it from the heat, and cool it in a water bath to room temperature. This prepolymer should contain approximately 15% free isocyanate. It will keep overnight or longer, if refrigerated.

CAUTION: Tolylene diisocyanate is extremely toxic, a strong irritant to the respiratory tract and skin, and a cancer-suspect agent. Neutralize spills with 5% aqueous ammonia solution. Wash from the skin with 30% aqueous isopropyl alcohol and then with soap and water. Handle prepolymer with as much care as the starting diisocyanate.

MICROSCALE

3. Polyurethane Foam

To prepare foamed polyurethane, in a small paper cup mix 0.175 g of buffered catalyst (prepared from 7 g of diethylaminoethanol, 3.32 mL of concentrated hydrochloric acid, and 6.6 mL of water), one drop of Dow Corning silicone oil, and 0.55 mL of water. To this, add 5 g of the prepolymer mix, and stir the mixture thoroughly with a metal spatula. Carry out this operation in the hood. Leave the cup with your name on it in the hood until the next lab period.

Fiberglass-Reinforced Polyester from Plastic Beverage Bottles

PET resin in the form of shredded particles of the clear plastic from beverage bottles is refluxed in the presence of a catalyst to effect an ester interchange reaction with propylene glycol. This glycolysis produces a mixture of three esters, each of which is terminated with a hydroxyl group.

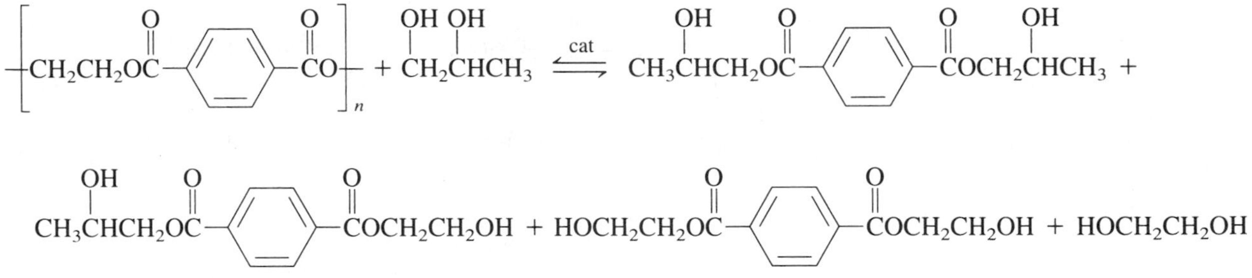

Alcohols react with anhydrides to give esters. In a very simple example, acetic anhydride reacts with ethyl alcohol to give the ester ethyl acetate and acetic acid.

If the anhydride is maleic anhydride, then the resulting ester terminates in a carboxyl group, which, in an equilibrium reaction, can react with another molecule of alcohol to give an unsaturated ester and water.

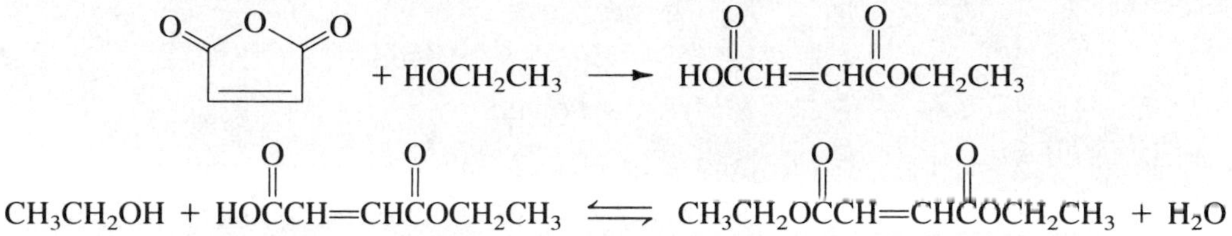

$$\text{(maleic anhydride)} + HOCH_2CH_3 \longrightarrow HOCCH=CHCOCH_2CH_3$$

$$CH_3CH_2OH + HOCCH=CHCOCH_2CH_3 \rightleftharpoons CH_3CH_2OCCH=CHCOCH_2CH_3 + H_2O$$

This reaction is employed with the mixture of diols produced in the glycolysis of PET. When this mixture reacts with maleic anhydride and water is allowed to distill from the reaction mixture, the result is an unsaturated polymeric ester:

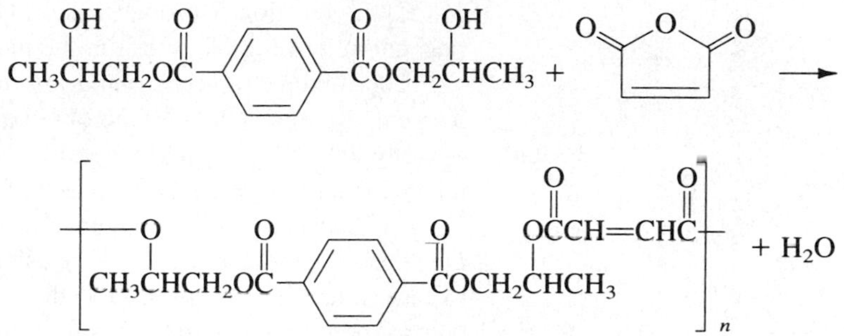

$$CH_3CHCH_2OC-\underset{OH}{\bigcirc}-COCH_2CHCH_3 + \text{(maleic anhydride)} \longrightarrow$$

$$\left[-O \quad CH_3CHCH_2OC-\bigcirc-COCH_2CHCH_3 \quad O \quad OCCH=CHC- \right]_n + H_2O$$

This unsaturated polyester is of fairly low molecular weight and thus is a liquid at room temperature. It can be crosslinked into a rigid, high molecular weight ester with styrene in a radical-catalyzed reaction:

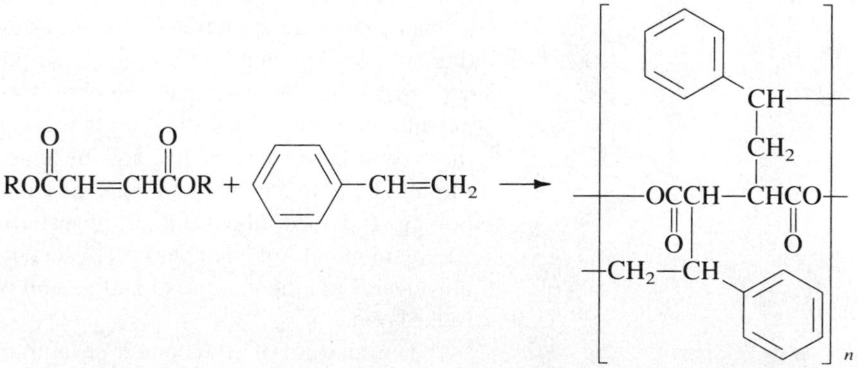

$$ROCCH=CHCOR + \bigcirc-CH=CH_2 \longrightarrow \left[\cdots \right]_n$$

This polyester, when reinforced with fiberglass, can be molded and polymerized in place to form such things as automobile bodies and boats ranging up to 100 ft in length.

4. Conversion of PET into Fiberglass-Reinforced Polyester

Microscale Procedure

To 3 g of PET cut into $\frac{1}{4}$-in. (6-mm) squares or material from a commercial shredder, add 1.4 g of propylene glycol, 5 mg of manganese acetate tetrahydrate, and 2 mg of manganese carbonate. Insert a thermometer into the mixture as far as possible, and heat the mixture on the sand bath until the propylene glycol boils (187°C). A rapid increase in the temperature is taken as an indication that the glycolysis reaction is over. When all the particles have disappeared and the temperature reaches 200°C, cool the mixture to about 150°C and add 1.56 g of maleic anhydride. Heat the mixture to boiling, and allow about 0.3 mL of water to escape (you probably will not see it). Again monitor the temperature. When it reaches 215°C, the reaction is probably over. Cool the mixture to 140°C, and with stirring, add to it 4.6 g of styrene in the hood.

The mixture of styrene and unsaturated polyester resin can now be polymerized. To do this, add 30 mg of cobalt octoate, stir well, and then add 100 mg of methyl ethyl ketone peroxide (the radical catalyst) with thorough stirring. The mixture can be poured onto aluminum foil to polymerize, or it can be poured onto 2.5 g of fiberglass on aluminum foil, the air squeezed out, and then allowed to polymerize in the hood. This fiberglass-reinforced polyester should be about $\frac{1}{8}$ in. (3 mm) thick. The original procedure says that these castings were allowed to cure at room temperature for 16 h, followed by a post-cure at 70°C for 1 h and a 100°C cure for 1 h. You could simply place the unpolymerized material in a well-ventilated oven at 80°C overnight.

Macroscale Procedure

Into a 20 × 150-mm test tube, place 6 g of PET cut into $\frac{1}{4}$-in. (6-mm) squares or material from a commercial shredder. Add 2.8 g of propylene glycol, 10 mg of manganese acetate tetrahydrate, and 4 mg of manganese carbonate. Insert a thermometer into the mixture as far as possible, and heat the mixture on the sand bath until the propylene glycol boils (187°C). A rapid increase in the temperature is taken as an indication that the glycolysis reaction is over. When all the particles have disappeared and the temperature reaches 200°C, cool the mixture to about 150°C and add 3.12 g of maleic anhydride. Heat the mixture to boiling, and allow about 0.6 mL of water to escape (you probably will not see it). Again monitor the temperature. When it reaches 215°C, the reaction is probably over. Cool the mixture to 140°C, and with stirring, add to it 9.2 g of styrene in the hood.

The mixture of styrene and unsaturated polyester resin can now be polymerized. To do this, add 60 mg of cobalt acetate (*Caution!*), stir well, and then add 0.20 g of methyl ethyl ketone peroxide (the radical catalyst) with thorough stirring. The mixture can be poured onto aluminum foil to polymerize or it can be poured onto 5 g of fiberglass on aluminum foil, the air squeezed out, and then allowed to polymerize in the hood. This fiberglass-reinforced polyester should

CAUTION: Cobalt acetate is a cancer-suspect agent and a mutagen. Handle with care.

be about $\frac{1}{8}$ in. (3 mm) thick. The original procedure says that these castings were allowed to cure at room temperature for 16 h, followed by a post-cure at 70°C for 1 h and a 100°C cure for 1 h. You could simply place the unpolymerized material in a well-ventilated oven at 80°C overnight.

Questions

1. What might the products be from an explosion of smokeless gunpowder? How many moles of CO_2 and H_2O would come from 1 mol of trinitroglucose? Does the molecule contain enough oxygen for the production of these two substances?

2. Write a balanced equation for the reaction of sebacoyl chloride with water.

3. In the final step in the synthesis of Bakelite, the partially polymerized material is heated at 85°C for several hours. What other product is produced in this reaction?

4. Give the detailed mechanism of the first step of the ring-opening polymerization of caprolactam.

Surfing the Web

Polyvinyl chloride:

http://www.basf-ag.basf.de/basf/html/e/produkte/gebiete/kunstst/products/pvc_1.htm

The history of Teflon:

http://www.dupont.com/teflon/100.html

For updated information visit:

www.mtholyoke.edu/courses/kwilliam/microscale.shtml
or
www.hmco.com/hmco/college/chemistry/Home.html

70

Qualitative Organic Analysis

Prelab Exercise: In the identification of an unknown organic compound, certain procedures are more valuable than others. For example, much more information is obtained from an infrared spectrum than from a refractive index measurement. Outline, in order of priority, the steps you will employ in identifying your unknown.

Identification and characterization of the structures of unknown substances are an important part of organic chemistry. It is often, of necessity, a micro process, e.g., in drug analyses. It is sometimes possible to establish the structure of a compound on the basis of spectra alone (IR, UV, and NMR), but these spectra must usually be supplemented with other information about the unknown: physical state, elementary analysis, solubility, and confirmatory tests for functional groups. Conversion of the unknown to a solid derivative of known melting point will often provide final confirmation of structure.

However, before spectra are run, other information about the sample must be obtained. Is it homogeneous (test by thin-layer, gas, or liquid chromatography)? What are its physical properties (melting point, boiling point, color, solubility in various solvents)? Is it soluble in a common NMR solvent? It also might be necessary to determine what elements are present in the sample and its percentage composition (mass spectroscopy).

Nevertheless, an organic chemist can often identify a sample in a very short time by performing solubility tests and some simple tests for functional groups, coupled with spectra that have not been compared with a database. Conversion of the unknown to a solid derivative of known melting point will often provide final confirmation of structure. This chapter provides the information needed to carry out this type of qualitative analysis of an organic compound.

Procedures

All experiments in this chapter can, if necessary, be run on two to three times the indicated quantities of material.

Physical State

Check for Sample Purity

Distill or recrystallize as necessary. Constant boiling point and sharp melting point are indicators, but beware of azeotropes and eutectics. Check homogeneity by TLC, gas, HPLC, or paper chromatography.

Note the Color

Common colored compounds include nitro and nitroso compounds (yellow), α-diketones (yellow), quinones (yellow to red), azo compounds (yellow to red), and polyconjugated olefins and ketones (yellow to red). Phenols and amines are often brown to dark-purple because of traces of air oxidation products.

CAUTION: Do not taste an unknown compound. To note the odor, cautiously smell the cap of the container and do it only once.

Note the Odor

Some liquid and solid amines are recognizable by their fishy odors; esters are often pleasantly fragrant. Alcohols, ketones, aromatic hydrocarbons, and aliphatic olefins have characteristic odors. On the unpleasant side are thiols, isonitriles, and low molecular weight carboxylic acids.

Make an Ignition Test

Heat a small sample on a spatula; first hold the sample near the side of a microburner to see if it melts normally and then burns. Heat it in the flame. If a large ashy residue is left after ignition, the unknown is probably a metal salt. Aromatic compounds often burn with a smoky flame.

Spectra

Obtain infrared and nuclear magnetic resonance spectra following the procedures of Chapters 12 and 13. If these spectra indicate the presence of conjugated double bonds, aromatic rings, or conjugated carbonyl compounds, obtain the ultraviolet spectrum following the procedures of Chapter 14. Interpret the spectra as fully as possible by reference to the sources cited at the end of the various spectroscopy chapters.

Explanation

Elementary Analysis, Sodium Fusion

This method for detection of nitrogen, sulfur, and halogen in organic compounds depends on the fact that fusion of substances containing these elements with sodium yields NaCN, Na_2S, and NaX (X = Cl, Br, I). These products can, in turn, be readily identified. The method has the advantage that the most usual elements other than C, H, and O present in organic compounds can all be detected following a single fusion, although the presence of sulfur sometimes interferes with the test for nitrogen. Unfortunately, even in the absence of sulfur, the test for nitrogen is sometimes unsatisfactory (nitro compounds in particular). Practicing organic chemists rarely perform this test. Either they know what elements their unknowns contain, or they have access to a mass spectrometer or atomic absorption instrument.

Rarely performed by professional chemists

　　Place a 3-mm cube of sodium[1] (30 mg, no more)[2] in a 10 × 75-mm Pyrex

Notes for the instructor

1. Sodium spheres $\frac{1}{16}''$ to $\frac{1}{4}''$ are convenient.

2. A dummy 3-mm cube of rubber can be attached to the sodium bottle to indicate the correct amount.

FIG. 70.1 Sodium fusion, just prior to addition of sample.

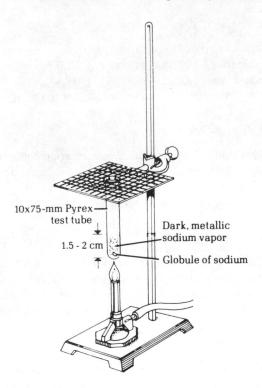

10x75-mm Pyrex test tube

1.5 - 2 cm

Dark, metallic sodium vapor

Globule of sodium

CAUTION: Manipulate sodium with a knife and forceps; never touch it with the fingers. Wipe it free of kerosene with a dry towel or filter paper; return scraps to the bottle or destroy scraps with methyl or ethyl alcohol, never with water. Safety glasses! Hood!

Do not use $CHCl_3$ or CCl_4 as samples in sodium fusion. They react extremely violently.

test tube, and support the tube in a vertical position (Fig. 70.1). Have a microburner with small flame ready to move under the tube, place an estimated 20 mg of solid on a spatula or knife blade, put the burner in place, and heat until the sodium first melts and then vapor rises 1.5–2.0 cm in the tube. Remove the burner, and at once drop the sample onto the hot sodium. If the substance is a liquid add 2 drops of it. If there is a flash or small explosion the fusion is complete; if not, heat briefly to produce a flash or a charring. Then let the tube cool to room temperature, be sure it is cold, add a drop of methanol, and let it react. Repeat until 10 drops have been added. With a stirring rod break up the char to uncover sodium. When you are sure that all sodium has reacted, empty the tube into a 13 × 100-mm test tube, hold the small tube pointing away from you or a neighbor, and pipette into it 1 mL of water. Boil and stir the mixture, and pour the water into the larger tube; repeat with 1 mL more water. Then transfer the solution with a Pasteur pipette to a 2.5-cm funnel (fitted with a fluted filter paper) resting in a second 13 × 100-mm test tube. Portions of the alkaline filtrate are used for the tests that follow.

(a) Nitrogen

The test is done by boiling a portion of the alkaline solution from the solution fusion with iron(II) sulfate and then acidifying. Sodium cyanide reacts with iron(II) sulfate to produce ferrocyanide, which combines with iron(III) salts,

Run each test on a known and an unknown.

inevitably formed by air oxidation in the alkaline solution, to give insoluble Prussian Blue, $NaFe[Fe(CN)_6]$. Iron(II) and iron(III) hydroxide precipitate along with the blue pigment but dissolve on acidification.

Place 50 mg of powdered iron(II) sulfate (this is a large excess) in a 10×75-mm test tube, add 0.5 mL of the alkaline solution from the fusion, heat the mixture gently with shaking to the boiling point, and then—without cooling—acidify with dilute sulfuric acid (hydrochloric acid is unsatisfactory). A deep-blue precipitate indicates the presence of nitrogen. If the coloration is dubious, filter through a 2.5-cm funnel and see if the paper shows blue pigment.

Cleaning Up   The test solution should be diluted with water and flushed down the drain.

(b) Sulfur

$Na_2(NO)Fe(CN)_6 \cdot 2H_2O$

Sodium nitroprusside

1. Dilute one drop of the alkaline solution with 1 mL of water, and add a drop of sodium nitroprusside; a purple coloration indicates the presence of sulfur.
2. Prepare a fresh solution of sodium plumbite by adding 10% sodium hydroxide solution to 0.2 mL of 0.1 M lead acetate solution until the precipitate just dissolves, and add 0.5 mL of the alkaline test solution. A black precipitate or a colloidal brown suspension indicates the presence of sulfur.

Cleaning Up   The test solution should be diluted with water and flushed down the drain.

(c) Halogen

Differentiation of the halogens

Do not waste silver nitrate.

Acidify 0.5 mL of the alkaline solution from the fusion with dilute nitric acid (indicator paper) and, if nitrogen or sulfur has been found present, boil the solution (hood) to expel HCN or H_2S. On addition of a few drops of silver nitrate solution, halide ion is precipitated as silver halide. Filter with minimum exposure to light on a 2.5-cm funnel, wash with water, and then with 1 mL of concentrated ammonia solution. If the precipitate is white and readily soluble in ammonium hydroxide solution it is AgCl; if it is pale yellow and not readily soluble it is AgBr; if it is yellow and insoluble it is AgI. Fluorine is not detected in this test since silver fluoride is soluble in water.

Cleaning Up   The test solution should be diluted with water and flushed down the drain.

> Run tests on knowns in parallel with unknowns for all qualitative organic reactions. In this way, interpretations of positive reactions are clarified and defective test reagents can be identified and replaced.

Beilstein Test for Halogens

A fast, easy, reliable test

Heat the tip of a copper wire in a burner flame until no further coloration of the flame is noticed. Allow the wire to cool slightly, then dip it into the unknown (solid or liquid), and again heat it in the flame. A green flash is indicative of chlorine, blue-green of bromine, and blue of iodine; fluorine is not detected because copper fluoride is not volatile. The Beilstein test is very sensitive; halogen-containing impurities may give misleading results. Run the test on a compound known to contain halogen for comparison with your unknown.[3]

Solubility Tests

Weigh and measure carefully.

Like dissolves like; a substance is most soluble in that solvent to which it is most closely related in structure. This statement serves as a useful classification scheme for all organic molecules. The solubility measurements are done at room temperature with 1 drop of a liquid, or 5 mg of a solid (finely crushed), and 0.2 mL of solvent. The mixture should be rubbed with a rounded stirring rod and shaken vigorously. Lower members of a homologous series are easily classified; higher members become more like the hydrocarbons from which they are derived.

If a very small amount of the sample fails to dissolve when added to some of the solvent, it can be considered insoluble; and, conversely, if several portions dissolve readily in a small amount of the solvent, the substance is obviously soluble.

If an unknown seems to be more soluble in dilute acid or base than in water, the observation can be confirmed by neutralization of the solution; the original material will precipitate if it is less soluble in a neutral medium.

If both acidic and basic groups are present, the substance may be amphoteric and therefore soluble in both acid and base. Aromatic aminocarboxylic acids are amphoteric, like aliphatic ones, but they do not exist as zwitterions. They are soluble in both dilute hydrochloric acid and sodium hydroxide, but not in bicarbonate solution. Aminosulfonic acids exist as zwitterions; they are soluble in alkali but not in acid.

The solubility tests are not infallible and many borderline cases are known. Carry out the tests according to the scheme of Fig. 70.2 and the following "Notes to Solubility Tests," and tentatively assign the unknown to one of the groups I–X.

Cleaning Up   Because the quantities of material used in these tests are extremely small, and because no hazardous substances are handed out as unknowns, it is possible to dilute the material with a large quantity of water and flush it down the drain.

---

3. http://odin.chemistry.uakron.edu/organic_lab/beil/
With seven very good color photos from the University of Akron, the Beilstein test is clearly demonstrated on this Web site. The dramatic differences among chlorine (green), bromine (blue-green), and iodine (blue) are quite clearly seen.

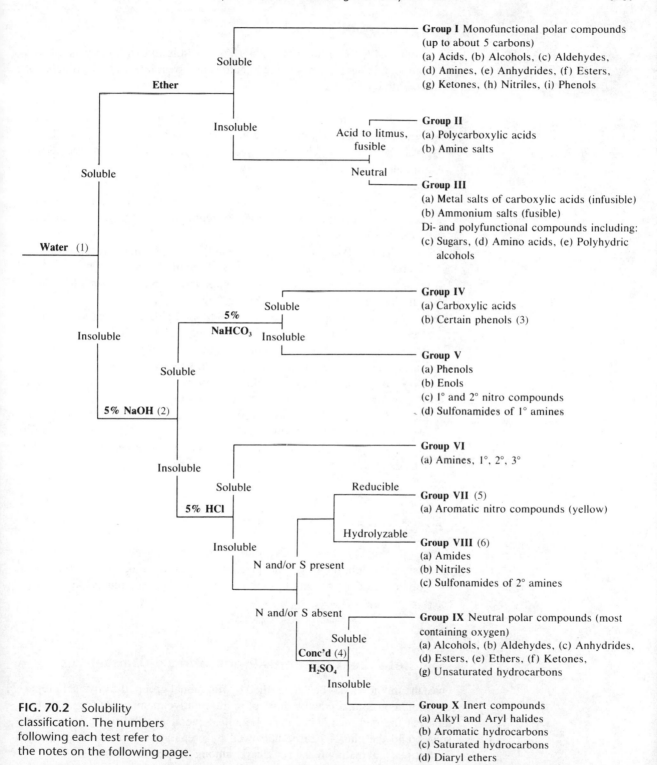

FIG. 70.2 Solubility classification. The numbers following each test refer to the notes on the following page.

Notes to Solubility Tests

See Fig. 70.2.

1. Groups I, II, III (soluble in water). Test the solution with pH paper. If the compound is not easily soluble in cold water, treat it as water insoluble but test with indicator paper.

2. If the substance is insoluble in water but dissolves partially in 5% sodium hydroxide, add more water; the sodium salts of some phenols are less soluble in alkali than in water. If the unknown is colored, be careful to distinguish between the *dissolving* and the *reacting* of the sample. Some quinones (colored) *react* with alkali and give highly colored solutions. Some phenols (colorless) *dissolve and then* become oxidized to give colored solutions. Some compounds (e.g., benzamide) are hydrolyzed with such ease that careful observation is required to distinguish them from acidic substances.

3. Nitrophenols (yellow), aldehydophenols, and polyhalophenols are sufficiently strongly acidic to react with sodium bicarbonate.

4. Oxygen- and nitrogen-containing compounds form oxonium and ammonium ions in concentrated sulfuric acid and dissolve.

5. On reduction in the presence of hydrochloric acid, these compounds form water-soluble amine hydrochlorides. Dissolve 250 mg of tin(II) chloride in 0.5 mL of concentrated hydrochloric acid, add 50 mg of the unknown, and warm. The material should dissolve with the disappearance of the color and give a clear solution when diluted with water.

6. Most amides can be hydrolyzed by short boiling with 10% sodium hydroxide solution; the acid dissolves with evolution of ammonia. Reflux 100 mg of the sample and 10% sodium hydroxide solution for 15–20 min. Test for the evolution of ammonia, which confirms the elementary analysis for nitrogen and establishes the presence of a nitrile or amide.

Classification Tests

After the unknown is assigned to one of the solubility groups (Fig. 70.2) on the basis of solubility tests, the possible type should be further narrowed by application of classification tests, e.g., for alcohols, or methyl ketones, or esters.

Complete Identification—Preparation of Derivatives

Once the unknown has been classified by functional group, the physical properties should be compared with those of representative members of the group (see tables at the end of this chapter). Usually, several possibilities present themselves, and the choice can be narrowed by preparation of derivatives. Select derivatives that distinguish most clearly among the possibilities.

Classification Tests

Group I. Monofunctional Polar Compounds (up to ca. 5 carbons)

(a) Acids

(Table 70.1; Derivatives, page 764)

No classification test is necessary. Carboxylic and sulfonic acids are detected by testing aqueous solutions with litmus. Acyl halides may hydrolyze during the solubility test.

(b) Alcohols

(Table 70.2; Derivatives, page 766)

Jones' Oxidation. Dissolve 5 mg of the unknown in 0.5 mL of pure acetone in a test tube, and add to this solution 1 small drop of Jones' reagent (chromic acid in sulfuric acid). A positive test is formation of a green color within 5 sec upon addition of the orange-yellow reagent to a primary or secondary alcohol. Aldehydes also give a positive test, but tertiary alcohols do not.

CAUTION: Cr^{+6} dust is toxic.

    Reagent: Dissolve/suspend 13.4 g of chromium trioxide in 11.5 mL of concentrated sulfuric acid, and add this carefully with stirring to enough water to bring the volume to 50 mL.

Cleaning Up   Place the test solution in the hazardous waste container.

Handle dioxane with care. It is a cancer-suspect agent.

Cerium(IV) Nitrate Test [Ammonium Hexanitratocerium(IV) Test]. Dissolve 15 mg of the unknown in a few drops of water or dioxane in a reaction tube. Add to this solution 0.25 mL of the reagent, and mix thoroughly. Alcohols cause the reagent to change from yellow to red.

    Reagent: Dissolve 22.5 g of ammonium hexanitratocerium(IV), $Ce(NH_4)_2(NO_3)_6$, in 56 mL of 2 N nitric acid.

Cleaning Up   The solution should be diluted with water and flushed down the drain.

(c) Aldehydes

(Table 70.3; Derivatives, page 767; Ch. 36)

2,4-Dinitrophenylhydrazones. All aldehydes and ketones readily form bright-yellow to dark-red 2,4-dinitrophenylhydrazones. Yellow derivatives are formed from isolated carbonyl groups and orange-red to red derivatives from aldehydes or ketones conjugated with double bonds or aromatic rings.

Dissolve 10 mg of the unknown in 0.5 mL of ethanol, and then add 0.75 mL of 2,4-dinitrophenylhydrazine reagent. Mix thoroughly and let sit for a few minutes. A yellow to red precipitate is a positive test.

Reagent: Dissolve 1.5 g of 2,4-dinitrophenylhydrazine in 7.5 mL of concentrated sulfuric acid. Add this solution, with stirring, to a mixture of 10 mL of water and 35 mL of ethanol.

Cleaning Up   Place the test solution in the hazardous waste container.

Schiff Test. Add 1 drop (30 mg) of the unknown to 1 mL of Schiff's reagent. A magenta color will appear within 10 min with aldehydes. Compare the color of your unknown with that of a known aldehyde.

Reagent: Prepare 50 mL of a 0.1% aqueous solution of *p*-rosaniline hydrochloride (fuchsin). Add 2 mL of a saturated aqueous solution of sodium bisulfite. After 1 h add 1 mL of concentrated hydrochloric acid.

Bisulfite Test. Follow the procedure in Chapter 36. Nearly all aldehydes and most methyl ketones form solid, water-soluble bisulfite addition products.

Destroy used Tollens' reagent promptly with nitric acid. It can form explosive fulminates.

Tollens' Test. Follow the procedure in Chapter 36. A positive test, deposition of a silver mirror, is given by most aldehydes, but not by ketones.

(d) Amides and Amines

(Tables 70.4, 70.5, and 70.6; Derivatives of amines, pages 769–771)

Hinsberg Test. Follow the procedure in Chapter 43, using benzenesulfonyl chloride to distinguish between primary, secondary, and tertiary amines.

(e) Anhydrides and Acid Halides

(Table 70.7; Derivatives, page 771) Anhydrides and acid halides will react with water to give acidic solutions, detectable with litmus paper. They easily form benzamides and acetamides.

Acidic Iron(III) Hydroxamate Test. With iron(III) chloride alone a number of substances give a color that can interfere with this test. Dissolve 2 drops (or about 30 mg) of the unknown in 1 mL of ethanol, and add 1 mL of 1 N hydrochloric acid followed by 1 drop of 10% aqueous iron(III) chloride solution. If any color except yellow appears you will find it difficult to interpret the results from the following test.

Add 2 drops (or about 30 mg) of the unknown to 0.5 mL of a 1 N solution of hydroxylamine hydrochloride in alcohol. Add 2 drops of 6 M hydrochloric acid to the mixture, warm it slightly for 2 min, and boil it for a few seconds. Cool the solution, and add 1 drop of 10% ferric chloride solution. A red-blue color is a positive test.

Cleaning Up   Neutralize the reaction mixture with sodium carbonate, dilute with water, and flush down the drain.

(f) Esters

(Table 70.8, page 772. Derivatives are prepared from component acid and alcohol obtained on hydrolysis.)

Esters, unlike anhydrides and acid halides, do not react with water to give acidic solutions and do not react with acidic hydroxylamine hydrochloride. They do, however, react with alkaline hydroxylamine.

Alkaline Iron(III) Hydroxamate Test. First test the unknown with iron(III) chloride alone. [See under Group I(e), Acidic Iron(III) Hydroxamate Test.]

To a solution of 1 drop (30 mg) of the unknown in 0.5 mL of 0.5 N hydroxylamine hydrochloride in ethanol, add 2 drops of 20% sodium hydroxide solution. Heat the solution to boiling, cool slightly, and add 1 mL of 1 N hydrochloric acid. If cloudiness develops add up to 1 mL of ethanol. Add 10% iron(III) chloride solution dropwise with thorough mixing. A red-blue color is a positive test. Compare your unknown with a known ester.

Cleaning Up   Neutralize the solutions with sodium carbonate, dilute with water, and flush down the drain.

(g) Ketones

(Table 70.14; Derivatives, page 776)

2,4-Dinitrophenylhydrazone. See under Group I(c), Aldehydes. All ketones react with 2,4-dinitrophenylhydrazine reagent.

Iodoform Test for Methyl Ketones. Follow the procedure in Chapter 36.

A positive iodoform test is given by substances containing the $CH_3\overset{\displaystyle O}{\overset{\displaystyle \|}{C}}-$ group or by compounds easily oxidized to this group, e.g., CH_3COR, CH_3CHOHR, CH_3CH_2OH, CH_3CHO, $RCOCH_2COR$. The test is negative for compounds of the structure CH_3COOR, CH_3CONHR, and other compounds of similar structure that give acetic acid on hydrolysis. It is also negative for $CH_3COCH_2CO_2R$, CH_3COCH_2CN, $CH_3COCH_2NO_2$.

Bisulfite Test. Follow the procedure in Chapter 36. Aliphatic methyl ketones and unhindered cyclic ketones form bisulfite addition products. Methyl aryl ketones, such as acetophenone, $C_6H_5COCH_3$, fail to react.

(h) Nitriles

(Table 70.15, page 777. Derivatives prepared from the carboxylic acid obtained by hydrolysis.)

At high temperature nitriles (and amides) are converted to hydroxamic acids by hydroxylamine:

$$RCN + 2\,H_2NOH \longrightarrow RCONHOH + NH_3$$

The hydroxamic acid forms a red-blue complex with iron(III) ion. The unknown must first give a negative test with hydroxylamine at lower temperature [Group I(f), Alkaline Iron(III) Hydroxamate Test] before trying this test.

Hydroxamic Acid Test for Nitriles (and Amides). To 1 mL of a 1 *M* hydroxylamine hydrochloride solution in propylene glycol add 15 mg of the unknown dissolved in the minimum amount of propylene glycol. Then add 0.5 mL of 1 *N* potassium hydroxide in propylene glycol, and boil the mixture for 2 min. Cool the mixture, and add 0.1 to 0.25 mL of 10% iron(III) chloride solution. A red-blue color is a positive test for almost all nitrile and amide groups, although benzanilide fails to give a positive test.

Cleaning Up   Since the quantity of material is extremely small, the test solution can be diluted with water and flushed down the drain.

(i) Phenols

(Table 70.17, page 778)

Iron(III) Chloride Test. Dissolve 15 mg of the unknown compound in 0.5 mL of water or water-alcohol mixture, and add 1 or 2 drops of 1% iron(III) chloride solution. A red, blue, green, or purple color is a positive test.

Cleaning Up   Since the quantity of material is extremely small, the test solution can be diluted with water and flushed down the drain.

A more sensitive test for phenols consists of dissolving or suspending 15 mg of the unknown in 0.5 mL of chloroform and adding 1 drop of a solution made by dissolving 0.1 g of iron(III) chloride in 10 mL of chloroform. (*Caution!* $CHCl_3$ is a carcinogen.) Addition of a drop of pyridine, with stirring, will produce a color if phenols or enols are present.

Group II. Water-Soluble Acidic Salts, Insoluble in Ether

Amine Salts

[Table 70.5 (1° and 2° amines), page 779; Table 70.6 (3° amines), page 771] The free amine can be liberated by addition of base and extraction into ether. Following evaporation of the ether, the Hinsberg test, Group I(d), can be applied to determine if the compound is a primary, secondary, or tertiary amine.

The acid iron(III) hydroxamate test, Group I(d), can be applied directly to the amine salt.

Group III. Water-Soluble Neutral Compounds, Insoluble in Ether

(a) Metal Salts of Carboxylic Acids

(Table 70.1, carboxylic acids; Derivatives, page 764)

The free acid can be liberated by addition of acid and extraction into an appropriate solvent, after which the carboxylic acid can be characterized by mp or bp before proceeding to prepare a derivative.

(b) Ammonium Salts

(Table 70.1, carboxylic acids; Derivatives, page 764)

Ammonium salts on treatment with alkali liberate ammonia, which can be detected by its odor and the fact that it will turn red litmus to blue. A more sensitive test utilizes the copper(II) ion, which is blue in the presence of ammonia [see Group VIII a(i)]. Ammonium salts will not give a positive hydroxamic acid test (Ih) as given by amides.

(c) Sugars

See Chapter 36 for Tollens' test and Chapter 63 for phenylosazone formation.

(d) Amino Acids

Add 2 mg of the suspected amino acid to 1 mL of ninhydrin reagent, boil for 20 sec, and note the color. A blue color is a positive test.

Reagent: Dissolve 0.2 g of ninhydrin in 50 mL of water.

Cleaning Up   Since the quantity of material is extremely small, the test solution can be diluted with water and flushed down the drain.

(e) Polyhydric Alcohols

(Table 70.2; Derivatives, page 766)

Periodic Acid Test for vic-Glycols.[4] Vicinal glycols (hydroxyl groups on adjacent carbon atoms) can be detected by reaction with periodic acid. In addition to 1,2-glycols, a positive test is given by α-hydroxy aldehydes, α-hydroxy ketones, α-hydroxy acids, and α-amino alcohols, as well as 1,2-diketones.

To 2 mL of periodic acid reagent add 1 drop (no more) of concentrated nitric acid and shake. Then add 1 drop or a small crystal of the unknown. Shake for 15 sec, and add 1 or 2 drops of 5% aqueous silver nitrate solution. Instantaneous formation of a white precipitate is a positive test.

---

4. R. L. Shriner, R. C. Fuson, D. Y. Curtin, and T. C. Morill. *The Systematic Identification of Organic Compounds,* 6th ed., John Wiley & Sons, Inc., New York, 1980.

Reagent: Dissolve 0.25 g of paraperiodic acid (H_5IO_6) in 50 mL of water.

Cleaning Up Since the quantity of material is extremely small, the test solution can be diluted with water and flushed down the drain.

Group IV. Certain Carboxylic Acids, Certain Phenols, and Sulfonamides of 1° Amines

(a) Carboxylic Acids

Solubility in both 5% sodium hydroxide and sodium bicarbonate is usually sufficient to characterize this class of compounds. Addition of mineral acid should regenerate the carboxylic acid. The neutralization equivalent can be obtained by titrating a known quantity of the acid (ca. 50 mg) dissolved in water-ethanol with 0.1 *N* sodium hydroxide to a phenolphthalein end point.

(b) Phenols

Negatively substituted phenols such as nitrophenols, aldehydrophenols, and polyhalophenols are sufficiently acidic to dissolve in 5% sodium bicarbonate. See Group I(i) for the iron(III) chloride test for phenols; however, this test is not completely reliable for these acidic phenols.

Group V. Acidic Compounds, Insoluble in Bicarbonate

(a) Phenols

See Group I(i).

(b) Enols

See Group I(i).

(c) 1° and 2° Nitro Compounds

(Table 70.16; Derivatives, page 717)

Iron(II) Hydroxide Test. To a small vial (capacity (1–2 mL) add 5 mg of the unknown to 0.5 mL of freshly prepared ferrous sulfate solution. Add 0.4 mL of a 2 *N* solution of potassium hydroxide in methanol, cap the vial, and shake it. The appearance of a red-brown precipitate of iron(III) hydroxide within 1 min is a positive test. Almost all nitro compounds give a positive test within 30 sec.

Reagents: Dissolve 2.5 g of ferrous ammonium sulfate in 50 mL of deoxygenated (by boiling) water. Add 0.2 mL of concentrated sulfuric acid and a piece of iron to prevent oxidation of the ferrous ion. Keep the bottle tightly stoppered. The potassium hydroxide solution is prepared by dissolving 5.6 g of potassium hydroxide in 50 mL of methanol.

Cleaning Up   Since the quantity of material is extremely small, the test solution can be diluted with water and flushed down the drain after neutralization with dilute hydrochloric acid.

(d) Sulfonamides of 1° Amines

An extremely sensitive test for sulfonamides (Feigl, *Spot Tests in Organic Analysis*) consists of placing a drop of a suspension or solution of the unknown on sulfonamide test paper followed by a drop of 0.5% hydrochloric acid. A red color is a positive test for sulfonamides.

The test paper is prepared by dipping filter paper into a mixture of equal volumes of a 1% aqueous solution of sodium nitrite and a 1% methanolic solution of *N*,*N*-dimethyl-1-naphthylamine. Allow the filter paper to dry in the dark.

CAUTION: Handle N,N-dimethyl-1-naphthylamine with care. As a class, aromatic amines are quite toxic and many are carcinogenic. Handle them all with care—in a hood if possible.

Cleaning Up   Place the test paper in the solid hazardous waste container.

Group VI. Basic Compounds, Insoluble in Water, Soluble in Acid

Amines

See Group I(d).

Group VII. Reducible, Neutral *N*- and *S*-Containing Compounds

Aromatic Nitro Compounds

See Group V(c).

Group VIII. Hydrolyzable, Neutral *N*- and *S*-Containing Compounds (identified through the acid and amine obtained on hydrolysis)

(a) Amides

Unsubstituted amides are detected by the hydroxamic acid test, Group I(h).

(1) Unsubstituted Amides. Upon hydrolysis, unsubstituted amides liberate ammonia, which can be detected by reaction with cupric ion [Group III(b)].

To 1 mL of 20% sodium hydroxide solution, add 25 mg of the unknown. Cover the mouth of the reaction tube with a piece of filter paper moistened with a few drops of 10% copper(II) sulfate solution. Boil for 1 min. A blue color on the filter paper is a positive test for ammonia.

Cleaning Up   Neutralize the test solution with 10% hydrochloric acid, dilute with water, and flush down the drain.

(2) Substituted Amides. The identification of substituted amides is not easy. There are no completely general tests for the substituted amide groups and hydrolysis is often difficult.

Hot sodium hydroxide solution is corrosive; use care.

Hydrolyze the amide by refluxing 250 mg with 2.5 mL of 20% sodium hydroxide for 20 min. Isolate the primary or secondary amine produced, by extraction into ether, and identify as described under Group I(d). Liberate the acid by acidification of the residue, isolate by filtration or extraction, and characterize by bp or mp and the mp of an appropriate derivative.

Cleaning Up   The test solution can be diluted with water and flushed down the drain.

Use care in shaking concentrated sulfuric acid.

(3) Anilides. Add 50 mg of the unknown to 1.5 mL of concentrated sulfuric acid. Carefully stopper the reaction tube with a rubber stopper, and shake vigorously. (*Caution!*) Add 25 mg of finely powdered potassium dichromate. A blue-pink color is a positive test for an anilide that does not have substituents on the ring (e.g., acetanilide).

Dichromate dust is carcinogenic, when inhaled. Cr^{+6} is not a carcinogen when applied to the skin or ingested.

Cleaning Up   Carefully add the solution to water, neutralize with sodium carbonate, and flush down the drain.

(b) Nitriles

See Group I(h).

(c) Sulfonamides

See Group V(d).

Group IX. Neutral Polar Compounds, Insoluble in Dilute Hydrochloric Acid, Soluble in Concentrated Sulfuric Acid (most compounds containing oxygen)

(a) Alcohols

See Group I(b).

(b) Aldehydes

See Group I(c).

(c) Anhydrides

See Group I(e).

(d) Esters

See Group I(f).

(e) Ethers

(See Table 70.9, page 773)
Ethers are very unreactive. Care must be used to distinguish ethers from those hydrocarbons that are soluble in concentrated sulfuric acid.

Fe[Fe(SCN)$_6$]

**Iron(III) hexathiocyanato-
ferrate(III)**

Ferrox Test. In a dry test tube grind together, with a stirring rod, a crystal of iron(III) ammonium sulfate (or iron(III) chloride) and a crystal of potassium thiocyanate. Iron(III) hexathiocyanatoferrate(III) will adhere to the stirring rod. In a clean tube place 3 drops of a liquid unknown or a saturated toluene solution of a solid unknown, and stir with the rod. The salt will dissolve if the unknown contains oxygen to give a red-purple color, but it will not dissolve in hydrocarbons or halocarbons. Diphenyl ether does not give a positive test.

Alkyl ethers are generally soluble in concentrated sulfuric acid; alkyl aryl and diaryl ethers are not soluble.

Cleaning Up   Place the mixture in the hazardous waste container.

(f) Ketones

(See page 776).

(g) Unsaturated Hydrocarbons

(Table 70.12, page 774)

Use care in working with the bromine solution

Bromine in Carbon Tetrachloride. Dissolve 1 drop (20 mg) of the unknown in 0.5 mL of carbon tetrachloride. Add a 2% solution of bromine in carbon tetrachloride dropwise with shaking. If more than 2 drops of bromine solution are required to give a permanent red color, unsaturation is indicated. The bromine solution must be fresh.

Cleaning Up   Place the mixture in the halogenated solvents container.

Potassium Permanganate Solution. Dissolve 1 drop (20 mg) of the unknown in reagent grade acetone and add a 1% aqueous solution of potassium permanganate dropwise with shaking. If more than one drop of reagent is required to give a purple color to the solution, unsaturation or an easily oxidized functional group is present. Run parallel tests on pure acetone and, as usual, a compound known to be an alkene.

Cleaning Up   The solution should be diluted with water and flushed down the drain.

Group X. Inert Compounds. Insoluble in Concentrated Sulfuric Acid

(a) Alkyl and Aryl Halides

(Table 70.10; Derivatives, page 773)

Do not waste silver nitrate.

Alcoholic Silver Nitrate. Add 1 drop of the unknown (or saturated solution of 10 mg of unknown in ethanol) to 0.2 mL of a saturated solution of silver nitrate. A precipitate that forms within 2 min is a positive test for an alkyl bromide, or iodide, or a tertiary alkyl chloride, as well as allyl halides.

If no precipitate forms within 2 min, heat the solution to boiling. A precipitate of silver chloride will form from primary and secondary alkyl chlorides. Aryl halides and vinyl halides will not react.

Cleaning Up   The quantity of material is extremely small. It can be diluted with water and flushed down the drain.

(b) Aromatic Hydrocarbons

(Table 70.13; Derivatives, page 775)
Aromatic hydrocarbons are best identified and characterized by UV and NMR spectroscopy, but the Friedel-Crafts reaction produces a characteristic color with certain aromatic hydrocarbons.

Keep moisture away from aluminum chloride.
CAUTION: Chloroform is carcinogenic. Carry out this test in a hood.

Friedel-Crafts Test. Heat a test tube containing about 50 mg of anhydrous aluminum chloride in a hot flame to sublime the salt up onto the sides of the tube. Add a solution of about 10 mg of the unknown dissolved in a drop of chloroform to the cool tube in such a way that it comes into contact with the sublimed aluminum chloride. Note the color that appears.

Nonaromatic compounds fail to give a color with aluminum chloride, benzene and its derivatives give orange or red colors, naphthalenes a blue or purple color, biphenyls a purple color, phenanthrene a purple color, and anthracene a green color.

Cleaning Up   Place the test mixture in the halogenated organic solvents container.

(c) Saturated Hydrocarbons

Saturated hydrocarbons are best characterized by NMR and IR spectroscopy, but they can be distinguished from aromatic hydrocarbons by the Friedel-Crafts test [Group X(b)].

(d) Diaryl Ethers

Because they are so inert, diaryl ethers are difficult to detect and may be mistaken for aromatic hydrocarbons. They do not give a positive Ferrox test (see

p. 757) for ethers and do not dissolve in concentrated sulfuric acid. Their infrared spectra, however, are characterized by an intense C—O single-bond, stretching vibration in the region 1270–1230 cm^{-1}.

Derivatives

1. Acids

(Table 70.1)

p-**Toluidides and Anilides.** Reflux a mixture of the acid (100 mg) and thionyl chloride (0.5 mL) in a reaction tube for 0.5 h. Cool the reaction mixture, and add 0.25 g of aniline or *p*-toluidine in 3 mL of toluene. Warm the mixture on the steam bath for 2 min, and then wash with 1-mL portions of water, 5% hydrochloric acid, 5% sodium hydroxide, and water. The toluene is dried briefly over anhydrous calcium chloride pellets and evaporated in the hood; the derivative is recrystallized from water or ethanol-water.

Caution: p-Toluidine is a highly toxic irritant.

Cleaning Up   The combined aqueous layers can be diluted with water and flushed down the drain. The drying agent should be placed in the hazardous waste container.

Amides. Reflux a mixture of the acid (100 mg) and thionyl chloride (0.5 mL) for 0.5 h. Transfer the cool reaction mixture into 1.4 mL of ice-cold concentrated ammonia. Stir until reaction is complete, collect the product by filtration, and recrystallize it from water or water-ethanol.

Thionyl chloride is an irritant. Use it in a hood.

Cleaning Up   Neutralize the aqueous filtrate with 10% hydrochloric acid, dilute with water, and flush down the drain.

2. Alcohols

(Table 70.2)

3,5-Dinitrobenzoates. Gently boil 100 mg of 3,5-dinitrobenzoyl chloride and 25 mg of the alcohol for 5 min. Cool the mixture, pulverize any solid that forms, and add 2 mL of 2% sodium carbonate solution. Continue to grind and stir the solid with the sodium carbonate solution (to remove 3,5-dinitrobenzoic acid) for about a minute, filter, and wash the crystals with water. Dissolve the product in about 2.5–3 mL of hot ethanol, add water to the cloud point, and allow crystallization to proceed. Wash the 3,5-dinitrobenzoate with water-alcohol and dry.

Note to instructor: Check to ascertain that the 3,5-dinitrobenzoyl chloride has not hydrolyzed. The mp should be >65°C. Reported mp is 68–69°C.

Cleaning Up   The aqueous filtrate should be diluted with water and flushed down the drain.

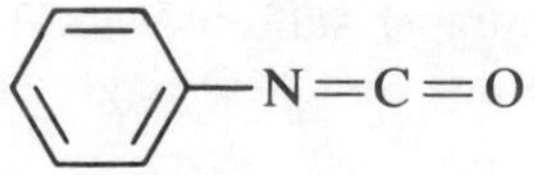

Phenyl isocyanate

CAUTION: *Lachrymator*

Phenylurethanes. Mix 100 mg of anhydrous alcohol (or phenol) and 100 mg of phenyl isocyanate (or α-naphthylurethane), and heat on the steam bath for 5 min. (If the unknown is a phenol add a drop of pyridine to the reaction mixture.) Cool, add about 1 mL of ligroin, heat to dissolve the product, filter hot to remove a small amount of diphenylurea which usually forms, and cool the filtrate in ice, with scratching, to induce crystallization.

Cleaning Up   The ligroin filtrate should be placed in the organic solvents container.

3. Aldehydes

(Table 70.3)

Semicarbazones. See Chapter 36. Use 0.5 mL of the stock solution and an estimated 1 mmol of the unknown aldehyde (or ketone).

2,4-Dinitrophenylhydrazones. See Chapter 36. Use 1 mL of the stock solution of 0.1 *M* 2,4-dinitrophenylhydrazine and an estimated 0.1 mmol of the unknown aldehyde (or ketone).

4. Primary and Secondary Amines

(Table 70.5)

Benzamides. Add about 0.25 g of benzoyl chloride in small portions with vigorous shaking and cooling to a suspension of 0.5 mmol of the unknown amine in 0.5 mL of 10% aqueous sodium hydroxide solution. After about 10 min of shaking the mixture is made pH 8 (pH paper) with dilute hydrochloric acid. The lumpy product is removed by filtration, washed thoroughly with water, and recrystallized from ethanol-water.

Cleaning Up   The filtrate should be diluted with water and flushed down the drain.

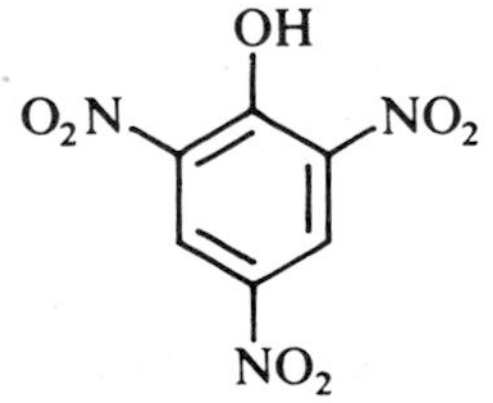

**Picric acid
(2,4,6-Trinitrophenol)**

Handle pure acid with care (explosive). It is sold as a moist solid. Do not allow to dry out.

Picrates. Add a solution of 30 mg of the unknown in 1 mL of ethanol (or 1 mL of a saturated solution of the unknown) to 1 mL of a saturated solution of picric acid (2,4,6-trinitrophenol, a strong acid) in ethanol, and heat the solution to boiling. Cool slowly, remove the picrate by filtration, and wash with a small amount of ethanol. Recrystallization is not usually necessary; in the case of hydrocarbon picrates the product is often too unstable to be recrystallized.

Cleaning Up   See page 508 for the treatment of solutions containing picric acid.

Acetamides. Reflux about 0.5 mmol of the unknown with 0.2 mL of acetic anhydride for 5 min, cool, and dilute the reaction mixture with 2.5 mL of water.

Acetic anhydride is corrosive. Work with this in a hood.

Initiate crystallization by scratching, if necessary. Remove the crystals by filtration, and wash thoroughly with dilute hydrochloric acid to remove unreacted amine. Recrystallize the derivative from alcohol-water. Amines of low basicity, e.g., *p*-nitroaniline, should be refluxed for 30–60 min with 1 mL of pyridine as a solvent. The pyridine is removed by shaking the reaction mixture with 5 mL of 2% sulfuric acid solution; the product is isolated by filtration and recrystallized.

Cleaning Up   The filtrate from the usual reaction should be neutralized with sodium carbonate. It can then be diluted with water and flushed down the drain. If pyridine is used as the solvent, the filtrate should be neutralized with sodium carbonate and extracted with ligroin. The ligroin/pyridine goes in the organic solvents container while the aqueous layer can be diluted with water and flushed down the drain.

5. Tertiary Amines

(Table 70.6)

Picrates. See under Primary and Secondary Amines.

$$R_3N + CH_3I$$
$$\downarrow$$
$$R_3\overset{+}{N}CH_3I^-$$

Methyl iodide is a cancer-suspect agent.

Methiodides. Reflux 100 mg of the amine and 100 mg of methyl iodide for 5 min on the steam bath. Cool, scratch to induce crystallization, and recrystallize the product from ethyl alcohol or ethyl acetate.

Cleaning Up   Since the filtrate may contain some methyl iodide, it should be placed in the halogenated solvents container.

6. Anhydrides and Acid Chlorides

(Table 70.7)

Acids. Reflux 40 mg of the acid chloride or anhydride with 1 mL of 5% sodium carbonate solution for 20 min or less. Extract unreacted starting material with 1 mL of ether, if necessary, and acidify the reaction mixture with dilute sulfuric acid to liberate the carboxylic acid.

Cleaning Up   Ether goes in the organic solvents container, and the aqueous layer should be diluted with water and flushed down the drain.

Amides. Since the acid chloride (or anhydride) is already present, simply mix the unknown (50 mg) and 0.7 mL of ice-cold concentrated ammonia until reaction is complete, collect the product by filtration, and recrystallize it from water or ethanol-water.

Cleaning Up   Neutralize the filtrate with dilute hydrochloric acid, and flush it down the drain.

Anilides. Reflux 40 mg of the acid halide or anhydride with 100 mg of aniline in 2 mL of toluene for 10 min. Wash the toluene solution with 5-mL portions each of water, 5% hydrochloric acid, 5% sodium hydroxide, and again with water. The toluene solution is dried over anhydrous calcium chloride and evaporated; the anilide is recrystallized from water or ethanol-water.

Cleaning Up   The combined aqueous layers are diluted with water and flushed down the drain. The sodium sulfate should be placed in the aromatic amines hazardous waste container.

7. Aryl Halides

(Table 70.11)

Nitration. Add 0.4 mL of concentrated sulfuric acid to 100 mg of the aryl halide (or aromatic compound) and stir. Add 0.4 mL of concentrated nitric acid dropwise with stirring and shaking while cooling the reaction mixture in water. Then heat and shake the reaction mixture in a water bath at about 50°C for 15 min, pour into 2 mL of cold water, and collect the product by filtration. Recrystallize from methanol to constant melting point.

Use great care when working with fuming nitric acid.

To nitrate unreactive compounds, use fuming nitric acid in place of concentrated nitric acid.

Cleaning Up   Dilute the filtrate with water, neutralize with sodium carbonate, and flush the solution down the drain.

Sidechain Oxidation Products. Dissolve 0.2 g of sodium dichromate in 0.6 mL of water, and add 0.4 mL of concentrated sulfuric acid. Add 50 mg of the unknown and boil for 30 min. Cool, add 0.4–0.6 mL of water, and then remove the carboxylic acid by filtration. Wash the crystals with water and recrystallize from methanol-water.

Cleaning Up   Place the filtrate from the reaction, after neutralization with sodium carbonate, in the hazardous waste container.

8. Hydrocarbons: Aromatic

(Table 70.13)

Nitration. See preceding, under Aryl Halides.

Picrates. See preceding, under Primary and Secondary Amines.

9. Ketones

(Table 70.14)

Semicarbazones and 2,4-dinitrophenylhydrazones. See preceding directions under Aldehydes.

10. Nitro Compounds

(Table 70.16)

Reduction to Amines. Place 100 mg of the unknown in a reaction tube, add 0.2 g of tin, and then—in portions—2 mL of 10% hydrochloric acid. Reflux for 30 min, add 1 mL of water, then add slowly, with good cooling, sufficient 40% sodium hydroxide solution to dissolve the tin hydroxide. Extract the reaction mixture with three 1-mL portions of *t*-butyl methyl ether, dry the ether extract over anhydrous calcium chloride pellets, wash the drying agent with ether, and evaporate the ether to leave the amine. Determine the boiling point or melting point of the amine and then convert it into a benzamide or acetamide as described under the section on Primary and Secondary Amines.

Cleaning Up   Neutralize the aqueous layer with 10% hydrochloric acid, remove the tin hydroxide by filtration, and discard it in the nonhazardous solid waste container. The filtrate should be diluted with water and flushed down the drain. Calcium chloride, after the ether evaporates from it, can be placed in the nonhazardous waste container.

11. Phenols

(Table 70.17)

α-Naphthylurethane. Follow the procedure for preparation of a phenyl-urethane under the Alcohols section.

Use great care when working with bromine. Should any touch the skin wash it off with copious quantities of water. Work in a hood and wear disposable gloves.

Bromo Derivative. In a reaction tube dissolve 160 mg of potassium bromide in 1 mL of water. *Carefully* add 100 mg of bromine. In a separate flask dissolve 20 mg of the phenol in 0.2 mL of methanol, and add 0.2 mL of water. Add about 0.3 mL of the bromine solution with swirling (hood); continue the addition of bromine until the yellow color of unreacted bromine persists. Add 0.6–0.8 mL of water to the reaction mixture, and shake vigorously. Remove the product by filtration, and wash well with water. Recrystallize from methanol-water.

Cleaning Up   Any unreacted bromine should be destroyed by adding sodium bisulfite solution dropwise until the color dissipates. The solution is then diluted with water and flushed down the drain.

TABLE 70.1   Acids

			Derivatives		
			p-Toluidide[a]	*Anilide*[b]	*Amide*[c]
bp	*mp*	*Compound*	*mp*	*mp*	*mp*
101		Formic acid	53	47	43
118		Acetic acid	126	106	79
139		Acrylic acid	141	104	85
141		Propionic acid	124	103	81
162		*n*-Butyric acid	72	95	115
163		Methacrylic acid		87	102
165		Pyruvic acid	109	104	124
185		Valeric acid	70	63	106
186		2-Methylvaleric acid	80	95	79
194		Dichloroacetic acid	153	118	98
202–203		Hexanoic acid	75	95	101
237		Octanoic acid	70	57	107
254		Nonanoic acid	84	57	99
	31–32	Decanoic acid	78	70	108
	43–45	Lauric acid	87	78	100
	47–49	Bromoacetic acid		131	91
	47–49	Hydrocinnamic acid	135	92	105
	54–55	Myristic acid	93	84	103
	54–58	Trichloroacetic acid	113	97	141
	61–62	Chloroacetic acid	162	137	121
	61–62.5	Palmitic acid	98	90	106
	67–69	Stearic acid	102	95	109
	68–69	3,3-Dimethylacrylic acid		126	107
	71–73	Crotonic acid	132	118	158
	77–78.5	Phenylacetic acid	136	118	156
	101–102	Oxalic acid dihydrate		257	400 (dec)
	98–102	Azelaic acid (nonanedioic)	164 (di)	107 (mono)	93 (mono)
				186 (di)	175 (di)
	103–105	*o*-Toluic acid	144	125	142
	108–110	*m*-Toluic acid	118	126	94
	119–121	DL-Mandelic acid	172	151	133
	122–123	Benzoic acid	158	163	130
	127–128	2-Benzoylbenzoic acid		195	165
	129–130	2-Furoic acid	107	123	143
	131–133	DL-Malic acid	178 (mono)	155 (mono)	
			207 (di)	198 (di)	163 (di)
	131–134	Sebacic acid	201	122 (mono)	170 (mono)
				200 (di)	210 (di)

(continued)

a. For preparation, see page 759.
b. For preparation, see page 759.
c. For preparation, see page 759.

TABLE 70.1   *continued*

			Derivatives		
			p-Toluidide[a]	*Anilide*[b]	*Amide*[c]
bp	*mp*	*Compound*	*mp*	*mp*	*mp*
	134–135	*E*-Cinnamic acid	168	153	147
	134–136	Maleic acid	142 (di)	198 (mono) 187 (di)	260 (di)
	135–137	Malonic acid	86 (mono) 253 (di)	132 (mono) 230 (di)	
	138–140	2-Chlorobenzoic acid	131	118	139
	140–142	3-Nitrobenzoic acid	162	155	143
	144–148	Anthranilic acid	151	131	109
	147–149	Diphenylacetic acid	172	180	167
	152–153	Adipic acid	239	151 (mono) 241 (di)	125 (mono) 220 (di)
	153–154	Citric acid	189 (tri)	199 (tri)	210 (tri)
	157–159	4-Chlorophenoxyacetic acid		125	133
	158–160	Salicylic acid	156	136	142
	163–164	Trimethylacetic acid		127	178
	164–166	5-Bromosalicylic acid		222	232
	166–167	Itaconic acid		190	191 (di)
	171–174	D-Tartaric acid		180 (mono) 264 (di)	171 (mono) 196 (di)
	179–182	3,4-Dimethoxybenzoic acid		154	164
	180–182	4-Toluic acid	160	145	160
	182–185	4-Anisic acid	186	169	167
	187–190	Succinic acid	180 (mono) 255 (di)	143 (mono) 230 (di)	157 (mono) 260 (di)
	201–203	3-Hydroxybenzoic acid	163	157	170
	203–206	3,5-Dinitrobenzoic acid		234	183
	210–211	Phthalic acid	150 (mono) 201 (di)	169 (mono) 253 (di)	149 (mono) 220 (di)
	214–215	4-Hydroxybenzoic acid	204	197	162
	225–227	2,4-Dihydroxybenzoic acid		126	228
	236–239	Nicotinic acid	150	132	128
	239–241	4-Nitrobenzoic acid	204	211	201
	299–300	Fumaric acid		233 (mono) 314 (di)	270 (mono) 266 (di)
	>300	Terephthalic acid		334	

a. For preparation, see page 759.
b. For preparation, see page 759.
c. For preparation, see page 759.

TABLE 70.2   Alcohols

| | | | Derivatives | |
| | | | 3,5-Dinitrobenzoate[a] | Phenylurethane[b] |
bp	mp	Compound	mp	mp
65		Methanol	108	47
78		Ethanol	93	52
82		2-Propanol	123	88
83		t-Butyl alcohol	142	136
96–98		Allyl alcohol	49	70
97		1-Propanol	74	57
98		2-Butanol	76	65
102		2-Methyl-2-butanol	116	42
104		2-Methyl-3-butyn-2-ol	112	
108		2-Methyl-1-propanol	87	86
114–115		Propargyl alcohol		63
114–115		3-Pentanol	101	48
118		1-Butanol	64	61
118–119		2-Pentanol	62	oil
123		3-Methyl-3-pentanol	96(62)	43
129		2-Chloroethanol	95	51
130		2-Methyl-1-butanol	70	31
132		4-Methyl-2-pentanol	65	143
136–138		1-Pentanol	46	46
139–140		Cyclopentanol	115	132
140		2,4-Dimethyl-3-pentanol	75	95
146		2-Ethyl-1-butanol	51	
151		2,2,2-Trichloroethanol	142	87
157		1-Hexanol	58	42
160–161		Cyclohexanol	113	82
170		Furfuryl alcohol	80	45
176		1-Heptanol	47	60(68)
178		2-Octanol	32	oil
178		Tetrahydrofurfuryl alcohol	83	61
183–184		2,3-Butanediol		201 (di)
183–186		2-Ethyl-1-hexanol		34
187		1,2-Propanediol		153 (di)
194–197		Linaloöl		66
195		1-Octanol	61	74
196–198		Ethylene glycol	169	157 (di)
204		1,3-Butanediol		122
203–205		Benzyl alcohol	113	77
204		1-Phenylethanol	95	92
219–221		2-Phenylethanol	108	78
230		1,4-Butanediol		183 (mono)
231		1-Decanol	57	59

(continued)

a. For preparation, see page 759.
b. For preparation, see page 759.

TABLE 70.2   *continued*

| | | | Derivatives | |
| | | | 3,5-Dinitrobenzoate[a] | Phenylurethane[b] |
bp	*mp*	*Compound*	*mp*	*mp*
259		4-Methoxybenzyl alcohol		92
	33–35	Cinnamyl alcohol	121	90
	38–40	1-Tetradecanol	67	74
	48–50	1-Hexadecanol	66	73
	58–60	1-Octadecanol	77(66)	79
	66–67	Benzhydrol	141	139
	147	Cholesterol	195	168

a. For preparation, see page 759.
b. For preparation, see page 759.

TABLE 70.3   Aldehydes

| | | | Derivatives | |
| | | | Semicarbazone[a] | 2,4-Dinitrophenylhydrazone[b] |
bp	*mp*	*Compound*	*mp*	*mp*
21		Acetaldehyde	162	168
46–50		Propionaldehyde	89(154)	148
63		Isobutyraldehyde	125(119)	187(183)
75		Butyraldehyde	95(106)	123
90–92		3-Methylbutanal	107	123
98		Chloral	90	131
104		Crotonaldehyde	199	190
117		2-Ethylbutanal	99	95(130)
153		Heptaldehyde	109	108
162		2-Furaldehyde	202	212(230)
163		2-Ethylhexanal	254	114(120)
179		Benzaldehyde	222	237
195		Phenylacetaldehyde	153	121(110)
197		Salicylaldehyde	231	248
204–205		4-Tolualdehyde	234(215)	232
209–215		2-Chlorobenzaldehyde	146(229)	213
247		2-Ethoxybenzaldehyde	219	
248		4-Anisaldehyde	210	253

(continued)

a. For preparation, see page 431.
b. For preparation, see page 429.

TABLE 70.3   *continued*

| | | | Derivatives | |
| | | | Semicarbazone[a] | 2,4-Dinitrophenylhydrazone[b] |
bp	*mp*	*Compound*	*mp*	*mp*
250–252		*E*-Cinnamaldehyde	215	255
	33–34	1-Naphthaldehyde	221	254
	37–39	2-Anisaldehyde	215	254
	42–45	3,4-Dimethoxybenzaldehyde	177	261
	44–47	4-Chlorobenzaldehyde	230	254
	57–59	3-Nitrobenzaldehyde	246	293
	81–83	Vanillin	230	271

a. For preparation, see page 431.
b. For preparation, see page 429.

TABLE 70.4   Amides

bp	mp	Name of Compound	mp	Name of Compound
153		*N,N*-Dimethylformamide	127–129	Isobutyramide
164–166		*N,N*-Dimethylacetamide	128–129	Benzamide
210		Formamide	130–133	Nicotinamide
243–244		*N*-Methylformanilide	177–179	4-Chloroacetanilide
	26–28	*N*-Methylacetamide		
	79–81	Acetamide		
	109–111	Methacrylamide		
	113–115	Acetanilide		
	116–118	2-Chloroacetamide		

TABLE 70.5   Primary and Secondary Amines

			Derivatives		
			Benzamide[a]	*Picrate*[b]	*Acetamide*[c]
bp	*mp*	*Compound*	*mp*	*mp*	*mp*
33–34		Isopropylamine	71	165	
46		*t*-Butylamine	134	198	
48		*n*-Propylamine	84	135	
53		Allylamine		140	
55		Diethylamine	42	155	
63		*s*-Butylamine	76	139	
64–71		Isobutylamine	57	150	
78		*n*-Butylamine	42	151	
84		Diisopropylamine		140	
87–88		Pyrrolidine	oil	112	
106		Piperidine	48	152	
111		Di-*n*-propylamine	oil	75	
118		Ethylenediamine	244 (di)	233	172 (di)
129		Morpholine	75	146	
137–139		Diisobutylamine		121	86
145–146		Furfurylamine		150	
149		*N*-Methylcyclohexylamine	85	170	
159		Di-*n*-butylamine	oil	59	
182–185		Benzylamine	105	199	60
184		Aniline	163	198	114
196		*N*-Methylaniline	63	145	102
199–200		2-Toluidine	144	213	110
203–204		3-Toluidine	125	200	65
205		*N*-Ethylaniline	60	138(132)	54
208–210		2-Chloroaniline	99	134	87
210		2-Ethylaniline	147	194	111
216		2,6-Dimethylaniline	168	180	177
218		2,4-Dimethylaniline	192	209	133
218		2,5-Dimethylaniline	140	171	139
221		*N*-Ethyl-*m*-toluidine	72		
225		2-Anisidine	60(84)	200	85
230		3-Chloroaniline	120	177	72(78)
231–233		2-Phenetidine	104		79
241		4-Chloro-2-methylaniline	142		140
242		3-Chloro-4-methylaniline	122		105
250		4-Phenetidine	173	69	137
256		Dicyclohexylamine	153(57)	173	103

(*continued*)

a. For preparation, see page 760.
b. For preparation, see page 760.
c. For preparation, see page 760.

TABLE 70.5 *continued*

bp	mp	Compound	Benzamide[a] mp	Picrate[b] mp	Acetamide[c] mp
			Derivatives		
	35–38	N-Phenylbenzylamine	107	48	58
	41–44	4-Toluidine	158	182	147
	49–51	2,5-Dichloroaniline	120	86	132
	52–54	Diphenylamine	180	182	101
	57–60	4-Anisidine	154	170	130
	57–60	2-Aminopyridine	165 (di)	216(223)	
	60–62	N-Phenyl-1-naphthylamine	152		115
	62–65	2,4,5-Trimethylaniline	167		162
	64–66	1,3-Phenylenediamine	125 (mono) 240 (di)	184	87 (mono) 191 (di)
	66	4-Bromoaniline	204	180	168
	68–71	4-Chloroaniline	192	178	179(172)
	71–73	2-Nitroaniline	110(98)	73	92
	97–99	2,4-Diaminotoluene	224 (di)		224 (di)
	100–102	1,2-Phenylenediamine	301	208	185
	104–107	2-Methyl-5-nitroaniline	186		151
	107–109	2-Chloro-4-nitroaniline	161		139
	112–114	3-Nitroaniline	157(150)	143	155(76)
	115–116	4-Methyl-2-nitroaniline	148		99
	117–119	4-Chloro-2-nitroaniline			104
	120–122	2,4,6-Tribromoaniline	198(204)		232
	131–133	2-Methyl-4-nitroaniline			202
	138–140	2-Methoxy-4-nitroaniline	149		
	138–142	1,4-Phenylenediamine	128 (mono) 300 (di)		162 (mono) 304 (di)
	148–149	4-Nitroaniline	199	100	215
	162–164	4-Aminoacetanilide			304
	176–178	2,4-Dinitroaniline	202(220)		120

a. For preparation, see page 760.
b. For preparation, see page 760.
c. For preparation, see page 760.

TABLE 70.6   Tertiary Amines

| | | Derivatives | |
| | | Picrate[a] | Methiodide[b] |
bp	Compound	mp	mp
85–91	Triethylamine	173	280
115	Pyridine	167	117
128–129	2-Picoline	169	230
143–145	2,6-Lutidine	168(161)	233
144	3-Picoline	150	92(36)
145	4-Picoline	167	149
155–158	Tri-*n*-propylamine	116	207
159	2,4-Lutidine	180	113
183–184	*N,N*-Dimethylbenzylamine	93	179
216	Tri-*n*-butylamine	105	186
217	*N,N*-Diethylaniline	142	102
237	Quinoline	203	133(72)

a. For preparation, see page 760.
b. For preparation, see page 760.

TABLE 70.7   Anhydrides and Acid Chlorides

| | | | Derivatives | | | |
| | | | Acid[a] | | Amide[b] | Anilide[c] |
bp	mp	Compound	bp	mp	mp	mp
52		Acetyl chloride	118		82	114
77–79		Propionyl chloride	141		81	106
102		Butyryl chloride	162		115	96
138–140		Acetic anhydride	118		82	114
167		Propionic anhydride	141		81	106
198–199		Butyric anhydride	162		115	96
198		Benzoyl chloride		122	130	163
225		3-Chlorobenzoyl chloride		158	134	122
238		2-Chlorobenzoyl chloride		142	142	118
	32–34	*cis*-1,2-Cyclohexanedicarboxylic anhydride		192		
	35–37	Cinnamoyl chloride		133	147	151
	39–40	Benzoic anhydride		122	130	163
	54–56	Maleic anhydride		130	181 (mono) 266 (di)	173 (mono) 187

(continued)

a. For preparation, see page 761.
b. For preparation, see page 761.
c. For preparation, see page 762.

TABLE 70.7   *continued*

					Derivatives		
					Acid[a]	*Amide*[b]	*Anilide*[c]
bp	*mp*	Compound		*bp*	*mp*	*mp*	*mp*
	72–74	4-Nitrobenzoyl chloride		241	201	211	
	119–120	Succinic anhydride		186	157 (mono)	148 (mono)	
					260 (di)	230 (di)	
	131–133	Phthalic anhydride		206	149 (mono)	170 (mono)	
					220 (di)	253 (di)	
	254–258	Tetrachlorophthalic anhydride		250			
	267–269	1,8-Naphthalic anhydride		274		250–282 (di)	

a. For preparation, see page 761.
b. For preparation, see page 761.
c. For preparation, see page 762.

TABLE 70.8   Esters

bp	mp	Compound	bp	mp	Compound
34		Methyl formate	169–170		Methyl acetoacetate
52–54		Ethyl formate	180–181		Dimethyl malonate
72–73		Vinyl acetate	181		Ethyl acetoacetate
77		Ethyl acetate	185		Diethyl oxalate
79		Methyl propionate	198–199		Methyl benzoate
80		Methyl acrylate	206–208		Ethyl caprylate
85		Isopropyl acetate	208–210		Ethyl cyanoacetate
93		Ethyl chloroformate	212		Ethyl benzoate
94		Isopropenyl acetate	217		Diethyl succinate
98		Isobutyl formate	218		Methyl phenylacetate
98		*t*-Butyl acetate	218–219		Diethyl fumarate
99		Ethyl propionate	222		Methyl salicylate
99		Ethyl acrylate	225		Dimethyl maleate
100		Methyl methacrylate	229		Ethyl phenylacetate
101		Methyl trimethylacetate	234		Ethyl salicylate
102		*n*-Propyl acetate	268		Diethyl suberate
106–113		*s*-Butyl acetate	271		Ethyl cinnamate
120		Ethyl butyrate	282		Dimethyl phthalate
127		*n*-Butyl acetate	298–299		Diethyl phthalate
128		Methyl valerate	298–299		Phenyl benzoate
130		Methyl chloroacetate	340		Dibutyl phthalate
131–133		Ethyl isovalerate		56–58	Ethyl *p*-nitrobenzoate
142		*n*-Amyl acetate		88–90	Ethyl *p*-aminobenzoate
142		Isoamyl acetate		94–96	Methyl *p*-nitrobenzoate
143		Ethyl chloroacetate		95–98	*n*-Propyl *p*-hydroxybenzoate
154		Ethyl lactate		116–118	Ethyl *p*-hydroxybenzoate
168		Ethyl caproate (ethyl hexanoate)		126–128	Methyl *p*-hydroxybenzoate

TABLE 70.9   Ethers

bp	mp	Compound	bp	mp	Compound
32		Furan	215		4-Bromoanisole
33		Ethyl vinyl ether	234–237		Anethole
65–67		Tetrahydrofuran	259		Diphenyl ether
94		*n*-Butyl vinyl ether	273		2-Nitroanisole
154		Anisole	298		Dibenzyl ether
174		4-Methylanisole		50–52	4-Nitroanisole
175–176		3-Methylanisole		56–60	1,4-Dimethoxybenzene
198–203		4-Chloroanisole		73–75	2-Methoxynaphthalene
206–207		1,2-Dimethoxybenzene			

TABLE 70.10   Halides

bp	Compound	bp	Compound
34–36	2-Chloropropane	100–105	1-Bromobutane
40–41	Dichloromethane	105	Bromotrichloromethane
44–46	Allyl chloride	110–115	1,1,2-Trichloroethane
57	1,1-Dichloroethane	120–121	1-Bromo-3-methylbutane
59	2-Bromopropane	121	Tetrachloroethylene
68	Bromochloromethane	123	3,4-Dichloro-1-butene
68–70	2-Chlorobutane	125	1,3-Dichloro-2-butene
69–73	Iodoethane	131–132	1,2-Dibromoethane
70–71	Allyl bromide	140–142	1,2-Dibromopropane
71	1-Bromopropane	142–145	1-Bromo-3-chloropropane
72–74	2-Bromo-2-methylpropane	146–150	Bromoform
74–76	1,1,1-Trichloroethane	147	1,1,2,2-Tetrachloroethane
81–85	1,2-Dichloroethane	156	1,2,3-Trichloropropane
87	Trichloroethylene	161–163	1,4-Dichlorobutane
88–90	2-Iodopropane	167	1,3-Dibromopropane
90–92	1-Bromo-2-methylpropane	177–181	Benzyl chloride
91	2-Bromobutane	197	(2-Chloroethyl)benzene
94	2,3-Dichloro-1-propene	219–223	Benzotrichloride
95–96	1,2-Dichloropropane	238	1-Bromodecane
96–98	Dibromomethane		

TABLE 70.11   Aryl Halides

| | | | Derivatives | | | |
| | | | Nitration Product[a] | | Oxidation Product[b] | |
bp	mp	Compound	Position	mp	Name	mp
132		Chlorobenzene	2, 4	52		
156		Bromobenzene	2, 4	70		
157–159		2-Chlorotoluene	3, 5	63	2-Chlorobenzoic acid	141
162		4-Chlorotoluene	2	38	4-Chlorobenzoic acid	240
172–173		1,3-Dichlorobenzene	4, 6	103		
178		1,2-Dichlorobenzene	4, 5	110		
196–203		2,4-Dichlorotoluene	3, 5	104	2,4-Dichlorobenzoic acid	164
201		3,4-Dichlorotoluene	6	63	3,4-Dichlorobenzoic acid	206
214		1,2,4-Trichlorobenzene	5	56		
279–281		1-Bromonaphthalene	4	85		
	51–53	1,2,3-Trichlorobenzene	4	56		
	54–56	1,4-Dichlorobenzene	2	54		
	66–68	1,4-Bromochlorobenzene	2	72		
	87–89	1,4-Dibromobenzene	2, 5	84		
	138–140	1,2,4,5-Tetrachlorobenzene	3	99		
			3, 6	227		

a. For preparation, see page 762.
b. For preparation, see page 762.

TABLE 70.12   Hydrocarbons: Alkenes

bp	Compound	bp	Compound
34	Isoprene	149–150	1,5-Cyclooctadiene
83	Cyclohexene	152	DL-α-Pinene
116	5-Methyl-2-norbornene	160	Bicyclo[4.3.0]nona-3,7-diene
122–123	1-Octene	165–167	(−)-β-Pinene
126–127	4-Vinyl-1-cyclohexene	165–169	α-Methylstyrene
132–134	2,5-Dimethyl-2,4-hexadiene	181	1-Decene
141	5-Vinyl-2-norbornene	181	Indene
143	1,3-Cyclooctadiene	251	1-Tetradecene
145	4-Butylstyrene	274	1-Hexadecene
145–146	Cycloctene	349	1-Octadecene
145–146	Styrene		

TABLE 70.13   Hydrocarbons: Aromatic

bp	mp	Compound	Melting Point of Derivatives		
			Nitro[a]		*Picrate*[b]
bp	*mp*	*Compound*	*Position*	*mp*	*mp*
80		Benzene	1, 3	89	84
111		Toluene	2, 4	70	88
136		Ethylbenzene	2, 4, 6	37	96
138		*p*-Xylene	2, 3, 5	139	90
138–139		*m*-Xylene	2, 4, 6	183	91
143–145		*o*-Xylene	4, 5	118	88
145		4-*t*-Butylstyrene	2, 4	62	
145–146		Styrene			
152–154		Cumene	2, 4, 6	109	
163–166		Mesitylene	2, 4	86	97
			2, 4, 6	235	
165–169		α-Methylstyrene			
168		1,2,4-Trimethylbenzene	3, 5, 6	185	97
176–178		*p*-Cymene	2, 6	54	
189–192		4-*t*-Butyltoluene			
197–199		1,2,3,5-Tetramethylbenzene	4, 6	181(157)	
203		*p*-Diisopropylbenzene			
204–205		1,2,3,4-Tetramethylbenzene	5, 6	176	92
207		1,2,3,4-Tetrahydronaphthalene	5, 7	95	
240–243		1-Methylnaphthalene	4	71	142
	34–36	2-Methylnaphthalene	1	81	116
	50–51	Pentamethylbenzene	6	154	131
	69–72	Biphenyl	4, 4′	237(229)	
	80–82	1,2,4,5-Tetramethylbenzene	3, 6	205	
	80–82	Naphthalene	1	61(57)	149
	90–95	Acenaphthene	5	101	161
	99–101	Phenanthrene			144(133)
	112–115	Fluorene	2	156	87(77)
			2, 7	199	
	214–217	Anthracene			138

a. For preparation, see page 762.
b. For preparation, see page 760.

TABLE 70.14 Ketones

bp	mp	Compound	Semicarbazone[a] mp	2,4-Dinitrophenylhydrazone[b] mp
56		Acetone	187	126
80		2-Butanone	136, 186	117
88		2,3-Butanedione	278	315
100–101		2-Pentanone	112	143
102		3-Pentanone	138	156
106		Pinacolone	157	125
114–116		4-Methyl-2-pentanone	132	95
124		2,4-Dimethyl-3-pentanone	160	88, 95
128–129		5-Hexen-2-one	102	108
129		4-Methyl-3-penten-2-one	164	205
130–131		Cyclopentanone	210	146
133–135		2,3-Pentanedione	122 (mono) 209 (di)	209
145		4-Heptanone	132	75
145		5-Methyl-2-hexanone	147	95
145–147		2-Heptanone	123	89
146–149		3-Heptanone	101	81
156		Cyclohexanone	166	162
162–163		2-Methylcyclohexanone	195	137
169		2,6-Dimethyl-4-heptanone	122	66, 92
169–170		3-Methylcyclohexanone	180	155
173		2-Octanone	122	58
191		Acetonylacetone	185 (mono) 224 (di)	257 (di)
202		Acetophenone	198	238
216		Phenylacetone	198	156
217		Isobutyrophenone	181	163
218		Propiophenone	182	191
226		4-Methylacetophenone	205	258
231–232		2-Undecanone	122	63
232		n-Butyrophenone	188	191
232		4-Chloroacetophenone	204	236
235		Benzylacetone	142	127
	35–37	4-Chloropropiophenone	176	223
	35–39	4-Phenyl-3-buten-2-one	187	227
	36–38	4-Methoxyacetophenone	198	228
	48–49	Benzophenone	167	238
	53–55	2-Acetonaphthone	235	262
	60	Desoxybenzoin	148	204
	76–78	3-Nitroacetophenone	257	228
	78–80	4-Nitroacetophenone		257
	82–85	9-Fluorenone	234	283
	134–136	Benzoin	206	245
	147–148	4-Hydroxypropiophenone		240

a. For preparation, see page 431.
b. For preparation, see page 429.

TABLE 70.15   Nitriles

bp	mp	Compound	bp	mp	Compound
77		Acrylonitrile	212		3-Tolunitrile
83–84		Trichloroacetonitrile	217		4-Tolunitrile
97		Propionitrile	233–234		Benzyl cyanide
107–108		Isobutyronitrile	295		Adiponitrile
115–117		*n*-Butyronitrile		30.5	4-Chlorobenzyl cyanide
174–176		3-Chloropropionitrile		32–34	Malononitrile
191		Benzonitrile		38–40	Stearonitrile
205		2-Tolunitrile		46–48	Succinonitrile
				71–73	Diphenylacetonitrile

TABLE 70.16   Nitro Compounds

bp	mp	Compound	bp	mp	Acetamide[a] mp	Benzamide[b] mp
					Amine Obtained by Reduction of Nitro Groups	
210–211		Nitrobenzene	184		114	160
225		2-Nitrotoluene	200		110	146
225		2-Nitro-*m*-xylene	215		177	168
230–231		3-Nitrotoluene	203		65	125
245		3-Nitro-*o*-xylene	221		135	189
245–246		4-Ethylnitrobenzene	216		94	151
	34–36	2-Chloro-6-nitrotoluene	245		157(136)	173
	36–38	4-Chloro-2-nitrotoluene		21	139(131)	
	40–42	3,4-Dichloronitrobenzene		72	121	
	43–50	1-Chloro-2,4-dinitrobenzene		91	242 (di)	178 (di)
	52–54	4-Nitrotoluene		45	147	158
	55–56	1-Nitronaphthalene		50	159	160
	83–84	1-Chloro-4-nitrobenzene		72	179	192
	88–90	*m*-Dinitrobenzene		63	87 (mono)	125 (mono)
					191 (di)	240 (di)

a. For preparation, see page 760.
b. For preparation, see page 760.

TABLE 70.17 Phenols

| | | | Derivatives | |
| | | | α-Naphthylurethane[a] | Bromo[b] |
bp	mp	Compound	mp	mp
175–176		2-Chlorophenol	120	48 (mono)
				76 (di)
181	42	Phenol	133	95 (tri)
202	32–34	p-Cresol	146	49 (di)
				108 (tetra)
203		m-Cresol	128	84 (tri)
228–229		3,4-Dimethylphenol	141	171 (tri)
	32–33	o-Cresol	142	56 (di)
	42–43	2,4-Dichlorophenol		68
	42–45	4-Ethylphenol	128	
	43–45	4-Chlorophenol	166	90 (di)
	44–46	2,6-Dimethylphenol	176	79
	44–46	2-Nitrophenol	113	117 (di)
	49–51	Thymol	160	55
	62–64	3,5-Dimethylphenol		166 (tri)
	64–68	4-Bromophenol	169	95 (tri)
	74	2,5-Dimethylphenol	173	178 (tri)
	92–95	2,3,5-Trimethylphenol	174	
	95–96	1-Naphthol	152	105 (di)
	98–101	4-t-Butylphenol	110	50 (mono)
				67 (di)
	104–105	Catechol	175	192 (tetra)
	109–110	Resorcinol	275	112 (tri)
	112–114	4-Nitrophenol	150	142 (di)
	121–124	2-Naphthol	157	84
	133–134	Pyrogallol	173	158 (di)

a. For preparation, see page 760.
b. For preparation, see page 763.

71

Searching the Chemical Literature

In planning a synthesis or investigating the properties of compounds, the organic chemist is faced with the problem of locating information on the chemical and physical properties of substances that have been previously prepared as well as the best methods and reagents for carrying out a synthesis. With over 13,000,000 compounds known and new ones being reported at the rate of more than 600,000 per year, the task might seem formidable. However, the field of chemistry has developed one of the best information retrieval systems of all the sciences. In a university chemistry library it is possible to locate information on almost any known compound in a few minutes. Even a modest library will have information on several million different substances. The ultimate source of information is the primary literature: articles written by individual chemists and published in journals. The secondary literature consists of compilations of information taken from these primary sources. For example, to find the melting point of benzoic acid, one would naturally turn to a secondary reference containing a table of physical constants of organic compounds. If, on the other hand, one wished to have information about the phase changes of benzoic acid under high pressure, one would need to consult the primary literature. If the piece of information, such as a melting point, is crucial to an investigation, the primary literature should be consulted because, for instance, transcription errors can occur in the preparation of secondary references.

Handbooks

For rapid access to information such as mp, bp, density, solubility, optical rotation, λ max, and crystal form, one turns first to the *Handbook of Chemistry and Physics,* in its 77th edition in 1997, where information is found on some 15,000 organic compounds, including the Beilstein reference (see below) to each compound. The compounds presented are well known and completely characterized. The majority are commercially available. The *Merck Index,* 12th edition in 1996 (also available on CD-ROM), contains information on more than 10,000 compounds, especially those of pharmaceutical interest. In addition to the usual physical properties, information and literature references to synthesis, isolation, and medicinal properties, such as toxicity data, are found. The last third of the book is devoted to such items as a long cross index of names (which is very useful for looking up drugs), a table of organic name reactions, an excellent section on first aid for poisons, a list of chemical poisons, and a listing of the locations of many poison control centers.

In every organic chemistry laboratory throughout the world can be found a copy of the *Aldrich Catalog Handbook of Fine Chemicals,* available without charge from the Aldrich Company. Not only does this catalog list the prices of the more than 31,000 chemicals (primarily organic) as well as the molecular weights, melting points, boiling points, and optical rotations, it also gives references to IR and NMR spectral data and to Beilstein, *Merck Index,* and Fiesers' *Reagents.* In addition, for each chemical the catalog provides the reference to the *Registry of Toxic Effects of Chemical Substances* (RTECS No.) and, if appropriate, the reference to Sax, *Dangerous Properties of Industrial Materials.* Special hazards are noted ("severe poison," "lachrymator," "corrosive"), and a reference is made to 1 of 30 different disposal methods for each chemical. The Aldrich catalog also gives pertinent information (uses, physiological effects, etc.) with literature references for many compounds. The catalog is on-line; it can be searched and orders placed on the Internet. A complete list of all chemicals sold commercially and the addresses of the companies making them is found in *Chem Sources,* published yearly.

After consulting these three single-volume references, one would turn to more comprehensive multivolume sources such as the *Dictionary of Organic Compounds,* 5th edition. This dictionary, still known as "Heilbron," the name of its former editor, now comprises seven volumes of specific information, with primary literature references, on the synthesis, reactions, and derivatives of more than 50,000 compounds. Rodd's *Chemistry of Carbon Compounds,* another valuable multivolume work with primary literature references, is organized by functional group rather than in dictionary form. Elsevier's *Encyclopedia of Organic Compounds* in about 20 volumes is an incomplete reference work on the chemical and physical properties of compounds. It is useful for those areas it covers. References to Elsevier are found in the *Handbook of Chemistry and Physics.*

Beilstein's *Handbuch der organischen Chemie* is certainly not a "handbook" in the American sense—it can occupy an entire alcove of a chemistry library! And the currently produced volumes must be among the most expensive contemporary works purchased by a library, for each book now costs more than $1000. However, with characteristic German thoroughness, this reference covers every well-characterized organic compound that has been reported in the literature up to 1949. Along with a main reference set come three supplements covering the periods 1910–1919, 1920–1929, and 1930–1949. Further supplements are bringing Beilstein ever closer to the present. Although written in German, this reference can provide much information even to those with no knowledge of the language. Physical constants and primary literature references are easy to pick out. A German–English dictionary for Beilstein is available on the Internet as well as a number of guides and leading references to its use.

Beilstein is organized around a complex classification scheme that is explained on the Internet and in *The Beilstein Guide* by O. Weissback, but the casual user can gain access through the *Handbook of Chemistry and Physics* and the *Aldrich Catalog.* In the *Handbook* the reference for 2-iodobenzoic acid is listed as $\mathbf{B9}^2$, 239, which means the reference will be found on p. 239 of Vol. (Band) 9 of the second supplement (Zweites Ergänzungwerk, EII). In the *Aldrich Catalog* the reference is given as Beil **9**, 363, which indicates information can be found in the main series on p. 363, Vol. 9. The *System Number* assigned to each compound can be traced through the supplements. The easiest access to Beilstein itself is through the formula index (*General Formel register*) of the second supplement. A rudimentary knowledge of German will enable one to pick out iodo (Jod) benzoic acid (Saure), for example. Beilstein itself is available, for a fee, on the Worldwide Web. Some 100 properties of seven million compounds are available for searching.

Chemical Abstracts

Secondary references are incomplete, some suffer from transcription errors, and the best—Beilstein—is 20 years behind in its survey of the organic literature, so one must often turn to the primary chemical literature. However, there are more than 14,000 periodicals where chemical information might appear. Chemists are fortunate to have an index, *Chemical Abstracts,* that covers this huge volume of literature very promptly and publishes biweekly abstracts of each article. The information in each abstract is then compiled into author, chemical substance, general subject, formula, and patent indexes. Since publication began in 1907, nomenclature has changed. Now no trivial names are used, so acetone appears as 2-propanone and *o*-cresol as benzene, 1-methyl, 2-hydroxy. It takes some experience to adapt to these changing names, and even now *Chemical Abstracts* does not follow the IUPAC rules exactly. The formula index is useful because it provides not only reference to specific abstracts but also a correct name that can be found in the chemical substances index.

Chemical Abstracts is also published on CD-ROM, which makes it easy to search. The disks come out five times yearly in order to cover the 600,000 articles that appear in the 8000 journals currently abstracted. In 1996 information on 1,220,000 compounds was added to *Chemical Abstracts.* Through the STN company the 12 million entries in *Chemical Abstracts* are searchable, for a fee, on line.

Current Contents

As long as a year can elapse between the time a paper appears and the time an abstract is published. *Current Contents* helps fill the gap by providing much of the information found in an abstract: a list of the titles, authors, and keywords that appear in titles of papers recently published—or, in some cases, papers about to be published.

Science Citation Index

Science Citation Index is unique in that it allows for a type of search not possible with any other index—a search forward in time. For example, if one wanted to learn what recent applications have been made of the coupling reaction first reported by Stansbury and Proops in 1962, one would look up their paper [*J. Org.,* **27,** 320 (1962)] in a current volume, say 1998, of *Science Citation*

Index and find there a list of those articles published in 1998 in which an author cited the 1962 work of Stansbury and Proops in a footnote.

Planning a Synthesis

In planning a synthesis the organic chemist is faced with at least three overlapping considerations: the chemical reactions, the reagents, and the experimental procedure to be employed. For students an advanced textbook such as Carey and Sundberg, *Advanced Organic Chemistry,* Vols. 1 and 2, 3rd ed., 1990, or March's *Advanced Organic Chemistry,* 4th ed., 1992, which have literature references, might be a place to start. Often a good lead is through named reactions, for which Mundy and Ellerd's *Name Reactions and Reagents in Organic Chemistry,* 1988, is a useful reference.

It is instructive for the beginning synthetic chemist to read about some elegant and classical syntheses that have been carried out in the past. For this purpose see Anand, Bindra, and Ranganathan, *Art in Organic Synthesis,* 2nd ed., 1988, for some of the most elegant syntheses carried out in recent years. Similarly, Bindra and Bindra, *Creativity in Organic Synthesis,* Vol. 1, and Fleming, *Selected Organic Syntheses,* are compendia of elegant syntheses. At a more advanced level see Cheng and Corey, *The Logic of Chemical Synthesis,* 1995. For natural product synthesis see the excellent series edited by ApSimon, *The Total Synthesis of Natural Products,* in nine volumes, the latest published in 1992. House's *Modern Synthetic Reactions,* 2nd ed., 1972, is a more comprehensive source of information with several thousand references to the original literature. It is brought further up to date at about the same level by Carruthers in *Some Modern Methods of Organic Synthesis,* 3rd ed., 1986.

An extremely useful reference is the *Compendium of Organic Synthetic Methods,* presently in eight volumes, the latest published in 1995. The material is organized by reaction type. In addition, *Organic Reactions,* published annually, should be consulted. Over 100 preparative reactions, with examples of experimental details, are covered in great detail in some 30 volumes. *Annual Reports in Organic Synthesis* is an excellent review of new reactions in a given year, organized by reaction type. *Stereoselective Synthesis* by Robert S. Atkinson, published in 1995, treats one of the newest and most important areas of organic chemistry.

The late Mary Fieser was responsible for publishing the 17 volumes of *Reagents for Organic Synthesis,* the last of which appeared in 1994. Started in 1967, this series critically surveys the reagents employed to carry out organic synthesis. Included are references to the original literature and suppliers of reagents, and to *Organic Syntheses* (see below), the critical reviews that have been written about various reagents. The index of reagents according to type of reaction is very useful when planning a synthesis. On a much larger scale is the comprehensive *Encyclopedia of Reagents for Organic Synthesis,* 1995, edited by Leo Paquette. It contains 3500 articles, giving a critical assessment of most of the reagents one would use in carrying out a synthesis.

To carry out a reaction on one functional group without affecting another functional group it is often necessary to put on a protective group. This rather specialized subject is admirably covered in Greene and Wuts' *Protective Groups in Organic Synthesis,* 2nd ed., 1991.

Before carrying out the synthesis itself one should consult *Organic Syntheses,* an annual series since 1921, which is grouped in six collective volumes with reaction and reagent indexes. *Organic Syntheses* gives detailed laboratory procedures for carrying out more than 1000 different reactions. Each synthesis is submitted for review, and then the procedure is sent to an independent laboratory to be checked. The reactions, unlike many that are reported in the primary literature, are carried out a number of times and therefore can be relied on to work. Laboratory techniques are covered in the 14 volumes of Weissberger *et al., Techniques of Organic Chemistry,* an uneven, multiauthor compendium, outdated in some respects. An excellent advanced text on the detailed mechanisms of many representative reactions is Lowry and Richardson's *Mechanism and Theory in Organic Chemistry,* 3rd ed., 1987.

The Modern Chemistry Library

Although the chemistry library as we now know it probably will not change markedly in the next decade, in one aspect it has already changed. It is now possible to search a large part of the chemical literature using a personal computer connected to a commercial database held in a large computer. By rapidly and efficiently calling into a database containing *Chemical Abstracts, Science Citation Index,* and several other indexes, a librarian trained to use the system can aquire information for the chemist much more rapidly than could be done by a manual search. In some cases, the information obtained would be impossible to get by conventional means. For example, to locate all references to the NMR spectra of insulin during the period 1976–1990 using the DIALOG system, one first asks for the number of references to NMR found in *Chemical Abstracts* during that period. The answer is typed out within a few seconds: 12,000. Next one asks how many references to insulin. The answer: 20,000. The next query asks the computer to cross the two lists for references common to both. The answer: 9. One can then ask that the references be typed out. At some expense, the entire abstract for each reference can be typed out. It is less expensive to have the abstracts typed off-line at the computer center and sent by mail.

Surfing the Web

http://www.library.upenn.edu/scitech/chemistry/guides/beilstei.html

A guide to Beilstein and references to other guides, from the University of Pennsylvania.

http://www-sul.stanford.edu/depts/swain/beilstein/bedict2.html

The editors of Beilstein have provided a German–English dictionary geared to the *Handbuch.*

http://stneasy.cas.org/english/help.html#about

Information about searching *Chemical Abstracts* on the Web, for a fee.

http://www.sigald.sial.com/aldrich/ald_e.htm

The searchable on-line Adrich catalog. A printed copy can be obtained from the Aldrich Chemical Co., 1001 West Saint Paul Ave., Milwaukee, WI 53233.

For updated information visit:

www.mtholyoke.edu/courses/kwilliam/microscale.shtml
or
www.hmco.com/hmco/college/chemistry/Home.html

MULTIPLES OF ELEMENTS' WEIGHTS

Elem	Wt	Elem	Wt	Elem	Wt	Elem	Wt	Elem	Wt	Elem	Wt
C	12.0112	C_{41}	492.457	H_{31}	31.2471	O_6	95.9964	I	126.904	P	30.9738
C_2	24.0223	C_{42}	504.468	H_{32}	32.2550	O_7	111.996	I_2	253.809	P_2	61.9476
C_3	36.0335	C_{43}	516.479	H_{33}	33.2630	O_8	127.995	I_3	380.713	P_3	92.9214
C_4	48.0446	C_{44}	528.491	H_{34}	34.2710	O_9	143.995			P_4	123.895
C_5	60.0557	C_{45}	540.502	H_{35}	35.2790	O_{10}	159.994	OCH_3	31.0345	P_5	154.869
C_6	72.0669	C_{46}	552.513	H_{36}	36.2869			$(OCH_3)_2$	62.0689	P_6	185.843
C_7	84.0780	C_{47}	564.524	H_{37}	37.2949	N	14.0067	$(OCH_3)_3$	93.1034		
C_8	96.0892	C_{48}	576.535	H_{38}	38.3029	N_2	28.0134	$(OCH_3)_4$	124.138	Na	22.9898
C_9	108.100	C_{49}	588.546	H_{39}	39.3108	N_3	42.0201	$(OCH_3)_5$	155.172	Na_2	45.9796
C_{10}	120.111	C_{50}	600.558	H_{40}	40.3188	N_4	56.0268	$(OCH_3)_6$	186.207	Na_3	68.9694
						N_5	70.0335	$(OCH_3)_7$	217.241		
C_{11}	132.123	H	1.00797	H_{41}	41.3268	N_6	84.0402	$(OCH_3)_8$	248.276	K	39.102
C_{12}	144.134	H_2	2.01594	H_{42}	42.3347			$(OCH_3)_9$	279.31	K_2	78.204
C_{13}	156.145	H_3	3.02391	H_{43}	43.3427	S	32.064	$(OCH_3)_{10}$	310.345	K_3	117.306
C_{14}	168.156	H_4	4.03188	H_{44}	44.3507	S_2	64.128				
C_{15}	180.167	H_5	5.03985	H_{45}	45.3587	S_3	96.192	OC_2H_5	45.0616	Ag	107.868
C_{16}	192.178	H_6	6.04782	H_{46}	46.3666	S_4	128.256	$(OC_2H_5)_2$	90.1231	Ag_2	215.736
C_{17}	204.190	H_7	7.05579	H_{47}	47.3746	S_5	160.320	$(OC_2H_5)_3$	135.185		
C_{18}	216.201	H_8	8.06376	H_{48}	48.3826	S_6	192.384	$(OC_2H_5)_4$	180.246	Cu	63.546
C_{19}	228.212	H_9	9.07173	H_{49}	49.3905			$(OC_2H_5)_5$	225.308	Cu_2	127.092
C_{20}	240.223	H_{10}	10.0797	H_{50}	50.3985	F	18.9984	$(OC_2H_5)_6$	270.369	Cr	51.996
						F_2	37.9968	$(OC_2H_5)_7$	315.431	Hg	200.59
C_{21}	252.234	H_{11}	11.0877	H_{51}	51.4065	F_3	56.9952	$(OC_2H_5)_8$	360.492	Pb	207.19
C_{22}	264.245	H_{12}	12.0956	H_{52}	52.4144	F_4	75.9936			Pt	195.09
C_{23}	276.256	H_{13}	13.1036	H_{53}	53.4224	F_5	94.992	$OCOCH_3$	59.045	Pt_2	390.18
C_{24}	288.268	H_{14}	14.1116	H_{54}	54.4304	F_6	113.99	$(OCOCH_3)_2$	118.090	Se	78.96
C_{25}	300.279	H_{15}	15.1196	H_{55}	55.4384	F_7	132.989	$(OCOCH_3)_3$	177.135	Th	204.37
C_{26}	312.290	H_{16}	16.1275	H_{56}	56.4463	F_8	151.987	$(OCOCH_3)_4$	236.180		
C_{27}	324.301	H_{17}	17.1355	H_{57}	57.4543	F_9	170.986	$(OCOCH_3)_5$	295.225		
C_{28}	336.312	H_{18}	18.1435	H_{58}	58.4623	F_{10}	189.984	$(OCOCH_3)_6$	354.270		
C_{29}	348.323	H_{19}	19.1514	H_{59}	59.4702			$(OCOCH_3)_7$	413.315		
C_{30}	360.334	H_{20}	20.1594	H_{60}	60.4782	Cl	35.453	$(OCOCH_3)_8$	472.360		
				H_{61}	61.4862	Cl_2	70.906	$(OCOCH_3)_9$	531.405		
C_{31}	372.346	H_{21}	21.1674	H_{62}	62.4941	Cl_3	106.359	$(OCOCH_3)_{10}$	590.450		
C_{32}	384.357	H_{22}	22.1753	H_{63}	63.5021	Cl_4	141.812				
C_{33}	396.368	H_{23}	23.1833	H_{64}	64.5101	Cl_5	177.265	$(H_2O)_{\frac{1}{2}}$	9.00767		
C_{34}	408.379	H_{24}	24.1913	H_{65}	65.5181			H_2O	18.0153		
C_{35}	420.390	H_{25}	25.1993			Br	79.904	$(H_2O)_{1\frac{1}{2}}$	27.0230		
C_{36}	432.401	H_{26}	26.2072	O	15.9994	Br_2	159.808	$(H_2O)_2$	36.0307		
C_{37}	444.413	H_{27}	27.2152	O_2	31.9988	Br_3	239.712	$(H_2O)_3$	54.0460		
C_{38}	456.424	H_{28}	28.2232	O_3	47.9982	Br_4	319.616	$(H_2O)_4$	72.0614		
C_{39}	468.435	H_{29}	29.2311	O_4	63.9976	Br_5	399.52	$(H_2O)_5$	90.0767		
C_{40}	480.446	H_{30}	30.2391	O_5	79.9970			$(H_2O)_6$	108.092		

BUFFER SOLUTIONS (0.2 M, except as indicated)

pH	Components	pH	Components
0.1	1 M Hydrochloric acid	8.0	11.8 g Boric acid + 9.1 g Borax
1.1	0.1 M Hydrochloric acid		($Na_2B_4O_7 \cdot 10H_2O$) per L
2.2	15.0 g D-Tartaric acid per L (0.1 M solution)	9.0	6.2 g Boric acid + 38.1 g Borax per L
3.9	40.8 g Potassium acid phthalate per L	10.0	6.5 g $NaHCO_3$ + 13.2 g Na_2CO_3 per L
5.0	14.0 g KH-Phthalate + 2.7 g $NaHCO_3$ per L	11.0	11.4 g Na_2HPO_4 + 19.7 g Na_3PO_4 per L
	(heat to expel carbon dioxide, then cool)	12.0	24.6 g Na_3PO_4 per L (0.15 M solution)
6.0	23.2 g KH_2PO_4 + 4.3 g Na_2HPO_4 (anhyd., Merck) per L	13.0	4.1 g Sodium hydroxide pellets per L (0.1 M)
7.0	9.1 g KH_2PO_4 + 18.9 g Na_2HPO_4 per L	14.0	41.3 g Sodium hydroxide pellets per L (1 M)